Elementary Cluster Analysis: Four Basic Methods that (Usually) Work

RIVER PUBLISHERS SERIES IN MATHEMATICAL AND ENGINEERING SCIENCES

Indexing: All books published in this series are submitted to the Web of Science Book Citation Index (BkCI), to SCOPUS, to CrossRef and to Google Scholar for evaluation and indexing.

Mathematics is the basis of all disciplines in science and engineering. Especially applied mathematics has become complementary to every branch of engineering sciences. The purpose of this book series is to present novel results in emerging research topics on engineering sciences, as well as to summarize existing research. It engrosses mathematicians, statisticians, scientists and engineers in a comprehensive range of research fields with different objectives and skills, such as differential equations, finite element method, algorithms, discrete mathematics, numerical simulation, machine leaning, probability and statistics, fuzzy theory, etc.

Books published in the series include professional research monographs, edited volumes, conference proceedings, handbooks and textbooks, which provide new insights for researchers, specialists in industry, and graduate students.

Topics covered in the series include, but are not limited to:

- Advanced mechatronics and robotics
- Artificial intelligence
- Automotive systems
- Discrete mathematics and computation
- Fault diagnosis and fault tolerance
- Finite element methods
- Fuzzy and possibility theory
- Industrial automation, process control and networked control systems
- Intelligent control systems
- Neural computing and machine learning
- Operations research and management science
- Optimization and algorithms
- Queueing systems
- Reliability, maintenance and safety for complex systems
- Resilience
- Stochastic modelling and statistical inference
- Supply chain management
- System engineering, control and monitoring
- Tele robotics, human computer interaction, human-robot interaction

For a list of other books in this series, visit www.riverpublishers.com

Elementary Cluster Analysis: Four Basic Methods that (Usually) Work

James C. Bezdek

Visiting Senior Fellow
University of Melbourne

River Publishers

Published, sold and distributed by:
River Publishers
Alsbjergvej 10
9260 Gistrup
Denmark

www.riverpublishers.com

ISBN: 978-87-70224-25-3 (Hardback)
978-87-70224-24-6 (Ebook)

For Arlene: my eyes, my ears, my heart

Contents

Preface

Clustering is an important component of the computational sciences. To some extent, this discipline powers Internet search engines such as Google and Yahoo, and clustering is also used for the analysis of data in application platforms such as Amazon, Facebook, and Twitter. Clustering underlies much of the data analysis that is done in fields such as information retrieval, data mining, knowledge engineering, image processing, medical information systems, document analysis, and machine learning. Many applications that depend on some form of pattern recognition revert to cluster analysis as a basic tool. Clustering is used to search for structure in virtually any kind of data and, thus, in the objects represented by that data.

Another book about clustering – Why? The availability of packaged clustering programs means that anyone with data can easily do cluster analysis on it. But ... many end users of this technology don't fully appreciate its many hidden dangers. In today's world of "grab and go algorithms," part of my motivation for writing this book is to provide (dare I say *naive*?) users with a set of cautionary tales about cluster analysis, for it is very much an art as well as a science, and it is easy to stumble if you don't understand its pitfalls. Indeed, it is easy to trip over them even if you do! Modern cluster analysis has become so technically intricate that it is often hard for the beginner or the non-specialist to appreciate and understand these hidden dangers. Here is how Yogi Berra put it, and he was right:

> *In theory there's no difference between theory and practice. In practice, there is*
>
> –Yogi Berra

What is in this book? This book is a step *backwards* to four classical methods for clustering in small, static data sets that have all withstood the tests of time. The youngest of the four methods is now almost 50 years old:

Gaussian mixture decomposition (GMD, 1898)
SAHN clustering (principally *single linkage* (SL, 1909))
Hard c-means (HCM, 1956, *also widely known as* (aka) "k-means")
Fuzzy c-means (FCM, 1973, reduces to HCM in a certain limit)

The dates shown are the first known writing (to me, anyway) about these four models. There are many different algorithms that attempt to optimize the objective function models (GMD, HCM, and FCM), which all define good clusters as part of extrema of optimization problems. The SAHN models are deterministic and operate in a very different way. This volume is sprinkled throughout with a style of emphasis that is designed to catch your eye and slow you down for what I regard as the most important ideas. This is the style:

> ☺ ***Forewarned is forearmed:*** Much of the confusion about the term "k-means" stems from the fact that this term is used quite indiscriminately for several sequential (local) and batch (global) versions for static data that optimize the same objective function in different ways.

And the use of the integers k and c also causes uncertainty and confusion. In common usage in different academic communities, both stand for the number of clusters. This integer is just a dummy index, and the two symbols are, in my mind, interchangeable. In other words, ***I regard Lloyd's batch k-means and basic hard c-Means (HCM algorithm 6.1, m = 1 in this volume) as synonyms, and I will freely switch from c-means to k-means and vice versa in an attempt to dispel the notion that these are different models and algorithms.*** But even this is a bit misleading since there are different ways to implement the batch algorithm other than the naïve version presented as Algorithm 6.1 in Chapter 6. ☺

The expansion of cluster analysis into every corner of our modern lives (big data, social networks, streaming video, wireless sensor networks, etc.,the list is endless) still rests on the foundation provided by these four models. I am (with apologies to Marvel Comics) very comfortable in calling HCM, FCM, GMD, and SL the *Fantastic Four*. By the way, everyone knows the original members of the FF: *The Thing*, *The Human Torch*, *The Invisible Woman*, and *Mr. Fantastic*. But other superheroes have served as full-fledged members of this exclusive team from time-to-time. Newsarama (2008) takes an ironic look at the all-time top 10 members of the Fantastic 4 (figure that out!); so if your favorite clustering algorithm is not in my FF, you have at least six-empty slots to accommodate it. Remember, these are (mathematical) models of our world, and be warned that

All models are wrong. Some models are useful

–Box and Draper (1987)

This volume bears the title "Elementary Cluster Analysis." A word that I quite like which can stand in for "elementary" in the context of educational literature is the word "primer." The (online) Free Dictionary gives these two definitions for the noun *primer*:

1. An elementary textbook for teaching children to read.
2. A book that covers the basic elements of a subject.

I chose the title of the book quite deliberately, in consonance with both definitions. The word *usually* in the title is very important because *all* clustering algorithms can and do fail from time to time. This also channels my admiration for an old but delightful book on numerical analysis by Acton (1970), which had the title *Numerical Methods that (Usually) Work*: the word *usually* was embossed on the cover of this book in very small letters after the word "that," but it is not part of the official title of the book. I was sorely tempted to use the longer subtitle "What can go *wrong* with methods that *usually* work?" but I decided that, while provocative and accurate, this subtitle would not really tell browsers much more about the contents of the book than the shorter one.

This definition also provides a nice and very apt analogy-for the first half of this book-which is about the basics of clustering, and it is written for newcomers. But ... the second half of this book is more demanding because I have tried to discuss the mathematics of the four classic methods in some detail. Even the basics of all four models have a surprising amount of depth; so the second half of this volume may seem more like an anti-primer to the novice.

As in many branches of computational science and engineering, cluster analysis is a vast topic. The overall picture in clustering is quite overwhelming; so any attempt to swim at the deep end of the pool in even a very specialized subfield requires a lot of training. But we all start out at the shallow end (or at least that is where we should start!), and this book is aimed squarely at teaching toddlers not to be afraid of the water. There is no section of this book that, if explored in real depth, can't be expanded into its own volume. So, if your needs are for an in-depth treatment of all the latest developments in any topic in this volume, the best I can do-what I will try to do anyway-is lead you to the pool and show you where to jump in. If you are a graduate student, professor, or professional in computational science or engineering, you may already know more than this book contains; so it will not be very useful to you. My hope is that this volume will be useful to the real novice, who thinks that clustering might be a good thing to try, but who knows very little about it. To close this part of the introduction, I repeat a cautionary statement made by Marriott (1974) about the dangers of believing what a computer tells you about clusters in your data:

> *If the results disagree with informed opinion, don't admit a simple logical interpretation, and don't show up clearly in a graphical presentation, they are probably wrong. There is no magic about numerical methods, and many ways in which they can break down. They are a valuable aid to the interpretation of data, not sausage machines automatically transforming bodies of numbers into packets of scientific fact.*

To summarize, Chapters 6–9 contain a fairly complete and extensive description of what are (arguably) the four clustering methods for static data that enjoy the most study and use around the world today. There are many, many worthy contenders that are not discussed here, and any one of them might move into the top 4 sooner or later. But that is in the future, and this volume concerns the past and the present. You know about the future?

> *The future is already here, it's just not very evenly distributed*
>
> – William Gibson

Numbering System The atomic unit for the numbering system is the chapter. Sections, figures, tables, examples, equations, algorithms, theorems, and exercises are all numbered consecutively within each chapter. Section 2.1 is section 1 of Chapter 2; Figure 3.5 is Figure 5 in Chapter 3; Table 4.1 is the first table in Chapter 4; Example 6.9 is the ninth example in Chapter 6; (2.2) is the second equation in Chapter 2.

Algorithms The focus of this book is a set of four basic models and algorithms for clustering in static data sets. These are the fantastic four:

Algorithm A6.1($m = 1$). Basic hard c-means (k-means): (HCM)
Algorithm A6.1($m > 1$). Basic fuzzy c-means: (FCM)
Algorithm A7.1. Basic EM algorithm for Gaussian mixture decomposition: (GMD/EM)
Algorithm A8.1. Basic SAHN Algorithm $\rightarrow$ A8.2 basic MST single linkage algorithm: (SL)

Beyond the Fantastic Four, there are six other algorithms that are given in pseudocode format in the text:

Algorithm A4.1. Visual assessment of tendency: (VAT/iVAT)
Algorithm A11.1. Maximin sampling algorithm: (MM)
Algorithm A11.2. Maximin directed random sampling algorithm: (MMDRS)
Algorithm A11.3. Visual assessment of tendency for big static data: (sVAT/siVAT)
Algorithm A11.4. Single linkage clustering in big static data: (clusiVAT)
Algorithm A12.1. Incremental update of within cluster dispersion: (WCD)

The *improved visual assessment of tendency* (iVAT) model and Algorithm A4.1 for visual assessment for clustering tendency appears early because it is a wonderful tool for understanding a bit more about data sets in their original space before taking a plunge into the icy waters of data partitioning. There are four algorithms in big data Chapter 11. Algorithm A11.1 describes an old, well-known method of sampling called *Maximin (MM) sampling* that, with a little freshening up, becomes useful for scaling up clustering algorithms for big (unloadable) data. The second algorithm, A11.2, is the extension of MM with *directed random sampling* (DRS), resulting in the MMDRS sampling algorithm. Algorithm 11.3 uses MMDRS sampling as a basis for extension of iVAT to siVAT (*sampled iVAT*) for arbitrarily large square relational data. This extension provides users with a way to visually estimate what cluster structure may or may not exist in big data sets with a small data heat map of an MMDRS sample of the big data set. In other words, siVAT provides a means for building core or skeletal cluster heat maps for big data sets that can't be directly assessed with iVAT. Maximin sampling and siVAT are married in A11.4, the clusiVAT algorithm, which is outlined in the last section of Chapter 11. This is a method that extends single linkage clustering to big data sets and, under certain circumstances, is exact single linkage. The final algorithm in the list above is A12.1, which provides a way to incrementally update a certain measure of within cluster scatter that can be used to derive a number of incremental stream monitoring functions.

I have tried to describe the algorithms in a (more or less) standardized format with pseudocode, and it is really pseudo in the sense that it is a mixture of three or four styles of representing programming languages. I am not a computer scientist-I am a computing scientist, so I have not been careful to observe all the latest trends and changes in computational techniques. My intent is to maximize clarity and minimize dependence on a particular language, operating system, compiler, host platform, and so on. You should be able to convert them into working programs with a minimum of trouble, and, of course, the fantastic four are all available in ready-to-use form in a variety of languages and formats with a quick jaunt to the Internet.

While these four clustering methods are readily available in MATLAB toolboxes (and in other, less well-known but readily available commercial and or professor packages), ...***you should implement them yourself***, from scratch. Why? Self-development is the only way to gain a deep understanding of the foundations, to learn the tricks, and appreciate how to tweak these models to extract the hopefully useful information you think your data contain. I already know your time is constrained, your financial support is limited, you operate under a lot of pressure, and you just want results. All of these are good excuses, but no more than that.

Build your own toolbox-it is infinitely more satisfying than using your neighbor's stuff. And, in this way, you will experience a bit of what Richard Feynman meant when he spoke of *the adventure of science*. He called it *the pleasure of finding things out*. Find out that you can write your own programs, make them work, and add your own improvements to them. You will really derive some pleasure from this. And you will discover new ideas of your own that can lead to methodological improvements or insights into the topic that interests you.

Numerical Examples: Almost all of the numerical examples were done in the MATLAB programming environment. All algorithms have parameters that affect their performance. Science is about quantitative models of our physical world, while art tries to express the qualitative content of our lives. You will encounter parameters that are user-defined, together with evasive statements like "pick a value for δ that is close to zero" or "don't choose a large value for m." What do instructions such as these mean? Lots of things are possible: (i) no one has better advice; (ii) the inventor of the algorithm tried lots of values, and values in the range mentioned produced the best results for her or him; (iii) 0.99 is closer to 1 than 0.95, and 22 is larger than

1.32, but you may never know which choice is better, and (unfortunately) this can make all the difference in the world; and, finally, (iv) sometimes we don't know why things work the way they do, but we should be happy if they work right this time-call it voodoo or luck, these are heuristic methods. My advice: if it works, be happy, but don't assume the same thing will work again on the next data set you analyze. Approach every problem as a new one.

Science is most often *NOT* very exact: it is a sequence of successively better approximations (to the things that happen) by models we invent to mirror the physical reality of processes we initiate, observe, or control. There is a lot of art in science, and this is nowhere more evident than in cluster analysis because, here, a human being remains in the loop, and the data always have the last word. And if the data are truly unlabeled and can't be examined directly (i.e., visually or manually in tabular form), there is no way for us to *ever* know if there even *is* a "best answer." We are always at the mercy of an unanticipated situation in the data: unusual structures, anomalous measurements, missing observations, improbable events that cause outliers, uncertainty about the interactions between variables, useless choices for numerical representation, sensors that don't respect our design goals, computers that lose bits, computer programs that have undetected flaws, and so on. When you read about an experiment with algorithmic parameters, keep an open mind-anything is possible.

The numerical examples have three objectives: (i) they are designed so that the underlying concepts are as transparent as possible; (ii) alternative solutions are considered when these are available; and (iii) failures of the methods involved are often highlighted. It is every bit as useful for you to know that something can't or will not work. For example, validation methods are often most useful for rejection of impossibly bad solutions as opposed to identifying the best of several pretty good ones. Sometimes methods fail because the scheme itself is weak, but, sometimes, they fail because of a peculiarity in the data. The examples are designed to ferret out explanations when clustering algorithms don't seem to produce the "right" answer.

I have tried to give you all the information you need to *replicate* the outputs reported in numerical examples. There are a few instances where this was not possible (for example, when an iterative procedure is initialized randomly or when the results were reported in someone's paper 15 or 20 years ago and the input data are lost, or when the authors of a paper simply could not supply more details). And it is always possible that the code used for the example implemented something other than the method described in the text, or it simply had undetected programming errors. Also, numerical results have been rounded off or truncated to make tables fit into the format of the book. Let me know if you find substantial differences between outputs you get (or got) and the results reported here.

Data: We are (quite literally) inundated with "big data." But the numerical examples in these volumes use very small ("toy") data sets. *Gasp! What?* In the age of big static data and streaming data, how useful can this be? Well, little data sets help us see what a formula means or what an algorithm does. There is much to be said for the pedagogical value of using a few points in the plane when studying and illustrating properties of various models. Moreover, small data sets often yield easy counterexamples for conjectures about a theoretical property or algorithmic performance. On the other hand, there are certain risks too.

Sometimes conclusions that are legitimate for small, specialized data sets become invalid in the face of large numbers of samples, features, and classes. And, of course, time and space complexity make their presence felt in very unpredictable ways as problem size grows. I don't subscribe to the idea that "if it doesn't work on big data, it is no good." But I certainly believe the corollary that it is better if it does! The next to last chapter in this volume, Chapter 11, is about clustering in big collections of static data. It is an attempt to mollify your fear that learning with little data leaves you disadvantaged in the race to analyze the big data sets that lie in your future.

Data get corrupted many ways, and in the electronic age, it should not surprise us to find (if we can) that this is a fairly common event. For example, on September 7, 1998, Bezdek *et al.* (1999b) fetched the ubiquitous Iris data from the anonymous FTP site "ftp.ics.uci.edu" under the directory "pub/machine-learning-databases," and discovered not one but *two* errors in it! This is not a criticism of this particular database. It simply alerts you to the fact that mistakes in data are inevitable. Again, don't be surprised: instead, be wary.

Acronyms and Abbreviations: Acronyms are like locusts. They spread unchecked across the technical landscape (mine anyway), leaving nothing in their wake but shreds of meaning that elude the most diligent reader. Why do we use them? Convenience. Laziness. This book has several hundred acronyms and abbreviations, and you would not remember what many of them mean for more than a few pages. There is a tabulation of the acronyms and abbreviations used in the text immediately preceeding the references.

Mathematics: This book contains a lot of math. How much rigor should be used when developing models and algorithms? There are a lot of well-known clichés about learning mathematics, not the least of which is nicely put in Simmons (1963):

> *It is a basic principle in the study of mathematics, and one too seldom emphasized, that a proof is not really understood until the stage is reached at which one can grasp it as a whole and see it as a single idea.*

Here is my own version: you have to be able to see the heart of a proof before the body is clear. On the other hand, my basic aim is to convey understanding in whatever form seems most effective and reasonable. I am again informed by Simmons (1981), who had this to say about proofs (18 years after he wrote the above) in his wonderful little book about pre-calculus mathematics:

> *For us the purpose of proof is to remove doubt and convey insight, not belabor the obvious. ... I use the word "proof" to mean an argument that I hope my intended audience will find convincing. A few mathematicians may object to this relativistic attitude. However, since students differ from logicians in their power of skepticism, and logicians differ among themselves from one generation to the next, it seems unlikely that any fixed, unalterable, absolute meaning can possibly be attached to the concept of proof. What a proof is depends on who and when you are.*

I agree emphatically and wholeheartedly with both of Simmons' ideas. When a formal proof seems desirable (this will not happen too often in this volume), I will advertise it as such. This volume does have definitions, proofs, and examples of the main mathematical concepts needed throughout the text, but I don't provide refresher notes or remedial material. A good working knowledge of linear algebra is essential. Useful, but less essential, mathematical background includes advanced calculus, numerical analysis, optimization, analysis of algorithms, and probability and statistics.

The material in Part 2 (Chapters 6–12) should be accessible to advanced undergraduates and graduate students majoring in engineering or science. This book is full of notation. I was sorely tempted to, but did not, include an appendix with all the notation. Why not? No good excuse, just being lazy. I assume readers are familiar with the basic mathematical symbols such as $\coprod$ $\prod$ $\therefore$ $\forall$ $\exists$ $\cup$ $\cap$ ∞, and so on. If you stumble on the math, ask your professor, your colleague, or go to the Internet-or ask me! Get some help.

I will not try to convince you that exercises cement ideas and concepts in mathematics that are complex and difficult to understand and retain-they do. I taught mathematics for 40 years, and I know a statement like this just causes eyes to roll. But please consider at least reading the problems that end each chapter, if only to broaden your overall grasp of the field itself.

Notes and Remarks: Each chapter ends with some notes of historical importance, bibliographic comments, and additional references. My experience is that many of the good and useful early papers and books about clustering have had a pretty short shelf life. Go to Yahoo! Answers, and enter the query "Why care about history?" You will find this query posted:

> *Why do people care so much about history? They're all dead?*

And the answer from the *arbiter of common sense*:

> *History defines us. Our language, our culture, our place on the earth is all defined by what went before. The lessons learned would all be pointless if we forgot why we learned them.*

So, the intent of this section is to extend the formal material of the text to include topics that, for whatever reason, eluded coverage in the body of the text; and to inform us all about the road that led us to this point in our pursuit of clusters. And, just to fix the importance of history in your mind, please consider Hegel's 1807 paradox:

> *Man learns from history that man learns nothing from history*

References: What type of referencing would serve you best? Because of the Internet, the answer to this question is at best a moving target. You will have access to search engines such as Yahoo, Google, and Bing. You can easily find out who wrote (and you can usually easily get in pdf form) the most cited papers on each topic, but this is a somewhat unreliable indication of their worth, for often, papers are cited heavily for all the wrong reasons.

Given this, providing references *almost* seems like a thing of the past! Why even have references anymore? Presumably, authors of books and papers know something about the topic, and they may be able to point you towards a few comprehensive references that enable you to enter their jungle with at least a rudimentary guide to help you locate the game you seek. This is the primary reason that texts and papers still provide written citations. Hopefully the references given will provide you with some good starting points for your journey into an area of interest. The references cited (I use the Harvard referencing system) appear in collected form at the end of the book.

There is another interesting fact that substantiates the impossibility of giving a really comprehensive set of references, summed up quite nicely this way in Aldenderfer and Blashfield (1984):

> *Cluster analysis methods have evolved from many disciplines and are inbred with the biases of these disciplines. This is important to note, because each discipline has its own biases as to the kinds of questions asked of the data, the types of data thought useful in building a classification, and the structure of classifications thought to be useful.*

This statement was made roughly 40 years ago; so just imagine the amplification of its impact today. References are inevitably biased toward what an author likes and knows (and doesn't like or know).

And the current size of the field precludes giving you a set of references that might be reasonably comprehensive on even the smallest subset of topics in this field. What I can do is indicate where some of these basic fields started, and (arguably, of course) who wrote the first paper, who wrote a survey worthy of that name, etc.

Exercises: It is not my intention to craft a set of problems worthy of a good textbook on these topics. But I do include a few exercises and problems at the end of most chapters. Some are just simple problems that involve definitions, theorems, proofs, and the like. Others are programming problems that will require a computer for solution. I have also included some much harder or more complicated exercises that may take you to a paper or book for explanation. As is always the case, the best way to use these problems is to sharpen your pencil and dig in. And, as Feynman said, you should enjoy solving problems!

Lastly: Changes in the technological landscape arrive so quickly these days that it is a never-ending task to stay on top of even a very small slice of the clustering pie. In the short time since the original version of this book was published in 2017 under the slightly different title "A Primer on Cluster Analysis: Four Methods that (Usually) Work," there have been quite a few important new developments in the clustering arena. Amongst the many possible additions and improvements to the original material in the first edition (including the elimination of some egregious typos and other mistakes!), I decided that the most useful updates would concern recent research about two topics: (i) acceleration and approximation schemes for clustering in big static data, now distributed throughout Chapters 6, 7, 8, and 11; and (ii) the rapidly growing field devoted to online algorithms for streaming data processing discussed in Chapter 12. Chapter 6 remains the heart of this volume: it has been extensively rewritten and includes a lot of new material. Chapters 8, 10, and 11 have also been updated and include some new examples and exercises. Chapter 11 has been renamed "Clustering in Static Big Data." The material on sampling has been expanded to include more theory and examples of approaches to approximation of clusters in Big Data based on a derivative of the MMDRS sampling strategy.

All of the algorithms discussed in Chapter 12 are called "streaming clustering" methods in the literature. However, most of the terminology used in this evolving field is co-opted from what I would call "classical cluster analysis," the topic of the first 11 chapters in both editions. But the semantics of classical cluster analysis is not readily adapted to stream processing algorithms. Chapter 12 (Structural Assessment in Streaming Data) emphasizes the distinction between static clustering and computations made by streaming algorithms. There will be some disagreement about my presentation in Chapter 12, but that is the way of science-eventually, a well-defined terminology will attach itself to this body of work, and we will all be the better for it. You may find my approach to the material in Chapter 12 controversial, but, at the same time, my hope is that it will stimulate your interest in pursuing research in this evolving and super-important field.

Acknowledgments: Since this is an elementary volume, most of the examples already exist in the open literature -many from papers I and/or my colleagues have published. One result of this is that the volume has a "homemade" look and feel. And, of course, it *is* homemade. But there are a few original examples which were coded and run by various friends and colleagues. Most notably, I want to thank Tim Havens (Michigan Technological University), Punit Rathore (MIT), and Dheeraj Kumar (IIT, Roorkee) for their assistance with several original examples. I would be remiss to not also mention several other contributors to the numerical work in some of the examples – namely, Thomas Runkler, Yang Lie, Masud Mostaghi, Mohammed Khalila, Wenrong Wu, Jeff Dale, Omar Ibrahim, Zahra Ghafoori, and Xiuyi Ye.

Let me offer special thanks to Dr. Nikhil Pal of the ISI in Calcutta. Nik offered to proofread the entire first edition (the Primer) of this book for me. He did. He found numerous typos and errors of fact. We had (and still have) different opinions about many of the non-factual items in this volume, but I often took Nik's advice and changed a statement or sentence to agree with him. I am profoundly grateful to him for his professional help and his personal friendship.

Many thanks are due to Nicki Dennis, the applied sciences publisher for River Publishing. She encouraged me to update and expand the first version of this book into its current form. Her help in the last two years has been pragmatic and inspirational. And editorial manager Junko Nakajima for River Publishing admirably shepherded the production of this volume through some difficult stages. My thanks to these two ladies.

And last, thanks are due and overdue to my wife Arlene, who puts up with me when it counts and who sacrificed a lot of hours together so I could finish this project. She is indeed my ears, my eyes, and my heart!

Disclaimer: As with all projects of this kind, the material presented is inevitably biased by the experience and knowledge of the author. Moreover, this is an enormous field, and the truth is that it is now far too large to even *know about* many important and useful methods that go unrecognized here. My writing is of course informed most heavily by my own work. I apologize for my bias and my ignorance and accept any and all blame for errors of fact and/or omission. How current is the material in this book? Well, you will certainly be able to find newer methods that have a modern ring and nice buzzwordy description with an Internet search-newer, but not necessarily better. That is how the science component of anything that is an art and a science works. Please send corrections, comments, and questions to: jcbezdek@gmail.com.

List of Figures

List of Tables

List of Abbreviations

Term	Definition
(↓)	Min-optimal validity function
(↑)	Max-optimal validity function
1-np	Nearest prototype
1G	First generation
AE	Autoencoder
AIC	Aikake information criterion
aka	"Also known as"
akFCM	Approximate kernelized FCM
AKM	Adapted k-means
AL	Average linkage (SAHN algorithm)
AM	Assessment model
AMM	Approximate Maximin (Sampling)
AO	Alternating optimization
ARI	Adjusted Rand Index (aka cRAND)
BI/E	Best internal/external (CVI)
BIRCH	Balanced iterative reducing and clustering using hierarchies
brFCM	Bit reduction fuzzy c-means
CEM	Classification EM
CF	Cluster footprint (summary statistics, aka Cluster Features)
CFA	Computational feature analysis
CL	Complete linkage (SAHN algorithm)
CLARA	Clustering large applications
CLARANS	Clustering large applications based on random search
CM	Clustering model
CNN	Computational neural network
CP	Candidate partition
CPOV	Computer point of view
cpu	Central processing unit
cRAND	Corrected Rand Index (aka ARI)
CS	Compact and Separated
CSF	Cerebro-spinal fluid
CWS	Compact and well-separated
CURE	Clustering using representatives

CVI	Cluster validity index
DAE	Deep autoencoder
DARPA	Defense advanced research procurement agency
DBI	Davies-Bouldin index
DBSCAN	Density-based spatial clustering of applications with noise
DCAD	Data capture anomaly detection
dec-iVAT	Decremental iVAT: sliding window model
dec-siVAT	Decremental iVAT: chunk-based model
DNN	Deep neural network
DI	Dunn's index
DO	Distinguished object
DS	Dempster-Shafer
EDM	Euclidean distance matrix
eFFCM	Extensible fast fuzzy c-means
EM	Expectation-maximization
EMST	Euclidean minimal spanning tree
EnFCM	Enhanced fuzzy c-means
eNERF	Extended NERFCM
EU	Existence-uniqueness
F4 or FF	Fantastic Four
ffIDCAD	Forgetting factor incremental data capture anomaly detection
FCM	Fuzzy c-means (model)
FCM/AO	FCM alternating optimization algorithm (A6.1, $m > 1$)
Fcn.	Function
FCQS	Fuzzy c-quadric shells
FCRM	Fuzzy c-regression models
FCV	Fuzzy c-varieties
FGFCM	Fast generalized fuzzy c-means
FONC	First-order necessary condition
GA	Genetic algorithm
GDI	Generalized Dunn's index
geFEM	Generalized fast EM
geFFCM	Generalized fast fuzzy c-means
geFHCM	Generalized fast hard c-means (k means)
Gen'l.	Generalized
GK	Gustafson-Kessel
GM	Gray matter
GMD	Gaussian mixture decomposition
GMD/EM	EM algorithm A7.1 for Gaussian Mixture Decomposition
GMG	Gradual model generator
GMM	Gaussian mixture model
GOFI	Goodness of fit index
GT	Ground truth

gpu	Graphics processing unit
HCM	Hard c-means (aka "k-means" model)
HCM/AO	HCM alternating optimization algorithm (A6.1, m = 1)
HFA	Human feature analysis
HPOV	Human point of view
IC	Information criterion
ICA	Independent component analysis
ICOMP	Information complexity
IDCAD	Incremental data capture anomaly detection
IDS	Intrusion detection system
IEEE	Institute of Electrical and Electronics Engineers
IFSA	International Fuzzy Systems Association
i.i.d.	Independent and identically distributed
iMMDRS	Incremental MMDRS
iNMMDRS	Incremental near MMDRS
inc-iVAT	Incremental iVAT: sliding window model
inc-siVAT	Incremental iVAT: chunk-based model
iSMF	Incremental stream monitoring function
ISODATA	Iterative self-organizing data analysis
iVAT	Improved VAT algorithm
k-nn	k-nearest neighbor (rule)
KCN	Potassium cynanide
kFCM	Kernelized fuzzy c-means
KKT	Karush-Kuhn-Tucker
LBG	Linde-Buzo-Gray
LDA	Linear discriminant analysis
LEM	Literal GMD/EM algorithm A7.1
LFCM	Literal FCM algorithm A6.1 (m>1)
LHCM	Literal HCM algorithm A6.1 (m=1)
LLE	Local linear embedding
LNERFCM	Literal non-Euclidean relational fuzzy c-means
LRT	Likelihood ratio test
LRLS	Linear regression with local sampling
LSE	Least squared error
LSL	Literal SL (Algorithm A8.2)
LVQ	Learning vector quantization
MDS	Multidimensional scaling
MEA	Multielectrode array
MF	Membership function
MI	Mutual information
ML	Maximum likelihood
MLE	Maximum likelihood estimation
MLP	Multilayered perceptron

MM	Maximin (sampling algorithm)
MMDRS	Maximin directed random sampling
MMDRS+	Approximate MMDRS sampling, full cycle RS
MMDRS*	Approximate MMDRS sampling, one cycle RS
MMP	Maximin principle
MOA	Massive online algorithms
mrFCM	Multi-stage random FCM
MRI	Magnetic resonance image
MS	Maximum sum (sampling algorithm)
MST	Minimal spanning tree
mvHCM	Multi-view hard c-means
NC	Necessary condition
NCOP	Non-linear constrained optimization problem
NERFCM	Non-Euclidean relational fuzzy c-means
NERHCM	Non-Euclidean relational hard c-means
NG	Neural gas
NIST	National Institute of Standards and Technology
NN	Neural network
NNR	Nearest neighbor rule
NPR	Nearest prototype rule
OEC	Online elliptical clustering
OEM	Online EM
oFCM	Online fuzzy c-means
oHCM	Online hard c-means
OG	Orthogonal
OKM	Online HCM (k-means)
p.s.d	Positive semi-definite
PA	Partition accuracy
PAM	Partitioning around medoids
PB-CVI	Pair-based cluster validity index
PC	Partition coefficient
PCA	Principal components analysis
PCE	Probably close enough
PCM	Possibilistic c-means
pD	p-dimensional, p = 1, 2, 3, etc.
PD	Positive-definite
PDF or pdf	Probability density function
PEM	Parametrized EM
PET	Positron emission tomography
PM	Prototype match
PR	Pattern recognition
PTA	Polynomial time approximation
PSO	Particle swarm optimization

RDI	Reordered dissimilarity image (aka cluster heat map)
RFCM	Relational fuzzy c-means
RHCM	Relational hard c-means
RHS	Right-hand side
RMS	Root mean square (error)
RP	Random projection
RS	Random sampling
RRP	Rogue random projection
rseFCM	Random sampling FCM = geFFCM
SAHN	Sequential, agglomerative, hierarchical, non-overlapping
SCA	Streaming cluster analysis
SC-MDS	Split and combine multidimensional scaling
sampKM	Random sampled HCM
scaleKM	Scaled (single pass) k-means
SDA	Streaming data analysis
sEM	Stepwise EM
SEM	Scalable EM
sHCM	Sequential hard c-means (MacQueen's k-means)
siVAT	Scalable iVAT
SL	Single linkage algorithm (A8.1; A8.2)
SMI	Standardized mutual information
SSQ	Sum of squared errors quality measure
specVAT	Spectral VAT
spFCM	Single pass fuzzy c-means
spHCM	Single pass hard c-means
SrEM	(Previous) short runs (of) EM
StEM	Stochastic EM
StreamsoNG	Streaming soft neural gas
sVAT	Scalable (sample) VAT
siVAT	Scalable (sample) iVAT
SVD	Singular value decomposition
SVM	Support vector machine
t-SNE	(t statistic) stochastic neighbor embedding
UCI	University of California at Irvine
UPGMA	Unweighted pair group mean average
VAT	Visual assessment of tendency
VEM	Vanilla EM (basic algorithm GMD/EM A7.1)
VFKM	Very fast k-means
VI	Variation of information index
VM	Validity model
VUI	Validating under the influence
wFCM	Weighted fuzzy c-means
wHCM	Weighted hard c-means

WGSS	Within group sum of squares
WM	White matter
WPGMA	Weighted pair group mean average
WPGMC	Weighted paired group median centroid
wrt	With respect to

Appendix A. List of Algorithms

Appendix D. List of Definitions

Appendix E. List of Examples

Appendix L. List of Lemmas and Theorems

Appendix V. List of Video Links

Part I

The Art and Science of Clustering

It's ok not to have all the answers in life ~ it's the willingness to search for the answers that counts

–Donna Gorman

Chapters 1–5 are devoted to basic topics and ideas that are shared by all classical clustering methods. Devising a useful solution that discovers cluster structure in a real problem requires some knowledge about a lot of things that might not be obvious to a casual user. There are some specific methods and algorithms in Part I, but they reside here mainly as examples which illustrate underlying notions about a topic that is important in the context of clustering. With this in mind, here is a preview of Part I.

Chapter 1 discusses the basic ideas underlying cluster analysis from the *human point of view* (HPOV). I firmly believe that most users of computational clustering algorithms (more or less) transfer what they think they know about clusters of real objects to computational counterparts built from data processing. But there are some very important differences between the human and computer points of view, and Chapter 1 paves the way for understanding them.

Chapter 2 discusses the foundations of fuzzy sets and models. The material in this chapter is quite basic and is needed to understand fuzzy clustering, which is covered in some detail in Chapter 6. Fuzzy models are compared to the other longstanding approach to modeling uncertainty – viz., probability theory. *Don't look for a winner* – they are both winners!

Chapter 3 is a counterpoint to Chapter 1 that discusses some ways to think about what it means to ask computers to look for clusters in numerical data; in short, how to think about clusters from the *computer point of view* (CPOV). Understanding the difference between the human and computer points of view is a great help when you are trying to make a computer find clusters.

Chapter 4 defines and illustrates the three main topics of batch (as opposed to streaming) cluster analysis: (tendency assessment, partitioning, and cluster validity). There is also a subsection about visual assessment of clustering tendency. Trying to understand what your data might contain before you look for clusters is much more useful than simply turning a clustering algorithm loose on your data and hoping for the best.

Chapter 5 is devoted to *feature analysis*, an important but somewhat under-represented part of the overall problem in the literature. Here, you will find a collection of techniques that attempt to condition unfriendly data so it is more amenable to CPOV cluster analysis. Many of the examples in this chapter use the ubiquitous Iris data, which is tiny, but has been analyzed so much in the past years that we have a good idea of how things will turn out with it. You will meet some very old techniques here, and a few new ones.

1

Clusters: The Human Point of View (HPOV)

An investment in knowledge pays the best interest

– Ben Franklin, Poor Richard's Almanac, c 1750

1.1 Introduction

There is a large difference between what humans and computers perceive as a set of clusters in either the real world or in the representations of it we build with data. Chapter 1 discusses the properties of a set of objects that people may see when describing them as a cluster. In this chapter, you will meet the seemingly awkward but very useful acronym HPOV, which stands for the "human point of view" (about clusters).

1.2 What are Clusters?

What does the word cluster mean to you? In English, cluster can be a noun or a verb. Here are some definitions:

The free dictionary (online) definition as a noun:

A group of the same or similar elements gathered or occurring closely together.

The Encarta® World English Dictionary definition as a verb:

To gather something into or form a small group.

The top Google hit for the query "What is a cluster?"

A grouping of a number of similar things.

These definitions all contain, either implicitly or explicitly, the three main ingredients that are used to define clusters: *similarity*, *closeness*, and *group*. Humans use the term "cluster" (and synonyms for it) in a naïve but completely understandable way. Examples of a *single cluster*: a bunch of grapes, a herd of cows, a school of fish, a cluster of stars. Examples of *sets of clusters* are as follows: political parties {conservative, liberal, and independent}; herds of animals {horses and cows}; soccer players for different teams {Manchester United, Juventus, and Bayern Munich}; stocks related to different fields {energy, computers, pharmaceuticals, and entertainment}.

Terminology : Clustering or classification? Here is the Encarta World English Dictionary definition of *classification* as a verb:

Classification is the allocation of items to groups according to type.

Much confusion about the difference between computational approaches to clustering and classification with numerical data stems from the obvious similarity between this definition and the ones just given for clusters. The Linnaean system of taxonomy, which has been with us for some 300 years or so, is called a classification system, but the "classes" of Linnaeus are what we call today a set of clusters. Most of the early work in computational cluster analysis was done by psychologists and numerical taxonomists, and both of these fields referred to the labeling of *finite* sets of objects as "classification." Well, providing labels for individual objects *is* classification. Indeed, the **Classification Society**, which is still quite active today, is devoted almost entirely to what most scientists and engineers now call clustering. No wonder we get confused! This confusion, which stems from competing uses of these two terms in natural language, will disappear when mathematical definitions are given for these two different activities.

What is the technical difference between computational clustering and classifier design? Classification involves the discovery of a classifier function, say **D**. The best function within a family of **D**-models is constructed or identified using a finite set of "training data." The value **D**(*) is a class label for any object (*) represented by the input data, and once **D** is known, this function provides a way to label *all* the (usually infinitely many) objects in its domain. Clustering, on the other hand, can only assign algorithmic labels to a *finite set* of objects and can't be directly generalized to produce labels of new objects that were not part of the processing. So, the real difference is one of scope -clustering finds some of the labels, and classification finds all of them.

Here are some synonyms for the three key words that characterize clusters:

similar (noun): resemblance, likeness, correspondence, match, relationship
close (adjective): compact, tight, concentrated, dense, packed
group (noun): collection, cluster, set

Similar: We say things are similar when they are identical or nearly identical -in the same political party, nearly the same religion, similar sweetness, matching textures, almost the same size, roughly the same color, etc. Similarity is the *degree of approximation of likeness* (more simply, the degree of likeness) between the objects being compared. Guitar notes may sound identical (the same), nearly the same, roughly the same, somewhat the same, not alike, etc. So, *similarity* (and its complement, *dissimilarity*) encompasses the idea of a *relationship* between (usually pairs of) things. Humans use the idea of similarity to identify *membership in clusters*: e.g., clusters of grapes, star clusters, clusters of admirers, and so on. The notion of similarity applies to a single pair of objects (one horse; one cow) but is easily extended to pairs of clusters of objects – flocks of ducks are more similar to flocks of geese than they are to herds of horses.

Close: In English, this term can be a verb (to close, or shut), adjective (dense), or noun (to end, or conclude); so it has many uses, and its relationship to our use of the term *similar* is often confusing. When we say objects are *close*, we almost always mean either (i) close in their relationship to each other or (ii) close with respect to some measure of distance. Thus, first cousins are closer to you than second cousins (in relationship), whereas Mexico is closer to the USA than Brazil (in physical distance). Distance between objects in a cluster is related to the measurement scale. For example, the stars in the cluster we call the "big dipper" look close to each other from our observation point on earth but are millions of miles apart from each other in space.

Don't confuse *proximity (or nearness) with similarity*. Similar objects can be so far apart in distance that they don't invite being called a cluster, e.g., herds of elephants in India and Africa. Conversely, objects of different types (dissimilar relationships) can be so close together in distance that it becomes hard to separate them into distinct clusters, e.g., horses and mules in a shared barnyard.

Group: Group may be a verb or a noun. As a verb, clustering means to form a group of similar objects. As a noun, a cluster specifies a group of objects. A *collection* of c clusters of a finite number n of real objects that are related in some way is a *c-partition* of the underlying set of objects. Without reference to any precise definition, we know that lions, tigers, and leopards comprise $c = 3$ clusters of predatory cats or that a handful of grapes and cherries are two clusters of small fruits which are somewhat similar but not the same.

Combining our definitions of cluster and similarity, we also know that the grapes in a cluster are more similar (or less dissimilar) to each other than they are to all the cherries. Moreover, you will probably agree that clusters of grapes and cherries are more like each other than clusters of, say, grapes and elephants.

You might mistake a cherry for a grape and place it in the "wrong" cluster, but this is an inconceivable mistake when the objects are cherries and elephants. The idea of collections of similar objects (clusters) matches our natural perception of the world. Later, we will see that the notion of "hybrids" may suggest modifications to this pleasant and very natural state of things.

Conflicts arise from different interpretations of the three words: similar, close, and group. Figure 1.1 emphasizes this for the word "group." According to our notion of a cluster, *one object should never be called a cluster* since the word "group" implies more than one object; so most observers would not call the single dot in view 1.1(a) a cluster[1].

Figure 1.1(b) is more like our intuitive idea of a cluster. And if a single point can't be a cluster, view 1.1(c) contains only one cluster. The observations in Figure 1.1(c) are usually described as one cluster with one "outlier" or "noise point" (the single blue dot). Implicit in the idea of a group is the idea of its count or cardinality (the number of objects in the set). *Cluster size* refers to the number of distinct objects in the set. By this definition, the size of the cluster in Figures 1.1(b) and 1.1(c) is 12. But there are other ways to define the size of a cluster, e.g., its diameter, its volume, and so on. When you encounter this term, be sure to know what definition of cluster size is being used.

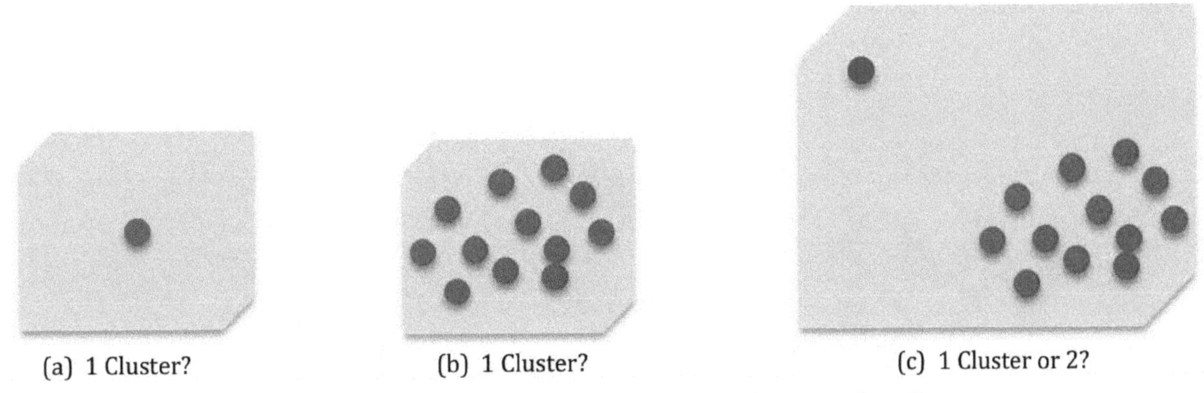

Figure 1.1 What is a group? When does a group become a cluster?

[1] Later, we will encounter a technical reason that requires singletons to be regarded as "[single linkage] clusters," but for now, we will use the literal definition.

Figure 1.2 illustrates closeness due to physical distance. The set of points in view 1.2(a) are close enough to each other that most observers would agree to call them a cluster of points. What about the points in view 1.2(b)? The distance between points here is sufficiently large to prevent many viewers from calling them a cluster. But this statement hides an implicit assumption, namely, that the points are not a cluster *at this scale*.

If these points were seen from a more distant vantage point, they would appear to be clustered like Figure 1.2(a) to us. A related concept that people use to characterize clustering is the *density* or *number of objects per unit of volume* of the collection.

Figure 1.3 illustrates how closeness arises due to relationships. The objects in set X_1 in Figure 1.3(a) form one cluster of animals – or is it two? The objects in set X_2 in Figure 1.3(c) form one cluster of birds – or is it two? Some observers will say there is only one cluster (8 birds); other observers will say there are two clusters (4 chickens and 4 ducks). So, HPOV clusters are hardly unique. There is a certain amount of subjectivity due to the bias and assumptions made by an individual about what constitutes a cluster.

If we wanted to relate object x in Figure 1.3(b) to one of these two groups, which would it be? Most observers would say that x has more in common with the animals in X_1 since all of those animals have 4 legs, 2 ears, and fur hides. But x is not one of the resident objects in X_1 – it is just more closely related to objects in X_1 than it is to those in X_2. In other words, a cow seems more similar (and hence, "closer") to rabbits and dogs than it does to chickens and ducks.

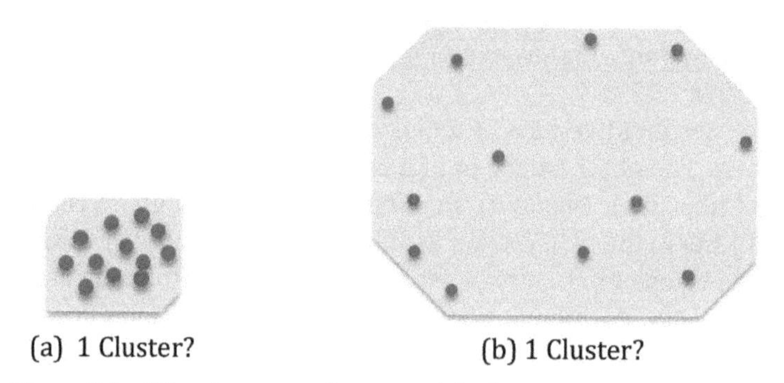

Figure 1.2 When is a group close enough in distance to become a cluster?

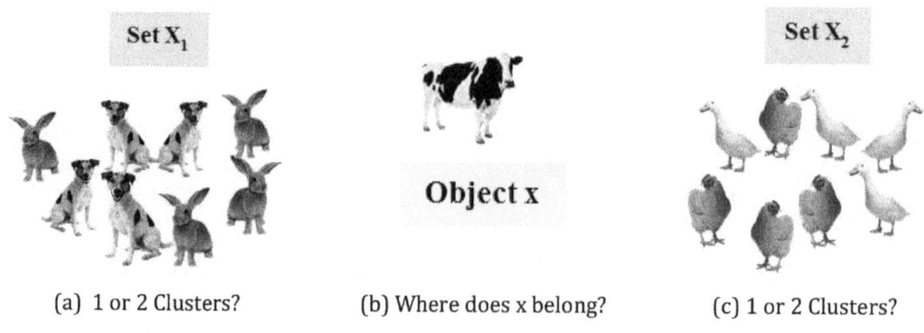

Figure 1.3 When is a group of closely related objects a cluster?

Another characteristic of points in a cluster is their arrangement (distribution, or shape). Figure 1.4 exhibits groups of similar, close objects that have the added distinction of differing shapes. This is a very useful property that humans easily perceive that computers struggle to recognize and imitate. How many clusters do you see in Figure 1.4(d)?

Groups of objects can be also be perceived as clusters due to a pattern in the arrangement. Consider some sequences of observations of coin flips that can land "heads" = H or "tails" = T on any toss. Figure 1.5 illustrates this for several sets of consecutive observations.

How many clusters do you see in view 1.5(a), left versus right? What about 1.5(b) and 1.5(c)? The least contentious views in Figure 1.5 are probably those in 1.5(c), where most people would agree that there are three clusters on the left, and two on the right (remember, a single observation can't be a group). These groups are more readily seen because they have some isolation (separation) from other clusters.

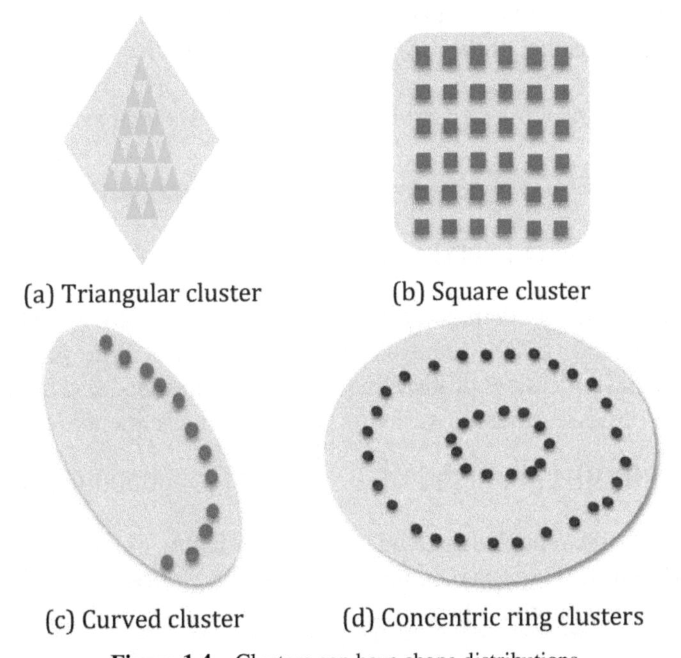

Figure 1.4 Clusters can have shape distributions.

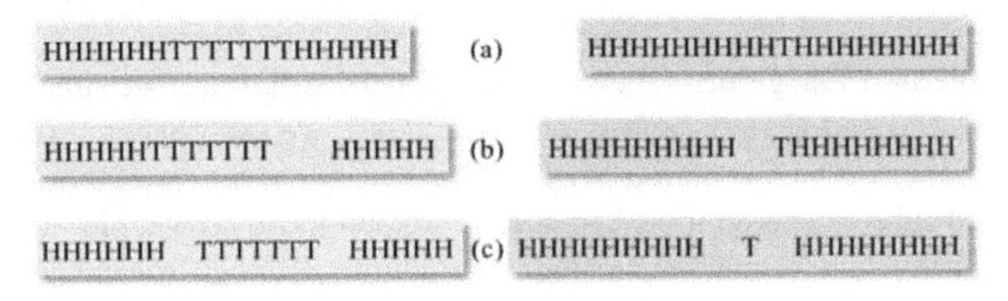

Figure 1.5 Clusters due to patterned arrangements.

Suppose the outcome of 20 coin tosses are the sequence

HHHHHHHHHHHHHHHHHHHT (A)

We are not interested in the *probability* of observing this sequence, but rather, in the visual perception it creates -whether it appears to be "structured." Sequence (A) can also be represented as a binary valued vector (say, H = 1 and T = 0); in this scheme, (A) becomes (11111111111111111110). Humans are left with the same visual impression from either representation; computers will almost always work on the bit string. Do you see a substructure? If someone asked you how many "clusters" were contained by this sequence, would you say c = 1 or c = 2? Or would you respond with the question – "what is a cluster?" In statistics, each subsequence of like symbols such as the 19 heads or 1 tail in (A) is called a *run* (or cluster). The number of adjacent like symbols is called the *length* of the run (size of the cluster). In the statistical context, (A) contains c = 2 clusters, and the boundary between them is easy for humans to see. The single "T" at the end of (A) seems at odds with our intuitive idea about a cluster being a group (more than one), but in the *experimental* context, it is not a statistical outlier. And, according to our discussion about what comprises a cluster, the "T" is not one.

If we rearrange the symbols in (A) while preserving their relative frequencies, the resultant visual effect is striking:

HHHHHHHHHTHHHHHHHHHH; (B)

11111111101111111111. (C)

In these sequences, you might decide that there are c = 3 clusters (or is it c = 2 with one inlier?) with sharply defined boundaries. In the statistical context sequence, (B) or (C) *seems* more probable than that in (A) (even though these two events have identical probabilities in a fair game) because of its visual appearance. Apparently, *context influences* our assessment of structure. Consider the sequence:

HHHHHHHHHHTTTTTTTTTT (or 11111111110000000000). (D)

Again, there seem to be c = 2 apparent clusters, and there will probably be much less disagreement about this than a similar statement about (A) because the boundary between the two runs of equal length is much more apparent in D. Finally, what about the sequence

HTHTHTHTHTHTHTHTHTHT (or 10101010101010101010)? (E)

Would you say (E) contains 20 clusters? Probably not, and if you agree that clusters comprise groups, you should say there are no clusters in (E). But (D) and (E) both contain ten observations from each of the two classes and are equiprobable in a fair game. What accounts for the difference in our visual perception in these two examples? While (D) and (E) are both highly structured, we see clusters in (D), but something more like *texture* in (E). As the length of a run relative to its background (the rest of the sequence) increases, the boundary between it and subsequent symbols becomes more and more noticeable.

Our perception of similarity, and, therefore, cluster substructure, *changes* as each of these sequences is presented. From this, we infer that a first principle for humans grouping objects involves *perceptual similarity* between pairs of objects *AND* their relationship to their background. Another point about the difference between (D) and (E) is that (D) exposes structure in (E) by simply rearranging the objects in the data.

This doesn't make sense in the context of sequences of coin flips, but the notion of rearrangement will be very important when we discuss visual assessment of clustering tendency in Chapter 4.

In summary, this chapter has been about seeing clusters from a *human point of view* (HPOV). The two most important characteristics of objects that humans describe as clusters are *compactness* and *separation.* In Chapter 3, the spotlight will be directed toward clusters in numerical data, found by computational means. Our goal there will be to describe clusters from the computer's standpoint, i.e., from a *computer point of view* (CPOV). Prior to that, however, we take a short diversion to discuss fuzzy sets and models, which will appear often in the remainder of the book.

1.3 Notes and Remarks

The main point of this chapter is that the basic ideas underlying the words *similar*, *close*, and *group* (*compactness* and *separation*) for humans must somehow be carried over to what we want them to mean for sets of data residing in computers. And the fact that we do disagree about what we see in the simple figures in this chapter tells us not to be very surprised if algorithms disagree about substructure in the data they process.

Starting this way focuses attention on the semantic meaning of the words used by the scientific community when they write about cluster analysis. Arguments abound in the scientific literature about what a "cluster" is; so it seems important to first understand how we use this term in the broader context of natural language.

Humans have always depended on their ability to process visual imagery from the world they live in and to cluster inputs into useful categories that help them in almost every imaginable way. We can certainly imagine that cave dwellers on the hunt in prehistoric times depended on recognizing outliers to a cluster of mammoths. In a more modern context, we immediately recognize clusters and patterns in sets of cars and trucks in the parking lot of a shopping mall, and we depend on our ability to classify (and thus find) our own vehicle in this apparent maze of visual input data.

It is hard to know when HPOV clustering was first coupled to computational schemes that might help humans with this task. A famous early example of HPOV clustering is the work reported by Petrie (1899), who used five methods to cluster 917 pieces of prehistoric Egyptian pottery from 4000 overlapping graves that spanned (approximately) the years 5800 to 4800 BC. His description of the five methods:

The next work is to place these as far as possible in the original order of the graves. For this, there are five methods, based on the following considerations:

1st. Actual superposition of graves or burials; but rarely found.
2nd. Series of development or degradation of form; very valuable if unimpeachable.
3rd. Statistical grouping by proportionate resemblance; the basis for classifying large groups.
4th. Grouping of similar types, and judgment by style; giving a more detailed arrangement of the result of the third method.
5th. Minimum dispersion of each type, concentrating the extreme examples.

Note especially that methods 3 and 5 employ computational approaches at a time when slide rules were the most advanced computing tools of the scientific community. Petrie's statement contains all the key words of HPOV clustering: similar, groups, and closeness (minimum dispersion). Figure 1.6 displays some of the reported results.

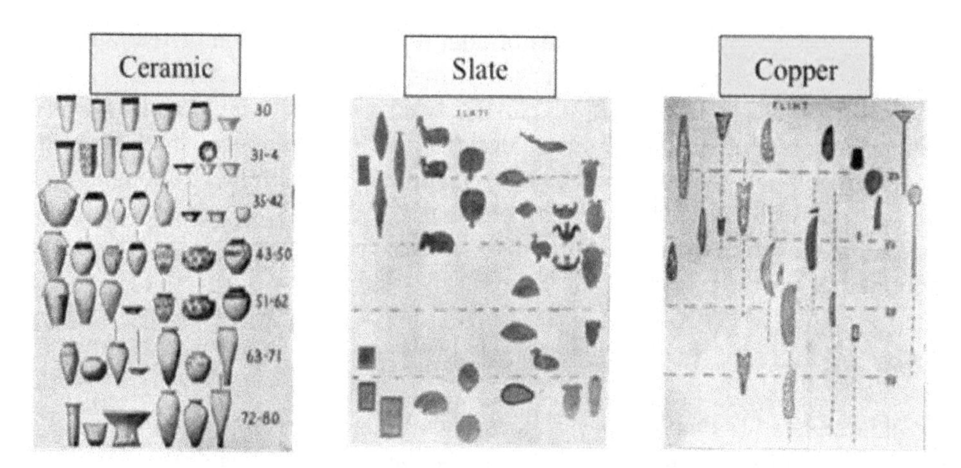

Figure 1.6 Petrie's arrangement of ancient Egyptian pottery.

This illustrates all of the elements of HPOV clustering discussed in Chapter 1 and leads us into the main question of Chapter 3: how to make computers do roughly the same thing using numerical measurements of the objects to be clustered?

1.4 Exercises

1. X is a simple deck of 52 cards, and integer c is the number of clusters in X. How many clusters are in this deck? Look at the possible questions about different properties for cluster membership.
2. Star clusters are groups of stars held together by a common gravitational bond. They vary greatly in size, shape, and age as well as the number of stars. Many of these clusters can be seen with the naked eye on a dark night. Astronomers have divided star clusters into open clusters or globular clusters according to their shape and number of stars. See Figure 4.3 for some examples.
   (a) Do you think there is a numerical representation of stars that would enable us to find clusters of them, which agree with astronomer-labeled clusters?
   (b) Comment on the possible utility of 2D vectors with components (visual intensity and distance from earth) as useful features to represent and analyze star clusters.
   (c) Are there other features that might make clustering stars easier for a computer?
3. (a), (b), and (c) are three sequences of 20 observations each of the results of flipping a coin (H = heads and T = tails). Do any of the sequences have natural groups (clusters)? Which sequence has the most structure? Which has the least? Try to justify your answers.
   (a) HHHHHHHHHTHHHHHHHHHH
   (b) HHHHHHHHHHTTTTTTTTTT
   (c) HTHTHTHTHTHTHTHTHTHT
4. Figure 1.7 is a data set that spells the name "Yang." We will cluster this data set with the fantastic four in Chapter 9. How many HPOV clusters do you see in this data set? Do the colors of the letters affect your answer?

5. Figure 1.8 is a cartoon rendering of 12 signals that an umpire in the game of Cricket makes to indicate decisions concerning bowling actions during a match. For example, (0) corresponds to "cancel ball," (1) to "dead ball," and so on, up to (11) for "wide."
   There are two visually evident HPOV clusters in this set of 12 gestures. What are they? Can you see any element of this set that might form a bridge between the two more distinct subsets? See Datta *et al.* (2019) for a CPOV solution.
6. Figure 1.9 is a real photograph of several types of African animals. This illustrates the idea of a *mixture* of data types. How many clusters do you see? If a singleton can't comprise a cluster, it is called an anomaly. Do you see an anomalous animal in this picture? This figure will pop up again in Chapter 12.

Figure 1.7 Scatterplot of a data set that spells the name "Yang."

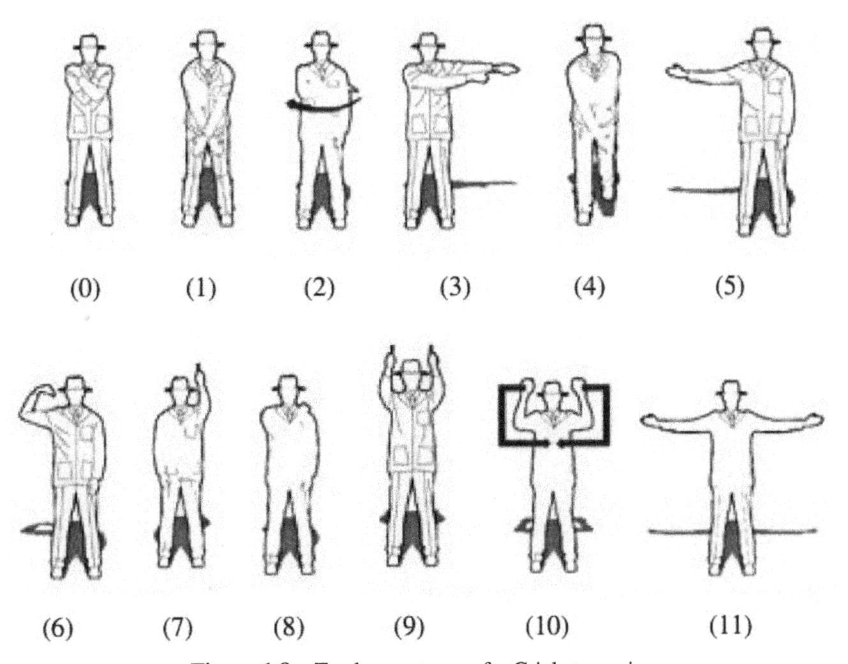

Figure 1.8 Twelve gestures of a Cricket umpire.

Figure 1.9 Clusters of African animals.

2

Uncertainty: Fuzzy Sets and Models

I can't keep from fooling around with our irrefutable certainties. It is, for example, a pleasure to knowingly mix up two and three dimensionalities, flat and spatial, and to make fun of gravity

–M. C. Escher

2.1 Introduction

There are three types of static clustering models and algorithms to optimize them in this book: deterministic (hard c-means (HCM) for object data and single linkage for relational data), probabilistic (Gaussian mixture decomposition (GMD) using the *expectation-maximization* EM algorithm), and fuzzy (fuzzy c-means (FCM) for object and relational data). Each of these four models has advantages and drawbacks. Many of you have learned (at least) univariate probability theory and mathematical statistics. Fewer readers have been exposed to fuzzy sets theory. Consequently, this chapter is devoted to a brief discussion of those topics in fuzzy sets and models that are germane to fuzzy clustering. Specific topics in probability will be discussed as they occur.

The importance of accounting for uncertainty in mathematical models is highlighted by the following quotes, excerpted from Feynman (1989) in his comments on the reliability analysis of the Challenger space shuttle:

> *A mathematical model was made of the erosion. . . . Uncertainties appear everywhere in this model. How strong the gas stream might be was unpredictable; .... Blowby showed that the ring might fail, even though it was only partially eroded. The empirical formula was known to be uncertain, for the curve did not go directly through the very data points by which it was determined. . . . Similar uncertainties surrounded the other constants in the formula, etc., etc.*

Feynman concluded:

> *When using a mathematical model, careful attention must be given to the uncertainties in the model.*

Feynman's assertion that "Blowby showed that the ring *might* fail, even though it was only *partially* eroded" (my italics) possesses uncertainty due neither to chance nor to incorrect measurements. Instead, this is fuzzy uncertainty due to imprecise natural language. Feynman was not advocating the use of a fuzzy model; the quote simply illustrates that scientists accustomed to precise calculations routinely use vague language, even during discussions about uncertainty in models. Sources of uncertainty that must be accounted for in mathematical models fall into at least three categories:

Imprecision in data collection: Data possess uncertainty due to errors, blunders, missing and incomplete measurements, and resolution limits due to the precision of the measuring device. It would be nearly impossible to measure your height to 0.001 cm using devices readily available in your home. Sensors always have a finite resolution, which bounds the certainty about the measurements it offers.

Outcomes that depend on chance: Statistical uncertainty arises when the outcome of a process is truly random – e.g., the result of flipping a fair coin or rolling an unbiased pair of dice.

Vagueness in natural language: Vague language causes a third type of uncertainty called linguistic imprecision. This type of uncertainty appears in almost every discourse between humans. "How much money do you have? About 4 dollars."

Fuzzy models are the primary tool for representation of linguistic imprecision. The basic idea of fuzzy sets is easy to grasp. Suppose, as you approach a red light, you must advise a driving student when to apply the brakes. Would you say "begin braking 31.24 meters from the crosswalk?" Or would your advice be more like "apply the brakes pretty soon?" The first instruction is too precise to be implemented; so precision in this case is quite useless, while vague directions such as commonsense advice can be interpreted and acted upon. Everyday language is one example of ways vagueness is used and propagated. Children quickly learn how to interpret and implement fuzzy instructions ("get to soccer practice early tomorrow"). Computational models of real systems should also be able to recognize, represent, manipulate, interpret, and use (act on) both fuzzy and statistical uncertainties.

2.2 Fuzzy Sets and Models

Crisp sets (conventional or hard sets) contain objects that satisfy precise properties required for membership. The set of real numbers H from 0 to 2 is crisp; we write $H = \{r \in \Re \mid 0 \leq r \leq 2\}$. Equivalently, H is described by its *membership function* (MF), also called a characteristic or indicator function, $m_H : \Re \rightarrow \{0, 1\}$,

$$m_H(r) = \begin{Bmatrix} 1; & 0 \leq r \leq 2 \\ 0; & \text{otherwise} \end{Bmatrix} \tag{2.1}$$

The crisp set H and the graph of m_H are shown in the left side of Figure 2.1. Every real number (r) either is in H or is not. Since m_H maps all real numbers onto the two points {0, 1}, crisp sets correspond to 2-valued logic – is or is not, on or off, black or white, 1 or 0.

In logic, values of m_H are called truth values with reference to the question "is r in H?" The answer is yes if and only if $m_H(r) = 1$ otherwise, no. If we know the set H, we know the function m_H, and conversely. The set theoretic and function theoretic descriptions of this set are *isomorphic*, indicated by $H \leftrightarrow m_H$.

How would you describe (mathematically) a set F of real numbers that are *close to one*? Fuzzy sets were introduced by Zadeh (1965) as a mathematical way to represent this type of linguistic vagueness. Since the property *close to 1* is vague, there is not a *unique* membership function for F. Rather, the modeler must decide, based on the potential application and properties desired for F, what m_F should be. Plausible properties for F include: (i) *normality* ($m_F(1) = 1$); (ii) *unimodality* (only $m_F(1) = 1$): (iii) *monotonicity* (the closer r is to 1, the closer $m_F(r)$ is to 1, and conversely); and (iv) *symmetry* (numbers equidistant to r from the left and right of r should have equal memberships).

Given these intuitive constraints, either of the functions shown on the right side of Figure 2.1 might be a useful representation of F. m_{F1} is continuous but not smooth at 1 (the triangular graph); m_{F2} is discrete (the staircase graph). You can construct an MF for F so that *every* number has some positive membership in F, but you would not expect numbers such as 1.34×10^{33}, for example, to have a membership value in F that is much larger than zero! One of the biggest differences between crisp and fuzzy sets is that crisp sets always have unique MFs, whereas *every fuzzy set has an infinite number of MFs* that may represent it. This is both a weakness and strength; uniqueness is sacrificed but gives a concomitant gain in terms of flexibility, enabling fuzzy models to be adjusted for maximum utility in a given situation.

In conventional set theory, sets of real objects such as the numbers in H are equivalent to, and isomorphically described by, a unique membership function such as m_H. However, there is no set-theoretic equivalent of real objects corresponding to a fuzzy set m_F. Fuzzy sets are always (and only) *functions*, from a universe of objects, say X, into [0,1]. This is depicted in Figure 2.2: *the fuzzy set is the function m* that carries X into [0,1]. The membership function is the key idea in fuzzy set theory; its values measure degrees to which objects satisfy imprecisely defined properties. Thus, $m(x_k) = 0.6$ in Figure 2.2 means that x_k shares the imprecise property defining m to the extent 0.6. Note especially that this value is interpreted as the *similarity* of x_k to the defining property of m. Specifically, $m(x_k)$ is *not* the probability or likelihood that x_k belongs to m.

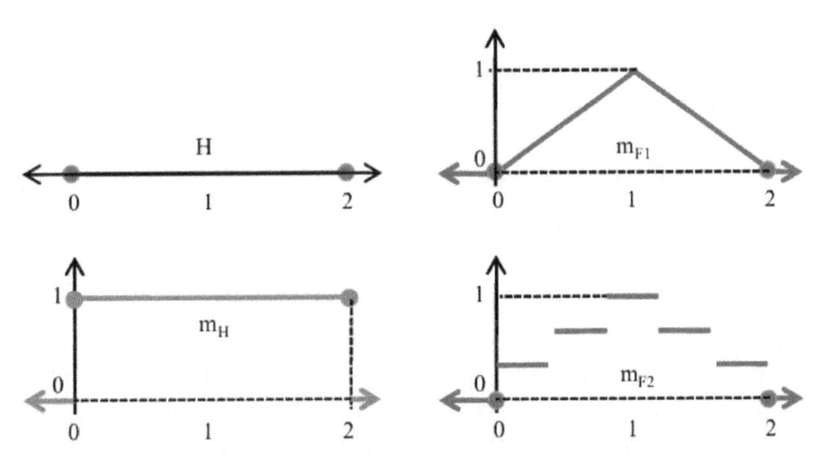

Figure 2.1 (Left) H = numbers between 0 and 2: (right) F = numbers near 1.

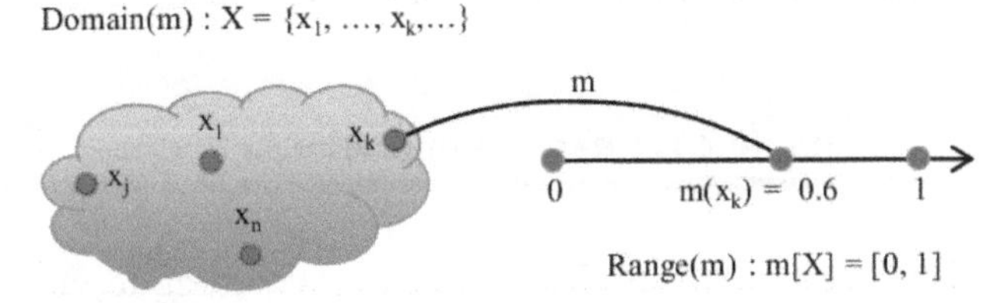

Figure 2.2 Fuzzy sets are membership functions $m : X \rightarrow [0, 1]$.

Definition 2.1. Fuzzy sets: Zadeh (1965).

X is any universe of discourse. A function $m : X \longmapsto [0, 1]$ is a fuzzy subset of U.

Every function $m : X \longmapsto [0, 1]$ is a fuzzy set in a *formal* mathematical sense. But many functions valued in [0,1] can't be interpreted as realizations of conceptual fuzzy sets. Functions that map X into the unit interval may be fuzzy sets but *are* fuzzy sets when and only when they match some intuitively plausible semantic description of imprecise properties of the objects in X. Zadeh (1965) proposed five operations that define an algebra for fuzzy sets (this was first, but there are now many different notions for these five operators).

Definition 2.2. Set operations for fuzzy sets: Zadeh (1965).

Let m_A and m_B be fuzzy subsets of a universe X. *Equality, containment, complementation, intersection, and union* for m_A and m_B are:

$$(=)\ \text{Equality}: A = B \Leftrightarrow m_A(x) = m_B(x); \tag{2.2a}$$

$$(\subset)\ \text{Containment}: A \subset B \Leftrightarrow m_A(x) \leq m_B(x); \tag{2.2b}$$

$$(\sim)\ \text{Complement}: m_{\widetilde{A}}(x) \leq 1 - m_A(x); \tag{2.2c}$$

$$(\cap)\ \text{Intersection}: m_{A\cap B}(x) = m_A(x) \wedge m_B(x); \tag{2.2d}$$

$$(\cup)\ \text{Union}: m_{A\cup B}(x) = m_A(x) \vee m_B(x). \tag{2.2e}$$

Since m_A and m_B are functions, these operations are defined pointwise – that is, they must hold for each x in the common domain of m_A and m_B. A common misunderstanding about fuzzy models is that they are offered as *replacements* for crisp models. But every crisp set is fuzzy by definition, and so fuzzy sets don't *replace* crisp sets; rather, they *enlarge* the axioms of conventional set theory. Most schemes that use fuzziness use it in this sense of *imbedding* crisp models. We work at preserving conventional structures, and letting them dominate the output whenever they can, or whenever they must. Verification that properties (2.2) retain their conventional meanings when A and B are crisp sets is left as an exercise. Example 2.1 shows how to interpret complementation, intersection, and union relative to the definitions given in (2.2c)–(2.2e).

Example 2.1. Fuzzy set operations.

Figure 2.3 is a photograph of the majestic Atlantic Bluefin Tuna. According to the Sea Shepherd Conservation Society, a mature adult has an average length ranging from 6.58 to 8.16 feet, with a maximum of 21 feet. They give the average weight as 1200 lbs. in the 1970s, which had dropped to 600 lbs. in 2013, up to a maximum of 1600 lbs.

Figure 2.3 The Bluefin Tuna.

Overfishing between 1970 and 1998 has resulted in a decrease of 70% in this fishery. Why? There are more people on our planet, which has increased pressure on all species of fish to supply us with food. Sushi restaurateur Kiyoshi Kimura paid $736,000 for a 593 pound Bluefin tuna in 2012 and $1.76 million a 488 pound Bluefin in 2013. That is around $3600 *per pound*. Ken Fraser caught a 1496 lb. Bluefin off the coast of Nova Scotia in 1979, long before its commercial price soared to present day values. According to the National Geographic magazine, the average length of the Atlantic Bluefin Tuna is now about 6.5 feet (the Paedocypris at 0.31 inches is perhaps the shortest fish known, while the oarfish is known to grow to 50 feet or so, but neither of these finds its way to sushi bars).

Let X = {Bluefin tuna}; x = length of any tuna f∈X; and RT = [0, 21] = range of lengths of all (known) tuna. Figure 2.4 shows membership functions for two fuzzy subsets of X : m_A = {lengths of fish close to 3 feet long} and m_B = {lengths of fish close to 7 feet long}. Notice how the membership functions chosen seem to mimic the physical situation. Neither is symmetric, both drop to 0 at 21', and both are normal at the nominal length at which the defining property is maximally satisfied. The graph of the membership function for "about 3 feet" dives to 0 faster than the graph for about 7 feet since 3 is further from 21 than 7.

Values for various operations are taken from the graph in Figure 2.4:

- $m_A(4) = 0.8$ = extent to which a 4′ fish is about 3′
- $m_B(4) = 0.6$ = extent to which a 4′ fish is about 7′

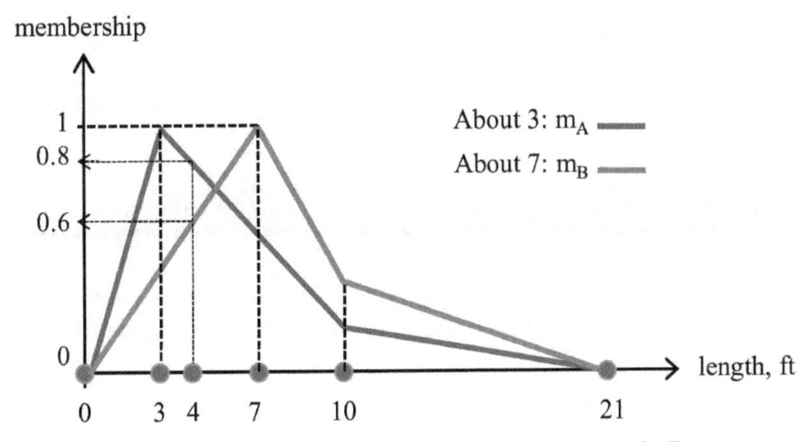

Figure 2.4 Membership functions m_A and m_B for a Bluefin Tuna.

- $m_{\hat{A}}(4) = 1 - 0.8 = 0.2$ = extent to which a 4′ fish is **NOT** about 3′
- $m_{A \cap B}(4) = \min\{0.6, 0.8\} = 0.6$ = extent to which a 4′ fish is about (3′ **AND** 7′)
- $m_{A \cup B}(4) = \max\{0.6, 0.8\} = 0.8$ = extent to which a 4′ fish is about (3′ **OR** 7′)

The fuzzy set operations of complementation, intersection, and union have their usual logical meanings of **NOT**, **OR**, and **AND**, as illustrated here. But as the membership functions change, so do the values obtained for **NOT**, **AND**, and **OR**. We leave the graphs of $m_{\hat{A}}$, $m_{A \cap B}$ and $m_{A \cap B}$ to the exercises.

Many writers have proposed other functions – in fact, infinite families of them – for, $(\hat{\bullet}, \cup, \cap)$ but the ones at (2.2) are used most often in real applications. Be careful about logical negation, which is *not* the same as the concept of a linguistic antonym. For example, *not old* is not necessarily the same (fuzzy set) as *young* (e.g., an alternative is new). This is an important distinction in models that depend on semantic definitions of imprecise properties. Typical questions about fuzzy sets are discussed next.

2.3 Fuzziness and Probability

Q1. What are membership function values? The value $m_A(4) = 0.8$ (Figure 2.4) tells us that 4 is *fairly close* (!) to 3. Thus, 0.8 is a measure of the *similarity* of the number 4 to others that possess the imprecise property that m_A represents. $m_A(4) = 0.8$ doesn't tell us anything about the *probability* of 4 being pretty close to 3.

Q2. Membership functions – where do they come from? Well, where do *probability density functions* (PDFs) come from? Two sources – data (objective) and people (subjective). Data, for example, are used to estimate the means and covariances of the component densities in a mixture of normal distributions. Similarly, the fuzzy c-means clustering algorithm discussed in Chapter 6 is used to find a fuzzy partition (whose rows are values of membership functions).

But what of the normal distribution itself? Where did *it* come from, and what is the justification for its use? Is using the normal distribution somehow different than adopting a fuzzy set because it seems to provide a reasonable and useful model of some process being described? ***NO***. The normal distribution originated with DeMoivre in 1744, and we use it because it often fits data from the physical world. *All* PDFs were introduced by people trying to model real processes, just as many MFs are introduced by humans to model imprecision. So, we *do* get PDFs and MFs in the same ways, and we choose them for the same reasons – as models of physical situations that help us understand, predict, and control some process of interest. Here is an example that shows how easily we can obtain membership functions from little people and big people too.

Example 2.2. Humans provide fuzzy membership functions.

Pediatric nurses use a device like the one shown in Figure 2.5 (there are dozens of them like this one – query "visual pain scales" on the Internet) to ascertain the amount of pain that a young child is experiencing. The child is asked to indicate how he or she feels by choosing one of the six faces on the scale. The numbers on the Wong-Baker (Wong, 2015) pain scale beneath the faces are used by caregivers/health professionals to quantify the pain.

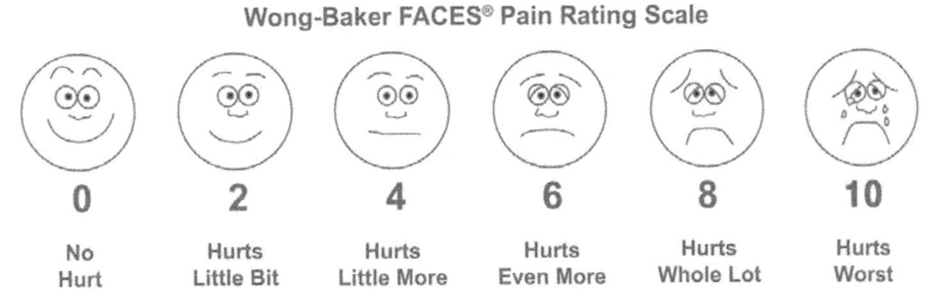

Figure 2.5 A discrete interactive pain scale for self-assessment of pain.

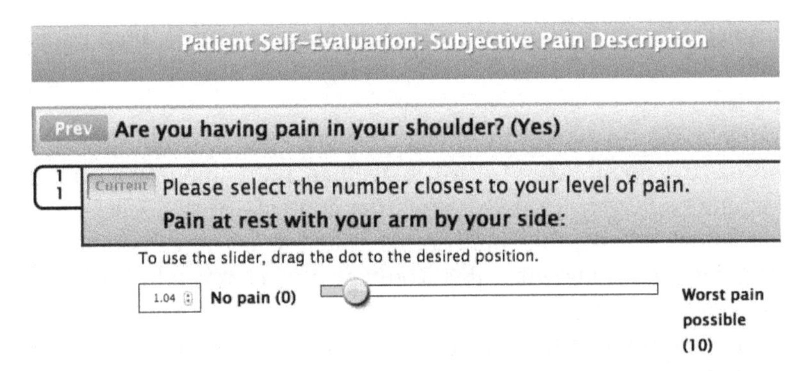

Figure 2.6 A continuous interactive pain scale for self-assessment of pain.

The leftmost face has the value 0, the second face has the value 2, and so on, up to the rightmost face, which has value 10. When divided by 10, these values represent a discrete fuzzy membership function for the fuzzy set F = {children with pain}. Note that the semantic descriptors assigned to each face are wonderful examples of exactly the type of linguistic imprecision that fuzzy sets are designed to capture: 0 is "no hurt"; 4 is "hurts little more"; 10 is "hurts worst." There are no faces for the integers 1, 3, 5, 7, and 9. This encourages young children who understand counting to use interpolation if they don't see a face they like but feel "somewhere in-between." An amusing coincidental fact about Figure 2.5 is that it is essentially mimics a clustering method called Chernoff faces (Chernoff, 1973).

Figure 2.6 shows an online version of this type of data collection used by many health care providers, which illustrates the collection of values of a continuous fuzzy membership function. Patients are asked to slide the cursor to the point that shows how much pain they are in. The mouse-controlled slider provides a means for self-assessment of pain to the nearest 0.01 – note the selection in the box, 1.04. As shown, 0 means "no pain," 10 means worst pain – "sedate me right now!" When normalized by division by 10 to the range [0, 1], each choice can be interpreted as a value of a continuous membership function for the fuzzy set F = {people in pain}.

The most important point of this example is that a question such as "how likely is it that you are in pain?" makes no sense whatsoever. That is, no probabilistic model would really accomplish what a health care provider is trying to learn by administering this self-assessment to a patient.

Q3. Isn't fuzziness just a clever disguise for probability? No! To see this, consider Example 2.3, which should convince you that the answer is definitely not. This example was first published in Bezdek and Pal (1992) and subsequently discussed by Woodall and Davis (1994) and Bezdek (1994).

Example 2.3. Memberships vs. probabilities -you decide what to drink.

Let U be the set of all liquids in the universe, and L = {all potable (= suitable for drinking) liquids}. L is a fuzzy subset of U. You have been abducted by aliens who have studied earth, and they know what *potability* means to humans. They recognize your need for food but not for water.

These aliens don't think humans pose a threat to them (well, do we?); so you are free to wander about in their ship. Exploring their ship, you come across pitchers A and B, marked as in Figure 2.7(a) (memb = membership, and prob = probability). Literally dying of thirst, you must drink – which pitcher would *you* choose to drink from?

Judging from the available information, you might conclude that A could contain, say, bath water, but it would not (discounting the possibility of a Machiavellian fuzzy modeler) contain liquids such as hydrochloric acid. That is, a *membership* of 0.96 means the contents of A are very *similar* to perfectly potable liquids. Ordinarily we would take, say, pure spring water, as the most potable or ideal liquid, and give it a membership of 1, assuming (or knowing) this would give you a much better idea of the meaning of 0.96 relative to a perfectly potable liquid.

On the other hand, the probability that B is potable = 0.96 means that over a long run of experiments, the contents of B are expected to be potable in about 96% of the trials; and in the other 4%? In these cases, the contents will be deadly – about 1 chance in 20. Do you like these odds well enough to gamble on B instead?

Suppose we examine the liquids in pitchers A and B and discover, as shown in Figure 2.7(b), that A contains *beer*, while B contains *potassium cyanide* (KCN). *After* observation, the membership value for A may be unchanged or it may be "tuned" (here, it is upgraded from 0.96 to 0.97 to account for this agreeable situation), but the probability value for B *must* drop from 0.96 to 0.0. This is illustrated in Figure 2.7(b).

before **observation**

Figure 2.7(a) Jugs with unknown contents.

after **observation**

Figure 2.7(b) Jugs with contents revealed.

Would changing the numerical information in this example affect your choice? *Maybe*. Suppose that the membership and probability values in Figure 2.7(a) were both 0.07. In this case, most observers would feel that no reasonable attempt to model the physical meaning of potability would assign a membership value this low to any liquid that was really potable. Consequently, most observers would switch to bottle B as their first choice since it (at least) offers a 7% chance of being drinkable. The point here is that decisions based on the numerical values in each model also depend on their actual magnitudes in the unit interval. The utility of the fuzzy model depends entirely on the plausibility of the fuzzy membership function for the fuzzy set L.

Example 2.3 shows that fuzzy and probabilistic models possess philosophically different kinds of information. Fuzzy memberships represent similarities of objects to imprecisely defined properties, while probabilities convey information about relative frequencies. This example doesn't imply that probability is in any way a defective or useless concept. What it does show is that choosing a model depends on the process being represented. When relative frequencies of chance occurrences are important, models should reflect this information. But when uncertainty is non-statistical, fuzzy models may offer a more natural and useful alternative.

Q4. Where do fuzzy models fit in with other models? Only one answer will do for this question – wherever they can provide either collateral or competitively better information about a physical process than other models. The binomial distribution seems intuitively optimal for modeling outcomes of coin flips. We could model this process with a fuzzy technique, but the results would almost certainly be nonsense.

On the other hand, "close to 7" can be represented by a statistical model, but this seems much less satisfactory than the fuzzy set m_B in Figure 2.4 because chance is absent from our naive description of the process. Fuzzy models are not really that much different from other approaches in the sense that sometimes they work, and sometimes they don't. This is the *only* criterion that should be used to judge any model. The most sensible approach to clustering is simple: use whatever model seems to get the job done best in the time you have to do it.

2.4 Notes and Remarks

Fuzzy sets, including the definitions at (2.2), were introduced by Lotfi Zadeh (1965). Some of the key ideas about the representation of vagueness were given about 30 years earlier by the philosopher Max Black (1937). There are now so many good textbooks and monographs on fuzzy sets and models that it is hard to provide a solid recommendation for just a few. The topic of interest to us, fuzzy clustering, is well represented in Bezdek (1981), Hoppner *et al.* (1999), Bezdek *et al.* (1999), Valente de Oliveira and Pedrycz (2007), Xu and Wunsch (2009), and Theodoridis and Koutroumbas (2009). The history of fuzzy clustering in an interview format (with Bezdek and Ruspini) appears in Seising (2015), and, more recently, in Ruspini *et al.* (2019). Chapter 5 in Celebri (2015) is a very nice survey of many aspects of fuzzy clustering by Vendramin, Naldi and Campello.

There were many amusing and usually quite contentious discussions about the relationship between probability and fuzziness in the first 30 years or so after Zaheh's paper appeared. Some of the material presented in this section about distinctions between the two approaches appeared in Chapter 1 of Bezdek and Pal (1992). Woodall and Davis (1994) objected strenuously to the jugs example (Example 2.3 in this section). You can see their argument and the author's (my) reply in Bezdek (1994). Readers interested in this aspect of fuzzy sets and models can jump into this fun but oh-so-serious fray with Stallings (1977), Gaines (1978), Tribus (1979), Lindley (1982), Cheeseman (1986), Cheeseman (1988), Kosko (1990), Elkan (1993), and Laviolette and Seaman (1994). Many historical anecdotes and updates to this debate appear in Seising *et al.* (2013).

It is hard to guess if the argument between the fuzzy and probabilistic camps will ever subside entirely. Successful use of fuzzy models in many real-world engineering applications has rendered this a moot debate for all but the most philosophically inclined and scientifically narrow-minded. If you are determined to use probability at all costs, Shafer (1992) admonishes you to remember that

> *The interpretation of belief functions is controversial because the interpretation of probability is controversial*

And Moulton (1939) still offers the best advice about such controversies to all of us:

> "*...every set of phenomena can be interpreted consistently in various ways, in fact, in infinitely many ways. It is our privilege to choose among the possible interpretations the ones that appear to us most satisfactory, whatever may be the reasons for our choice. If scientists would remember that various equally consistent interpretations of every set of observational data can be made, they would be much less dogmatic than they often are, and their beliefs in a possible ultimate finality of scientific theories would vanish.*"

2.5 Exercises

1. "Membership values have nothing to do with probability." Comment and/or justify with an example.
2. Propose a probabilistic model for the fuzzy set "close to 6." Comment on its utility. Does this model convey what you want it to mean about a statement like "I have about 6 dollars in my pocket?"
3. Propose a fuzzy model for the flip of a coin. Comment on its utility. Does your model capture what you expect to happen if you flip such a coin 100 times?
4. Show that Equations (2.2) reduce to their conventional meanings when A and B are crisp sets.

5. Give an example of a fuzzy set m_A which has this property: $m_A = m_{\tilde{A}}(x) = m_{A\cap E} = m_{A\cup E}$ for all x in the domain X of the membership function. The function you find proves that fuzzy sets can't provide the same measure-theoretic structure as probability theory. Ans. $m_A(x) = 1/2 \ \forall \ x \in X$.
6. Prove that $m_{A\cap B}$ is the largest fuzzy set contained in both m_A and m_B.
7. Prove that $m_{A\cup B}$ is the smallest fuzzy set containing both m_A and m_B.
8. Propose a membership function m for the set of numbers that are either close to 6 or close to 12, that is: normal (in the fuzzy sets sense) and bimodal at 6 and 12, monotone decreasing from 6 and 12 in both directions, symmetric about 9, and that assigns positive membership to every real number. Give the equation for m(r), and plot the graph of m.
9. Plot the membership functions for the fuzzy sets $m_{\tilde{A}}$, $m_{A\cap B}$, and $m_{A\cup B}$ in Example 2.1.
10. Two fuzzy sets m_A and m_B are defined on the interval [0,10] as follows: $m_A(X) = \frac{1}{1+x^2}$; and $m_B(X) = 2^{-x}$. Plot the membership functions of these fuzzy sets along with the graphs of their complement, union, and intersection.
11. Suppose in Example 2.3, the values for Figure 2.7(a) were memb = 0.5 and prob = 0.5. Which jug would you drink from? Are the two models equivalent or equally useful when their numerical values are equal?
12. Assume $x \geq 0$: Consider two fuzzy sets $m_A(x) = 1/(1+20x)$ and $m_B(x) = (1/(1+10x))^{1/2}$. Show that both of DeMorgan's laws hold for the fuzzy sets A and B. That is, prove that $(A \cap B)^c = A^c \cup B^c$ and $(A \cup B)^c = A^c \cap B^c$. Here superscript "c" stands for the complement of the set.
13. Devise and graph reasonable membership functions for these fuzzy sets.

    (a) The set of long fish
    (b) The set of heavy fish
    (c) The set of fish that are heavy and long
    (d) The set of fish that are heavy or long
    (e) The set of fish that are not long
    (f) The set of fish that are not heavy
    (g) The set of very heavy fish
    (h) The set of very long fish
    (i) The set of fish that are very heavy and very long
    (j) The set of fish that are very heavy or very long
    (k) The set of fish that are not very heavy and very long

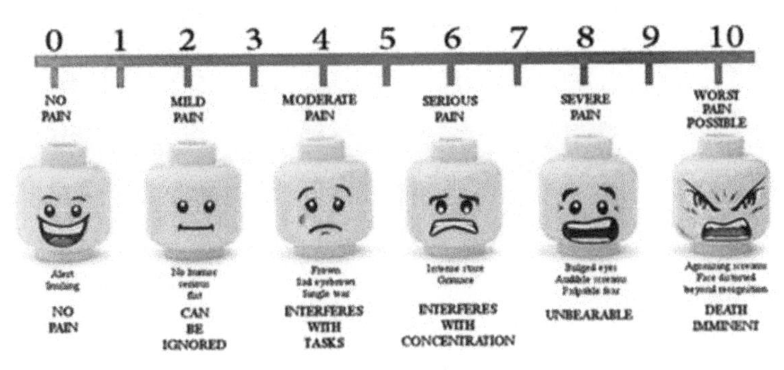

Figure 2.8 The Smith (2001) faces of pain.

14. Figure 2.8 is another version of the Wong-Baker faces of pain shown in Figure 2.5. I put it here mostly for fun. Do you like this one better than Figure 2.5? There are literally dozens of figures like this on the Internet: query "visual pain scales" to see some of them.

Give an example of a device you can probably find in your bathroom at home that collects values of a continuous fuzzy membership function. Fuzzy sets are everywhere!

3

Clusters: The Computer Point of View (CPOV)

No great discovery was ever made without a bold guess

–Sir Isaac Newton

3.1 Introduction

Clusters of real objects in the natural world seem unremarkable, but when we want to talk about clusters in *numerical data* (we really mean clusters in the objects represented by the data), we must first describe the objects in terms of a mathematical model that enables us to measure properties that are related to possible groupings of the objects.

Repeat: "Clusters in data" are well-defined only through a mathematical model.

This chapter describes components of the overall process of finding clusters from static sets of objects that are represented by numerical data. Models provide mathematical properties that define CPOV clusters in data. Algorithms are methods that find the CPOV clusters in static data by optimizing a model. Several different algorithms may be associated with a single model, and they can produce different "best" results. Because HPOV clusters are an intuitively simple concept, the idea that different algorithms can find different clusters in the *same data* with the *same model* can be difficult to accept.

Two kinds of numerical data, defined in the next section (the "variables" alluded to above), are used in computational approaches to clustering. From now on, computational clustering in static (batch) data will simply be referred to as clustering. To describe the various components in the clustering process, we need some notation.

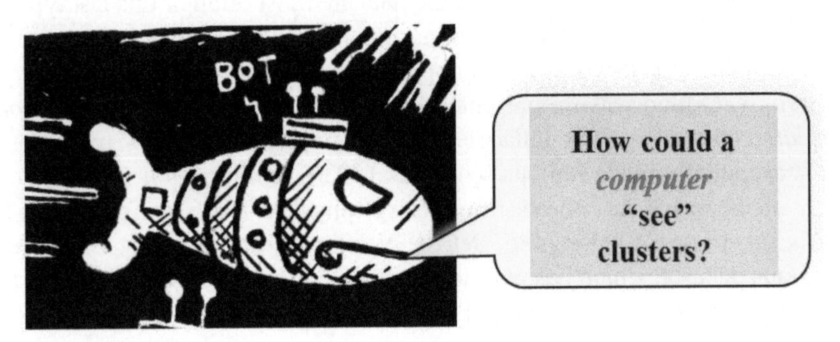

Figure 3.1 What are computer-point-of-view (CPOV) clusters, anyway?

3.2 Label Vectors

The *real numbers* are denoted by $\Re$. The *Cartesian product* of $\Re$ with itself p times is $\Re^p = \underbrace{\Re \times \cdots \times \Re}_{\text{p times}}$.

Column vectors $\mathbf{x} = \begin{pmatrix} x_1 \\ \vdots \\ x_p \end{pmatrix}$ are $p \times 1$, denoted by boldface type. The *transpose* of $\mathbf{x}$ is the $1 \times p$ *row vector* $\mathbf{x}^T = (x_1, \ldots x_p)^T$; and $\mathbf{0}^T = (0, 0, \ldots, 0)$ is the zero vector in $\Re^p$. Matrices with p rows and n columns are in $\Re^{pn} = \Re^p \times \Re^n$. The vectors in a $p \times n$ matrix (A) can be regarded as either column vectors in $\Re^p$ or row vectors in $\Re^n$ or a single vector in $\Re^{pn}$. These alternate representations are shown in Equation (3.1).

$$\underbrace{\mathbf{A} = [a_{ij}]_{p \times n}}_{\text{matrix}} = \underbrace{\begin{bmatrix} \mathbf{A}_{(1)} \\ \vdots \\ \mathbf{A}_{(p)} \end{bmatrix}}_{\text{rows in } \Re^n} = \underbrace{[\mathbf{A}^{(1)} \ldots \mathbf{A}^{(n)}]}_{\text{columns in } \Re^p} = \underbrace{\begin{pmatrix} a_{11} \\ \vdots \\ a_{pn} \end{pmatrix}}_{\text{vector in } \Re^{pn}} \tag{3.1}$$

Alternate terms for $\Re^p$: p-space or p-dimensional (pD, e.g., 1D, 2D, etc.) Cartesian product. When $\Re^p$ is equipped with the usual vector addition, scalar multiplication, and Euclidean inner product, it is called "Euclidean space."

Let $O = \{o_1, \ldots, o_n\}$ denote n objects (fish, cigars, motorcycles, guitars, beers, etc.). When object o_k in O is represented by (column) vector $\mathbf{x}_k$ in $X = \{\mathbf{x}_1, \ldots, \mathbf{x}_n\} \subset \Re^p$, X is an *object data representation* of O. These vectors have a variety of names in the literature, the most common being feature vectors, pattern vectors, or object vectors. For $1 \le k \le p$, the *k*th component of the *i*th feature vector (x_{ki}) is the value of the *k*th feature measurement or attribute (e.g., height, weight, length, color, etc.) of the *i*th object.

Alternatively, when each *pair* of objects in O is represented by a relationship between them, we have *relational data*. There are many names for a relation that begets relational data values; for example, measures of proximity, agreement, concordance, resemblance, distance, etc. Let $R = [r_{ij}]$ be an $n \times n$ matrix of relational values on $O \times O$, r_{ij} being the relation between o_i and o_j. Relational data is often *dissimilarity* data, denoted as $D = [D_{ij}]$, where D_{ij} is the pairwise dissimilarity (usually a distance) $D_{ij} = d(o_i, o_j)$ between o_i and o_j, for $1 \le i, j \le n$. R can be also be a matrix of similarities, denoted as S, based on a variety of measures (Theodoridis and Koutroumbas, 2009). Finally, R may be a relation specified by a person observing a process involving pairs of objects. *Rectangular* relational data are most often of this last type. For example, information retrieval problems often use rectangular relational data, where the columns are keywords and the rows are publications in which those words might appear.

When each object in $o_k \in O$ (and when represented by a feature vector, $\mathbf{x}_k \in X$) has a *physical label*, O is a set of *labeled data*; otherwise, O is unlabeled. For example, Anderson's Iris data, collected by Anderson (1935), and subsequently made famous by Fisher (1936), comprises $n = 150$ feature vectors in $p = 4$ dimensions. Each vector in Iris has one of three (crisp) physical labels assigned to it by Anderson, corresponding to the Iris subspecies it belongs to: Setosa, Versicolor, or Virginica. Examples 1.8 and 1.9 introduce you to this famous data set, which has probably appeared in more clustering papers than any other data set on the planet. Papers that discuss Iris could fill a few years worth of issues of an *IEEE Transactions on the Iris Data* – it is almost a rite of passage for cluster analyzers.

☺ ***Forewarned is forearmed:*** Subsets with physical ("*ground truth* GT") labels may or may not be CPOV "clusters" according to any model and algorithm that detects substructure in numerical data. Consequently, data which are provided with physical labels by some collection process (including synthetic generation of points with "known" class labels) will always be called *"labeled subsets,"* as opposed to subsets after labels have been attached to the objects by a clustering algorithm. Then, and only then, labeled subsets are CPOV clusters and will be called (CPOV) *"clusters."* ☺

Measured features can be continuous real variables (length, weight, etc.), discrete-valued n-ary variables (true or false, 1 = blue, 2 = yellow, 3 = green, etc.), qualitative values (textures, etc.), or any combination of these types. Another fairly common problem with feature vector data is missing and/or corrupted values. Real data suffer from these problems more often than you might imagine. Ways that clustering can and must be adjusted to accommodate mixed data types and missing or incomplete data will be discussed as they arise.

There are four types of class labels but only three mathematical structures that represent them: *crisp (or hard), fuzzy, probabilistic, and possibilistic.* Non-crisp labels are often called *soft labels.*

Definition 3.1. Three label vector structures for four types of labels.

Let integer c be the number of classes, $1 < c < n$ (in many papers, the integer used is k: if you are more comfortable with k, use it instead!). The *crisp, fuzzy/probabilistic, and possibilistic label vectors* in $\Re^c$ are

$$N_{pc} = \{\mathbf{y} \in \Re^c : y_i \in [0,1] \, \forall \, i; y_i > 0 \, \exists \, i\}; \tag{3.2a}$$

$$N_{fc} = \{\mathbf{y} \in N_{pc} : \sum_{i=1}^{c} y_i = 1 \}; \tag{3.2b}$$

$$N_{hc} = \{\mathbf{y} \in N_{fc} : y_i \in \{0,1\} \forall i\}. \tag{3.2c}$$

Figure 3.2 depicts these three sets for c = 3. N_{hc} is the canonical (unit vector) basis of $\Re^c$. The *i*th vertex of N_{hc}, $\mathbf{e}_i = (0, \ldots, 1, \ldots, 0)^T$, where the 1 occupies the *i*th place, is the *crisp* label for class i, $1 \leq i \leq c$. The set N_{fc} is a piece of a hyperplane, and, as will be proved in Section 6.1 (cf., Equation (6.4a)), is the *convex hull* of N_{hc}. For example, the vector $\mathbf{y} = (0.1, 0.6, 0.3)^T$ is a label vector in N_{f3}; its entries lie between 0 and 1, and sum to 1. There are two interpretations for the elements of N_{fc}. If $\mathbf{y}$ comes from a method such as maximum likelihood estimation in mixture decomposition, $\mathbf{y}$ is a (usually posterior) *probabilistic* label, and y_i is interpreted as the probability that, given $\mathbf{x}$, it is in, or came from, class or component i of the mixture (cf. section 7.2). On the other hand, if $\mathbf{y}$ is a label vector for an object (o) generated by, say, the fuzzy c-means clustering model (Bezdek, 1973), $\mathbf{y}$ is a *fuzzy* label for (o), and y_i is interpreted as the membership of (o) in class i. An important point, then, is that N_{fc} has the same mathematical structure for probabilistic and fuzzy labels. Finally, $N_{pc} = [0,1]^c - \{\mathbf{0}\}$ is the unit (hyper)cube in $\Re^c$, *excluding the origin.* Vectors such as $\mathbf{z} = (0.7, 0.2, 0.7)^T$ in N_{p3} are called *possibilistic* label vectors, and in this case, z_i is interpreted as the possibility that $\mathbf{x}$ is in, or came from, class i. Labels in N_{pc} are produced, e.g., by possibilistic clustering algorithms (Krishnapuram and Keller (1993)). Evidently, $N_{hc} \subset N_{fc} \subset N_{pc}$.

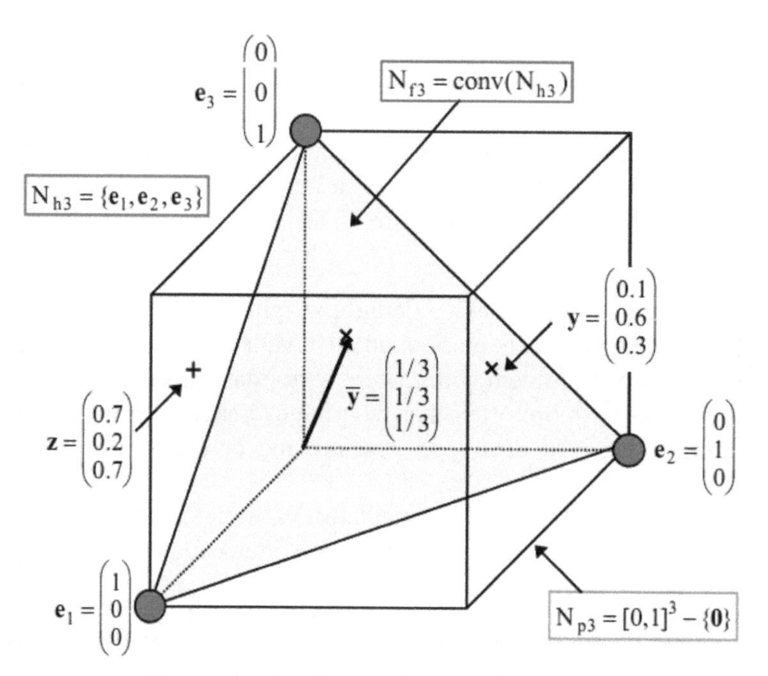

Figure 3.2 Label vectors for c = 3 classes in $\Re^c$.

3.3 Partition Matrices

Clustering (unsupervised learning) in an unlabeled data set representing a finite set of objects $O = \{o_1, \ldots, o_n\}$ is the assignment of one of the four types of *labels* to the objects in O. The label vectors at Equation (3.2) become the columns of three types of *c-partitions* of O, which are sets of (cn) values $\{u_{ik}\}$ that are most conveniently arrayed as c × n matrices, say $U = [u_{ik}]$. Let $\mathbf{U}^{(k)}$ denote the *k*th column of U (it is a label vector in $\Re^c$).

Definition 3.2. Crisp and soft partition matrices.

The non-degenerate -(this means no zero rows, corresponding to empty clusters) *crisp, fuzzy/probabilistic, and possibilistic c-partitions* of n objects are

$$M_{pcn} = \{U \in \Re^{cn} : \mathbf{U}^{(k)} \in N_{pc}\ \forall k;\ 0 < \sum_{i=1}^{c} u_{ik} < n\ \forall i\}; \tag{3.3a}$$

$$M_{fcn} = \{U \in M_{pcn} : \mathbf{U}^{(k)} \in N_{fc}\ \forall k\}; \tag{3.3b}$$

$$M_{hcn} = \{U \in M_{fcn} : \mathbf{U}^{(k)} \in N_{hc}\ \forall k\}. \tag{3.3c}$$

There are two special cases that must be discussed, namely, c = 1 and c = n. Both of these cases arise in Chapters 8 and 11 even though O may not contain cluster structure in any well-defined sense. The case c=1 is represented uniquely by the crisp *1-partition* $\mathbf{1}_n = [1\ 1 \cdots 1]$, which asserts that all n objects belong to a single cluster. At the other extreme, c = n is represented uniquely by the n-partition $U = I_n$, the $n \times n$ identity matrix, up to a permutation of columns. In this case, each object is in its own singleton cluster, a situation that disagrees with our basic understanding of what constitutes a cluster as a group of similar objects. Thus, choosing c = n or c = 1 rejects the hypothesis that O contains clusters. The agglomerative SAHN clustering algorithms discussed in Chapter 8 begin and end with these two special cases.

Evidently, $M_{hcn} \subset M_{fcn} \subset M_{pcn}$. There are four types of clustering algorithms corresponding to the four types of partitions but, again, only three distinct mathematical structures. The *fuzzy/probabilistic* partitions share a common mathematical structure, but the assumptions about data underlying fuzzy and probabilistic clustering models are very different. The reason these matrices are called *partitions* follows from the interpretation of their entries.

If the labels are crisp, we hope the values $\{u_{ik} \in \{0, 1\}\}$ identify c natural subgroups in O. If U is crisp or fuzzy, u_{ik} is the *membership* of o_k in the *i*th partitioning crisp or fuzzy subset cluster. Ruspini (1969) introduced M_{fcn}. Bezdek (1981) contains more details about the algebraic and geometric nature of these partition sets (cf., Chapter 6). If $P(\sim U)$ in M_{fcn} is probabilistic, $p_{ik} \sim (u_{ik})$ is usually the (posterior) probability $p_{ik} = Pr(i|o_k)$ that, given o_k, it came from class i. The values $\{u_{ik} = p_{ik} \in [0, 1]\}$ comprising probabilistic partitions found by Gaussian mixture decomposition (GMD) (cf., Chapter 7) arise through Bayes rule. Historically, it is not clear when this interpretation of these values was first associated with clustering. If U in M_{pcn} is possibilistic, it has entries between 0 and 1 that don't necessarily sum to 1 over every column, and the values are usually called *typicalities* to distinguish them from memberships. In this case, u_{ik} is interpreted as the possibility that o_k belongs to the *i*th possibilistic subset (Krishnapuram and Keller (1993)).

An alternative characterization of any crisp U in M_{hcn} is as a *cluster list vector*. This is the vector $\mathbf{u} \in \Re^n$ whose *j*th entry gives the label in $\{1, 2, \ldots, c\}$ of the *j*th object partitioned by U. For example, $U = \begin{bmatrix} 1 & 0 & 1 & 0 & 0 \\ 0 & 1 & 0 & 1 & 1 \end{bmatrix} \leftrightarrow \mathbf{u} = (1, 2, 1, 2, 2)$. The double arrow indicates that we can go back and forth between these two representations: given either, the other is easily constructed. The cluster list is a much more efficient representation of U for implementation in the crisp case. The matrix form of crisp partitions finds its value in the theory.

Another way to represent any crisp U in M_{hcn} is in terms of the c crisp subsets of object set O that are defined by the rows of U. Specifically, we may write O as the union of the c subsets $\{O_j\}$, where $O_i \cap O_j = \emptyset$, $i \neq j$. The *i*th row of U contains a 1 in column k when o_k is in crisp subset i, and $\sum_{k=1}^{n} u_{ik} = n_i = |O_i|$.

Ultimately, soft partitions are usually converted to crisp ones. The most common method for doing this is to convert each soft column $\mathbf{U}^{(k)} \in N_{pc} - N_{hc}$ of c-partition U into a crisp label by replacing the maximum element of the soft label vector $\mathbf{U}^{(k)}$ with a 1, and setting the remaining (c-1) elements to 0. Here is a formal definition:

Definition 3.3. Hardening a soft c-partition of the data.

$U \in (M_{pcn} - M_{hcn})$. $\mathbf{H}(U) = [h(U^{(1)}) \cdots h(U^{(n)})]$ is a *hardening of U* by the function $h : (N_{pc} - N_{hc}) \mapsto N_{hc}$, defined as

$$h(\mathbf{U}^{(k)}) = \mathbf{e}_i \Leftrightarrow u_{ik} > u_{jk} \quad j \neq i;\ \forall k. \tag{3.4}$$

When ties occur, the usual practice is to assign the membership of 1 arbitrarily to any one y of the winners and treat the other maximums like the non-winners.

When U is fuzzy, $\mathbf{H}(\mathrm{U})$ is its *maximum membership* partition. When P is a probabilistic c-partition, $h(\mathbf{P}^{(k)})$ at Equation (3.4) is nothing other than Bayes rule for posterior probability: (decide $o_k \in$ class $i \Leftrightarrow pr(i|o_k) > pr(j|o_k) \ \forall j \neq i$). Here is an example of what the three types of partitions look like.

Example 3.1. Fuzzy memberships: representation of hybrids.

Let $X = \{o_1 = \text{domestic cat}, o_2 = \text{African Serval}, o_3 = \text{Savannah cat}\}$. The Savannah cat is the result of breeding an African Serval with a domestic cat, i.e., it is a hybrid animal. Table 3.1 shows three 2-partitions of these three objects.

Table 3.1 An example of the three partition types at c = 3.

	$U_1 \in M_{h23}$			$U_2 \in M_{f23}$			$U_3 \in M_{p23}$		
Object	o_1	o_2	o_3	o_1	o_2	o_3	o_1	o_2	o_3
Domestic cat	1.0	0.0	0.0	1.0	0.2	0.4	1.0	0.2	0.5
African Serval	0.0	1.0	1.0	0.0	0.8	0.6	0.0	0.8	0.7

The Savannah cat, o_3, is labeled by the last column of each partition, and, in the crisp case, it must be (erroneously) given full membership in one of the two crisp subsets partitioning this data. In U_1, o_3 is labeled "African Serval." Non-crisp partitions enable models to (sometimes!) avoid such mistakes. The last column of U_2 allocates more (0.6) of the membership of o_3 to the Serval class; but it also assigns a lesser membership (0.4) to o_3 as a domestic cat. U_3 illustrates possibilistic label assignments for the objects in each class. This can seem like a more realistic assignment than the fuzzy partition because it indicates typicalities of the Savannah cat to each of the participating parents. Finally, observe that hardening each column of U_2 and U_3 with Equation (3.4) in this example makes them identical to U_1, i.e., $U_1 = \mathbf{H}(U_2) = \mathbf{H}(U_3)$. Crisp partitions don't possess the information content that can be used to suggest fine details of infrastructure such as hybridization or mixing that are available in U_2 and U_3. Consequently, if this sort of knowledge is useful, extract it before you harden U!

3.4 How Many Clusters are Present in a Data Set?

The examples in Chapter 1 should convince you that people can disagree about clusters in sets of objects they observe: why should computers be able to do better? Even worse: different algorithms can find different "good" clusters in the same data. (More on what a *good* cluster is later.)

Repeat: Different algorithms for the same model can find different "good" clusters in the same data.

Example 3.2. The "best" clusters in data depend on the question you ask of that data.

Let O = {Father, Mother, Daughter, Son}. Suppose that the father has brown eyes, while the mother, son, and daughter have blue eyes. Moreover, the mother has blood type A, the father and daughter have blood type AB, and the son has blood type B. As the crisp criterion selected to define clusters varies, we get

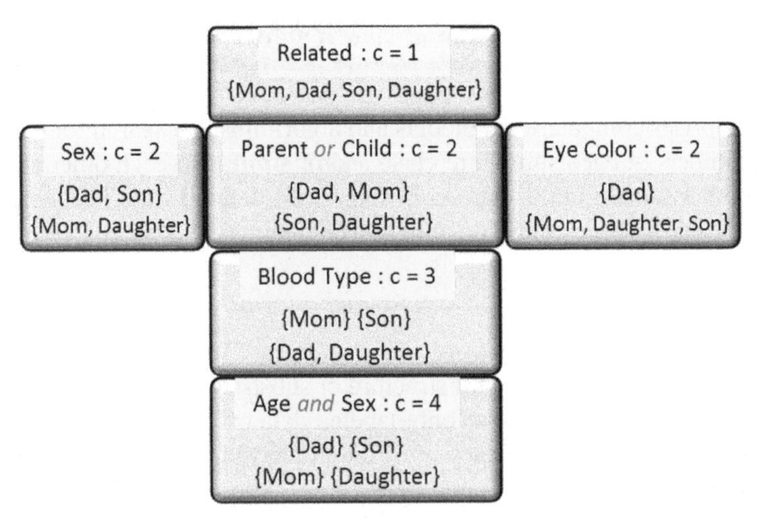

Figure 3.3 Some natural groups in O = {Father, Mother, Daughter, Son}.

various hard subsets or natural groups in O, some of which are shown in Figure 3.3. Note especially that there are different and yet correct (for the criterion chosen) clusters at c = 1, 2, 3, and 4; and there are also three different correct sets of clusters at c = 2. This illustrates that there may be many different, useful interpretations of substructure in a fixed data set even at the same value of c. Why? ***As the question varies, so does the answer.***

This illustrates that the best choice for c always depends on the properties specified for membership of objects in the clusters. The criterion (or criteria) that defines the clusters is all-important in determining the optimality of a solution. Also recognize that there are three different types of properties for different partitions of O: *simple* (related, sex, blood type, and eye color); *compound union* (age and sex); and *compound intersection* (parent or child).

Finally, note that the criteria in this example are *precise* properties of the objects in O, but there are many examples of clusters defined by imprecise object properties. For example, we might ask – who is in the cluster of tall people in O? This cluster is a *fuzzy subset* of O, which is not a "real" set, but certainly a set we might like to describe to a basketball coach using the techniques discussed in Chapter 2.

3.5 CPOV Clusters: The Computer's Point of View

You can lodge an Internet search for virtually any term beginning "clusters of xxxx," and you will find (hundreds of) references to computational approaches to clustering sets of objects (xxxx) represented by data. For example, one first return for the query "clusters of patients" is

> *A Cluster Analysis of Patients With Schizophrenia in Community Care ...Using variables that are widely discussed in the literature on community care (1), we identified different clusters of patients on the basis of the severity ...*

Now we can vaguely define (computational) cluster analysis.

Definition 3.4. Computer point of view (CPOV) cluster analysis.

CPOV cluster analysis comprises computational models and algorithms that search for **groups** (c-partitions) in *unlabeled data*. The members of each subset are **close** and/or **similar** to each other and, simultaneously, not so close or similar to other subsets in the data set in some well-defined way. The words "close," "not so close," and "similar" render this a vague definition at best. The main point is that CPOV clusters are defined *only* by models and algorithms.

Please consider why there is emphasis on the words *unlabeled data* in Definition 3.4. The basic tenet of clustering with numerical data is to discover useful structure or substructure in unlabeled data, BUT ... there will be many labeled data sets in the text because only labeled data offer a means for evaluating clustering models and algorithms, and comparing the efficacy of different algorithms for discovering cluster structure in O. After all, with unlabeled data, we don't – and never will – know if a CPOV clustering is "right" in any sense of the word. The relationship between physical labels for subsets of labeled data to CPOV labels assigned it by clustering algorithms is discussed in detail in the sequel.

In Chapter 1, it was noted that humans can and do make a clear distinction between the concepts of similar and close. We regard African and Asian elephants, for example, as very similar because they have very similar attributes, but these objects are far apart in their physical locations. But for computers, faced only with vectors in p-space or numerical relationships between pairs of objects, the terms similar and close are almost always synonyms. Vectors close to each other in distance are (more) similar to each other because of this than vectors that are far apart in distance and, hence, not very similar.

> ☺***Forewarned is forearmed:*** A super-important point: *all* clustering algorithms do their job, i.e., they generate CPOV solutions. That is, they all identify (possibly very different) "optimal" partitions based on (possibly very different) well-defined mathematical properties.☺

It is our hope, of course, that mathematically optimal groupings are *in some sense* an accurate portrayal of groups in the physical process represented by the data. This is worth repeating. ***Buyer beware – every clustering algorithm will do its job***: it will give CPOV clusters on demand. And even when there are HPOV clusters in data, different models and algorithms often offer different CPOVs about labels for the data.

One final remark about *singleton clusters*. We argued in Chapter 1 that from the HPOV, we would never identify a single object as a "cluster." At the other extreme, one cluster of grapes, for example, is regarded as an HPOV cluster. But what about the computer? There are algorithms that happily return singleton and whole set clusters, quite unaware of our semantic objections to this point. Single linkage in Chapter 8, for example, begins with n singletons "clusters" (c = n) and terminates when the whole set is connected at c = 1 "cluster." So from a CPOV, we have two choices. Either (i) we can reject CPOV singletons and whole sets as clusters by adding and enforcing this constraint to clustering software, thereby precluding these two events or (ii) we can accept singleton and whole set clusters from computer programs, thereby affirming in yet another way that the CPOV differs from the HPOV. For technical convenience, we choose the latter option: we will allow CPOV singletons and whole sets to be clusters when the need arises.

3.6 Notes and Remarks

The main purpose of this chapter is to get you to think about the relationship between our perceptions of clusters in the real world versus what we try to make computers recognize as clusters in numerical data. It would be maddeningly tiresome – but correct -to continually stress that when we look at a data set such as X_{30} in Figure 4.11(a) in the next chapter, it seems to have c = 3 clusters from a *human point of view* (HPOV) and Figure 4.11(b) confirms this from a *computer point of view (CPOV) relative to the model/algorithm = HCM/alternating optimization*. The tedium of saying all this every time we want to speak of algorithmic clusters overrides our desire to describe them with technical accuracy.

Consequently, almost everyone who writes about CPOV clusters, including this author, lapse into the sloppy habit of simply saying "this data set has c clusters" or "that validity index indicates c clusters," without specifying the model and algorithm that recognizes them as such. I will try to be pretty careful about this. But when the context of discussion leaves no doubt that the clusters in question are CPOV clusters seen by a certain model and algorithm, I will often lapse into this conventional form of sloppiness by not stating the explicit facts exactly.

The question "what is a (CPOV) cluster in data?" has been addressed by many authors over the years. Every book on cluster analysis answers this question in one way or another, and there have been many attempts to frame an objective answer to this question in the literature. For example, McQuitty (1967), Van Rijsberger (1970), and Hubert (1974) all propose different notions of "perfect clusters." For Van Rijsberger, a perfect cluster is one in which the least similar pair of data in the cluster is more similar than the most similar pair with one object in the cluster and the other object outside of it. McQuitty's definition of a perfect cluster is one in which each datum is more similar to all others in its cluster than it is to any datum outside the cluster. Three definitions of the perfect CPOV cluster, all different, and at odds with one another. And surely there are dozens of others. But finding good CPOV clusters is important; so attempts to characterize them have continued throughout the years. For example, Keller and Sledge (2007) assign a "degree of clusterness" to single subsets based on possibilistic c-means validity models; and on a visual approach based on iVAT images such as those in Chapter 4.

There are only two answers to the question "what is a cluster?" You can make an HPOV answer like those discussed in Section 1.2 about any set of objects based on your understanding of the objects themselves. When the question is more specifically "what is a cluster *in data*?" you can make an HPOV answer if you can see the data. But the only defensible answer if you can't see the data is that *a [computationally obtained] cluster is whatever your model defines it to be*. The distinction between HPOV and CPOV clusters made in this volume emphasizes that "natural clusters" don't exist in high-dimensional spaces. You can *believe* in the existence of such clusters based on your knowledge and/or intuition about the objects, but the only *evidence* you can ever have that they exist is computational evidence from some model and algorithm. There is nothing "natural" about such evidence.

3.7 Exercises

1. Give an example of a physical situation with three classes for which possibilistic label vectors in M_{pcn} such as the label vector $(0.6, 0.5, 0.4)^T$ are physically plausible.
2. Prove that the number of crisp c-partitions of n distinct points is given by $|M_{hcn}| = (1/c!)\left[\sum_{j=1}^{c}\binom{c}{j}(-1)^{c-j}j^n\right]$. Use this to show that for data sets where c is small compared to n (c<<n), $|M_{hcn}| \approx c^n/c!$ (cf., Feller (1959)).

3. You have two algorithms A_1 and A_2. A_1 produces probabilistic labels while A_2 generates fuzzy labels. In a two class problem for an input vector **x**, suppose both classifiers generate the same label vector $\binom{0.8}{0.2}$. In practice, a hard label is needed; the maximum membership or maximum probability scheme will both assign **x** to class 1. From the practical point of view, does it matter which solution you choose? Justify your answer.

4

The Three Canonical Problems

Theory attracts practice as the magnet attracts iron

–Carl Friedrich Gauss (attributed)

4.1 Introduction

This chapter is about the three important components of CPOV cluster analysis in static unlabeled data. The overall problem comprises the three operations depicted in Figure 4.1, which shows different models (AM, CM, and VM -***assessment, clustering,*** and ***validity models***) for each of the three phases. This somewhat formal representation provides a means for understanding how the three components fit together. When AM, CM, and VM are truly different, passing from one activity to the next often results in a mismatch that can lead to very surprising and often disappointing results.

In addition to the confusion caused by disagreements about the meaning of the terms clustering and classification discussed in Section 1.2, confusion can exist about what the term "clustering" means within the clustering community. Why? Because the term "cluster analysis" is usually shortened to the simpler term "clustering," which can encompass all three problems in Figure 4.1 or just the second one. It would be more accurate to always refer to activity 2 as partitioning, but I will lapse into the commonly accepted usage by occasionally calling it clustering as shown in Figure 4.1. Some approaches to clustering don't fit well into the detailed breakdown shown in Figure 4.1. For example, it is hard to separate the clustering model from the (only) algorithm that optimizes it in the single linkage scheme presented in Chapter 8.

4.2 Tendency Assessment – (Are There Clusters?)

The discussion of tendency assessment is nicely prefaced by this quote from J. W. Tukey (1977).

It is important to understand what you CAN DO before you learn to measure how WELL you seem to have done it.

Tukey's remark points squarely at the distinction between tendency assessment and validation: (i) assessment of clustering *tendency* (should you look for clusters at all -what CAN be done?); (ii) validation of clustering *outputs* (can you confirm the results -how WELL did you do it?). It is pretty clear, however, that the intermediate activity – *finding* the clusters (channeling Tukey, you might say, DOING IT) – is the main activity in cluster analysis.

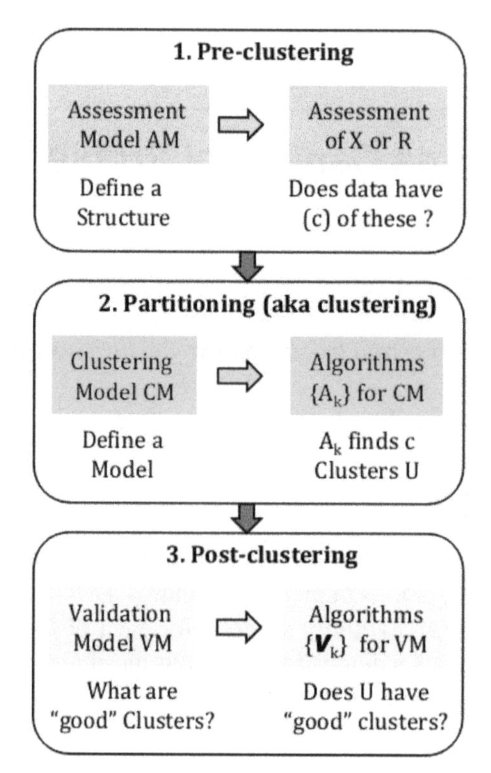

Figure 4.1 The three canonical problems comprising cluster analysis in static data.

4.2.1 An Overview of Tendency Assessment

Models for *tendency assessment* (AM in Figure 4.1) usually define what clusters are by instead defining what they are not. This description of assessment points out that you can't assess a data set for cluster tendency without a working definition for what clusters are, and AM supplies that definition. Most AMs are statistical, often basing tendency assessment on a hypothesis test against an assumed distribution that either is, or is not, one with clusters. A typical hypothesis is that the data are uniform or random over the sampling window; so a failed test suggests that the data have some sort of cluster structure.

Should you subject your data to a statistical assessment test for the existence of structure before proceeding to partitioning? Well, what do you learn if a statistical hypothesis test asserts that the data possess cluster structure in the sense of assessment model AM, but the criteria that define the AM are not the same as those of CM, your CPOV clustering model? The assessment model AM, for example, may simply tell you that the data are not random, while CM is designed to find compact, well-separated hyperspherical clouds of object data vectors.

This mismatch between AM and CM leaves a lot of room for misinterpretation of any algorithmic CPOV clusters discovered in the data during the partitioning step. Readers interested in the statistical approach to tendency assessment can get a start in this direction in the text by Jain and Dubes (1988). The fact that AMs don't usually match up very well with CMs probably explains the generally unenthusiastic reception of any suggestion toward doing assessment at all. Can we really know when data have cluster structure? Let us assume that we have some idea about what clusters are and see where this takes us.

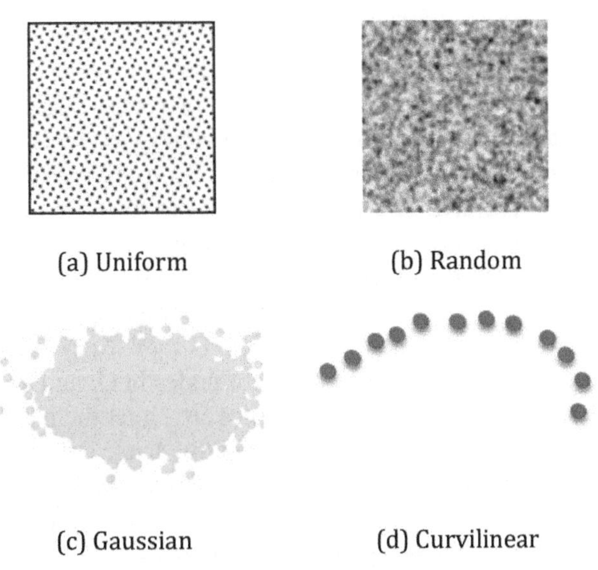

Figure 4.2 Some data sets that don't have (more than one) cluster.

Some examples of data that don't have clusters are shown in Figure 4.2. Uniformly (Figure 4.2(a)) and randomly (Figure 4.2(b)) distributed data definitely don't have HPOV clusters. There are also situations where data are not uniformly or randomly distributed; yet, there is no interesting clustering structure. For example, if all the data come from a single normal distribution as in Figure 4.2(c), then they will most likely not be uniformly or randomly distributed; yet, there is no HPOV cluster substructure within this single cluster. The data has an interesting *shape*, perhaps, but not clusters. And the set of points in Figure 4.2(d) is a single cluster; it doesn't have cluster substructure. But . . . if you apply *any* clustering model and algorithm to one of the data sets in Figure 4.2 at, say, $c > 1$, it will do its job – it will return a partition of these data sets into c (non-existent) clusters. How many clusters? However many you instruct the algorithm to find.

Definition 4.1. Assessment of clustering tendency.

(Cluster) tendency assessment comprises computational assessment (AM) models and algorithms that attempt to answer the question: does the data set (X or R) have computationally obtainable cluster structure?

4.2.2 Minimal Spanning Trees (MSTs)

We will need *minimal spanning trees* (MSTs) for two applications. First, the VAT and iVAT visual tendency assessment methods in the next subsection find a MST and use the sequence of edge additions required to build it to reorder the rows and columns of the input dissimilarity matrix. This implicitly rearranges the ordering of the underlying objects in O, much in the same way that Petrie (1899) did for the artifacts pictured in Figure 1.6. And, second, the single linkage model in Chapter 8 finds clusters by cutting edges of the MST.

Many readers will have enough background in graph theory to feel comfortable with the casual way this idea arrives in these pages. But since this is an elementary excursion, let us interrupt the main safari and take a short side trip in this direction for the benefit of readers with somewhat less background on this topic.

It is really easy to feel intimidated by the jargon associated with graph theory because there are a lot of technical definitions for the meaning of ordinary words in this field (e.g., simple, connected, complete, cyclic, trees, paths, and so on). Hopefully, you will find just enough about this topic here to keep on swimming without shouting, "throw me a life preserver." Any good text on discrete mathematics or graph theory can take you on deep-sea trips – for example, Rosen (2007) is one of the best (and the historical vignettes in this book are really fun to read too).

A finite binary relation on $O \times O$ can be viewed as a graph $G = (V, E, W)$ where $V = \{o_j\}$ is a set of n vertices (or nodes) of G; E is a set of m edges between pairs of nodes in G; and W is a set of weights (often distances) along the edges in E. In many applications, the nodes are either the n objects $O = \{o_1, \ldots, o_n\}$ or n vectors $X = \{\mathbf{x}_1, \ldots, \mathbf{x}_n\} \subset \Re^p$ that represent the objects: in this latter case, the vector $\mathbf{x}_i$ is regarded as the *i*th vertex of G. Typically, the matrix of weights is the distance matrix between pairs of vectors in $X \times X$ so that $W = D$ and $D = D^T$ is symmetric. In this case, G is *undirected* (which means that edges in E go both ways), *connected* (there is a path connecting each pair of distinct nodes in G), and *complete* (there is exactly one edge between each pair of distinct nodes).

A *path of length t* in G from node $\mathbf{x}_1$ to node $\mathbf{x}_{t+1}$ is a set of t edges $\{e_1, \ldots, e_t\} \subset E$ that pass through the intermediate vertices $\{\mathbf{x}_2, \ldots, \mathbf{x}_t\}$, forming the sequence $\mathbf{x}_1 \xrightarrow{e_1} \mathbf{x}_2 \cdots \rightarrow \cdots \mathbf{x}_t \xrightarrow{e_t} \mathbf{x}_{t+1}$. A non-trivial path is a circuit if $\mathbf{x}_1 = \mathbf{x}_{t+1}$. A *path or circuit is simple* if it contains each edge just once. G is *simple* if each edge connects two different vertices, and no pair of vertices is connected by more than one edge. A *tree* T in G is a connected, undirected graph with no *simple circuits*. A *spanning tree* is a tree that contains every node in V. An MST is a spanning tree with a minimum sum of weights along its edges. Figure 4.3 illustrates some of these ideas using standard views of several constellations of stars.

Example 4.1. Graphs and trees in the stars.

Figure 4.3(a) shows a standard representation of the c = 2 HPOV star clusters that are popularly known as the Big and Little Dippers. Each cluster contains seven nodes (stars), seven edges (lines connecting the nodes), and a simple circuit (the closed loops). In an abstract setting, if the 14 stars are the vertices of G, n = m = 14, and the two line drawings of the big and little dippers would be called subtrees of G.

The subtrees, in our context, correspond to c = 2 clusters of 7 stars each, and this is exactly how the single linkage algorithm in Chapter 8 determines clusters – by cutting edges in an MST on all of the nodes if one exists. Making a visual estimate (i.e., not based on computing any kind of distance using coordinates of the stars), the minimum distance between the two subtrees looks to be the edge shown by the red line connecting Mizar to Kochab. This is the edge that single linkage would cut to produce two clusters *on the way down* from a single cluster consisting of all the vertices if there were no circuits in the two subtrees (i.e., there is no MST on the 14 nodes because of the circuits). And it is also the edge that an MST algorithm would add *on the way up* when growing the tree from an initial state beginning with all the vertices in singleton clusters to bridge the gap between the two subtrees, assuming again that no cycle had been created. The terminology – *on the way down* and *on the way up* – used in this example refers to the way the agglomerative implementation of single linkage described in Chapter 8 is implemented. There is also a divisive version of single linkage, for which the terms would be reversed.

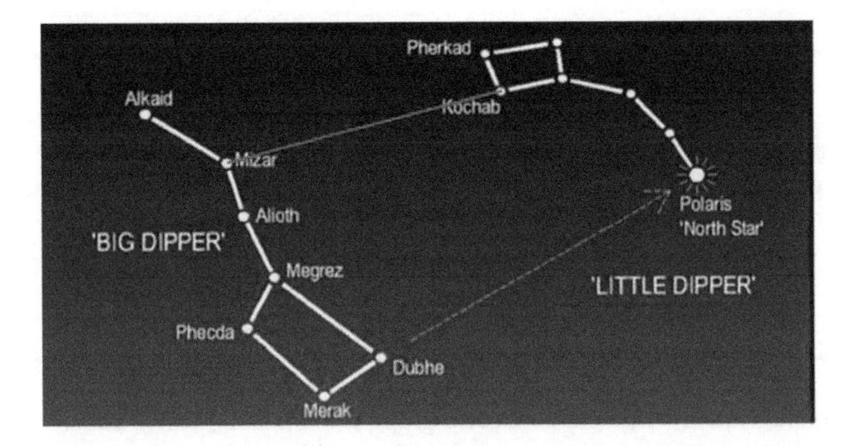

(a) (Sub) trees for the Big and Little Dippers have simple circuits.

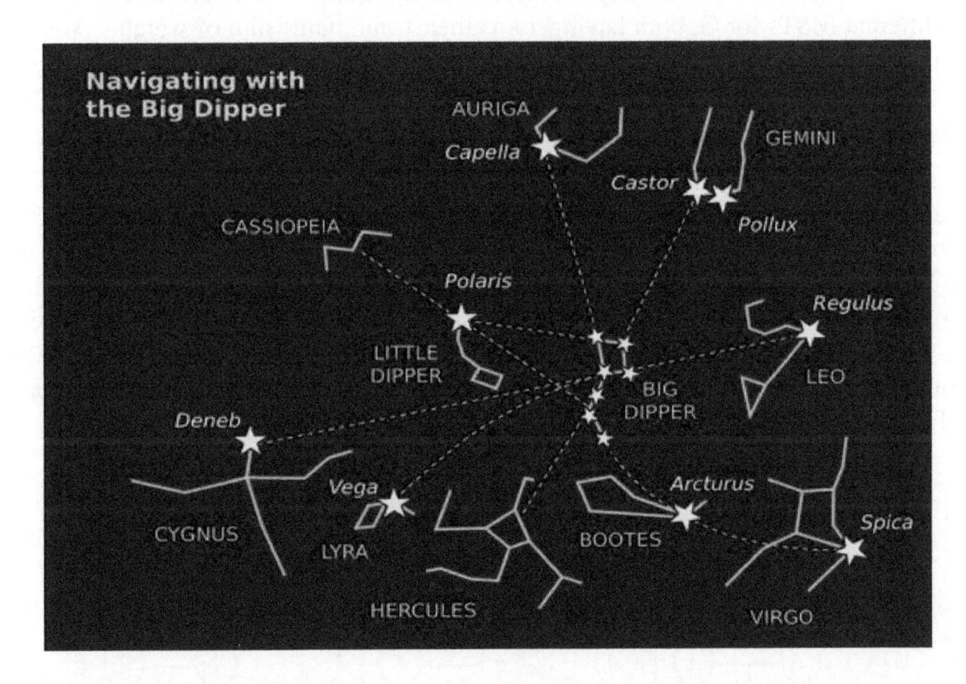

(b) Eleven star clusters in the neighborhood of the Big and Little Dippers.

Figure 4.3 HPOV clusters of stars represented as (sub)trees in a graph.

Figure 4.3(b) is a view of c = 11 named HPOV star clusters (the names are all capital letters) in the neighborhood of the Big and Little Dippers. Exercise 3 at the end of the chapter asks some questions about these graphs. Problem 2 in Chapter 1 and Section 5.6 contain discussions of problems associated with identification of CPOV star clusters like these using numerical coordinates such as distance and light intensity.

Definition 4.2. Minimal spanning tree (MST).

A *minimal spanning tree* (MST) T for a graph G = (V, E, D) is a connected weighted subgraph of an undirected weighted graph G that is a spanning tree of G with the smallest possible sum of weights (or distances) along its edges. The MST for G is unique if the edge weights in D are all different. Otherwise, G may have more than one MST. It is common to call T an MST of either G or D.

Example 4.2. Minimal spanning trees in graphs.

Figure 4.4(a) shows an undirected, weighted, connected graph G = (V, E, D) which has n = 6 vertices in V, and m = 8 edges in E. Each edge has an integer weight, which will be called a distance. Since more than one of the 8 weights are equal to 3, there may be more than one MST for this graph. Figures 4.4(b) and 4.4(c) show two different MSTs for G, both having 14 as their (minimum) sum of weights. We will return to this example in Chapter 8, where the single linkage clustering algorithm is applied to G. Single linkage will produce different sequences of clusters in G (a cluster in G is a subset of V) depending on whether it constructs MST1 or MST2.

Some exercises at the end of the chapter should firm up your understanding of these concepts. So, given G = (V, E, D), how do we find an MST of G? There are two standard algorithms that guarantee finding an MST in G due to Kruskal (1956) and Prim (1957). Both methods start with an empty graph and add edges to it one at a time so that cycles are never formed. Both methods are greedy algorithms[2] that add a next unused edge of smallest weight to the current set of minimal weight edges in different ways (Rosen (2007)). The visual assessment method listed below as Algorithm 4.1 uses a slight modification of Prim's approach which will be discussed in the next subsection.

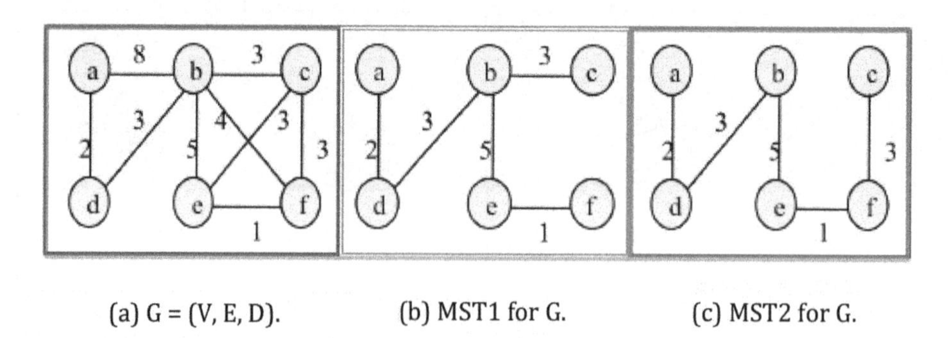

(a) G = (V, E, D). (b) MST1 for G. (c) MST2 for G.

Figure 4.4 Non-unique MSTs in a graph.

---

[2]Greedy algorithms find an optimal solution at each step, without looking ahead to see how the current choice affects optimality of the overall solution. This strategy doesn't always produce an optimal solution of the whole problem, but an optimal solution is guaranteed by these two algorithms.

4.2.3 Visual Assessment of Clustering Tendency

And what is the use of a book, "thought Alice," without pictures or conversation?

–Lewis Carroll (1866), Alice in Wonderland

The drawbacks of statistical assessment of clustering tendency are ameliorated to some extent by an approach based on data visualization. Visual clustering is the search for structure in graphical representations of numerical data. Three important functions determine different aspects of visual clustering: display functions, reordering functions, and clustering functions. Visual methods have been used for all three of the canonical problems shown in Figure 4.1. There are two branches of visualization according to the type of input data. Feature vector methods make visual displays such as scatterplots of projections of the input vectors: this topic is covered in Sections 5.4 and 5.5.

When the input is relational data, displays are often made in the form of images of the input data matrix. Since $X = \{x_1, \ldots, x_n\} \subset \Re^p$ can always be converted to a square, symmetric distance matrix $D(X) \in \Re^{nn}$ using any vector norm, visual displays of relational (usually dissimilarity) data are a bit more general than direct visualization of $X \subset \Re^p$ by projection into a viewable space.

Visual methods in cluster analysis, including approaches for the tendency assessment problem, capitalize on the ability of humans to recognize similarity and form HPOV clusters by inspection of visual stimuli (cf., Figure 1.6). For example, early humans quickly learned the difference between clusters of short-faced bears and prairie dogs and acted accordingly. Visual clustering was and still is an evolutionary advantage for all of the early species that have survived to the present day. The common denominator in many visual imaging methods is the *reordered dissimilarity image* (RDI), often called a *cluster heat map*.

Let's assume that D is a square, symmetric dissimilarity matrix and that D* is a reordering of D by any well-defined way to permute its rows and columns. Let I(D) and I(D*) be gray-level images of D and D*. The intensity of each pixel in I(D) and I(D*) corresponds to the similarity or dissimilarity between the addressed row and column objects in the two matrices. In gray-level RDIs, it is customary to use black for zero distances (so, the diagonal pixels will always be black) and white for very large distances. Example 4.3 illustrates how the ordering of the objects affects the distances on which I(D) and I(D*) depend.

Example 4.3. Reordering can improve visual cluster assessment.

Sometimes a set of objects contains meaningful groups, but the order of presentation of the objects in the input data doesn't equip the matrix of distances D between pairs of objects with the information about cluster structure that enables I(D) to expose the structure. For example, consider the set of seven cherries $\{c_i\}$ and four pineapples $\{p_j\}$ in Figure 4.5(a), which are mixed together randomly, $O = \{c_1, c_2, p_1, c_3, p_2, c_4, p_3, c_5, c_6, c_7, p_4\}$. The crisp 2-partition U of O corresponding to this ordering is also shown this view.

Measure the weight of each object, and let $d_{ij} = |\text{weight}(o_i) - \text{weight}(o_j)|$. Now form n × n matrix of distances $D = [d_{ij}]$, with indices corresponding to this ordering, and, from this, have a look at the image I(D). Don't expect I(D) to suggest much structure in O. Why not? Now imagine rearranging the fruits so that the cherries and pineapples are adjacent as in the lower view of Figure 4.5. This is exactly what Petrie did with the physical archeological objects shown in Figure 1.6. View 4.5(b) offers a much clearer idea about HPOV clusters in the 11 fruits than the mixture seen in the (a) panel.

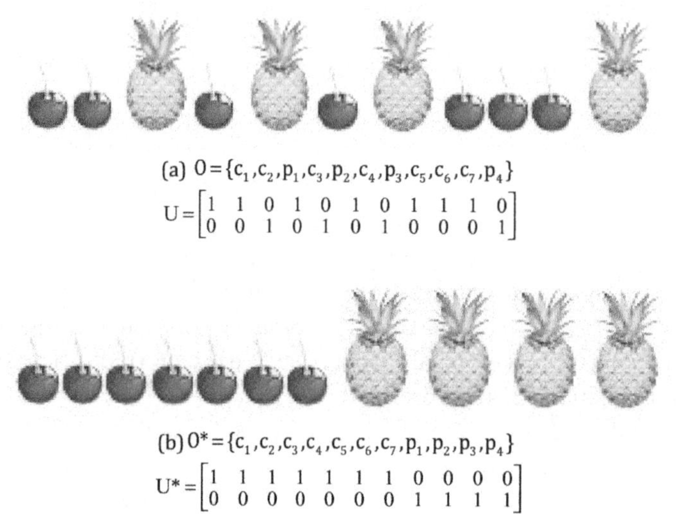

Figure 4.5 How rearrangements affect I(D) and I(D*).

The reordered object set $O^* = \{c_1, c_2, c_3, c_4, c_5, c_6, c_7, p_1, p_2, p_3, p_4\}$ is shown in Figure 4.5(b), with its crisp 2-partition U*. Now recompute the distance matrix using the ordering in O*, obtaining D*, and imagine the change in the image from I(D) to I(D*). All of the intracluster distances between pairs of cherries will be small, similar to each other and adjacent. The intracluster distances between pairs of pineapples will also be small, similar to each other and adjacent. There will be a third set of distances – the intercluster distances between the cherries and pineapples, and these distances will be much larger than the intracluster distances, and not adjacent. What do you expect to see in the reordered image I(D*)? Note that the reordering of O to O* is mirrored by the reordering of U to U*.

To see how reordering U to U* affects the images I(D) and I(D*), imagine computing the 11×11 matrices $D(U) = 1_{11\times11} - U^T U$ and $D(U^*) = 1_{11\times11} - \{U^*\}^T U^*$. What do you think they will look like? Your mind's eye should see two diagonal blocks of zeros in D(U*): a 7×7 block, and a 4×4 block, corresponding to the two HPOV clusters in O*.These two blocks will be black squares on the diagonal. Problem 4.6 asks you to compute these matrices. This little example exhibits the basis of a large family of techniques for visual assessment of relational data: the key to success is reordering of the input data.

Many times, a simple rearrangement of the objects, realized by a permutation of the rows and columns of D to D*, enables I(D*) to present a much more informative idea about structure in the data than I(D) does. An RDI is "useful" if I(D*) highlights potential clusters as a set of "dark blocks" along its diagonal. Each dark block represents a group of objects that are fairly similar.

There are a lot of models and algorithms in the literature that do the reordering. According to Wilkinson and Friendly (2009), RDIs had appeared in well over 4000 biological or biomedical publications in the period 1998–2009. In this volume, we will use only one of them, the *visual assessment* of (clustering) *tendency* (VAT) algorithm described by Bezdek and Hathaway (2002). (Actually, it is the *improved VAT* (iVAT) algorithm that makes several appearances throughout the book.) Before specifying the details of this model and algorithm, let us have a look at another example that further illustrates the cherries/pineapples discussion of Example 4.3.

Example 4.4. Reordering D → D* improves visualization.

Figure 4.6(a) shows 20 (x, y) pairs of points in the plane, which are labeled by color for visual clarity. You can imagine them as clusters of grapes (blue), apples (red), kiwi fruits (green), and a single orange (orange). According to our earlier agreement, the orange will be called a (singleton) cluster, even though it doesn't constitute a group. Figure 4.6(b) depicts a random sequence of Euclidean edge distances between pairs of 2D points, beginning at the arrow, which are then stored in the input data matrix D in this same random order. Figure 4.6(c) shows I(D), the image corresponding to D. You can see that even though the feature vectors in this data quite clearly form four HPOV clusters, the image I(D) fails to reveal this structure. Figure 4.6(c) looks like a random jumble of discordant pixels. This apparent randomness reflects the random ordering of the distances, which corresponds to the order in which the edges were collected, shown by the lines in Figure 4.6(b).

Figure 4.6(d) depicts edges in the MST of the data, beginning at the same point as in Figure 4.6(b). In this simple example, the MST is built by beginning at the arrow, first connecting all the blues, then jumping to the closest red, connecting all the reds, then jumping to the closest green, connecting them, and finally, moving down to connect the single orange. Now use the ordered sequence of edge additions to reorder (permute) the 20 objects from their mixed arrangement in the random edge sequence in Figure 4.6(b) to the ordered arrangement of MST edges represented by Figure 4.6(d). This is essentially the way that VAT operates – it finds an MST on the data and uses the sequence of edge additions to reorder D.

The VAT image I(D*) in Figure 4.6(e) agrees exactly with the HPOV of the clusters in Figure 4.6(a). Recall that black is the color used for minimum distance (zero) and white is the color used for maximum distance. The largest dark diagonal block is 9×9, representing the intracluster distances in the blues, which are all relatively close to each other. The next two 5×5 dark diagonal blocks represent the reds and the greens, and the lone pixel at the lower right in Figure 4.5(b) corresponds to the orange. So, the four dark blocks are essentially a pictorial representation of the intracluster (or within groups) distances, while the lighter colored off-diagonal pixels having larger intensities correspond to the larger intercluster

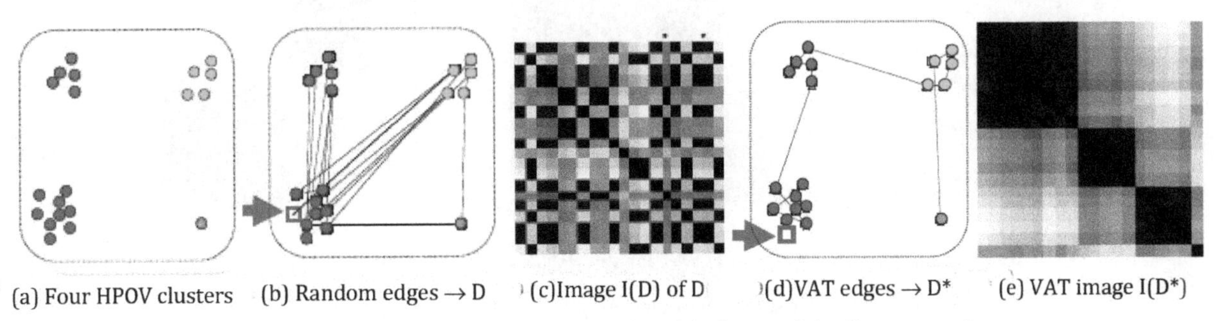

Figure 4.6 Reordering improves visual acuity of the image of the distance matrix.

(or between cluster) distances. Figure 4.6(e) presents a very clear idea of the cluster structure in the input data. This type of visualization is independent of the dimension of the input data. The matrix D is not a function of p; so images like this can be made from data with very large dimensionality (but, of course, the size of p will certainly affect the time it takes to make D).

> ☺ ***Forewarned is forearmed.*** A VAT image *suggests* that there may be c = 4 clusters in the input data in Figure 4.6(a), but VAT doesn't *find* the clusters in the data. VAT displays RDIs of dissimilarities in O -it doesn't partition O. ☺

The best way to use VAT (or any other cluster heat map of your raw data) is to form an estimate about substructure by visual inspection of the image. Counting dark diagonal blocks in a VAT image of "good" data should suggest the most likely number of CPOV clusters in the data. This helps you direct the partitioning effort toward a likely solution. But, be a bit careful. VAT can and does fail. Example 4.4 is a pretty nice illustration of how well VAT can work in an ideal case, but this ideal is rarely realized in practice. Fortunately, there is a pretty good upgrade to VAT – viz., *improved VAT* (iVAT), illustrated in Example 4.5.

Example 4.5. Visual assessment of cluster tendency with iVAT.

The data in Figure 4.7(a) appears to have c = 3 HPOV clusters of concentric points in the plane. Converting the 2D coordinates of the data in view 4.7(a) to Euclidean distances yields the dissimilarity data matrix $D = [d_{ij}] = [(x_i - x_j)^T(x_i - x_j)]$. The VAT image $I(D^*)$ of the VAT reordered data, $D \rightarrow D^*$, appears in panel 4.7(b). Do you see any hint of the structure of the data in this view? Probably not.

This happens because, unlike the four HPOV clusters in Figure 4.6(a), there are, in this data set, a few inliers between the three rings, and when VAT forms the MST of this data, it jumps back and forth from one ring to another, instead of calmly collecting all of the edges in each ring before moving to the next one. So, even with rearrangement by MST ordering, the VAT image is quite deceiving – it suggests that the input data has very little cluster structure, when, in fact, it has quite a lot.

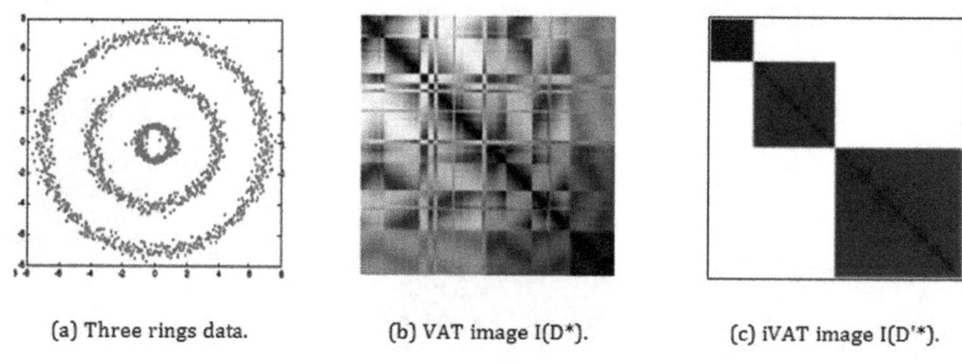

(a) Three rings data. (b) VAT image $I(D^*)$. (c) iVAT image $I(D'^*)$.

Figure 4.7 Visual assessment of clustering tendency with an iVAT image.

Applying the *improved visual assessment of tendency* (iVAT; Havens and Bezdek (2011)) algorithm to D results in a new reordered matrix called D'^*, which underlies the iVAT image $I(D'^*)$ in Figure 4.7(c). The three dark blocks along the diagonal of $I(D'^*)$ strongly suggest that the input data may have c = 3 clusters. And, in fact, the sizes of the three sub-blocks accurately reflect the sizes of the three rings in the input data: the smallest block corresponds to the innermost ring, while the largest block is associated with the outermost ring. But again, this image merely suggests that the input data are structured like this: iVAT doesn't *find* any partition of the object data in Figure 4.7(a). Since we can see this HPOV cluster structure in the input data, this visual assessment of clustering tendency seems unnecessary. But it is quite simple to make a data set just like this in, say, four-dimensional space, and then the HPOV fails. VAT and iVAT are still able to produce images such as these from data of arbitrary dimensions, and this enables us to at least have some idea about what the input data might contain: in effect, this scheme lets us take a peek at structure in the input space, even though we can't actually see things there.

---

iVAT transforms $D \to D'^*$ by first replacing the input distance d_{ij} between nodes (i, j) with the minimum of the maximum of all the path-based distances d'_{ij} that connect them, when $d_{ij} > d'_{ij}$. This amounts to performing feature extraction on the relational data set D, and this transformation is important enough in various clustering applications to warrant a separate discussion of it which is given in Section 5.5.4.

The iVAT method was introduced in Wang *et al.* (2010), and the transformation $D \to D'^*$ was done iteratively there in $O(n^3)$ time. The recursive version given in Algorithm 4.1 reduces the time complexity to $O(n^2)$. iVAT images are almost always superior to VAT images; so recursive iVAT will be used exclusively to produce RDIs in the sequel.

Example 4.6. Clusters of ellipses with two dissimilarity measures.

There are many interesting types of relational dissimilarity that are not induced by measuring distance between pairs of vectors in $X \subset \Re^p$. In this example, there are n = 40 objects. The *i*th object is a sensor, represented mathematically by an ellipsoidal summary of data collected over a specified time window, ellipse e_i, comprising the sample mean and covariance matrix of sensor measurements taken across a specified window in time. The 40 ellipsoids are called E40 in Figure 4.8(a). There are 10 ellipses in the upper right cluster, and 15 ellipses in each of the two clusters in the left, lower part of this view. Even though most of the sample means in the group of 30 are quite close, the orientation of the two sets of 15 ellipses makes the c = 3 HPOV clusters very easy to see. That is, *we* can see them, but can the computer? Should the primary interpretation of E40 be c = 2 clusters, or should it be c = 3 clusters?

Surprisingly, there is no standard way to measure the distance between pairs of ellipsoids. Moshtaghi *et al.* (2011) devised three ways to construct dissimilarities $D_\varphi = [d_{ij}] = [\varphi(e_i, e_j)]$, where φ was one of three functions called the *focal distance* D_f, the *transformation energy* D_{te}, or the *compound normalized* distance D_c between pairs of ellipsoids. These three ways to measure the distance between pair of ellipsoids produce very different results when submitted to iVAT. The images based on the distance matrices D_f and D_{te} are shown in Figures 4.8(b) and 4.8(c).

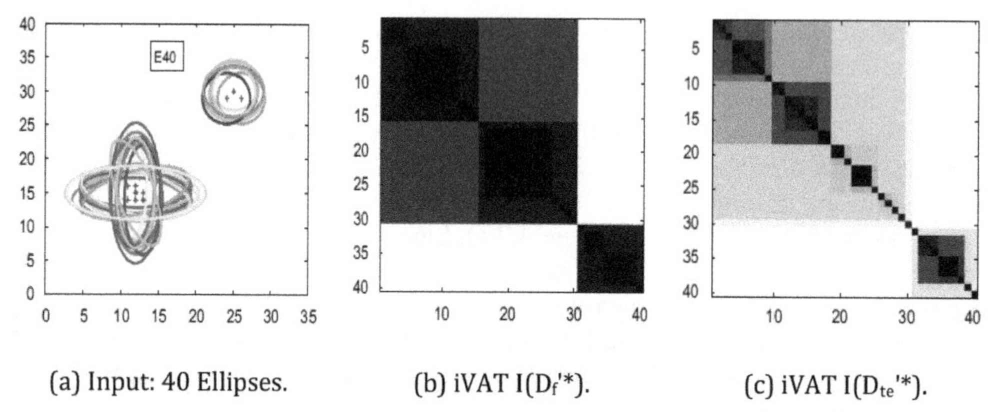

(a) Input: 40 Ellipses. (b) iVAT $I(D_f'^*)$. (c) iVAT $I(D_{te}'^*)$.

Figure 4.8 Two iVAT images of c = 3 subsets of ellipses.

Figure 4.8(b) depicts the structure possessed by E40 quite well. The 30 × 30 upper dark diagonal block corresponds to the mixed set of 30 ellipses in the input data, and the lower 10 × 10 dark block on the diagonal of this image represents the well-separated group of 10. If you agree that first and foremost, E40 should be interpreted as having c = 2 clusters, this view is most satisfying. But very importantly, please note that the larger 30 × 30 block contains two well-defined 15 × 15 sub-blocks that represent the two clusters of 15 ellipses that have horizontal and vertical orientations. This is an important feature of iVAT RDIs: they often depict primary cluster structure and, within it, secondary and even tertiary substructure.

As you can see in Figure 4.8(c), changing the input matrix sent to iVAT from D_f to D_{te} dramatically affects the image you get – for better, if passing from view (c) to view (b), or for worse, if passing from view (b) to view (c). The gross (primary) c = 2 structure of E40 is faintly visible in Figure 4.8(c), but the two primary clusters are broken into many subclusters with the primary blocks. So, another takeaway lesson for this example is that the iVAT image you produce from a distance matrix can be severely altered by changing the way distances are measured. It is an important point. If Figure 4.8(c) is the view of E40 that you rely on to infer something about cluster structure in the input data, you will be badly misled.

4.2.4 The VAT and iVAT Reordering Algorithms

Examples 4.2–4.6 should convince you that VAT and iVAT are useful additions to every clustering toolbox. Algorithm A4.1 contains pseudocode for VAT (Bezdek and Hathaway (2002)) and for recursive iVAT (Havens and Bezdek (2011)).The term VAT/iVAT will often appear in the text because both algorithms are shown together in A4.1. In practice, VAT is rarely used because iVAT images are almost always superior to VAT images; so recursive iVAT will be used exclusively to produce RDIs in the sequel. Thus, iVAT sometimes appears alone as if it is independent of VAT, but keep in mind that VAT must be run in order for iVAT to follow; so the term VAT/iVAT is technically a bit more correct.

Algorithm A4.1. Visual assessment of tendency (VAT/iVAT).

		Algorithm VAT (Bezdek and Hathaway, 2002)
1	**In**	D, an $n \times n$ matrix of dissimilarities:
		$D = D^T;\ d_{ij} \geq 0\ \forall\ i, j\ ;\ d_{ii} = 0\ \forall i$
2	**Set**	$K = \{1, 2, \ldots, n\}: I = J = \emptyset$:
3		Select $(i, j) \in \arg\max\{D_{st} : s \in K, t \in K\}$
4		$P(1) = i :\ I = \{i\} :\ j = K - \{i\}$
5		% Initialize MST at either end of edge with largest weight in D
6	**For** m = 2 to n **do**:	
7		Select $(i, j) \in \arg\min\{D_{st} : s \in I, t \in J\}$
8		$P(m) = j :\ I = I \cup \{j\} :\ J = J - \{i\} :\ d_{m-1} = d_{ij}$
9	**For** $1 \leq i, j \leq n$ **do**:	
10		$[D*]_{ij} = [D]_{P(i)P(j)}$
11	**Out**	VAT reordered dissimilarities D*: arrays **P**, **d**
12		% Create VAT RDI I(D*) using D*
		Algorithm iVAT (Havens and Bezdek, 2011)
13	**In**	D* = VAT reordered dissimilarity matrix: $D'^{*} = [0]$
14	**For** k = 2 to n **do**:	
15		$j = \underset{r=1,\ldots,k-1}{\operatorname{argmin}} \{D^{*}_{kr}\}$
16		$D'^{*}_{kc} = D^{*}_{kc}\ ;\ c = j$
17		$D'^{*}_{kc} = \max\{D^{*}_{kj}, D'^{*}_{jc}\};\ c = 1, \ldots, k-1; c > j$
18	**For** $j = 2,\ n/2,$ while $i < j$:	
19		$D'^{*}_{ji} = D'^{*}_{ij}$
20		**Out iVAT** Reordered dissimilarities D'^{*}: arrays **P**, **d**
		% Create iVAT RDI $I(D'^{*})$ using D'^{*}

Here are four notes about the implementation of A4.1.

A4.1(i): The input matrix D for VAT in line 1 is positive-definite and symmetric. There are symmetric distance matrices, however, that are not : the 2x2 matrix with 0's on the diagonal and 1's in the off-diagonal addresses is indefinite, but is a legitimate distance matrix. And the size of D can be an issue. This basic version is only useful for fairly small values of n (say, $n \approx 10,000$ or so). Extensions to rectangular, asymmetric, and big data inputs are covered in the notes and remarks for this chapter. There are several versions of VAT-based algorithms for streaming data which will be discussed in Chapter 12.

A4.1(ii): Prim's MST algorithm usually starts at either end (i.e., vertex) of a smallest weight edge. Initialization at line 3 starts at the opposite extreme – either end of a largest weight edge. This prevents VAT from a certain type of off-course deviation that is discussed in Bezdek and Hathaway (2002).

A4.1(iii): The argmax and argmin function calls in lines 3, 7, and 15 may produce sets, not single values. For example, in A4.1 $(i, j) \in \arg\max\{D_{st} : s \in K, t \in K\}$ is the *set of all ordered pairs* (i, j) that have a maximum distance. In case of ties, use a vertex from either end of any one edge in the set.

A4.1(iv): The arrays **P** and **d** will not be seen again until Chapter 11. These two arrays are used by clusiVAT Algorithm A11.4 to extend single linkage clusters from a small data matrix D_n to an intractably big data matrix D_N. The **P** array contains the order of insertion of the edges in the MST; the **d** array contains the corresponding edge weights in the reordered list.

To summarize, VAT reorders the input dissimilarity matrix $D \to D^*$ and enables display of a gray-scale image $I(D^*)$ whose *ij*th element is a scaled dissimilarity value between objects o_i and o_j. (Many cluster heat maps are rendered in color.) Each element on the diagonal of a VAT image is zero (colored black). Off the diagonal, the values range from 0 to 1. If an object is a member of a cluster, then it also should be part of a submatrix of "small" values, whose diagonal is superimposed on the diagonal of the image matrix. There is a very strong relationship between VAT/iVAT images and single linkage clustering which will be made explicitly in Section 8.5.

Most clustering methods have parameters to tinker with which enable you to tune them up with good choices (or wreck them with bad choices), but A4.1 is completely user-friendly in the sense that it has no parameters to choose. You will always get the same output every time A4.1 runs on the same input when the MST of D is unique.

While it may seem unscientific, pictorial representations such as iVAT images of multidimensional data are a useful precursor to formal analysis with partitioning techniques. Images such as these help us understand what *may be true* about clusters in data sets. But they never *prove* anything about the data, as noted by Ronald Fisher nearly 100 years ago:

> *The preliminary examination of most data is facilitated by the use of diagrams. Diagrams prove nothing, but bring outstanding features readily to the eye; they are therefore no substitute for critical tests as may be applied to the data, but are valuable in suggesting such tests, and in explaining conclusions founded upon them.*
>
> –Ronald Fisher (1924)

After we decide that a data set contains cluster substructure by any method whatsoever, our next task is to *find* the clusters. That is the job of clustering algorithms (the partitioning step in Figure 4.1), the topic we turn to next.

4.3 Clustering (Partitioning the Data into Clusters)

The second problem shown in Figure 4.1 – partitioning the data – is the heart of cluster analysis. First, you need to choose a clustering model CM that defines CPOV clusters. Your choice depends, to some extent, on the numerical representation (X or R) you have of the objects O. Once you have a model in hand, you must choose an algorithm (a member of the set $\{A_k\}$ in Figure 4.1) capable of finding clusters that are *optimal in the sense of the model.*

Much of the clustering literature ignores the distinction between a model that defines clusters and the algorithms that are used to find the optimal clusters defined by it. This is unfortunate because the *mathematical properties that define CPOV clusters in the data* are used by algorithms associated with the model while seeking a U for O. These properties *reside in the mathematical model* that underlies the algorithm. We illustrate the distinction with a discussion of *hard c-means*[3] (HCM) clustering, which begins with the objective function

$$J_1(U, \mathbf{V}; X) = \sum_{k=1}^{n} \sum_{i=1}^{c} u_{ik} \|\mathbf{x}_k - \mathbf{v}_i\|_A^2. \tag{4.1a}$$

In Equation (4.1a), $X = \{\mathbf{x}_1, \ldots, \mathbf{x}_n\} \subset \Re^p$ is a set of object data, $U \in M_{hcn}$ is a crisp c-partition of X, the c vectors $\mathbf{V} = \{\mathbf{v}_1, \ldots, \mathbf{v}_c\} \subset \Re^{cp}$ are *cluster centers* (prototypes, templates, code vectors, dictionary elements, centroids, etc.) of the crisp clusters that U imposes upon X, and the matrix $A \in \Re^{pp}$ is a positive-definite weight matrix that induces the inner product norm $\|\mathbf{x}_k - \mathbf{v}_i\|_A = \sqrt{(\mathbf{x}_k - \mathbf{v}_i)^T A(\mathbf{x}_k - \mathbf{v}_i)}$. What does the function $J_1(U, \mathbf{V}; X)$ measure? Let X_i denote the n_i vectors in X that correspond to the n_i 1's in row i of U. For any $\mathbf{V} = \{\mathbf{v}_1, \ldots, \mathbf{v}_c\} \subset \Re^{cp}$, an alternative way to write Equation (4.1a) is

$$J_1(U, \mathbf{V}; X) = \sum_{i=1}^{c} \left(\sum_{\mathbf{x}_k \in X_i} \|\mathbf{x}_k - \mathbf{v}_i\|_A^2 / n_i \right). \tag{4.1b}$$

This form shows that J_1 is sum of c values, each an average of the sum of squared distances between cluster center $\mathbf{v}_i$ and the points "attached to it," i.e., in X_i. This is a geometric interpretation of J_1. We will explain the statistical meaning of Equation (4.1a) in Chapter 6. How shall we choose a pair (U, **V**) that points to "good" clusters in X? J_1 is the HCM objective function, but it is not yet a clustering model. A well-defined model of clusters in X requires another step, and that is using J_1 to define the non-linear optimization problem

$$\underset{(U,\mathbf{V}) \in M_{hcn} \times \Re^{cp}}{\text{minimize}} \left\{ J_1(U, \mathbf{V}; X) = \sum_{k=1}^{n} \sum_{i=1}^{c} u_{ik} \|\mathbf{x}_k - \mathbf{v}_i\|_A^2 \right\}. \tag{4.2}$$

While this looks like a very common type of optimization problem, we will discover in Chapter 6 that it is not readily solved. Equation (4.2) is legitimately called a clustering model, and it has a number of names in the literature. For example, many authors refer to Equation (4.2) as the *with in group sum of squares* (WGSS) clustering criterion, while others refer to it as the minimum variance partitioning problem. Optimal CPOV clusters of X are the U part of optimal pairs for J_1. It is not hard to show that, for a given $U^* \in M_{hcn}$, the optimal (and unique) choice for the unconstrained "half-problem" in **V**, namely, $\min\{J_1(U^*, \mathbf{V}; X)\}$, is given by

$$\mathbf{v}_i^* = \sum_{k=1}^{n} u_{ik}^* \mathbf{x}_k \Big/ \sum_{k=1}^{n} u_{ik}^* = \sum_{\mathbf{x}_k \in X_i} \mathbf{x}_k / n_i, 1 = i = c. \tag{4.3}$$

The c cluster centers at Equation (4.3) are the centroids of the c clusters defined by the rows of U. Legendre (1805, p. 75) closed his introduction to the method of least squares by describing this result as follows:

> *We see, therefore, that the method of least squares reveals, in a manner of speaking, the center around which the results of observations arrange themselves, so that the deviations from the center are as small as possible.*

[3]Many writers call this the k-means model (k = c). More discussion on this appears in the notes and remarks for this chapter.

The update formula for U given $\mathbf{V}^*$ that solves the other half problem when $\mathbf{V}^*$ is fixed, $\min\{J_1(U, \mathbf{V}^*; X)\}$, is the well-known formula which is usually attributed in principle to Lloyd (1957) when A = I induces the Euclidean norm on the input space:

$$u_{ik}^* = \left\{ \begin{array}{ll} 1; & \|\mathbf{x}_k - \mathbf{v}_i^*\|_A^2 \le \|\mathbf{x}_k - \mathbf{v}_j^*\|_A^2, j \ne i \\ 0; & \text{otherwise} \end{array} \right\}; \left\{ \begin{array}{c} 1 \le i \le c \\ 1 \le k \le n \end{array} \right\}. \tag{4.4}$$

Many writers interpret Equations (4.3) and (4.4) as paired *first-order necessary conditions* (FONCs) for extrema of J_1. However, this is not exactly true. Equation (4.3) is a first-order necessary condition, obtained by setting the gradient of $J_1(U^*, \mathbf{V})$ equal to $\mathbf{0}$ and solving for $\mathbf{V}^*$. But Equation (4.4) can't be obtained this way since $J_1(U^*, \mathbf{V})$ is not differentiable in U. Chapter 6 contains the derivation for Equation (4.4), which is necessary, but not in the same sense as Equation (4.3). We examine this aspect of the HCM model in more detail in Chapter 6.

How should we look for "good" U*'s that solve the optimization problem at Equation (4.2)? You might think there is only one way to go about this, but there are about half a dozen different algorithms in the literature that attempt to efficiently find solutions for Equation (4.2). The most naive approach is exhaustive search. Since M_{hcn} is a finite set, it is possible, in principle at least, to simply enumerate all the U's, find the **V**'s from Equation (4.3) that are paired with them, and pick the $(U^*, \mathbf{V}^*)$ pair that yields the smallest value of J_1. For tiny data sets, this can be done, but for data sets where c is small compared to n, $|M_{hcn}| \approx c^n/c!$ so exhaustive search quickly becomes impossible (Duda and Hart (1973)). For example, there are roughly $3^{150}/6 \approx 6.16(10^{70})$ crisp 3-partitions of the Iris data – *way* too many to look at all of them! Nonetheless, exhaustive search is one algorithm which might be used in an attempt to solve Equation (4.2) – that is, it is one HCM clustering algorithm.

A second, much more popular, HCM algorithm is based on calculating U and **V** using formulae (4.3) and (4.4), which are the basis for approximately minimizing J_1 with *alternating optimization* (AO *aka* (also known as) grouped coordinate descent, block optimization, etc.). Chapter 10 presents an in-depth look at this optimization technique. In a few words, just guess U (or **V**), compute **V** (or U), compute a new U (or **V**), compare the old and new estimates, and stop when successive iterates produce little change. We will discuss this approach to HCM in much greater detail in Chapter 6. Here, it suffices to observe that this second algorithm is the most common approach to solving Equation (4.2), and we identify it in the remainder of Chapter 4 as the HCM/AO algorithm (Lloyd's algorithm when the model norm is Euclidean).

Are there other ways to approximately solve Equation (4.2)? Sure. Various authors have proposed minimization of J_1 using, for example, *genetic algorithms* (GAs) (Hall *et al.*, 1999; Pacheco and Valencia, 2003). The GA approach basically ignores Equations (4.3) and (4.4) and, instead, generates populations of possible solutions $\{(U, \mathbf{V})\}$ of Equation (4.2) that are candidates for minimizing J_1, which in this context is called a fitness function. Exercising the GA identifies the "fittest" (U, **V**) – i.e., the one that produces the smallest value of J_1 arising from the candidates on hand. There are many different models in the HCM/GA family, depending on the specifics of the GA in use.

Problem (4.2) has also been attacked with *particle swarm optimization* (PSO), for example, in Deng *et al.* (2005). This constitutes a fourth family of HCM algorithms, the HCM/PSO family. So, now we have HCM/Search, HCM/AO, HCM/GA, HCM/PSO, etc. All of these algorithms are legitimately called "HCM clustering algorithms" because they all attempt to approximately optimize the HCM clustering model at Equation (4.2). Some other approaches are briefly discussed in Section 6.9. Figure 4.9 illustrates this point:

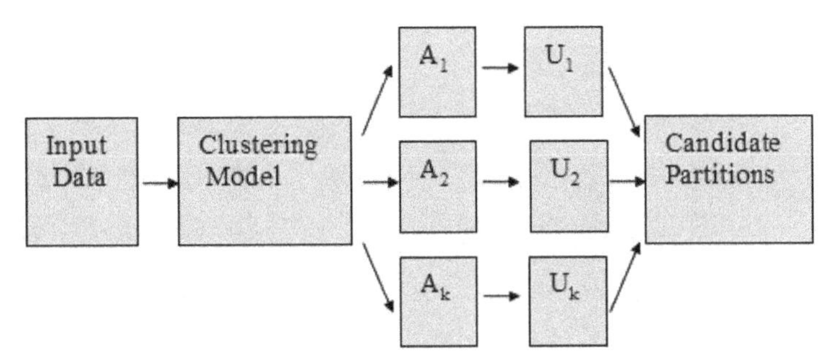

Figure 4.9 Different $\{A_k\}$ and $\{U_k\}$ for a fixed model and input data.

for a fixed data set and fixed clustering model, it is possible to discover competing, non-equal candidate partitions of the data with various algorithms.

By far, the most common approach to optimization problem (4.2) is the basic HCM/AO algorithm, which will be used exclusively in the remainder of this book. The AO approach to solving Equation (4.2) is used so often, compared to other schemes for minimizing J_1, that either part, the CM (4.2) or the AO algorithm {(4.3), (4.4)}, is called "hard c-means." This terminology is incomplete.

Clustering models should always be separated from the algorithms used to exploit or optimize them because – as depicted in Figure 4.9 -different algorithms can and do produce very different clusters for the same data and clustering model. Even more disturbing, perhaps, is that the same algorithm and model can produce the "wrong" clusters, as is illustrated in the next example.

Example 4.7. The global HCM minimum can be misleading.

Figure 4.10(a) shows X_{29}, a data set of 29 points in the plane, that are partitioned into c = 2 HPOV subsets, comprising 4 points in the left subset (X_1), and 25 points in the right subset (X_2). Let the hard partition $X_{29} = X_1 \cup X_2$ with $X_1 \cap X_2 = \emptyset$ be represented by the crisp partition matrix $U_1 \in M_{h2,29}$. The points in each subset are separated in (Euclidean) distance by one unit in the two orthogonal directions. The set X_1 has centroid $\mathbf{v}_1$ (shown as a red triangle in the center of the 4 points): $\mathbf{v}_1$ is not in X_1. The sum of squared Euclidean distances between the 4 points in X_1 to $\mathbf{v}_1$ is $\sum_{x \in X_1} \|\mathbf{x} - \mathbf{v}_1\|^2 = 2$. The set X_2 has centroid $\mathbf{v}_2$, which is the center point of X_2. The sum of squared Euclidean distances between the 25 points in X_2 to $\mathbf{v}_2$ is $\sum_{x \in X_2} \|\mathbf{x} - \mathbf{v}_2\|^2 = 100$.

So, the sum of squared errors for the 2-partition U_1 of X_{29} shown in Figure 4.10(a) relative to the pair of centroids $\mathbf{V}_1 = (\mathbf{v}_1, \mathbf{v}_2)$ is $J_1(U_1, \mathbf{V}_1) = \sum_{x \in X_1} \|\mathbf{x} - \mathbf{v}_1\|^2 + \sum_{x \in X_2} \|\mathbf{x} - \mathbf{v}_2\|^2 = 2 + 100 = 102$. To emphasize the meaning of (the value of) the k-means objective function J_1, look at Figures 4.10(b) and 4.10(c), which show the sets X_1 and X_2 repositioned in different parts of the plane. The two sets don't intersect in either view, i.e., $X_1 \cap X_2 = \emptyset$ is still true, and, hence, the partition U_1 is intact. What *has* changed? The coordinates of the 29 points and the two centroids are different, but, importantly, the sum of squared errors from the points in each subset to its centroid remains the same. So, the value of the objective function doesn't change either, that is, $J_1(U_1, \mathbf{V}_1) = 102$ for all three sets of points. Do you see that X_1 and X_2 can be placed *anywhere* in the plane and the value of $J_1(U_1, \mathbf{V}_1)$ will never change if the positions of the individual

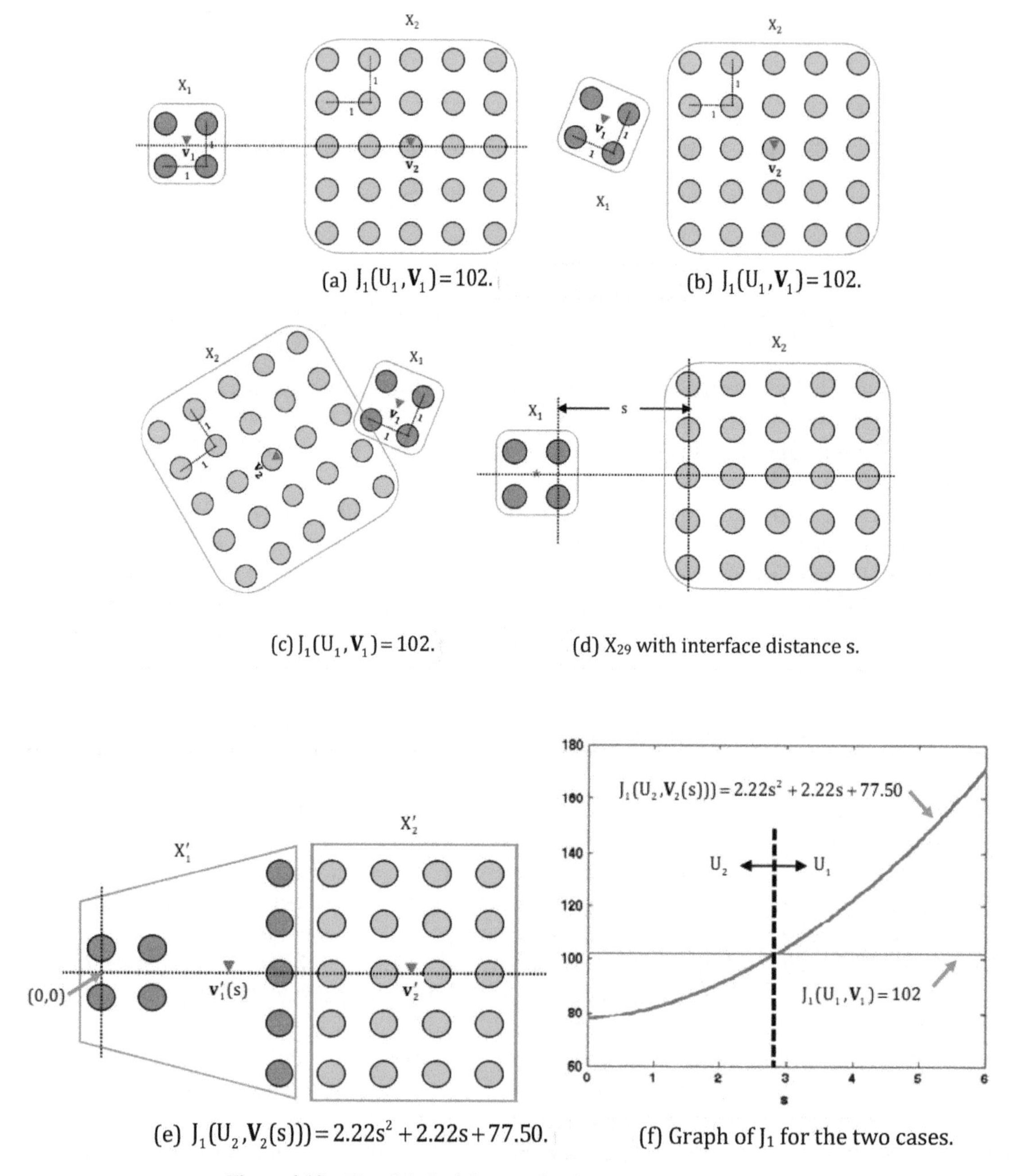

Figure 4.10 The global minimum of J_1 for two 2-partitions of X_{29}.

points relative to each other are fixed? Since U_1 agrees with our HPOV intuition about CPOV clusters in X_{29}, it is natural to suppose that 102 is the global minimum of J_1. Let us look into that more closely.

The value of J_1 will change only if the 29 points are assigned to a different pair of subsets, say $X_{29} = X'_1 \cup X'_2$ with $X'_1 \cap X'_2 = \emptyset$ and $X_1 \neq X'_1$, $X_2 \neq X'_2$. In Figure 4.10(d), the two subsets $\{X_1, X_2\}$ are aligned so that their centroids are on the horizontal line connecting them and are separated by an interface distance s. (The points in all the views in Figure 4.10 look like they have area, but this is just so you can see them.) We know that this arrangement of the two subsets will yield the value $\mathbf{J}_1(\mathrm{U}^*, \mathbf{V}^*) = 102$ for any value of s for which the two subsets are disjoint. Is there another 2-partition of X_{29} which could yield a smaller value of J_1 than 102? Now you can see the idea in Equation (4.2): find the pair $(\mathrm{U}, \mathbf{V}) \in M_{h2,29} \times \Re^2$ that makes J_1 as small as possible. For this little example, there are about $(2^{29}/2)$ partitions of X_{29} into two disjoint subsets; so it is not a trivial undertaking to decide whether 102 is the global minimum of J_1 for $X_{29} = X_1 \cup X_2$. And, in fact, as we shall see shortly, it is not.

Now consider the partition $U_2 \in M_{h2,29}$ shown in Figure 4.10(e). The centroid $\mathbf{v}_2'$ of the 16 points called X_2' on the right side in Figure 4.10(e) is fixed, and it is easy to show that $\sum_{\mathbf{x} \in X_2'} \|\mathbf{x} - \mathbf{v}_2'\|^2 = 65$. But the position of the centroid $\mathbf{v}_1'(s)$ for the 9 points in X_1' is a function of the interface distance s; so the vector of centroids is too, $\mathbf{V}_2(s) = (\mathbf{v}_1'(s),\ \mathbf{v}_2')$. Consequently, J_1 is, in turn, a function of s, $J_1(U_2, \mathbf{V}_2(s))$. It is likely that for most values of s, $J_1(U_2, \mathbf{V}_2(s)) \neq J_1(U_1, \mathbf{V}_1)$. If we choose the coordinate system so that the origin (0,0) is as shown in view (e), the value of the y coordinate of $\mathbf{v}_1'(s)$ is zero. The value of the x coordinate of $\mathbf{v}_1'(s)$ depends on s.

Straightforward but tedious algebra for the coordinate of $\mathbf{v}_1'(s)$ leads to $J_1(U_2, \mathbf{V}_2(s)) = 2.22s^2 + 2.22s + 77.50$ as a function of the interface distance. Equating this quadratic to 102 yields two intersections, (–3.86, 102) and (2.86, 102). Since s is positive, we take the second root. Thus, $s > 2.86 \Rightarrow J_1(U_1, \mathbf{V}_1) < J_1(U_2, \mathbf{V}_2(s))$; so J_1 points to the HPOV clusters as the better partition. But when interface distance $s < 2.86 \Rightarrow J_1(U_2, \mathbf{V}_2(s)) < J_1(U_1, \mathbf{V}_1)$, J_1 points to the partition U_2 in Figure 4.10(e) as the more desirable solution. Why? Because the overall sum of squared errors for the configuration U_2 is less than it is for U_1. Figure 4.10(f) illustrates this graphically: to the left of the vertical line at 2.86, J_1 prefers the partition U_2. This shows that 102 is definitely not the global minimum of J_1 over $(\mathrm{U}, \mathbf{V}) \in M_{h2,29} \times \Re^2$. This is the theory. But in practice, what happens depends on the initialization and termination criteria for the k-means algorithm used to generate 2-partitions of X_{29}. Most trials of Lloyd's HCM algorithm A6.1 will terminate at the HPOV solution U_1 for at $s > 4$. However, for $3<s<4$, the situation becomes unpredictable. And for $s <3$, sometimes, two columns of the larger subset will be cut off and joined with the 4 points called X_1.

This example has two main points. First, failure to find the "right" partition of X_{29} is not a failure of any HCM *algorithm*. No alternate way to find the partitions will avoid this. This is an example of *model failure*. Minimizing J_1 tends to "equalize" cluster sizes by allocating points that seemingly belong to bigger clusters to clusters having a smaller number of points (cf., Duda and Hart (1973) for another example). Second, like all objective function methods, a smaller value of J_1 doesn't always point to a "better" partition, even though it is a better solution to an optimization problem such as Equation (4.2); so we almost always need some additional machinery to separate the wheat from the chaff. That machinery is the heart of the third canonical problem in Figure 4.1, namely, cluster validity.

Forewarned is forearmed: For any c, $1 < c < n$, the set of all feasible values $\{J_1(\mathrm{U}, \mathbf{V}; X) : \mathrm{U} \in M_{hcn}; \mathbf{V} \in \Re^{cp}\}$ is finite because there are only finitely many values of $\mathrm{U} \in M_{hcn}$, so this set always has a global minimum that is greater than or equal to zero. But ... since this set is a discrete subset in the joint solution space contained in $M_{fcn} \times \Re^{cp}$, there can't be any "local minimums" of J_1 in the classical sense for the error surface defined by $J_1(U, \mathbf{V}; X)$. See Figure 10.3(b) for a graphic illustration of this fact. We return to this aspect of HCM in Chapters 6 and 10. ☺

It is hard (ok, impossible) to build a taxonomy of clustering models that separate them into mutually exclusive, exhaustive sets. Table 4.1 lists a few of the many ways to classify clustering models. The columns in Table 4.1 are *not* coupled. For example, most probabilistic clustering models, such as the one discussed in Chapter 7, are (U, **V**, +) models. A different way to organize approaches to clustering is shown in Table 4.2. Each of the model types from Table 4.1 is represented by various approaches that fit into the categories in Table 4.2. For example, Chapter 7 describes the (U, **V**, +) *Gaussian mixture decomposition* (GMD) model, which is a statistical model for object data defined by a global maximum likelihood objective function. This model allows overlapping partitions (probabilistic clusters), is non-adaptive *with respect to* (wrt) the number of clusters, and is usually optimized by an iterative (AO) approach called the expectation-maximization (EM) algorithm.

Tables 4.1 and 4.2 offer different ways to classify clustering models and algorithms, but the most useful way to classify models is by the parameters they estimate (the first column in Table 4.1). U is a c-partition generated by a U-only model such as the single linkage model in Chapter 8. **V** is a set of c cluster centers that might be obtained, for example, by the *learning vector quantization* (LVQ) model of Kohonen (1989). (U, **V**) is a (partition, prototype) pair that arises, for example, from any of the c-means models. And finally, (U, **V**, +) is a clustering model such as GMD, which produces a partition, a set of means, and the added (+) parameters that are the prior probabilities and covariance matrices for the observed data.

Table 4.1 Some different ways to classify clustering models.

Parameters	Partition	Strategy	Model Structure	Implementation
U only	Crisp	Hierarchical	Objective function	Batch
V only	Fuzzy	Partitional	Graph theoretic	Incremental
(U, **V**)	Probabilistic	Overlapping		Iterative
(U, **V**, +)	Possibilistic	Ordinal		Streaming

Table 4.2 A taxonomy of clustering methodologies.

Data type	Object Data/Relational Data
Direction	Agglomerative/divisive
Nesting	Hierarchical/non-hierarchical
Partition	Overlapping/non-overlapping
Criterion	Objective function/graph-theoretic/heuristic
Domain of criterion	Local/global
# of clusters	Adaptive/non-adaptive
Model type	Crisp/fuzzy/probabilistic/possibilistic
Architecture	Sequential/simultaneous/parallel
Algorithm	Direct/iterative/decompositional

Here is a definition for computational clustering, the second activity shown in Figure 4.1:

Definition 4.3. CPOV clustering.

CPOV clustering (partitioning the data set X or R) comprises computational models and algorithms that identify *candidate partitions* $U \in M_{pcn}$ of a given input data set[4].

The correct way to refer to CPOV clusters in data is to *always* identify them with reference to a particular model and algorithm. For example, the fact that X contains compact, well-separated HCM/AO clusters may not guarantee that it contains identifiable clusters using some other model, or even some other algorithm that attempts to optimize the HCM objective function, such as HCM/PSO. Unfortunately, when writing about clusters in data, there is a tendency to get lazy and ignore this crucial fact: that they are clusters only with reference to a well-defined mathematical model *and* algorithm. You may read that the four-dimensional Iris data "has 2 or 3 clusters" as if Iris really does have two or three (HPOV) clusters, independent of the way a mathematical model defines the clusters. This, of course, is impossible.

4.4 Cluster Validity (Which Clusters are "Best"?)

Assuming that we have some ways to *get* clusters (i.e., partitions U of input data set X or R), we turn to the last problem in Figure 4.1 – how to validate them? Figure 4.1 shows yet a third model, VM, a *validity model*, brought in to evaluate the validity of the clusters found by the CM. Basically, there are two general approaches to validation: *visual* and *non-visual*. The iVAT algorithm A4.1 can be used in conjunction with other techniques for visual validation. Non-visual methods almost always depend on a *cluster validity index* (CVI). Why would we need such a validation model?

To understand the role of CVIs for validation of U's in CP, we begin with a discussion of what – if anything -can be deduced about candidates when you have crisp c-partitions of an unlabeled object set O, and nothing else -not even access to the data. Are there ANY internal CVIs – functions of U alone -in this category? Let us consider the following example.

Example 4.8. Evaluating crisp partitions without any help – needles in a haystack.

Your professor sends you the four CPOV crisp partitions, $\{U_1, U_2, U_3, U_4\}$ shown at (4.5) of a set $O = \{o_1, o_2, o_3, o_4\}$ comprising n = 4 objects , and asks you: "which one is best"?

$$U_1 = \begin{bmatrix} 1 & 1 & 1 & 1 \end{bmatrix}, U_2 = \begin{bmatrix} 0 & 0 & 1 & 1 \\ 1 & 1 & 0 & 0 \end{bmatrix}, U_3 = \begin{bmatrix} 1 & 1 & 0 & 0 \\ 0 & 0 & 1 & 0 \\ 0 & 0 & 0 & 1 \end{bmatrix}, U_4 = \begin{bmatrix} 1 & 0 & 0 & 0 \\ 0 & 1 & 0 & 0 \\ 0 & 0 & 1 & 0 \\ 0 & 0 & 0 & 1 \end{bmatrix}. \quad (4.5)$$

You don't know what the objects are, how the U's were obtained, or what property of U would serve to define a best partition among them. (And worse, it is impossible for you to guess what "best" means to your

[4]For completeness, this definition includes the case of possibilistic cluster analysis. But M_{pcn} will be an infrequent visitor in the sequel.

professor – for him or her, it probably means "can we publish this?") This amounts to randomly selecting four vertices in four different partition sets, $\{M_{hc4} : c = 1, 2, 3, 4\}$, which, on the face of it, seem equally likely to be "best" in some as yet undefined sense.

You can propose any number of functions that will define a "best" partition. For example, you could decide that the best partition should use the minimum storage space; so you would choose U_1, which puts all 4 objects together in 1 cluster. But how is this choice related to the four objects? It is not. Or, you might decide that the partitions for c = 1 and c = 4 should be ruled out on the general principle that they are in some sense trivial. It is mighty hard to imagine a function that chooses a best partition that is actually related to the set O in any meaningful way, and even harder to imagine what you will tell your professor. I don't think there are any functions in this category. As for your professor? You are on your own.

Suppose we know a little more about the objects. Specifically, suppose we have a data set for which the HPOV clusters can be seen visually. In this case, we could tell if a candidate CVOV clustering of the objects was useful, or "best." But, please consider what happens in Example 4.9.

Example 4.9. The simplest data sets can fool k-means clustering.

Figure 4.11(a) shows data set X_{30} comprising 30 points in the plane that have three visually apparent HPOV clusters of 10 points each. The coordinates of the 30 points are listed in Table 7.2. View 4.11(b) shows the crisp CPOV 3-partition U* of X_{30} found by HCM/AO ("k-means") from a random initialization at c = 3. The points in the three terminal clusters are numbered 1–10, 11–20, and 21–30, as shown in Figure 4.11(b). Later examples will refer to these three groups using this labeling of the data, and U* will be called the *ground truth partition* of X_{30}. U* is the terminal state for this data set using the HCM/AO model and

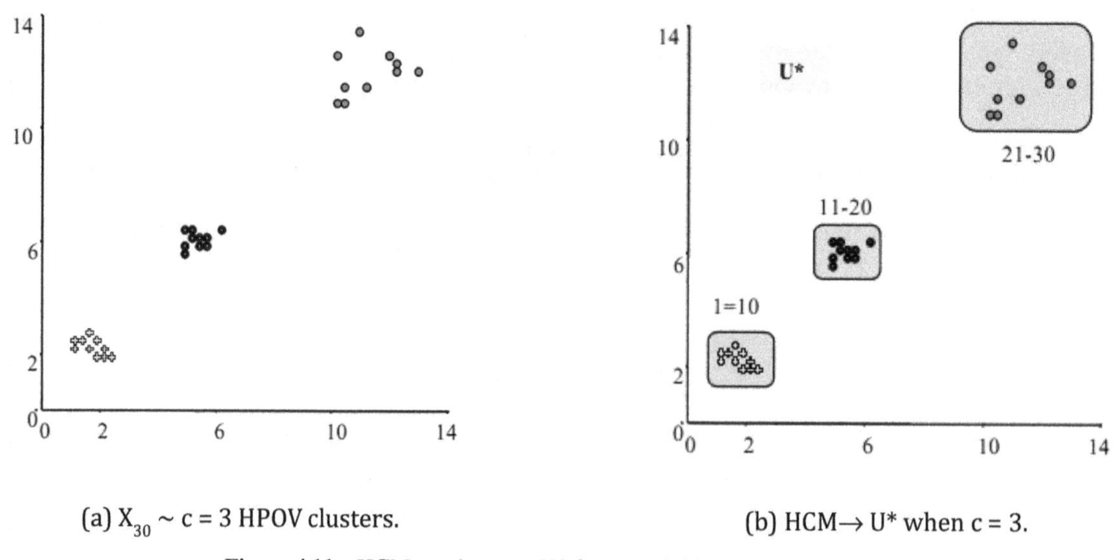

Figure 4.11 HCM terminates at U* from any initialization when c = 3.

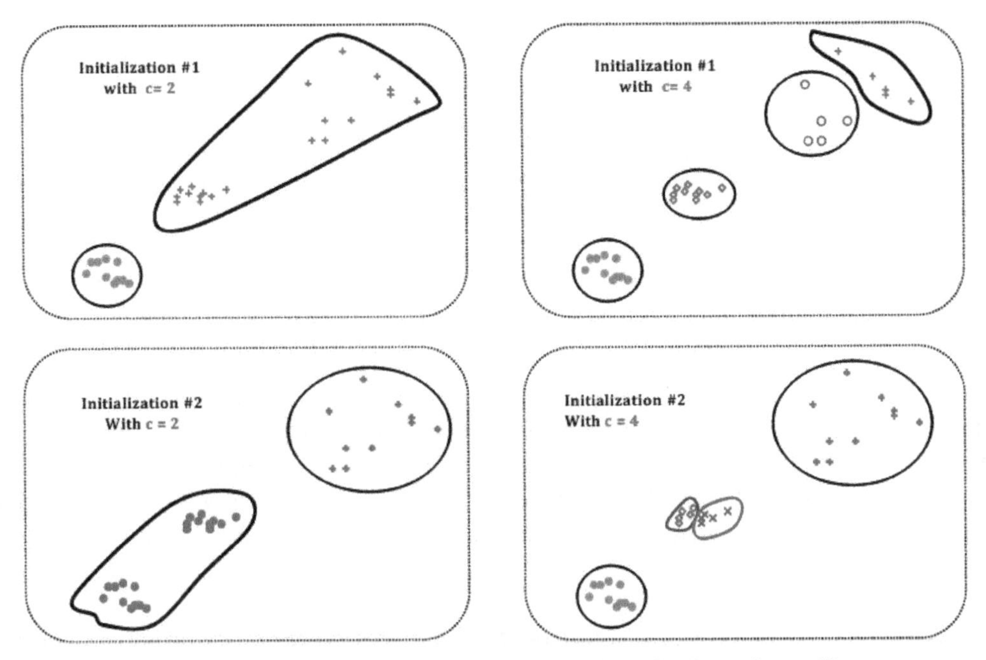

Figure 4.12 HCM terminations from different initializations when $c \neq 3$.

clustering algorithm from any initialization, but ... only as long as we instruct k-means (HCM) to look for c = 3 clusters.

Figure 4.12 shows four other partitions of X_{30} found by requesting HCM/AO to seek either c = 2 or c = 4 clusters in the data. On the left, Figure 4.12 contains two different partitions of X_{30} into c = 2 clusters. The right views in Figure 4.12 are two different sets of HCM/AO clusters at c = 4. The only variable for the processing at c = 2 and c = 4 was the initialization of U for the AO loop. This demonstrates two things. First, if we don't look in the right solution set (that is, $M_{h3,30}$), it is impossible to obtain the HPOV result. And, second, even though the data seem to have compact, well-separated clusters that are well-matched by a mathematical model (HCM here), the algorithm used to find the "good" clusters can produce very undesirable results – results that are sensitive to implementation parameters such as the initial guess for U.

So, Figure 4.12 shows that HCM/AO can generate solutions we want to reject and that values of J_1 alone are not necessarily capable of telling us anything about these unwanted results (incidentally, this tells us that successively close values of J_1 are not a very good way to terminate the k-means iteration). Perhaps, some validity model (VM) can help.

You may wonder: if the global minimum of the HCM model J_1 can't identify the clusters you want (and we already know from Example 4.2 that this can happen), then why not directly optimize a validity model? First, no model can capture all the properties that "good" clusters might possess, and this of course includes any particular VM we might propose. For example, we seek, from data set to data set, clusters with quite different properties: compactness, isolation, maximal crispness, density gradients, particular distributions, etc. And, more importantly, many of the most common CVIs don't fit naturally into a well-behaved framework for mathematical optimization. So, we use validity measures as an "after the fact" way to gain further confidence in a particular clustering solution.

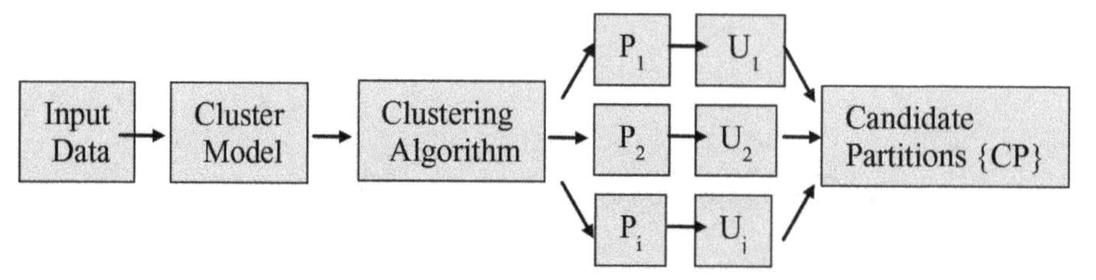

Figure 4.13 Different parameters can produce different clusters.

Suppose you had the five partitions for X_{30} shown in Example 4.8, but $X_{30} \subset \Re^p$ with $p > 3$. One partition might be correct (most useful in some well-defined sense) or, at least, best amongst the members of the five solutions you have, and the other four, like those in Figure 4.12, might be very misleading. But since you can't see the data, there is no HPOV that will help you sort through these solutions to find the right or most attractive partition. How would you know which to choose? And what would "correct" or "best" even mean? This is the crux of the validation problem. The situation in Figure 4.12 is specific to different runs of the HCM/AO algorithm. This can happen with *any* clustering algorithm. Figure 4.13 is a companion to Figure 4.9 that illustrates a different aspect of collecting candidate partitions.

In Figure 4.13, the input data (X or R), *clustering model* (CM), and *clustering algorithm* (A) are all fixed. The only variables are the algorithmic parameters $\{P_j\}$, which may be either (i) model parameters such as c, the number of clusters, or (ii) implementation parameters such as the termination criterion for the algorithm. Partition U_i is the partition obtained by CM/A using parameters P_i. The set of *candidate partitions* (CP) contains more than one possible solution.

Definition 4.4. The set of CPOV candidate partitions (CP).

For a given data set (X or R), any finite set of CPOV partitions obtained by clustering in X or R is a set of *candidate partitions*, CP = $\{U \in M_{fcn}: c_m \leq c \leq c_M\}$, where c_m and c_M are the minimum and maximum values for c represented amongst the matrices in CP.

There is still another way to create the set CP that complicates things even more. We may have different clustering models in each of Figures 4.9 and 4.13. This leads us to the most general case, which is shown in Figure 4.14. In this figure, X or R is the input data; CM_i is the *i*th clustering model i; A_{ij} is clustering algorithm j for clustering model i; P_{ijk} are parameters k for model (i) and algorithm (ij); and U_{ijk} is the resultant partition of the data.

We have already discussed the possibility that there can be more than one "best c," depending on the nature of the question being asked about the data. Sometimes CP contains only one U at each value of c in $\{c_m, \ldots, c_M\}$. In this case, "choosing the best U" in CP is often taken to mean "choosing the best c." But Example 3.2 illustrates that there can be more than one (best U and best c) pair for a fixed data set. You can see that there are a *LOT* of possibilities for populating CP, the set of candidate partitions.

Another source of confusion about cluster validity is due to the very fine line that may exist between the three problems of assessment, clustering, and validation. The canonical problems can be quite distinct for some approaches to cluster analysis but practically impossible to separate for others. For example, some

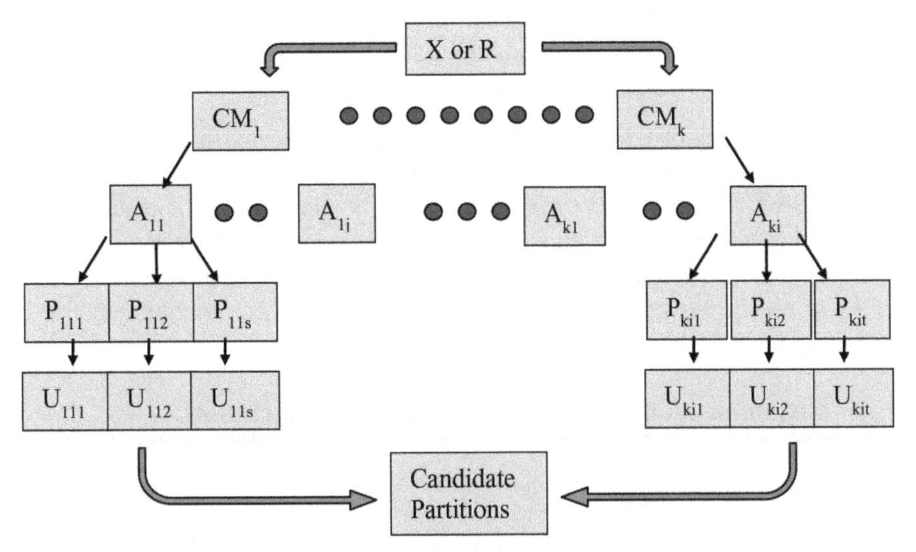

Figure 4.14 The set CP can contain partitions built from many variations.

Table 4.3 (Some) different ways to classify validation models.

Data	U	VM	VM	VM	VM assesses
X	M_{hcn}	Visual	Internal	Direct	One cluster
R	M_{fcn}	Non-visual	External	Indirect	All clusters

methods solve the assessment problem implicitly by finding clusters, while some validation methods are part of the clustering model itself.

Just as there are a lot of ways to characterize clustering models, there are many ways to divide the landscape of validity models. One of the most important distinctions is whether the partition U is crisp or soft. A second distinction can be drawn based on the type of data that is clustered: *object data* (X) and *relational data* (R). A third characteristic will be whether the validation method is *internal* (uses only the information available from the algorithmic outputs) or *external* (uses "outside" information such as a reference or ground truth partition). And finally, we may call a validation model *direct* if it assesses crisp CPOV clusters in the data; otherwise, it is *indirect*. Table 4.3 summarizes some (but not all) ways to classify validation models. The columns of Table 4.3 are independent.

Definition 4.5. The cluster validity problem.

Cluster validity (validation) comprises computational models and algorithms that identify a "best" member amongst a set of *candidate partitions* $CP = \{U \in M_{fcn} : c_m \leq c \leq c_M\}$. Scalar measures of cluster validity are called *cluster validity indices* (CVIs). We call an index whose minimum value over a set of candidate partitions (CP) points to the "best one" as *min-optimal*, indicated by the symbol (↓). Similarly, indices that indicate a preferred choice of U in CP by their maximum value are *max-optimal* (↑). Internal CVIs will have the notational form $V(U)$; external CVIs will look like $V(U|Q)$, where Q is called the reference partition. Almost always, $Q = U_t$ or U_{gt} (cf. Definition 4.6). Here and hereafter, *cluster validity indices* (CVIs) will be denoted by the letter *V*. These *V*'s are functions that map various domains into the reals (yeah, it's another *V* to confuse you, but this one is the name of a function, unbolded, in a different font).

While any of the characteristics of CVIs in Table 4.3 provides a means for classifying them, there is an overriding consideration that elevates the distinction between internal and external CVIs to the top spot in our list. And that consideration is whether or not the data to be clustered is labeled ("fake clustering") or unlabeled ("real clustering"). Specifically, in the context of *real* clustering (with *unlabeled* data sets), internal CVIs are the *only* way to validate candidate partitions. So, the obvious question is: how shall we choose an internal CVI to use in the real case? Well, we try to develop some confidence in one or more internal CVIs by evaluating them with labeled data.

Definition 4.6. Crisp true and ground truth partitions.

There are two types of crisply labeled data:

(i) *Synthetic data* has – *by construction*-a known number of (in principle, HPOV clusters) labeled subsets, say, with a "true" crisp partition matrix $Q = U_t \in M_{hc_tn}$.

(ii) *Real data* (e.g., measured or collected) that possesses c_{gt} physically labeled subsets, in which case the data have the crisp "ground truth" partition $Q = U_{gt} \in M_{hcn}$.

For convenience, both of these cases are sometimes represented with a common notation, viz., $U_{GT} = U_r$ or U_{gt}.

☺ ***Forewarned is forearmed.*** Many writers call both types of partitions in Definition 4.6 "ground truth" partitions, but there is a good reason to make the distinction. We have emphasized often that real labeled data (such as the Iris data) has labeled subsets, which may or may not be CPOV clusters. But synthetic data (such as X_{30} in Example 4.8) has -by construction – (at the least, mathematical) HPOV clusters. This is why the two cases are separated in this volume. ☺

There are three ways to evaluate internal and external CVIs with labeled data, all of which have been studied extensively. It is important to understand that these three methods (I like to call them CVI races) are used for two purposes: (i) to evaluate the relative quality of different clustering algorithms used to generate sets of candidate partitions (U's in CP); and (ii) to evaluate competing internal CVIs. In the case of real clustering, the data are always unlabeled; so internal CVIs are the only choice, and the Best I/E method is the preferred way to choose one.

Best c: evaluation of internal CVIs: $V(U)$
Best U: evaluation of external CVIs: $V(U|Q)$
Best I/E: agreement of $V(U)$ with $V(U|Q)$

Any attempt to provide an in-depth treatment of this topic would require a considerable volume. Instead, examples of each method are scattered throughout Chapters 6, 7, 8, and 9. In the next chapter, we turn to the data itself – is it adequate to support the cluster analyses we intend to make? If not, what can we do about it?

4.5 Notes and Remarks

Assessment strategies often ignore the question of how a clustering model defines the CPOV clusters it seeks. This approach implicitly assumes that the assessment model defines clusters in a way that complements the definition of clusters implied by the clustering model. But to emphasize this point again, unless the models used to define clusters in steps 1 and 2 of Figure 4.1 are the same (AM = CM), it is hard to see how CM can deliver the same clusters that AM predicts we should be able to find. Jain and Dubes (1988) is one of the earliest references that discusses the problem of cluster assessment using formal, statistical methods.

Tukey (1977) and Everitt (1978) are early entry-level texts that discuss statistical and informal graphical methods (visual displays) for deciding what -if any -substructure is in unlabeled object vector data. Cleveland (1993) offers a very thorough discussion of the use of graphical methods for one-, two-, and three-dimensional data. There are many simple techniques such as histograms, Chernoff faces, box and whisker plots, radar graphs, trees and castles, and Andrews plots that are attractive for the analysis of small data sets in a few dimensions. Friendly and Denis (2005) have the following to say about the scatterplot:

> *Indeed, among all the forms of statistical graphics, the humble scatterplot may be considered the most versatile, polymorphic, and generally useful invention in the entire history of statistical graphics. Tufte (1983) estimated that between 70 and 80 percent of graphs used in scientific publications are scatterplots.*

A classic reference for the principles of effective visual display is the updated version of Tufte (2001). Many useful data mining and visualization methods are covered in Witten and Frank (2005). Soukup and Davidson (2002) is a nice reference for the practical application of some of the most well-known techniques. Speaking of classics, Tufte (1996) made this impossible-to-beat statement about visualization in 1996:

> *As for a picture, if it isn't worth a thousand words, the hell with it*
>
> –E. R. Tufte

The history of the minimal spanning tree is fairly well-known. According to Graham and Hell (1985), the MST problem and an algorithm for solving it were reported in 1926 by Boruvka (1926) (cf., Nesetril (2001) for translation of the 1926 paper). The two MST algorithms discussed in Section 4.2.2 that are now used almost universally were apparently discovered within a year of each other, Kruskal's (1956) and Prim's (1957) algorithms. However, Jarnik (1930) published a Czech language paper, which is acknowledged to be the same as Prim's algorithm, and many authors also refer to Prim's method as the *Jarnik-Prim* (JP) algorithm (and, in fact, even as the DJP algorithm, the "D" referring to Edsger Dykstra, who "rediscovered" it independently in 1959).

Prim's method has time complexity $O(\mathrm{m}\log\mathrm{n})$, whereas Kruskal's algorithm takes $O(\mathrm{m}\log\mathrm{m})$ operations ($|O| = |V| = n$, $|E| = m$). So, Kruskal's algorithm is preferable when G is sparse, i.e., when $m << n(n-1)/2$, but in most clustering problems, $m \approx n(n-1)/2$; so the time complexity of using either method is roughly the same. See Rosen, (2007) for most of the applications discussed in this book.

The visual representation of structure in unlabeled dissimilarity data has a long and rich history, which apparently began with Loua (1873), who presented a hand-shaded dissimilarity matrix image describing 20 neighborhoods in Paris. You can see Loua's image in Wilkinson and Friendly (2009). The earliest use of imaging techniques such as iVAT dates back to the 1909 work by Czekanowski (1909). This paper seems to have been the first to discuss a method for clustering (without reordering) in dissimilarity data using a visual approach.

Tryon (1939) paved the way for a different approach to visual clustering using visual assessment and aggregation of hand-rendered profile graphs for all three canonical problems discussed in this chapter. He and his wife clustered a group of 170 12-year-old boys into c = 5 groups that had different personality traits (such as "fights," "bossy," "unkempt," etc.). This important work was all done by hand, and the clusters were formed – much like Petrie's Egyptian pottery arrangements shown in Figure 1.6 -by visually grouping profile graphs of the 20 rows of the correlation matrix for the boys.

Reordering dissimilarity data to discover groups within it exemplified by models such as the VAT/iVAT images in Examples 4.1–4.4 apparently began with Cattell (1944). He first depicted clusters in pairwise dissimilarity data about the objects in O as an n $\times$ n image showing each distance as a hand-shaded pixel with one of three possible "intensities," reproduced here as Figure 4.15.

Compare this to Figure 4.6 to see the origin of VAT/iVAT in a very primitive state. I believe that Cattell (1944) also introduced the term "single linkage" into the clustering lexicon for the SAHN model and algorithm that occupies most of Chapter 8, as evidenced in the following quote that describes the construction of Figure 4.15:

> *With or without* ***single linkage*** *lists he then manipulates the order of the variables in an attempt to bring linkage correlations alongside the diagonal or as near to it as possible. If the process is successful, the resultant clustering is clearly and strikingly recorded, as in Diagram 1.*
>
> –Cattell

Important advances in visual clustering include Sneath (1957), who was the first person to generate dissimilarities with a modern computer. Floodgate and Hayes (1963) reported the first use of single linkage clustering with a computer, but their RDI was still hand drawn. I believe that Robert Ling (1973) was the first person to completely automate a computational method for reordering dissimilarity data and producing

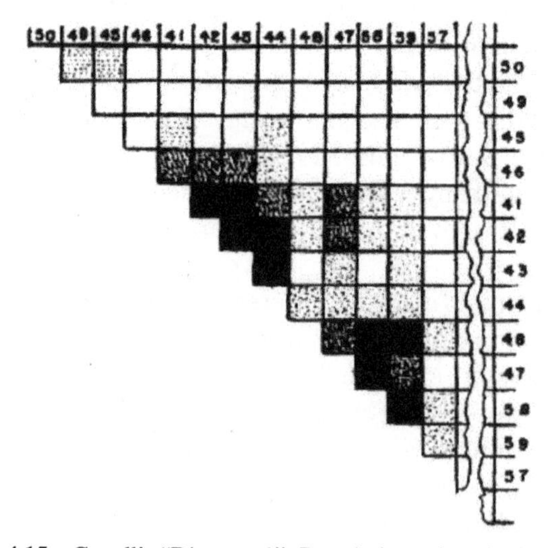

Figure 4.15 Cattell's "Diagram 1": Reordering using single linkage.

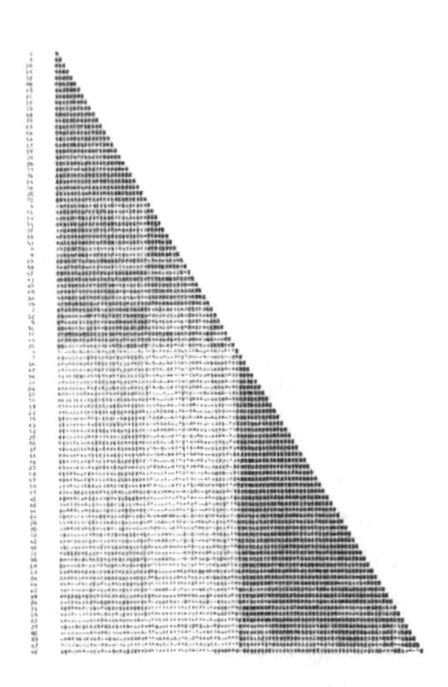

Figure 4.16 Ling's RDI of 99 points in Iris, made with the SHADE algorithm.

an image from it. Ling's algorithm, called SHADE, used complete-linkage hierarchical clustering (cf., Chapter 8) to reorder the data, and the "image" was created by overstriking standard printed characters by backspacing the carriage to produce 15 halftone intensities.

Figure 4.16 is an example of his technique, applied to the first 99 of the 150 points in Anderson's (1935) Iris data. I am not certain, but I believe this was done because the punch cards used by the computer for his program (implemented in FORTRAN IV) could only accommodate 99 characters per line. Only half of this image was created to save time and space, but it is based on a symmetric dissimilarity matrix; so it has the same general appearance as more modern digital RDIs such as iVAT images. The two dark blocks along the diagonal of this image represent two CPOV clusters in the 99 points found by complete linkage clustering algorithm.

Later methods for reordering and visualizing dissimilarity data include the "graphical method of shading" proposed by Johnson and Wichern (2007). Tran-Luu (1996) searched for the most "acceptable" block-diagonal matrix by optimizing an objective function that measures "blockiness." Other reordering methods include graph-based schemes, such as depth first search and schemes specific to unweighted connected graphs, such as the Reverse Cuthill-Mckee scheme described in George and Liu (1981), King's (1970) algorithm, and Sloan's (1986) algorithm. A recent innovation for weighted connected graphs is spectral ordering, West (2001), which reorders the connection matrix according to an Eigen-based decomposition of the graph's Laplacian matrix. This method is very effective and stable but suffers from a very high computational complexity $-\boldsymbol{O}(\mathrm{n}^6)$ for fully connected graphs. Siirtola and Makin (2005) present a very readable description of cluster analysis using matrix reordering based on the well-known barycentric heuristic. Figures 5 and 6 in their paper are visually quite similar to Figures 4.6 and 4.7 in this chapter. Reordering is also known as *seriation* in some academic circles. Liiv (2008) gives a very useful survey of seriation methods for social network analysis. Wang *et al.* (2010) describe an algorithm called specVAT which uses a different form of reordering than VAT and iVAT. An example using this type of ordering is given in Chapter 11.

See Zhang *et al.* (2009) for a commercial application of VAT in the context of role-based access control in industrial networks. The relationship between VAT and single linkage cluster analysis is discussed at length in Havens *et al.* (2009) and will be considered in some detail in Chapter 11.

Havens and Bezdek (2012) developed two methods for extending VAT/iVAT images for assessment of *co-clusters* in rectangular input data called coVAT/coiVAT. In particular, these algorithms are relevant to visual assessment of clustering tendency for the four clustering problems associated with co-clustering. The four problems involve finding row clusters, column clusters, clusters in the union of the row and column objects, and pure co-clusters within the set of clusters in the union.

The use of VAT type cluster heat maps for tendency assessment is usually based on visual assessment of the reordered dissimilarity image. This puts a human in the loop, and it becomes a matter of subjective judgment as to how many dark blocks appear along the diagonal of an RDI. For compact, well-separated subsets of objects, iVAT images are crisp and clear. But as the subsets mix and overlap, the images become less and less amenable to meaningful interpretation. This leads to the idea that there should be some way to automate the process of extracting the number of dark blocks. Much time and effort has been expended on this problem without much success (Keller and Sledge (2007), Hu and Hathaway (2008), Havens *et al.* (2008), Pakhira and Dutta (2011)). An extensive family of relatives of VAT/iVAT is surveyed in a recent paper by Kumar and Bezdek (2020), who discuss this aspect of visual tendency assessment in more detail.

Brian Everitt nicely sums up the basic utility of visual methods as follows:

> *[visual methods] should in no way be regarded as methods to be used to the exclusion of other types of multivariate analysis; indeed, they will, in general, be most helpful when used alongside, and in association with, other forms of analysis*

–Brian Everitt (1978)

This volume discusses (arguably) the four most heavily used models for cluster analysis. It is useful to understand where this topic fits into the overall structure in the larger enterprise known as *pattern recognition* (PR). Duda and Hart (1973) characterized *pattern recognition* (PR) as "a field concerned with machine recognition of meaningful regularities in noisy or complex environments." Bezdek (1981) said "pattern recognition is the search for structure in data." Theodoridis and Koutrombas (2009) state that" Pattern recognition is the scientific discipline whose goal is the classification of *objects* into a number of categories or classes."

All three of these definitions of PR include cluster analysis (indeed, the third one *almost* mirrors Definition 4.3 of CPOV clustering!). But pattern recognition also includes many other areas of investigation -most notably classifier design in all its forms, which will not be discussed in this book, and feature analysis of every kind, a few methods of which will be covered in Chapter 5. The fields commonly called machine learning and data mining nowadays also fall under the umbrella of pattern recognition methods.

The PR literature contains a lot of overlap, duplication, confusing jargon, and, in some instances, open hostility between "rival factions" that might well be called "scientific tongs." For example, you will find enormous literatures now that deal with data mining, machine learning, and knowledge discovery, and it is easy to come away with the impression that these fields have nothing, or at most little, in common with cluster analysis. But in fact, all of them largely (but not wholly) deal with exactly this topic. These literatures are so diverse and non-intersecting that it is impossible to imagine unifying the notation, models, algorithms, and examples that abound in each camp, much less making peace between the camp dwellers themselves. Just be alert, as you read books and papers about clustering, that it exists in many forms and under many different names.

There are many *really* good texts and papers about virtually every facet of clustering, and you can easily find hundreds of them with a quick Internet search. Several of these are mentioned here because of either their historical value or because they seem outstanding as general introductions to the subject. Tryon (1939) has already been mentioned in connection with his method for visual clustering.

I believe that the term "cluster analysis" (and, thus, the first book about it) was introduced in Tryon (1939). Tryon apparently coined the term to distinguish his orthometric approach to factor analysis from (standard) factor analysis, which was the analytic "tool du jour" of many psychologists of his day. Tryon proposed both analytic and visual methods for discovering cluster structure in the data he studied. We will return to some of Tryon's work in a later chapter.

The HCM model (4.2) and the AO algorithm based on Equations (4.3) and (4.4) that accompany it are studied in (excruciating?) detail in Chapter 6. The notes and remarks following Chapter 6 contain a fairly extensive account of the history of this most famous of all clustering models and algorithm, which is listed as one of the Top 10 algorithms in data mining by Wu and Kumar (2009). For now, it suffices to note that the first work (in English, anyway) that presaged the hard or classical c-means models appears in the 1957 paper by Lloyd (1957).

Probabilistic clustering based on GMD is discussed in Chapter 7. Statisticians started using probabilistic models to look for clusters in the late 19th century. Karl Pearson (1894) seems to have been the first author to address object vector clustering with a probabilistic model. His 1894 paper decomposes a mixture of two univariate Gaussian distributions with an algorithm based on sample moments. Seven years later, Pearson (1901) introduced principal components analysis as a means for extracting new features that made clusters more apparent in a new feature space. There are many excellent texts on probabilistic clustering with mixture models, including Titterington *et al.* (1985), which is perhaps still the definitive reference on mixtures, and McLachlan and Basford (1988), which contains a number of applications of mixture models to clustering. Devijver and Kittler (1982) is still a standard reference for nearest neighbor models, of which there are now an almost uncountable number.

Aldenderfer and Blashfield (1984) is a nice little book that amplifies the view of clustering espoused by Sneath and Sokal (1973), the classic reference for sequential, nested, hierarchical algorithms (such as the linkage algorithms of Chapter 8). While it is pretty old now, Aldenderfer and Blashfield (1984) have worthwhile insights about many issues in the practical use of clustering that are still timely today. Hartigan (1975) and Jain and Dubes (1988) continue to be useful texts.

Two fundamentally different approaches to cluster validity are validation of a single cluster vs. validation of a whole c-partition of the data. The former approach is represented by papers such as Keller and Sledge (2007), McQuitty (1967), Van Rijsberger (1970), and Hubert (1974). But most researchers concentrate exclusively on the second approach, which contains many more methods – almost too many to document but much less to discuss and illustrate.

You can find evidence reaching back to the earliest days of cluster analysis to support the assertion that of the three problems (assessment, clustering, and validation), validation is by far the most difficult. Thorndike observed that:

> *The approximation procedure for arriving at the optimum definition of clusters for a specified value of k [k = c in this book], the number of clusters, seems moderately satisfying. Now we must face the much nastier problem of determining the appropriate value for k.*

–Thorndike (1953)

A good deal of time has been spent discussing why validation is the most important and difficult of the three canonical problems. In the first place, there can be more than one "right answer," as in Example 3.2. Furthermore, even if there is a "right answer" at, say, c*, but you are looking for it in M_{fcn} with $c \neq c^*$, you can't arrive at this right solution. And finally, the ultimate problem for all validation models is that every clustering algorithm will impose structure on the data even when there is none.

Most of the books that have been mentioned contain at least one chapter on cluster validity. Surveys on crisp *cluster validity indices* (CVIs) that compare various validation schemes in one way or another began to appear in the early 1980s; cf., Dubes and Jain (1980). Milligan and Cooper (1985) compared 30 validity tests (which they called "stopping rules") using partitions generated by four hierarchical clustering methods. Gurrutxaga *et al.* (2011) present a very thorough critique of the "best c" methodology first championed by Milligan and Cooper. A highly specialized, nicely written survey of 15 internal CVIs was made in 2002 by Dimitriadou *et al.* (2002). These authors generated 162 12-dimensional data sets with each of the 12 variables restricted to having the value 0 or 1. Each data set had n=1000 samples, and the authors arranged for the data to have ground truth values of c* = 4, 5, or 6 clusters with varying degrees of probability. Nguyen *et al.* (2010) present a best c study of similarity measures and distance-based CVIs that compare pairs of crisp partitions using a variety of information-theoretic measures. They identify a total of 26 measures that are subdivided into 10 similarity measures and 16 distance measures.

Arbelaitz *et al.* (2013) published a very ambitious comparison of 30 internal min-optimal or max-optimal CVIs for crisp c-partitions. They avoid internal indices such as the modified Hubert index that requires subjective decisions (where is the "knee" in a graph?), and they also eschew CVIs for soft partitions. Their data sets were divided into two categories: 10 instances each of 72 synthetic data sets having c^*_{HPOV} HPOV clusters with a minimum 100 points per cluster and 20 real-world data sets having c_{PL} physically labeled subsets. Three crisp clustering algorithms were used to populate CP: it would be misleading to assert that the conclusions in this paper are right or wrong. Indeed, the terms "good CVIs" and "bad CVIs" are perhaps oxymoronic in the field of cluster validity. But this is a strong, well thought out, comprehensive approach to the business of CVI comparisons that should be studied by anyone with serious intentions to work in the field of cluster validation.

Validation of fuzzy partitions began with Ruspini (1969), who introduced the use of entropy in the context of cluster validity. This method was the basis for the partition coefficient discussed in Bezdek (1974). Pal and Bezdek (1995) present an extensive analysis of CVIs devoted to validation of FCM partitions. Bezdek *et al.* (1997) compare a number of validation models for probabilistic clusters generated by the EM algorithm A7.1 for GMD. Examples from both of these papers will appear in Chapters 6 and 7. See Wang and Zhang (2007) for a recent survey of many indices for validation of fuzzy partitions.

4.6 Exercises

1. Compute $D(U) = \mathbf{1}_{11\times 11} - U^T U$ and $D(U^*) = \mathbf{1}_{11\times 11} - \{U^*\}^T U^*$ in Example 4.3 to confirm that D(U*) contains two diagonal blocks of zeros: a 7×7 block and a 4×4 block, corresponding to the two clusters in O*.
2. The average weight of a pineapple is 1000 grams, and the average weight of a cherry is 5 grams. Using this numerical information for the fruits in Example 4.3, compute the matrices U, U*, D, and D*, where D is the Euclidean distance between each pair of fruits. If you have MATLAB, make images of D and D*: does this help you understand how reordering the input data improves your visualization of it?

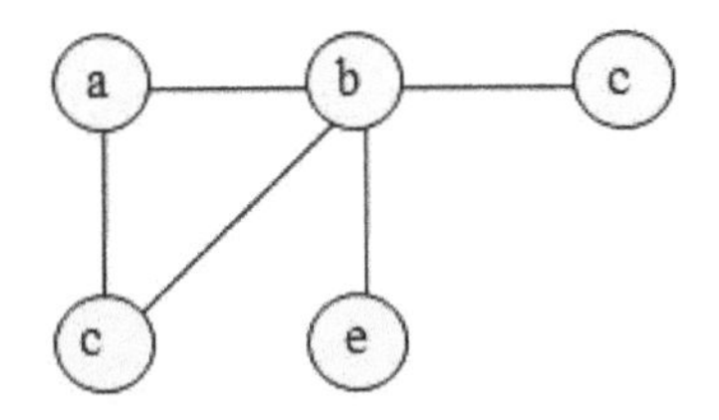

(a) Unweighted graph G1.

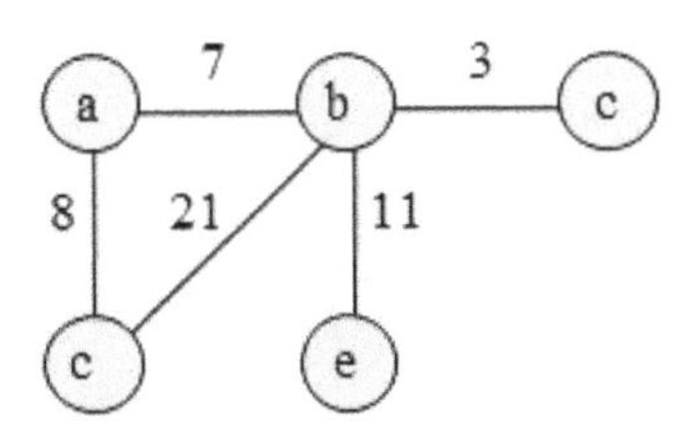

(b) Weighted graph G2.

Figure 4.17 Spanning trees and minimal spanning trees.

3. Find the crisp 4-partition U* of the HPOV clusters in Figure 4.5(a). Compute the 19×19 matrix $D(U^*) = \mathbf{1}_{19\times 19} - \{U^*\}^T U^*$ and show that $I(D(U^*)) = I(D^*)$ as in Example 4.2.

4. Which star clusters in Figure 4.3(B) contain cycles? If you had to reduce the number of HPOV clusters in this figure from 11 to 10, which pair of clusters would you merge? This amounts to reversing the idea we depend on for single linkage clustering.

5. Find three different spanning trees for the *unweighted* graph G1 in Figure 4.17(a).

6. Find the sum of weights for each of the spanning trees found in Exercise 5 for the weighted graph G2 shown in Figure 4.17(b). Which one is the MST? Why is it unique?

7. Show that the value of J_1 in Example 4.7 is $J_1(U_1, \mathbf{V}_1) = 102$ for the subsets $\{X_1, X_2\}$ in Figure 4.10(a) and is $J_1(U_2, \mathbf{V}_2(s)) = 2.22s^2 + 2.22s + 77.50$ for the subsets $\{X_1', X_2'\}$ in Figure 4.10(e).

5

Feature Analysis

...in order to compare and classify the immense variety of shapes, structures and phenomena around us we can't take all their features into account, but have to select a few significant ones.

–Fritjof Capra, 1975.

5.1 Introduction

So far, our discussion of clustering has barely considered the most important element for any computational analysis – viz., the data. The unfortunate truth is that all clustering models and algorithms are at the mercy of the data they process. We discuss this aspect of clustering more carefully in this chapter before plunging into the icy waters of the four basic models and algorithms.

An input data set (X or R) can come to us in many ways, and it can have properties that make it amenable to or difficult for CPOV cluster analysis. Often, we cluster the data as soon as we can just to see what happens. If there is any reason to suspect that the data possess a certain type of cluster structure, we may be surprised or disappointed when this initial effort fails to reveal it. What can go wrong? We have discussed many possible problems already, such as having the wrong model, using the wrong algorithm, and choosing the wrong parameters. Remember, "wrong" here often means "didn't work for me as expected" ... but some related method might.

But what about the data itself? We may have chosen everything discussed in Chapter 4 perfectly but still fail to find those elusive clusters because we have the wrong type of data. That is, the data don't represent the structure we expect to find adequately for discovery by any computational cluster analysis. Far worse, of course, is that the process generating the data simply has no cluster structure, so data representing the objects don't either – a situation we often refuse to acknowledge. Instead, we often persist in the belief that there *must be* some clusters in the data anyway. We are all guilty of belief, and in cluster analysis in particular, here is what often happens:

We decide what we believe first, and then we go looking for facts that support what we believe.

–Barbara Kingsolver

This section discusses some ways we can try to fix the data so that it works harder to satisfy our belief that there are clusters lurking in it somewhere, if only we can improve it enough. This is what feature analysis is all about.

5.2 Feature Nomination

Many writers of papers and books about clustering and classifier design give the impression that the data sets we use arrive almost as if by magic, and there is not much we can do about them. For academics, this has a ring of truth. But someone, somewhere, had to collect or build every data set, and it is very important, if you are designing the data collection procedure, to get this part of the analysis right. Section 5.2 discusses problems and issues associated with choosing what features or relationships to measure.

If you are lucky enough to have some input to the data collection process, you may also have the responsibility for *feature nomination*. Nomination is the right word for this procedure: think about the features you need, propose them, and then find out if you can measure them. Not everyone will have this luxury of course. But usually, the best data are collected after the data collection team has asked the people responsible for analyzing the data a number of pertinent questions about what the objective of data collection is: how to measure the data? what type of information is being sought with the data?, and so on. Some of these questions should resonate with you even if you never get near an actual sensor, and thinking about them will almost always lead you to a fruitful method of feature analysis. Here are some of the questions you should ask:

Q1. What problem is to be solved?
Q2. What variables are important for this problem?
Q3. Which variables can be measured?
Q4. Can I afford to measure them?
Q5. Is there time to measure them?
Q6. What analysis will the data support?
Q7. Are more features always better?

Question Q1 seems so obvious that it shouldn't need listing, but there are many examples of projects where a clear definition of the objective is never articulated. Without it, the overall effort will probably be wasted. Presuming a clearly defined problem, Q2 and Q3 ask you about the variables. You may have a pretty good idea about which ones are important (Q2), but looking ahead to Q3 may prejudice your answer. For example, texture is a very useful feature for distinguishing apples from lemons. You can separate these two fruits with (almost) 100% accuracy while blindfolded by just touching them. But ..., how will you measure this property? What is the cost of building a sensor capable of measuring the skin texture of these two fruits (Q4)? Is it crucial for the problem you are working on right now (Q5)? Are the features you nominate good ones for the analysis you have to do (Q6)?

For us, the answer to Q6 is most assuredly "to find CPOV clusters in the data," but before you can provide sensible suggestions for what to collect, you need to understand what you can and can't expect CPOV clustering to discover. The last question in this list (Q7) is perhaps the most interesting one. It is natural to think that adding more features to the vectors in a given object data set will increase the amount of information the data possess about the problem to be solved. But this is not necessarily the case.

An early study by Highleyman (1962) showed that adding more features to a set of object data vectors could sometimes *decrease* their potential for yielding correct classifications in various circumstances. A related phenomenon is adding more independent features. As parameter p, the number of features, increases, it becomes more and more difficult to add new features that are not somehow correlated to previous ones. Our next example, an abridged version of Example 2.19 in Bezdek *et al.* (1999b), illustrates feature nomination, measurement, and selection of object vector data.

Example 5.1. Nominating and selecting features for three fruits.

Figure 5.1 displays an apple, a tomato, and an orange. To ask and answer questions about these objects by computational means, we need data. Let us look for object vectors to represent each fruit (tomatoes are fruits to scientists, vegetables to cooks). What to measure? Thinking about what to measure, we nominate features that seem capable of representing object properties that will be useful in solving an automatic sorting problem. Consider these four candidates: weight, diameter, color, and texture. Each choice presents questions. For example, it is easy and cheap to measure weight; so this is a natural choice. But is it a good one? Probably not, since the weights of typical pieces of these three fruits are pretty close, but it is so cheap to measure weight, let us regard it as a good candidate. What about color? Not so hot either because apples and tomatoes can be red, yellow, and green. Oranges can be green, yellow, orange, and so on. Diameter doesn't seem much better than weight in terms of its ability to discriminate between the three fruits. Texture seems like a natural choice since humans can often discern the three fruits this way by touch. Let us settle for weight, diameter, and texture.

Once the features (weight, diameter, and texture) are nominated, sensors are needed to measure their values for each object in some mixture of samples. The weight of each fruit is readily obtainable, but the diameter and texture values require more thought, more time, more work, and will be more expensive to collect. Shape (diameter) will be an expensive feature to measure, and it may not capture a property of these classes that is useful for the purpose at hand. Texture can be represented by a binary variable, say 0 = "smooth" and 1 = "rough." It may not be easy or cheap to acquire a texture value for each fruit, but it could be done. After setting up the measurement system, each fruit passes through it, generating an object vector of measurements. In Figure 5.1, each feature vector is in 3-space.

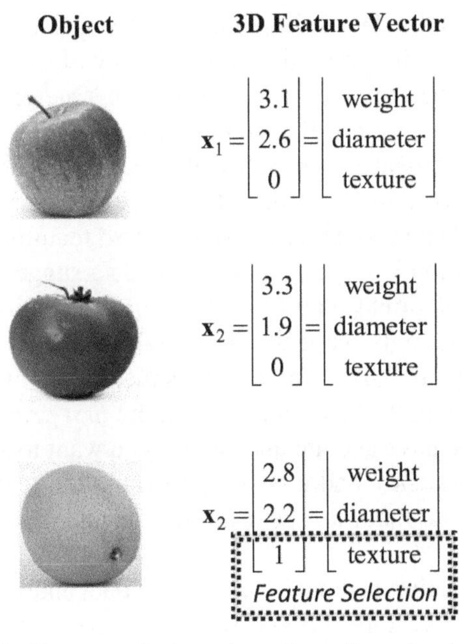

Figure 5.1 Feature analysis produces three-dimensional object data.

Suppose we want to separate the citrus from the non-citrus fruits. The only feature value we need to inspect is the third one (texture). Apples and tomatoes are not citrus fruits and usually have very smooth skins. Oranges are citrus fruits, and they will (almost) always have rough texture. Thus, as shown in Figure 5.1, we may *select* the feature of texture and disregard the first and second features, when separating the citrus from the other two fruits. This selection is made by a human who has knowledge of the question to be answered. Channeling the use of the terms HPOV and CPOV for clusters, this might be called *human feature analysis* (HFA), not to be confused with *computational feature analysis* (CFA).

The feature selection function ϕ that formally accomplishes the selection of the texture feature is the *orthogonal* (OG) projection of $\Re^3$ onto its third axis, $\forall k$, $\phi(x_{k1}, x_{k2}, x_{k3}) = x_{k3}$. This reduces p from 3 to 1 and makes the computational solution simpler and possibly more accurate, since calculations which include all three features use measurements of the other variables that may make the data less separable. It is certainly possible for an anomaly to fool the system because real data exhibits noise. In this example, noise corresponds to, say, a very rough-skinned apple.

Suppose the data in Example 5.1 contained a pineapple. This fruit has a much rougher texture than an orange but is *not* a citrus fruit. One way to deal with the new problem is to modify the resolution of texture. In this case, the texture measurement could be modified to, perhaps, a ternary variable; 0 = "smooth," 1 = "rough," and 2 = "very rough." Texture alone would still suffice since the solution is now characterized as oranges: texture = 2. Although it is easy to say this, remember that the system under design must convert the texture of each fruit into one of the numbers 0, 1, or 2. This is possible but may be expensive. An attractive alternative would be to use mass and binary-valued texture, both of which are already measured and available. Texture would separate oranges and pineapples from apples and pears, and weight would separate oranges from pineapples.

If the *question* we intend to pose to the data in Example 5.1 changes, the feature selected in Figure 5.1 might not be so useful. For example, if the problem were to remove from a conveyor belt all fruits that were too small for a primary market, then texture would be useless. One or both of the first two variables would work better for this problem. However, the diameter and weight of each fruit are probably fairly well-correlated; so collecting either measurement might be roughly equivalent to having them both. Analysis might even yield an approximate functional relationship between these two features. One of the primary uses of feature analysis is to remove redundancies of this type in measured features. In Example 5.1, the physical meaning of the variables and the question being asked about the data suggested the selection illustrated in Figure 5.1. In more subtle cases, computational analysis of the available features, of which there may be very many (thousands, perhaps!), is often the only recourse. If possible, the level of redundancy in the features should be controllable so that CFA is guided by a way to assess measurement error.

The intention in this subsection is to get you thinking about the *first principle of feature analysis*: what is the relationship between the data you have and the question(s) you want to answer? Careful thought about what *might have been most useful to measure* had it been up to you often leads to a useful alteration of the features that have *already been measured* that must be used by you, and this may help your ensuing cluster analysis a lot.

To summarize, "supervised" nomination is done by a (human) data analyst, who, for example, might be asked by data collection experts: "what features do you want us to measure if we want you to process the data to find abnormalities in this patient's brain?" You, the data analyst, might suggest a blood test, but more likely, an imaging procedure such as a magnetic resonance image. Nominations of the data analyst are based on prior knowledge of the physical process that generates the data and the questions that the data analyst will try to answer.

But in many instances, the data analyst is not involved in the decision about what to collect. For example, you might be given a set of *positron emission tomography* (PET) scans on a group of patients collected by a neurologist who says "see if you can find signatures (markers of apparent clusters) for necrosis, white matter, gray matter, cerebral spinal fluid, and tumors in these patients' brain scans." This might be called "unsupervised nomination" because the analyst has not been consulted prior to data collection. Deriving useful information from any cluster analysis is vastly improved when there is an active interface between the data analyst and the data collector. If you are in a position to influence what features to measure, do it. This will make the ensuing computational effort much more efficient and useful.

5.3 Feature Analysis

There are many ways to condition the data so that your computer may have a better chance at finding clusters in it. Over the years, researchers have accumulated a bag of tricks for conditioning data that we collectively lump together as *feature analysis*. The use of this term is not standard. As used here, the term feature analysis consolidates many fields that are often presented as separate topics (some of them are really cottage industries in their own right). To make this point, for example, we find (in Wikipedia, with apologies) this definition for *data cleansing*.

Definition 5.1. Data cleansing.

Data cleansing or data scrubbing is the process of detecting and correcting (or removing) corrupt or inaccurate records from a record set, table, or database. Used mainly in databases, the term refers to identifying incomplete, incorrect, inaccurate, irrelevant, etc., parts of the data and then replacing, modifying, or deleting this dirty data.

Many feature analysis methods are special cases of *feature extraction*, a term that can include, for example, feature selection, normalization, standardization, scaling, filtering, smoothing, and noise suppression. There are many techniques in this field, but the terminology and coverage of feature analysis is quite specialized in most textbooks because most of the tricks we use to improve data are purpose-built. That is, they are designed for a particular type of data or application and, hence, don't generalize well to other kinds of problems, nor do they lend themselves well to general prescriptions for how to find them. With this disclaimer about terminology, we discuss some common methods that are often used for data conditioning when the objective is cluster analysis.

To this point, the term "feature" has been used to describe the attributes of object data vectors in $X = \{x_1, \ldots, x_n\} \subset \Re^p$. The *k*th component of the *i*th feature vector (x_{ki}) is the value of the *k*th feature measurement or attribute of the *i*th object. At first glance, the form of relational data doesn't lend itself well to using the word "feature" in this way. But when $D = [d_{ij}]$ is a dissimilarity matrix, d_{ij} can be regarded as a *joint feature* of objects i and j. We will meet several methods that transform D into a new matrix D^*, and the entries of the new matrix, $D^* = [d^*_{ij}]$, can be regarded as new (extracted) joint features of pairs of objects in the relational data.

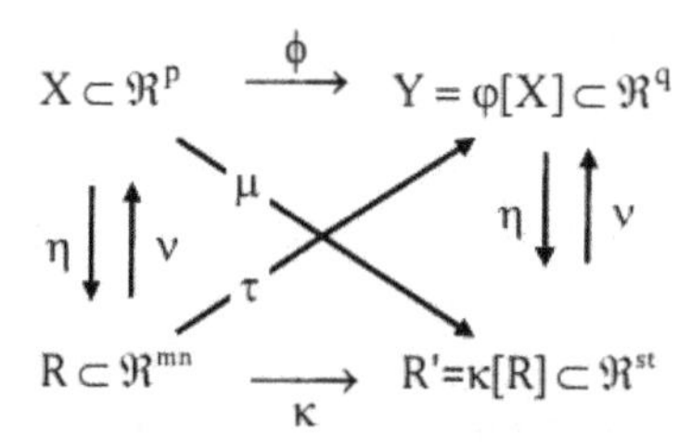

Figure 5.2 Feature extraction for input data X or R.

Figure 5.2 depicts the general idea in *feature extraction*. (Depending on the application context, this operation is sometimes called *data compression*.) The function $\phi : \Re^p \rightarrowtail \Re^q$ performs feature vector extraction by creating new q-dimensional features from the given p-dimensional originals. For object data, ϕ is often a projection of X onto a manifold of some type. The manifolds usually lie in Euclidean spaces, but the range of ϕ can be quite a bit more general.

The set $Y = \phi[X] = \{y = \phi(x) \in \Re^q\}$ is called the *image of X under* ϕ or the projection of X onto its range. Two kinds of functions are used: analytic (or closed form) expressions that can be evaluated exactly in a finite number of operations. For example, functions such as $y = \phi(x) = 2x + 1$; and *algorithms* that transform input data into another form via some set of computational steps that are hard or impossible to describe in a compact analytic form. For example, Sammon's algorithm constructs features non-linearly by an iterative procedure that can't be readily described in closed form (Sammon, 1969).

The function κ in Figure 5.2 transforms relational data (usually dissimilarity data) from one relation matrix to another relation matrix of possibly different size. The VAT and iVAT transformations in Algorithm 4.1, $D \rightarrow D^*$ and $D \rightarrow D'^*$, are examples of matrix transformations that preserve size. In the more general context of the word "feature", this operation is also feature extraction. The functions η and ν in Figure 5.2 convert vectors to matrices and conversely. An example of η is the construction of VAT matrix D from input data X, $D = \eta\,[X]$. The operation of *multidimensional scaling* (MDS) is an example of the function ν, which converts square dissimilarity matrices into sets of (abstract) feature vectors (Cox and Cox, 2000).

The relationship between the dimensions (p, q) or (mn, st) is not fixed. For object data set X, the typical case is that $p \geq q$. When $p > q$, we call X the *upspace*, and $Y = \phi\,[X]$ the *downspace* determined by applying ϕ to X. If p is "big"[5], it is often desirable to look for Y in a much smaller space so that $p >> q$, thereby reducing both time and space requirements for downstream processing. Selection by orthogonal projection is of this type. Orthogonal projection of $\mathbf{x}_k \in \Re^p$ to the *j*th axis in $\Re^p$ is done by the projection function $\phi(\mathbf{x}_k) = \phi(x_{k1}, \ldots, x_{kj}, \ldots, x_{kn}) = x_{kj}$. And, conversely, there are instances for which $p << q$. The leading example of this idea is the kernel *support vector machine* (SVM; Theodoridis and Koutroumbas (2009)), which, *in principle*, maps p-space into a much higher dimensional q-space where the extracted features are (hopefully) linearly separable. An example for which $p = q$ is normalization of X by converting each of its vectors to a unit vector, $\phi(\mathbf{x}_k) = \mathbf{x}_k/\|\mathbf{x}_k\|$.

[5]The meaning of the word big is a moving target: 50 years ago big meant p = 10 or so, but, today, big p's run into the hundreds of thousands of measured features. This is another axis (besides n, the number of objects beneath the data) of today's "BIG Data" volume debate. More on this in Chapter 11.

5.4 Feature Selection

Feature extraction functions alter the initial representation of objects by the input data by creating a new set of data that are (hopefully) more amenable to the computations that will follow. The most common form of extraction is the special case called feature selection highlighted in Figure 5.1. That is, {feature selection functions} ⊂ {feature extraction functions}. In simple terms, feature selection amounts to ignoring selected "bad" features (but don't throw the bad ones away – they might be super useful for different queries in the same data). But how do we choose the features to be eliminated?

There are many methods for identifying "good features" and "bad features." To name a few, orthogonal projection onto a subset of the coordinate axes, orthogonal projection onto a new coordinate system (e.g., *principal components analysis* (PCA), *random projection* (RP), etc.), oblique projection, correlation, *neural networks* (NN), and even clustering (!) have all been used to select "good" features. The usual aim is to reduce the dimension of the input space, thereby reducing both time and space complexity for subsequent processing, without losing too much "useful" information about cluster structure.

Figure 5.3 illustrates several feature selection and extraction schemes for object vector data. The input data are depicted in p-space, $X \subset \Re^p$. Five different feature extraction functions are illustrated. Images of X under the first four functions are subsets of linear varieties (manifolds and subspaces) of $\Re^p$. The fifth image, $\phi[X] \subset \Re^u$, lies on a non-linear manifold (the blue curve) in u-space.

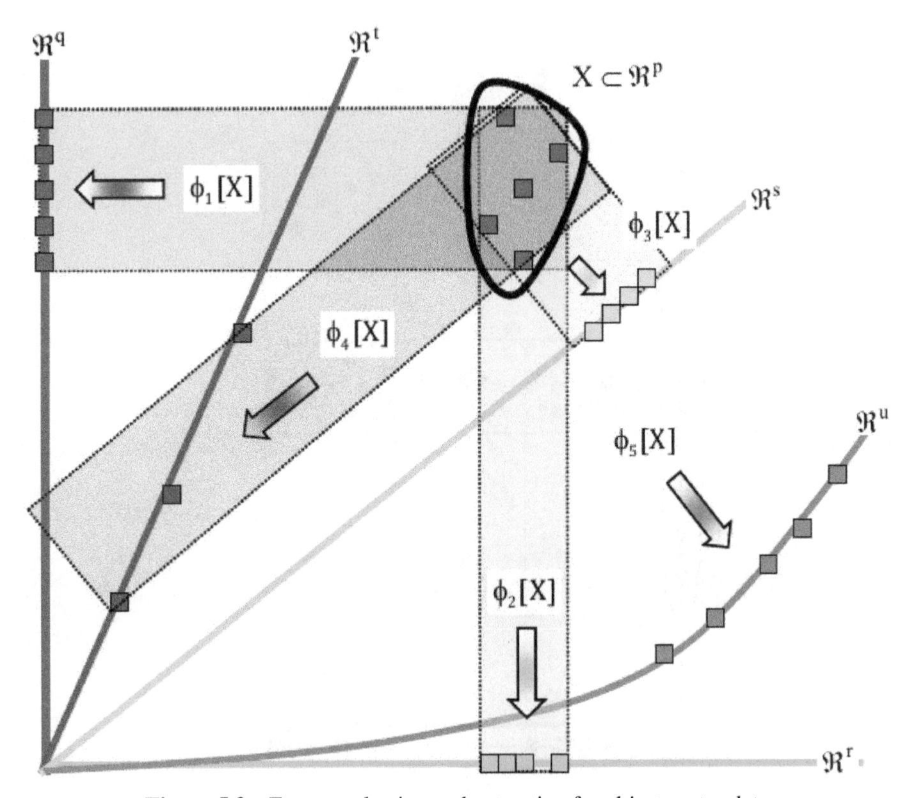

Figure 5.3 Feature selection and extraction for object vector data.

The first set $\phi_1[X] \subset \Re^q$, shown as a projection onto the vertical axis, shows the *selection* of $q < p$ features by orthogonal projection of X onto $\Re^q$. The set $\phi_2[X] \subset \Re^r$, projection of X onto the horizontal axis, illustrates the same thing, *selection* of $r < p$ features by orthogonal projection of X onto $\Re^r$. There are two views with $q \neq r$ because there are a lot of possibilities when choosing subsets of input features by orthogonal projection onto coordinate axes. How many? That is left as an exercise. Example 5.2 demonstrates that this type of feature selection is fraught with danger.

Example 5.2. Orthogonal projection can obscure structure.

Figure 5.4 depicts a small set X of points in the (x_1, x_2) plane, constructed so that X has c = 4 visually apparent HPOV clusters. The points in each of the four clusters are colored, so the 2D data are labeled. Projection of X onto the horizontal axis with ϕ_1 creates the set Y_1, which seems to contain c = 2 clusters, while projection of X onto the vertical axis with ϕ_2 produces the set Y_2, which apparently has c = 3 clusters. The projections are shown *without* colors -unlabeled, as are the data for all true clustering problems. In particular, you would not be able to see the "mixing" of differently labeled subsets in the projections. For example, the rightmost cluster in Y_1 has points in it from the three clusters in 2D that are *physically labeled* as red, blue, and green, but you would not be able to detect this by inspecting the projection onto the x_1 axis if the data were unlabeled.

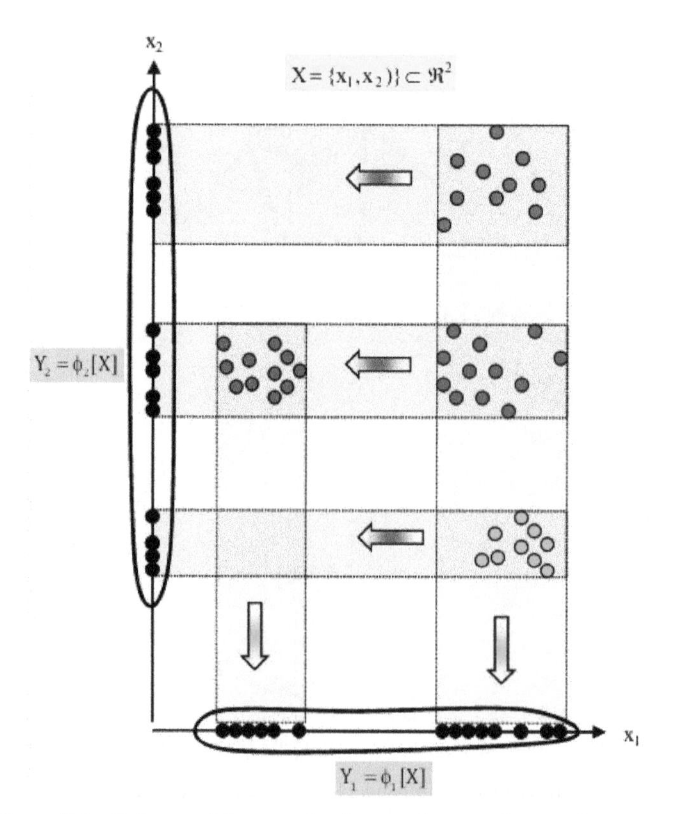

Figure 5.4 Orthogonal feature selection can destroy cluster substructure.

Figure 5.4 conveys three messages. First, visual examination of feature subsets can be misleading to both the HPOV and the CPOV: Y_1 and Y_2 both suggest the wrong number of clusters to seek. Second, feature selection can't *add* information about clusters to the original data, but it can certainly *eliminate* information possessed by the original data that is needed to discover the clusters. Third, feature selection can be quite useful for answering *some* questions but not *all* questions. Here, for example, if we wanted to isolate all the olive colored points in X (perhaps, they are samples of the edible fish among a mixed sample of edible and inedible fish species), feature 1 would suffice. So, feature selection can be a blessing or a curse – seductive and, yet, valuable.

Assuming that a set of *labeled data* with representatives of c physically labeled classes must contain c CPOV clusters is a very common mistake. Our next example illuminates the relationship between algorithmically assigned cluster labels (CPOV labels) and physical labels that may come with data (HPOV labels). A reasonable question to ask at this point is – why would you perform cluster analysis on labeled data? After all, clustering proposes to assign labels to unlabeled data, and labeled data already have them. We answer this question rhetorically. How will you ever know (and perhaps, more importantly, convince others) that your clustering model and algorithm "work" if you don't process some labeled data sets that seem to find the right answers? Well, you won't, and you can't.

Example 5.3. Feature selection in Iris – the unlabeled case.

This example concerns the ubiquitous Iris data, which was collected by the botanist Anderson (1935). He measured four numerical features for each flower: x_1 = sepal length, x_2 = sepal width, x_3 = petal length, and x_4 = petal width. And he collected measurements for 50 flowers from each of three physically distinct subspecies: Iris Setosa, Iris Versicolor, and Iris Virginica. Consequently, the Iris data is a set of n = 150 feature vectors in $\Re^4$ that are separated into three groups of 50 points each by *physical labels*. As advertised in Chapter 3, these sets of 50 points are *HPOV subsets* – it remains to be seen if they can also be construed as *CPOV clusters*. Anderson collected these data but did not publish their values. Fisher (1936) obtained the data from Anderson and was apparently the first author to publish the data values. Hence, the Iris data are often (erroneously) referred to as "Fisher's Iris Data."

While Iris contains observations for 50 plants from each of three different subspecies of Iris flowers, the numerical representation in $\Re^4$ of these objects doesn't seem to support the hypothesis that Iris contains three geometrically well-defined CPOV clusters. Anderson stated that the 100 samples for Versicolor and Virginica were collected from a mixed stand in the field, while the third set of measurements (Setosa) was taken on flowers that were well-separated from the other two subspecies. And, indeed, the first two of the three classes (Versicolor and Virginica) have substantial overlap in the measurement space, while the subset of 50 Setosa plants seems well-separated from the mixed subset of Versicolor and Virginica measurements.

Many authors (including this one) argue that there are only c = 2 CPOV clusters in Iris, and, hence, "good" clustering algorithms and validity models should indicate that c = 2, with a corresponding 2-partition of Iris that separates the data into a well-defined subset of 50 points for Setosa and 100 points for the other two subspecies. But, there are also many papers that proclaim that Iris has c = 3 CPOV clusters, agreeing with the physical labels attached to the plants and displaying a method that finds this to be the case! These authors subscribe to the theory that if X has c labeled subsets, these are necessarily c CPOV clusters as well.

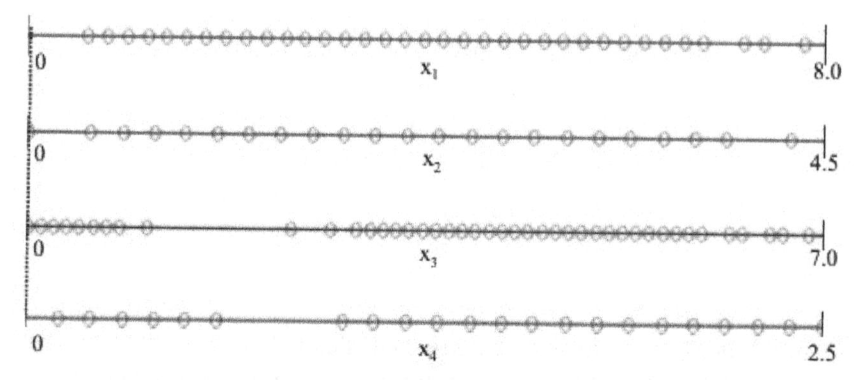

Figure 5.5 OG projections of unlabeled Iris onto each feature axis.

Such is not the case. Remember, saying that Iris has 2 or 3 [CPOV] clusters without reference to a particular model and algorithm is like saying that pickup trucks have either 2 or 3 body lengths or cab styles without specifying the makes and models of the trucks.

Let's see what we can learn about Iris from feature selection. Figure 5.5 contains (not-to scale) plots of the four sets of one-dimensional measurements (that is, the four views are the orthogonal projections of Iris onto each of its four coordinate axes). The data are shown without labels. In the spirit of Chapter 1, have a look at these four 1D views of Iris and decide how many HPOV clusters each view supports. Plots of the values of the first two features (x_1 and x_2) show very little cluster structure and might lead your HPOV assessment of Iris based on these two graphs to the conclusion that Iris has no HPOV clusters. But the plots of the third and fourth features (x_3 and x_4) suggest that you might declare that there are two HPOV clusters in Iris. But remember, you are looking at 1D projections of 4D data, and we are discussing HPOV clusters in data that seemingly have three HPOV "clusters." Temper your judgment of what to infer about Iris from these views with the knowledge provided by Figure 5.4. The only concrete (but not inconsequential) knowledge we can gain from Figure 5.5 is this: if we knew the labels, we would know that the 50 Setosa points are the cluster of points on the left side of the third and fourth plots. So, perfect separation of Setosa from Versicolor and Virginica could be achieved with just one feature; either x_3 or x_4.

Figure 5.6 contains all six orthogonal projections of Iris onto pairs of coordinate axes. The selected feature pairs are again shown without labels so that you can decide for yourself what you learn about Iris by looking at these 2D scatterplots. Look at them now and ask yourself – "how many clusters do I see?" You will most likely see two clusters pretty clearly in five of the six views, viz., views (b)–(f).

If we add labels to Figure 5.6, the smaller cluster in all these views corresponding to the 50 Setosa plants is not so distinct in view 5.6(a), but there is some separation if you know where to look for it. If these views showed the labels with different symbols or colors, you would probably decide that at least five of the six 2D projections support the idea that Iris contains c = 2 HPOV clusters. But again, remember what we learned about inferences of this sort from Figure 5.4. It would be pretty adventurous to assert from an inspection of these 2D views that a particular CPOV clustering model and algorithm would identify c = 2 as its best guess solution using the 4D data without trying it.

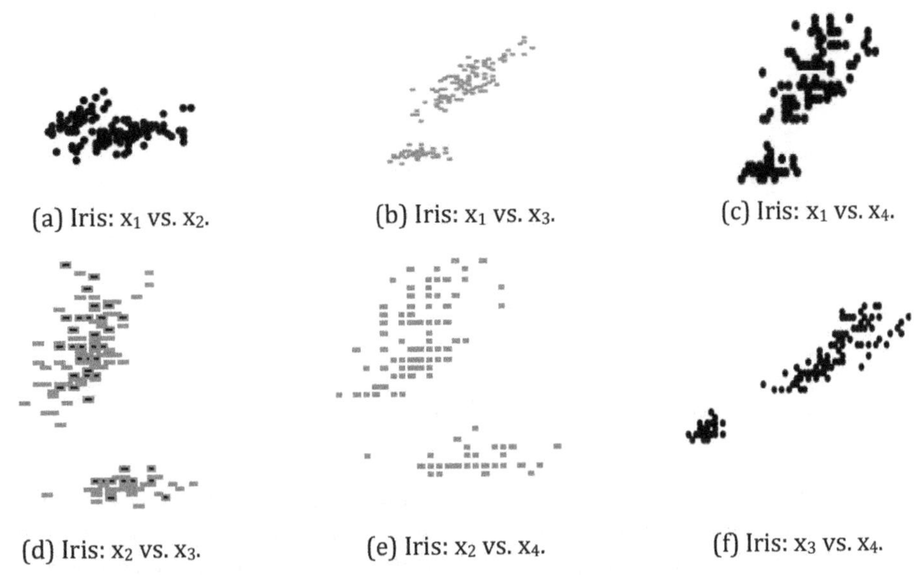

Figure 5.6 OG projections of unlabeled Iris onto all six pairs of 2D coordinate axes.

Example 5.4. Feature selection in Iris – the labeled case.

Figure 5.7(a) shows the orthogonal projection of Iris onto the third and fourth coordinate axes, the same scatter plot as Figure 5.6(f) at a different scale but with labels of the data shown by different symbols and colors. This is the view that is usually invoked by authors who state that Iris has only two clusters, even though there are three physically labeled subsets. But we know from Figure 5.4 that what we see in 2D is not necessarily what we would see in 4D if we could look there.

Figure 5.7(a) suggests that (we should believe that) Iris has two HPOV clusters, but what about CPOV clusters? Usually, no reference is given to a specific model and algorithm. Two HCM/AO clusters? Two single linkage clusters? Two suitably hardened GMD/EM clusters? Would a hardened FCM/AO partition agree? Without specifying a model and algorithm that identifies what *kind* of CPOV clusters, the statement that Iris "has two clusters" is vacuous. That is, the existence of CPOV clusters in data is explicitly dependent on each particular model and algorithm that seeks them.

A finite set of points $X = X_1 \cup X_2$ is said to be *linearly separable* when there is a hyperplane H in the input space for which all of the points from X_1 are on one side of H, while all the points in X_2 fall on the opposite side. The two mixed classes (Versicolor and Virginica) in Figure 5.7(a) are not linearly separable from each other in these two dimensions, but their union is clearly separable from the 50 Setosa points in this view. Figure 5.4 warns us that we can't conclude that Versicolor and Virginica will not be distinct from each other in all four dimensions either. The safest conclusion we can draw from visual inspection of Iris in Figures 5.6 and 5.7(a) is that some (perhaps many) clustering models and algorithms will probably identify $c = 2$ as "best," and others will suggest that the "best" interpretation of CPOV cluster structure is found at $c = 3$. It will be the job of validation strategies to help us decide which interpretation of the data is the "best." Unfortunately, we will learn there that no validation method will always lead us to the right choice.

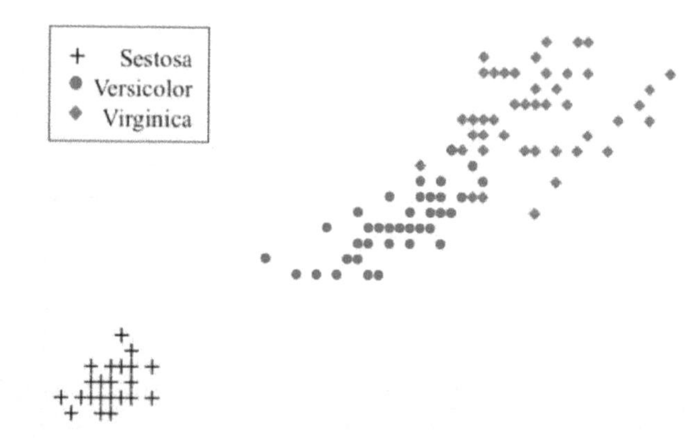

Figure 5.7(a) OG projection of Iris (labels shown) onto features $\{x_3, x_4\}$.

Hmmm..... so, can we learn *anything* from Figure 5.7(a)? Sure. If all we want is a way to separate Setosa from the other two subspecies, the HPOV clusters in these two coordinates suggest, just as in Figure 5.6, that all we really need is feature 3 or feature 4 to do a perfect job for the given data and, probably, an almost 100% reliable indicator for any subsequent data of the same type. If Figures 5.5(c, d) and 5.6(b)–(f) were labeled this way, they all would tell us exactly the same thing, as we shall see next.

In the early days of data visualization, an unknown author introduced a method called the *scatterplot matrix* that offers a way to compare sets of 2D projections of multivariate static data having four or more dimensions. The method is easy to describe. If the data are p-dimensional, there are $p(p-1)/2$ two-dimensional projections of it. Construct a $p \times p$ matrix whose diagonals identify the p features. The *i*th row of the scatterplot matrix has as its p entries views of the $[(i, 1), (i, 2), \ldots, (i, i) = (\text{ID of feature i}), \ldots, (i, p)]$ two-dimensional scatterplots. Each (i, j) pair will appear in the (i, j) and (j, i) addresses of the scatterplot matrix. The (j, i) pair will appear as a reflection of the (i, j) pair about the line y = x because the two variables are plotted in reverse order in the two panels.

Figure 5.7(b) (cf., Figure 16, Friendly and Denis, 2005) is a modern rendition of this matrix which shows the six panels in Figure 5.6, each appearing twice (in flipped pairs) in a 4×4 scatterplot matrix of the Iris data. The labels of the three subsets are shown by different symbols. The ellipses in each panel capture 68% of the data points in each labeled subset, and the lines are regression lines for each of the 50 points in each subset. For example, the unlabeled pair (1, 3) in Figure 5.6(b) appears in the (1, 3) cell of Figure 5.7(b), and in the (3,1) cell, it is reflected about the line y = x. The ellipses for Versicolor and Virginica in Figure 5.7(b) appear to be tangent in four views and slightly separated in the (4, 2) panel. But these are 68% ellipses; so 32% of the points from each subset lie outside of them. This is pretty convincing evidence that these 100 points belong to a single cluster - but what kind of cluster? Not an HPOV cluster - we can't see the 4D data. So, a CPOV cluster?- yes but, again, only with respect to a specified model and algorithm.

Figure 5.7(c) (cf., Figure 3, Rivera-Galicia, 2013) is a different rendition of this same matrix. This is a more traditional presentation of a scatterplot matrix because the range of each feature of the Iris data is plotted along the outside of the panels. The points in Figure 5.7(c) are also labeled, but in this diagram, by different colors instead of different symbols. The unlabeled pair (1, 3) in Figure 5.6(b) also appears in the

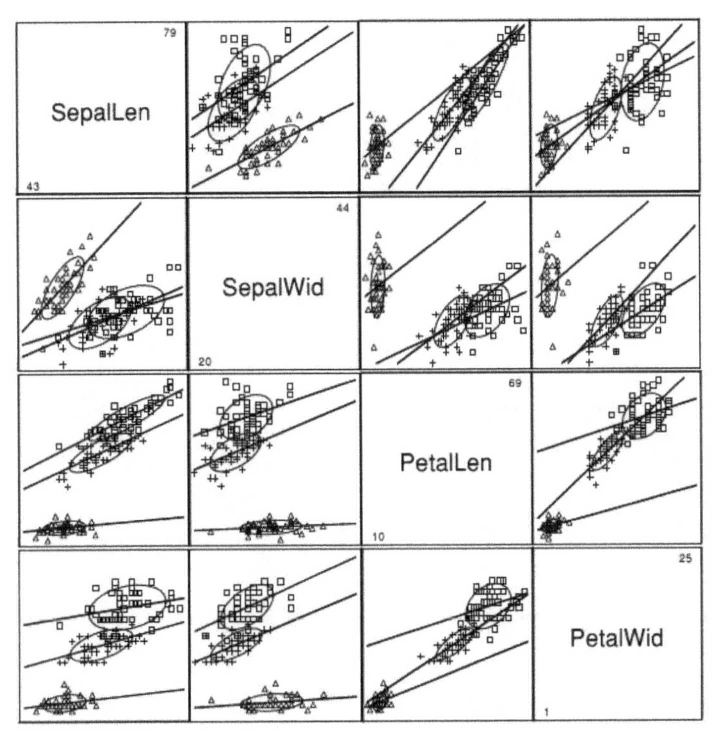

Figure 5.7(b) A scatterplot matrix of labeled Iris Pairs: △ = Setosa; + = Versicolor; = □ Virginica.

(1, 3) cell of Figure 5.7(c). Are there any panels in Figure 5.7(c) which show that Setosa *can't* be linearly separable from the other 100 points in Iris in its upspace?

According to Cleveland (1993), the important idea of this type of presentation is that it affords you with a chance to visually link the features in one scatterplot with features in another. The idea is that you can scan a row or column to link effects on different scatterplots. In the present context, do either of these scatterplot matrices contribute more to our belief about CPOV clusters in the 4D Iris data than earlier unlabeled views? Well, the ellipses in Figure 5.7(b) and red dots in Figure 5.7(c) certainly suggest that Setosa is a separate cluster.

We can't make truly representative plots of three-dimensional orthogonal projections of Iris, and, of course, looking at all four features (as vectors) is impossible. There are many methods available for making other types of visual displays of Iris (iVAT, for example, as shown in Figure 5.20 in problem 5.8 at the end of this chapter), and some of them are quite informative about what Iris might or might not contain, but, for now, we have learned what we can about it via feature selection.

5.5 Feature Extraction

Figure 5.3 shows the extraction of s new features $\phi_3[X]$ by orthogonal projection of X onto the linear manifold $M_3 \subset \Re^s$. The most common way to do this is a very popular method called *principal components analysis* (PCA, Johnson and Wichern (2007)). It is popular because it is fast, the extraction technique is

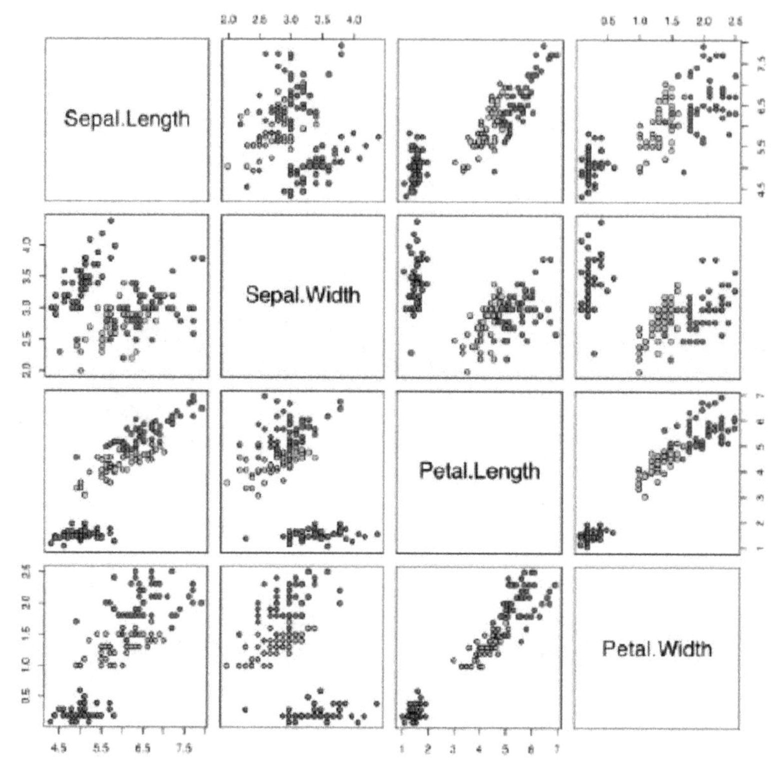

Figure 5.7(c) A scatterplot matrix of labeled Iris pairs: • = Setosa; • = Versicolor; • = Virginica.

analytic (PCA can be done with an explicitly known linear transformation of X), and it has a clear statistical meaning (it maximizes the variance of the projected features in relation to the variance of the original features).

5.5.1 Principal Components Analysis

The *principal components* (PCs) of $X = \{x_1, \ldots, x_n\} \subset \Re^p$ are the eigenvectors of the sample covariance matrix of X, ordered by its eigenvalues. These vectors are the unique solution of the optimization problem: find the linear manifold M in p-space upon which the orthogonal projection of the sample variance is minimum. The grand sample mean and (unbiased) sample covariance matrix of X are computed as usual:

$$\bar{x} = \sum_{k=1}^{n} x/n \tag{5.1}$$

$$S = \sum_{k=1}^{n} (x_k - \bar{x})(x_k - \bar{x})^T/(n-1) \tag{5.2}$$

Assuming S to be positive-definite, PCA extracts and orders its p eigenvalues, say, $\lambda_1 \geq \lambda_2 \geq \cdots \geq \lambda_p > 0$. For i = 1 to p, let $\mathbf{v}_i$ be the unit eigenvector corresponding to λ_1; i.e., $S\mathbf{v}_1 = \lambda_1\mathbf{v}_1$ with $\mathbf{v}_1^T\mathbf{v}_1 = 1$.

Let P_k denote the $p \times k$ matrix whose columns are the first k (ordered) eigenvectors of S, and define, for $q = 1,\ldots,k$, $\phi_{PC,q} : \Re^p \rightarrow \Re^q$ as $\mathbf{y}_q = \phi_{PC,q}(\mathbf{x}) = P_q^T\mathbf{x}$. The vector $\mathbf{y}_q$ is called the *qth order principal component* of **x**, and $y_{qi} = \mathbf{v}_i^T\mathbf{x}$ is called the *i*th score or loading of **x**. Importantly, the function $\phi_{PC,q}$ is linear, $\phi_{PC,q}(\alpha\mathbf{x} + \beta\mathbf{y}) = \alpha\phi_{PC,q}(\mathbf{x}) + \beta\phi_{PC,q}(\mathbf{y})$.

For a chosen value of q, $Y_{PC,q}$ comprises the first q eigenvectors of S. This choice provides the unique q-dimensional projection of X onto a vector subspace through the origin which maximizes the fraction of sample variance in X accounted for by the vectors in $Y_{PC,q}$. The amount of variance in $Y_{PC,q}$ is simply the ratio of sums of the q eigenvalues used to the p eigenvalues of S, i.e., $0 \leq \left(\sum_{i=1}^{q} \lambda_i / \sum_{i=1}^{p} \lambda_i\right)$. The time complexity of PCA depends on the number of eigenvectors used. If all p eigenvalues are computed, the time complexity of PCA is $O(np^2)$.

To summarize, the linear manifold, say, M_q, defined by $\phi_3 = \phi_{PC,q}$ as shown in Figure 5.3, is determined by the first q PCs of X. This manifold is the linear span of the first q eigenvectors of S, $M_q = \text{span}\{\mathbf{v}_1, \ldots, \mathbf{v}_q\}$. When $q = 1$, M_q is a line; when $q = 2$, M_q is a plane, and so on, up to $q = (p-1)$, for which M_q is a hyperplane. The orthogonal projection of X onto M_q completes the PCA feature extraction. The set $Y_{PC,q} = \phi_{PC,q}[X] \subset \Re^q$ is the set of new q-dimensional feature vectors extracted from the original set X of p-dimensional feature vectors. When PCA is used for visualization of high dimensional feature vector data, the usual choice is to take $q = 2$.

Example 5.5. Feature extraction in Iris with PCA.

Figure 5.8 shows the features extracted by PCA using the first pair of principal components of Iris. Labels of the points in each class are again shown by different symbols and colors; Setosa is class 1, Versicolor is class 2, and Virginica is class 3. Compare this 2D data set to the data scatterplotted in Figure 5.7(a) to see the similarity of $Y_{PC,2}$ for Iris with the orthogonal projection of Iris onto its third and fourth coordinate axes.

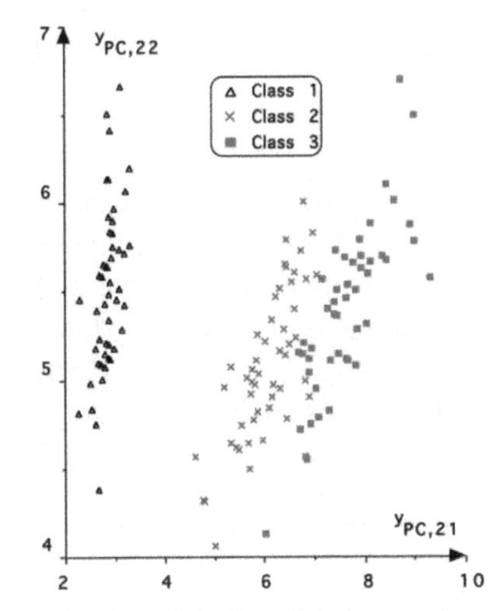

Figure 5.8 PCA projection of $X = X_{Iris}$ (labels shown) onto $Y_{PC,2} = \phi_{PC,2}[X]$.

Bear in mind as you look at Figure 5.8 that this visual representation of Iris is based on replacing the original four features by two new ones. You might be tempted to interpret this 2D scatterplot of 4D data as a "true picture" of what the upspace 4D data contains. From this viewpoint, you would conclude, if these points were not colored (labeled), that Iris has only two CPOV clusters, but these would be HPOV clusters in this 2D data set, not in the 4D Iris data. There is no assurance that the HPOV clusters you can see in $Y_{PC,2}$ correspond to CPOV clusters in the 4D Iris data. Remember, clusters in Iris are necessarily CPOV clusters. What *kind* of CPOV clusters? That depends entirely on the model and algorithm you use to find them.

Do you think that the data in Figure 5.8 are "better" than those in Figure 5.7 for cluster analysis? One reason sometimes given to dislike this scheme is that the original features have physical meaning (here, the 4D data are physical measurements on individual flowers), while the derived PCA features in Figure 5.8 don't have any real meaning (the extracted features are physically "dimensionless"). For example, the coordinates of the points in Figure 5.7(a) represent real measurements (the third and fourth features of each flower, viz., petal width and petal length). But the coordinates of the 2D points in Figure 5.8 have no physical meaning - they are simply numerical features built by PCA from the measured data. This aspect of PCA underlies a continuing argument about its utility - the math camp versus the physical dimensions gang. Just be aware this argument is out there.

Nonetheless, we see that Setosa is again linearly separable from the other 100 data points and that the two mixed classes are non-linearly separable, but the "mixing" between Versicolor and Virginica seen in Figure 5.7(a) is a bit less severe in the PCA data. So, soft (fuzzy or probabilistic) clustering this extracted data might produce a more interesting or useful interpretation of substructure in Iris than clustering the selected 2D data in Figure 5.7(a).

Several other methods of linear feature extraction that are popular include Fisher's *linear discriminant analysis* (LDA) and *independent component analysis* (ICA). Please refer to Theodoridis and Koutroumbas (2009) for more functions of this type. Next, we depart from the well-traveled paths of classical linear feature extraction and discuss a method called random projection that has received a lot of recent attention.This is an attractive method which has some really nice mathematical structure and that seems to have applications in many areas of computer science. But as we shall see, looks can be deceiving.

5.5.2 Random Projection

PCA is based on the idea of finding an orthogonal projection Y of X that maximizes the amount of projected variance. There are many other criteria that have been used to define good projections of X. One of the most natural requirements that can be imposed on a feature extraction function is to require that it preserves distances between pairs of points in the upspace and downspace. This idea made its way into the feature extraction literature early, in Sammon's (1969) non-linear method, which is discussed in Section 5.5.3. The mathematical apparatus needed for understanding this concept is contained in the following.

Definition 5.2. Lipschitz continuity, isometry, and contraction.

Let (X, d) and (Y, d') be metric spaces. A function $f : X \mapsto Y$ is an *isometry* of X onto Y if, and only if, $d(x_i, x_j) = d'(f(x_i), f(x_j)) \forall (x_i, x_j) \in X \times X$. We indicate that X and Y are isometric by writing $X \leftrightarrow Y$. More generally, if $d(x_i, x_j) = \kappa d'(f(x_i), f(x_j)) \forall (x_i, x_j) \in X \times X$ for some constant $\kappa > 1$, f is *Lipschitz continuous* with *Lipschitz constant* κ. Evidently, f is an isometry if and only if $\kappa = 1$. When $0 \leq \kappa < 1$, f is called a *contraction* mapping because it shrinks all the pairwise distances in $X \times X$ by the uniform amount κ.

When $X = \Re^p$ and $Y = \Re^q$ are equipped with vector norms, and $\mathbf{y} = f(\mathbf{x}) \forall \mathbf{x} \in X$, the general definitions reduce to

$$f : X \rightarrowtail Y \text{ is an isometry } \Leftrightarrow \|\mathbf{x}_i - \mathbf{x}_j\| = \|\mathbf{y}_i - \mathbf{y}_j\| \ \forall\, i,\, j; \tag{5.3a}$$

$$f : X \rightarrowtail Y \text{ is a contraction } \Leftrightarrow \|\mathbf{x}_i - \mathbf{x}_j\| = \kappa\|\mathbf{y}_i - \mathbf{y}_j\|;\ 0 \leq \kappa < 1,\ \forall\, i,\, j. \tag{5.3b}$$

The essential fact about isometries is that X and its image $Y = f[X]$ are regarded as (topologically) identical, differing only in the types of objects in the two sets.

One of the most interesting but often confounding linear feature extraction techniques is based on random projection. Johnson and Lindenstrauss (1984) published a paper about extending Lipschitz continuous maps from metric spaces to Euclidean spaces in a way that bounded the Lipschitz constant κ by a prescribed amount. The paper began with a lemma which was needed for subsequent proofs in the paper. Now, the lemma is almost universally regarded as the most important part of the paper (by applications-oriented researchers anyway). This lemma and many of its immediate relatives are all called the "JL Lemma," even though some extensions of it seem quite different from the original.

Let $T : \Re^p \rightarrowtail \Re^q$ be a linear transformation. In the context of this chapter, $\Re^p$ is the upspace and $\Re^q$ is a downspace. T linearly transforms $X = \{\mathbf{x}_1, \ldots, \mathbf{x}_n\} \subset \Re^p$ into the extracted set $Y = T[X] = \{\mathbf{y}_1, \ldots, \mathbf{y}_n\} \subset \Re^q$. Let $R_{p \times q} = [r_{ij}]$ be the matrix representation of T, i.e., $\mathbf{y}_j = T(\mathbf{x}_j) = R\mathbf{x}_j, j = 1, \ldots, n$. T is called a *random projection* operator if the entries of R are chosen according to a specified random distribution. Here is one form of the JL lemma:

Theorem 5.1. The Johnson-Lindenstrauss (JL) lemma.

Let $0 < \varepsilon < 1$. For any integer $q \geq q_0 = \lceil \frac{4\ln(n)}{\varepsilon^2/2 - \varepsilon^3/2} \rceil$ and any set $X = [\mathbf{x}_1, \ldots, \mathbf{x}_n] \subset \Re^p$, there is a linear map $T : \Re^p \rightarrowtail \Re^q$ which can be found in (randomized) polynomial time so that $\forall (\mathbf{x}_i, \mathbf{x}_j) \varepsilon X \times X$ with probability $1 - (1/n^2)$

$$\sqrt{(1-\varepsilon)}\, \|\mathbf{x}_i - \mathbf{x}_j\| \leq \|T\mathbf{x}_i - T\mathbf{x}_j\| \leq \sqrt{(1+\varepsilon)}\, \|\mathbf{x}_i - \mathbf{x}_j\|. \tag{5.4}$$

What does it mean? This theorem says that a linear map T exists (but is not unique), which satisfies the inequalities in Equation (5.4). The construction of a matrix that represents T is not specified in Theorem 5.1, but one choice that works is given in Dasgupta and Gupta (2002), and this is the one used in Example 5.6. T is an isometry if and only if $\varepsilon = 0$. When $\varepsilon > 0$, T is not a contraction because each image distance $\|T\mathbf{x}_i - T\mathbf{x}_j\|$ is bounded above and below. Condition (5.4) essentially defines an envelope that bounds each distance. In Figure 5.9, the (horizontal in the diagram) distance $\|\mathbf{y}_i - \mathbf{y}_j\| = \|T\mathbf{x}_i - T\mathbf{x}_j\|$ lies somewhere in the orange trapezoid, which is a function of the pair $(\mathbf{y}_i, \mathbf{y}_j) = (T\mathbf{x}_i, T\mathbf{x}_j)$.

Proof. This proof requires a certain degree of sophistication in statistics and a pretty reasonable background in probability theory. It is not super hard, but it is long and a bit out of our way; so please refer to Dasgupta and Gupta (2002, pp. 61, 62) for the details.

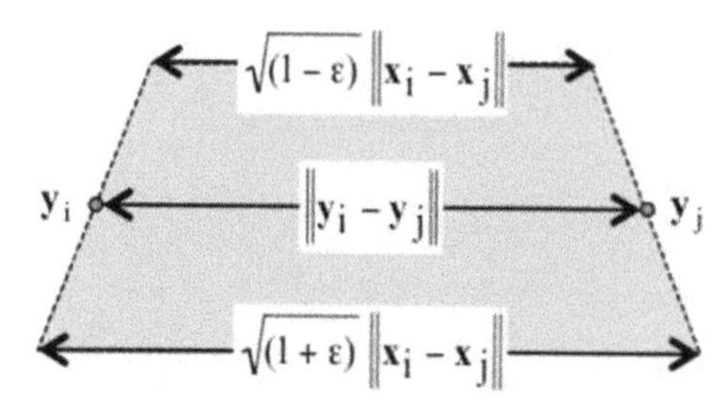

Figure 5.9 The geometric implication of Theorem 5.1 for the pair $(\mathbf{y}_i, \mathbf{y}_j)$.

It is important to see the role of uncertainty in this theory. The JL guarantee – that T preserves *all* of the $n(n-1)/2$ pairwise distances in X × X in the projected vectors in Y = T[X] to within $(1\pm\varepsilon)$-is not an ironclad guarantee; it is probabilistic. But the probability $1-(1/n^2)$ is so close to 1 for even very small values of n (for example, the guarantee for n = 10 exists with probability 0.99) that it is easy to overlook this uncertainty. Writers anxious to press on with an application often ignore this aspect of the lemma, but *this probability is never one.* The term random projection is an accurate one. If you do this projection 10 times with different projection matrices on a fixed data set X, you will almost certainly get 10 very different sets of projected features in downspace $\Re^q$. On the other hand, if you repeat PCA 10 times on X using fixed parameters for PCA, you will get the same projected features all 10 times since the optimization problem that defines PCA has a unique solution.

The norm in Equation (5.4) is the Euclidean norm. The integer q_0 is sometimes called the *target dimension* of the projection. When all $n(n-1)/2$ pairwise distances in Y do satisfy Equation (5.4) for some q, Y is said to possess a $JL(\varepsilon)$ *certificate*. Projections of $X \subset \Re^p \to Y \subset \Re^q$ for $q \in \{2, 3, \ldots, q_0 - 1\}$ are *not* covered by the JL lemma, but it is possible, albeit unlikely, that a projection into one of these "unauthorized" downspaces carries a $JL(\varepsilon)$ certificate. Bezdek *et al.* (2016b) call these "out of bounds" projections *rogue random projections* (RRPs). Example 5.6, abstracted from their paper, shows that RRP can be as effective as PCA for linear feature extraction. But it also shows that RRP can produce less than optimal results. After all, "rogue" means unorthodox and unpredictable – look it up – and that is exactly what to expect when you do feature extraction this way.

Example 5.6. RRP feature extraction in Iris with random projection.

Dasgupta and Gupta (2002) prescribe the random projection mapping R (the matrix representation of T in Theorem 5.1) that was used in this example as $\sqrt{q}\mathbf{y}_i = R\mathbf{x}_i; i = 1, \ldots, n$. Specifically, the (qp) entries of R are chosen randomly and are *independently and identically distributed* (i.i.d.), drawn from the univariate Gaussian distribution with zero mean and unit variance, $N(0, 1)$. That is, $R_{q\times p} = [r_{ij}] = [\text{rand}(N(0,1))]$

For the Iris data, n = 150, Bezdek *et al.* (2016) chose $\varepsilon = 0.9$. So, by Theorem 5.1, the JL target dimension is $q_0 = \left\lceil \frac{4\ln(n)}{\varepsilon^2/2 - \varepsilon^3/2} \right\rceil = 124$. This means that transforming the Iris data from p = 4 to q = 2 with random projection using this value of ε is way outside the JL theory, and, hence, using RP on Iris to produce 2D features is a case of *rogue random projection* (RRP). Figure 5.10 contains scatterplots of the best (Figure 5.10(a)) and worst (Figure 5.10(b)) extractions in 100 trials of RRP on Iris. Here, best and worst mean largest and smallest values of the Pearson correlation coefficient between the Euclidean distance

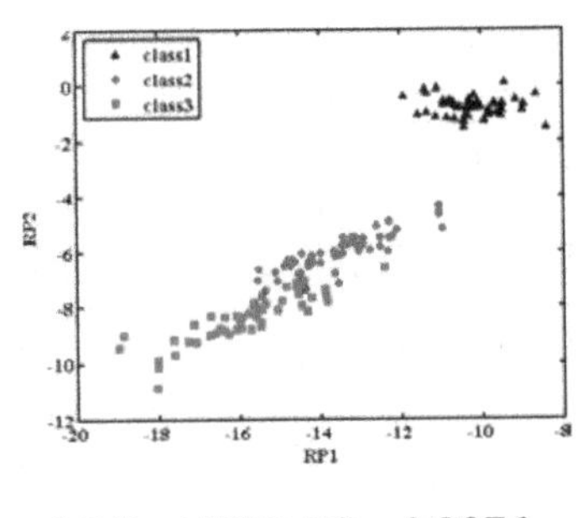

(a) Best RRP: CC_p =0.9976.

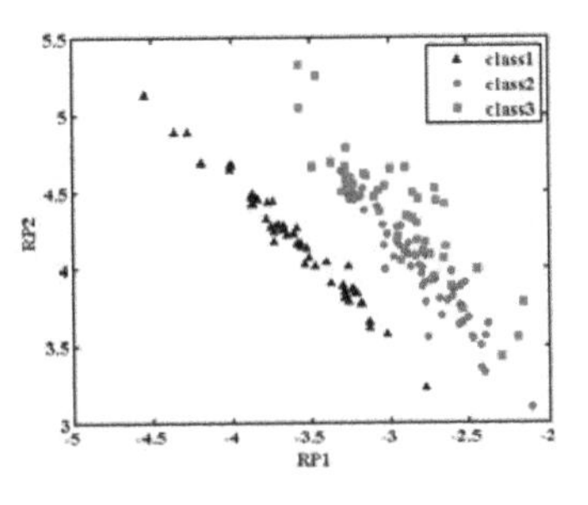

(b) Worst RRP: CC_p = 0.2461.

Figure 5.10 RRP projection of Iris (labels shown) onto $Y = q^{1/2}R[X]$.

matrices D_X in the upspace and D_Y in the downspace. Specifically, Pearson's correlation can be written in terms of the Euclidean inner product and norm for matrices as

$$CC_p(D_X, D_Y) = \langle D_X, D_Y \rangle / \|D_X\| \bullet \|D_Y\| \tag{5.5}$$

Equation (5.5) has the usual meaning of correlation. If the projection map is an isometry, then $D_X = D_Y \Rightarrow CC_p(D_X, D_Y) = 1$, which indicates perfect agreement between D_X and D_Y (an optimal RRP). On the other hand, when $CC_p = 0$, there is maximal disagreement between the upspace and downspace distance matrices, which is taken as indication that there is no good reason to trust the corresponding RRP of X.

Comparing Figure 5.10(a) to Figures 5.7(a) (feature 3 vs. feature 4 of Iris) and 5.8 (PCA extraction) shows that when RRP is good (note that CC_p is *almost* 1), it produces 2D features that are quite comparable to the two previous methods. The red and blue points in view 5.9(a) are not linearly separable, but the mixing of these two subsets is about the same as it is in the data (Figure 5.7(a)) or in the PCA extraction (Figure 5.8). On the other hand, look at panel 5.10(b), which shows the worst RRP extraction (the CC_p value 0.2461 is a pretty low correlation) in 100 trials. Here, the red and blue points are thoroughly mixed, and the red points are quite dispersed. Clustering an unlabeled version of Iris in this extracted data would be rather unsatisfactory compared to CPOV clusters that we might find in the other three 2D representations of Iris.

5.5.3 Sammon's Algorithm

We remarked in connection with Figure 5.2 that there were two forms of feature analysis: closed-form functions and algorithmic maps. The three methods discussed above (orthogonal projection, PCA, and RRP) all used closed-form (known) linear functions. The general approach to non-linear feature extraction, represented by ϕ_5 in Figure 5.3, is often an algorithmic map that defies description in closed form (e.g., some computational neural networks). This subsection discusses an algorithm of this type.

Random projection brings probability into the mix with linear functions that preserve metric topology in a well-specified sense. The criterion in Equation (5.4) is a step along the way toward trying to extract a set of downspace features that are actually an isometric image of their upspace parent. One of the most popular early non-linear extraction methods, and, perhaps, the first one that used isometry as a criterion of goodness, is due to Sammon (1969).

Sammon's model looks for a set $Y_S \subset \Re^q$, $q < p$, whose downspace elements all have the same pairwise distances as their upspace pre-images in X. As above, let $D_X = [d_{ij}]$ be a distance matrix for $X \subset \Re^p$ and $D_{Y_s} = [d'_{ij}]$ be a distance matrix whose *ij*th entry, $d'_{ij} = d'(\mathbf{y}_i, \mathbf{y}_j) = d'(A_s(\mathbf{x}_i), A_s(\mathbf{x}_j)); 1 \le i \neq j \le n$, is the distance between the (unknown) vectors $\mathbf{y}_i = A_s(\mathbf{x}_i)$ and $\mathbf{y}_j = A_s(\mathbf{x}_j)$. The function $A_s : \Re^p \rightarrowtail \Re^q$ is Sammon's algorithm, which attempts to solve Sammon's optimization problem defined with Sammon's objective function J_S:

$$\underbrace{\min}_{Y \subset R^q} = \left\{ J_s(D_Y|D_X) = \left(\frac{1}{\sum\limits_{i<j} d_{ij}} \right) \sum_{i<j} \frac{(d_{ij} - d'_{ij})^2}{d_{ij}} \right\} \tag{5.6}$$

D_X is fixed by X, while $Y_S \leftrightarrow D_{Y_s}$ is variable.

$J_s(D_{Y_s}|D_X) = 0 \Leftrightarrow D_{Y_s} = D_X \Leftrightarrow X \leftrightarrow Y_S$. That is, J_S is zero if and only if all n(n–1)/2 distances in X are preserved exactly in Y_S. Thus, A_S attempts to find a Y_S that is isometric to X by approximating solutions to optimization problem (5.6). Sammon began by guessing an initial $Y_0 \mapsto D_{Y_0}$ and then used the method of steepest descent to iteratively minimize J_S. Since there are different ways to optimize the model, it is a bit of a misnomer to speak of Sammon's algorithm without reference to his choice for the optimization scheme. To be clear, Equation (5.6) is Sammon's model, and steepest descent is the algorithm Sammon (1969) chose to optimize his model. There are other ways to approximate solutions to problem (5.6). Example 5.7 replaces steepest descent with Newton's method.

Example 5.7. Feature extraction in Iris with Sammon's algorithm and PCA.

Figure 5.11 compares the scatterplots of projections of Iris into two dimensions with the first two principal components of PCA to that made by running Sammon's model with q = 2 and a random initialization. Newton's method was used to minimize Sammon's model for the output shown in view 5.11(a). The scatterplot in Figure 5.11(b) is the same as the plot in Figure 5.8, but made at a different scale, with the red points in Figure 5.8 colored green in Figure 5.11(b).

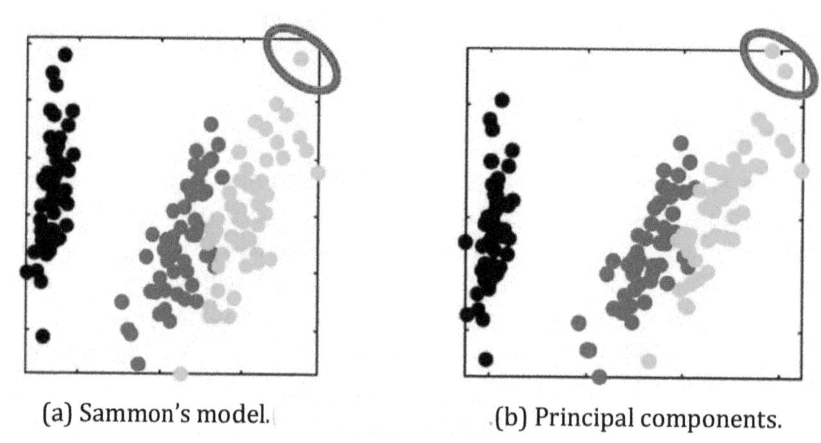

(a) Sammon's model. (b) Principal components.

Figure 5.11 Projections of Iris. (a) Typical Sammon result. (b) Unique PCA.

These scatterplots are not identical, but they certainly contain roughly the same information about apparent CPOV cluster structure in the 2D views. The most noticeable difference is highlighted by the red ellipses in the upper right-hand corner of both views. One of the two points in the PCA view has gone elsewhere under the Sammon mapping. We will see that this is not always the case: Sammon's model appears to do a better job of feature extraction than PCA in Example 7.2.

Another point worth mentioning here is that the points extracted by Sammon's method are in all likelihood not unique. Since this algorithm can begin with different initializations and is iteratively optimized with some scheme such as Newton's method or gradient descent, Sammon's algorithm may offer different sets of projected points upon termination from different starting points. In other words, the points seen in Figure 5.11(a) are one output of Sammon's method, but there may well be others. On the other hand, the PCA projection in Figure 5.11(b) is unique.

The main drawback to Sammon's method is *time* – it usually works fine for a few hundred points that have a small number of features, but in the crazy world of big data today, Sammon's method takes too much time to be of much practical value. On the other hand, its *ambition* – to find a downspace projection Y that is isometric to X -is still of great interest, as we have seen in the RRP attempts to do so.

5.5.4 Autoencoders

Sammon's model is based on a well-defined objective function that defines "good" extracted features as those which solve problem (5.6). There are many approaches to feature extraction that are purely computational – i.e., the model and the feature extraction function ϕ in Figure 5.3 that implements it are both computer programs. The model will have a conceptual basis, but it is realized algorithmically, not as a closed-form function such as the principal components model. *Computational neural networks* (CNNs) have long been a fertile breeding ground for feature extraction of this kind. There is no lack of good references to NNs, and this book will not discuss their many uses in pattern recognition. However, in the context of feature analysis, a short discussion of their use for feature extraction seems warranted.

Perhaps the simplest feature extraction technique based on CNN construction is the *autoencoder* (AE). This network is a special case of the *multilayered perceptron* (MLP), which becomes an AE when the output layer has the same number of nodes as the input layer and is trained to reconstruct the labels of the inputs. This idea was apparently first advanced by Rumelhart and McClelland (1986). Elman and Zipser (1988) is a classic application of this technique in speech recognition, and Cottrell *et al.* (1989) were among the first to use it for image compression.

The basic architecture of the *first generation* (1G) scheme is illustrated in Figure 5.12. The inputs are p-dimensional vectors $(\mathbf{x}_k \in \Re^p)$ which enter on the LHS of the network, one at a time. The functions f and F at each computing node are the integrator and transfer functions. The integrator function f is usually linear, a dot product function that realizes a hyperplane. A typical choice for the transfer function F is a non-linear sigmoid function.

The layers to the left of the extracted output vectors $(\mathbf{y} \in \Re^q)$ in the middle of the network are called A in this diagram, and the layers to the right of the extraction layer are called A^{-1} since their job is to (approximately) invert the feature extraction done by algorithm A. The overall mapping in Figure 5.12 is **CNN**: $\Re^p \rightarrowtail \Re^p$. The aim is for the CNN to function like the identity map, $\mathbf{CNN}(\mathbf{x}_k) = (\mathbf{x}'_k) \approx (\mathbf{x}_k)$. The idea is that if **CNN** does function as the identity, then the data flowing through every layer of the network will, by and large, possess the same "information" as the inputs themselves. And, in particular, the vectors

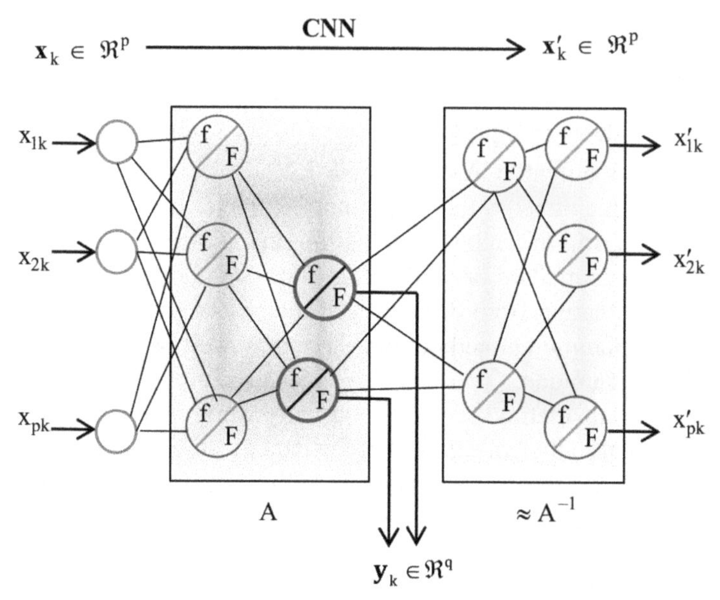

Figure 5.12 A basic first generation autoencoder for feature extraction.

$\{\mathbf{y}_1, \ldots, \mathbf{y}_n\} \subset \Re^q$ that are copied from the output of the central hidden layer should be good extracted features in this loosely defined sense. Here is an example that extracts two features from the Iris data with a first generation auto encoder.

Example 5.8. Feature extraction in Iris with a 1G autoencoder.

The architecture of the network is as follows. There are p = 4 input nodes, one hidden layer with q = 2 computing nodes, and an output layer with four computing nodes. All six computing nodes use $y = f(\mathbf{x}) = \langle \mathbf{x}, \mathbf{w} \rangle + \alpha$ for the integrator function and $F(y) = (1 + e^{-y})^{-1}$ for the transfer function. Since the range of the Iris data is [0.1, 7.9], each logistic function was scaled by 10, enabling the output of each node to range over the interval [–10, 10]. Thus, the specific form of each node function is $10(F \circ f)$.

The parameters acquired during learning are the weight vectors $\{\mathbf{w}_i\}$ and bias constants $\{\alpha_i\}$ of the six hyperplanes at the computing nodes. All 150 points in Iris were fed sequentially through this network during training to acquire the network weight vector. After termination, each point in X_{Iris} is fed through the network one more time, generating $\mathbf{Y} = \{\mathbf{y}_1, \ldots, \mathbf{y}_{150}\} \subset \Re^2$, a labeled set of 2D vectors that can be used to represent the 4D Iris data. Figure 5.13 is a scatter plot of the 150 points in Y found by this technique using the parameters specified in Bezdek *et al.* (1999a). Each $\mathbf{y}_k$ inherits the label of $\mathbf{x}_k$ in the original data set; so the class labels of the 50 points in each of the three clusters can be illustrated by different symbols.

The dashed vertical lines at $y_1 = -1.03$ and $y_2 = 2$ represent a linear classifier that separates the extracted data into three subsets. The 4 blue x's to the left of the vertical line $y_1 = -1.03$ are the resubstitution errors committed by this classifier (i.e., the error rate is $4/150 = 2.66\%$). Compare these features with the representations of Iris made by previous methods to see that these extracted features are *qualitatively* similar to all of the other approaches and will probably support the same types of CPOV clusters in Iris that previous sets of features do.

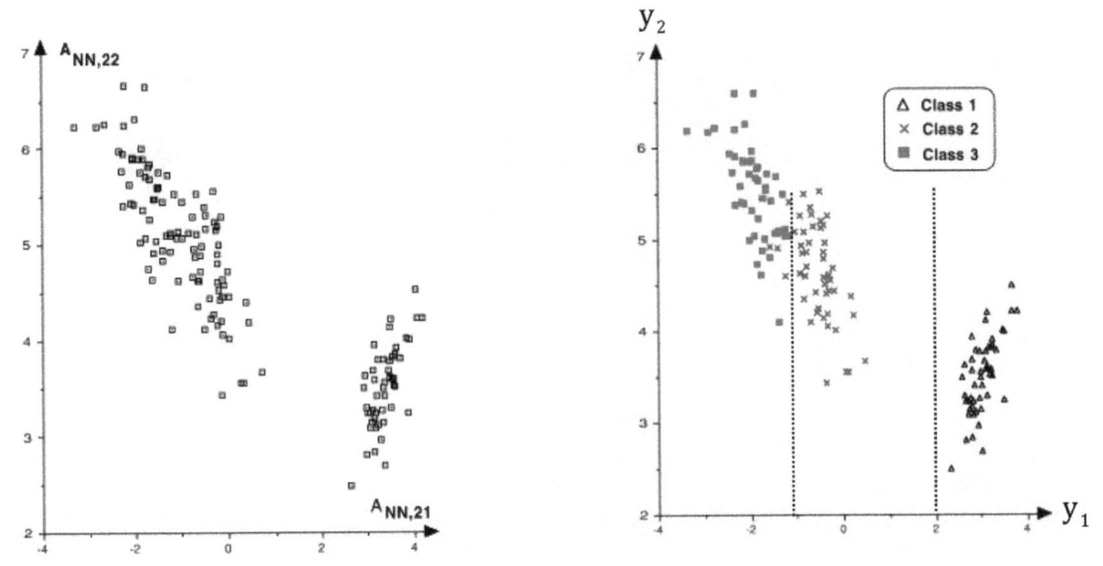

Figure 5.13 A 1G autoencoder projection (labels shown in right view) of the Iris data.

All the usual activation functions for NNs can be used in autoencoders; if linear activations are used, or there is only a single hidden layer of sigmoid functions, then the optimal solution to the autoencoder is strongly related to PCA (Boulard and Kamp (1988)). The computer vision community has a long-standing interest in image compression, and this is a technique that enjoys great popularity in that community. Recent developments in AEs based on the idea of "deep learning" neural networks have emerged in the last decade or so. See the material in Section 5.7 on "deep autoencoders" for a quick look into this emerging field.

5.5.5 Relational Data

We close this subsection with another iVAT example. This one illustrates the function κ in Figure 5.2, which performs feature extraction for relational data. As promised, immediately following Example 4.5, this example explains how iVAT feature extraction is responsible for the clear display of the apparent clusters in the concentric rings data in Figure 4.7 and the four lines data in Figure 5.14 which appears in Example 5.9.

Example 5.9. Feature extraction in distance data for iVAT.

Shown in Figure 5.14(a) is a set of 2D data that form c = 4 visually apparent HPOV linear clusters. This data appeared in Wang *et al.* (2010). They converted this data to dissimilarity data D using the Euclidean (2-norm) function $\eta(x_i, x_j) = d_{ij} = \|x_i - x_j\|_2$. Figure 5.14(b) shows the VAT image of this data matrix made by Algorithm A4.1 with D as input. The VAT image doesn't contain much evidence that hints at substructure in D (and hence X). You would conclude from this view of the data that clusters are quite mixed or, more likely, not present at all.

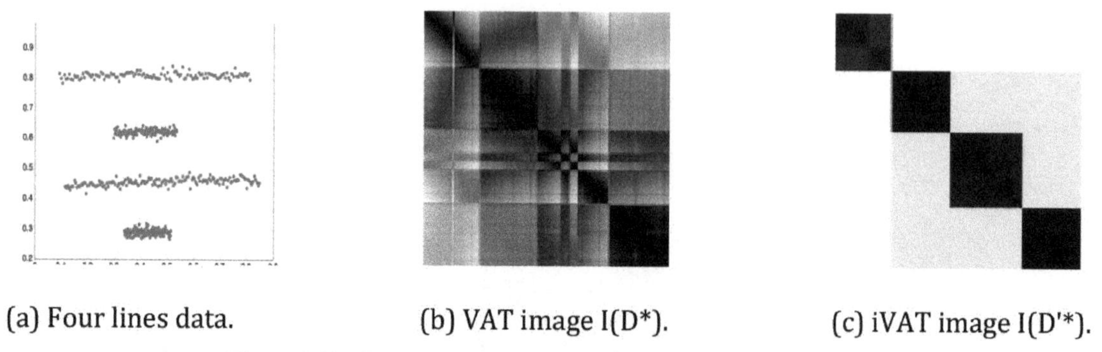

(a) Four lines data. (b) VAT image I(D*). (c) iVAT image I(D'*).

Figure 5.14 Feature extraction (D*→D′*) improves visual display.

The four dark diagonal blocks in the iVAT image shown in Figure 5.14(c), however, make the very clear suggestion that the input data may have c = 4 clusters. What makes the difference? A transformation of D using minimax distance. We regard D as the symmetric weight matrix of a simple, undirected graph G(V, E). Recall from Chapter 4 that a *path of length k* in G from node i to node j is a set of k edges $\{e_1, \ldots, e_k\} \subset E$ that connect vertex i to vertex j. Let P_{ij} denote the set of paths from i to j in G (V, E). Different paths in P_{ij} may have different path lengths; let n_k denote the length of path $p_k \in P_{ij}$. Let d_{kq} be the distance (weight) of edge q along $p_k \in P_{ij}$; q runs from 1 to n_k for path p_k. Then compute

$$d'_{ij} = \underbrace{\min}_{p_k \in P_{ij}} \{\underbrace{\max\{d_{kq}\}}_{1 \leq q < n_k}\} \; \forall \, (i > j). \tag{5.7}$$

If $d'_{ij} \leq d_{ij}$, replace $d_{ij} \leq d'_{ij}$. Applying this operation to the $n(n-1)/2$ upper triangular elements of D and then reflecting the values to the lower triangular portion of D result in the new symmetric matrix D′. Now apply VAT to D′, which builds a different MST than the VAT MST on the objects in the data. The improved visual display of possible cluster structure in the data created by this new MST is evident in Figure 5.14(c), which clearly suggests that the input data contain c = 4 HPOV clusters. This simple process of feature extraction (on D) provides a great improvement to the ability of iVAT to assess possible cluster structure in the input data, and it is independent of p, the dimension of the input data set X.

5.6 Normalization and Statistical Standardization

If you read a dozen papers on clustering, you will discover that ideas about why, when, and how to use normalization (called "standardization" in some texts and papers, especially those with a statistical orientation) is very diverse. Some authors insist on normalizing X or R immediately, in all situations. Others use this type of conditioning more judiciously, presenting an argument for or against normalizing a particular data set. And yet others simply ignore normalization altogether. The most common normalizations are unit vectorization for object vector data X, and unit normalization for relational data matrix R:

$$x'_k = \phi(x_k)/\|x_k\|, \; x_k \in \mathbf{X}; \tag{5.8}$$

$$[r'_{ij}] = [\kappa(r_{ij})] = [r_{ij}/\max\{r_{ij}\}], \; r_{ij} \in \Re \tag{5.9}$$

These normalizations don't change the dimensions or physical meaning of the input data. The next example shows that normalization (5.8) can be counterproductive when you want to find groups in object vector data.

Example 5.10. Normalization can obscure cluster structure.

Let $\mathbf{0} \in \Re^p$ be the zero vector. The set $\partial B(\mathbf{0}, 1) = \{\mathbf{x} \in \Re^p : \sqrt{\mathbf{x}^T\mathbf{x}} = 1\}$ is the surface of the (Euclidean) hypersphere $B(\mathbf{0}, 1) = \{\mathbf{x} \in \Re^p : \sqrt{\mathbf{x}^T\mathbf{x}} \leq 1\}$ centered at $\mathbf{0}$ having unit radius with respect to a chosen vector norm. For p = 2, the Euclidean norm (the "2-norm") results in the real circle $\partial B_2(\mathbf{0}, 1) = \{\mathbf{x} \in \Re^p : \sqrt{\mathbf{x}^T\mathbf{x}} = 1\}$, as seen in both views in Figure 5.15. The left view in Figure 5.15(a) contains two HPOV clusters X_1 and X_2 (the data are red squares and blue dots; so you can see them clearly). Let's normalize these points with Equation (5.8) and plot the results.

The result of transforming $X_1 \cup X_2$ with Equation (5.8), where each vector in X is simply divided by its length in the Euclidean norm, is shown in the right view of Figure 5.15. The set $Y = \phi_2[X]$ of unit vectors extracted from X by projecting them onto the unit circle lies on the boundary of the circle. Before normalization, the HPOV clusters exist and the corresponding CPOV clusters can be found pretty easily by, for example, HCM/AO with c = 2. To see the impact of this projection, imagine the red and blue points without colors, say, all black points. After normalization, the HPOV clusters are gone (undetectable by either visual or computational clustering) since the points in X_1 become interspersed with those in X_2 on the unit circle, as shown on the right in Figure 5.15. This illustrates that normalization is not necessarily beneficial prior to clustering in X.

Another way to interpret Figure 5.15 is to notice that $Y = \phi_2[X]$ is an example of feature extraction by non-linear projection onto the manifold $\partial B_2(\mathbf{0}, 1)$ as in Figure 5.3 with the general non-linear map shown there as $Y = \phi_5[X]$. Seen this way, destruction of cluster substructure in X by normalization (5.8) is exactly like the loss of cluster structure illustrated by orthogonal projection of data in Figure 5.4.

Unit vectors at Equation (5.8) can be built with any vector norm from two infinite families of norms; so normalization (5.8) can be "customized" for a particular type of data or problem. Specifically, we can choose an *inner product norm* $\| \mathbf{x} - \mathbf{v} \|_A = \sqrt{(\mathbf{x} - \mathbf{v})^T A(\mathbf{x} - \mathbf{v})}$; $\mathbf{x}, \mathbf{v} \in \Re^p$ parameterized by the positive-definite weight matrix A. When $A = I_p$ the p × p identity matrix, we get the Euclidean norm with circular unit

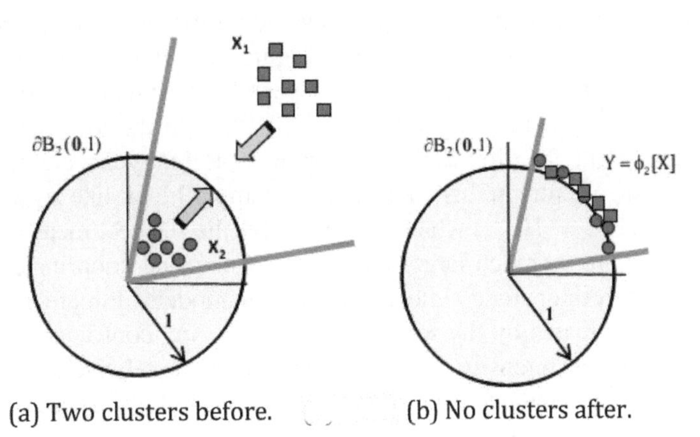

Figure 5.15 Two clusters disappear under unit normalization (5.8).

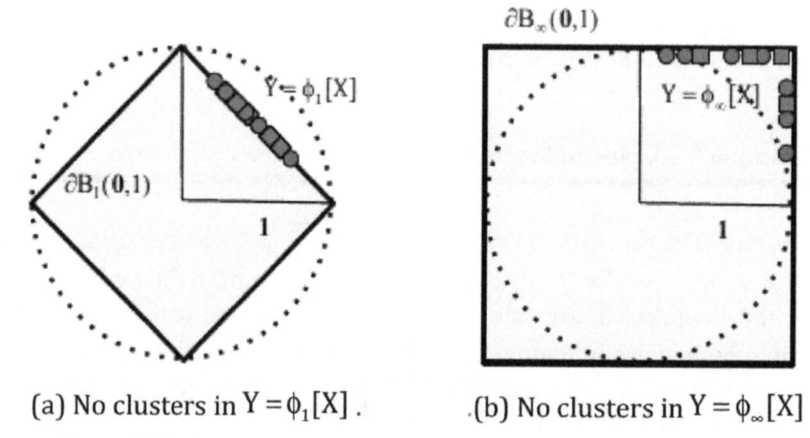

Figure 5.16 One and sup norm normalizations can destroy HPOV clusters.

balls as in Figure 5.15. The other option for Equation (5.8) is a Minkowski *norm* from the q-norm family $\|\mathbf{x}\|_q = \left(\sum_{j=1}^{p} |x_j|^q\right)^{1/q}$; $q \geq 1$.

The Minkowski 1-norm (q = 1) has as its unit ball the diamond $\partial B_1(\mathbf{0}, 1)$ as shown in Figure 5.16(a). Projection of the data set $X_1 \cup X_2$ in Figure 5.15(a) onto $\partial B_1(\mathbf{0}, 1)$ results in the set $Y = \phi_1[X]$ as shown in Figure 5.16(a). Using the Minkowski sup norm ($q \to \infty$) for the projection, $Y = \phi_\infty[X]$, results in the square unit ball $\partial B_\infty(\mathbf{0}, 1)$, and projection of the data X in Figure 5.13 results in $Y = \phi_\infty[X]$ as shown in Figure 5.16(b). The apparent HPOV clusters in Figure 5.15 are essentially destroyed by either of these normalizations of this small fictional data set.

Don't confuse one-time normalization of X *prior* to clustering with (re)normalization *during* processing, which is often recommended for some clustering algorithms. For example, some implementations of the self-organizing map and various support vector machine designs require renormalization during iteration to control the growth of the data as processing proceeds (Kohonen (1989)). And there are algorithms that *require* some type of normalization, perhaps, to satisfy a model constraint of some kind.

Sometimes the features in an object vector data set X have very different dynamic ranges. For example, we might represent each star in several constellations (cf., Figure 4.3) by a pair of measurements such as (apparent intensity magnitude, distance from earth in light years). The range of Pogson intensity values currently used to indicate visual magnitude is about $[-26.74, 30]$, with the lowest value (our Sun) being brightest and the highest values being the dimmest stars. Distances in light years are hard to comprehend: 1 light year is 9.46073×10^{12} km. Vectors in this coordinate system will look like $\mathbf{x}_{\text{Sun}} = (-26.74, 1.44 \times 10^8)$, $\mathbf{x}_{\text{Zeta Leporis}} = (3.55, 6.64143 \times 10^{14})$, which is (currently) the furthest measurable star from earth. Values of the second coordinate are so much larger than those of the first coordinate that any calculation involving distances between these vectors (and almost all clustering models ultimately use distance to define (dis)similarity) will be totally dominated by the second component. Any contribution to separation of star clusters that might be offered by visual intensity is simply swamped by the relative disparity of the numerical distance values. But we do use visual intensity for HPOV clustering in the stars; so normalization of each feature seems desirable for CPOV data analysis.

Do we really need to account for this? In this particular example, if the unit of the second feature is changed from km to light years, the range of distances is something like [0, 71]. So, in this little example, we seemingly find a way out of feature disparity by changing units – or do we? Now we have $(-26.74, 0.000015)$ for our Sun and (3.55. 70.2) for Zeta Leporis. The disparity between second coordinates is still very large even though the range is now more similar to that for light intensity, but, now, the intensity values might dominate the calculations. What else can we do?

We can equalize features in a rough and tumble way by simply scaling each of them by an amount that limits their ranges. If all of the features are observations of continuous variables, a popular and often effective means for equalizing the weight of disparate feature values is a statistically oriented linear transformation of each feature. Specifically, if $X = \{x_1, \dots, x_n\} \subset \Re^p$ is a set of n p-vectors, the *j*th feature has sample mean (m_j) and sample standard deviation s_j computed as

$$m_j = \sum_{k=1}^{n} x_{jk}/n, j = 1, \dots, p \tag{5.10}$$

$$s_j = \sqrt{\sum_{k=1}^{n} (x_{jk} - m_j)^2/(n-1)}, j = 1, \dots, p \tag{5.11}$$

We level the playing field, so to speak, using the feature means and standard deviations for normalization:

$$y_{jk} = \phi_{msd}(x_{jk}) = \frac{x_{jk} - m_j}{s_j}; \quad j = 1, \dots, p; \quad k = 1, \dots, n \tag{5.12}$$

Each feature in the transformed data set $Y = \phi_{msd}[X]$ extracted by Equation (5.12) from X has zero mean and unit standard deviation; so this transformation is sometimes called *statistical normalization*. When should you do this? If the problem suggests that the features should have equal influence in the decision process, this transformation makes sense. But if the context suggests that certain features should have higher weights than others, this might be a bad idea.

There are many other functions and algorithms used to normalize data in various circumstances. Perhaps, the most popular non-linear normalization is *softmax*. This normalization looks a lot like Equation (5.12) because it uses a scalar multiple of the statistical transformation as its first step, viz.,

$$x'_{jk} = \phi_{sm1}(x_{jk}) = (x_{jk} - m_j)\omega s_j; \quad j = 1, \dots, p; \quad k = 1, \dots, n \tag{5.13}$$

$$y_{jk} = \phi_{sm2}(x'_{jk}) = 1/(1 + e^{-x'_{jk}}), \quad j = 1, \dots, p; \quad k = 1, \dots, n \tag{5.14}$$

The overall softmax feature extraction function is the composition $\phi_{sm} = \phi_{sm2} \circ \phi_{sm1}$. The factor ω in Equation (5.13) is a user-defined parameter that controls how much "squashing" of the input data is done by the sigmoid function at Equation (5.14), which changes the range of all of the input data to the half-open unit interval (0, 1]. This normalization emphasizes central tendency in the data. Values close to each feature mean are given more weight, while values that are far from their mean rapidly approach zero.

5.7 Notes and Remarks

Years ago (yeah, *many* years ago!), it was hard to find a comprehensive *chapter* on the topic of feature analysis, but many early books had one or more sections devoted to the topics that were introduced in this section. For example, Duda and Hart (1973), Tou and Gonzalez (1974), and Devijver and Kittler (1982) provided good early discussions of some aspects of feature extraction and selection.

As computational pattern recognition has grown, so has the literature on this subject. Lui (1998) was perhaps the first book that was entirely devoted to feature analysis. This edited volume contains 24 chapters on feature analysis written by many experts in the field. Liu and Motoda (2007) is an update of Liu's previous volume that contains 20 newer chapters on various selection and extraction techniques. Another very extensive and excellent book of 31 chapters presents topics in a huge volume edited by Guyon *et al.* (2006). Chapters 6 and 7 in Theodoridis and Koutroumbas (2009) contain perhaps the most extensive integrated treatment of feature analysis in recent textbooks. These authors provide a very nice exposition of many aspects of this topic. There is no shortage of references for older techniques and statistical procedures such as principal components and factor analysis. One of the clearest discussions on PCA is presented in Johnson and Wichern (2007).

There have been many techniques for feature selection based on various NN architectures. See Pal and Chintalapudi (1997) for a method based on attenuation of less useful features by tuning a standard *multi-layered perceptron* (MLP). Simultaneous feature selection and functional approximation using a five-layer feed-forward NN are discussed in Chakraborty and Pal (2001). Later, using both MLP and radial basis function networks, Chakraborty and Pal (2008) extend this idea to sensor selection, which might be viewed as a generalized form of feature selection, where each sensor selected corresponds to choosing a set of features.

Removing correlated features can also be achieved by clustering the *columns* of an object data matrix. Clustering can sometimes be used to extract or select features. Bezdek and Castelaz (1977) illustrate how to use terminal point prototypes from fuzzy c-means to select subsets of triples from a set of 11 (binary-valued) features for 300 stomach disease patients.

Vempala (2004) offers an excellent summary of the state of the art of random projection in 2004. Applications discussed by Vempala include: NP-hard combinatorial optimization (maximum cut, minimum coloring, and VLSI layouts); Euclidean, volume preserving, and minimum distortion graph embeddings; robust concept learning models; approximate nearest neighbor classifiers; latent semantic indexing; feature extraction (dimensionality reduction) for pattern recognition; low rank approximations of matrices.

The use of random projection for feature extraction is presented in Chapter 5 as a method to extract "good" features to support clustering in lower dimensional spaces rather than in the input data space. The most heavily represented clustering algorithm that has been exercised on randomly projected data is the ubiquitous hard c-means model and algorithm (HCM, aka k-means, where $k = c$ is the number of clusters involved; cf., Chapter 6). Another model and algorithm that appears frequently in this context is the expectation-maximization (EM) algorithm for probabilistic clustering (cf., Chapter 7).

Dasgupta (1999) has an excellent discussion of the theory of learning Gaussian mixtures using JL projection, and Dasgupta (2000) is a follow-up study using the EM algorithm for approximate estimation of the mixing parameters. Dasgupta's focus is on parametric estimation, not soft clustering, but one of the four sets of estimates produced by the EM algorithm is a probabilistic partition; so these two papers are an excellent place to begin a study of this topic.

There are some surprising and somewhat counterintuitive results in Dasgupta (1999) that are important for cluster analysis. Of these, perhaps, the most important is that the random projection of two p-variate Gaussian distributions can be accomplished with minimal loss of a certain well-defined measure of separation between the two Gaussian (2G) distributions for $q_{2G} \geq \boldsymbol{O}(\ln c))$. Compare this to the generic requirement for a $\mathrm{JL}(\varepsilon)$ certificate that $q_0 = \boldsymbol{O}(\ln(n))/\varepsilon^2$ for random projection of an n-sample from an unspecified distribution. Dasgupta asserts that $q_{2G} = \lceil 10(\ln c) \rceil$ works nicely for p-variate Gaussians. For example, this means that for a mixture of c Gaussians in p-space, random projection into $\Re^{\lceil 10(\ln(c)) \rceil}$ approximately preserves his measure of pairwise Gaussian separation between the density functions in $\Re^p$ and $\Re^q$. However, Dasgupta's measure doesn't provide any index of separation between two or more finite subsets of samples.

Fern and Brodley (2003) note that a JL(ε) certificate probably guarantees $(1 \pm \varepsilon)$ pairwise distance preservation, but it says little about cluster preservation. Their experiments show that random projections are very unstable in the sense that projections of X based on different projection matrices can be quite different. This effect is also noted in Figure 3 of Ventkatasubramanian and Wang (2011). So, for clustering in randomly projected downspaces, it seems that an additional guarantee beyond the JL(ε) certificate is needed – one concerning separation of the clusters under projection. Fern and Brodley conjecture that the minimum downspace dimension required to achieve good projected separability is q = 5 (which is much lower than the JL target dimension in most real data situations). These authors don't specify any measure of separation that might be used to confirm or deny their conjecture. They add the disclaimer that "to our knowledge, it is still an open question how to choose the dimensionality for a random projection in order to preserve separation among clusters in general clustering applications." Their clustering approach combines an ensemble of EM clusters on different projections with a final pass using complete linkage to find the clusters.

To summarize, there are two issues in play when randomly projected data are used for clustering: (i) distortion of pairwise distances; (ii) separation of clusters. Many papers are available that address (i) in great detail for projections in the JL range, but measures that assess the preservation of cluster separation between general types of cluster structure under random projection are almost non-existent in the current literature.

A non-imaging method for visualization of feature vector data that doesn't use the idea of reordering is *multidimensional scaling* (MDS), which creates feature vector data in low-dimensional spaces, much like Sammon's algorithm, but MDS begins with relational data instead of feature vectors. The basic idea in MDS is to find a (realization) set $Y = \{y_1, \ldots, y_n\} \subset \Re^q$ (with specified dimension q) so that the found dissimilarities $\{d'(y_1, \ldots, y_n)\}$ between pairs of vectors in Y, arrayed as D′ (approximately) match those in D, i.e., $D \approx D'$. When q = 1, the problem is called *unidimensional* scaling. MDS is, in some sense, inverse to the operation represented by the downward, vertical transformation on the LHS of Figure 5.2 that carries feature vectors X into dissimilarities $R = \eta(X)$. In the notation of Figure 5.2, MDS is also feature extraction, viz., transformation of an $n \times n$ input data set R (as D) to the new form $Y = \{y_1, \ldots, y_n\} \subset \Re^q$ shown by the dashed upward vertical line on the RHS of Figure 5.2.

Schoenberg (1935) and Young and Householder (1938) planted the seeds of MDS, and Torgerson (1952) germinated the idea in his classic 1952 paper. Rapid growth into a mature plant over the next 30 years is nicely summarized in Mardia *et al.* (1979), Coxon (1982), and Davison (1983). This field is still expanding. Distinctions can be made between metric and non-metric MDSs, between weighted and unweighted MDSs, between single matrix and multiple matrices, and between deterministic and probabilistic matrices. Many relatives of MDS are presented in the excellent text by Cox and Cox (2000). We return to the use of MDS as a means for scaling up k-means (HCM) in Chapter 11.

The arrival of interest in "big data" (cf., Chapter 11) coincides with a groundswell of new developments in and other techniques for data visualization. Ferreira de Oliveira and Levkowitz (2003) provide a very useful survey of the state of the art at that time, which they call "visual data mining." These authors describe and illustrate six basic visualization styles: geometric projection, icon based, pixel-oriented, hierarchical, graph-based, and hybrid. The VAT/iVAT algorithm A4.1 is a pixel-oriented method based on a hierarchical reordering principle.

More recently, Maaten and Hinton (2008) introduced an interesting competitor to methods such as PCA, random projection, and Sammon's algorithm. Their technique is really a combination of two methods. It begins with projection from the upspace $X = \{\mathbf{x}_1, \ldots, \mathbf{x}_n\} \subset \Re^p$ to $Z = \{\mathbf{z}_1, \ldots, \mathbf{z}_n\} \subset \Re^s$ in an intermediate downspace (typically, s = 30) using PCA. The second step in their approach, called t-SNE (student's t statistic conjoined with stochastic neighbor embedding), then optimizes a probabilistically defined objective function that seeks an optimal set in the target projection space. This results in projection from $Z = \{\mathbf{z}_1, \ldots, \mathbf{z}_n\} \subset \Re^s$ to $Y = \{\mathbf{y}_1, \ldots, \mathbf{y}_n\} \subset \Re^2$. There are some very persuasive scatterplots of real data sets in this paper, but it uses PCA to find an intermediate projection space before stochastic optimization takes over.

Roweis and Saul (2000) present another very popular type of dimensionality reduction called *local linear embedding* (LLE). This method is similar to spline approximation, but, in LLE, small pieces of the data (actually each point) are embedded in manifolds. Rather than providing more details on this method, let me refer you to their wonderful homepage on LLE, which will redirect you to tutorials, papers, examples, code, and related work: (*https://www.cs.nyu.edu/ ~ roweis/lle*). Here is a quote from their website that differentiates their method from, say, random projection per the JL theory:

> *LLE is doing something very different than finding an embedding which preserves distances from each point to its nearest neighbours. Imagine 3-dimensional (or higher) data distributed on a 2-dimensional square lattice with unit spacing. Colour the points red/black according to a checkerboard pattern. An embedding which places all the red points at the origin and all the black points one unit away exactly preserves the distance from each point to its 4 nearest neighbours but completely destroys the essential structure of the manifold.*

Valdes *et al.* (2016) present a very insightful analysis and examples of several visualization techniques, including linear and nonlinear, classical and computational intelligence-based methods: principal components, Sammon mapping, Isomap, locally linear embedding, spectral embedding, t-distributed stochastic neighbor embedding (t-SNE), generative topographic mapping, neuroscale, and genetic programming. Valdes *et al.* use a virtual reality modeling language to view 3D visualizations that may soon become a standard tool for visualizing big data. This wide-ranging paper deserves a careful reading.

Data can also be visualized using interactive software systems such as the IBM Open Visualization Data Explorer ™ (http://www.research.ibm.com/dx/). And at the website https://lvdmaaten.github.io/drtoolbox/ you will find the description of a MATLAB toolbox developed by Laurens van der Maaten that contains 34 dimensionality reduction algorithms, many of which can be used to project data from $\Re^q \rightarrowtail \Re^2$ for visualization of possible cluster structure in the raw data as well as extraction of low-dimensional representations of the raw p-dimensional data.

Looking at the 2D scatterplots of Iris in Figures 5.6, 5.7, 5.8, 5.10, 5.11, and 5.12 might lead you to think that 2D visualizations of the 4D Iris data all look pretty much alike. And from this, you might conclude that visualization methods always produce similar results. This is a bad thing to believe because each and every view of upspace data made by a method that produces downspace visualizations depends not only on the method but on the data itself.

Figure 5.17 shows the results (with the points unlabeled) of applying four methods of data projection to Iris using van der Maaten's toolbox that are not discussed in the text. A quick look at these 2D scatterplots

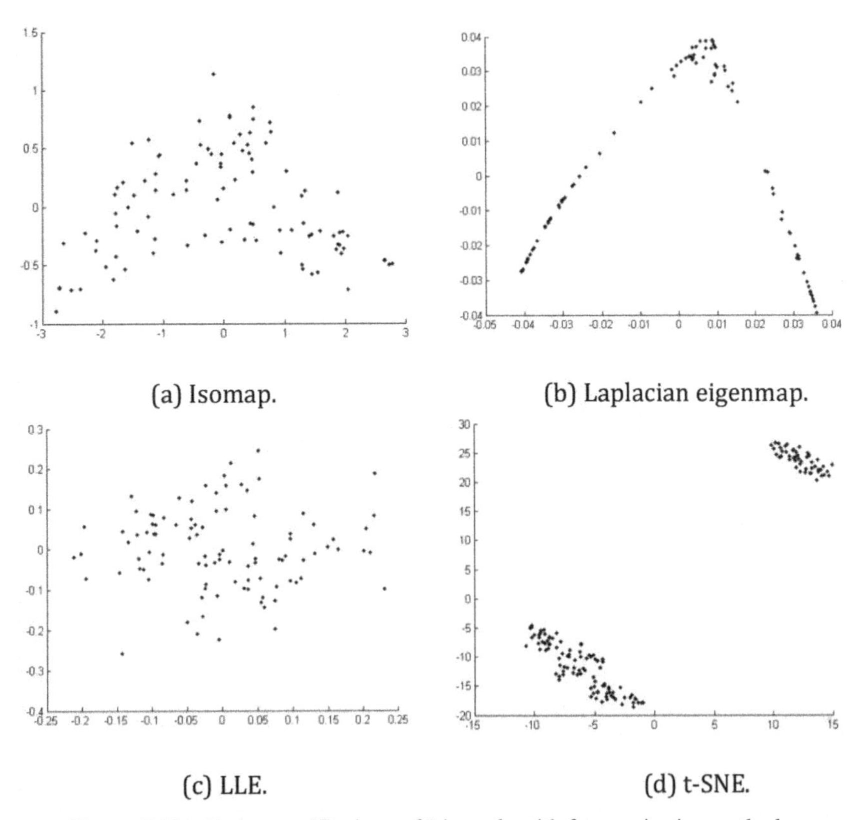

(a) Isomap. (b) Laplacian eigenmap.

(c) LLE. (d) t-SNE.

Figure 5.17 Four more 2D views of Iris made with four projection methods.

is all you need to see to understand that different projection methods can produce wildly dissimilar 2D (downspace) views of upspace data. Figures 5.17(a) and 5.17(c) would suggest to you that Iris contains just one cluster. Figure 5.17(b) is a truly bizarre looking scatterplot that might suggest that Iris contains c = 2 clusters.

Figure 5.17(d), the t-SNE projection of Iris from $\Re^4 \rightarrowtail \Re^2$, offers the clearest *suggestion* that this data set contains c = 2 clusters even though there are three labeled subsets, and this is consistent with other evidence that supports this idea. But please notice my deliberate use of the word "suggestion," for that is all it is -it is not *evidence* that supports this assertion; it is just a view that may strengthen your degree of belief that Iris only contains two CPOV clusters.

For a second example, let us have a quick look at some visualizations of the MNIST image data, which is a set of labeled data containing n = 70,000 points. Each point is a p = 784-dimensional feature vector that represents an image of one of the 10 digits from 0 to 9. Figure 5.18 reproduces four 2D visualizations of 6000 samples (600 of each of the 10 digits) of the MNIST image data that appeared in van der Maaten and Hinton (2008). As noted above, this paper introduced t-SNE.

The center panel, Figure 5.18(b), shows a color key for the 10 labels. The other four views in Figure 5.17 are 2D representations of the 6000 vectors in upspace $\Re^{784}$ made by the four methods identified in the captions. There is clearly no way to know what is actually "up there" in $\Re^{784}$, but the 10 labeled subsets in

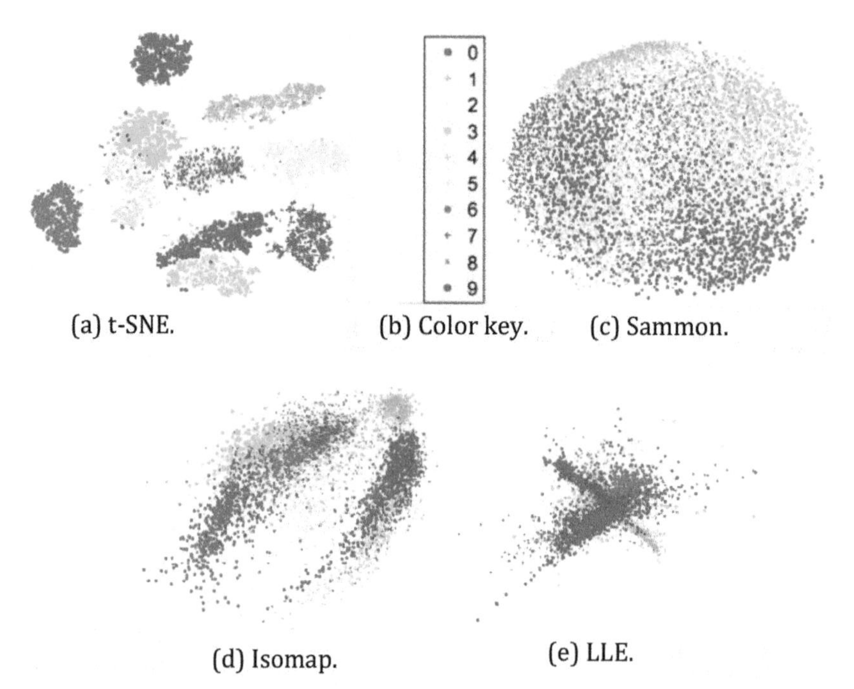

Figure 5.18 Four 2D views of 6000 MNIST samples made with four projection methods.

the data suggest that a good projection of the upspace data might show c = 10 subsets in the downspace view. Certainly, t-SNE provides the most agreeable snapshot of the input data in Figure 5.18. Sammon's algorithm produces an almost circular visualization, which is a very typical result for this method - not much separation between downspace classes. Isomap and LLE fail to show the presumed structure either. So, in this example, t-SNE seems like a superior result. This scheme is a very popular method for visualization of high-dimensional data and has been used in many applications. See Ghafoori *et al.* (2020) for a version of this method called ln-SNE based on the log-normal distribution instead of Student's t.

Another recent development that is worth mentioning is the deployment of *deep autoencoders* (DAEs) for feature extraction and clustering. The idea of "deep learning," using multilayered or *deep neural networks* (DNNs) (Bengio (2009)), has become the basis for a new generation of deep embedded autoencoders that jointly optimize extracted features with the objective of a clustering algorithm. Figure 5.19, taken from Guo *et al.* (2018), conveys the basic idea. The left side is a set of encoder layers that begin with many nodes at the input layer to accommodate high-dimensional (upspace) input data vectors (**x**). Successive layers on the left scale down the numbers of nodes in each additional layer until a "flattening layer" is reached (this is the larger layer that sandwiches the downspace at the central layer). The data travels back up to the input space on the right side, with output ($\mathbf{x}'$). The encoder-decoder cycle attempts to minimize the sum of squared errors, $\sum \|\mathbf{x} - \mathbf{x}'\|^2$, between input/output pairs. After satisfactory replication of the input data is achieved, the extracted features in the downspace are sent to a clustering algorithm.

The central layer in the DAE architecture of Figure 5.19 produces a low-dimensional set of extracted features (typically q = 10 features). The function L shown in Figure 5.19 is an overall loss function having two terms: L_1 is the reconstruction loss incurred by the autoencoder; L_c is the clustering loss incurred by the embedded clustering algorithm; λ is a coefficient that essentially controls the amount of distortion in

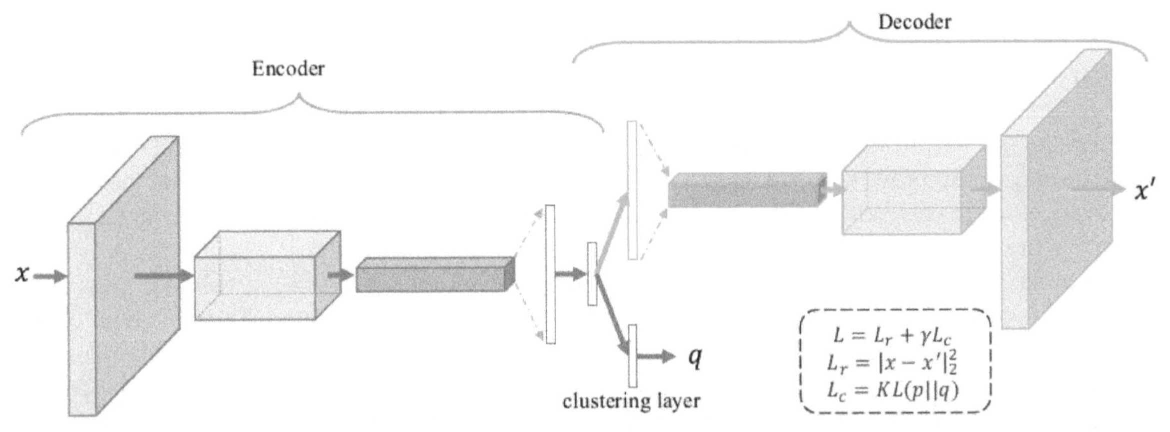

Figure 5.19 A deep embedded clustering model (after Figure 2 in Guo *et al.* (2017)).

the extracted features. This architecture is interesting because it tries to simultaneously optimize a set of extracted features and the clusters that can be discovered in them using a specified clustering scheme. The most heavily represented clustering algorithm in this new area is, of course, k-means (HCM). Feng *et al.* (2019) have applied a variation of this architecture to fuzzy c-means. If the clustering algorithm is changed (no trivial task to accomplish), the extracted features will probably change too; so this is akin to customizing a set of features for a particular clustering algorithm. See Yang *et al.* (2017) or Xie *et al.* (2016) for more on this topic.

No distinction was made in this chapter between feature analysis for labeled and unlabeled data since the basic assumption for "real" clustering is that the data are always unlabeled. Because of this, all the methods presented in Chapter 5 might be called "unsupervised" feature analysis. However, it is worth mentioning that there is another category of feature analysis methods for *labeled data*. As you might expect, methods in this category are to some extent context-dependent, but the basic idea is that the class labels are used to guide you toward a set of features that enhances separation of the classes. This would be the case, for example, if the labeled data were being used train a classifier. Better separation of the classes in the extracted data should provide a better basis for training and tested the learning model and algorithm.

For example, principal components analysis extracts features from unlabeled data toward the end of securing a better set of data for clustering. The method of *common spatial patterns* is similar in spirit to PCA but uses class-labeled covariance matrices to extract features that hopefully improve class separation. Canonical correlation analysis (Hotelling, 1936) is another example of supervised feature analysis that has many applications to learning methods. See Hardoon *et al.* (2004) for an excellent survey that clearly explains the distinction between the two types of feature analysis.

5.8 Exercises

1. *Feature extraction*, as opposed to *feature selection*, is transformation of the "raw" features into a new set of values, presumably better for some pattern recognition problem than the original data. Propose a linear transformation of the features in Figure 5.1 that reduces p from 3 to 2 that *might* be a good set for sorting the unlabeled fruits into three separate bins. Discuss the plausibility or failure of the features your transformation produces.

2. Do you think that the weight and diameter of an orange are correlated? How would you establish this? Does this affect your opinion of the usefulness of these two features as a basis for clustering, i.e., does correlation justify elimination of features?
3. You are designing an automatic medical diagnostic system. Data collection provides measurements of six features: x_1, x_2, x_3, x_4, x_5, and x_6. The use of redundant features adds to the cost of data collection and may or may not improve subsequent processing. Analysis of measured data shows that the sample correlation coefficient between x_1 and x_3 is 0.94 and that between x_3 and x_6 is 0.96. As a designer, would you use all six features or some specific subset(s)? Justify your answer clearly.
4. To design a classifier for different metallic substances, you decide to measure the volume, mass, and coefficient of reflection of each sample. The coefficient of reflection lies between 0 and 1; volume and mass can have any positive values. Discuss problems in using these three features for classifier design; suggest a set of transformed features that might be better.
5. You are required to obtain a set of feature vectors that represent individual samples of each of the following fruits:

apple	blueberry	grapefruit	orange	prune
apricot	cantaloupe	kiwi	peach	pineapple
banana	coconut	lemon	pear	tangerine
blackberry	grape	nectarine	plum	watermelon

For each of the following properties, discuss the most plausible number of HPOV groups amongst the 20 fruits and nominate features that you can measure and/or compute with that might allow you to find the groups by clustering the feature vectors so obtained:

(a) Individual fruits
(b) Citrus fruits
(c) Type of plant
(d) Shape
(e) Color of skin
(f) Hardness of skin
(g) Texture of skin

6. The *resolution* of an object property plays an important role in determining its usefulness for cluster analysis. For example, a pineapple has skin texture that possesses the fuzzy, compound property *hard and rough and sharp*. The path through the binary tree from skin texture to pineapple is shown in Figure 5.20. Choose a fruit from the list in problem 5 that best matches the compound property along each of the other seven paths to leaves in the tree in Figure 5.20. If no match seems possible, indicate with "none."
7. How many different subsets of features can you select from $X = \{x_1, \ldots, x_n\}$?
8. Let D_E be the Euclidean distance matrix for the Iris data. Use whatever software package you have for matrix visualization to display the iVAT image $I(D_E'^*)$ of X_{Iris}. Compare your image to Figure 5.21, the iVAT image of Iris made with MATLAB. Are they the same? Can you see which block in the image corresponds to the 50 Setosa points? Is this block 50×50? If your image is inverted, does it matter? Does this image, based on all four features, suggest that Iris will contain $c = 2$ CPOV clusters? Make iVAT images of X_{Iris} using different distance matrices. For example, try the Minkowski 1-norm and sup norm. Compare iVAT views of Iris with Figure 5.21. Do any of these images alter your belief about possible cluster structure in Iris? Does the t-SNE view of Iris in Figure 5.17(b) help you decide?

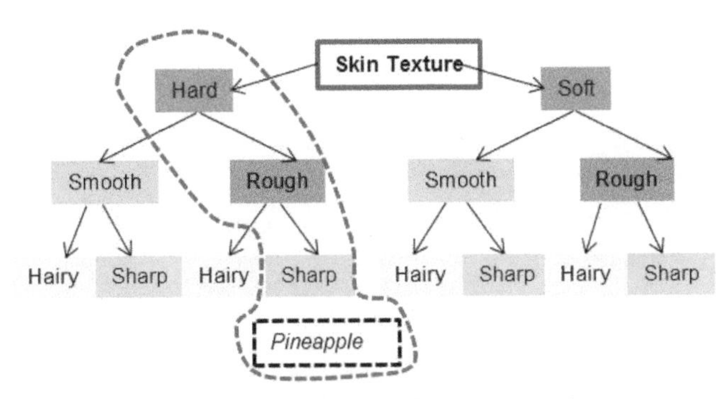

Figure 5.20 Resolution of object properties.

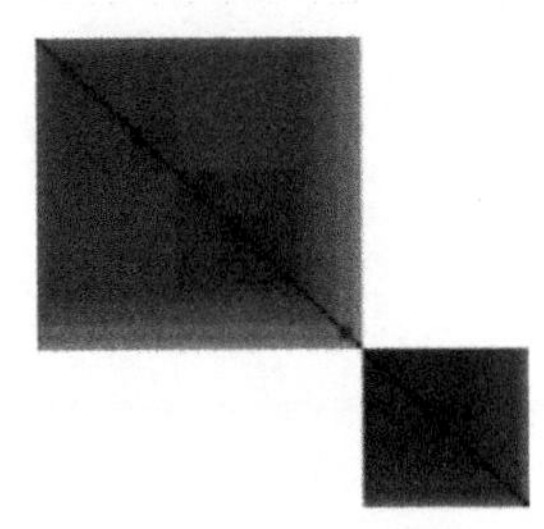

Figure 5.21 iVAT image of the Iris data for the Euclidean distance matrix.

Conclusion for Part 1: Preamble for Part 2

Ok, this concludes the discussion of the basic problems in cluster analysis. Here is the plan for the remainder of the book. Chapter 6 describes the hard and fuzzy c-means (aka "k-means") models and AO algorithms for optimizing them for object data. Chapter 7 develops the Gaussian mixture model and the EM/AO algorithm for probabilistic clustering in object vector data. Chapter 8 is about the SAHN model and algorithms, with special emphasis on crisp single linkage clustering in relational data. Chapter 9 presents a comparative analysis and some examples of the four basic models. Chapter 10 covers the theory of local and global numerical convergence of AO, including corollaries that cover FCM and GMD/AO. Chapter 11 discusses some ways to approach problems associated with scaling up the Fantastic Four to "big data" -i.e., clustering in sets of static data that are too large to mount in your computer for all-at-once cluster analysis. And, finally, Chapter 12 contains an introduction to the increasingly important topic of streaming analysis.

Part II

Four Basic Models and Algorithms

A mathematician who is not something of a poet will never be a complete mathematician.

–Karl Weierstrass

The title of this book advertises it as an elementary introduction to clustering. Consequently, the second part of the text covers only four basic "blue chip" methods for partitioning data into clusters. There are literally thousands upon thousands of new or advanced models and algorithms that are offshoots, relatives, and generalizations of these four methods and many others that can't be sheltered under this umbrella. But these four methods are still at the heart of many – if not most – of the applications of cluster analysis reported by various research and application communities under banners such as pattern recognition, vector quantization, mixture decomposition, machine learning, data mining, and the like. Many writers call the implementations I will present "naïve," presumably meaning "original" or unrefined. To avoid misconceptions, I will call them basic methods, but it is ok for you to think "naïve" instead.

The first three methods covered in this part (crisp-HCM/AO, fuzzy-FCM/AO, and probabilistic-GMD/EM) are arguably the three best known and most heavily used models and algorithms for object (feature vector) data. These three models are based on approximate optimization of an objective function, which serves to define what "good clusters" are. The algorithms given here to optimize these three models rely on *alternating optimization* (AO) to approximate solutions for their constrained non-linear optimization problems. Well, almost. Hard c-means (k-means) uses an iterative method that is essentially AO but with an important theoretical qualifier.

The fourth method, the SAHN model, and, in particular, *single linkage* (crisp-SL), is used for square *relational data.* Since any object data set $X \subset \Re^p$ can be converted to dissimilarity data matrix $D(X) \in \Re^{nn}$ using any vector norm, SL can be used for both types of numerical data. Thus, the SAHN model is, in some sense, more general than the first three methods.

Chapter 6 covers crisp (*hard c-means*, HCM, aka *k-means*) and fuzzy (*fuzzy c-means*, FCM). This chapter is long but efficient in that much of what can be said about the theory and implementation of HCM/FCM carries over to Chapter 7 with very little substantive change. Several algorithms that generalize or are otherwise related to FCM are also covered.

Chapter 7 presents probabilistic clustering based on a special case of the mixture model. The algorithm is an AO scheme called the *expectation-maximization* (EM) algorithm for *maximum likelihood estimation* (MLE) when the mixture components are normally distributed; so this model and algorithm are called *Gaussian mixture decomposition* (GMD). This is by far the most popular way to generate probabilistic clusters for object data.

Chapter 8 develops a completely different approach to crisp clustering in square dissimilarity data – the *sequential, agglomerative, hierarchical, non-overlapping* (SAHN) model. The general case of this method is examined briefly, but the lion's share of attention is given to its most important member, the *single linkage* (SL) case. At first glance, SL is a heuristic model, but a deeper look reveals that it has a graph-theoretic interpretation based on making cuts in a minimal spanning tree of the input data.

Chapter 9 offers a comparison of the four clustering models across a number of criteria, including model type, algorithmic structure, time and space complexity, scalability for big data, etc. Several applications that illustrate various aspects of the four models are discussed with a view toward indicating the breadth and depth of clustering in its many forms.

Chapter 10 contains a more or less complete mathematical development of the general theory of local and global convergence for alternating optimization. The exact theories for the basic FCM and GMD/EM algorithms appear here as corollaries to the general theory. There is no corresponding convergence theory for HCM because it is not an alternating optimization scheme in the strict sense of this chapter. You will need to have some background in the calculus of several variables to follow the examples and proofs in this chapter. But even if you don't want the details, please skim through the material in this chapter so that you will have some insight into this part of numerical pattern recognition.

Chapter 11 discusses some of the many ways that have been tried to carry offshoots of the Fantastic Four into the world of clustering in ***BIG STATIC DATA***. My aim in this chapter is to whet your appetite for this topic (you will not be able to escape it if your path continues along the clustering highway) in a way that lets you consume what you want without getting too full.

Chapters 6–11 all rest on the underlying assumption that the clustering will be done on static sets of batch data. Chapter 12 deviates from this way of thinking about data. This chapter is concerned with the analysis of streaming data – that is, data that passes an observer (the computational model) one piece at a time; is processed for information about its relationship to previous data in the stream; and is then (almost always) discarded. Many of the models and algorithms discussed in this chapter are offshoots of the Fantastic Four that their authors describe as "streaming clustering." I will argue that this terminology is very misleading since all three canonical problems of static clustering that are defined in Chapter 4 are non-existent in the streaming case. Nonetheless, this is a significant new field which will get more and more important in the future as our technology embraces more and more data from the streaming environment. This chapter concludes with a discussion of two emerging techologies for streaming data analysis: incremental stream monitoring functions and visualization of data streams.

6

The c-Means (aka k-Means) Models

Of all the principles that can be proposed for this purpose [distributing errors among the equations], I think there is none more general, more exact, or easier to apply, than that which we have used in this work; it consists of making the sum of the squares of the errors a ***minimum****. By this method, a kind of equilibrium is established among the errors which, since it prevents the extremes from dominating, is appropriate for revealing the state of the system which most nearly approaches the truth*

–Legendre, 1805, pp. 72–73.

6.1 Introduction

This chapter is about the basics of the hard and fuzzy c-means models. The history of these two models and clustering algorithms is pretty well-known and will be discussed in some detail in the notes and remarks at the end of the chapter (including its roots in the least squares method of Legendre alluded to in his quote above). We appended "/AO" to HCM throughout Chapter 4 to emphasize that a complete specification of CPOV clusters found in data requires the identification of both a model and an algorithm that extracts approximate clusters that are defined by the model. Since the optimization approach for the three objective function models used in the remainder of this book is exclusively AO, the notational burden this causes will often be reduced by dropping "/AO." But don't forget that while AO is perhaps the most popular and effective approach for finding approximate extrema of these objective functions, it is never the only choice. Some of the important issues and questions raised by these two c-means models will be further discussed in Chapter 10.

6.2 The Geometry of Partition Spaces

This section contains some important and useful facts about the sets M_{hcn} and M_{fcn} defined at Equations (3.3b, c) that will help us when describing and evaluating various clustering models and validity methods. Let us begin by recalling Definition 3.2, which specifies the sets of non-degenerate crisp and fuzzy c-partitions of any n objects $O = \{o_1, \ldots, o_n\}$. In contrast to the coordinate-free notation in Equations (3.3b, c), the notation here is in explicit coordinate form, which might make the definitions a bit more transparent:

Definition 6.1. Non-degenerate crisp and fuzzy c-partitions.

$$M_{hcn} = \left\{ U \in \Re^{cn} : u_{ik} \in \{0,1\} \forall i,k; \sum_{i=1}^{c} u_{ik} = 1 \forall k; 0 < \sum_{k=1}^{n} u_{ik} < n \forall i \right\}; \quad (6.1a)$$

$$M_{fcn} = \left\{ U \in \Re^{cn} : 0 \le u_{ik} \le 1 \forall i,k; \sum_{i=1}^{c} u_{ik} = 1 \forall k; 0 < \sum_{k=1}^{n} u_{ik} < n \forall i \right\}. \quad (6.1b)$$

Here are some non-degenerate crisp partitions: $U_1 = \begin{bmatrix} 1 & 1 & 0 & 0 & 0 \\ 0 & 0 & 1 & 1 & 1 \end{bmatrix}$: $U_2 = \begin{bmatrix} 1 & 0 & 0 \\ 0 & 1 & 0 \\ 0 & 0 & 1 \end{bmatrix}$: $U_3 = \begin{bmatrix} 1 & 0 & 0 & 0 \\ 0 & 1 & 1 & 0 \\ 0 & 0 & 0 & 1 \end{bmatrix}$.

Here are some non-degenerate fuzzy partitions: $U_4 = \begin{bmatrix} 1 & 0.1 & 0.85 & 0 & 0.5 \\ 0 & 0.9 & 0.15 & 1 & 0.5 \end{bmatrix}$: $U_5 = \begin{bmatrix} 0.33 & 0.1 & 0.2 \\ 0.67 & 0.3 & 0.7 \\ 0 & 0.6 & 0.1 \end{bmatrix}$.

Note that columns 1 and 4 of U_4 are crisp label vectors; but the converse is not possible – there can't be a fuzzy or probabilistic label vector for any column of a crisp partition. Exercise 6.3 asks you to prove that the cardinality of M_{hcn} is $|M_{hcn}| = \left(\frac{1}{c!}\right) \sum_{j=1}^{c} \binom{c}{j} (-1)^{c-j} j^n$. A useful approximation when c is much less than n is $M_{hcn} \approx \frac{c^n}{c!}$. For example, for the Iris data, we have roughly $|M_{h3,150}| \approx \left(\frac{3^{150}}{6}\right)$ (or) $|M_{h2,150}| \approx \left(\frac{2^{150}}{2}\right)$, depending on whether you think the Iris data has 2 or 3 CPOV clusters. In either case, there are far too many crisp c-partitions of Iris for exhaustive search to find a "good partition" under the crisp constraint on the partitions.

The constraint $0 < \sum_{k=1}^{n} u_{ik}$; $i = 1, \ldots, c$ precludes a row of zeroes; and $\sum_{k=1}^{n} u_{ik} < n$; $i = 1, \ldots, c$ prevents a row full of ones because the only crisp 1-partition of n objects is the *1-partition* $1_n = \begin{bmatrix} 1 & 1 & \cdots & 1 \end{bmatrix}$. At the other end of the spectrum for c, the only crisp *n-partition* of O is the $n \times n$ identity matrix, $U_{n \times n} = \mathrm{diag}[1] = I_n$. These two partitions will be important in Chapter 8 since they are the initial and final partitions in nested sequences found by the SAHN clustering models. Certain technical details about M_{fcn} necessitate consideration of the larger set defined in the following.

Definition 6.2. Degenerate crisp and fuzzy/probabilistic c-partitions.

For $2 \le c \le (n\text{-}1)$, the sets of *degenerate* crisp and fuzzy c-partitions of n objects are

$$M_{fcn_0} = \{U \in \Re^{cn} : \mathbf{U}^{(k)} \in N_{fc}\ \forall k;\ \forall i,\ 0 \le \sum_{i=1}^{c} u_{ik} \le n\} \quad \text{(fuzzy/probabilistic)} \quad (6.2a)$$

$$M_{hcn_0} = \{U \in \Re^{cn} : \mathbf{U}^{(k)} \in N_{hc}\ \forall k;\ \forall i,\ 0 \leq \sum_{i=1}^{c} u_{ik} \leq n\}\ \text{(crisp)}. \tag{6.2b}$$

$M_{fcn} \subset M_{fcn_0}$, so M_{fcn_0} also imbeds the crisp c-partitions M_{hcn} and their degenerate superset. Overall then, we have $M_{hcn} \subset M_{hcn_0} \subset M_{fcn_0}$ and $M_{hcn} \subset M_{fcn} \subset M_{fcn_0}$. The degenerate crisp partitions M_{hcn_0} seem uninteresting since they don't represent CPOV clusters in the sense of Definition 4.3, but this set plays an important part in understanding the mathematical structure of crisp and soft partitioning algorithms.

Here are some degenerate crisp c-partitions: $U_6 = \begin{bmatrix} 0 & 0 & 0 & 0 & 0 \\ 1 & 1 & 1 & 1 & 1 \end{bmatrix}$: $U_7 = \begin{bmatrix} 0 & 0 & 0 \\ 0 & 0 & 0 \\ 1 & 1 & 1 \end{bmatrix}$:

$$U_8 = \begin{bmatrix} 0 & 0 & 0 & 0 \\ 1 & 1 & 1 & 1 \\ 0 & 0 & 0 & 0 \end{bmatrix}.$$

Please note that the columns of degenerate partitions must still satisfy the column constraint. Are there any non-trivial fuzzy degenerate 2-partitions? Well, for c = 2, if one row is all zeroes, the column constraint necessitates that the other row has only 1's. Thus, $U_9 = \begin{bmatrix} 0 & 0 & 0 & 0 & 0 \\ 1 & 1 & 1 & 1 & 1 \end{bmatrix}$ is a crisp 2-partition, and, hence, by definition, it is fuzzy as well, even though all of its entries are either 0 or 1. For c > 2, we can have truly fuzzy columns in a degenerate soft partition, such as the first and third columns in

$$U_{10} = \begin{bmatrix} 0 & 0 & 0 \\ 0.3 & 0 & 0.65 \\ 0.7 & 1 & 0.35 \end{bmatrix}.$$

Figure 6.1 is a diagram that illustrates how the *crisp* degenerate and non-degenerate partitions might look if we could see each partition as a vector in $\Re^{cn}$, the vector space of all real c × n matrices. Looking ahead, we will discover that interpreting the points in Figure 6.1 as vertices of a certain geometric solid enables us to understand the constraints on crisp and soft clustering algorithms in a very nice way.

To understand the foundational structure of c-partitions and, subsequently, how this structure constrains clustering algorithms that look for them, we need some additional machinery from linear algebra. If you are rusty on this topic, Noble and Daniel (1987) provide an excellent discussion of many topics that will arise in Part II.

Definition 6.3. Convex hull of a finite set.

The *convex hull* of a set a finite set $X = \{\mathbf{x}_1, \ldots, \mathbf{x}_n\} \subset \Re^p$ is the set of all linear combinations of the vectors in X that have convex coefficients (positive values that sum to 1)

$$\text{conv}(X) = \left\{ \mathbf{x} \in \Re^p : \mathbf{x} = \sum_{k=1}^{n(x)} \alpha_k \mathbf{x}_k; \sum_{k=1}^{n(x)} \alpha_k = 1; \alpha_k \geq 0; \mathbf{x}_k \in X \right\}. \tag{6.3}$$

The number of terms $n(\mathbf{x})$ needed for the expansion $\mathbf{x} = \sum_{k-1}^{n(x)} \alpha_k \mathbf{x}_k$ in Equation (6.3) is a function of $\mathbf{x}$. An equivalent way to characterize the convex hull of X is to say that conv(X) is the minimal convex superset of

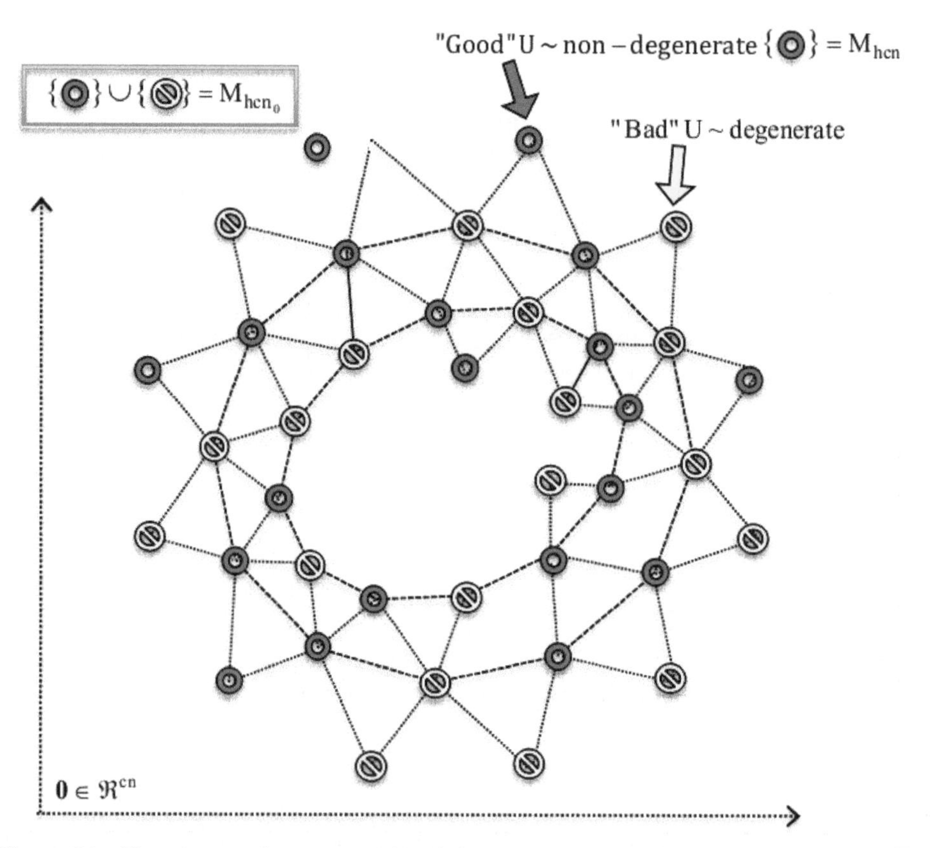

Figure 6.1 The crisp non-degenerate (red) and degenerate (yellow) c-partitions as vectors in $\Re^{cn}$.

X; i.e., conv(X) is the intersection of all the convex supersets of X. An *extreme point* of a convex set X is one that doesn't lie on any line segment with distinct endpoints in X. Let ext(X) denote the extreme points of the convex hull of X. When a convex set X is closed and bounded, it is the closed convex hull of ext(X). In other words, every point in the convex hull of such a set can be written as a convex combination of its extreme points.

The next result has profound implications for understanding the mathematical structure underlying all crisp, fuzzy, and probabilistic clustering algorithms. The statement of this result appeared in Dunn (1973) without proof. The proof appeared in Bezdek (1973) but has never been published in the open literature, and so will be given here in complete detail. For readers unfamiliar with the notation used in Theorem 6.1, if X and Y are any sets, the set of all functions from X to Y is denoted as $Y^X = \{f \mid f : X \mapsto Y\}$.

Theorem 6.1. The convex hull of the set M_{hcn_0}.

Let $\text{conv}(M_{hcn_0})$ be the convex hull of M_{hcn_0}. Recall that the set $N_{hc} = \{y \in N_{fc} : y_i \in \{0,1\} \forall i\}$ is the standard basis of $\Re^c$. Then

$$N_{fc} = \text{conv}(N_{hc}); \tag{6.4a}$$

$$(N_{fc})^X = \text{conv}((N_{hc})^X); \tag{6.4b}$$

$$M_{fcn} = \text{conv}(M_{hcn_0}). \tag{6.4c}$$

What does it mean? Equation (6.4a) shows that every fuzzy/probabilistic label vector has at least one convex decomposition by a finite number of crisp label vectors: you have already seen this, but not explicitly stated this way. Look at Figure 3.2 again. The extreme points of the shaded triangle are the vectors $N_{h3} = \{\mathbf{e}_1, \mathbf{e}_2, \mathbf{e}_3\}$, and the triangle is their convex hull, $N_{f3} = \text{conv}\{\mathbf{e}_1, \mathbf{e}_2, \mathbf{e}_3\}$. Equation (6.4a) says that any vector $\mathbf{x} \in \Re^c$ can be expressed as $\mathbf{x} = \sum_{k-1}^{c} \alpha_k \mathbf{e}_k$ where $\{\mathbf{e}_k\}$ is the canonical basis of $\Re^c$. Equation (6.4b) is needed to get to Equation (6.4c), which is much more interesting. This equation asserts that any non-degenerate fuzzy/probabilistic c-partition has at least one convex decomposition by its extreme points; that is, a finite number of crisp c-partitions (which may or may not include some degenerate terms).

Proof. Let X and Y be any sets. The proof uses three general lemmas. All three lemmas involve showing that sets are equal; so the proofs are made in the usual way (two sets are equal iff each is contained in the other).

Lemma 6.1. A set S is convex if and only if the set of functions from any X into S is too.

Let X be any set, and let S be any set in a vector space V, $S \subset V$;

Then S is convex in V $\Leftrightarrow$ S^X is convex in V^x.

($\Rightarrow$) Assume that S is convex, and let $f, g \in S^X$; so $\forall x \in X$, $f(x), g(x) \in S$. Since S is convex, $\alpha f(x) + (1 - \alpha)g(x) \in S \; \forall \alpha \in [0, 1]$. This means the function $\alpha f + (1-\alpha)g \in S^X \; \forall \alpha \in [0, 1]$; so S^X is convex in V^X.

($\Leftarrow$) Let $x, y \in S$. The constant functions $f(z) = x$, $g(z) = y \; \forall z \in S$ are in S^X; so $\alpha f + (1 - \alpha)g \in S^X \; \forall \alpha \in [0, 1]$ is in S^X because S^X is convex. Thus, $(\alpha f + (1 - \alpha)g)(z) = \alpha f(z) + (1 - \alpha)g(z) = \alpha x + (1 - \alpha)y$ is in S whenever x and y are in S; so S is convex in V. ■

Lemma 6.2. A set B is a convex hull if and only if the set of functions from any X into B is also.

Let X be any set and let $A, B \subset V$ be any sets in a vector space V;

then $B = \text{conv}(A) \Leftrightarrow B^X = \text{conv}\left(A^X\right)$.

($\Rightarrow$) Assume that $B = \text{conv}(A)$. It is easy to check that $A \subset B \Leftrightarrow A^X \subset B^X$. Suppose there is a convex set $\left(T^X \subset V^X\right) \ni \left(A^X \subset T^X \subset B^X \subset V^X\right)$. By Lemma 6.1, T is convex, and $A \subset B \Leftrightarrow A^X \subset B^X$ implies that $A \subset T \subset B \subset V$. But B is the minimal convex superset of A in V; so this inclusion relation is impossible. Thus, the assumption about T^X is wrong; B^X is the minimal convex superset of A^X in V^X; i.e., $B^X = \text{conv}(A^X)$.

($\Leftarrow$) If $B^X = \text{conv}(A^X)$, then B is convex by Lemma 6.1. Suppose that there is a set T, $T \subset V$, T is convex, and that $A \subset T \subset B \subset V$. T^X must be convex since T is; so $\left(A^X \subset T^X \subset B^X \subset V^X\right)$. Lemma 6.1 shows that this is impossible; no such T can exist. Therefore, $B = \text{conv}(A)$. ■

Lemma 6.3. Vector space isomorphisms preserve convexity and convex hulls.

V, W are vector spaces, $L : V \mapsto W$ is an isomorphism, $A \subset B \subset V$, $S = L[A] \subset T = L[B] \subset W$.

Then (i) B is convex $\Leftrightarrow T = L[B]$ is convex; and (ii) $B = \text{conv}(A) \Leftrightarrow L[B] = \text{conv}(L(A))$.

(i) Let $\alpha \in [0, 1]$ and $\mathbf{x}, \mathbf{y} \in B$. Since L is an isomorphism, there are unique vectors $\mathbf{u}, \mathbf{v} \in L[B]$ so that $\mathbf{u} = L(\mathbf{x})$, $\mathbf{v} = L(\mathbf{y})$. L is linear and invertible; so $\alpha\mathbf{u} + (1-\alpha)\mathbf{v} \in L[B] \Leftrightarrow \alpha\mathbf{x} + (1-\alpha)\mathbf{y} \in B$.
(ii) Equation (6.3) gives

$$B = \text{conv}(A) \Leftrightarrow B = \{\mathbf{x} \in V : \mathbf{x} = \textstyle\sum_{k=1}^{n(x)} \alpha_k \mathbf{z}_k : \sum_{k=1}^{n(x)} \alpha_k = 1;\ \alpha_k \geq 0;\ \mathbf{z}_k \in A\}$$
$$\Leftrightarrow L[B] = \{\mathbf{u} \in W : \mathbf{u} = \textstyle\sum_{k=1}^{n(x)} \alpha_k L(\mathbf{z}_k) : \sum_{k=1}^{n(x)} \alpha_k = 1;\ \alpha_k \geq 0;\ \mathbf{z}_k \in A\}$$

Since $\mathbf{z}_k \in A \Leftrightarrow L(\mathbf{z}_k) \in L[A]$, L[B] is the convex hull of L[A]. ■

Lemmas 6.2 and 6.3 are now used to complete the proof of Theorem 6.1. First, let $\mathbf{x} \in N_{fc} \Rightarrow \sum_{i=1}^{c} x_i = 1;\ 0 \leq x_i \leq 1\ \forall\ i$. Let $\mathbf{e}_i = (0, \ldots, 1, \ldots, 0)^T \in \Re^c$, where the 1 is in the *i*th place. The set $\{\mathbf{e}_1, \mathbf{e}_2, \ldots, \mathbf{e}_c\}$ is the canonical basis of $\Re^c$ and coincides with the crisp label vectors N_{hc} up to an arrangement. Now put $\alpha_i = x_i\ \forall\ i$ so that $\mathbf{x} = \sum_{i=1}^{c} \alpha_i \mathbf{e}_i$ is a convex combination of the vectors $\{\mathbf{e}_1, \mathbf{e}_2, \ldots, \mathbf{e}_c\}$. This shows that $N_{fc} = \text{conv}(N_{hc})$. Put $A = N_{hc}$, $B = N_{fc}$, and $V = \Re^c$ in Lemma 6.2, obtaining Equation (6.4b).

Finally, let $(\mathbf{U}^{(1)}, \ldots, \mathbf{U}^{(n)})$ denote any n label vectors in $N_{fc} = \text{conv}(N_{hc})$. Let $\mathbf{L} : N_{fc} \mapsto \Re^{cn}$ be the function that creates a matrix U whose n columns are the vectors $(\mathbf{U}^{(1)}, \ldots, \mathbf{U}^{(n)})$; thus, $L((\mathbf{U}^{(1)}, \ldots, \mathbf{U}^{(n)})) = U \in \Re^{cn}$. Each $\mathbf{U}^{(k)}$ is a label vector in N_{fc}; so the image of $\mathbf{L}$ is a partition matrix. But the n label vectors $(\mathbf{U}^{(1)}, \ldots, \mathbf{U}^{(n)})$ are not coupled; so it is possible that when arrayed as in matrix form, there may be a zero row in U; so $U \in M_{fcn_0}$. For any real α, $\beta \in \Re$ and $U, W \in \Re^{cn}$, we compute $L(\alpha(\mathbf{U}^{(1)}, \ldots, \mathbf{U}^{(n)}) + \beta(\mathbf{W}^{(1)}, \ldots, \mathbf{W}^{(n)})) = L(\alpha\mathbf{U}^{(1)} + \beta\mathbf{W}^{(1)}, \ldots, +\alpha\mathbf{U}^{(n)} + \beta\mathbf{W}^{(n)}) = \begin{bmatrix} \alpha u_{11} + \beta w_{11} & \cdots & \alpha u_{1n} + \beta w_{1n} \\ \alpha u_{c1} + \beta w_{c1} & \cdots & \alpha u_{cn} + \beta w_{cn} \end{bmatrix} = \alpha U + \beta W = \alpha L((\mathbf{U}^{(1)}, \ldots, \mathbf{U}^{(n)}) + \beta L((\mathbf{W}^{(1)}, \ldots, \mathbf{W}^{(n)}))$. Moreover, it is easy to check that $\mathbf{L}$ is bijective (1-1 and onto); so $\mathbf{L}$ is an isomorphism. Apply Lemma 6.3 to conclude that $M_{fcn} = \text{conv}(M_{hcn_0})$. ■

Here is an example that shows how Theorem 6.1 can be used in the context of fuzzy/probabilistic partitioning. This example uses two different convex decomposition algorithms to write the same soft partition as the convex sum of crisp partitions.

Example 6.1. Convex decomposition of a fuzzy/probabilistic 2-partition of five objects.

Consider the fuzzy (or probabilistic) partition $U = \begin{bmatrix} 0.9 & 0.8 & 0.3 & 0.4 & 0.05 \\ 0.1 & 0.2 & 0.7 & 0.6 & 0.95 \end{bmatrix}$. The minimax algorithm (Bezdek and Harris, 1979) decomposes U into a sum of five terms, each with a convex coefficient which can be interpreted as the "weight" of the contribution of the corresponding hard partition in the expression.

$$U = \begin{bmatrix} 0.9 & 0.8 & 0.3 & 0.4 & 0.05 \\ 0.1 & 0.2 & 0.7 & 0.6 & 0.95 \end{bmatrix}$$

$$= 0.6\begin{bmatrix} 1 & 1 & 0 & 0 & 0 \\ 0 & 0 & 1 & 1 & 1 \end{bmatrix} + 0.2\begin{bmatrix} 1 & 1 & 1 & 1 & 0 \\ 0 & 0 & 0 & 0 & 1 \end{bmatrix} + 0.1\begin{bmatrix} 0 & 0 & 0 & 1 & 0 \\ 1 & 1 & 1 & 0 & 1 \end{bmatrix}$$

$$+ 0.05\begin{bmatrix} 1 & 0 & 1 & 1 & 1 \\ 0 & 1 & 0 & 0 & 0 \end{bmatrix} + 0.05\begin{bmatrix} 1 & 0 & 1 & 1 & 0 \\ 0 & 1 & 0 & 0 & 1 \end{bmatrix}$$

The coefficients of this expansion are always monotone decreasing, and each coefficient can be interpreted as the "weight" or importance of the contribution of the corresponding crisp partition in the expression. The partition corresponding to the largest weight, $U_1 = \begin{bmatrix} 1 & 1 & 0 & 0 & 0 \\ 0 & 0 & 1 & 1 & 1 \end{bmatrix}$, is just the hardened version of U, found by applying the hardening function $\mathbf{h} : (N_{pc} - N_{hc}) \mapsto N_{hc}$ at Equation (3.4) to U which replaces the maximum value in every column by a "1." The coefficient (0.6) of this term is the minimum maximum value. If U is the result, for example, of Gaussian mixture decomposition with the EM algorithm (GMD/EM; Dempster, Laird and Rubin, 1977), this crisp first term partition corresponds to hardening U using Bayes rule (replace each maximum posterior probability by "1"). This can be viewed as a sort of first-order approximation to a soft U by a non-soft partition.

Since the first hard partition extracted by the minimax decomposition always has the largest convex weight (analogous to the first principal component, which accounts for the largest projected variance), it is tempting to regard the minimax decomposition as a sort of "principal components analysis" of U, with the first term being the most important component, and so on.

There is another algorithm in Bezdek and Harris (1979) for convex decomposition that they called the F-decomposition of U. Here is the result of applying it to the same U:

$$U = \begin{bmatrix} 0.9 & 0.8 & 0.3 & 0.4 & 0.05 \\ 0.1 & 0.2 & 0.7 & 0.6 & 0.95 \end{bmatrix}$$

$$= 0.05\begin{bmatrix} 0 & 0 & 0 & 0 & 1 \\ 1 & 1 & 1 & 1 & 0 \end{bmatrix} + 0.05\begin{bmatrix} 0 & 0 & 0 & 0 & 0 \\ 1 & 1 & 1 & 1 & 1 \end{bmatrix} + 0.1\begin{bmatrix} 1 & 0 & 0 & 0 & 0 \\ 0 & 1 & 1 & 1 & 1 \end{bmatrix}$$

$$+0.3\begin{bmatrix} 1 & 1 & 1 & 1 & 0 \\ 0 & 0 & 0 & 0 & 1 \end{bmatrix} + 0.1\begin{bmatrix} 1 & 1 & 0 & 1 & 0 \\ 0 & 0 & 1 & 0 & 1 \end{bmatrix} + 0.4\begin{bmatrix} 1 & 1 & 0 & 0 & 0 \\ 0 & 0 & 1 & 1 & 1 \end{bmatrix}$$

The F-decomposition works, but note that the second term on the right side uses the degenerate 2-partition $\begin{bmatrix} 0 & 0 & 0 & 0 & 0 \\ 1 & 1 & 1 & 1 & 1 \end{bmatrix}$ of the five objects. The order of appearance of the six terms in this decomposition is an artifact of the F algorithm. The minimax partition ($U_1 = \begin{bmatrix} 1 & 1 & 0 & 0 & 0 \\ 0 & 0 & 1 & 1 & 1 \end{bmatrix}$ above) appears in this expansion, but the convex coefficient for it drops from 0.6 to 0.4. The proportion of terms with a degenerate crisp partition tends to increase with c. There is little to like about the F-decomposition because it offers no clear explanation or interpretation of the soft partition it represents. But this second decomposition does exhibit the fact that degenerate partitions are needed in Theorem 6.1 to get a complete picture of crisp, fuzzy, and probabilistic partitioning. And, it also demonstrates that convex decompositions are not necessarily unique.

Clustering by convex decomposition was not vigorously pursued when it appeared in 1979. This was due, at least in part, to the fact that the computer power available for exploration of the minimax decomposition was not sufficient to make it a practical idea forty years ago. This method could and should probably be explored in the current era of seemingly unlimited computational power.

Equation (6.4c) tells us that M_{fcn} is the convex hull of the very large finite set M_{hcn_0}. The convex hull of a finite set of points in $\Re^p$ is called a *convex polyhedron* (or *convex polytope*). Thus, the set of soft fuzzy/probabilistic partitions of n objects into c clusters is a convex polytope in $\Re^{cn}$.

What do you think about the partition $\bar{U}= [1/c]$, which assigns an equal membership (or probability) $u_{ik} = 1/c$ to all n objects in each of the c clusters in the underlying objects? Well, certainly there is only one of them. The centroid $\bar{U}= [1/c]$ plays a key role in fuzzy/probabilistic clustering. It is this matrix we move to the origin to secure the vector subspace that has the dimension of M_{fcn}.

Proposition 6.1. The soft c-partition $\overline{U}$ is the centroid of M_{hcn_0}.

$\bar{U}= [1/c]$ is the *centroid of* M_{hcn_0} and M_{fcn} for $2 \leq c \leq (n-1)$.

What does it mean? This simple proposition will find many uses in clustering. It asserts that the soft partition $\overline{U} = [1/c]$ which assigns all n objects the membership (or probability) $1/c$ is the geometric center (aka barycentric center) of the extreme points of M_{hcn_0} and its convex hull M_{fcn}. The restriction on c eliminates the impossible cases.

Proof. Let S be any set in a vector space V, $S \subset V$. The centroid of S is the point in V that minimizes the sum of squared Euclidean distances between the vectors in S to any point in V. In our case, $S = M_{hcn_0}$. The squared Euclidean norm of any $c \times n$ matrix in $U \in \Re^{cn}$ is $\|U\|^2 = \sum_{k=1}^{n}\sum_{i=1}^{c} u_{ik}^2$. Every matrix in M_{hcn_0}, non-degenerate or degenerate, has n ones and $n(c-1)$ zeroes, while $\overline{U}$ has nc values equal to $1/c$. Compute the squared distance from $U \in M_{fcn_0}$ to $\overline{U}$:

$$\begin{aligned}\|U-\overline{U}\|^2 &= \sum_{k=1}^{n}\sum_{i=1}^{c}(u_{ik}-(1/c))^2 = n\left((1-(1/c)\right)^2 + n(c-1)/c^2 = n\left(1-2/c+1/c^2\right) + n(c-1)/c^2 \\ &= n\left((c-2)/c+1/c^2\right) + n(c-1)/c^2 = \left(n\left(c^2-2c+1\right)/c^2\right) + \left(\left(n(c-1)/c^2\right)\right. \\ &= \frac{n}{c^2}\left(c^2-2c+1+c-1\right) = \frac{n\left(c^2-c\right)}{c^2} = \frac{n(c-1)}{c}\end{aligned} \tag{6.5}$$

Thus, the squared distance from $\overline{U}$ to any point in M_{hcn_0} (the crisp c-partitions on n objects) is constant, $\|U-\overline{U}\|^2 = \frac{n(c-1)}{c}$; so the sum of squared distances is $\frac{n(c-1)}{c}\left(|M_{hcn_0}|\right)$.

To see that $\overline{U}$ yields the minimum distance, it suffices to show that one term in this sum will increase for any other soft partition. Suppose that there is another soft partition, say $U_\varepsilon \neq \overline{U}$, so that $\|U-U_\varepsilon\|^2 < \|U-\overline{U}\|^2$. Create U_ε by adding and subtracting any $\varepsilon \in (0,1)$ to a pair of elements in the same column of $\overline{U}$. Then U_ε has (nc-2) elements equal to $(1/c)$ and two altered values $(1/c) \pm \varepsilon$. The two altered values must be in the same column, say column $\mathbf{U}_\varepsilon^{(k)}$ of U_ε, to satisfy the column constraint. There are two cases: (i) $(1/c) \pm \varepsilon$ are both subtracted from 0 in column $\mathbf{U}^{(k)}$ of U (ii) or one of the values is matched to a 0 in $\mathbf{U}^{(k)}$ and the other one is matched to the 1 in column $\mathbf{U}^{(k)}$.

For case (i), $\|U-U\|^2$ has $(nc-2)$ terms as before, plus the two terms $(0-((1/c)-\varepsilon))^2+(0-((1/c)+\varepsilon))^2$ instead of the two terms $(0-(1/c))^2+(0-(1/c))^2 = 2/c^2$. Since $-((1/c)-\varepsilon)^2+\left(-((1/c)+\varepsilon)^2\right) = \left(2/c^2\right)+2\varepsilon^2 > 2/c^2$, the sum or squared errors increases by $2\varepsilon^2$; i.e., the squared distance from U_ε to any point in M_{hcn_0} increases and, hence, so does the overall sum of squares to all the points in M_{hcn_0}.

For case (ii), without loss of generality, we assume that $(1/c)+\varepsilon$ is matched to 0, while $(1/c)-\varepsilon$ is matched to 1. So, the pair of altered terms in the distance calculation is $(1-((1/c)-\varepsilon))^2+(0-((1/c)+\varepsilon))^2$. After a little algebra, the affirmative answer depends on whether $\left(1+2\varepsilon+2\varepsilon^2-\frac{2}{c}\right) > \frac{2}{c^2}$. It is left as an exercise to show that this inequality holds for $2 \le c \le (n-1)$; so this case also increases the sum of squared distances. Combining the two cases gives the result: $\overline{U}$ is the centroid of M_{hcn_0}.

To see that $\overline{U}$ is the centroid of M_{fcn}, suppose, to the contrary, that there is $U^* \in M_{fcn}$, $U^* \neq \overline{U}$, and that U^* is the centroid of M_{fcn}. By definition of the centroid, the squared distance from U^* to all the points in M_{fcn} must be minimum, including the points in M_{hcn_0}, since $M_{hcn_0} \subset M_{fcn}$. But $\overline{U}$ is the centroid of M_{hcn_0}; so $\|U-U^*\|^2 \ge \|U-\overline{U}\|^2 \ \forall\, U \in M_{hcn_0}$. This contradicts the assumption that $U^* \neq \overline{U}$. ■

Putting all these facts together gives us an idea of what the set $M_{fcn} = \text{conv}\left(M_{fcn_0}\right) \subset \Re^{cn}$ might look like if we could see this solid in $\Re^{cn}$. Figure 6.2(a) is a rendition of the Archimedean solid called an isododecahedron made by a woodworker (https://www.youtube.com/watch?v=8UeH1aTBy7k). Figure 6.2(b) shows part of the roof surface of a real 3D convex polytope, which is a photograph of (part of) the roof of AAMI Park, a sports stadium for soccer and rugby in Melbourne, Australia. Figure 6.2(c) is a stylized depiction of what we might see if we could look into $\Re^{cn}$ at the convex polytope M_{fcn} for $2 \le c \le$ (n-1).

Please go back to Figure 6.1 for a moment. The imaginary solid shown in Figure 6.2(c) is built by simply connecting all the points (recall that each point is a crisp c-partition on n objects) in Figure 6.1 with constant length edges. The set $M_{fcn} = \text{conv}(M_{hcn_0})$ is multi-faceted, akin to a soccer ball. Each face of this convex polytope is "flat." The points along the edges (in between vertices), on the flat faces, and all of the interior points are soft partitions. In the sequel, this structure is sometimes called the *cn-soccer ball*. All of the extreme points of this polytope, that is, the vertices in M_{hcn_0}, are either good (red = non-degenerate) or bad (yellow = degenerate) crisp c-partitions on n objects, and they are circumscribed by (that is, imbedded in the surface of) the hypersphere $B(\bar{U}, \sqrt{n(c-1)/c})$ centered at $\bar{U}$ with radius $\sqrt{n(c-1)/c}$.

When c = 1, the soccer ball collapses to a point, the *1-partition* $U_{1\times n} = \begin{bmatrix} 1 & 1 & \dots & 1 \end{bmatrix}$, and, likewise, at the other extreme, c = n, it collapses to the n-partition $U_{n\times n} = I_n = \text{diag}[1]$. It may help you to imagine the ball being initially flat at c = 1. As air is pumped in (c increases), it gets bigger and bigger up to c = (n-1), but when c becomes n, it pops and collapses back to a single point. It is a fetching analogy, but be careful – its validity rests with the value of n.

Forewarned is forearmed: Let's be clear about this: Figure 6.2(c) is an imaginary rendition of ***ALL*** the partitions that can be found by *every* crisp, fuzzy, or probabilistic clustering algorithm. Possibilistic partitions are not imagined in this diagram but would correspond to all of the points inside a hyperbox in $\Re^{cn}$ whose cn faces are tangent to the circumscribing hypersphere of the cn soccer ball is contained.

(a) 19" Geodesic sphere in $\Re^3$.

(b) Roof of AAMI Park in $\Re^3$.

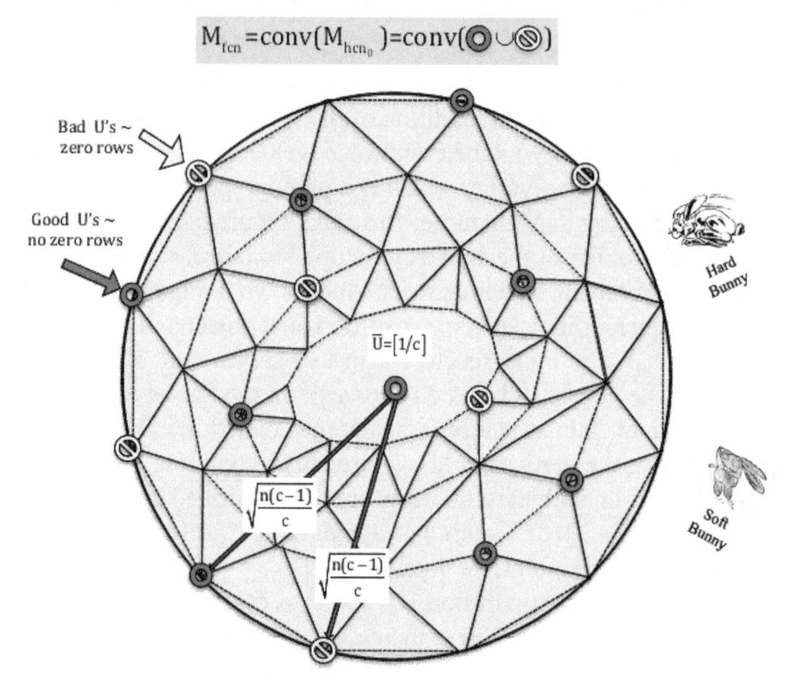

(c) The convex polytope of fuzzy/probabilistic partitions, $M_{fcn}=conv(M_{hcn_0}) \in \Re^{cn}$.

Figure 6.2 Convex polytopes: (a) and (b) in $\Re^2$ are real; (c) in $\Re^{cn}$ "in the mind's eye."

Example 6.2. A tale of two rabbits in c-partition land.

Did you notice those two bunnies in Figure 6.2(c)? What are they doing there? Well, they put us in a position to better understand one of the biggest difficulties encountered by many crisp clustering algorithms – the *local trap state*. And they can also help us understand why soft clustering can sometimes be superior to classical crisp approaches. Here is the tale of the two bunnies.

Hard bunny

Hard bunny is black and white (only two colors, 0 and 1). Think of yourself as k-means (HCM) bunny, at the start of a trip through the terrain in Figure 6.1. Your adventure starts at an initial crisp partition – a good (red) vertex in the land of crisp c-partitions. How will you find the "best" vertex? Well, in the case of HCM, the best vertex is the one which, when paired with its companion FONC for **V**, yields the global minimum of the (generalized) k-means objective function, i.e.,

$$J_1(U, \mathbf{V}; X) = \sum_{i=1}^{k} \left(\sum_{k=1}^{n} u_{ik} \left\| \mathbf{x}_k - \mathbf{v}_i \right\|_A^2 \right).$$

During the search for this best crisp CPOV partition of the input data, you can only hop from vertex to vertex during each iteration in your quest; always hoping to land on one that is good enough to end your trip (cf., Figures 9.7 and 10.1(b)). You know that the cn-soccer ball has a finite number of vertices; so why not hop to every one of them and find the global minimum by exhaustion? In a word, because YOU will be REALLY exhausted!

How many vertices are available? The finite cardinality of M_{hcn} is roughly $c^n/c!$ (the exact value has been given, the proof is left as an exercise). You might be lucky and land on a really good one according to some other criterion right away. But it is more likely that you will get tired; so your search will stop at an approximation that is accepted by your constraints. Those constraints are the control parameters of your crisp clustering algorithm. Unfortunately, you can't look ahead, and since you are tired already, you may decide to stop at a very unsatisfactory vertex (a local trap state – it *can't be a local minimum* – why not?). *Hopping from vertex to vertex is all every crisp clustering algorithm can **ever** do.* Different models set up the search over the

Soft Bunny

Soft bunny has colors ranging from black to white (any value from 0 to 1). Soft bunny can hop just like hard bunny, springing from vertex to vertex. But soft bunny can do more because he or she has a very important additional capability – soft bunny can *burrow*. Fuzzy clustering methods such as FCM, and probabilistic algorithms such as the EM scheme for GMD, allow soft bunny to hop around on these vertices (because $M_{hcn} \subset M_{fcn}$), but he or she can also travel into the *interior* of the cn-soccer ball depicted at Figure 6.2(c) while seeking an approximate solution that satisfies the optimization constraints. Now there are *infinitely many solutions* to examine. So, it sounds like soft bunny has a much harder problem than hard bunny, but, to the contrary, this imbedding makes many aspects of the search much easier. For example, soft objective functions are often differentiable in the matrix U (impossible in the crisp case), which allows us to employ the machinery of calculus and optimization theory when seeking an optimal partition. Not all soft clustering algorithms behave like FCM bunny, but the EM scheme for GMD certainly does.

And, importantly, a tunnel through the interior sometimes enables soft bunny to avoid a vertex that hard bunny finds irresistible. Of course, some tunnels eventually lead to the surface; so soft bunny may end up very close to a crisp vertex – perhaps, the same one that hard bunny chooses, if the data are strong enough to demand it. We will see this happen shortly.

Finally, soft bunny may have a home at $\bar{U}$, the centroid of M_{fcn}. As soft bunny travels up one of these tunnels toward the surface from $\bar{U}$, the uncertainty about memberships of the objects in the clusters decreases until, at the surface, it emerges at a crisp c-partition of the objects. No uncertainty about vertices in different ways, and some of them (like k-means) involve detours to look for good companion parameters (like the cluster centers **V**), but none of them can ever do more. Not all crisp clustering algorithms visit the vertices of Figure 6.1 in this particular way. We will see that single linkage bunny in Chapter 8 follows a very different path.

membership remains (but this doesn't guarantee the crisp memberships are "right"). So, moving toward the center means increasing uncertainty, and conversely.

We will return to this aspect of the c-means models again in Chapter 9, informed further by the theory of alternating optimization developed in Chapter 10.

The *dimension* of M_{fcn} is the dimension of the vector subspace in $\Re^{cn}$ obtained by translating it to the origin. Bezdek and Harris (1979) proved that $\dim(M_{fcn}) = n(c-1)$. You can intuit this result by simply thinking about the columns of U: there are n of them, and each one has (c-1) independent values. The last value in each column is determined by the column sum constraint.

6.3 The HCM/FCM Models and Basic AO Algorithms

There are quite a few crisp clustering algorithms that are called k-means in various books and papers. The two crisp clustering models that are most often known as, and sometimes confusingly both called, hard k-means, are the *batch* (global) model, which was introduced in Section 4.3 as HCM, and *sequential* (local or incremental) *hard c-means* (sHCM). The sHCM algorithm is discussed in the notes and remarks for this chapter and appears again in Chapter 12, where it has become something of a staple for clustering in streaming data.

The confusion is exacerbated by many authors who refer to either of these models as "k-means" without identifying which version, global or sequential, is meant. To make matters worse, there are different versions of each of these basic approaches. And, finally, there is the apparently never-ending discussion about whether to use (c) or (k) to denote the number of clusters[6].

The HCM model begins with the objective function at Equation (4.1), repeated here for convenience as

$$J_1(U, \mathbf{V}; X) = \sum_{k=1}^{n} \sum_{i=1}^{c} u_{ik} \left\|\mathbf{x}_k - \mathbf{v}_i\right\|_A^2, \tag{6.6}$$

where $X = \{\mathbf{x}_1, \ldots, \mathbf{x}_n\} \subset \Re^p$ is a set of object data, $U \in M_{hcn}$ is a crisp c-partition of X, and the c vectors $\mathbf{V} = \{\mathbf{v}_1, \ldots, \mathbf{v}_c\} \subset \Re^{cp}$ are *cluster centers* (prototypes, templates, code vectors, etc.) of the crisp clusters imposed on X. The matrix A in Equation (6.6) is a positive-definite weight matrix that induces the inner product norm.

Definition 6.4. Positive-definite matrices: inner product norms.

A real $p \times p$ matrix $A \in \Re^{pp}$ is *positive-definite* (PD) if and only if (i) it is symmetric, and (ii) $\mathbf{x}^T A \mathbf{x} > 0, \mathbf{x} \neq \mathbf{0} \in \Re^p$. Let $PD^{pp} \subset \Re^{pp}$ denote the set of real positive-definite p×p matrices. It is left as an exercise

[6]This integer is a dummy variable, right? You make the call.

to show that $A \in PD^{pp} \Leftrightarrow A^{-1} \in PD^{pp}$. We will also encounter matrices that are *positive semi-definite* (p.s.d, $x^T Ax \geq 0$), *negative-definite* (n.d., $x^T Ax < 0$), or *negative semi-definite* (n.s.d., $x^T Ax \leq 0$). Every matrix $A \in PD^{pp}$ induces the unique inner product A-norm

$$d_{ikA} = \|x_k - v_i\|_A = \sqrt{(x_k - v_i)^T A (x_k - v_i)} = \sqrt{\sum_{s=1}^{p}\sum_{t=1}^{p} x_{ks} a_{st} v_{it}}. \tag{6.7}$$

It is called the *inner product* A-norm because $\langle x, x\rangle_A = x^T Ax$ is an inner product (dot product and scalar product) on $\Re^p \times \Re^p$ (cf., Exercise 6.13). In other contexts, the function $Q_A(\mathbf{x}) = \mathbf{x}^T A\mathbf{x}$ is also called a *quadratic form* in A.

The fuzzy c-means model begins with the generalization of the HCM objective function at Equation (6.6):

$$J_m(U,V;X) = \sum_{k=1}^{n}\sum_{i=1}^{c} (u_{ik})^m \|x_k - v_i\|_A^2;\ m \geq 1. \tag{6.8}$$

In the sequel, we may refer to $J_m(U, \mathbf{V}; X)$ more briefly as $J_m(\mathbf{Q}; X)$, where $\mathbf{Q} = (U,\mathbf{V})$. In Equation (6.8), $U \in M_{fcn}$ is a non-degenerate fuzzy c-partition of X and $m \geq 1$ is a weighting exponent (m is usually called the *fuzzifier*) applied to the memberships of the data in the c fuzzy clusters in X. At m = 1, Equation (6.8) reduces to the HCM objective function at Equation (6.6), a case that must be handled separately. The objective function J_2 was discussed by Dunn (1974) for the special case m = 2. Bezdek (1973) generalized J_2 to any value of $m > 1$. The optimization problem associated with Equation (6.8) is the family of hard and fuzzy c-means models.

Definition 6.5. The hard and fuzzy c-means models (HCM/FCM).

$X = \{x_1, \ldots, x_n\} \subset \Re^p$ is a set of object data. $(U, \mathbf{V}) \in M_{fcn} \times \Re^{cp}$, $A \in PD^{pp}$ induces the inner product A-norm $d_{ikA} = \|x_k - v_i\|_A = \sqrt{(x_k - v_i)^T A (x_k - v_i)}$. The *hard (m = 1, aka "k-means") and fuzzy (m >1) c-means models* are the non-linear constrained optimization problems

$$\underset{U\in M_{hcn}, \mathbf{V}\in\Re^{cp}}{\text{minimize}} \left\{ J_1(U, V; X) = \sum_{k=1}^{n}\sum_{i=1}^{c} (u_{ik}) \|x_k - v_i\|_A^2 \right\} \text{(hard c -means, HCM = "k-means")}. \tag{6.9}$$

$$\underset{U\in M_{fcn}, \mathbf{V}\in\Re^{p}}{\text{minimize}} \left\{ J_m(U, V; X) = \sum_{k=1}^{n}\sum_{i=1}^{c} (u_{ik})^m \|x_k - v_i\|_A^2 \right\} \text{(fuzzy c-means, FCM, } m \geq 1). \tag{6.10}$$

Note that the weighting parameter (m) for memberships in the FCM model is stated as $m \geq 1$. It turns out that problem (6.10) has problem (6.9) as its limit at m = 1. That is, problem (6.9) is a special case of problem (6.10). They are written separately here to make explicit the distinction between the crisp and fuzzy cases. As an aside, we remark that for m = 1, papers that talk about "convergence to a local minimum" of the k-means (HCM) algorithm are incorrect: the function J_1 at Equation (6.9) is not continuous in U; so it can't have local minimums.

According to Drineas *et al.* (2004), the classical k-means optimization problem at Equation (6.9) is NP-hard, even at c = 2. These authors discuss a relaxation method for approximating solutions to Equation (6.9) at m = 1. We will discuss a somewhat more straightforward method here and, eventually, will show that the HCM objective function in Equation (6.9) at m = 1 can be formulated in at least three other ways. Derivation of theoretical conditions for extrema of J_m in the *joint* variables (U, **V**) is intractable. Solutions for Equations (6.9) and (6.10) in the separated block variables U and **V** by AO are discussed next.

Case 1: Fuzzy c-means, m > 1. Because there are both equality ($\sum_{i=1}^{c} u_{ik} = 1$) and inequality ($0 \leq u_{ik} \leq 1$) constraints on the elements of U, optimization problem (6.10) is a *Karush–Kuhn–Tucker* (KKT) problem (Luenberger (1984)). However, applying the full KKT theory to the optimization of J_m is difficult. Instead, Dunn and Bezdek both used the simpler Lagrange multiplier theory to find *first order necessary conditions* (FONCs) for extreme points (U, **V**) of J_2 and J_m, respectively, by ignoring the inequality constraints and then verifying that the solution obtained by Lagrange multipliers satisfied the inequality constraints anyway. This approach begins for the fuzzy case by fixing m > 1 and then reducing Equation (6.10) into two smaller "half problems" involving optimization over just one of the two sets of variables. Specifically, consider the paired (half) problems

$$\underset{\mathbf{V}\in\mathfrak{R}^{cp}}{\text{minimize}}\left\{J_m(U^*,\mathbf{V};X)=\sum_{k=1}^{n}\sum_{i=1}^{c}\left(u_{ik}^*\right)^m\|\mathbf{x}_k-\mathbf{v}_i\|_A^2\right\}; \tag{6.11a}$$

$$\underset{U\in M_{fcn}}{\text{minimize}}\left\{J_m(U,\mathbf{V}^*;X)=\sum_{k=1}^{n}\sum_{i=1}^{c}\left(u_{ik}\right)^m\|\mathbf{x}_k-\mathbf{v}_i^*\|_A^2\right\}. \tag{6.11b}$$

U^* is fixed in Equation (6.11a); $\mathbf{V}^*$ is fixed in Equation (6.11b). This pair of reduced problems can be solved using calculus. The problem at Equation (6.11a), minimization of $J_m(U^*, \mathbf{V}; X)$, is unconstrained; so solving $\nabla_{\mathbf{v}_i} J_m(U^*,\mathbf{V};X) = \mathbf{0}$ for $\mathbf{v}_i$ leads directly to first-order necessary conditions on the cluster centers for m > 1, viz.,

$$\mathbf{v}_i^* = \sum_{k=1}^{n}\left(u_{ik}^*\right)^m \mathbf{x}_k \Big/ \sum_{k=1}^{n}\left(u_{ik}^*\right)^m;\ 1 \leq i \leq c. \tag{6.12}$$

As long as the norm in Equation (6.10) is an inner product norm, the function $\|\mathbf{x}_k - \mathbf{v}_i\|_A^2$ is differentiable with respect to $\mathbf{v}_i$, and Equation (6.12) follows. But, if, for example, the model norm used in Equation (6.10) is a Minkowski norm (other than the Euclidean norm, for which A is the p × p identity matrix), solutions of Equation (6.11a) can't be approached via calculus.

Now consider the half-problem at Equation (6.11b). This is a KKT problem in U when $\mathbf{V}^*$ is fixed. The problem can be treated as a Lagrange multiplier problem by ignoring the inequality constraints on the values of $\{u_{ik}\}$ – but later, we will have to double back and check the solution to make sure it does satisfy the inequality conditions. Consider one term in the all-positive sum over k in Equation (6.11b). We seek a solution for the reduced problem based on just one column, say $\mathbf{U}^{(i)}$, the *i*th column of U:

$$\underset{U\in N_{fc}}{\text{minimize}}\left\{g\left(\mathbf{U}^{(i)},\mathbf{V}^*;X\right)=\sum_{i=1}^{c}\left(u_{ik}\right)^m\|\mathbf{x}_k-\mathbf{v}_i^*\|_A^2\right\}. \tag{6.13}$$

Forming the Lagrangian of this function for any value of $m > 1$ and setting its gradient equal to the zero vector leads to

$$u_{ik}^* = \left(\|\mathbf{x}_k - \mathbf{v}_i^*\|_A^2\right)^{\frac{1}{m-1}} / \sum_{j=1}^{c} \left(\|\mathbf{x}_k - \mathbf{v}_j^*\|_A^2\right)^{\frac{1}{m-1}} ; 1 \le i \le c; m > 1. \tag{6.14}$$

Equation (6.14) is the FONC for solutions to Equation (6.13). Since the n columns $\{U^{(k)}\}$ of U are independent and $J_m(U^{(k)}, \mathbf{V}^*; X) \ge 0 \ \forall \ k$, solutions of the n problems represented by Equation (6.13) for one i, can be written as $\sum_{k=1}^{n} \{\min\{J_m(U^{(k)}, \mathbf{V}^*; X)\}\} = \min\left\{\sum_{k=1}^{n} J_m(U^{(k)}, \mathbf{V}^*; X)\right\} = \min\{J_m(U, \mathbf{V}^*; X)\}$. Thus, we are entitled to add "for all k" to Equation (6.14), yielding the FONCs for Equation (6.11b):

$$u_{ik}^* = \left(\|\mathbf{x}_k - \mathbf{v}_i^*\|_A^2\right)^{\frac{1}{m-1}} / \sum_{j=1}^{c} \left(\|\mathbf{x}_k - \mathbf{v}_j^*\|_A^2\right)^{\frac{1}{m-1}} ; 1 \le i \le c; 1 \le k \le n; m > 1. \tag{6.15}$$

The inequality constraints on the memberships in U* have been ignored; so it remains to be shown that the cn values in U* all lie in [0, 1]. If all n columns of U* are computed with Equation (6.15), it is clear that $u_{ik} \in (0, 1) \forall i, k$ if none of the data points coincide with any of the cluster centers. To handle the singular case when $\exists k, j \ni \|\mathbf{x}_k - \mathbf{v}_j\|_A^2 = 0$, define an index set for the row distances in column (k) that are zero

$$I_k = \left\{1 \le i \le c : \|\mathbf{x}_k - \mathbf{v}_i^*\|_A^2 = 0\right\}; \tag{6.16}$$

$$\widetilde{I}_k = \{1, 2, \ldots, c\} - I_k. \tag{6.17}$$

For k's such that $I_k = \varnothing$, the c values $\{u_{ik}^* \in (0, 1) : i = 1, \ldots, c\}$ are computed with Equation (6.15). For any k such that $I_k \neq \varnothing$, Equation (6.15) is replaced by the singularity condition

$$I_k \neq \varnothing \Rightarrow u_{ik}^* = 0 \, \forall i \in \widetilde{I}_k; \sum_{i \in I_k} u_{ik}^* = 1. \tag{6.18}$$

It is left as an exercise for you to show that if some of the columns in U* are computed using Equation (6.18), that U* is still a non-degenerate c-partition of X. Equations (6.16) and (6.17) may seem obscure, but they are easy to understand: when singularity occurs for column $(\mathbf{U}^*)^{(k)}$, zero out the terms for all rows where the distances vanish and distribute the membership of $\mathbf{x}_k$ arbitrarily over the remaining rows for this column. The collective experience for nearly 50 years with formula (6.15) is that the occurrence of FCM singularity is exceedingly rare in practice but must be accounted for in both the theory and in implementations that use the FONCs for J_m. This completes a sketch of the proof of FCM theorem 6.2.

It is useful when describing implementation details to write Equations (6.12) and (6.15) more compactly. Suppressing the (*) indicating optimal, half passes through X for cluster center updates are conveniently written as $\mathbf{V} = G_m(U; X)$. In other words, G_m is the function that produces $\mathbf{V}$ from any pair (U; X) using Equation (6.12). In like manner, the other half of these conditions is represented as $U = F_m(\mathbf{V}; X)$ – that is, F_m is the function that produces U from any pair ($\mathbf{V}$; X) with Equations (6.15) and possibly (6.16). It is worth emphasizing that given any U, computing $\mathbf{V}$ with Equation (6.12) yields a pair (U, $\mathbf{V}$) that satisfies the FONCs for J_m. And conversely, starting with any $\mathbf{V}$, computing U from Equation (6.15) also gives an optimal pair. Thus, (6.12) and (6.15) yield pairs of companion conditions for the FONCs that define solutions to problem (6.10).

Theorem 6.2. FONCs for basic AO solutions of the FCM model ($m > 1$).

Let $\|x_k - v_i\|_A^2$ be any inner product norm, and let $X = \{x_1, \ldots, x_n\} \subset \Re^p$ have $c < n$ distinct points. For all k, define the *singularity sets* $I_k = \left\{1 \le i \le c : \|x_k - v_i^*\|_A^2 = 0\right\}$, $\tilde{I}_k = \{1, 2, \ldots, c\} - I_k$. For $m > 1$, $(U^*, V^*) \in M_{fcn} \times \Re^{cp}$ *may be* an extreme (or stationary) point for $J_m(U, V; X) = \sum_{k=1}^{n}\sum_{i=1}^{c}(u_{ik})^m \|x_k - v_i\|_A^2$ *only if* $V^* = G_m(U^*; X) = (v_1^*, \ldots, v_c^*)$, or in coordinate form,

$$v_i^* = \sum_{k=1}^{n}\left(u_{ik}^*\right)^m x_k \Big/ \sum_{k=1}^{n}\left(u_{ik}^*\right)^m, 1 \le i \le c; \text{and} \tag{6.19a}$$

$U^* = F_m(V^*; X)$, or in coordinate form,

$$I_k = \varnothing \Rightarrow u_{ik}^* = \frac{\left(\|x_k - v_i^*\|_A^2\right)^{\frac{1}{m-1}}}{\sum_{j=1}^{c}\left(\left\|x_k - v_j^*\right\|_A^2\right)^{\frac{1}{m-1}}}; \begin{cases} 1 \le i \le c \\ 1 \le k \le n \end{cases} ; \text{or} \tag{6.19b1}$$

$$I_k \neq \varnothing \Rightarrow u_{ik}^* = 0 \forall i \in \tilde{I}_k, \sum_{i \in I_k} u_{ik}^* = 1; 1 \le k \le n. \tag{6.19b2}$$

What does it mean? Conditions (6.19) are the basis for the AO algorithm that is used to approximately minimize J_m. The functional form of AO is simple Picard iteration in the joint variables, which begins with an initial guess for U_0 (or for V_0) and then loops through the FONCs at Equations (6.19a) and (6.19b) in half steps, for example, as $\cdots U_t = F_m(V_{t-1}; X) \rightarrow V_t = G_m(U_t; X) \rightarrow U_{t+1} = F_m(V_t; X) \cdots$, where (t) is the iteration counter and the functions G_m and F_m yield the right-hand sides of Equations (6.19a) and (6.19b1/6.19b2), respectively. Each full step of this procedure endows J_m with the descent property, i.e., $J_m(U_{t+1}, V_{t+1}; X) \le J_m(U_t, V_t; X)$ for each iterate t; so looping through the conditions at Equations (6.19a) and (6.19b1/6.19b2) do lead J_m to an equal or smaller value at each step. Termination of this looping structure is discussed after A6.1 is recorded. The extreme points referred to in the statement of Theorem 6.2 may be local minima or saddle points but can't be local maxima (Bezdek *et al*., 1987).

Proof. All of the essential steps for the proof have been sketched out above. A complete, explicit proof of Theorem 6.2 appears in Section 11 of Bezdek (1981). We will record the algorithm that depends on Theorem 6.2 shortly, but, first, let us take up the special case of problem (6.9).

Case 2: Hard c-means (aka "k-means"), m = 1. Solutions for the HCM model in Definition 6.5 are also approached as two half-problems. We begin again with the reduced problem at Equation (6.11a) but require that the fixed U variables for this problem are $U^* \in M_{hcn}$ (a crisp partition), so the half problem at hand is

$$\underset{V \in \Re^{cp}}{\text{minimize}}\left\{J_1(U^*, V; X) = \sum_{k=1}^{n}\sum_{i=1}^{c}(u_{ik}^*)\|x_k - v_i\|_A^2\right\}. \tag{6.20}$$

It will be shown below that this is equivalent to fixing m = 1 in the FCM model. As long as the norm in Equation (6.9) or ((6.11a) at m = 1) is differentiable with respect to $\mathbf{v}_i$, obtaining FONCs for $\mathbf{V}^*$ with U* fixed proceeds exactly as in the fuzzy case, yielding

$$\mathbf{v}_i^* = \sum_{k=1}^{n} u_{ik}^* \mathbf{x}_k \Big/ \sum_{k=1}^{n} u_{ik}^* = \left[\sum_{\mathbf{x}_k \in X_i} \mathbf{x}_k / n_i \right], \quad 1 \leq i \leq c, \tag{6.21}$$

which was exhibited as Equation (4.3) in Chapter 4. The second form of Equation (6.21) in square brackets shows that $\mathbf{v}_i^*$ is just the centroid of the points currently assigned to cluster (i). Turning to the other half-problem (6.11b) for m=1,

$$\underset{U \in M_{hcn}}{\text{minimize}} \left\{ J_1(U, \mathbf{V}^*; X) = \sum_{k=1}^{n} \sum_{i=1}^{c} u_{ik} \|\mathbf{x}_k - \mathbf{v}_i^*\|_A^2 \right\}, \tag{6.22}$$

we fix $\mathbf{V}^*$ and set out to find FONCs for U, but, here, we hit a bump in the road. Possible solutions to Equation (6.22) lie amongst the finite but very large number of "good" vertices on the cn soccer ball (M_{hcn} in Figure 6.2). The good news is that since M_{hcn} is finite, problem (6.22) certainly has a global solution that is bounded below by zero. The bad news is that J_1 is not differentiable in the variables $\{u_{ik}\}$, so, we can't use calculus to establish FONCs for extrema of (6.22), and this means that the function $J_1(U, \mathbf{V}^*)$ *can't have local minima* in the ordinary sense. On the other hand, because U is crisp, we are able to derive, for any $\mathbf{V}^*$, a unique "companion *necessary condition* (NC)" for U, and this forms the basis for an AO algorithm that approximates solutions of Equation (6.22).

The rationale underlying the U update for HCM (k-means) is shown in Figure 6.3. From a theoretical point of view, the solution of Equation (6.22) doesn't need an inner product A-norm for the distances. All we need is a set of positive real numbers that can be ordered – any dissimilarity function would satisfy this part of the argument. But we do need a differentiable norm for the other half of the problem (finding $\mathbf{V}^*$); so we leave it in place here.

Having $\mathbf{V}^*$, we compute $d_{ikA}^2 = \|\mathbf{x}_k - \mathbf{v}_i^*\|_A^2$; $i = 1, \ldots, c$. Consider the choice for $\mathbf{U}^{(k)}$, the crisp label vector that is to be assigned to $\mathbf{x}_k$ as shown in Figure 6.3. This illustration is for the Euclidean norm, but the concept is theoretically correct for any distance measure on $\Re^p$.

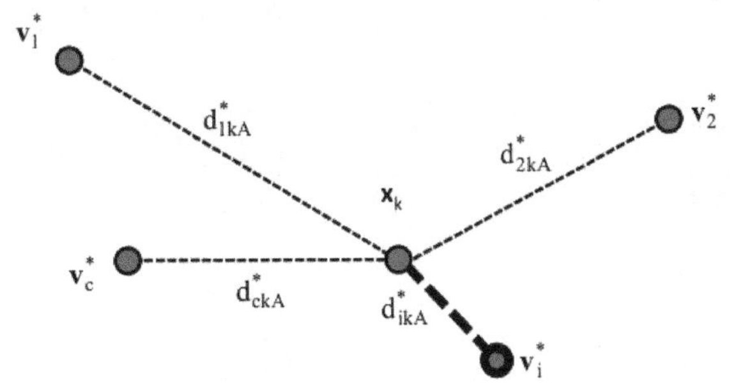

Figure 6.3 The nearest prototype assignment of crisp label vector $U^{(k)}$.

The c terms in $J_1(U, \mathbf{V}^*)$ corresponding to the *k*th column are

$$J_1\left(\mathbf{U}^{(k)}, \mathbf{V}^* : X\right) = u_{1k}d^*_{1kA} + u_{2k}d^*_{2kA} + \cdots + u_{ck}d^*_{ckA}. \tag{6.23}$$

Suppose we set $u_{jk} = 1$ in Equation (6.23). Then the column constraint for $\mathbf{U}^{(k)}$ forces us to assign 0's to the remaining (c-1) membership function multipliers, $\{u_{ik} = 0 : i \neq j\}$. The chosen value 1 and the (c-1) 0's are the coordinates of $\mathbf{U}^{(k)}$. If we place the 1 at position j of $\mathbf{U}^{(k)}$, the value of J_1 increases by the amount d^*_{jkA}. The minimum for (6.23) occurs for the minimum value of d^*_{jkA}, which is d^*_{ikA} in Figure 6.3 because $\mathbf{v}^*_i$ is the closest prototype to $\mathbf{x}_k$. Continuing until all n columns of U are chosen, we end up with the condition for U* when $\mathbf{V}^*$ is fixed, shown at Equation (4.4) and again here as Equation (6.24):

$$u^*_{ik} = \begin{Bmatrix} 1; & \|\mathbf{x}_k - \mathbf{v}^*_i\|^2_A \le \left\|\mathbf{x}_k - \mathbf{v}^*_j\right\|^2_A, j \neq i \\ 0; & \text{otherwise} \end{Bmatrix}; \begin{Bmatrix} 1 \le i \le c \\ 1 \le k \le n \end{Bmatrix}. \tag{6.24}$$

This is an intuitively satisfying result: it simply assigns each $\mathbf{x}_k$ to the crisp cluster with the closest prototype. In classifier design, this is called the *nearest prototype* (1-NPR) rule. While the argument for updating U this way is based on minimally *increasing* J_1, we expect that the overall value of J_1 *decreases* when comparing successive values of J_1 for (U, **V**) pairs. Singularity for HCM/AO based on conditions (6.24) is not as dramatic as it is for FCM. Here, the only problem that can occur is when the set of c distances $\{d^*_{jkA}: 1 \le j \le c\}$ doesn't have a unique minimum. This simply requires an arbitrary tie-breaking procedure in the HCM/AO algorithm.

Theorem 6.3. Necessary conditions for the HC (k-means) model (m = 1).

Let $\|\mathbf{x}_k - \mathbf{v}^*_i\|^2_A$ be any inner product norm, and let $X = \{\mathbf{x}_1, \ldots, \mathbf{x}_n\} \subset \Re^p$ have $c < n$ distinct points partitioned by $U \in M_{hcn} \leftrightarrow X = \bigcup_{i=1}^{c} X_i; \varnothing = \underset{i \neq j}{X_i \bigcap X_j} : u_{ik} = \begin{Bmatrix} 1; & \mathbf{x}_k \in X_i \\ 0; & \text{otherwise} \end{Bmatrix}$, $\sum_{k=1}^{n} u_{ik} = n_i = |X_i|;\ i = 1, \ldots, c$. Then $(U^*, \mathbf{V}^*) \in M_{hcn} \times \Re^{cp}$ ***may*** be a fixed point for $J_1(U, \mathbf{V}; X) = \sum_{k=1}^{n}\sum_{i=1}^{c} u_{ik}\|\mathbf{x}_k - \mathbf{v}_i\|^2_A$ *only if* $\mathbf{V}^* = G_1(U^*; X) = (\mathbf{v}^*_1, \ldots, \mathbf{v}^*_c)$, or in coordinate form,

$$\mathbf{v}^*_i = \sum_{k=1}^{n} u^*_{ik}\mathbf{x}_k \Big/ \sum_{k=1}^{n} u^*_{ik} = \sum_{\mathbf{x}_k \in X_i} \mathbf{x}_k / n_i;\ 1 \le i \le c;\ \text{and} \tag{6.25a}$$

$U^* = F_1(\mathbf{V}^*; X)$, or in coordinate form,

$$u^*_{ik} = \begin{Bmatrix} 1; & \|\mathbf{x}_k - \mathbf{v}^*_i\|^2_A \le \left\|\mathbf{x}_k - \mathbf{v}^*_j\right\|^2_A, j \neq i \\ 0; & \text{otherwise} \end{Bmatrix}; \begin{Bmatrix} 1 \le i \le c \\ 1 \le k \le n \end{Bmatrix}. \tag{6.25b}$$

What does it mean? Conditions (6.25) are the basis for the iterative algorithm (almost always called Lloyd's algorithm for hard k-means) that is used to approximately minimize J_1. Iteration is a simple loop beginning with an initialization for U_0 (or for $\mathbf{V}_0$), and then looping back and forth through the necessary conditions, $\cdots U_t = F_1(\mathbf{V}_{t-1}; X) \rightarrow \mathbf{V}_t = G_1(U_t; X) \rightarrow U_{t+1} = F_1(\mathbf{V}_t; X) \cdots$. The term "fixed point" used in the statement of this theorem deserves an explanation which will be given shortly. Termination of this sequence is discussed after A6.1 is recorded.

Proof. There are many ways to formalize a proof for this theorem. The basic steps have been outlined here. There is a fairly long history associated with the development of Equations (6.25), which began in 1956–1957, much of which is covered in Section 6.9.

Should the iterative procedure outlined in Theorem 6.3 be called *alternating optimization* (AO)? Since the notion of a local extremum of a function is a neighborhood-based concept, but crisp non-degenerate U's don't have neighborhoods (they are the good vertices among the set of vertices in Figure 6.1), the idea that J_1 can *have* local extrema is not well-defined in the classical sense. J_1 is not even jointly continuous in the variables $(U, \mathbf{V})$. So, iteration through conditions (6.25) can't be *formally* viewed through the lens of classical AO, but there is little harm in regarding HCM (Lloyd's k-means) as alternating optimization as long as the distinction about what the term means is made clear. The half conditions for $\mathbf{V}^*$ at Equation (6.25a) for a fixed U *are* FONCs, and the half-conditions at Equation (6.25b) are *necessary conditions* (NCs) on U for a fixed $\mathbf{V}^*$, but they are not "first-order" conditions in the conventional sense of satisfying necessary conditions based on zeroes of first derivatives. It is not wrong to call the conditions (6.25) necessary conditions for problem (6.9), and since each half step are NCs, it is easy to slip into the habit of simply calling both halves of Equation (6.25) FONCs for optimization of J_1, but this is technically incorrect. Section 10.4 sheds some additional light on this question when HCM is captured in the spotlight of the convergence theory of alternating optimization.

Since the set $S_1 = \{(U^*, \mathbf{V}^*)\}$ of feasible solutions for optimization of J_1 is finite, S_1 must contain at least one pair that are a global minimum for J_1. But finding such a pair by exhaustive search is out of the question because of the huge number of vertices in M_{hcn} that must be examined. Each full step of Lloyd's k-means procedure does endow J_1 with the descent property. That is, for each iterate t, $J_1(U_{t+1}, \mathbf{V}_{t+1}; X) \leq J_1(U_t, \mathbf{V}_t; X)$, so looping back and forth through the conditions at Equations (6.25a) and (6.25b) do lead J_1 toward an equal or smaller value at each step. When the iteration is terminated because $J_1(U_{t+1}, \mathbf{V}_{t+1}; X) = J_1(U_t, \mathbf{V}_t; X)$, some authors describe $(U_t, \mathbf{V}_t)$ as a local minima, but this is not correct: $(U_t, \mathbf{V}_t)$ is a fixed point of the Picard operator. The best we can do is to require that any $(U^*, \mathbf{V}^*)$ pair offered as an approximate solution to Equation (6.9) at m = 1 satisfies the necessary conditions for U^* and $\mathbf{V}^*$ shown at Equation (6.25). This might seem a bit disappointing from the theoretical standpoint, but, historically, this pair of slightly mismatched conditions have been used as a basis for crisp clustering more than any other method on planet earth.

There is an additional relationship between the FONCs for FCM and the NCs for HCM that is derived explicitly in Bezdek (1981). If all n minimums in Equation (6.25b) for HCM are unique, and all n singularity sets for FCM in Equation (6.19b1) are empty, then taking limits as m approaches 1 from above yields

$$\underset{m \xrightarrow{+} 1}{\text{limit}} \{J_m(U^*, \mathbf{V}^*; X)\} = J_1(U^*, \mathbf{V}^*; X) \tag{6.26a}$$

if, and only if, the cn values in U* are computed with Equation (6.25b). This proves that the FCM model is a direct generalization of the HCM (k-means) model. Moreover, the FONCs for FCM have the NCs used by Lloyd's algorithm as their limits as m approaches 1 from above. It is easy to see that for i = 1 to c,

$$\underset{m \xrightarrow{+} 1}{\text{limit}} \left\{ \frac{\sum_{k=1}^{n} (u_{ik}^*)^m x_k}{\sum_{k=1}^{n} (u_{ik}^*)^m} \right\} = \frac{\sum_{k=1}^{n} u_{ik}^* x_k}{\sum_{k=1}^{n} u_{ik}^*} = \sum_{x_k \in X_i} \frac{x_k}{n_i}. \tag{6.26b}$$

Less obvious, but true, is that the FONCs for FCM U at m > 1 converge to the NCs for HCM at m = 1 as m approaches 1 from above; so for all i and k,

$$\underset{m \xrightarrow{+} 1}{\text{limit}} \left\{ \frac{\left(\|x_k - v_i^*\|_A^2\right)^{\frac{1}{m-1}}}{\sum_{j=1}^{c} \left(\left\|x_k - v_j^*\right\|_A^2\right)^{\frac{1}{m-1}}} \right\} = \left\{ \begin{array}{ll} 1; & \|x_k - v_i^*\|_A^2 \le \left\|x_k - v_j^*\right\|_A^2, j \neq i \\ 0; & \text{otherwise} \end{array} \right\}. \tag{6.26c}$$

This amounts to an analytic proof that computing U* for HCM must be done with the 1-np rule. Because it is "necessary" to compute U this way, there is an added temptation for calling Equations (6.25) FONCs for problem (6.9) at m = 1. At the other extreme, as m grows without bound,

$$\underset{m \to \infty}{\text{limit}} \{J_m(U^*, V^*; X)\} = 0. \tag{6.27}$$

This suggests that J_m decreases monotonically on optimal pairs, and it does. See Pal and Bezdek (1995) for more discussion about limit properties of HCM and FCM. J_1 also has an interesting and important statistical interpretation. To exhibit it, we need to define the within cluster and between cluster scatter matrices for a set of vectors in $\Re^p$.

Definition 6.6. The scatter matrices of Wilks (1962).

$X = \{x_1, \ldots, x_n\} \subset \Re^p$, $U \in M_{hcn}$, $\bar{x}_i = \sum_{k=1}^{n} u_{ik} x_k / n_i$ is the sample mean of the vectors in cluster X_i, $\sum_{k=1}^{n} u_{ik} = n_i = |X_i|$; $i = 1, \ldots, c$. The *scatter matrix* S_i of the *i*th cluster and the overall *within cluster scatter matrix* S_W associated with U are

$$S_i = \sum_{k=1}^{n} u_{ik}(x_k - \bar{x}_i)(x_k - \bar{x}_i)^T = \sum_{x_k \in X_i} (x_k - \bar{x}_i)(x_k - \bar{x}_i)^T, 1 < i < c, \tag{6.28}$$

$$S_W = \sum_{i=1}^{c} S_i = \sum_{i=1}^{c}\sum_{k=1}^{n} u_{ik}(x_k - \bar{x}_i)(x_k - \bar{x}_i)^T = \sum_{i=1}^{c} \left[\sum_{x_k \in X_i} (x_k - \bar{x}_i)(x_k - \bar{x}_i)^T \right]. \tag{6.29}$$

Using the Euclidean norm (set A = I_{pxp} in Definition 6.4) allows us to write J_1 at Equation (6.9) as the trace of S_W:

$$J_1(U, V: X) = \sum_{i=1}^{c} \left(\sum_{x_k \in X_i} \|x_k - v_i\|^2 \right) = \sum_{i=1}^{c} \left(\sum_{x_k \in X_i} \|x_k - \bar{x}_i\|^2 \right) = tr(S_W). \tag{6.30}$$

Since the trace of S_W is proportional to the sum of variances in the p coordinate directions, minimizing J_1 for the special case of the Euclidean norm also minimizes this measure of variance of the c clusters in U. This statistical connection to HCM justifies calling it the *minimum variance partitioning problem*[7], and is, perhaps, yet another reason why HCM (k-means) is such a popular clustering model.

Now we are ready to state the naive iterative algorithm based on Theorems 6.2 and 6.3, which describes a basic version of HCM/AO and FCM/AO. Recall our terminology for the looping structure established earlier, that $\cdots U_t = F_m(V_{t-1}; X) \to \mathbf{V}_t = G_m(U_t; X) \to U_{t+1} = F_m(\mathbf{V}_t; X) \cdots$.

Algorithm A6.1. Basic HCM (k-means)/FCM AO for the batch HCM and FCM models

	Algorithm HCM/FCM
1	**In** Unlabeled feature vector data set $X = \{\mathbf{x}_1, \ldots, \mathbf{x}_n\} \subset \Re^p$
2	***% User choices for model parameters***
3	**Set** Number of clusters: $c \in \{2, \ldots, n-1\}$,
4	Fuzzifier for J_m: m = 1 for HCM, m > 1 for FCM,
5	$A \in PD^{pp}$ for *model norm* $d_{ikA} = \lVert \mathbf{x}_k - \mathbf{v}_i \rVert_A$.
6	***% User Choices for Implementation Parameters***
7	Termination threshold: $\epsilon \in (0, \infty)$:
8	Termination norm: $\lVert * \rVert_{err}$:
9	Iterate limit T: $t \in \{2, \ldots, T\}$:
10	Initial guess: choose $\mathbf{V}_0 = (\mathbf{v}_{1,0}, \ldots \mathbf{v}_{c,0}) \in \Re^{cp}$
11	$U_0 = F_m(\mathbf{V}_0)$: $\mathbf{V}_1 = G_m(U_0)$: t = 1
12	***While*** $(t \leq T \text{ and } \lVert \mathbf{V}_t - \mathbf{V}_{t-1} \rVert_{err} > \varepsilon)$
13	$U_t = F_m(\mathbf{V}_t)$
14	% If m = 1: if $\min_i\{d_{ikA}\}$ is not unique, set first = 1
15	% If m > 1: if $\min_i\{d_{ikA}\}$ is zero, compute $\mathbf{U}_t^{(k)}$ with (6.19b2)
16	$\mathbf{V}_{t+1} = G_m(U_t) : t = t + 1$
17	***Endwhile***
18	$U^* = U_t : \mathbf{V}^* = \mathbf{V}_t$
19	**Out** $(U^*, \mathbf{V}^*)$

The argument that establishes Equation (6.24) only uses the notion of nearest prototype, which, in turn, depends only on a way to measure "nearest," i.e., on distance. So, the model norm in line 5 of A6.1 must be an A-norm in order to use the FONCs for FCM at m > 1, but it could be any metric on the input space for HCM in A6.1(m = 1). The computation of cluster centers at each $^1/_2$ step remains the same in all cases. To avoid continually repeating the same runtime parameters that are used in many of the numerical examples, here is a list of default choices that are used for A6.1 in most of the examples presented below. Deviations from the default values will be stated explicitly.

[7] Be careful to distinguish between the batch and sequential versions of hard c-means (k-means). Duda and Hart (1973) refer to the sequential version as the basic minimum squared error partitioning problem.

Default Choices for A6.1: The model Norm is Euclidean (also called the 2-norm), $d_{ik,A=I} = \|x_k - v_i\|_2$; m = 1 for HCM, m = 2 for FCM; $\varepsilon = 0.001$; T = 100. ***Initialization Rule***: Choose c, then randomly select c *distinct* points in X as $\mathbf{V}_0$. ***Termination rule***: The termination norm is the *squared* Euclidean norm for vectors in $\Re^{cp}$, $\|\mathbf{V}_t - \mathbf{V}_{t-1}\|_{err} = \|\mathbf{V}_t - \mathbf{V}_{t-1}\|_2^2$.

Several examples that highlight some of the advantages and drawbacks of A6.1 will be given shortly. But we need to take a short detour to discuss a topic that will help us with evaluation of sets of CPOV clusters found by A6.1.

6.4 Cluster Accuracy for Labeled Data

Throughout this section, please remember that while there are two types of ground truth, U_{GT} stands for *either* U_t or U_{gt}. Example 5.3 asserts that *labeled data* containing representatives of c physically labeled subsets need not contain c (CPOV) clusters in any HPOV sense. As a simple example, we may have 10^6 seismographs labeled "normal," and 1 seismograph labeled "earthquake"; so there are two physically labeled classes but only one HPOV cluster that has one anomalous observation. Or, consider a set of 10 objects comprising 5 apples and 5 oranges, which has 2 HPOV clusters. Suppose you measure the weights of these 10 objects and call the resultant measurements X, say $X = \{1.1, 1.5, 1.3, 0.98, 1.0, 1.6, 1.4, 0.95, 1.05, 1.22\}$. It is hard to imagine a clustering model and algorithm that would return two CPOV clusters from X as the "best" partition of the data. Don't forget that we have agreed to call the labeled data *subsets* – they are never CPOV clusters.

In Example 5.3, you were asked – why cluster labeled data at all? There are two good reasons. First, how will you ever know (and, perhaps, even more importantly, convince others!) that your clustering model and algorithm "work" if you don't process some labeled data sets that find CPOV clusters which seem to be the right HPOV answers? Well, you can't, and you won't. Second, you will have to choose and use an internal CVI for real clustering, and the only way to evaluate candidates for this job is by fake clustering, as discussed in connection with Definition 6.1. The clustering literature (and this text) are replete with examples that do use labeled data for precisely these reasons.

Several metrics can be used to compare the CPOV clusters recovered from a labeled data set to U_{GT}. Two of the most common measures are discussed here. One of the primary tools that is used to compare *crisp* candidate partitions U to U_{GT} is the confusion matrix.

Definition 6.7. Confusion matrix of $U_{GT} \in M_{hcn}$ against a CPOV partition $U \in M_{hcn}$

Let $U_{GT} \in M_{hcn}$ be a *ground truth partition* for any labeled data set (X or R), and let $U \in M_{hcn}$ be a c-partition of (X or R). The entries of the $c \times c$ *confusion matrix* $C(U_{GT}, U)$ between the crisp labels in the two partitions are defined as

$$[C(U_{GT}, U)]_{ij} = [\#\text{labeled } j \in U | \#\text{labeled } i \in U_{GT}]. \tag{6.31}$$

The *j*th column of $C(U_{GT}, U)$ counts the instances in the *j*th cluster in U, while the *i*th row counts the instances in the *i*th ground truth cluster. Thus, the *ij*th entry of the confusion matrix is the Euclidean dot product of the *i*th row of U_{GT} with the *j*th column of U^T. In the notation of Equation (3.1), an alternative way to write

Equation (6.31) is

$$[C(U_{GT}, U)]_{ij} = \left\langle (U_{GT})_{(i)}, (U^T)^{(j)} \right\rangle \Rightarrow C(U_{GT}, U)] = U_{GT}U^T. \tag{6.32}$$

The diagonal entries of the confusion matrix indicate how many examples from class i are labeled "correctly" in U – that is, how many cluster labels match the ground truth labels for this class. The *i*th diagonal values of C, C_{ii} is the number of (label) matches for class i. Evidently the trace of $C(U_{GT}, U)$ counts the *overall* number of correctly matched labels between U_{GT} and U:

$$tr(C(U_{GT}, U)) = \sum_{i=1}^{c} C_{ii} = tr(U_{GT}U^T) = \langle U_{GT}, U \rangle. \tag{6.33}$$

The last form in Equation (6.33) asserts that this function is an inner product on pairs of matrices, which can be interpreted as vectors in $\Re^{cn}$; the proof is left as an exercise. From this, it follows that the off-diagonal entries of the confusion matrix contain information about label mismatches class by class. Normalizing the trace in Equation (6.33) by n, the number of objects in O provides a simple interpretation of the overall match between U and U_{GT}:

Definition 6.8. Partition accuracy of $U \in M_{hcn}$ wrt $U_{GT} \in M_{hcn}$.

Let $U_{GT} \in M_{hcn}$ be a ground truth partition for any labeled data set (X or R), and let $U \in M_{hcn}$ be a CPOV c-partition of (X or R). Let C_{ii} denote the number of labels that are matched for class i in U and U_{GT}. The *partition accuracy* (PA) of U with respect to U_{GT} is

$$PA(U|U_{GT}) = \frac{tr(C(U_{GT}, U))}{n} = \frac{\sum_{i=1}^{c} C_{ii}}{n} = \frac{\langle U_{GT}, U \rangle}{n} = \left(\frac{\#\text{ matched}}{\#\text{ tried}} \right). \tag{6.34}$$

$PA(U|U_{GT})$ is defined only for crisp partitions $U \in M_{hcn}$. To extend this metric for evaluation of soft (fuzzy or probabilistic) partitions $U \in M_{fcn} - M_{hcn}$, the soft partition must be hardened using Equation (3.4) to convert $U \mapsto \mathbf{H}(U)$ before $PA(U|U_{GT})$ can be computed. We call $PA(\mathbf{H}(U)|U_{GT})$ the *hardened partition accuracy* of U.

When the returned CPOV labels agree perfectly with the ground truth labels, $PA(U|U_{GT}) = 1$. When none of the labels in U match those in U_{GT}, $PA(U|U_{GT}) = 0$; so

$$0 \leq PA(U|U_{GT}) \leq 1 \;\; \forall \;\; U \in M_{hcn}; \text{ or} \tag{6.35a}$$

$$0\% \leq 100 \bullet PA(U|U_{GT}) \leq 100\% \; \forall \; U \in M_{hcn}. \tag{6.35b}$$

Equation (6.35b) states the partition accuracy as a percentage. This form of PA is the one referred to when the term "percent correct" is used. The PA is our first example of a cluster validity index (CVI), and Equation (6.35) tells us that the "best" partitions in the sense of PA are ones that maximize this index. When the preferred value of any CVI is its maximum, we will call it a *max-optimal* (↑) index. Similarly, indices that are to be minimized are *min-optimal* (↓). Hence, the partition accuracy is (↑).

$PA(U|U_{GT})$ is more heavily used (and, perhaps, more easily justified) in the field of classifier design than it is in cluster analysis. After a classifier is trained with some or all of the ground truth samples, $PA(U|U_{GT})$ becomes a predictor of its future performance on new unlabeled samples. Other names for $PA(U|U_{GT})$ appear in classifier design literature, depending on the relationship of the training data to the ground truth. If all of the samples are used for both training and testing, $PA(U|U_{GT})$ is often called the *resubstitution error* of the classifier. Other names for this number are used if only part of the ground truth is used for training. The remaining samples are sometimes divided in test and validation sets, jackknifing through the various sets is employed to reduce estimation bias, and so on.

We don't need many of the concepts related to this idea (training set, test set, validation set, error bias, etc.) for our work in clustering. When the input data are labeled, the primary use of $PA(U|U_{GT})$ will be to compare candidate partitions obtained by any clustering algorithm to ground truth labels and, hence, to each other. When we have candidate partitions from more than one method, this affords us a "seat of the pants" comparison between clustering algorithms too. Finally, it is worth pointing out that a poor value of $PA(U|U_{GT})$ doesn't necessarily indicate that U is a "bad" set of clusters: it simply means that U and U_{GT} are not well-matched.

Assume the objects $O = \{o_1, \ldots, o_n\}$ are labeled and, hence, crisply partitioned, for example, like this: $O = O_1 \cup O_2 \cup \cdots \cup O_c$. Let $|o_i| = n_i; 1 \leq i \leq c$ so that $\sum_{i=1}^{c} n_i = n$. The n_i objects in O_i are the ones labeled *class i* in the data set. Recall that these labeled objects are not called *cluster i* for the reasons given in Example 5.3. The *natural ordering* of the ground truth partition U_{GT} that corresponds to this arrangement for O is an aligned c-partition (cf. Definition 11.2) like this:

$$U_{GT} = \begin{bmatrix} [1]_{n_1} & 0 & \cdots & 0 \\ 0 & [1]_{n_2} & \cdots & 0 \\ \vdots & 0 & \ddots & \vdots \\ 0 & & & [1]_{n_c} \end{bmatrix}_{c \times n} . \tag{6.36}$$

Usually, the data are presented to an algorithm (all at once, not sequentially) and are indexed in their natural order. Users sometimes assume that the order of the numerical labels assigned to the CPOV clusters by their clustering method of choice corresponds to the same order as the labels that come with the ground truth data. But it is a mistake to assume that the cluster labels for a returned CPOV partition are in 1-1 correspondence with the physical labels of the subsets in the ground truth inputs. The computer can't know which CPOV cluster "belongs to" which subset of HPOV labels; so some type of intervention is necessary to guard against the possibility that measures such as $PA(U|U_{GT})$ that compare the ground truth labels to a set of CPOV cluster labels provide a false report. This is known as the *correspondence (or registration) problem.* We will present a means for solving this problem shortly, but, first, let's look at its effect on a simple problem.

Example 6.3. The correspondence problem between U and U_{GT}.

Recall that the Iris data has 50 labeled vectors in $\Re^4$ in each of three physical classes, Setosa, Versicolor, and Virginica. The natural order of the input vectors is usually: first 50 = Setosa; next 50 = Virginica; last 50 = Versicolor. The ground truth partition of Iris corresponding to this natural ordering is

$$U_{GT,Iris} = \begin{bmatrix} [1]_{1\times 50} & [0]_{1\times 50} & [0]_{1\times 50} \\ [0]_{1\times 50} & [1]_{1\times 50} & [0]_{1\times 50} \\ [0]_{1\times 50} & [0]_{1\times 50} & [1]_{1\times 50} \end{bmatrix} . \tag{6.37}$$

Changing the order of the input data will change U_{GT} as well. For example, suppose we interleave the three subsets with a permutation function π so that the inputs alternate between the three subspecies: Setosa = $\{\mathbf{x}_1, \mathbf{x}_4, \mathbf{x}_7, \ldots, \mathbf{x}_{148}\}$; Versicolor = $\{\mathbf{x}_2, \mathbf{x}_5, \mathbf{x}_8, \ldots, \mathbf{x}_{149}\}$; Virginica = $\{\mathbf{x}_3, \mathbf{x}_6, \mathbf{x}_9, \ldots, \mathbf{x}_{150}\}$. The ground truth partition corresponding to this new input order becomes

$$U_{\pi(GT,Iris)} = \begin{bmatrix} 1 & 0 & 0 & 1 & 0 & 0 & \cdots \\ 0 & 1 & 0 & 0 & 1 & 0 & \cdots \\ 0 & 0 & 1 & 0 & 0 & 1 & \cdots \end{bmatrix}. \quad (6.38)$$

This shows how important it is to know the input order, even when the clustering method operates globally (i.e., in batch mode). This problem arises from reordering the object indices. A second type of reordering problem is associated with the algorithmic labels assigned to (the rows of) a partition by any clustering algorithm. Suppose we run a crisp clustering method on Iris with inputs in the natural order corresponding to the labels shown in Equation (6.37) and suppose that this method produces this partition U_1 of the data :

$$U_1 = \begin{bmatrix} [0]_{1\times50} & [1]_{1\times50} & [0]_{1\times50} \\ [1]_{1\times50} & [0]_{1\times50} & [0]_{1\times50} \\ [0]_{1\times50} & [0]_{1\times50} & [1]_{1\times50} \end{bmatrix}. \quad (6.39)$$

Equation (6.34) gives us $PA(U_1|U_{GT,Iris}) = 0.33$, i.e., there are 33% correct matches between U_1 and $U_{GT,Iris}$ due to the last 50 points in the data. It is possible that the algorithm producing U_1 made, as this computation indicates 100 mistakes. But there is another much more probable explanation. The computer doesn't know that the first 50 points are labeled Setosa in the ground truth; so U_1 could be correct, differing only in the correspondence between the algorithmically assigned labels and the ground truth labels (that is, between the rows of U_1 and $U_{GT,Iris}$). This is the situation shown in Table 6.1.

Table 6.1 Registration of algorithmic labels to GT labels in Iris.

Subspecies	$U_{GT,Iris}$	$* = U_1$		$* = \pi(U_1)$
Setosa	1	2		1
Versicolor	2	1	→	2
Virginica	3	3		3
$100 \bullet PA(*|U_{GT})$		33%		100%

If we examine the partition U_1 cluster by cluster and discover that the assignments do agree with the physical labels but in the wrong order, we would reorder the algorithmic labels the computer gives us, arriving at the column $\pi(U_1)$ in Table 6.1, leading to $100 \bullet PA(\pi(U_1)|U_{GT,Iris}) = 100\%$.

☺ ***Forewarned is forearmed:*** Be careful when you compare the accuracy of your crisp partitions to a set of true or ground truth labels. Before you believe the percentage suggested by $100 \bullet PA(U)|U_{GT})$, you should always check that the order of the true or ground truth subset labels in U_{GT} matches the order of the CPOV cluster labels in U assigned to the objects by a clustering algorithm. ☺

When the correspondence problem arises, how do we relabel U so that its clusters are "aligned" with the reference clusters in U_{GT}? For object vector data, several relabeling methods based on cluster prototypes are discussed in Bezdek *et al.* (1999b). Various algorithms exist for relabeling clusters in relational data. For example, an accuracy metric that relies on permuting the cluster labels to register U to U_{GT} is developed and illustrated in Cai *et al.* (2005). A brief description of this method follows.

Let L_{GT} be the vector of ground truth labels in U_{GT} for n objects. Assume without loss that L_{GT} is arranged so that it has the form $L_{GT} = (\underbrace{1,1,\ldots 1}_{n_1}, \underbrace{2,2,\ldots 2}_{n_2}, \ldots, \underbrace{c,c,\ldots c}_{n_c})$, where n_i is the number of objects with label i, $1 \leq i \leq c$. Let L_A denote the cluster list vector of labels in U obtained by clustering the n objects with clustering algorithm A, $L_A \in \{1,\ldots,c\}^n$. U is either a crisp partition or a hardened version of a soft partition of the data. It is possible that U replicates the ground truth clusters in U_{GT} exactly, but U doesn't present the labels in ground truth order – this is exactly the case for U_1 and U_{GT} in Table 6.1.

Let Π denote the set of all permutation functions on n symbols. Define $\delta(a.b) = \left\{ \begin{matrix} 1; & a = b \\ 0; & a \neq b \end{matrix} \right\}$, and consider the optimization problem

$$AC(L_{GT}, L_A) = \underbrace{\max}_{\pi \in \Pi} \left\{ \sum_{i=1}^{n} \delta([L_{GT}]_i, \; [\pi(L_A]_i)/n \right\}. \tag{6.40}$$

The global solution of Equation (6.40) identifies the permutation of L_A that is the best match to L_{GT}. A method for solving this problem with dynamic programming based on the Kuhn–Munkres algorithm is given in Lovasz and Plummer (1986).

The partition accuracy $PA(U|U_{GT})$ is computable as long as (i) U has the same size as U_{GT} and (ii) U is crisp. There is a second less general measure of cluster quality that is often useful when U is derived from an object vector clustering algorithm and the clustering algorithm also produces point prototypes (or cluster centers) $\mathbf{V} = (v_1, \ldots v_c) \in \Re^{cp}$. For example, the pairs (U,**V**) from HCM, (**H**(U), **V**) from FCM, and (**H**(P), **M**) from EM in Chapter 7 all satisfy this qualification.

When $O = \{o_1, \ldots, o_n\}$ is labeled data represented by a set of n feature vectors $X \subset \Re^p$, the labels crisply partition X. It is a simple matter to compute the sample mean $\bar{m}_i$ of the vectors labeled *class i* in X. The c sample means $\mathbf{M}_{GT} = \{\bar{m}_1, \ldots, \bar{m}_c\} \subset \Re^{cp}$ can be interpreted as the *true* or *ground truth prototypes* according as X is labeled synthetically or by physical means.

Definition 6.9. Prototype matching of $\mathbf{V} \subset \Re^{cp}$ wrt $\mathbf{M}_{GT} \subset \Re^{cp}$.

Let $U_{GT} \in M_{hcn}$ be a GT partition of $X = \{x_1, \ldots, x_n\} = X_1 \cup X_2 \cup \cdots \cup X_c$, $|X_i| = n_i$, $1 \leq i \leq c \Rightarrow \sum_{i=1}^{c} n_i = n$. Let the sample mean of the vectors in X_i be $\bar{m}_i = \sum_{x \in X_i} x/n_i$ and define $\mathbf{M}_{GT} = \{\bar{m}_1, \ldots, \bar{m}_c\}$. If $\mathbf{V} = (v_1, \ldots, v_c) \in \Re^{cp}$ is any set of c CPOV point prototypes associated with clustering X, the *prototype match* (PM) of **V** with respect to M_{GT} is

$$PM(\mathbf{V}|\mathbf{M}_{GT}) = \|\mathbf{V} - \mathbf{M}_{GT}\|_{PM}. \tag{6.41}$$

It is easy to see that $0 \leq PM(\mathbf{V}|\mathbf{M}_{GT}) < \infty$; so this cluster validity index is min-optimal (↓) – that is, the smaller, the better. You need to make sure that the indices for the two sets of prototypes are aligned because the PM measure can also suffer from the registration problem. In cases like HCM, FCM, or GMD/EM, the registration problem is the same for both U and **V**. Thus, the same reordering needed to match clusters in U to the label ordering in U_{GT} is the right reordering for the clusters centers **V** to match the order of the means $\mathbf{M}_{GT}$.

Unlike the partition accuracy, the prototype matching measure is usually not normalized since division by the number (cp) might make this a very small value[8]. The norm $\|*\|_{PM}$ in Equation (6.41) is not specified. The arguments **V** and $\mathbf{M}_{GT}$ can be interpreted either as vectors of length (cp) or as $c \times p$ matrices; so there are many choices for $\|*\|_{PM}$. The basic idea, regardless of the choice for this norm, is to compare the approximation of the ground truth mean vectors to the CPOV cluster prototypes.

As the norm in Equation (6.41) varies, the degree to which the approximation matches the sample means changes. For example, if $\|*\|_{PM}$ is the sup norm $\|*\|_{\infty}$ at Equation (6.50e), the comparison is based on just one value, viz., the maximum deviation between the two sets of (cp) values. On the other hand, the Euclidean norm (the usual choice), $\|*\|_{PM} = \|*\|_2$ at Equation (6.50a), compares the distance between the two vectors in $\Re^{cp}$. Smaller values of Equation (6.41) indicate better matches to ground truth sample means.

$PA(U|U_{GT})$ and $PM(\mathbf{V}|\mathbf{M}_{GT})$ measure the partition accuracy and prototype matching of CPOV clusters with respect to the ground truth of the labeled data. These two measures find their main application in the comparison of different clustering methods, and both are employed in various examples throughout the text.

Example 6.4. Partition accuracy/prototype matching: c-means on X_{Iris}.

X_{Iris} appeared in Examples 5.3 and 5.4, where it was used to illustrate feature selection and extraction, and in Example 6.2 in connection with the registration problem. Results similar to this example have been published many times, but there is a lot of pedagogical value in using the same data for many different examples. Partitioning X_{Iris} with HCM and FCM illustrates how to extract a "best CPOV" partition from a set of candidate partitions using these two measures of CPOV quality.

The ground truth partition $U_{GT,Iris}$ is used in its natural order, Equation (6.37). However, we have argued that there is no *a priori* reason to expect CPOV clusters to agree with $U_{GT,Iris}$. And when the data are beyond human examination (as they are here, with $p = 4$), there are no HPOV clusters either – just those *oh-so-seductive* physical class labels suggesting that we regard $c = 3$ as the correct number of CPOV clusters to seek.

If the numerical representation chosen for Iris by Anderson was perfect, almost any clustering model and algorithm would be able to find CPOV clusters that agreed with $U_{GT,Iris}$ at $c = 3$. In particular, c-means (HCM), which produces crisp partitions of X_{Iris} on demand, would find $U_{GT,Iris}$ if the data truly represented the physical structure corresponding to the labels. To see if this happens, HCM and FCM were each applied to X_{Iris} 100 times at $c = 3$, with all parameters being the default choices given for A6.1. The *same* initialization was used *for each run* of HCM and FCM; so the only variable in the 100 runs is the initial guess, which for this example was U_0. Over the 100 runs, HCM and FCM both stopped at more than one terminal pair; so J_1 had more than one fixed point, and J_2 had more than one local extreme point.

The partition accuracy and prototype matching indices for HCM and FCM were then averaged over the hundred runs, taking care in all cases to register the labels of $U_{GT,Iris}$ and M_{GT} to the algorithmic CPOV results, so that mismatch counts are based on labels for the same objects represented by the data. The Euclidean norm was used in Equation (6.41), $\|*\|_{PM} = \|*\|_2$. Table 6.2 shows the results.

[8]Bradley *et al.* (1998) suggest normalization of Equation (6.41) with just c, not cp. If c is small relative to p, this makes sense.

Table 6.2 Partition accuracy (PA) and prototype matching (PM) on X_{Iris}.

	$\overline{PA}(*\|U_{GT,Iris})$ (↑)	$\overline{PA}(H(*)\|U_{GT,Iris})$ (↑)	$\overline{PM}(*\|M_{GT,Iris})$(↓)
HCM	0.88	☠	0.35
FCM	☠	0.89	0.23

Table 6.3 Sample means and 100-run average prototype estimates of X_{Iris}.

Subspecies	Feature	Sample Means $\bar{m}_{i,GT,Iris}$	Average Prototypes $\bar{v}_{i,HCM}$	Average Prototypes $\bar{v}_{i,FCM}$
Setosa	SL	5.006	5.025	5.008
	SW	3.418	3.429	3.402
	PL	1.464	1.481	1.495
	PW	0.244	0.255	0.256
Versicolor	SL	5.936	5.985	5.907
	SW	2.770	2.777	2.767
	PL	4.260	4.518	4.390
	PW	1.326	1.493	1.411
Virginica	SL	6.588	6.732	6.736
	SW	2.974	3.035	3.040
	PL	5.552	5.562	5.593
	PW	2.026	1.983	2.028

The average partition accuracy of the crisp HCM and hardened FCM partitions is virtually the same, indicating that these two algorithms committed, on average, the same number of label mismatches to the ground truth labels. Since $0.88(150) = 132$, the average number of mismatches in 100 trials of HCM was 18 label "mistakes." FCM was only very slightly better, $0.89(150) = 133.5$; so the average number of mismatches in 100 trials of FCM was 17.5 label "mistakes." In most single runs, FCM will commit 17 mismatches, HCM will commit 18.

The value 0.23 for FCM prototype matching, $\overline{PM}(*|M_{GT,Iris})$ in Table 6.2, shows that the FCM cluster centers were somewhat better at replicating the sample means than the HCM cluster centers, whose error was 0.12 higher on average (remember, smaller values of the PM indicate better matches to the labeled prototypes). You can see this explicitly by comparing the columns for (100 run average) estimates of the coordinates of the labeled subsample means of X_{Iris} produced by HCM and FCM shown in Table 6.3 to the subsample means of Iris. Abbreviations in Table 6.3 for the four features are: SL = sepal length; SW = sepal width; PL = petal length; PW = petal width. Since $U_{GT,Iris}$ has three physical labels, you might conjecture that we have found the "best" average CPOV clusters of X_{Iris} at $c = 3$, at least for these two models and algorithms. Do you think this is the best answer? Please review Examples 5.3 and 5.4 before you decide.

6.5 Choosing Model Parameters (c, m, $||*||_A$)

This section contains some facts (and fictions) related to each of the model parameters needed to implement the crisp and fuzzy k-means Algorithm A6.1. This important aspect of both clustering algorithms has received continuous attention since the inception of these two models – these choices are a veritable fount of new papers about HCM (k-means) and FCM (fuzzy k-means).

6.5.1 How to Pick the Number of Clusters c

If you are processing data which contain instances of c physically labeled subsets, there is a natural inclination to look for c CPOV clusters in the data, ignoring the labels, and then compare the CPOV cluster labels to the labels of the data using the partition accuracy at Equation (6.34). Aside from the obvious question – why look for clusters in labeled data at all? – we have illustrated that physically labeled subsets may not coincide with CPOV clusters found by any model and algorithm. If the data are unlabeled (as they will be in a real cluster analysis), the general question of how to choose the number of clusters (either before clustering or after you have done it) is the most important choice you have to make. Why? If the solution you hope to find lies in M_{fc^*n}, but you are looking in $M_{fc'n}$, $c' \neq c^*$, the solution you seek is not in the search space.

Let us begin with an informal method. The lower limit for c is, of course, c = 1, i.e., rejection of the idea that the data have (more than one) clusters. The only way you can computationally justify this value for c is by a tendency assessment that says "don't look for clusters." A common rule of thumb for the largest number of clusters to seek is to limit c by the square root of n, $c \leq \lfloor \sqrt{n} \rfloor$. Is this practical? For the Iris data, this rule suggests $c \leq \lfloor \sqrt{150} \rfloor = 12$, which seems ok, and we assume that this estimate is high. But suppose $n = 10^{12}$? There is a vanishingly small chance that there are $c = 10^6$ interesting CPOV clusters in such data.

In most problems, you will have some indication or intuitive idea of a reasonable upper limit on c from traits and properties of the objects generating the data. For example, *magnetic resonance image* (MRI) slices of the healthy human brain usually don't have more than 8 or 10 substantially different types of tissue; so a reasonable upper limit is something like $c < 10$ or so. Many published studies suggest that overestimates of c are better than underestimates because merging CPOV clusters is more straightforward than splitting them. Merger is often accomplished by joining the objects in CPOV clusters corresponding to very similar prototypes or very similar sets of memberships. See Bensaid *et al.* (1996), Strehl and Ghosh (2002), Xiong *et al.* (2004), or Hore *et al.* (2009b) for representative studies.

The question of what c to seek lies at the heart of the tendency assessment problem that was introduced in Section 4.2. Visual representation of the data by methods such as VAT/iVAT Algorithm A4.1 often provide a pretty accurate estimate for a good *range* of values for this parameter, but cluster heat maps also have their limits. Most notably, large numbers of objects require very large dissimilarity matrices, which can make a basic method awkward, if not intractable. Nonetheless, many visual methods are truly useful prior to clustering (pre-clustering), whereas much of the current literature is preoccupied with selecting a best value of c *after* clustering – the process called cluster validity in this volume. Several validation schemes are illustrated in the sequel.

There have been many proposals over the years about how to find a good value for c once the candidate partitions are in hand. For example, Tibishirani *et al.* (2001) introduce the idea of testing the output of any clustering algorithm against a gap statistic that forms a null hypothesis for improvement of the within cluster dispersion. Hamerly and Elkan (2003) offer a modification of HCM they call "G-means" which is applied to c-means (k-means) partitions on the fly. The G-means algorithm is based on a statistical test for the hypothesis that a subset of data follows a Gaussian distribution. G-means runs HCM with increasing c in a hierarchical fashion until the test accepts the hypothesis that the data assigned to each c-means center are Gaussian.

But this is really what most users will do. Run HCM or FCM at a number of selected values of c, thereby generating a set $CP = \{U \in M_{fc_kn} : c_m \leq c_k \leq c_M\}$ of candidate partitions as in Definition 4.4. Then choose a "best partition" from CP, where "best" takes its meaning using any one of hundreds of validation methods (the gap statistic, for example), which identifies the best partition in some well-defined sense. Choosing U this way automatically also chooses c, the number of CPOV clusters your computer thinks reside in the data. Of course, if you repeat this experiment with new initializations, you may end up with a different choice for c. Then what?

6.5.2 How to Pick the Weighting Exponent m

There are a number of studies about how to choose the weighting $m \in (1, \infty)$ for FCM, but no method has emerged that points the way to a general answer for making the best choice. The best thing to know when faced with choosing this parameter is the effect it has on FCM partitions. We know what happens to $J_m(U^*, \mathbf{V}^*)$, U^*, and $\mathbf{V}^*$ as m approaches 1 from above: FCM/AO becomes Lloyd's c-means (m = 1 in A6.1), HCM/AO. What happens to U as m approaches infinity? Take the limit of Equation (6.19b1):

$$\lim_{m\to\infty}\{u^*_{ik}\} = \lim_{m\to\infty}\left\{\frac{\left(\|\mathbf{x}_k - \mathbf{v}^*_i\|^2_A\right)^{\frac{1}{m-1}}}{\sum_{j=1}^{c}\left(\left\|\mathbf{x}_k - \mathbf{v}^*_j\right\|^2_A\right)^{\frac{1}{m-1}}}\right\} = [1/c] = \bar{U}. \tag{6.42a}$$

Now take the limit of Equation (6.19a) to find that, for all i, $1 \le i \le c$:

$$\lim_{m\to\infty}\{\mathbf{v}^*_i\} = \lim_{m\to\infty}\left\{\sum_{k=1}^{n}(u^*_{ik})^m\mathbf{x}_k \Big/ \sum_{k=1}^{n}(u^*_{ik})^m\right\} = \sum_{k=1}^{n}\mathbf{x}_k/n = \bar{\mathbf{V}}. \tag{6.42b}$$

So, as m approaches infinity, $\lim_{m\to\infty}\{(U^*, \mathbf{V}^*)\} = (\bar{U}, \bar{\mathbf{V}})$; that is, U becomes the centroid $\bar{U}$ of M_{fcn} and $\mathbf{V}$ becomes the centroid (or grand mean) $\bar{\mathbf{V}} = \bar{\mathbf{m}}$ of the data set X. Thus, lower and lower values of m push the partition toward crispness (and FCM toward HCM), and, conversely, raising m pushes all the memberships toward each other. Finally, $\lim_{m\to\infty}\{(1/c)^m\} = 0$ shows that $\lim_{m\to\infty}\{J_m(U^*, \mathbf{V}^*)\} = 0$.

The most common choice of m, of course, is m = 1, for which A6.1 becomes HCM (Lloyd's c-means). For $m > 1$, the most common choice is m = 2. Exercise 11 asks you to show that $J_2(U^*, \mathbf{V}^*)$ is the average harmonic mean of the squared errors of the n data vectors in X.

There is an electrical analog of the meaning of the objective function J_m for the values m = 1, 2, that follows by interpreting each distance d_{ikA} as the electrical resistance of the connection between $\mathbf{x}_k$ and $\mathbf{v}_i$. The circuit of series resistors built by connecting all n points this way has minimum resistance $J_1(U^*, \mathbf{V}^*)$. The circuit corresponding to a network of n sets of c resistances $\{d_{ikA} | 1 < i < c\}$, each of the n sets of c resistors hooked up in parallel, has $J_2(U^*, \mathbf{V}^*)$ as its resistance (Bezdek, 1976).

Another piece of theoretical information about m comes from the derivative of J_m with respect to m at an optimal pair for J_m. Suppressing the (*) for optimality

$$\frac{dJ_m}{dm}(U, \mathbf{V}) = \sum_{k=1}^{n}\sum_{i=1}^{c} u^m_{ik}\log(u_{ik})d^2_{ikA}. \tag{6.43}$$

It is easy to show that J_m is strictly monotone decreasing in m on any finite interval [1, b]. McBratney and Moore (1985) combined this derivative with the square root of c and studied the function

$$\phi(m, c) = -\sqrt{c}\left(\frac{dJ_m}{dm}(U, V)\right) = -\sqrt{c}\left(\sum_{k=1}^{n}\sum_{i=1}^{c} u_{ik}^m \log(u_{ik}) d_{ikA}^2\right), \tag{6.44}$$

with a view toward understanding the tradeoff between choices for m and c. They used a very small ($n = 9, p = 2, c = 3$) artificial data set and plotted values of $\phi(m, c)$ versus m at terminal pairs for J_m. Their conclusion was that larger values of c should be coupled with smaller values for m. This heuristic might have merit when optimal pairs are computed for a range of values of c, and trial m's are computed and evaluated with Equation (6.43) at each fixed c. This may or may not provide a reliable method for choosing m at each c, but it, at least, gives users a start.

Attempts to theoretically justify a choice for m are typified by work such as Yu *et al.* (2004). These authors review a number of theorems about tests for local optimality of (U^*, $\mathbf{V}^*$) for J_m and then derive a result that underlies two heuristic rules. One rule is based on the parameters of the input data vectors $\{\mathbf{x}_1, \ldots, \mathbf{x}_n\} \subset \Re^p$:

$$\text{Rule } \alpha : [\text{choose}]\ m \leq \left(\frac{\min(p, n-1)}{\min(p, n-1) - 2}\right) \text{ if } \min(p, n-1) \geq 3. \tag{6.45}$$

The second heuristic for choosing m is based on an analysis of the eigenvalues of the matrix $H^T H/n$, where $\mathbf{H}$ is the row vector

$$\mathbf{H} = \left(\frac{\mathbf{x}_1 - \bar{\mathbf{V}}}{\|\mathbf{x}_1 - \bar{\mathbf{V}}\|_2}, \frac{\mathbf{x}_2 - \bar{\mathbf{V}}}{\|\mathbf{x}_2 - \bar{\mathbf{V}}\|_2}, \ldots, \frac{\mathbf{x}_n - \bar{\mathbf{V}}}{\|\mathbf{x}_n - \bar{\mathbf{V}}\|_2}\right); \ \bar{\mathbf{V}} = \sum_{k=1}^{n} \mathbf{x}_k \Big/ n. \tag{6.46}$$

Let λ_M denote the maximum eigenvalue of $\mathbf{H}^T\mathbf{H}/n$. Yu *et al.* give their second rule in terms of this eigenvalue,

$$\text{Rule } \beta : [\text{choose}]\ m \leq \left(\frac{1}{1 - 2\lambda_M}\right) \text{ if } \lambda_M < 0.5. \tag{6.47}$$

Yu *et al.* offer these observations about the choice of m for FCM using rules α and β:

(i) Rule α is an approximation of Rule β
(ii) Rule β "must be obeyed" if its condition is satisfied, for otherwise the most probable terminal point is the grand mean $\bar{\mathbf{V}}$ for all c cluster centers.
(iii) $m = 2$ is not a reasonable choice if $\lambda_M < 0.25$.
(iv) When $\lambda_M \geq 0.5$, rules α and β are invalid, and the selection of m is left to choice by the user.

Neither of these rules, nor any of these comments, tell us how to select m, but they do add some fuel to the fire of discussion about how m affects and, in turn, is affected by other parameters of the FCM model. Attempts to link the value of m to characteristics of the input data include the papers of Dembele and Kastner (2003) and Futschik and Carlisle (2005).

A beguiling analysis about choosing m is offered by Schwammle and Jensen (2010), who strongly reject the conventional default choice $m = 2$. They suggest that an optimal fuzzifier m can be obtained as the value of

$$m_{opt,SJ}(p, n) = 1 + \left(\frac{1418}{n} + 22.05\right) p^{-2} + \left(\frac{12.33}{n} + 0.243\right) p^{-\kappa}, \tag{6.48}$$

where $\kappa = (0.0406\ln(n) - 0.1134)$ – a pretty bizarre looking and uncommonly precise constant for a parameter of a fuzzy clustering algorithm! The function at Equation (6.48) relies only on the dimension p and number of input vectors n in X, and its closed form is indeed tempting.

Their examples with 10 real data sets choose optimal m's ranging from 1.13 to 2.15 and never land on the default value m = 2. However, their analysis is based entirely on the use of the Euclidean norm for J_m, and their synthetic clusters are confined to Gaussian clouds; so the term "optimal m" should be approached with some degree of caution. But, since there is no *a priori* reason NOT to use this as a starting point – why not try it? Our next example studies the relationship of m to HCM and FCM partitions of a small, two-dimensional data set in a different way.

Example 6.5. The effect of parameter m in fuzzy c-means.

We return to the data set X_{29} shown in Figure 4.10(d), repeated in the window on the right in Figure 6.4. Recall that the c = 2 HPOV subsets $\{X_1, X_2\}$ in X_{29} produce the value $J_1(U_1, V_1) = 102$, independent of the interface distance s in Figure 4.10(d), and that the breakpoint at which J_1 switches from U_1 to U_2 is s = 2.86. Recall also from Example 4.7 that the value of J_1 for the partition U_2 and the companion cluster centers $\mathbf{V}_2(s)$ that accompany it is $J_1(U_2, \mathbf{V}_2(s)) = 2.22s^2 + 2.22s + 77.50$.

$J_1(U_2, \mathbf{V}_2(s))$ has a minimum less than 102 for $s < 2.86$, leading k-means A6.1(m = 1) to *usually* terminate at the visually undesirable solution with five visually apparent mistakes (the left column of the right subset) as shown in Figure 4.10(e). Numerical experiments have shown that, depending on the way the iterative loop is initialized, HCM *may* terminate at this undesirable solution even when $s > 2.86$. And for most trials of HCM, when $s \geq 4$, HCM will terminate at the HPOV solution. It was noted in Example 4.7 that between s = 3 and s = 4, the behavior of HCM is unpredictable. And when $s < 3$, sometimes two columns of the larger cluster on the right (seen as "10 mistakes" in Figure 6.5) will be assigned to the left cluster instead of the right cluster.

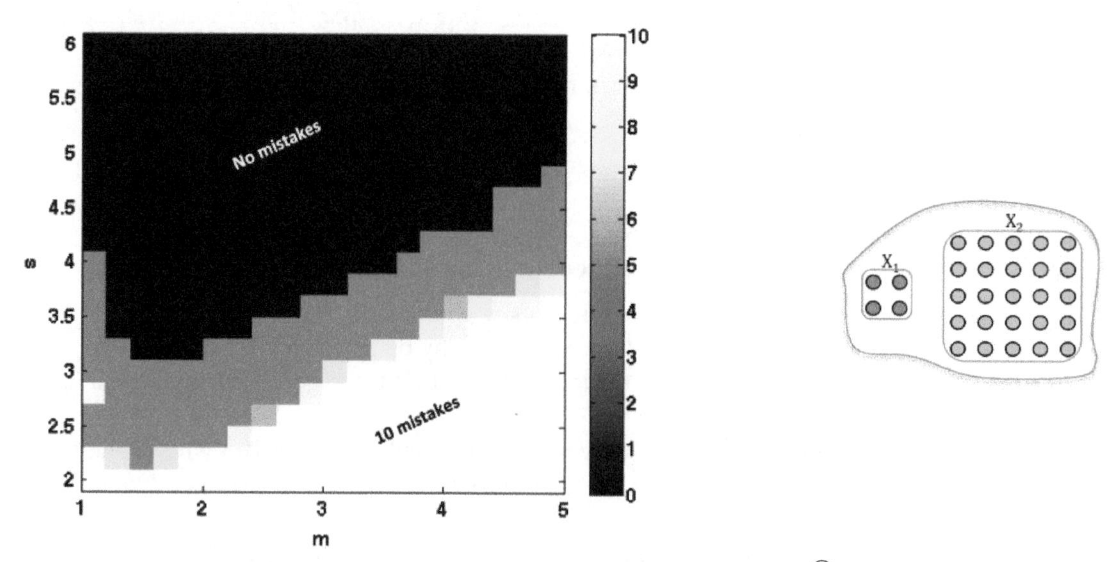

Figure 6.4 Misclassifications in right cluster (X_2) of X_{29} after hardening FCM $\widehat{U}$'s as a function of (m, s).

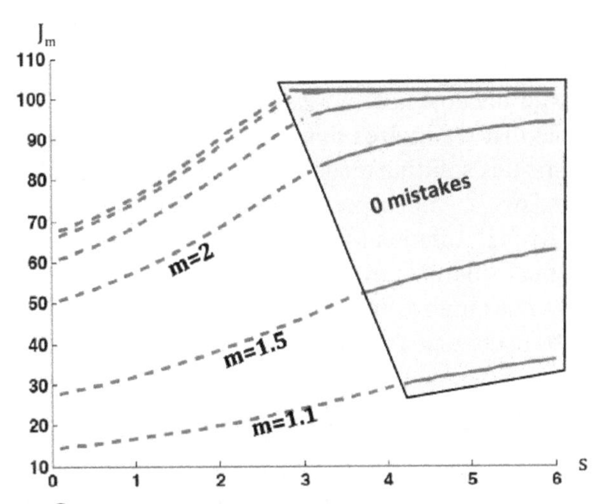

Figure 6.5 $\mathbf{H}(\widehat{U})$ mismatches to HPOV labels for X_2 of X_{29} as a function of $(s, \mathbf{J}_m)$.

What happens when we use FCM with $m > 1$ to cluster the points in X_{29}? To make a comparison with the behavior of HCM as a function of s, terminal fuzzy partitions $\widehat{U}$ found by A6.1 using its default parameters are hardened to $\mathbf{H}(\widehat{U})$ with Equation (3.4). Figure 6.4 compares the number of label mismatches (median over 15 runs, called mistakes here) of $\mathbf{H}(\widehat{U})$ to the crisp labels in the right cluster of the HPOV partition U_1 for various values of the parameter pair (m, s) with c = 2 fixed. For $s > 4$, the hardened version of FCM on the 25-point subset coincides with the c-means labels for subset X_2 at all values of m between 1 and 3.5. For $s > 5$, the two sets of labels are the same at all values of m.

Figure 6.5 shows the relationship between $\mathbf{H}(\widehat{U})$ and J_m in the form of a nomogram, each curve representing one value for m. Here we can see how the label mistakes between the HPOV partition U_1 and the undesirable partition U_2 vary against values of J_m. The blue portion of each curve (blue curves in the beige colored trapezoid) is the range for which no mistakes occur. The dashed red portion of each curve represents the range for which the first column of the right cluster is "stripped" from the right group and instead attached to the left group upon hardening.

This example has three take-away lessons. First, in common with all hill-climbing algorithms, initialization is very important for A6.1 (k-means and fuzzy k-means); so repeated trials are always recommended. Second, termination of AO iteration on successive values of the objective function is a bad idea because these values don't always reflect the relative quality of the CPOV clusters found by HCM and FCM. And, finally, the value of m in A6.1 for the fuzzy case importantly controls the quality of partitions found by FCM.

Another facet of clustering with HCM/FCM that is illustrated by this example is that *both* of these models attempt to equalize the numbers of points in clusters found with A6.1. When the distribution of the points in X is known to have one or more large HPOV clusters and several small ones, A6.1 (for *any* m ≥ 1) tries to split points away from the larger clusters and assign them to the smaller ones as shown in Figures 6.4 and 6.5.

This happens because the overall value of J_m can be reduced by replacing many smaller intracluster distances needed to retain HPOV integrity by a few larger intercluster distances to the prototypes in the CPOV solution. This tendency is well-documented in many papers and texts. Figure 6.13, p. 220 in Duda and Hart (1973) is an early example that recognizes this tendency in J_1. There are many heuristic and *ad hoc* methods that attempt to overcome this splitting tendency (see Mahmood *et al.* (2007) for one approach), and it is good to be aware that it exists. If you suspect that your data have imbalanced cluster sizes, take precautions against being led to a "wrong" solution by A6.1.

Most studies that result in practical solutions from an FCM clustering of data use m in the range $1 < m \leq 3$. There are examples of using m as high as m = 30 in an attempt to filter out noise (cf., Example 4 in Bezdek *et al.* (1981)). Yu *et al.* (2004) note that a proper choice of m depends on the input data.

6.5.3 Choosing the Weight Matrix (A) for the Model Norm

Theorems 6.2/6.3 require that the distance norm used in basic HCM/FCM is induced by an inner product, which ensures that the norm is differentiable with respect to cluster center $\mathbf{v}_i$ in the hard and fuzzy cases. Definition 6.4 asserts that every positive-definite matrix $A \in PD^{pp} \subset \Re^{pp}$ induces a unique inner product norm. Here is a formal statement of that assertion.

Theorem 6.4. Inner products, vector A-norms, and A-distances in $\Re^p$.

Let $A \in PD^{pp}$. Then $\forall\ \mathbf{x}, \mathbf{y} \in \Re^p$, $\langle \mathbf{x}, \mathbf{y} \rangle_A = \mathbf{x}^T A \mathbf{y}$ is an inner product on $\Re^p$. Conversely, if $\langle *, * \rangle$ is an inner product on $\Re^p$, there is a unique $A \in PD^{pp}$, called the weight matrix of the inner product, so that $\forall\ \mathbf{x}, \mathbf{y} \in \Re^p$, $\langle \mathbf{x}, \mathbf{y} \rangle_A = \mathbf{x}^T A \mathbf{y}$. Moreover, each inner product induces the unique *A-norm* $\|\mathbf{x}\|_A$ on $\Re^p$ and *A-distance* (or norm metric) $d_A(\mathbf{x} - \mathbf{y})$ on $\Re^p \times \Re^p$ shown in

$$\langle \mathbf{x}, \mathbf{y} \rangle_A = \mathbf{x}^T A \mathbf{y}; \tag{6.49a}$$

$$\|\mathbf{x}\|_A = \sqrt{\langle \mathbf{x}, \mathbf{x} \rangle_A} = \sqrt{\mathbf{x}^T A \mathbf{x}}; \tag{6.49b}$$

$$d_A(\mathbf{x} - \mathbf{y}) = \|\mathbf{x} - \mathbf{y}\|_A = \sqrt{\langle \mathbf{x} - \mathbf{y}, \mathbf{x} - \mathbf{y} \rangle_A} = \sqrt{(\mathbf{x} - \mathbf{y})^T A (\mathbf{x} - \mathbf{y})}. \tag{6.49c}$$

What does it mean? Theorem 6.4 provides us with a rich supply of inner products, norms, and distances (metrics), one for every $A \in PD^{pp}$. There are infinitely many ways to choose an inner product, A-norm, and A-norm metric as a basis for the topology of open and closed sets in $\Re^p$. Line 5 of A6.1 requires a choice for the norm-inducing weight matrix $A \in PD^{pp}$; so the basic HCM/FCM algorithms are really infinite families of clustering algorithms, one for each choice of $A \in PD^{pp}$.

Proof. This is a standard theorem which can be found in most linear algebra textbooks. The proofs of Equation (6.49) are left to the exercises.

A *canonical story* is one that scholars consider "great" or "major." In the present context, a canonical story is a description of a model that shows its equivalence to some major physical or mathematical property of something in the real world. For example, to get from one corner of a city block to the diagonally opposite corner, a taxicab must go around the block along two adjacent streets (cf., the green diamond in Figure 6.6, which corresponds to such a block). The popular alternate name of the 1-norm (the taxicab or Manhattan metric) reminds us of its canonical story.

While there are infinitely many inner product norms, only three inner product induced norms have "canonical stories": the Euclidean, Mahalanobis, and diagonal norms. And there are only two non-inner product norms with canonical stories – the 1-norm and sup norm (max norm). Since the 1 and sup norms are not inner product (induced) norms, they can't be used directly in A6.1. See Bobrowski and Bezdek (1991) for one way to alter A6.1 when the 1-norm or sup-norm is used in the formulation of problem (6.9). These remarks are collected in the following definition.

Definition 6.10. The five good norms.

Let $\bar{m} = \sum_{k=1}^{n} x_k/n$ and $S = \left(\sum_{k=1}^{n}(x_k - \bar{m})(x_k - \bar{m})^T/(n-1)\right)$ be the grand mean and sample covariance matrix of $X = \{x_1, \ldots, x_n\} \subset \Re^p$. Let $\Delta = \text{diag}(S)$ be the diagonal matrix obtained from S by zeroing its off-diagonal entries. For $\mathbf{x}, \mathbf{y} \in \Re^p$, the five good norms are

$$\|\mathbf{x}-\mathbf{y}\|_2 = \sqrt{(\mathbf{x}-\mathbf{y})^T(\mathbf{x}-\mathbf{y})} = \left(\sum_{j=1}^{p} |x_j - y_j|^2\right)^{\frac{1}{2}} : \text{Euclidean (or 2 – norm) } (A = I_{p \times p}); \quad (6.50a)$$

$$\|\mathbf{x}-\mathbf{y}\|_{S^{-1}} = \sqrt{(\mathbf{x}-\mathbf{y})^T S^{-1}(\mathbf{x}-\mathbf{y})} : \text{Mahalanobis norm, } (A = S^{-1}); \quad (6.50b)$$

$$\|\mathbf{x}-\mathbf{y}\|_{\Delta^{-1}} = \sqrt{(\mathbf{x}-\mathbf{y})^T \Delta^{-1}(\mathbf{x}-\mathbf{y})} : \text{Diagonal norm, } (A = \Delta^{-1}); \quad (6.50c)$$

$$\|\mathbf{x}-\mathbf{y}\|_1 = \sum_{j=1}^{p} |x_j - y_j| : \text{1 – norm (Manhattan or Taxicab norm)}; \quad (6.50d)$$

$$\|\mathbf{x}-\mathbf{y}\|_\infty = \max_{1 \le j \le p} \{|x_j - y_j|\} : \text{sup – norm (max norm)}. \quad (6.50e)$$

The Mahalanobis and diagonal norms shown here are sample based and are often called statistical distances. In the context of Chapter 7, we will have population-based versions of these two measures based on the mean and covariance of an underlying probability distribution.

These five "good norms" appear in perhaps 99% of all literature on clustering. These five have canonical stories and, hence, are attractive choices for measuring distances between pairs of feature vectors in $\Re^p$. We will gain a deeper understanding of what CPOV clusters may mean by studying some properties of the five good norms. The Euclidean, Mahalanobis, and Diagonal norms are the most distinguished members of the infinite family of inner product norms, parameterized by the positive-definite weight matrix A. The 1-norm and sup norm are *Minkowski norms* from the q-norm family $\|\mathbf{x}\|_q = \left(\sum_{j=1}^{p} |x_j|^q\right)^{\frac{1}{q}}$, $q \ge 1$. Equation (6.50e) for the sup norm is realized by taking the limit of the q-norm as q approaches infinity.

The only member in the intersection of the two infinite families is the Euclidean norm, for which $A = I_p$ and/or $q = 2$. Because of this, the Euclidean norm has the alternate name "2-norm" and the alternate notation $\|*\|_2$. The Euclidean norm is by far the most popular model norm for J_m (and for almost all other clustering models as well), but this mathematical property is probably not responsible for its popularity. There are compelling geometric and statistical reasons for using the 2-norm, but, perhaps, the most compelling of all reasons is the canonical story of the Euclidean norm: it measures distances directly from point to point, just as we do in our real world - Euclidean 3-space.

Figure 6.6 shows the shape of the unit balls $B_{(*)}(\mathbf{0}, 1) = \{\mathbf{x} \in \Re^2 : \|\mathbf{x}\|_{(*)} = 1\}$ (that is, the sets of unit vectors) in the norm-induced topology for each of the five good norms. Please be clear about what this figure shows. You are used to seeing all the unit vectors with the same length (which is Euclidean length), as they will for the vectors that lie along the black circle which represents the unit ball in the Euclidean norm. But for the other norms, the vectors look like they have different lengths (because your eye sees them from the Euclidean point of view). Thus, there are only four vectors (the ones along the coordinate axes) in the infinite set of unit vectors for the green diamond corresponding to the 1-norm that will "look like 1." The rest of the unit vectors on this diamond look shorter, but they all yield the value $\|\mathbf{x}\|_1 = 1$.

Figure 6.6 shows us why the norm used by J_m may impose an implicit shape on (all c) possible CPOV clusters of points in feature vector data. The circular shape of the unit vectors for the 2-norm seen in Figure 6.6 generalizes to a hypersphere in $\Re^p$. This explains why classical k-means is good at recognizing and capturing spherical clusters when the model norm is the Euclidean norm and X contains subsets of points that are roughly this shape. On the other hand, in the absence of hyperspherical substructure in X, Euclidean distance does its best to *impose* this type of geometry on the clusters. Figure 6.7 illustrates this effect.

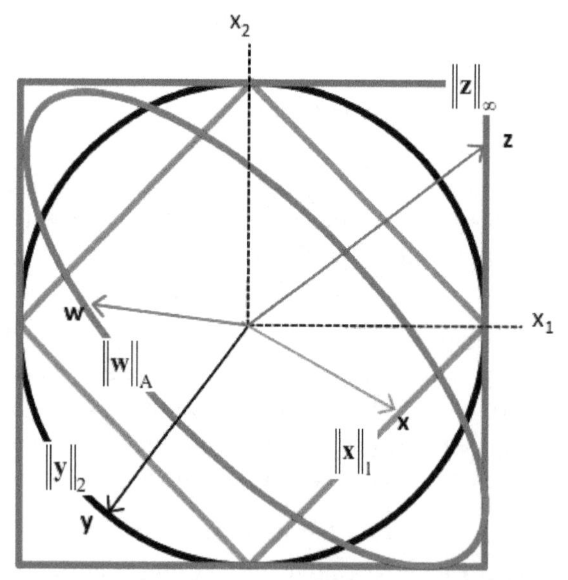

Figure 6.6 Unit balls for the five good norms in $\Re^2$: $\|\mathbf{x}\|_1 = \|\mathbf{y}\|_2 = \|\mathbf{z}\|_\infty = \|\mathbf{w}\|_A = 1$.

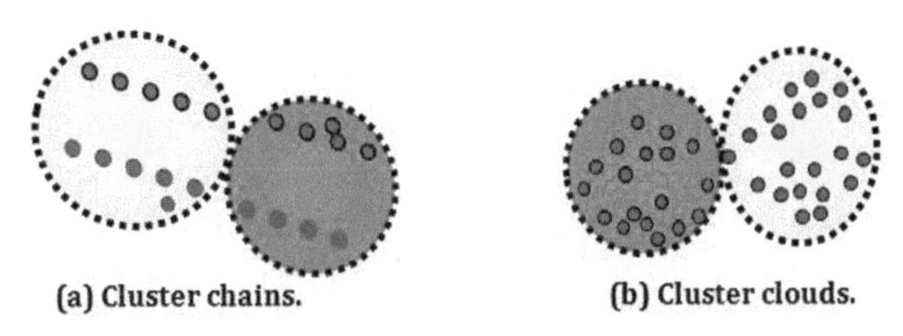

Figure 6.7 The Euclidean (2-norm) imposes circular structure even when none exists.

The left view (a) in Figure 6.7 shows two more or less linear chains (the red and blue data points), but when HCM or FCM using the Euclidean norm is applied to such data, the likely result is the incorrect (from the HPOV) 2-partition of the data shown in view 6.7(a). In view 6.7(b), there are two roughly circular HPOV clouds of data; so any of the c-means algorithms as well as the EM/GMD algorithm in Chapter 7 will probably retrieve CPOV clusters like the ones shown in view 6.7(b) when using the Euclidean norm (actually, this will be true for c-means models using almost any inner product norm that is not too stretched in one direction).

This effect can sometimes be alleviated by a change of model norm. Consider the red ellipse in Figure 6.6 corresponding to the unit vectors associated with an inner product norm induced by some weight matrix A, $\|\mathbf{x}-\mathbf{y}\|_A = \sqrt{(\mathbf{x}-\mathbf{y})^T A(\mathbf{x}-\mathbf{y})}$. As matrix A varies, it distorts the shape of Euclidean geometry in an infinite number of ways, even stretching the ellipse to (in the limit) a set of lines (a degenerate ellipse). This illustrates the importance of understanding the topology of the open and closed balls induced on $\Re^p$ by a particular choice of distance that is illustrated in Figure 6.6. And it also highlights a very important limitation of the c-means models – viz., that all c clusters found by A6.1 have an implicitly defined single type of shape which is determined by the model norm chosen. We will discuss some alterations to J_m that attempt to accommodate various geometric features of the data such as different cluster shapes in Chapter 9. And we will discuss an algorithm in Chapter 8 – the *single linkage* (SL) method, which has just the opposite geometric preferences to those of c-means. SL will do well on data with chains or strings such as that shown in Figure 6.7(a). Conversely, SL will be confused by points in the region of cloud overlap: it may “jump the gap” too early in the clustering process and produce undesirable results with clouds of data such as shown in Figure 6.7(b).

We have seen that each of the three model parameters c, m, and $\|\|_A$ can and do affect partitions found by HCM/FCM. But this is only part of the story about A6.1. Now we turn to the rest of the story – the implementation parameters for these two families of algorithms.

6.6 Choosing Execution Parameters (V_0, ε, $||*||_{err}$, T)

The second most important choice to be made (as said above, the most important is choosing c, the number of clusters to seek) is the initial guess for $\mathbf{V}_0$ (or U_0) because the objective functions underlying the c-means models almost always have many attractive termination states (local extrema for $J_{m>1}$ or fixed points for J_1).

6.6.1 Choosing Termination and Iterate Limit Criteria

Termination vs. Convergence. Please be careful to differentiate between the meanings of the words convergence and termination. *Convergence* refers to the theoretical limit of an infinite sequence or series. Chapter 10 examines the theory of convergence for A6.1. *Termination* means cutting off the sequence or series before it gets to its theoretical limit. Terminal solutions should always be regarded as approximations to any theoretically guaranteed point of convergence. In this context, the theory of Newton's method is similar to A6.1. Both iteration sequences have convergence theories, but we can't run the iteration to infinity; so these methods are usually terminated computationally before a point of convergence is actually achieved. In the case of A6.1, the three parameters $(T,\ \varepsilon,\ \|*\|_{err})$ all play a role in termination. Local and global convergence theory for FCM is given in Chapter 10. There is no corresponding convergence theory for HCM because it is not alternating optimization in the sense of Theorem 10.1. But Lloyd's algorithm does provide A6.1 (m = 1) with the descent property; so it is AO in this more restricted sense.

The termination criterion $\|\mathbf{V}_{t+1} - \mathbf{V}_t\|_{err} \leq \varepsilon$ requires choices for the norm $\|*\|_{err}$ and threshold ε, which clearly depend on each other. This tradeoff is often approached by first choosing the norm and then experimenting a bit with values for ε until you are happy with the results[9]. Choosing $\varepsilon = 0.5$, for example, usually leads to rapid termination in just a few iterations, but the cost is pretty coarse estimates. On the other hand, don't choose $\varepsilon = 0.00000001$ unless you have plenty of time and/or there is a specific need for very fine estimates. Many writers report that $\varepsilon = 0.001$ is usually a good compromise between speed and accuracy.

Termination on U or **V** will eventually lead to the same (U*, **V***), but the actual stopping point and number of iterations needed to achieve the same result is almost always a function of which set of startup and termination variables is used. With all other parameters fixed, it can happen that $U_0 \rightarrow (U_1^*,\ \mathbf{V}_1^*)$ while $\mathbf{V}_0 \rightarrow (U_2^*,\ \mathbf{V}_2^*)$ and $(U_1^*,\ \mathbf{V}_1^*) \not\equiv (U_2^*,\ \mathbf{V}_2^*)$. This is a matter of fineness or granularity of the pairs of successive estimates. Because there are far fewer comparison parameters for the cp elements of **V** than there are for the cn parameters needed for U termination are used, it is usually faster (that is, occurs at a smaller iterate counter t) to satisfy the termination criterion on successive values of **V** than on successive values of U. If ε is fairly small (say, $\varepsilon \sim 0.0001$ or so), it will be the case that $(U_1^*,\ \mathbf{V}_1^*) \cong (U_2^*,\ \mathbf{V}_2^*)$. If you want equality, just keep running both schemes until it occurs. This aspect of termination – quality of the successive estimates – is usually not worth worrying about, but certainly the amount of time and space it takes for a given quality of estimate depends on your choice of this pair of termination criteria. For most applications, estimates to two decimal places will suffice.

Another aspect of termination is the choice of the error norm. The usual choice (the default choice for A6.1) is the Euclidean matrix norm, which is essentially the *root mean squared* (RMS) error between successive pairs of estimates. The 1-norm, sup norm, and Mahalanobis norm have all been used for termination on successive pairs in both U or **V**. This choice clearly affects both the quality of the estimates of (U*, **V***) and the time it takes to get them. Some writers advocate normalizing the error norm by the number of values it examines, i.e., by computing $\|\mathbf{V}_{t+1} - \mathbf{V}_t\|_{err}/cp \leq \varepsilon$ or $\|U_{t+1} - U_t\|_{err}/cn \leq \varepsilon$. These choices are not as crucial as other things you have to pick; so the default choice for the error norm in A6.1 will be the Euclidean norm (squared since there is no reason to compute square roots unless values overflow).

---

[9]By the way, the MATLAB implementation of FCM and HCM doesn't use $\|\mathbf{V}_{t+1} - \mathbf{V}_t\|_{err}$ for termination of A6.1. Their implementation, in common with many others, uses the absolute value of successive differences in the objective function.

Finally, you should want to ask: why *not* use successive values of J_m as the termination criterion, $|J_m(U_{t+1}, \mathbf{V}_{t+1}) - J_m(U_t, \mathbf{V}_t)| \leq \varepsilon$, instead of either $\|U_{t+1} - U_t\|_{err} \leq \varepsilon$ or $\|\mathbf{V}_{t+1} - \mathbf{V}_t\|_{err} \leq \varepsilon$? After all, we are relying on $J_m(U, \mathbf{V};\ X)$ to lead us to good CPOV clusters; don't we believe it? Sure we do, but with reservations. For starters, its value is not a very good measure of fit for the $c(p+n)$ variables being estimated by A6.1. In the absence of any certain knowledge about what value of J_m we are shooting for, it is impossible to set a realistic estimate of what amount of change in this function is indicative of a good terminal solution. But more importantly, using J_m for this purpose is similar to computing the mean of many numbers – it is a useful summary for some purposes but, probably, not for this one. Since we seek the U's and **V**'s (and not the J_m's), termination on U or **V** seems much closer to the overall objective of the cluster analysis in the sense that many more of the estimated parameters have to be close in order to stop the iteration with successively close U's or **V**'s.

What about Lloyd's classical k-means (A6.1, m = 1)? First, recall that, in Example 4.7, we saw that a smaller value of J_1 than 102 could be achieved by the 2-partition $(U_2, \mathbf{V}_2)$ of X_{29} that was, from the HPOV, less desirable than the partition $(U_1, \mathbf{V}_1)$ associated with the value 102. Many authors *do* use this as their termination criterion, as does MATLAB. And some authors claim even more if Lloyd's algorithm arrives at a point where another iteration of A6.1 $(m = 1)$ doesn't change the previous set of hard labels determined by Equation (6.25b). Let us see what this amounts to.

A fixed point of a function $\mathbf{f} : \Re^p \mapsto \Re^p$ is a point in the domain of f that maps to itself. So, $\mathbf{x}^*$ is a fixed point of **f** if, and only if, $x^* = f(x^*)$. It is not difficult to cast HCM and FCM as fixed-point iterations in either set of variables U or **V** by simply writing the output of both 1/2 steps of A6.1 as a composition of two functions. Recall the characterization of iterates in A6.1: $\cdots U_t = F_m(\mathbf{V}_{t-1}; X) \rightarrow \mathbf{V}_t = G_m(U_t; X) \rightarrow U_{t+1} = F_m(\mathbf{V}_t; X) \cdots$. To illustrate, we can write the next 1/2 step for $\mathbf{V}_t$ in terms of the previous pair of 1/2 iterates as $\mathbf{V}_t = G_m(F_m(\mathbf{V}_{t-1})) = G_m \circ F_m(\mathbf{V}_{t-1})$. Termination of hard k-means occurs when $\mathbf{V}^* = G_m \circ F_m(\mathbf{V}^*)$ because, at this juncture, repeated iteration will now remain at the fixed point. Some authors incorrectly interpret this as stopping at a local minimum of J_1, but as we have shown, this function can't have local minima. It is correct to say that the termination point in this case is a fixed point in the joint variables $(U, \mathbf{V})$: this situation is called a *trap state* in the sequel.

For the special case of J_1, there are many theorems about "provable approximations" to the true but unknown global minimum of k-means. It is hard to find much to like about the practical value of this kind of assertion, knowing that Lloyd's algorithm often stops at a fixed point of J_1 whose relationship to the global minimum is unknown.

Turning now to the iteration limit T, it is rare but not unknown that A6.1 will enter a limit cycle, just looping back and forth between several states without producing further improvement in the descent of J_m toward a local extrema. In this case, your program may run forever unless an iteration limit T is prespecified. And, perhaps, more importantly, the basic c-means algorithms at A(6.1) can be *oh so slow*. You may want to cut off the iteration before arriving at a fixed point or extrema of the objective function; this is the job of termination threshold T. The choice of T is somewhat arbitrary, and it can interfere with the march toward a reasonable estimate of a local solution. A small value for T may cut off the iteration before J_m approaches a close enough set of successive values for the termination criterion to stop the looping.

Multiple Runs. $J_m(U, \mathbf{V};\ X)$ has $c(n + p)$ unknowns. For the Iris data (nowadays, a tiny data set), this amounts to the estimation of 3(150+4) = 462 unknown values using the 600 measured values of the data. Aside from the problem that there is not enough data to reliably estimate the unknowns, there is the problem of how many local extrema the objective function may have. We know from Example 6.3 that A6.1 has more than one local stopping point for both the hard and fuzzy cases.

For data of any real size, say $n \geq 1{,}000{,}000$, it is not hard to imagine that $J_{m>1}$ has quite a few local minima and, possibly, saddle points as well. The situation for J_1 is comparable: there may be many U's at good vertices on the cn-soccer ball that, when paired with their companion **V**'s, will stall AO for HCM at a local trap state.

Because HCM and FCM can and do terminate at different CPOV solutions when initialized at different starting points, many authors conduct several (or even better, *lots of*) trials with different initializations when using A6.1. Measures of validation and vector prototypes can be meaningfully averaged over a number of trials; averaging sets of output partitions is far more difficult. So, a good reason for multiple runs is to acquire some level of confidence about the CPOV solution obtained by A6.1. If, for example, 10 runs of c-means from different starting points all terminate at the same (U^*, $\mathbf{V}^*$), you can be fairly confident that the algorithmic CPOV clusters offered are not too dependent on the initialization. But if there are four or five sets of visibly different outputs, something bad has happened, and more experiments and/or different clustering models and algorithms may be required. This can be taken as an indication that the data being processed: (i) don't have CPOV cluster structure; or (ii) much more likely, that HCM/FCM can't find it; or (iii) you have just been really unlucky in your choice of initializations. When this happens, try another approach.

6.6.2 How to Pick an Initial V_0 (or U_0)

Perhaps, the earliest example of a numerical scheme for estimating the zeroes of a real function is Newton's method, which uses gradient descent. This method will converge in theory to a desired root – provided the iteration starts "close enough" to the "right" solution (i.e., the one you want). FCM (both steps) and the **V** half of HCM are based on zeroing gradients, but the AO scheme in A6.1 is *not* classical gradient descent. However, just like Newton's method, HCM and FCM need "good" initializations to avoid stopping at "bad" clustering solutions. What constitutes a good initialization? There is no theoretical answer to this question. The topic of how to initialize these two algorithms occupies much space in the pattern recognition literature, but the fact is that every data set has its own trap states, unusual geometry, special conditions, unruly noise, etc.; so there is no hope of specifying a general scheme for initialization that guarantees success. But there are some general considerations that bear on this problem. In many of these schemes, the idea of acceleration (cf., Section 6.6.3) is a criterion for good initialization: indeed, the ideas of acceleration and good initialization are, in some sense, inseparable, and are certainly not independent. For example, a good initialization might be defined as one that accelerates the algorithm toward a terminal state. And, conversely, an ideal acceleration might be defined as one that arrives at a desirable set of CPOV clusters from any initialization.

The first thing to realize is that the AO sequence $\cdots \mathbf{V}_{t-1} \rightarrow U_{t-1} \rightarrow \mathbf{V}_t \rightarrow U_t \cdots$ can be started with either U or **V**, and the two choices are theoretically equivalent. Moreover, it is not necessary to start and stop on the same sets of unknowns. You can initialize on U or **V** and terminate on successive estimates of the other set of variables. However, initializing and terminating with **V** reduces the time and space complexity used by A6.1 dramatically for all but the smallest data sets. Specifically, if you use **V**, the termination step requires storage and comparison (in the chosen error norm) of two sets of values for **V** of size cp. Compare this to using two sets of values for U of size cn to see that the advantage lies with starting and stopping on

V. In all non-trivial applications, c << n; so time and storage are both reduced by using **V** instead of U for initialization and termination. Consequently, initialization with **V** is almost universally accepted as the better choice.

Many schemes for initializing A6.1 with **V** have been used, some of which are discussed by Hung and Yang (2001) and, more recently, by Celebi and Kingravi (2015). There are several popular, very simple ways to construct $\mathbf{V}_0$. For example, you might choose the first c distinct points in X; choose c distinct points randomly in X; or distribute an initial set $\mathbf{V}_0 = \{\mathbf{v}_{1,0}, \ldots, \mathbf{v}_{c,0}\}$ of computed cluster centers along the diagonal of the hyperbox containing X, and so on. These simple schemes are still used, but there has been an ongoing effort to improve initialization of c-means for many years. Most of the newer schemes involve some type of sampling. Only a few of them are discussed below.

One of the earliest papers that suggest sampling as a means toward better c-means clustering (better meaning faster, and toward smaller values of J_1) is Bradley and Fayyad (1998). These authors present a scheme for initialization called "refinement" that begins with $K \neq c$ small subsamples of the input data. The aim is to select K samples that are close to the modes (local maxima) of the joint probability distribution function that presumably generated the data. HCM is applied to each of the K subsamples, and, at termination, the clustering results are combined. Then c-means is applied to the resultant set K times, initialized with each component set, the best solution is chosen, K is adjusted, and iteration continues to termination.

Maximin (MM) sampling (cf., Section 11.3.4) to obtain $\mathbf{V}_0$ has been advocated by many authors. For example, Elkan (2003, HCM) and Hathaway *et al.* (2006, HCM and FCM) both use this method to begin A6.1, but they use different methods to initialize MM sampling itself (cf., Section 11.7 for different ways to initialize MM). Celebi and Kingravi (2015) discuss MM and three variations of it for initialization of hard c-means as well as two approaches based on principal components analysis. They compared initializing Lloyd's algorithm using the 6 methods on 24 data sets from the UCI database. Their conclusions for a typical run: (i) MM initialized at the grand mean of the data provides the slowest termination of c-means, whereas the KK method (MM initialized at the $\mathbf{x} \in X$ which has the largest Euclidean norm) yields the fastest termination; and (ii) from an optimization point of view, the VP and PP initializations are superior to the other four methods (PP uses a divisive hierarchical approach based on principal component analysis (PCA), and VP is an approximation to PP).

Arthur and Vassilvitskii (2007) discuss a very popular algorithm called c-means++ (aka k-means++) that initializes HCM in a slightly different random fashion. In the seeding step, initial cluster centers are found using an adaptive sampling scheme called D^2-sampling. This approach begins by drawing an initial center from X at random. Then, $(c-1)$ additional centers are sequentially added to the previously sampled centers using D^2-sampling. The c-means++ seeding scheme requires c full passes through the data. There is a provable guarantee in this paper which is that the expected value of J_1, when Lloyd's algorithm is applied to X using c-means++ seeding, is $E[J_1(U, \mathbf{V}; X)] \leq 8(\ln c + 2)J_1(U^*, \mathbf{V}^* : X)$, where $J_1(U^*, \mathbf{V}^* : X)$ is the global minimum of J_1. In common with many subsequent papers, a theoretical analysis such as this ties "provably good" samples to $J_1(U^*, \mathbf{V}^* : X)$. But the global solution $(U^*, \mathbf{V}^*)$ of problem (6.9) can't be found without examining all of the good vertices of M_{hcn}; so these theorems are not particularly useful in either fake (labeled data) or real (unlabeled data) cluster analysis. Nonetheless, these authors demonstrate good results on four data sets, and the c-means++ algorithm has many adherents. For example, Bachem *et al.* (2016) give a useful review of this method and provide their own provably good theory for an updated version of c-means++ seeding based on a Markov Chain Monte Carlo method.

Ostrovsky *et al.* (2012) prescribe a method for initialization of HCM by selecting the $\{\mathbf{v}_{i,0}\}$ with a probabilistic seeding scheme that is similar in some ways to the maximin sampling scheme (Algorithm A11.1 in this volume). The crux of this approach is to select the first pair of centers as far from each other as possible and then continue to add next seeds as far from the set of current seeds as possible. Newling and Fleuret (2017) describe a way to use the medoids associated with the Clarans algorithm of Ng and Han (1994, 2002) to initialize c-means. They claim that their Clarans-based method outperforms other algorithms on 23/23 data sets with a mean decrease over c-means ++ of 30% for initialization mean squared error (MSE) and 3% for final MSE.

There are many other variations on selecting initial cluster centers for A6.1, but none seems superior for a wide variety of different data sets. So, the default initialization rule shown beneath A6.1 – viz., just pick any c distinct points in X as $\mathbf{V}_0 = \{\mathbf{v}_{1,0}, \ldots, \mathbf{v}_{c,0}\}$, is used in the sequel for most of the numerical examples. See Section 11.3 for more on this topic.

☺ ***Forewarned is forearmed:*** By the way, if you choose an initialization of A6.1 for which two or more rows of U_0 or more than one vector in $\mathbf{V}_0$ are identical, these values will never change (in theory) since updates to U and **V** proceed row-wise in U and from cluster to cluster for **V**. For example, if rows i and j of U_0 are equal, then $\mathbf{U}^{(0,i)} = \mathbf{U}^{(0,j)} \Rightarrow \mathbf{v}_{0,i} = \mathbf{v}_{0,j} \Rightarrow \mathbf{U}^{(1,i)} = \mathbf{U}^{(1,j)} \ldots$. So, however you initialize, make sure you don't inadvertently make this mistake. It is unwise to rely on roundoff error to perturb either half step enough to dodge this problem; so just avoid it by making a wiser choice when you start. ☺

6.6.3 Acceleration Schemes for HCM (aka k-Means) and (FCM)

Naïve implementation of Algorithm A6.1 (HCM, m = 1; FCM, m > 1) has at least time complexity O(tcpn), where t is the number of iterations to termination and the model norm is Euclidean. (See Section 9.3 if you need to refresh your memory about the big-oh notation.) This means that the time consumed by either of these methods grows linearly in the number of clusters (c), number of dimensions (p), and number of samples (n). Linear growth is the basis for the notion of scalability, which can be roughly summarized as meaning something like "twice as many samples (clusters, dimensions) will take about twice as much CPU time, etc." So, the basic k-means algorithms are scalable. But … they can be mighty slow for large values of n and/or p. The choice of model norm also affects runtime, but the most important factor in this regard is the calculation of the (cn) point to prototype distances needed to update the membership matrix at each half-iterate. Acceleration becomes more and more desirable as we process vector data with higher and higher numbers of input features. This subsection discusses acceleration of hard and fuzzy k-means. Acceleration for GMD/EM is discussed in Chapter 7: acceleration for single linkage is discussed in Chapter 8.

Acceleration approaches for iterative algorithms can be roughly grouped into data transformations, approximate versions of literal equations ("literal" means exact implementation of the naïve algorithm), improvement in initial starting positions, clever implementations, and different computational styles (serial, parallel, mounted on *graphics processing units* (GPUs), etc.). Methods that aim for faster run time while producing exactly the same result as a literal scheme are called *exact acceleration schemes.*

Three of the four basic algorithms covered in this volume (Algorithms A6.1 ~HCM (c-means), FCM (fuzzy c-means), and A7.1 ~ GMD/EM) are Picard (or fixed point) iteration schemes that seek local extrema of non-linear objective functions. Acceleration of successive approximations to solve non-linear equations has a long, rich history in numerical analysis (cf., Aitken, 1926); so it should not surprise you to discover that there have been many studies of this kind for these three methods. Most (but not all) of the tricks that work for FCM will probably work for c-means (HCM) too. In fact, many of them also work for GMD/EM since all three of these are based on alternating through uncoupled necessary conditions. On the other hand, there are some clever tricks to speed up HCM that are not available to FCM or GMD/EM because of the differences with the update structure of Equation (6.24). We begin with some ways to accelerate hard c-means.

Hard c-Means (A6.1, m = 1). There is an extensive literature discussing various methods to accomplish exact acceleration of hard c-means (aka k-means). The computation at Equation (6.24) is based on the geometry shown in Figure 6.3, which invokes the nearest prototype rule to determine the cluster assignment for each individual **x** in X. This calculation appears in A6.1 (m = 1) as Equation (6.25b), where the NPR computation is made for all c clusters and n input points in each ½ step of HCM. Most of the time consumed in the iterative loop for HCM is spent on determining possible cluster reassignments based (6.25b), i.e., on the (cn) distances $\left\{\|\mathbf{x}_k - \mathbf{v}_i\|_A^2\right\}$ between the data points and the current cluster centers. Hamerly and Blake (2015) observe that "The implementation of Lloyd's algorithm spends the majority of its time here, and much of the computation done here is unnecessary; the information needed for this step can be derived using some caching and geometry." This subsection explores a few techniques based on this general idea. It is worth noting at the outset that most of the methods in this subsection are NOT applicable to either FCM or GMD-EM because the ½ step membership updates for $\{u_{ik}\}$ or $\{p_{ik}\}$ done by these soft partitioning procedures necessarily adjusts all n values of $\{u_{ik}\}$ for each i from 1 to c.

Pelleg and Moore (1999) discuss a method for accelerating hard c-means (A6.1, m = 1) using an implementation based on kd-trees. Their method depends on the fact that nodes in the tree can represent many points in the data. Acceleration is realized by limiting the iterative update for HCM through Equations (6.25) to hyperboxes on the nodes, instead of simply running the updates for every $\mathbf{x}_k \in X$. They give a numerical example for n = 433, 208 two-dimensional samples of real star data that shows speedups ranging from 26 to 136. A fun fact about this example is that c = 5000 centers were requested – that's a lot of centroids! These authors also discuss an example using synthetic data, c = 72 Gaussian subsets of n = 20,000 points in spaces with dimensions from 2 to 8. Their algorithm is faster than HCM A6.1 for $2 < p < 5$, but for higher numbers of features (p = 6, 7, and 8), the naive k-means algorithm A6.1 is somewhat faster than both of their acceleration methods. A similar method for exact acceleration based on kd-trees appears in Kanungo *et al.* (2002).

Moore (2000) introduces the use of metric trees in conjunction with cached sufficient statistics to accelerate HCM, arguing that kd-trees do well on low-dimensional input data but fare less well as the dimensionality of the data grows. His scheme uses the concept of an anchors hierarchy to quickly establish nodes in the metric tree. This paper explicitly shows how to employ the triangle inequality to good advantage for exact acceleration of hard k-means. His definition of speedup is the ratio of the number of distance calculations needed by A6.1(m = 1) to the number needed by the metric tree method. Numerical experiments on 14 data sets and three values of c are discussed (see Table 6.4(c) for a summary). Moore observes that there is wild variation in the speedup realized in various experiments using metric trees due to the choice of thresholds used by the algorithms tested.

Phillips (2002) introduces two methods called "Compare-means" and "Sort-means" that yield computational economies using the triangle inequality to avoid all (cn) calculations at Equation (6.25b) at each half step of HCM. He presents experiments on three quite different data sets: Landsat image data (2.4 mb in 6D), the Forest Cover data (541 k in 44D), and border gateway patrol data (141 k in 29D). He summarizes the speedup of HCM by his two methods this way: "Algorithm Sort-means is significantly faster than algorithm Compare-means (except when the number of means is very large), and the difference is most pronounced when the number of dimensions of the data set is small."

One of the very best papers on this topic is by Elkan (2003), who provides an excellent survey of much of the early effort expended on accelerating hard k-means. Elkan states that there are at least three general approaches for accelerating the update calculations at Equation (6.25b) used in each $1/2$ loop of basic HCM: *locality sensitive hashing* (Indyk and Motwani, 1998); *subtrees of nearby points*, beginning with (Ramasubramanian and Paliwal, 1990); and *triangle inequality methods* that enable a user to avoid making all (cn) computations in Equation (6.25b) at each $1/2$ step, beginning with (Hodgson, 1988). Elkan gives many references besides these three; interested readers may consult Elkan (2003) for further details.

Elkan asserts that his acceleration method for k-means satisfies three basic properties: (i) it can be initialized with random cluster centers; (ii) it produces exactly the same results as Lloyd's batch c-means algorithm A6.1 $(m = 1)$; and (iii) it works with *any* metric on the input space. The last property makes this method more general than updates to the memberships made by Equation (6.25b) in the basic HCM algorithm A6.1, which is restricted to inner product induced norm metrics. Acceleration is achieved by abandoning any (distance to cluster center) computation for which the bounds show that the considered center is either too far or too near to the current cluster center to alter the label assignment that will be made by the (NPR) necessary condition at Equation (6.25b). The metric requirement guarantees the triangle inequality, which is used to derive two lemmas about distance bounds between data points and cluster centers. Here are Elkan's lemmas:

Lemma 6.4. Triangle inequality bounds (Elkan, 2003).

Let (X, d) be any metric space, $x, a, b \in X$.

(a) (Upper bound). $d(a, b) \geq 2d(x, a) \Rightarrow d(x, b) \geq d(x, a)$.
(b) (Lower bound). $\max\{0, d(x, a) - d(a, b)\} \leq d(x, b)$.

What does it mean? Elkan shows how to use these two lemmas to accelerate HCM when $X = \Re^p$ and the points a and b are k-means cluster centers. The lemmas essentially establish upper and lower "zones of influence" for the points a and b. Since every A-norm on $\Re^p$ induces a norm metric on it (cf., problem 12 in the exercises for this chapter), Elkan's lemmas apply directly to the case of A6.1 $(m = 1)$, i.e., to HCM.

Proof. **(a)** The triangle inequality gives $d(a, b) \leq d(a, x) + d(x, b)$; so $[d(a, b) - d(x, a)] \leq d(x, b)$. Apply the hypothesis to the left side of this inequality:$[d(a, b) - d(x, a)] \geq [2d(x, a) - d(x, a)] = d(x, a)$; so $d(x, a) \leq d(x, b)$. **(b)** By the triangle inequality, we have $d(x, a) \leq d(x, b)+d(b, a) \Rightarrow [d(x, a) - d(a, b)] \leq d(x, b)$and $0 \leq d(x, c)$. ■

Switching to the notation of A6.1, let the metric d be $d(x, y) = \|x - y\|_A \forall\, x, y \in \Re^p$, $a = v_i$, $b = v_j$. Suppose a point **x** is currently assigned to cluster j with center v_j. Using squared distances without loss of generality, Equation (6.24) will now move **x** from cluster j to cluster i with center v_i iff $\|x - v_i\|_A^2 \leq \|x - v_j\|_A^2 \;\; \forall\, j \neq i$. If $\|v_j - v_i\|_A^2 \geq 2\|v_j - x\|_A^2$, then Lemma 6.4(a) guarantees that $\|v_i - x\|_A^2 \geq \|v_j - v\|_A^2$; so we don't need to compute $\|v_i - x\|_A^2$ because the test at Equation (6.24) will fail. Now suppose that $\|v_j - x\|_A^2$ is not known exactly but is bounded above, say $\|v_j - x\|_A^2 \leq B(x)$. If $2B(x) >$

$\|v_i - v_j\|_A^2$, distances $\left(\|x - v_j\|_A^2 \text{ and } \|x - v_i\|_A^2\right)$ are not needed. And if $2B(x) < \min_{i \neq j}\left\{\|v_i - v_j\|_A^2\right\}$, then **x** must remain with cluster j, and the $(c - 1)$ distance calculations for this **x** can be skipped. In order to use this method, one upper bound is maintained for each point in the data set X.

To see how part b of Lemma 6.4 is used, suppose **x** is currently labeled j and that this point was closest to v_j' in the previous iteration. Let v_j be the current version of the same center (i.e., the *j*th cluster center has changed from $v_j' \rightarrow v_j$). Suppose that there was a lower bound $b_{(j)}'$ for which $\left\|v_j' - x\right\|_A^2 \geq b_{0j}'(x)$. If the *j*th center has moved only a short distance in the HCM update of cluster center j via Equation (6.25a), Elkan estimates a new lower bound for the current center as $\|x - v_j\|_A^2 \geq \max\left\{0, \left\|x - v_j'\right\|_A^2 - \left\|v_j - v_j'\right\|_A^2\right\} \geq \max\left\{0, b_j' - \left\|v_j - v_j'\right\|_A^2\right\} \doteq b_j$. The rationale offered for this is that in the later stages of HCM iteration, most centers will change but little, and the bounds will carry this information to the next iteration to avoid unnecessary distance calculations.

Suppose that **x** is currently assigned to cluster (j), that $\|v_j - x\|_A^2 \leq B(x)$, and that $\left\|v_j' - x\right\|_A^2 \geq b_j'(x)$. If $B(x) \leq b_j'(x)$, then $\|v_j - x\|_A^2 \leq B(x) \leq b_j'(x) \leq \left\|v_j' - x\right\|_A^2$; so the two bounds suffice. In other words, it is not necessary to calculate either of the distances $\|v_j - x\|_A^2$ or $\left\|v_j' - x\right\|_A^2$ in this case. Elkan provides a detailed pseudocode for this accelerated version of k-means and then gives numerical examples on six real data sets. The data range in size from (***Random***: $n = 10,000, p = 1000$) to (***BIRCH***: $n = 100,000, p = 2$). Tabulated results show the least speedup of 1.50:1 for $c = 3$ clusters in the ***Random*** data; up to 351:1 for $c = 100$ clusters in the ***BIRCH*** data. An interesting aspect of this method is that the speedup factor increases with increasing c (more calculations are avoided) and decreases with increasing p (higher dimensional input vectors require more computation per distance needed).

As noted in the previous subsection, Arthur and Vassilvitskii (2007) propose accelerating k-means by choosing initial cluster centers from the data points in X in a random but specific way. Their procedure, called c-means++ (aka *k-means ++*), picks a first center and then chooses successive centers by weighting the data points according to their squared distance from the closest center already chosen. The rationale for this method is philosophically similar to the reasoning behind the Maximin random sampling algorithm A11.1 given in Chapter 11. These authors compare k-means to k-means++ using two synthetic and two real data sets using 20 run averages. Average speedups are positive in all but one instance, ranging from −0.9 (k-means is faster) to 9.6. Their average speedup on the 10% KDD intrusion data, $n = 494,019$, samples in $p = 35$ dimensions with $c = 50$ clusters is 2.7:1.

Hamerly (2010) introduced a modification of Elkan's (2003) method that eliminates all but one of the c lower bounds per point. In Elkan's algorithm, each data point **x** has one upper bound $B(x)$, which bounds the distance between x and its closest center, and c lower bounds $\{b_j(x) : 1 \leq j \leq c\}$. Rather than keeping c lower bounds for each data point, Hamerly's algorithm keeps only one lower bound, say $b(x)$, per data point, which is a lower bound on the distance between the data point and its second-closest cluster center. As long as $B(x) \leq b(x)$, the center assignment of the point **x** can't change. Hamerly asserts that Elkan's algorithm is superior for high-dimensional data sets, while his is better for low-dimensional data sets, and suggests combining the two methods at the end of his paper. Seven data sets of various sizes and dimensionalities are used to illustrate the method. See Table 6.4c for a summary of the results.

Drake and Hamerly (2012) introduce the idea of having a variable number of lower bounds per point on the c-means distances. Their algorithm affords a compromise between the approaches of Elkan (2003) and Hamerly (2010). They use the same data sets as these two earlier papers and conclude that Elkan's algorithm performs best for high-dimensional data sets, while Hamerly's algorithm dominates speedup for low-dimensional inputs [3]. These authors state that the variable bounds algorithm successfully blends the two earlier algorithms for better efficiency in medium dimensional input spaces. They assert that the strengths of these three algorithms are complementary and suggest the possibility of constructing a piecewise algorithm using Hamerly's algorithm for $p < 20$, their algorithm for $20 \leq p \leq 120$, and Elkan's algorithm for $p > 120$.

Hamerly and Drake (2015) provide a really nice, very extensive compendium of much of work in this area. These authors review several approaches to acceleration of hard c-means that don't use triangle inequality bounds, but their forte is concentrated on improved approaches to acceleration based on Elkan's idea. They motivate their approach with the following observation: "We investigated the source code of c-means implementations for the software packages ELKI, graphlab, Mahout, MATLAB, MLPACK, Octave, OpenCV, R, SciPy, Weka, and Yael, and found that none of them use the triangle inequality bound acceleration techniques." This interesting fact begs the question posed by Ostrovsky *et al.* (2012), who questioned why practitioners continue to use the basic or naïve versions of Lloyd's AO algorithm A6.1 ($m = 1$), along with EM and hierarchical methods, which they describe as "old standbys with no known performance guarantees."

Hamerly and Drake (2015) itemize four ways that lower and upper bounds can be used to avoid some distance calculations that Equation (6.24) requires for implementation of the basic c-means algorithm A6.1 ($m = 1$):

1. To prove that $\mathbf{v}_j'$ is closer to $\mathbf{x}$ than $\mathbf{v}_j$, given only distances $\left\|\mathbf{x} - \mathbf{v}_j'\right\|_A^2$ and $\left\|\mathbf{v}_j - \mathbf{v}_j'\right\|_A^2$.
2. To prove that $\mathbf{v}_j'$ is closer to $\mathbf{x}$ than $\mathbf{v}_j$, given only norms $\|\mathbf{x}\|_A^2$ and $\|\mathbf{v}_j\|_A^2$and distance $\left\|\mathbf{x} - \mathbf{v}_j'\right\|_A^2$.
3. To maintain an upper bound on $\|\mathbf{x} - \mathbf{v}_j\|_A^2$ when $\mathbf{v}_j$ is moving.
4. To maintain a lower bound on $\|\mathbf{x} - \mathbf{v}_j\|_A^2$ when $\mathbf{v}_j$ is moving.

Hamerly and Drake's (2015) Table 2.2, replicated here as Table 6.4(a), itemizes some acceleration methods that use sorting, distance bounds, and the triangle inequality to speed up various versions of the hard c-means algorithm. The columns in Table 6.4(a) identify the basis of different approaches, and the symbol ☺ means that the algorithm in that row uses the method identified by that column. Here is a key to the columns of Table 6.4a:

1. Distance from center $\mathbf{v}_j$ to its closest other center.
2. Distance from center $\mathbf{v}_j$ to all other c-1 centers.
3. Upper bound on $\left\{\|\mathbf{x} - \mathbf{v}_j\|_A^2 : 1 \leq j \leq c\right\}$.
4. Lower bound on $\left\{\|\mathbf{x} - \mathbf{v}_j\|_A^2 : 1 \leq j \leq c\right\}$ per data point.
5. For each $\mathbf{v}_j$, sort $\{\mathbf{v}_i : 1 \leq i \neq j \leq c\}$ by distance to $\mathbf{v}_j$.
6. Sort $\{\mathbf{v}_j : 1 \leq j \leq c\}$ by $\left\{\|\mathbf{v}_j\|_A^2 : 1 \leq j \leq c\right\}$.

Table 6.4a Triangle inequality methods for accelerating literal hard c-means (HCM A6.1 (m = 1)).

Algorithm	1	2	3	4	5	6
Lloyd (1957) Batch c-means						
Phillips (2002) Compare-means	☺					
Phillips (2002) Sort-means		☺			☺	
Elkan (2003)	☺	☺	☺	c		
Hamerly (2010)	☺		☺	1		
Drake/Hamerly (2012)	☺		☺	b < c		
Hamerly/Drake (2015) Annular	☺		☺	1		☺
Hamerly/Drake (2015) Heap	☺		☺	0		

Please note that the bounds shown in column 4 of Table 6.4(a) are the number of lower bounds needed *per data point*. Specifically, Elkan's and Hamerly's algorithms keep (c) lower and one upper bound, respectively. Drake and Hamerly (2012) bridged the gap between these two extremes by using $1 < b < k$ lower bounds on the b closest centers to each point. According to Blake and Hamerly, the value of b can be selected in advance or adaptively learned while the algorithm runs. Since Drake's algorithm uses one upper bound per clustered point, his method incurs an overhead of $n(b + 1)$ total distance bounds.

After a comprehensive review of many methods, Hamerly and Drake (2015) introduce two new methods (called annular and heap) for acceleration of HCM. The annular method depends on sorting the centers by their vector norms, a strategy that can eliminate many centers whose norms are too large or too small to be closest to $\mathbf{x}$. The heap algorithm orders all the points by the likelihood that each will need updating, indicated by the value of $b(\mathbf{x}) - B(\mathbf{x})$. Numerical experiments basically use the same seven data sets used by Elkan (2003), Hamerly (2010), and Drake and Hamerly (2012). Their first conclusion is that the best acceleration is achievable with lower dimensional, "naturally clustered" input data. This paper goes on to describe several versions of accelerating kernelized c-means and also presents a very interesting discussion about the numbers of distance calculations done by various algorithms. Overall, this chapter is arguably the best recent survey of acceleration methods for the naïve HCM algorithm and is well worth reading if you intend to use the hard c-means algorithm A6.1 (m = 1) in any form.

A different approach, called the PK method, is espoused by Poteras *et al.* (2014). These authors assert that after a certain number of iterations, only a few of the data points in X are reassigned. Their method, instead of skipping calculations of some of the distances at each $\mathbf{x} \in X$, skips looking at some of the $\mathbf{x}$'s altogether. The way this is accomplished is to impose a limit on the points that need updating via an interval based on a threshold (they call it a *constant width*) that divides the data into points that need to be visited versus points that are omitted from further consideration in A6.1(m = 1). They illustrate the use of this approach on seven synthetic data sets of two-dimensional data points that are subsequently divided into c = 4, 8, and 12 clusters. They verify that their approach does produce the same literal partition as Lloyd's c-means with reduction of the run time ranging from 49% to 72% or about 2:1 to 3.6:1.

Jain and Buksh (2016) offer a derivative of the PK method for excluding some of the $\mathbf{x}$'s discussed by Poteras *et al.* (2014). These authors claim that the constant width threshold used by the PK method is not related to any particular data set, and so its effectiveness is limited. Jain and Buksh propose basing the width of the exclusion interval on the data set that is to be clustered. They find the value of the range for each input feature and use 10% of the minimum range as the width for the PK algorithm. Their numerical examples use seven data sets, including three well-known, very small, real data sets (Iris, Wine, and Ionosphere) and four instances of a synthetically generated data set in two dimensions. Like Poteras *et al.* (2014), these authors give no other information about the structure of these synthetic data. Curiously, they report acceleration in terms of improvement to the PK method. Their tabulation shows a speedup of the PK method itself of something less than 2:1; so, by extrapolation, this scheme realizes a speedup of HCM of something like 3:1 to 7:1. The approach espoused by the authors of these two papers may have merit, but the quality of acceleration realized is not evident in the numerical evidence.

To wrap up this subsection, here is one more very recent approach to acceleration of Lloyd's HCM. This one uses an old technique from numerical optimization theory called *Anderson Acceleration* (AA; Anderson, 1965). The basic idea of this method is to increase the *rate of convergence* (see Definition 10.1) of c-means algorithm A6.1(m = 1) by reducing the overall number of iterations required to achieve termination. This is a general method in many areas of numerical optimization that applies to fixed point iteration schemes.

Recall the characterization of iterates in A6.1: $\cdots U_t = F_m(\mathbf{V}_{t-1}; \mathbf{X}) \rightarrow \mathbf{V}_t = G_m(U_t; \mathbf{X}) \rightarrow U_{t+1} = F_m(\mathbf{V}_t; \mathbf{X}) \cdots$. Zhang *et al.* (2018) write the next $1/2$ step for $\mathbf{V}_t$ in terms of the previous pair of $1/2$ iterates as $\mathbf{V}_t = G_m(F_m(\mathbf{V}_{t-1})) = G_m \circ F_m(\mathbf{V}_{t-1})$. Termination of HCM occurs when $\mathbf{V}^* = G_m \circ F_m(\mathbf{V}^*)$ because, at this juncture, repeated iteration will now remain at the fixed point. Zhang *et al.* incorrectly describe this as convergence to a local minimum of $J_1(U, \mathbf{V}; X)$, when it is really stopping at a local trap state of the c-means objective function. Be that as it may, these authors then describe applying AA to the function $G_m \circ F_m$. What this means is that instead of the next value of $\mathbf{V}_t$ depending on only the previous value $\mathbf{V}_{t-1}$, the last (η_{aa}) iterates $(\{\mathbf{V}_{t-1}, \mathbf{V}_{t-2}, \ldots, \mathbf{V}_{t-\eta_{aa}}\})$ are examined with a view toward determining an accelerated iterate.

Zhang *et al.* make two observations of importance about this procedure: (i) the choice of (η_{aa}) is a tradeoff between accuracy and computational efficiency; and (ii) using a fixed value for (η_{aa}) doesn't always guarantee the descent property for $J_1(U, \mathbf{V}; X)$. To overcome these problems, these authors introduce a way to achieve dynamic adjustment to the number of iterates examined. Their procedure has three options: (a) if the current iterate increases J_1 or the decrease in the current iteration is small compared to the previous iteration, then reduce (η_{aa}); (b) if the decrease in J_1 is large enough compared to the previous iteration, then increase (η_{aa}); (c) otherwise, (η_{aa}) is not altered.

Zhang *et al.* support their scheme using numerical examples on 20 data sets whose parameters range between the pairs of values shown in Table 6.4(b). They compare Lloyd's A6.1(m = 1) to their method but with several twists. First, the assignment step (Equation (6.25b)) in original c-means is replaced by Hamerly's (2010) assignment step. Second, comparisons are made using four different types of initialization, viz., Init 1 = c-means++ (Arthur and Vassilvitskii, 2007), Init 2 = afk-mc2 (Bachem *et al.*, 2016), Init 3 = bf (Bradley and Fayyad, 1998), and three versions of the CLARANS initialization, Init 4, Init 5, and Init 6 (Newling and Fleuret, 2017). The numerical examples compare the total CPU time taken by this modified version of c-means with and without AA. In all 120 tests (20 data sets $\times$ 6 initialization methods), the terminal value of mean squared error, i.e., MSE final = $J_1(U_f, \mathbf{V}_f; X)$ is the same with and without AA. Table 6.4(b) (from Table 3 in Zhang *et al.* 2018) exhibits the results shown for data set 20, which is the full KDD '99 Cup data.

The bold entries in Table 6.4(b) indicate the winner of each test. The values labeled (η_{aa}) denote the (dynamically adjusted ranges of the) number of previous iterates used by AA of c-means for each of the six initializations. For example, the (η_{aa}) entry for Init 5, viz., 38↔53, indicates that during this iteration, (η_{aa}) ranged between 38 and 53. For this data set, the use of AA results in reduced CPU time in five of the

Table 6.4b Six runs on the KDD '99 Cup data with and without Anderson Acceleration.

KDD	Init 1 c = 10		Init 2 c = 10		Init 3 c = 10	
	W^0 AA	(η_{aa})	W^0 AA	(η_{aa})	W^0 AA	(η_{aa})
# Iter	13	**11↔12**	20	**13↔15**	**8**	9↔10
cpu (s)	11.09	**11.09**	14.83	**13.37**	**5.84**	6.80
MSE	3.92		4.07		3.89	

KDD	Init 4 c = 10		Init 5 c = 10		Init 6 c = 10	
	W^0 AA	(η_{aa})	W^0 AA	(η_{aa})	W^0AA	(η_{aa})
# Iter	14	**8↔10**	83	**38↔53**	163	**58↔72**
cpu (s)	7.72	**6.04**	221	**191**	10408	**7212**
MSE	3.91		1.86		0.77	

six tests, the best speedup being 10408/7212 = 1.44:1 for Init 6. But note that for Init 3, AA takes longer to terminate with AA than without it; in other words, AA *decelerated* the version of c-means being tested in this run, about 0.85:1, reported as (−0.85). Zhang *et al.* state that their version of AA outperforms the unaccelerated version using the four initializations (Inits. 4, 5, and 6 are lumped together) in 15, 20, 18, and 19 cases, respectively, for the 20 data sets.

Fuzzy c-Means (A6.1, m>1). The earliest method for acceleration of FCM is apparently due to Cannon *et al.* (1986), who discussed a way to approximate some of the calculations in the necessary equations at (6.19) for literal implementation of FCM with six lookup tables. They called this method *approximate FCM* (AFCM). They compared FCM to AFCM by segmenting three 256 × 256 test images which were made from nine-band flight data showing a region of Oklahoma into c = 10 regions. This method afforded a modest speedup of about 6:1 for FCM. In 1986, this seemed promising: with today's computational power, it has little practical value. But this paper affords a pretty interesting insight about the meaning of the term big data by stating that: "Large values of N, say 1mb, are easily encountered in applications." More on this in Chapter 11.

Kamel and Selim (1994) and Kolen and Hutcheson (2002) improved the time and space complexity of FCM by implementing the Picard loop through Equations (6.19) cluster by cluster (column by column in the former, and row by row in the latter) on U and **V**, instead of the batch style shown in A6.1. This "hidden partition" trick eliminates the need for storage of the c × n partition matrix U, which reduces the per iteration complexity of naïve FCM from $O(c^2np)$ to $O(cnp)$ and thereby improves the ability of memory to accommodate more data. While these two papers limit their discussion to improving FCM, the same implementation trick works for both HCM and GMD/EM because all of the necessary conditions for A6.1 and A7.1 are uncoupled across the cluster index. Borgelt and Kruse (2003) discuss five variations of a scheme based on the difference between two consecutive steps of the alternating optimization scheme in A6.1 as a gradient that can be used for modification similar to back propagation updating in neural networks.

Cheng *et al.* (1998) propose a *multistage random FCM* (mrFCM) that purports to accelerate FCM by getting better initial guesses for the cluster centers on bigger and bigger subsets of loadable small data sets. But the last step of mrFCM is a run of LFCM on the entire data set; so this method is really aimed at reaching the literal solution more rapidly by guiding FCM toward $U_n^{literal} \in M_{fcn}$. Since the final run is a pass of LFCM through the entire data set, mrFCM doesn't address the problem of clustering in unloadable data. See Altman (1999) for a one-step version of the mrFCM procedure.

Eschrich *et al.* (2003) develop an algorithm called *bit reduction FCM* (brFCM), which is a way to reduce the number of distinct vectors which must be clustered by FCM. This very different approach to clustering in static big data essentially reduces $N \to n$ by replacing many similar or identical copies of the same instance by a single example. More specifically, data quantization forces different continuous values into the same quantization bins, creating identical feature vectors from "similar" ones. In brFCM, aggregation then creates a single exemplar representing the quantization bin. The value of this exemplar is the mean value of all full-precision feature vectors quantized to this bin, weighted by the number of replicants in the bin.

Eschrich *et al.* compare brFCM to LFCM by segmenting 32 magnetic resonance images into different tissue types and 172 infrared images into trees, grass, and target regions. Average speedups ranging from 59 to 290 times as fast as LFCM are achieved without significant loss in partition quality. This method can be viewed as an accelerator for FCM/HCM or as a way to cluster with FCM/HCM in big static data. The natural domain for this approach is image segmentation of single gray-level images whose intensities take only a limited number of integer values.

Cai *et al.* (2007) develop an acceleration scheme for image segmentation with clustering using a modification of FCM they call *fast generalized fuzzy c-means* (FGFCM). This approach builds on four previous modifications of FCM for image segmentation that all utilize a fixed parameter (α), by replacing (α) with a new parameter (s_{ij}) which is a function of pixel location (i, j) within the image. The crux of their approach is to replace the input image, which has q gray levels, with a new image that reflects the spatial and gray-level variation measure by (s_{ij}) in small (usually 3×3 or 5×5) windows of the input image centered at (i, j). FGFCM has two steps: (i) in the first step, the input image is replaced by a linearly weighted sum image ξ which is generated using (s_{ij}); (ii) in the second step, the fast segmentation method called *enhanced fuzzy c-means* (EnFCM) introduced by Szilágyi *et al.* (2003) is performed on the gray-level histogram of ξ.

An important point about FGFCM is that the CPU time for image segmentation is dependent on the number of gray levels q rather than the size of the image. Specifically, for an image of size $N = n \times m$, the time complexity is reduced from $\boldsymbol{O}(cNI_{FCM})$ to $\boldsymbol{O}(cqI_{FGFCM})$, where c is the number of the clusters, $N = n \times m$, and I_{FCM} and I_{FGFCM} are the numbers of iterations, respectively, used by FCM A6.1(m>1) and FGFCM, to achieve termination for the 8-bit images $(q = 256)$ discussed by Cai *et al.* The typical spatial resolution of an image in 2021 is 16 bits, for which $q = 65,526$. But this is still usually much less than N; so the acceleration of FCM using this idea for images is a useful one. These authors give an example using an $(N = 181 \times 181 = 32,761)$ 8-bit MRI image; so $q = 256$ against $N = 32,761$. They report speedups in the range 141 to 156 to 1 for FGFCM against the basic FCM algorithm A6.1$(m = 2)$.

Anderson *et al.* (2008) implement FCM on a GPU instead of the traditional implementation in cpu-based software. Speedup is realized in their scheme by changing the way the literal algorithm is implemented in its computational environment. This method of acceleration should be useful for any clustering algorithm, but GPU implementations demand exacting programming and intimate knowledge of the hardware involved. With ongoing improvements to programming languages that support GPU implementations, this is an idea whose time is just around the corner.

Hore *et al.* (2009a) study acceleration of both HCM and FCM using several methods that begin by chunking the data. Each chunk is processed separately, clusters are summarized by their cluster centers, and the sets of cluster centers are then merged by ensemble techniques, leading to an overall approximation of the literal results. Their experiments don't give speedup in terms of the literal algorithms. Instead, they compare speeds of several techniques for merging ensembles to the method advocated by Strehl and Ghosh (2002) and show speedups as high as 600,000:1 on image data with 4 million samples.

Havens *et al.* (2011) present an acceleration study of *kernelized FCM* (kFCM). Their method, called *approximate kernelized FCM* (akFCM), is similar in spirit to Cannon *et al.*'s (1986) AFCM in the sense that it provides approximations to exact solutions for the literal version of kFCM. Assuming an input of n feature vectors, they derive an approximation to the exact kernel distance used by kFCM, viz., $d_\kappa(i,j) = [\mathbf{U}^{(i)}]^T K \mathbf{U}^{(i)} - 2([\mathbf{U}^{(i)}]^T K)_j + K_{jj}$, where $\mathbf{U}^{(i)}$ is the *i*th column vector in U, normalized with the 1-norm, and K is a chosen $n \times n$ Kernel matrix. The approximate kernel distance is $d_\kappa(i,j) \cong \alpha_i^T K_{\xi\xi} \alpha_i - 2(K_\xi \alpha_i)_j + K_{jj}$, where $K_{\xi\xi}$ and K_ξ are $s \times s$ and $n \times s$ submatrices of Kernel matrix K. Acceleration is due to the reduction in the size of K, which in turn reduces the complexity of the kernel distance calculation from $\boldsymbol{O}(n^2)$ to $\boldsymbol{O}(sn)$.

Havens *et al.* (2011) offer two examples of akFCM using a small synthetic data set of $n = 7500$ two-dimensional points having $c = 50$ visually apparent HPOV clusters; and the MNIST image data, $n = 70,000$ points in $p = 784$ features with $c = 10$ presumed CPOV clusters. Speedup depends, as expected, on the sample size s chosen. Small submatrix sizes ($s = 3$ or so) lead to large speedups: an average speedup of about 19:1 for the small data and about 1000:1 for MNIST. Larger values of s result in larger kernel submatrices and, hence, smaller speedups: somewhere around 3:1 for both data sets at $s = 1500$. These two examples don't show an appreciable difference in cluster quality between the approximate and exact versions of kFCM. Two additional examples on data sets (Quadraped Mammals and the Forest Cover data) which are too big to be tractable for kFCM are offered to show that approximate clustering with akFCM is feasible. These examples are very much in the spirit of the approximate single linkage clusters found by clusiVAT, discussed in Chapter 11; they show how to approximate FCM clustering for data sets that are too big for a literal analysis. Havens *et al.* (2012) compare acceleration methods for FCM that are based on three ideas: (i) sampling followed by noniterative extension; (ii) incremental techniques that make one sequential pass through subsets of the data; and (iii) *kernelized FCM* (kFCM) that provides approximations based on sampling. This paper will be discussed in detail in Chapter 11.

Table 6.4(c) summarizes the average speedup of each of the literal hard and fuzzy k-means clustering algorithms realized by the acceleration methods reviewed in this section. The citation given is the first named author of the paper discussed. When the number of samples (n) is given in mb, the data sets are usually images. Some of the values in Table 6.4(c) are averages or estimates: see the papers for more precise information. Some larger values of n have been rounded off to fit into the table. For example, 597,971 becomes 0.6 mb in Table 6.4(c). The notation #X indicates the *number of data sets* used in numerical experiments. The notation "ns" means *not specified.* The notation "gls" means *gray levels.* The speedup for both hard and fuzzy c-means is expressed as a ratio, L:M, computed as [CPU time of literal implementation of c-means algorithm A6.1]/[CPU time of the acceleration method]. Some speedup times are estimated from other measures used in several of the papers.

Table 6.4c Average speedup by some acceleration methods for literal algorithm A6.1.

Acceleration of Fuzzy k-Means: FCM Algorithm A6.1(m>1)					
Ref.	# X	p	c (or k)	n	Speedup L:M
Cannon (1986)	3	10	10	0.25 mb	6:1
Kamel (1994)	4	ns	ns	ns	1.2: 1
Cheng (1998)	13	3, 6	10	0.4 mb	3:1
Altman (1999)	1	3	3	1 mb	3–10:1
Kolen (2002)	1	9	10	20 mb	9:1
Borgelt (2003)	6	8–13	2, 3	≤ 4177	2:1
Eschrich (2003)	2	2,3	5,7	0.4 mb	59–290:1
Cai (2007)	4	1 (256 gls)	8	32 kb	141–156:1
Anderson (2008)	1	4–32	4–64	64, 8192	10–100:1
Hore (2009a)	8	3–617	3-12	150, 4 mb	Up to 600,000:1
Havens (2011)	2	2, 784	10, 50	7500, 70 kb	3–1000:1
Acceleration of Hard k-Means: HCM Algorithm A6.1(m = 1)					
Ref.	# X	p	c (or k)	n	Speedup L:M
Pelleg (1999)	2	2	5000	39 kb, 150 kb	26–136:1
Moore (2000)	14	2–10,000	3, 20, 100	10 kb–150 kb	(−0.6)–20,000:1
Phillips (2002)	3	3, 6, 54, 29	10–5000	2.4 mb, 581 kb, 141 kb	12–16:1
Elkan (2003)	6	2–1000	3, 20, 100	60 kb–150 kb	1.5–351:1
Arthur (2007)	4	5–35	5–50	10 kb, 500 kb	1–9.6:1
Hamerly (2010)	7	2–56	3, 20, 100, 500	60 kb–1.25 mb	2.1–17:1
Drake (2012)	7	2–784	50, 200	60 kb–580 kb	4–16:1
Poitras (2014)	7	2	4, 18, 12	100 kb–5 mb	2–3.6:1
Hamerly (2015)	7	2–784	2, 8, 32, 128	60 kb–1 mb	2–50:1
Jain (2016)	7	2–34	2, 4, 13	150–750 kb	3–7:1
Zhang (2018)	20	2–561	10, 100, 1000	7 kb–4.90 mb	(−0.6)–3.54:1

☺ ***Forewarned is forearmed:*** First, this subsection discusses only a tiny fraction of the papers in the literature that address acceleration for the basic hard and fuzzy c-means algorithms represented at A6.1. Another important thing to remember when looking at Table 6.4(c) is that classical k-means (HCM A6.1, m = 1, aka c-means) is a special case of fuzzy c-means (A6.1, m> 1). So, some of the acceleration or approximation schemes that have been tested explicitly for FCM *automatically* apply, possibly with slight modifications to accommodate the crisp case, to hard c-means as well. But, and perhaps surprisingly, the converse is not true. For example, the triangle inequality methods shown in Table 6.4(a), which can be used to avoid all (cn) computations at Equation (6.24) for HCM don't apply to FCM (nor to GMD-EM A7.1). ☺

What we can conclude from Table 6.4(c)? Acceleration usually results in at least an order of magnitude speedup, which can get quite a bit better for special kinds of data or with novel implementations. This doesn't sound like a worthwhile endeavor unless you think carefully about what it means. For example, there are today data sets for which these algorithms take a week (168 hours) for one run to termination. An order of magnitude improvement cuts this to about 17 hours[10]. But that still sounds like a long time to us in 2021 and beyond. So, in Chapter 11, we will turn to other ways to approach the problems caused by big static data sets.

To end this subsection, let me offer this quote from Moore (2000): "If there is no underlying structure in the data (e.g., if it is uniformly distributed) there will be little or no acceleration in high dimensions no matter what we do. This gloomy view, supported by recent theoretical work (Indyk, Amir, Efrat and Samet (1999)), means that we can only accelerate [clustering algorithms for] data sets that have interesting internal structure. Resorting to empirical results with real data sets, however, there is room for some cautious optimism for real world use."

6.7 Cluster Validity With the Best c Method

In real cluster analysis, $X = \{x_1, \ldots, x_n\} \subset \Re^p$ is unlabeled; so you will not know the number of clusters to seek (indeed, for $p > 3$, you will never know if there *are* subsets in X that you might regard as HPOV clusters in X!). But X *always* contains CPOV clusters; so with your favorite clustering algorithm, you will probably want to generate a set of candidate partitions $CP = \{U \in M_{fc_k n} : c_m \leq c_k \leq c_M\}$ as specified by Definition 4.4. Once the candidates are found, what happens next? Cluster validity happens next. In this section, we will introduce a few of the very many internal CVIs that can be used to evaluate sets of candidate CPOV partitions found with various clustering algorithms.

Recall from Definition 4.5 that *cluster validity indices* (CVIs) are denoted by the letter *V*. The heuristic for many of these CVIs is that their maximum value over a set of candidate partitions indicates the "best choice." We called such indices *max-optimal* (↑). Similarly, many CVIs are designed to minimize at the "best choice," i.e., they are *min-optimal* (↓) validity indices. Cluster validity indices are often normalized in one of three ways. The basic assumption is that a given *raw* CVI may need to be adjusted to account for various problems that are either index-specific (e.g., variable range) or conceptually deficient (e.g., uncorrected for chance).

[10] It sounds a lot better if you say that an order of magnitude acceleration cuts the time of a literal 60 second run to about 6 seconds, doesn't it?

6.7.1 Scale Normalization

The general idea of normalization for scale appeared at the beginning of Section 5.6, where we discussed various normalizations of feature vectors. The range of many raw validation measures depends on one or more problem parameters such as n, p, or c. For example, the raw *variation of information* (VI) index introduced by Meila (2003) is valued in [0, log(n)]. This index is defined as $V_{VI}(U,W) = V_H(U,W) - V_{MI}(U,W)$, where $V_H(U,W)$, $V_{MI}(U,W)$ are the joint entropy of and mutual information between two crisp partitions $U, W \in M_{hcn}$. So, V_{VI} has a maximum value that increases monotonically with the number of samples (n) in a crisp c-partition. This makes it hard to ascertain what should be regarded as a "good" value of VI. To correct this, Kraskov *et al.* (2005) defined a normalized VI as $\widehat{V_{VI}}(U,W) = 1 - [V_{MI}(U,W)/V_H(U,W)]$, which has the range [0, 1], independent of the number of samples being partitioned.

Scale normalization removes some uncertainty about how to interpret indices across various candidates in CP. The most common intervals for this type of adjustment are [−1, 1] for correlation type measures and [0,1] for positive-definite functions. We will call this type of adjustment to any CVI *normalization of range.*

6.7.2 Statistical Standardization

This type of normalization was discussed in the context of feature extraction in Section 5.6. Equation (5.12) for statistical standardization was introduced as a means for equalizing the effects of scale variation in different features in object vector data. The numerator in this linear transformation centers the variate in question (i.e., zeroes it mean value), and division by the standard deviation imposes unit variance upon it. Validity functions are sometimes normalized in the same way by regarding them as random variables with an assumed distribution over their domains.

Statistical standardization of CVIs dates back to at least 1974, when Brennan and Light (1974) applied the general transformation at Equation (5.12) to an agreement index that was subsequently discussed in Hubert and Arabie (1985). This type of correction follows the general formula:

$$\widehat{V} = \frac{V - \mu_V}{\sqrt{\mathrm{var}(V)}}, \quad \mu_V = \mathrm{E}[V]; \ \mathrm{var}(V) = \mathrm{E}[(V - \mu_V)^2], \tag{6.51}$$

where E denotes the expectation operator of the random variable. Romano *et al.* (2014) used Equation (6.51) to adjust the *mutual information* (MI) index $V_{MI}(U,W)$ (Cover and Thomas, 1991), which is the joint entropy between two crisp partitions $U, W \in M_{hcn}$. Romano *et al.* compute the mean and variance of $V_{MI}(U,W)$ under the assumption that it follows the hypergeometric distribution[11] and then adjust it per Equation (6.51), arriving at a *standardized mutual information* (SMI) index $\widehat{V_{MI}}(U,W)$. Romano *et al.* regard $\widehat{V_{MI}}(U,W)$ as a correction for selection bias.

6.7.3 Stochastic Correction for Chance

The third type of normalization is a constant baseline correction for chance based on the formula

$$V^* = \frac{V - \mu_V}{\max(V) - \mu_V}, \quad \mu_V = \mathrm{E}[V]. \tag{6.52}$$

[11]The hypergeometric distribution assumes partitions U and V, both having c rows and n columns, are chosen randomly.

The calculation of Equation (6.52) requires a statistical null hypothesis for the computation of $\mu_V = E[V]$ and applies to the case where V is max-optimal (↑). In this case, V^* will be bounded above by 1, will be 0 when the index is equal to its expected value, and may take negative values otherwise. The earliest discussion of this type of correction appears to be in Hubert and Arabie (1985) who provide a very lucid account of problems associated with choosing an appropriate form for $\max\{V\}$. Their choice of the form of the maximum leads to the *adjusted rand index* (ARI) that appears in Chapter 9.

6.7.4 Best c Validation With Internal CVIs

The simplest type of evaluation (of the CVIs, which themselves are used to evaluate the candidate partitions) is made with internal CVIs. This type of experiment is often called "best c" validation but might be more accurately called "best (U, c)" validation. This method was popularized in 1985 by Milligan and Cooper (1985) and has many adherents. Suppose we have a set of (T) internal cluster validity indices $\mathrm{CVI_{int}} = \{V_j : 1 \le j \le T\}$. Different indices in this set will have different domains. For example, the Xie-Beni index V_{XBm} has three arguments $(U, \mathbf{V}; X)$, while the partition coefficient V_{PC} has U's in CP as its only argument. The following method can be used to evaluate internal CVIs.

Definition 6.11. Best c validation with internal CVIs using labeled data.

Given a set X or R of *labeled* data and any crisp clustering model and algorithm. Assume a set of T internal CVIs is being evaluated, say $\{V_j : 1 \le j \le T\}$:

(i) generate CPOV partitions $\mathrm{CP} = \{U \in M_{fcn} : c_m \le c \le c_M\}$; (Bc.1)

(ii) compute $\{V_j(U) : 1 \le j \le T;\ U \in \mathrm{CP}\}$; (Bc.2)

(ii) for each V_j, determine a "winner" partition $U_{w_j} \in M_{fc_{w_j}n}$ that has c_{w_j} CPOV clusters; (Bc.3)

(iv) if $c_{w_j} = c_t$ (synthetic) or $c_{w_j} = c_{gt}$(real), declare the validation test for V_j a success; (Bc.4)

(v) repeat steps (1)–(4) for N different sets of CPs generated by any means. Let N_j denote the number of successes at (Bc.4) for V_j in N trials; (Bc.5)

(vi) rank the $\{V_j : 1 \le j \le T\}$ based on $\{N_j : 1 \le j \le T\}$; i.e., their recovery of the designated number of labeled subsets in U_t or U_{gt}. (Bc.6)

This scheme is based on choosing a best internal CVI by letting each CVI pick a best U from CP without recourse to the structural information possessed by the labels in the partition U_t or U_{gt}. The only parameter that counts for success is whether or not the recovered value c_{w_j} matches the target value (c_t or c_{gt}) of the labeled data. While it seems wasteful to ignore the structural information possessed by the labels of the data, in the context of *real* clustering (with *unlabeled* data sets), this is the *only* way to validate candidates; so the underlying motivation for this type of validation is actually *to choose a "good" CVI* – which means, one you are willing to trust with unlabeled data[12]. So, the main point of this type of experiment is to develop some confidence in one or more internal CVIs for use in real clustering by comparing their ability to identify the best (U, c) pair without recourse to the labels on common sets of candidate partitions.

6.7.5 Crisp Cluster Validity Indices

How many validation methods for crisp partitions are there? In 1985, Hubert and Arabie (1985) began a paper on this topic as follows: "We will not try to review this literature comprehensively since that task would require the length of a monograph." It is hard to guess how many new CVIs have emerged in the 35 years or so since Hubert and Arabie made this statement; there are surely hundreds more now. Since this is an introductory text, we will illustrate a few basic CVIs that have been used successfully off and on through the years. Generally speaking, CVIs are validity functionals (V) that map some domain D into the real line $\Re$, $V{:}\mathrm{D} \mapsto \Re$. The domain of the function varies: for crisp partitions, it can be as simple as M_{hcn} but will usually have more arguments.

The objective here is to simply introduce you to the idea of how to interpret values of internal CVIs when applied to labeled data sets. And how they can (and can't) be employed in the service of cluster analysis. Consult one of the many good books on clustering for an in-depth treatment – e.g., Chapter 16 in Theodoridis and Koutrombas (2009).

> ☺ ***Forewarned is forearmed:*** No CVI is capable of identifying what seems to be a "best" CPOV partition of an arbitrary input data set X or D as the data vary across the infinite variety of structures they can and do possess. So, academicians are free to just keep inventing them, and so, they do. And, of course, new CVIs always work on the data sets used to introduce them; so why not? Each index is another paper for some lucky vita! Too cynical? Maybe not. The point is simple: when you enter the jungle of cluster validity, watch out. ☺

Crisp CVIs usually attempt to measure two geometric properties that are considered essential for "good" CPOV clusters, viz., *compactness* and *separation*. Let $X = \{\mathbf{x}_1, \ldots, \mathbf{x}_n\} \subset \Re^p$ be partitioned as $X = \bigcup_{i=1}^{c} X_i;\ X_i \cap X_j = \varnothing,\ i \neq j,\ |X_i| = n_i$ and $\bar{\mathbf{v}}_i = \sum_{\mathbf{x} \in X_i} \mathbf{x}/n_i\ \forall i$. Denote the set of cluster means as $\bar{\mathbf{V}} = \{\bar{\mathbf{v}}_1, \ldots, \bar{\mathbf{v}}_c\}$. Davies and Bouldin (1979) introduced a crisp internal CVI which is a function of the ratio of sums of pairwise within cluster scatter (a measure of compactness) to between cluster separation. Since scatter matrices depend on the geometry of the clusters, this index has both a statistical and geometric rationale,

$$V_{DB_{qt}}(U, \bar{\mathbf{V}}; X) = \left(\frac{1}{c}\right) \sum_{i=1}^{c} \left[\underbrace{\max}_{j, j \neq i} \left\{ (\alpha_{i,t} + \alpha_{j,t}) \Big/ \left(\left\| \bar{\mathbf{v}}_i - \bar{\mathbf{v}}_j \right\|_q \right) \right\} \right], \tag{6.53}$$

where $\alpha_{i,t} = \left(\sum_{\mathbf{x} \in X_i} \|\mathbf{x} - \bar{\mathbf{v}}_i\|^t / n_i \right)^{1/t}$. Here, t is a positive integer, $\|*\|^t$ is the *t*th power of the Euclidean norm, and $q \geq 1$ defines the Minkowski q-norm, $\|\bar{\mathbf{v}}\|_q = \left(\sum_{j=1}^{p} |\bar{v}_j|^q \right)^{1/q}$. Parameters q and t can be selected independently, but it is rare not to see $q = t = 2$ in the literature. The *Davies–Bouldin index* (DBI) at Equation (6.53) essentially averages c ratios of pairwise (compactness/separation) measures

[12]Think of using labeled data this way as a sort of validation chaperone. You (the labeled data) monitor your kid (a CVI) at a few dances and build up enough confidence to finally let him or her (the unlabeled data – sans chaperone) go it alone.

$(\alpha_{i,t} + \alpha_{j,t}) / \left(\|\bar{\mathbf{v}}_i - \bar{\mathbf{x}}_j\|_q\right)$. Since minimum within cluster dispersion and maximum between class separation are both desirable, low values of $V_{DB_{qt}}$ are taken as indicants of good cluster structure. In other words, $V_{DB_{qt}}$ is *min-optimal* (↓).

Dunn (1973) defined an internal CVI for crisp partitions based on roughly the same geometrical rationale as the Davies-Bouldin index $V_{DB_{qt}}$ in that they both characterize "good" partitions as sets of crisp clusters that are compact and well-separated. To describe this index, let X_i and X_j be non-empty subsets of $\Re^p$, and let $d : \Re^p \times \Re^p \to \Re^+$ be any metric on $\Re^p \times \Re^p$. Assume that $X = \{\mathbf{x}_1, \ldots, \mathbf{x}_n\} \subset \Re^p$ is partitioned as $U \in M_{hcn} \leftrightarrow X = \bigcup_{i=1}^{c} X_i;\ X_i \cap X_j = \varnothing,\ i \neq j$. The standard definitions of the *diameter* Δ of X_i and the *set distance* δ between X_i and X_j are

$$\Delta(X_i) = \underbrace{\max}_{\mathbf{x},\mathbf{y} \in X_i} \{d(\mathbf{x}, \mathbf{y})\} \text{ (diameter); and} \tag{6.54}$$

$$\delta(X_i, X_j) = \delta_{SL}(X_i, X_j) = \min_{\substack{\mathbf{x} \in X_i \\ \mathbf{y} \in X_j}} \{d(\mathbf{x}, \mathbf{y})\}. \tag{6.55}$$

The set distance between X_i and X_j at Equation (6.55) is often called the *single linkage* (SL) distance because it is used by the SL algorithm discussed in Chapter 8 (see Figure 8.4 for a depiction of this distance in the Euclidean case). For any partition $U \leftrightarrow X = X_1 \cup \cdots \cup X_c$, Dunn defined the separation index of U (universally known as *Dunn's index* (DI)) as

$$V_{DI}(U) = \frac{\underbrace{\min}_{i \neq j} \{\delta(X_i, X_j)\}}{\underbrace{\max}_{1 \leq k \leq c} \{\Delta X_k\}}. \tag{6.56}$$

Evidently, the geometric objective of Dunn's index is to maximize intercluster distances (big numerators = good separation) whilst minimizing intracluster distances (small biggest diameter = more compact clusters). The quantity $\delta(X_i, X_j)$ in the numerator of V_{DI} is analogous to $\|\bar{\mathbf{v}}_i - \bar{\mathbf{v}}_j\|_q$ in the denominator of $V_{DB_{qt}}$; the former measures the distance between clusters directly on the points in the clusters, whereas the latter uses the distance between their cluster centers for the same purpose. The use of ΔX_k in the denominator of Equation (6.56) is analogous to $\alpha_{k,t} = \left(\sum_{\mathbf{x} \in X_k} \|\mathbf{x} - \bar{\mathbf{v}}_k\|^t / n_i\right)^{1/t}$ in the numerator of $V_{DB_{qt}}$; both are measures of the scatter volume of cluster X_k. Thus, extrema of V_{DI} and $V_{DB_{qt}}$ share roughly the same geometric objective: maximizing intercluster distances whilst minimizing intracluster distances.

The measures of separation and compactness in Equation (6.56) occur "upside down" from their appearance in Equation (6.53); consequently, *large* values of V_{DI} correspond to good clusters. Hence, the number of clusters c* that maximizes V_{DI} is taken as the optimal value of c, i.e., V_{DI} is *max*-optimal (↑). Dunn's index is not defined on $\mathbf{1}_n$ when c = 1 or on I_n when c = n. The next definition describes a property related to V_{DI} that is important for some theoretical aspects of single linkage (Chapter 8) and Maximin sampling (Chapter 11).

Definition 6.12. Compact Separated (CS) Clusters (Dunn, 1973)

A set of crisp clusters $U \leftrightarrow X = X_1 \cup \cdots \cup X_c$ are *compact and separated* (CS) if, and only if : $\forall$ p, q, r with $q \neq r$, any pair of points x, y in X are closer together (as measured by d) than any pair of ponts u, v with u in X_q and v in X_r. Dunn asserted that this property of a data set was completely determined by the following equivalent criterion: X is CS $\Leftrightarrow \underbrace{\max}_{U \in M_{hcn}} \{ V_{DI}(U)\} > 1$.

While Dunn's index is easy to compute for a given c-partition of X, it is not so easy to ascertain whether a given data set X has a CS partition, because this requires (possibly) examining all of the good vertices of the cn soccer ball, which is impossible for data sets with all but the smallest cardinalities. But in the context of evaluating a set of candidate partitions, all the U's are known, so Dunn's index finds many applications in the literature. Dunn defined a more stringent index of separation he called *compact and well separated* (CWS) clusters:

$$V_{DI,CWS}(U;X) = \frac{\min\limits_{i \neq j}\{\mathrm{dist}(X_i, \mathrm{conv}(X_j))\}}{\underbrace{\max}_{1 \leq k \leq c}\{\Delta X_k)\}}. \tag{6.57}$$

Analogous to the CS case, Dunn stated that X can be partitioned into CWS clusters relative to (d) if and only if $\underbrace{\max}_{U \in M_{hcn}} \{ V_{DI,CWS}(U;X)\} > 1$. Evidently, $V_{DI,CWS}$ is also *max-optimal*. This index has a very attractive geometrical interpretation. However, estimation of the convex hull $\mathrm{conv}(X_j)$ from numerical input data for even p = 2 is computationally difficult; so $V_{DI,CWS}$ finds little use in practice and will not be considered further here.

Dunn's index is a function of three parameters: (d, δ, Δ). The metric is quite general; the set distance and diameter functions are much more limited. Bezdek and Pal (1998) proposed a set of 18 generalizations of Dunn's index using (a) = six measures of the distance between sets and (b) = three definitions for the diameter of a set. Let Δ_b be any positive semi-definite (diameter) function on $P(\Re^p)$, the *power set* of $\Re^p$ (the set of all subsets of $\Re^p$) and let δ_a denote any positive semi-definite, symmetric (set distance) function on $P(\Re^p) \times P(\Re^p)$. The *generalized Dunn's indices* (GDIs) are, for $U \in M_{hcn} \leftrightarrow X = \bigcup_{i=1}^{c} X_i$; $X_i \cap X_j = \varnothing,\ i \neq j$,

$$V_{GDI_{ab}}(U) = \frac{\underbrace{\min}_{i \neq j}\{\delta_a(X_i, X_j)\}}{\underbrace{\max}_{1 \leq k \leq c}\{\Delta_b(X_k)\}};\ 1 \leq a \leq 6;\ 1 \leq b \leq 3. \tag{6.58}$$

The choice a = b = 1 corresponds to Dunn's original index at Equation (6.56), $V_{GDI_{11}}(U) = V_{DI}(U)$. The set distances for a = 2 and a = 3 are the complete and average linkage distances shown in Figure 8.4. Diameter $\Delta_2(X_k)$ is the average of the distances in subset X_k; $\Delta_3(X_k)$ is the average distance between the points in X_k and its centroid $\bar{v}_k = \sum_{x \in X_k} x/n_k$. This diameter is relatively insensitive to the addition or deletion of a few aberrant points and is a measure of the scatter volume of X_k.

The GDIs are all basically ratios of pairwise (separation/compactness); so in some sense, they all measure the quality of crisp partitions in a way that is inverse to $V_{DB_{qt}}$ – good partitions of X have high values of $V_{GDI_{ab}}$, i.e., they are all max-optimal (↑) indices. Experimental evidence in several studies suggests that the most reliable combinations of a and b in Equation (6.54) correspond to $V_{GDI_{33}}$ and $V_{GDI_{53}}$.

For example, in Arbelaitz *et al.* (2013), $V_{GDI_{53}}$ is rated fourth best overall in a set of 30 internal CVIs, which were extensively tested over many synthetic and real data sets. The index $V_{GDI_{53}}$ plays an important role in incremental stream monitoring functions in Ibrahim *et al.* (2020). Bezdek and Pal (1998) give an extensive discussion of the 18 GDIs, and Rathore *et al.* (2019a) develop some approximation algorithms for the GDIs in the big data case.

Example 6.6. Crisp cluster validity for HCM on Iris.

The input data set is the Iris data. Before we discuss post-clustering evaluation of HCM partitions of Iris, let us have a look at a pre-clustering visual assessment of the CPOV structure that Iris might possess (cf., problem 8 of Chapter 5). The Euclidean distance matrix $D_{E,\ Iris}$ is computed on pairs of vectors in Iris.

The VAT image of Iris in Figure 6.8(a) is somewhat blurry and has disrupted intensity patterns, but the diagonal blocks do suggest that Iris is primarily composed of c = 2 clusters. The suggestion of the iVAT image in Figure 6.8(b) is unequivocal: it tells us to expect c = 2 CPOV clusters found by clustering Iris with most algorithms. The sizes of the two dark blocks – 50 × 50 pixels in the lower right and 100 × 100 pixels in the upper left – correspond exactly to the numbers of vectors in Setosa and {Versicolor}∪{Virginica}.

Please compare Figures 5.7(a) and 6.8. Figure 5.7(a) suggests that this is the case based on two of the four input features: Figure 6.8 is stronger evidence, since the images shown here are based on all four input features. Table 6.5 shows typical values of $V_{DB_{22}}(U)$, $V_{GDI_{11}}(U)$, and $V_{GDI_{33}}(U)$ obtained by exercising HCM (k-means) algorithm A6.1 (m = 1) with its default parameters on Iris to get partitions for nine values of c in the integer range $2 \leq c \leq 10$. These are typical (unaveraged) values that depend on the initializations used, but they suffice to illustrate the idea of using internal CVIs to assess sets of candidate partitions.

The best value indicated by each index is shown in boldface and highlighted. Apparently, Dunn's original index $V_{GDI_{11}} = V_{DI}$ weakly chooses c = 3, the number of physical subsets in the labeled data. Note that $V_{GDI_{11}} = V_{DI}$ also maximizes at c = 7 and experiences a variation of at most 0.04 over the tested values for c. The other two CVIs point more emphatically to c = 2, the number of CPOV clusters that HCM prefers in Iris for the initialization used. It would, of course, be much more informative to repeat these calculations 100 times with different initializations for HCM, average the results, and report the averages.

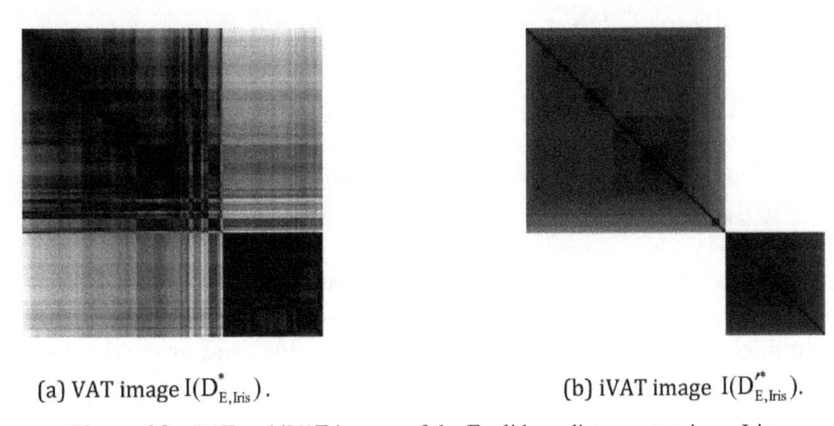

(a) VAT image $I(D^{*}_{E,Iris})$. (b) iVAT image $I(D'^{*}_{E,Iris})$.

Figure 6.8 VAT and iVAT images of the Euclidean distance matrix on Iris.

Table 6.5 Typical values of three internal crisp CVIs on HCM Partitions of X_{Iris}.

c	Type	2	3	4	5	6	7	8	9	10
$V_{DB_{22}}$	(↓)	**0.47**	0.73	0.84	0.99	1.00	0.96	1.09	1.25	1.23
$V_{GDI_{11}} = V_{DI}$	(↑)	0.08	**0.10**	0.08	0.06	0.09	**0.10**	0.08	0.06	0.06
$V_{GDI_{33}}$	(↑)	**2.00**	1.32	0.99	0.63	0.53	0.53	0.39	0.34	0.31

Would the conclusion be different? No, but HCM will likely have several different local trap states for 100 initializations; so there probably will not be 100% agreement. The main lesson you should take away from this example is that different CVIs will vote for different best partitions residing in *the same set of candidates*. This is not an aberration: it happens – *a lot* – in the jungle of cluster validity.

6.7.6 Soft Cluster Validity Indices

When the U's in CP are soft (fuzzy or probabilistic), there are two fundamentally different ways to evaluate them. First, the function **H** at Equation (3.4) can be employed to harden each column of $U \in M_{fcn} \to \mathbf{H}(U) \in M_{hcn}$. Then $\mathbf{H}(U)$ can be evaluated with any direct internal or external method available, as in Example 6.6, for instance. Other defuzzifications of U (e.g., α-cuts at different levels) can produce different crisp partitions and, hence, different values for validity indices when soft partitions are converted to crisp ones.

The alternative to hardening U is to conduct an evaluation of U using some function of the soft partitions. We will discuss a couple of these here, just to convey the flavor of this second approach. Ruspini (1969) opined that an entropy function would be useful for assessing the quality of fuzzy partitions. Bezdek (1974) broadened Ruspini's idea with a more formal approach to validity methods associated with soft partitions. Here are his definitions of the *partition coefficient (PC) and partition entropy (PE)* :

$$V_{PC}(U) = \frac{1}{n}\left(\sum_{k=1}^{n}\sum_{i=1}^{c} u_{ik}^2\right) = \frac{\|U\|_2^2}{n} = \frac{tr(UU^T)}{n}. \tag{6.59}$$

$$V_{PE}(U) = -\frac{1}{n}\left(\sum_{k=1}^{n}\sum_{i=1}^{c}[u_{ik}\ \ln_a(u_{ik})]\right);\ a \in (1,\infty). \tag{6.60}$$

Properties of these internal CVIs are well-known:

$$V_{PC}(U) = 1 \Leftrightarrow V_{PE}(U) = 0 \Leftrightarrow U \in M_{hcn}(\text{crisp}); \tag{6.61}$$

$$V_{PC}(U) = 1/c \Leftrightarrow V_{PE}(U) = \ln_a(c) \Leftrightarrow U = \bar{U} = [1/c]. \tag{6.62}$$

V_{PC} and V_{PE} are examples of a class of internal CVIs that are functions of $U \in M_{fcn}$ alone. A strong and well-justified criticism of indices like these is that, while U is a function of the data set X, functionals such as V_{PC} and V_{PE} are only implicitly functions of X. In other words, they don't use or see the data itself as part of the validation process. Since the PC takes its maximum (and the PE its minimum) on every hard c-partition, these CVIs can't discriminate between the various good vertices of the cn-soccer ball in Figure 6.2. At the other extreme, the PC takes its unique minimum (and PE its unique maximum) at the centroid $\bar{U} = [1/c]$ of the cn-soccer ball shown in Figure 6.2.

The heuristic for these CVIs is that V_{PC} is *max-optimal* (↑), and V_{PE} is *min-optimal* (↓); that is, higher values of the PC (lower values of PE) are assumed to indicate "better" partitions of the data. This heuristic is clearly open to question since there are an infinite number of different partitions that produce the same value of the PC (or of the PE) for every value in their ranges.

It is not clear that, for example, as $V_{PC}(U)$ approaches 1, the corresponding CPOV partition U is getter "better" as it becomes harder (closer to a crisp c-partition or vertex in M_{hcn}). Since $V_{PC}(U) = 1$ for *every* U in M_{hcn}, we can't assert that just because the PC is close to 1, U is a good clustering of X. Empirical studies vary: some have shown that maximizing the PC (or minimizing the PE) over $CP \subset M_{fcn}$ sometimes (but not always) leads to a good interpretation of the data. But there are also studies where the PC and/or PE are dead last in validity races. At the end of the day, the PC and PE and their relatives are essentially just measures of fuzziness. But these two indices have enjoyed on-again, off-again success over the years and are still actively included in many validation experiments.

At the other extreme, however, it is clear that when an algorithm terminates at a partition that is close to $\bar{U}$, that algorithm is not finding very distinct CPOV clusters in X. This may be the fault of the algorithm for any of the reasons we have enumerated (bad features, bad initialization, bad parameters, wrong algorithm, and so on). Or ... the data may actually lack enough structure to satisfy the model underlying the analysis. In any case, as the PC or PE approaches the centroid of the cn-soccer ball, its unique minimum (or maximum) helps us see that CPOV structure is ***not*** being found, and this is useful. So, the most common use of these soft CVIs is to eliminate bad CPOV partitions as ones that are just "too soft" to be of much value in terms of interpreting data structure.

Yet another aspect of these CVIs is their specific relationship to FCM and, in particular, to the weighting exponent m of the memberships in A6.1. To see this, consider the relationship of the partition coefficient to this parameter. Let $\{U_m\} \subset M_{fcn}$ be a sequence of partitions corresponding to $\{J_m(U_m, \mathbf{V}_m; X)\}$. Then

$$\lim_{m \xrightarrow{+} 1} \{V_{PC}(U_m)\} = 1; \tag{6.63a}$$

$$\lim_{m \to \infty} \{V_{PC}(U_m)\} = 1/c. \tag{6.63b}$$

The PC is not a function of m, but the PC is often used evaluate FCM partitions, which are functions of m. The limits in Equation (6.63) show two things. When the PC arrives at 1 in (6.63a), its value is constant on all crisp U's for every c, so it can't show a preference for a particular choice of c. At the other extreme, when m becomes large, the PC will select c = 2 even when the FCM candidates range across c because of the second limit in Equation (6.63b). For example, the PC will maximize at 1/2 because as m approaches infinity with c = 2, the second limit at Equation (6.63a) yields 1/2, and larger values of c will yield smaller values of the PC. What to do? Add more ingredients to the CVI.

Many algorithms generate collateral information in the form of computed parameters – e.g., the cluster centers **V** from the c-means models – that are useful for validation, as we have seen, for example, in the case of the Davies–Bouldin index. One of the most highly regarded measures of fuzzy cluster validity that uses all three c-means variables (U, **V**; X) is the well-known internal CVI of Xie and Beni (1991). Let $U \subset M_{fcn}$

and $\mathbf{V} = \{\mathbf{v}_1, \ldots, \mathbf{v}_c\} \subset \Re^{cp}$ be a vector of c distinct points (for us, these will be cluster centers from FCM), and let $\|\mathbf{x}\|^2 = \mathbf{x}^T\mathbf{x}$ be the Euclidean norm. The *extended Xie–Beni index* for FCM (Pal and Bezdek, 1995) is

$$V_{XB_m}(U, \mathbf{V}; X) = \frac{\sum_{i=1}^{c}\sum_{k=1}^{n} u_{ik}^m \|\mathbf{x}_k - \mathbf{v}_i\|^2}{n\left(\underbrace{\min}_{i \neq j}\left\{\|\mathbf{v}_i - \mathbf{v}_j\|^2\right\}\right)} = \frac{J_m(U, \mathbf{V}; X)}{n\left(\underbrace{\min}_{i \neq j}\left\{\|\mathbf{v}_i - \mathbf{v}_j\|^2\right\}\right)}; \; 1 \leq m < \infty. \tag{6.64}$$

The original version of $V_{XB_m}(U, \mathbf{V}; X)$ was defined in Xie and Beni (1991) for the case m = 2, $V_{XB_2}(U, \mathbf{V}; X)$). Xie and Beni pointed out that while $V_{XB_2}(U, \mathbf{V}; X)$ is functionally related to the FCM objective function for m = 2, it could also be used on fuzzy partition pairs (U, **V**) found by any clustering algorithm that produces $U \subset M_{fcn}$ and $\mathbf{V} = \{\mathbf{v}_1, \ldots, \mathbf{v}_c\} \subset \Re^{cp}$. In particular, V_{XB_1} becomes a worthwhile internal CVI for k-means at m = 1. While the model norm specified by Xie and Beni was Euclidean, any inner product norm can be used in Equation (6.64).

Xie and Beni explained the original index by writing it as a ratio of the (renamed) *total generalized variation* $\sigma_2(U, \mathbf{V}; X)$ of (U, **V**), and sep(**V**), a measure of the separation of the prototype vectors **V**:

$$\sigma_2(U, \mathbf{V}; X) = \sum_{i=1}^{c}\left(\sum_{k=1}^{n} u_{ik}^2 \|\mathbf{x}_v - \mathbf{v}_i\|^2\right) = J_2(U, \mathbf{V}; X); \tag{6.65}$$

$$\text{sep}(\mathbf{V}) = \underbrace{\min}_{i \neq j}\left\{\|\mathbf{v}_i - \mathbf{v}_j\|^2\right\}. \tag{6.66}$$

According to this interpretation, a good (U, **V**) pair should produce a small value of σ_2 (and, more generally, σ_m) because that is what small values of the objective function J_m are supposed to indicate. And well-separated $\mathbf{v}_i$'s will produce a high value of sep(**V**). More generally, when $V_{XB_m}(U_1, \mathbf{V}_1; X) < V_{XB_m}(U_2, \mathbf{V}_2; X)$ for either of these reasons (or both), U_1 is presumably a better partition of X than U_2. Consequently, the *minimum* of V_{XB_m} over a set of U's $\in CP \subset M_{fcn}$ is taken as the most desirable partition of X (i.e., V_{XB_m} is *min-optimal* (↓)). Here are some limit properties of Xie –Beni CVIs (cf., Pal and Bezdek (1995) for proofs):

$$\underset{m \xrightarrow{+} 1}{\text{limit}}\{V_{XB_m}(U, V; X)\} = \underset{m \xrightarrow{+} 1}{\text{limit}}\{V_{XB_2}((U, \mathbf{V}; X)\} = \frac{J_1(U, \mathbf{V}; X)}{n\left(\underbrace{\min}_{i \neq j}\left\{\|\mathbf{v}_i - \mathbf{v}_j\|^2\right\}\right)}; \tag{6.67}$$

$$\underset{m \to \infty}{\text{limit}}\{V_{XB_2}(U, \mathbf{V}; X)\} = \frac{\sum_{k=1}^{n}\|\mathbf{x}_k - \bar{\mathbf{m}}\|^2 \Big/ nc}{0} = \infty; \tag{6.68a}$$

$$\underset{m \to \infty}{\text{limit}}\{V_{XB_m}(U, \mathbf{V}; X)\} = \frac{0}{0} = \text{indeterminate}. \tag{6.68b}$$

In Equation (6.68a), the vector $\bar{m} = \sum_{x \in X} x/n$ is the grand mean of the input data. The reason that the limits in Equations (6.68a) and (6.68b) are different is because the membership exponent for V_{XB_2} has a fixed exponent (m = 2), whereas the memberships for V_{XB_m} have m as a variable. These limit properties show that when the set $CP \subset M_{fcn}$ is generated by FCM algorithm A6.1 (m > 1), validation of partitions can be indirectly influenced by the model parameter m which is used to find the U's in CP. Interestingly, if the candidates come from a different clustering model (e.g., the GMD/EM algorithm discussed in Chapter 7), these limit properties don't affect the quality of evaluations made by the Xie–Beni CVIs as much as they do for fuzzy c-means.

Example 6.7. Fuzzy cluster validity for FCM on Iris.

The input data is again the Iris data. This example studies the use of three soft internal CVIs: $V_{PC}(U)$, $V_{XBm}(U)$, and $V_{XB2}(U)$. Candidate fuzzy partitions $U \in CP \subset M_{fcn}$ are generated over the integer range $2 \leq c \leq 10$ as in Example 6.6. The parameters of FCM for this example: T = 100, ε = 0.00001, the model norm $\|\mathbf{x}\|^2 = \mathbf{x}^T\mathbf{x}$ is the Euclidean norm, $\|*\|_{err} = \|\mathbf{V}_{t+1} - \mathbf{V}_t\|_1$ is the 1-norm on $\Re^{cp}$, and the default choice for $\mathbf{V}_0$ is a set of c randomly chosen distinct points in X. For a particular c, the same initial centroids were used for runs at m = 1.2 and m = 7.

Table 6.6 lists values of the three indices on FCM candidates for two values of m. For m = 1.2, all three indices behave well and point to c = 2. The behavior of both Xie–Beni indices as functions of c for m = 7 is very peculiar. For example, V_{XB2} is 0.10 for c = 2, grows to 808,502 at c = 8, and then plummets back to nearly 0 (2.54 *is* nearly zero relative to the values in this row) at c = 10. The limiting behavior of V_{XBm} shown at Equation (6.68b) is on display in the third row for m = 7, viz., it is really unpredictable. Looking at the trend of values, leaves you with the very correct impression that there is a very strong and unpredictable interaction between c and m for both of these CVIs.

Table 6.6 Typical values of three internal soft CVIs on FCM partitions of X_{Iris}.

m=1.2

c		2	3	4	5	6	7	8	9	10
V_{PC}	(↑)	**0.99**	0.98	0.96	0.95	0.95	0.95	0.95	0.96	0.94
V_{XB_2}	(↓)	**0.01**	0.12	0.18	0,48	0.72	1.12	2.05	1.40	1.16
V_{XB_m}	(↓)	**0.01**	0.13	0.19	0.51	0.77	1.20	2.19	1.50	0.96

m=7

c		2	3	4	5	6	7	8	9	10
V_{PC}	(↑)	**0.54**	0.37	0.27	0.22	0.18	0.16	0.14	0.13	0.14
V_{XB_2}	(↓)	**0.10**	0.62	1.13	7.98	26326	32470	808502	17599	2.54
V_{XBm}	(↓)	0.0013	0.0016	0.0006	0.0015	2.0163	1.0905	14.4105	0.1645	**0.0000**

To summarize, very low or high values of m may influence CVIs used to evaluate candidate partitions produced by FCM. Moreover, indices that also use **V** from FCM may experience additional problems.

And, finally, indices that are *explicit* functions of m and (U, **V**) may be very unreliable and/or unpredictable for small or large values of m[13].

☺ ***Forewarned is forearmed.*** Here is the takeaway lesson from Example 6.7. Some validity indices have surprising and sometimes alarmingly unpredictable dependency on elements of the clustering model and algorithm that seem at first glance to be rather unrelated to their job – which is to tell you whether or not to believe the outputs. This type of behavior should serve as a strong warning about the care you need to exercise when using internal CVIs, which are sometimes "VUI" – *validating under the influence* of hidden variables (such as m in Example 6.7). ☺

Another example of hidden influences on CVIs is offered by looking more carefully at the PC and PE indices. The ranges of the partition coefficient V_{PC} and partition entropy V_{PE} depend monotonically on c. Specifically, for every $U \in CP \subset M_{fcn}$, $V_{PC}(U) \in [1/c, 1]$, $V_{PE}(U) \in [0, \ln_a(c)]$. As discussed in Section 6.7.1, dependence of these CVIs on the problem parameter c makes interpretation of the relative quality of U's difficult to evaluate. To address this problem, Roubens (1979) defined a normalization of scale for V_{PC} (which he called a "Non-fuzziness Index" in his paper[14])

$$V_{PC,R}(U) = \left(\frac{cV_{PC}(U) - 1}{c - 1}\right). \tag{6.69}$$

And Bezdek (1974) defined a similar normalization of the partition entropy

$$V_{PE,B}(U) = V_{PE}(U)/\ln_a c. \tag{6.70}$$

Both of these indices are valued in [0, 1] for all values of c and have the following bounds:

$$U \in M_{hcn} \Rightarrow V_{PC,R}(U) = 1\, ;\, V_{PE,B}(U) = 0\, ; \tag{6.71}$$

$$\bar{U} = [1/c] \Rightarrow V_{PC,R}(\bar{U}) = 0\, ;\, V_{PE,B}(\bar{U}) = 1. \tag{6.72}$$

$V_{PC,R}$ is max-optimal (↑) and $V_{PE,B}$ is min-optimal (↓). These simple normalized CVIs also have a checkered history in the literature but still find successful applications for specific data sets. For example, Friedman *et al.* (2015) employ $V_{PC,R}$ to decide how many clusters of individual neurons exist in a set of neuron spike trains. Here is an example excerpted from Moshtaghi *et al.* (2015) to illustrate.

Example 6.8. Best c validation for 2 data sets with 15 synthetic clusters.

Figure 6.9(a) and (b) shows scatterplots of two synthetic data sets of n = 5000 points in $\Re^2$ that have c = 15 HPOV clusters. These data were introduced by Fränti and Virmajoki (2006) and have become somewhat of

---

[13] Perhaps you are wondering what a low or high value of m is? Most users of FCM choose values of m in the range (1,10]. One or two studies have tested FCM with m up to about 30, but values of m < 5 or so are more usual, and m = 2 is by far the most common choice. For most users, m = 1.05 is a small value of m, and 5 is a large value of m.

[14] Dave (1996) reinvented Rouben's index and called it the modified partition coefficient. This appears to be a case of innocent and independent duplication.

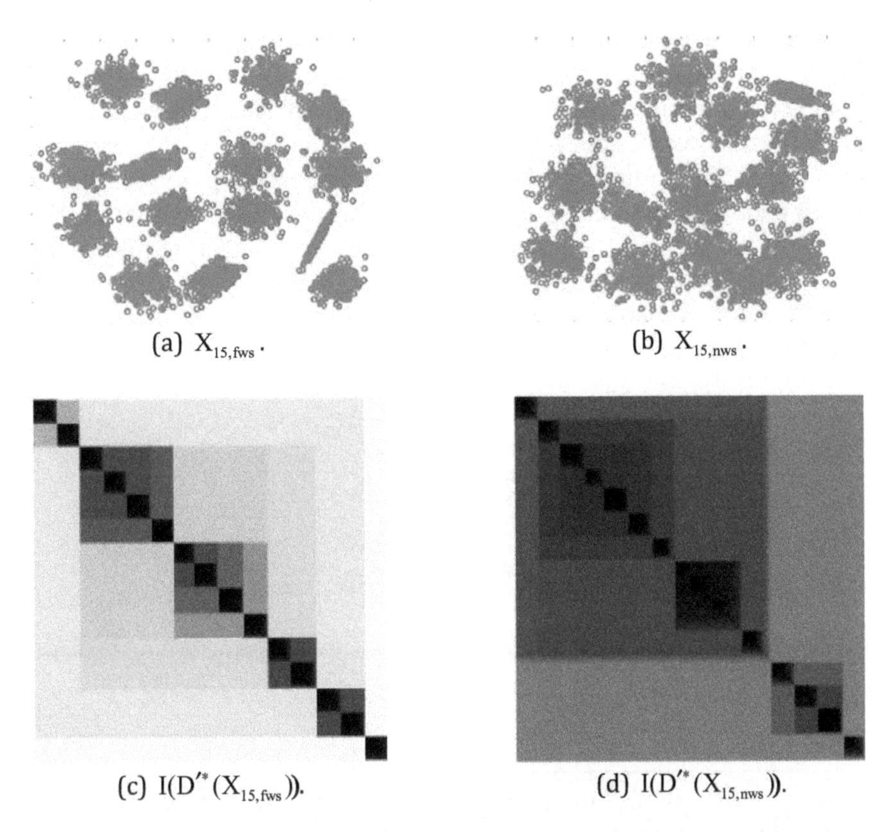

(a) $X_{15,fws}$. (b) $X_{15,nws}$.

(c) $I(D'^{*}(X_{15,fws}))$. (d) $I(D'^{*}(X_{15,nws}))$.

Figure 6.9 Scatterplots and iVAT images of two sets of HPOV Gaussian clusters.

a favorite in the clustering literature. View 6.9(a) shows $c = 15$ *fairly well separated* (fws) HPOV clusters ($X_{15,fws}$) that are samples of 15 2D Gaussian distributions of varying shapes. View 6.9(b) is a second data set ($X_{15,nws}$) with much less clarity and separation between the 15 subsets; these clusters are *not well separated* (nws). These two data sets and their "true" crisply labeled HPOV partitions of $c_t = 15$ subsets can be downloaded at http://cs.joensuu.fi/?isido/clustering/. Dunn's index at Equation (6.56) is 0.0084 for $X_{15,fws}$ and 0.0021 for $X_{15,nws}$. So, the ground truth partitions for both sets are not compact and separated in the sense of Dunn's definition. (This doesn't necessarily imply there is no CS partition of either one.)

Figures 6.9(c) and 6.9(d) are iVAT images of the Euclidean distance matrices of the two data sets. The primary structure in the iVAT image in view (c) clearly suggests that $X_{15,fws}$ contains 15 fairly distinct clusters. There are five secondary structures that apparently subsume 14 of the individual clusters in larger groups of 2, 4, 4, 2, and 2 subsets, reading from top to bottom, left to right. The structure suggested by the iVAT image of $X_{15,nws}$ in view 6.9(d) is much less well-defined. Fifteen individual clusters are still recognizable in this image, but the boundaries of the diagonal sub blocks are almost all "fuzzy" (in the visual sense). Comparing these two images certainly conveys an accurate idea of the difference between the two data sets and leads us to expect that most validity indices will point to $c_t = 15$, which is the known number of HPOV subsets as constructed in each of these two data sets.

Fuzzy c-means A6.1 was applied to these two data sets using the default choices for all parameters, except that a finer termination threshold, viz., $\varepsilon = 0.00001$ was used. Each data set was processed

N = 10 times. Initialization of each run was done by randomly selecting c distinct vectors in the input data as initial cluster centers. Five internal CVIs are used in this example: the original and scaled partition entropies, the original and scaled partition coefficients, and the original Xie–Beni index. Table 6.7 shows the best c's selected by the 10 run averages of each index.

The normalized versions of the partition coefficient and partition entropy select the target value of c ($c_w = c_t = 15$) for both data sets while the original unnormalized CVIs are wrong in three of the four sets of 10 trials. Interestingly, the average Xie–Beni index $\bar{V}_{XB,2}$ selects the target value for the noisy data but underestimates c by 1 for the more well-separated data shown in Figure 6.9(a). The unscaled partition coefficient behaves oppositely, and the unscaled partition entropy (horribly) selects c = 2 for both input sets. Table 6.8 affords a closer look at the average values over 10 trials of the five CVIs for each value of c in the study range, c = 2 to c = 20.

Table 6.7 Best c choice of five CVIs on $X_{15,fws}$ and $X_{15,nws}$: 10 run average.

	c_t	$\bar{V}_{PE}$	$\bar{V}_{PE,B}$	$\bar{V}_{PC}$	$\bar{V}_{PC,R}$	$\bar{V}_{XB,2}$
$X_{15,fws}$	**15**	2	**15**	**15**	**15**	14
$X_{15,nws}$	**15**	2	**15**	2	**15**	**15**

Table 6.8 Average values of CVIs over 10 runs of FCM on $X_{15,fws}$ and $X_{15,nws}$.

	$X_{15,fws}$					$X_{15,nws}$				
c	$\bar{V}_{PE}$	$\bar{V}_{PE,B}$	$\bar{V}_{PC}$	$\bar{V}_{PC,R}$	$\bar{V}_{XB,2}$	$\bar{V}_{PE}$	$\bar{V}_{PE,B}$	$\bar{V}_{PC}$	$\bar{V}_{PC,R}$	$\bar{V}_{XB,2}$
2	**0.46**	0.67	0.67	0.40	0.30	**0.45**	0.65	**0.71**	0.42	0.25
3	0.68	0.62	0.62	0.42	0.14	0.63	0.58	0.64	0.47	0.15
4	0.80	0.58	0.58	0.44	0.11	0.78	0.56	0.60	0.46	0.12
5	0.87	0.54	0.58	0.47	0.11	0.91	0.56	0.55	0.44	0.16
6	0.93	0.52	0.57	0.48	0.11	0.97	0.54	0.54	0.45	0.16
7	0.97	0.50	0.56	0.49	0.14	1.01	0.52	0.54	0.46	0.13
8	0.96	0.46	0.58	0.52	0.12	1.03	0.49	0.55	0.49	0.12
9	0.97	0.44	0.59	0.54	0.15	1.03	0.47	0.56	0.51	0.12
10	0.93	0.40	0.62	0.57	0.12	1.01	0.44	0.58	0.53	0.13
11	0.90	0.37	0.64	0.61	0.11	1.00	0.42	0.59	0.55	0.13
12	0.85	0.34	0.67	0.64	0.09	0.95	0.38	0.62	0.59	0.10
13	0.79	0.31	0.70	0.67	0.07	0.92	0.36	0.64	0.61	0.10
14	0.71	0.27	**0.73**	0.71	**0.06**	0.89	0.34	0.66	0.63	0.09
15	0.66	**0.24**	**0.75**	**0.74**	0.19	0.83	**0.31**	0.69	**0.66**	**0.07**
16	0.71	0.26	0.73	0.71	0.82	0.87	0.32	0.67	0.65	0.36
17	0.74	0.26	0.72	0.70	0.90	0.92	0.32	0.65	0.63	0.48
18	0.77	0.27	0.71	0.69	0.83	0.97	0.33	0.63	0.61	0.66
19	0.81	0.27	0.69	0.68	0.84	1.00	0.34	0.62	0.60	0.68
20	0.85	0.28	0.68	0.66	0.87	1.04	0.35	0.60	0.58	0.75

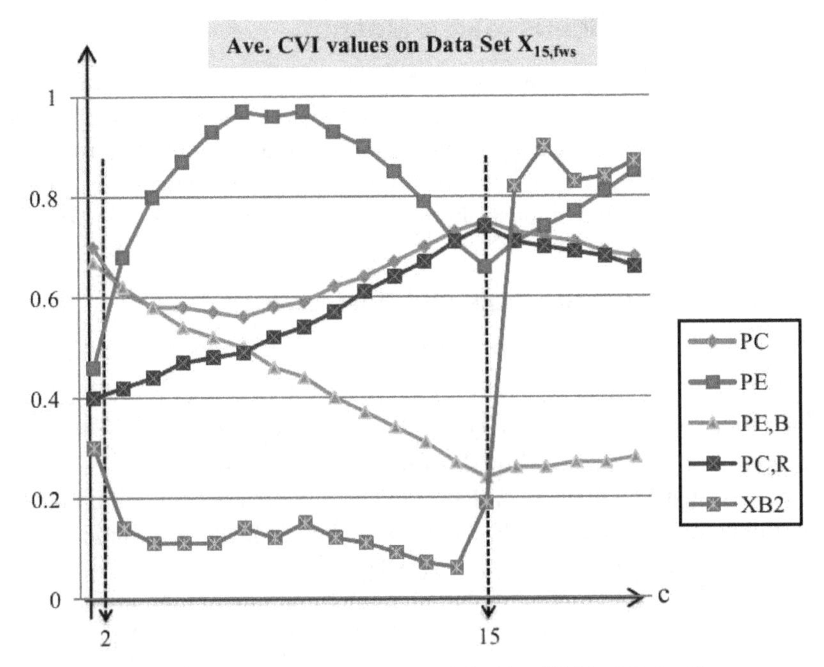

Figure 6.10 Graphs of the CVIs (the left side of Table 6.8) for $X_{15,fws}$.

Sometimes it's hard to learn a lot by staring at tables of numbers. Let's have a look at graphs of the values in the columns on the left side of this table. Figure 6.10 plots the values of the first five columns in Table 6.8 corresponding to average values of the CVIs over 10 runs on data set $X_{15,fws}$. The red curve for partition entropy has a global minimum (0.46) at c = 2, rises to a global maximum (0.97) at c = 7 and again at c = 9, then drops to a well-defined local minimum (a "knee") at c = 15, and, finally, rises steadily until c = 20. The green curve – the normalized partition entropy – is much more well-behaved. This function declines steadily until it reaches a global minimum (0.24) at c = 15 and then gradually increases for c from 16 to 20. So, in this instance anyway, scale normalization has been quite effective in rectifying a misleading interpretation of candidate partitions offered by the partition entropy.

Both forms of the partition coefficient take global maximums at c = 15 and present a very tidy idea about the candidates presented for evaluation. The most interesting and somewhat puzzling graph is the blue one – the Xie–Beni index, which looks quite erratic and takes a fairly well-defined global minimum (0.06) at c = 14 ... one stop too early! This is typical for the Xie–Beni index, which can be quite effective or very much off in its analysis of candidate partitions from experiment to experiment.

The strongest conclusion you should draw from Examples 6.5–6.7 is that internal CVIs can be quite unpredictable. If you had to pick one (or several) to evaluate candidate partitions in real applications with unlabeled data, how would you do that? The short answer is that we can evaluate competing internal CVIs using the best c method with labeled data as was done here and simply adopt the internal CVI that seems to win the most internal CVI races. But a much better way is to use the best match *internal/external* (BI/E) strategy that is defined and illustrated in Chapter 9. This approach enables you to choose an internal index that retrieves U_{GT} as assessed and agreed upon by an external helper criterion.

6.8 Alternate Forms of Hard c-Means (aka k-Means)

So many papers about crisp c-mean (k-means) appear every year that it is hard to decide what else should be included about it this volume. There seem to be an unending number of interesting new developments about A6.1 for Lloyd's case, $m = 1$. This section discusses several models that present the c-means objective function appearing at Equation (6.9) in alternate forms. We begin with a theorem that appears in Boutsidis (2010), which purports to bound the value of J_1 that can be obtained on a randomly projected subset of input data.

6.8.1 Bounds on k-Means in Randomly Projected Downspaces

The objective function identified as the k-means objective function by Boutsidis *et al.* (2010) is somewhat different than the formulation of $J_1(U, V; X) = \sum_{k=1}^{n} \sum_{i=1}^{c} u_{ik} \|x_k - v_i\|_A^2$ at Equation (6.9). They formulate the k-means objective function as $J_{1B}(W; X) = \|X - WW^T X\|_2^2$, where W is an $(n \times c)$ indicator (membership) matrix which is related, but not equal, to the crisp c-partition matrix $U^T \in M_{hcn}$, and data set X is regarded as an $(n \times p)$ matrix. Their formulation of the c-means optimization problem at Equation (6.9) calls for finding an indicator matrix $W^* \in \Psi$, where Ψ is the set of all such indicator matrices, which minimizes $J_{1B}(W; X)$. That is, $J_{1B}(W^*; X) = \min_{W \in \psi} \left\{ \|X - WW^T X\|_2^2 \right\}$.

Please note that the function $J_{1B}(W; X)$ doesn't appear to contain the means (the cluster centers **V**), and the algorithm shown by Boutsidis *et al.* (***Algorithm 1****: A random projection algorithm for k-means clustering*) doesn't return cluster centers; so it is hard to see the equivalence of their formulation with the one given at Equation (6.9). Nonetheless, their numerical examples used MATLAB's *k-means* function, and they state that they initialize the cluster centers with points chosen from the input data in a specific way.

The main objective in Boutsidis *et al.* (2010) is a theorem that relates the optimal value $J_1(W_p^*; X)$ discovered by running their algorithm on the upspace data $X = \{x_1, \ldots x_n\} \subset \Re^p$, to the value $J_1(W_q^*; X)$, where the partition W_q^* is discovered by running c-means on the randomly projected downspace data $Y = R[X] = \{y_1, \ldots, y_n\} \subset \Re^q$ as described in Theorem 5.1[15]. Recall that in Example 5.6, the transformation $R_{q \times p} = [r_{ij}] = [\text{rand}(N(0, 1)]$ was the basis for the random projection shown there. Achlioptas (2003) proved that a simpler transformation could be used to secure a JL certificate. His projection matrix was constructed by choosing each entry as $R_{ij} = \pm \left(1/\sqrt{q}\right)$ with probability 1/2. With this construction for the downspace data, Boutsidis *et al.* (2010) proved the following theorem.

Theorem 6.5. (2+ϵ) bound for optima of $J_{1B}(W; X)$ in a downspace with a JL certificate.

Let the upspace data $X = \{x_1, \ldots, x_n\} \subset \Re^p$ be represented as an $n \times p$ data matrix. Extract a randomly projected downspace data set $Y = R[X] = \{y_1, \ldots, y_n\} \subset \Re^q$, using the Achlioptas (2003) projection matrix $R_{ij} = \pm \left(1/\sqrt{q}\right)$ where $q \geq q_0 = Kc/\varepsilon^2$, $c < \min\{p, n\}$, and $0 < \varepsilon < 1/3$. K is a "sufficiently large constant" that insures a JL certificate for Y. Let $J_1(W_p^*; X) = \min_{W \in \psi} \left\{ \|X - WW^T X\|_2^2 \right\}$ and $J_1(W_q^*; Y) = \min_{W \in \psi} \left\{ \|Y - WW^T Y\|_2^2 \right\}$. Then with Prob $\geq (0.97 - \delta)$ and $\gamma = 1 + \varepsilon$

$$J_1(W_q^*; X) \leq (2 + \varepsilon) J_1(W_p^*; X), \text{ i.e., } \left\|X - W_q^* W_q^{*T} X\right\|_2^2 \leq (2 + \varepsilon) \left\|X - W_p^* W_p^{*T} X\right\|_2^2. \quad (6.73)$$

[15] Evaluating R at each element of X in its domain produces a set called the image of X under R, denoted here as Y = R[X].

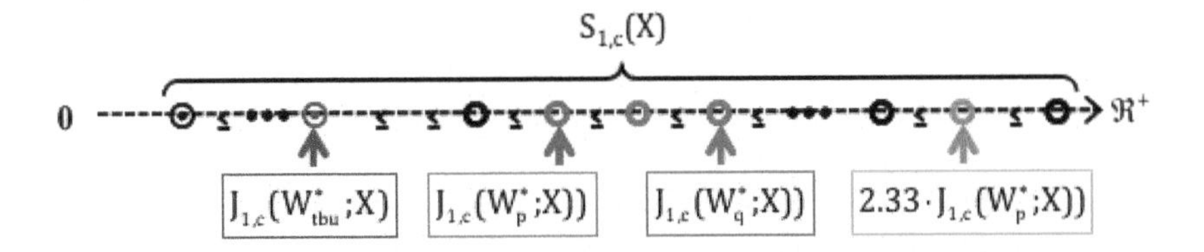

Figure 6.11 Graphical representation of Theorem 6.5 when $c \leq \min\{p, n\}$.

Proof. See Theorem 1, Section 4 of Boutsidis *et al.* (2010).

What does it mean? Please note that the indicator matrices W_p^* and W_q^* are of the same size, i.e., they are both c-partitions that can be applied to the n data points in X or Y. W_p^* is found by running c-means on $X \subset \Re^p$, and W_q^* is found by running c-means on $Y \subset \Re^q$, both initialized at the same set of input indices (in the upspace and downspace). In words, the optimal value of the c-means objective function on the randomly projected downspace data Y of dimension q in the worst case ($\varepsilon = 1/3$) is not worse than 2.33 times its optimal value on X of dimension p in the upspace when the same partition is used for both X and Y. This result in our previous notation reads $J_1(U_q^*, \mathbf{V}(U_q^*); X) \leq (2 + \varepsilon)J_1(U_p^*, \mathbf{V}(U_p^*); X)$, where the cluster centers $\mathbf{V}(U_q^*)$ and $\mathbf{V}(U_p^*)$ are the necessary condition companions of the two c-partitions of X computed with Equation (6.25a). Figure 6.11 depicts result (6.73) graphically.

The set $S_{1,c}(X)$ along the horizontal axis in Figure 6.11 is the set of real positive values of the objective function $J_{1,c}(W; X)$, i.e., $S_{1,c} = \{J_{1,c}(W; X) \geq 0 : W \in \Psi\}$. The integer c is attached to the subscripts in Figure 6.11 because c is restricted; the theorem is true only for every $c < \min\{n, p\}$. Since we don't know (and can't know, unless we examine all the good vertices of M_{hcn}) the *true but unknown* (tbu) global minimum $J_{1,c}(W_{tbu}^*; X)$ of the proposed objective function, it is almost certainly less than the three computed values in Figure 6.11. Since $J_{1,c}(W; X)$ is not a continuous function of W, none of these three computed values occur at "local minima." Theorem 6.5 assumes that the optimal partitions W_p^* and W_q^* are known, but running c-means on X or Y produces only computed approximations to these two partitions; so none of the four values in Figure 6.11 are actually known.

Theorem 6.5 provides some insight into the relationship between running Boutsidis *et al.*'s version of c-means on X in the input space as opposed to using it on the randomly projected subset Y in a downspace that satisfies the JL lemma. However, insight is pretty much all we get. Why? Well, there is a lot of uncertainty in the theory which underlies Equation (6.73).

Ⓢ ***Forewarned is forearmed:*** There is little practical value in Theorem 6.5 because we don't know the globally optimal minimum $J_{1,c}(W_{tbu}^*; X)$; so we can't know how good the computed approximations to any of the terms in Equation (6.73) are[16]. Moreover, the result holds with "$\text{Prob} \geq 0.97 - \delta$"; so it is not guaranteed that the random projection even satisfies the hypothesis for a JL certificate. And, of course, we don't know a best value of δ, nor do we know the value of the "sufficiently large" constant K; so we don't know the specific dimension q of the downspaces.

[16]These values are available if you are willing to spend the time and computing resources to evaluate $J_{1B}(W; X)$ on all of the good (red) vertices of the cn-soccer ball in Figure 6.2. But for any reasonable number (n) of samples, this seems impractical: plan a nice long vacation after you begin such a run.

Moreover, if you did these calculations 100 times, using 100 different random projections of X, the partition W_q^* would vary quite a bit. But all other things being equal, it would still be pinched in between the two values shown in Figure 6.11 that depend on W_p^*. ☺

Given all these forewarnings, what is the value of this theorem? Well, it is a psychological crutch of sorts for users of random projection and c-means. And there is value in having this kind of prosthetic. Theorem 6.5 asserts, under proper circumstances (and with a little luck!), that using Lloyd's algorithm on randomly projected downspaces doesn't run off the rails. The necessary conditions on U and **V** for HCM at Equation (6.25) are *apparently* abandoned in this formulation, but the clustering problem is still solved by alternating between estimates of W and **V**, as can be seen by the following quote from Boutsidis *et al.* (2010).

> *In practice, the Lloyd algorithm is very popular and although it doesn't admit a worst case theoretical analysis, it empirically does well. We thus employ the Lloyd algorithm for our experimental evaluation of our algorithm in Section 5. Note that, after using the proposed dimensionality reduction method, the cost of the Lloyd heuristic is only* $O(n^2/\varepsilon)$ *per iteration. This should be compared to the cost of* $O(ncp)$ *per iteration if applied on the original high dimensional data.*

6.8.2 Matrix Factorization for HCM for Clustering

The form of the c-means functional discussed by Kim and Park (2008) is similar to $J_{1B}(W;X) = \|X - WW^TX\|_2^2$, but Kim and Park incorporate the c-means cluster centers explicitly in their formulation. By viewing c-means as a lower rank *non-negative matrix factorization* (NMF) with special constraints rather than a clustering method, they derive constraints to impose on the NMF formulation so that it behaves as a variation of c-means. In other words, their method interprets the c-means objective function as that of a lower rank approximation with special constraints.

In general, unweighted non-negative matrix factorization of X asks for a pair of matrices C and B^T whose product provides an approximation for X in the form $X \cong CB^T$. This leads quite naturally to the *least squared errors* (LSE) optimization problem $\min_{C,B}\left\{\|X - CB^T\|_2^2\right\}$. Note that $\|(X - CB^T)^T\| = \|X^T - BC^T\|$. Since the problem is symmetric in its unknowns, the distinction between the roles of C and B is (mathematically) unimportant. In this context, $C = V^T$ is called a codebook or dictionary, and the aim is to use it as a lower dimensional representation of the input data set X. This is similar to the role played by the cluster centers in the c-means model.

Let the input data $X = \{x_1, \ldots x_n\} \subset \Re^p$ be arrayed as the $p \times n$ data matrix X and the c cluster centers $\mathbf{V} = \{v_1, \ldots v_c\} \subset \Re^p$ be arrayed as the $p \times c$ matrix V. Let $U \in M_{hcn} \leftrightarrow X = \bigcup_{i=1}^{c} X_i;\ X_i \cap X_j = \varnothing,\ i \neq j;\ |X_i = n_i|\ \forall i$ be a crisp c-partition of X. Kim and Park write Equation (6.9) (their Equation (4)) as

$$\min_{U,\mathbf{V}}\left\{J_1(U,\mathbf{V};X) = \sum_{k=1}^{n}\sum_{i=1}^{c} u_{ik}\|\mathbf{x}_k - \mathbf{v}_i\|_2^2\right\} = \min_{U,V}\left\{J_1(U,V;X) = \|X^T - VU\|_2^2\right\}. \tag{6.74}$$

For optimization, Kim and Park use the approach of Patero and Tapper (1994), who call their basic AO algorithm alternating regression or alternating non-negative least squares. The gist of this procedure is: initialize the cluster center matrix V with non-negative values; scale its columns to unit vectors using the

Euclidean norm; fix this matrix and solve the half problem $\min_{U}\left\{\|X^T - VU^T\|_2^2\right\}$; then freeze U and solve the other half problem $\min_{V}\left\{\|X^T - VU^T\|_2^2\right\}$; continue Picard iteration until termination. Kim and Park give a number of numerical examples that show the NMF-based approach is often superior to the naïve HCM Algorithm A6.1. Another interesting aspect of their paper is the presentation of cluster assessment using images of consensus matrices. The images shown for some of the examples are very similar in flavor to the VAT/iVAT images shown elsewhere in the sequel.

This is equivalent to the AO procedure in A6.1 for HCM with m = 1. Why have this alternative? Well, it leads to two very interesting extensions of the theory that have so far eluded attempts for generalization to the fuzzy case: (i) computable lower bounds for the c-means objective function; and (ii) lower dimensional representation of the input data. Both of these developments require a nodding acquaintance with the singular value decomposition; so we make a brief detour to record the salient facts about this well-known technique. Almost any good text on linear algebra will provide a much more detailed discussion of the theory. Noble and Daniel (1987) is excellent, and the review of SVD in Cox and Cox (2000) is eminently readable.

6.8.3 SVD: A Global Bound for $J_1(U, V; X)$

Unfortunately, the almost universal notation associated with the *singular value decomposition* (SVD) of any matrix (our interest is in the data matrix X) uses the notation U and V for the matrices whose columns are the left and right eigenvectors of the decomposition. We will use $\boldsymbol{U}$ and $\boldsymbol{V}$ in this different font in an attempt to minimize the confusion over an already badly overworked pair of symbols.

Definition 6.13. Singular values of a rectangular matrix.

Let $X_{p\times n}$ be a p × n matrix. The *singular values* (SVs) of X are the square roots of the eigenvalues of X^TX. These numbers are denoted by $\{\sigma_i : 1 \le i \le n\}$ and indexed so that $\sigma_1 \ge \sigma_2 \ge \cdots \ge \sigma_n \ge 0$.

Some books refer to the SVs of X as the square roots of the *non-zero* eigenvalues of X. If the rank of X is r, then X has r non-zero singular values. The singular value decomposition of X is

$$X_{p\times n} = \boldsymbol{U}_{p\times p}\Sigma_{p\times n}\boldsymbol{V}^T_{n\times n}. \tag{6.75}$$

The columns $\{\boldsymbol{U}^{(1)}, \boldsymbol{U}^{(2)}, \ldots, \boldsymbol{U}^{(p)}\}$ of $\boldsymbol{U}$ are orthonormal *left singular vectors* of X; the columns $\{\boldsymbol{V}^{(1)}, \boldsymbol{V}^{(2)}, \ldots, \boldsymbol{V}^{(n)}\}$ of $\boldsymbol{U}$ are orthonormal *right singular vectors* of X. $\boldsymbol{U}$ and $\boldsymbol{V}$ are orthogonal matrices, $\boldsymbol{U}^T\boldsymbol{U} = \boldsymbol{V}^T\boldsymbol{V} = I$. The $\{\boldsymbol{V}^{(i)}\}$ are the eigenvectors of the symmetric matrix X^TX, and the $\{\boldsymbol{U}^{(i)}\}$ are the eigenvectors of the symmetric matrix XX^T. The p × n matrix Σ has the r non-zero SVs of X on its diagonal, and all other entries are zero.

Every matrix has a singular value decomposition with at most $r = \text{rank}(X) \le \min\{p, n\}$ non-zero singular values. Assuming rank(X) = r, an equivalent way to write the SVD at Equation (6.75) is $X_r = \sum_{j=1}^{r}\sigma_j\boldsymbol{U}^{(j)}(\boldsymbol{V}^{(j)})^T$. Since p is usually much less than n, most data sets can be decomposed into a very few terms. For example, the Iris data has p = 4; so (as a data matrix) it can't have more than four non-zero singular values.

The r terms in a SVD of X provide an r-rank approximation to X. A well-known theorem from linear algebra (cf., Theorem 8.2, Vempala (2007)) guarantees that the SVD $X_r = \sum_{j=1}^{r} \sigma_j U^{(j)}(V^{(j)})^T$ is the best r-rank approximation to X in the least squares sense. Let $R(r) = \{Y \in \Re^{pn} : \mathrm{rank}(Y) = r\}$. Then

$$\|X - X_r\|_2^2 = \left\| X - \sum_{j=1}^{r} \sigma_j U^{(j)}(V^{(j)})^T \right\|_2^2 = \min_{Y \in R(r)} \left\{ \|X - Y\|_2^2 \right\}. \tag{6.76}$$

Moreover, the sum of squares of the singular values *not used* in the expression $X_r = \sum_{j=1}^{r} \sigma_j U^{(j)}(V^{(j)})^T$ yields the error in the approximation. For $c \leq r$, the error incurred by using only the first c terms in $X_r = \sum_{j=1}^{r} \sigma_j U^{(j)}(V^{(j)})^T$ is

$$\|E_c\|_2^2 = \left\| X - \sum_{j=1}^{c} \sigma_j U^{(j)}(V^{(j)})^T \right\|_2^2 = \sum_{j=c+1}^{n} \sigma_j^2. \tag{6.77}$$

It is this result in the context of HCM clustering that provides an estimate for the global minimum of the c-means objective function. We know that $\mathrm{rank}(X) \leq \min\{p, n\}$, $\mathrm{rank}(U) \leq \min\{c, n\}$. As long as $c \leq p$, we can replace VU by the best approximation to X by a matrix of rank c, i.e., $X_c = \sum_{j=1}^{c} \sigma_j U^{(j)}(V^{(j)})^T$, and, from this, we get a lower bound on the global minimum of the HCM objective function in terms of the residual singular values of the input data for all crisp c-partitions of X, viz.,

$$J_1(U, V; X) \geq \|X^T - X_c^T\| \geq \sum_{j=c+1}^{n} \sigma_j^2. \tag{6.78}$$

Equation (6.78) provides a *computable* benchmark that tells us how small successive values of the c-means objective function need to be in order for the iteration in A6.1 to approach the global minimum. Let us recognize this result formally as follows.

Theorem 6.6. A global bound for J_1 (U, V; X).

Let $X = \{\mathbf{x}_1, \ldots \mathbf{x}_n\} \subset \Re^p$ have matrix form $X = [x_{ij}]_{p \times n} = \begin{bmatrix} \uparrow & & \uparrow \\ \mathbf{x}_1 & \cdots & \mathbf{x}_n \\ \downarrow & & \downarrow \end{bmatrix}$, with $r = \mathrm{rank}(X) \leq \min\{p, n\}$. Let c be an integer, $c < \min\{p, n\}$. Let $J_{1,c}(U_p^*, \mathbf{V}_p^* : X)$ denote the global minima of $J_{1,c}$ with model norm induced by A = I in the upspace (the Euclidean norm). Let $X_{p\times n} = U_{p\times p} \Sigma_{p\times n} V_{n\times n}^T$ be the SVD of X at Equation (6.75), and let the rank c approximation to X be $X_c = \sum_{j=1}^{c} \sigma_j U^{(j)}(V^{(j)})^T$. Then

$$\sum_{j=c+1}^{\min\{p,n\}} \sigma_j^2 = \|X - X_c\|_2^2 \leq J_{1,c}(U_p^*, \mathbf{V}_p^* : X). \tag{6.79}$$

Proof. See Zha *et al.* (2001) or Kim and Park (2008) for proofs that use different representations of the HCM objective function. The form of $J_{1,c}$ used by Zha *et al.* (2001) is $J_{1,c}(U, \mathbf{V}; X) = \mathrm{tr}(X^T X) - \mathrm{tr}(P^T X^T X P)$, where P is the orthonormal matrix $P = \mathrm{diag}\left(1/\sqrt{n_i}\right)$ and, as usual, $n_i = |X_i|$. Result (6.79) then follows after showing that the problem $\min\{\mathrm{tr}(X^TX) - \mathrm{tr}(P^TX^TXP)\}$ is equivalent to the problem $\max\{\mathrm{tr}(P^TX^TXP)\}$.

What does it mean? This is a pretty remarkable but somewhat limited result. It tells us that we can bound the true but unknown global minimum of the k-means objective function (that uses the Euclidean norm) for a given input data set X in the upspace by finding the SVD of X, extracting the c largest singular values and then summing the squares of the remaining singular values to obtain the bound. Please make sure that you see the hypothesis $1 \le c \le \min\{p, n\}$, because this restricts the result to a limited number of possible clusters. For example, the Iris data has $p = 4$, $n = 150$; so this theorem is valid for Iris only when c is less than or equal to 4.

Theorem 6.6 is represented graphically in Figure 6.12, which adds the lower bound in Equation (6.79) to the illustration in Figure 6.11. Note again that the values represented in Figure 6.12 depend on the parameter c and that Theorem 6.6 holds only for values of c that are no larger than $\min\{p,n\}$.

Figures 6.11 and 6.12 have a similar appearance, but there is a very great difference between the bounds shown in Figure 6.12 based on Theorems 6.5 and 6.6 and the additional lower bound illustrated in Figure 6.12. The three bounds from Theorem 6.5 are not computable in theory since you don't have a value for the global minimum of $J_{1,c}$ in either the upspace or downspace (unless you find them by exhaustion). But the lower bound from Theorem 6.6 at the left side of Figure 6.12 *is* computable for each value of $c \le \min\{p, n\}$ because it only depends on having the input data set X and its c largest singular values, which are (in principle at least) obtainable through the SVD.

However, the tightness of the bound, i.e., the difference between the unknown global minimum $J_{1,c}(U^*_{p,tbu}, V^*_p : X)$ and the bound $\sum_{j=c+1}^{\min\{p,n\}} \sigma_j^2$ is not known. Intuition suggests that as the rank of X, and, hence, the potential number of non-zero singular values, increases, the tightness of the bound decreases because there will be more terms in the residual sum $\sum_{j=c+1}^{\min\{p,n\}} \sigma_j^2$. If the non-zero singular values are greater than 1, squaring them may make the bound quite large. The difference also depends on c, the number of clusters sought; but generally speaking, c will be less than rank(X), particularly in the case of high-dimensional input data. Examples 6.8, 6.9, and 6.10 explore this.

Theorems 6.5 and 6.6 can't be discovered with the representation (6.9) of the c-means objective function as a point of departure. It is the alternate formulations of c-means in this subsection that lead to these aspects of the theory. Using one of these matrix factorization representations seemingly limits the model norm to

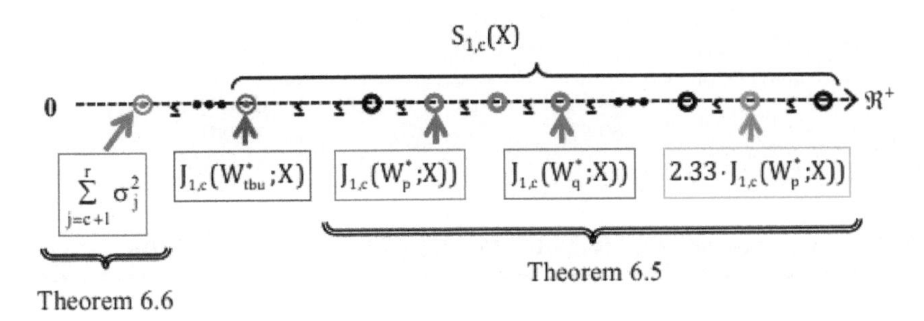

Figure 6.12 Lower bound on $J_{1,c}(U, V; X)$ when $c \le \min\{p, n\}$.

Euclidean (this *seems* to be the case, but it may well be true that these two theorems can be extended to any inner product A-norm). On the other hand, Equation (6.9) leads to the rich and satisfying theory of earlier sections that provide a means for developing the unique soft generalization of HCM provided by FCM. And looking at HCM by its representation at Equation (6.9) easily allows the model norm to be any inner product norm on the input space. Both points of view are useful. The various representations of the c-means objective functions lead to complementary facts and a deeper understanding of c-means clustering.

Example 6.9. Lower bounds on HCM/FCM using Theorem 6.6 with the Iris data.

By now you are well-acquainted with the Iris data. Written as a data matrix, X_{Iris} is $p \times n = 4 \times 150$, so its rank is less than or equal to 4, which limits the number of singular values to 4, and $c \leq \min\{p, n\} = \min\{4, 150\} = 4$. It turns out that X_{Iris} does have four non-zero singular values. One run of HCM and FCM, A6.1 with m = 1 and 2, using the default parameters (the model norm is the Euclidean norm, indicated here with the second subscript 2), starting both algorithms from the same initial guess at c = 2 and c = 3, resulted in the lower bound values shown in Table 6.9. Terminating HCM (aka k-means) on successive values of its objective function without the guidance of Theorem 6.6 is a bit like flying blind. You would obtain a value of $J_{1,c}(U, \mathbf{V}; X_{Iris}))$ for c = 2 or 3 similar to the ones in Table 6.9 but have no idea if you were closing in on the global minimum. But the values in the first column, for the lower bound, are independent of the parameters chosen for A6.1: these values depend only on the SVs of X itself.

On the other hand, the bound in Equation (6.79) is not very tight; so you might be closer to the global minimum than the values in Table 6.9 suggest. At the end of the day, this is what is wrong with terminating HCM/FCM and GMD/EM on successive values of their objective functions. You never really know how close you are to the global minimum (or maximum, in the case of GMD/EM), and even if you did know, there is no guarantee that when you got there, the most satisfactory partition it points to is the best clustering solution (revisit Figure 4.10 to refresh your memory on this point). The SVs of Iris don't change. The only way to establish the quality of the lower bound in this example would be to find the global solution for c-means on Iris (by exhaustion) and compare it to the values in Table 6.9.

There is no theory that provides a satisfactory matrix factorization for the FCM objective function and, hence, no fuzzy version of Theorem 6.6 (yet). But the values in Table 6.9 are arranged so that $J_{m=2,c}(U, \mathbf{V}; X_{Iris})$, which is less than $J_{m=1,c}(U, \mathbf{V}; X_{Iris})$ for c = 2 and 3, is to the left (the FCM values are below the HCM values). Does this tempt you to think that this might always be true?

Table 6.9 Lower bounds for $J_1(U, \mathbf{V}; X_{Iris})$ using Theorem 6.6.

c	$\sum_{j=c+1}^{4} \sigma_j^2$	Fuzzy c-means $J_{m=2,c}(U, \mathbf{V}; X_{Iris})$	Hard c-means $J_{m=1,c}(U, \mathbf{V}; X_{Iris})$
2	5.348	128.923	152.369
3	1.879	60.576	78.941

Example 6.10. Lower bounds on HCM/FCM using Theorem 6.6 with the GM_{25} data.

Let us explore the value of $\left[J_1(U, \mathbf{V} : GM_{25}) - \sum_{j=c+1}^{\min(p,n)} \sigma_j^2\right]$ with the GM_{25} data set in Rajasegarar *et al.* (2010). These data are sets of vectors arranged in 26 clusters of ellipsoids. Each cluster has 10 ellipsoids.

Table 6.10 Lower bounds for $J_1(U, \mathbf{V}; GM_{25})$ using Theorem 6.6.

c	$\sum_{j=c+1}^{25} \sigma_j^2$	FCM : $J_{m=2,c}$	HCM : $J_{m=1,c}$
2	450	4820	9330
3	430	3220	9030
4	409	2410	8700
5	389	1930	8370
6	369	1610	8090
7	348	1380	7730
8	328	1210	7480
9	308	1070	7120
10	288	965	6780
11	268	877	6550
12	248	804	6170
13	228	742	5880
14	209	689	5490
15	189	643	5190
16	169	603	4930
17	150	568	4700
18	131	536	4280
19	111	508	4000
20	92.1	482	3630

Each ellipsoid has 100 vectors drawn from Gaussian distributions. So, GM_{25} has $n = 26,000$ data vectors drawn from $c = 26$ labeled Gaussian distributions in $\Re^{25}$. Hence, as a data matrix GM_{25} has size $25 \times 26,000$; so $\text{rank}(GM_{25}) \leq 25$. One run of HCM and FCM, A6.1 with m = 1 and 2, using the default parameters (the model norm is the Euclidean norm), starting both algorithms from the same initial guess at each value of c, resulted in the (rounded off) lower bound values shown in Table 6.10.

Table 6.10 shows the same type of calculations as in Table 6.9 for values of c from 2 to 20 for this data set, which turns out to have 25 non-zero singular values. The lower bound in column 2 decreases monotonically as c increases, as it must, since there are fewer terms in the sum of squares of the SVs of X. The values of J_2 and J_1 also do this. And again, the FCM values are always less than the HCM values. The smallest value of the difference between the bound and the obtained value of J_1 occurs at c = 20, which has the least number of residual SVs in the experiment, viz., $\left[J_1(U, \mathbf{V} : GM_{25}) - \sum_{j=21}^{25} \sigma_j^2\right] = 3630 - 92.1 = 3537.9$. This is a fairly big number. Perhaps, there are other initializations that would lead to lower values in this experiment. Let's try an even bigger data set.

Example 6.11. Lower bounds on HCM/FCM using Theorem 6.6 with the MNIST data.

MNIST (http://yann.lecun.com/exdb/MNIST/) has n = 70,000 images, each of which is represented by a single vector with $p = 784$ dimensions (cf. Figure 5.18). This data set is a subset of the collection of handwritten digits that are available from the *National Institute of Standards and Technology* (NIST); so there are $c = 10$ labeled subsets in the data. Each of the 70,000 vectors begins as a 28×28 pixel image of one of the digits 0 to 9. Each pixel has an integer value between 0 and 255. These gray-level values are normalized to the interval [0,1] by dividing all the pixel values by 255. Finally, the values in each image are concatenated to produce a 784-dimensional feature vector.

Table 6.11 Lower bounds for $J_1(U,\mathbf{V};X_{MNIST})$ using Theorem 6.6.

c	$\sum_{j=c+1}^{784}\sigma_j^2$	FCM : $J_{m=2,c}$	HCM : $J_{m=1,c}$
2	6.25E+06	1.20E+11	2.25E+11
3	6.12E+06	8.00E+10	2.14E+11
4	6.00E+06	6.00E+10	2.05E+11
5	5.88E+06	4.80E+10	1.98E+11
6	5.78E+06	4.00E+10	1.92E+11
7	5.69E+06	3.43E+10	1.88E+11
8	5.61E+06	3.00E+10	1.85E+11
9	5.52E+06	2.67E+10	1.83E+11
10	5.45E+06	2.40E+10	1.78E+11
11	5.38E+06	2.18E+10	1.77E+11
12	5.31E+06	2.00E+10	1.74E+11
13	5.24E+06	1.85E+10	1.72E+11
14	5.18E+06	1.71E+10	1.70E+11
15	5.12E+06	1.60E+10	1.67E+11
16	5.06E+06	1.50E+10	1.66E+11
17	5.00E+06	1.41E+10	1.65E+11
18	4.94E+06	1.33E+10	1.63E+11
19	4.89E+06	1.26E+10	1.61E+11
20	4.84E+06	1.20E+10	1.60E+11

Thus, X_{MNIST} is a $784 \times 70,000$ data matrix; so $\text{rank(MNist)} \leq 784$. This data set has 714 singular values greater than 1, and the remaining SVs are very close to zero. Table 6.11 exhibits the lower bounds, and values of the HCM and FCM objective functions obtained by running A6.1 just once beginning with the same initialization and using the default parameters for both algorithms as in Examples 6.9 and 6.10. The most notable difference between this example and the previous two is the difference between the lower bound and the value of J_1 which occurs again at c = 20, $\left[J_1(U,\mathbf{V}:X_{MNIST}) - \sum_{j=21}^{784}\sigma_j^2\right] = 1.6E+11-4.84E+06 = 185995160000$. This is a really big number because (714 − c) of the residual singular values are greater than 1, and the remaining SVs are very close to zero. The values in these three experiments confirm that as the size of the data set (and, hence, the number of residual singular values) increases, so does the looseness of the lower bound at Equation (6.79).

6.9 Notes and Remarks

The HCM and FCM optimization problems are very clearly *least squared error (LSE)* estimation problems. Both problems are well-defined in the sense that $J_m(U,\mathbf{V})$ always has a global minimum, and the literature is replete with approaches for finding local extrema of $J_m(U,\mathbf{V})$ for $m > 1$. Optimizers that attempt to seek local minima of $J_1(U,\mathbf{V})$ are stymied by the fact that it is not continuous in the U variables. Each $(U,\mathbf{V}) \in M_{hcn} \times \Re^{cp}$ is isolated – in other words, there is no neighborhood about each point; so $J_1(U,\mathbf{V})$ can't have joint local minima. The LSE families of models have many favorable mathematical properties, some of which are important in the development of FCM. But much of the strong theory of LSE, in particular, the orthogonal projection theorem, is unavailable for the analysis of HCM because M_{hcn} is finite. There will be more about this in Chapter 10.

It is impossible to overestimate the popularity of the HCM model – "k-means." There are literally thousands of papers that study its theory and many thousands more that use it (perhaps, indiscriminately at times) as the "go-to algorithm" or "clustering method du jour," quite independently of its good and bad points. It would take a not-so-small monograph (and an army of writers!) to record the important steps and applications in the historical evolution of HCM (aka "k-means, batch form"). The trail through the history of least squares leading up to HCM is an interesting one; so we will take a short detour to explore this path on our safari through the jungle that is cluster analysis.

In common with many past events, the history of least squared error estimation is cloudy. Bell (1966) credits Gauss with the invention of the method of least squares for parameter estimation, but it is beyond dispute that Legendre published the first formal exposition of it in 1806 (Bell, 1966, p. 259). (The original publication is in French – Legendre (1805)). Gauss subsequently asserted in 1809 that he had used the method as early as 1795 but did not make a formal presentation of it until 1809. The ensuing argument about which of these eminent mathematicians "invented" the method of least squares is one of the more fascinating storied and somewhat acerbic rivalries in the history of Science.

Plackett (1972) takes issue with parts of Bell's account of the events but seems to agree that Gauss probably knew about least squares earlier than 1805. He writes: "The method of least squares was discovered independently by Gauss and Legendre during the period which followed [1783–1789]. Although Gauss had been using the method since about 1795, the first explicit account was published in 1805 by Legendre. Four years later Gauss gave his version, in the course of which he referred to his earlier work." So, Plackett agrees with Bell about the mathematics but has a very different interpretation of the social aspects of the dispute.

Stigler (1981) investigates the details of the method through the eyes of a modern statistician and goes to great computational lengths to discover whether or not Gauss did indeed use least squares to calculate his published values for the French data pertaining to the Meridian arc. Stigler writes: "Examination of the French arc data must, I feel, be taken as supportive of the view that Gauss treated these data with least squares in 1799, more than five years before Legendre published on the subject. The case is not unchallengeable, but considering the complexity of the problem, it is as strong as we have a right to expect." And what of Gauss and Legendre? Well, they traded very civilized but, nonetheless, very sharp barbs pointed at each other, several of which are reprinted in Plackett and Stigler. If this part of the LSE story interests you, read the accounts rendered by Bell (1966), Plackett (1972), and Stigler (1981). We are hardly immune to this sort of thing in modern science today, nor will we ever be. Reading a bit of history about it may help us understand each other a little better, and that is always a good thing.

The first written work (at least in English) describing a batch version of an AO-like algorithm for minimizing signal loss in the context of pulse code modulation appeared in an unpublished technical report written by Lloyd (1957). Cox (1957) also discussed this approach in a less formal way the same year. However, the first discussion of J_1 as a clustering criterion is widely credited to Steinhaus (1956). Moreover, Lloyd acknowledged that an even earlier appearance of his method apparently was discussed in a paper written by Lukaszewicz and Steinhaus (1955). The LBG (Linde, Buzo and Gray, 1980) algorithm that first appeared in 1980 is widely accepted as the first paper that introduced this method to the field of vector quantization; their algorithm is basically an extension of Lloyd's scheme.

One of the earliest methods for automating the choice of c using HCM is due to Ball and Hall (1967). They used conditions (6.25) as a basis for their crisp *iterative self-organizing data analysis* (ISODATA) algorithm, which combines k-means A6.1 ($m = 1$) with a number of heuristic procedures for (dynamic) determination of the "correct" number of clusters to seek. Duda and Hart (1973) call the version of HCM given in A6.1 *basic ISODATA* to distinguish it from the more sophisticated version of Ball and Hall (1967). They credit their version to a 1965 program developed by Singleton and Kautz at SRI (Duda and Hart, p. 250). Early papers by Dunn and Bezdek incorrectly called the basic FCM/AO algorithm ($m > 1$ in A6.1) "fuzzy ISODATA," even though there were no heuristics attached to it for discovery of a best c. Later papers replaced the term fuzzy ISODATA by fuzzy c-means, but the incorrect use of fuzzy ISODATA still occurs now and then. Be aware of this change if you dive into the early waters of FCM.

Thorndike (1953) was perhaps the first to describe a sequential procedure that purports to define a good partition as one that minimizes a sum of distances (but not squared errors as used by J_1). Much more well-known is a *sequential* version of *k-means* (sHCM) that is arguably the first competitive learning algorithm, another branch of the clustering family which now has its own large and active congregation – the *learning vector quantization* (LVQ) crowd, headed up by Kohonen and his disciples.

Sequential hard c-means (sHCM, most often called "k-means" in the literature) is usually attributed to MacQueen (1967), who began his paper this way: "The main purpose of this paper is to describe a process ... which is called k-means ... that gives partitions which are reasonably efficient in terms of within-class variance." Here is MacQueen's description of sHCM:

> "Stated informally, the k-means procedure consists of simply starting with k groups each of which consists of a single random point, and thereafter adding each new point to the group whose mean the new point is nearest. After a point is added to a group, the mean of that group is adjusted in order to take account of the new point. Thus, at each stage the k-means are, in fact, the means of the groups they represent (hence the term k-means). ... When the sample points are rearranged in a new random order, there is some variation in the grouping which is obtained."

Evidently, MacQueen's algorithm is sequential but is designed to find a prespecified number of clusters in static data. It ends in one pass through a batch data set X with estimates for c crisp clusters $\{X_i\}$ and c cluster centers (the c-means) $V = \{v_1, \ldots, v_c\} \subset \Re^{cp}$. MacQueen provided a highly detailed analysis of convergence in probability under pretty strong statistical assumptions for his version of sHCM. Some variations of sHCM which are variously called "adaptive" sHCM are presented, for example, by Darken and Moody (1990) and Chinrungrueng and Sequin (1995), but none of these derivative algorithms are able to determine changes to c during processing.

MacQueen himself credits Sebestyen with the invention of sHCM, saying at the end of his introduction "Sebestyen (1962) has described a procedure called "adaptive sample set construction," which involves the use of what amounts to the k-means process. This is the earliest explicit use of the process with which the author is familiar." Sebestyen's procedure differed slightly from sHCM. Here is a paraphrased version from his book:

> "When the first input is introduced, it is assigned to class 1, and this input becomes the mean vector of class 1. The second input arrives. If it is within a threshold distance T of the first mean, it is assigned to class 1, and the mean is updated. Otherwise, it becomes the mean of a new class 2. Continuing, this procedure results in a set of points which represent the locations of clusters of inputs, whose approximate sizes are T."

Sebestyen called his algorithm "sample set construction." A slightly modified version of it which imposes an upper limit on the number of clusters that can be created is the well-known *leader algorithm* (Hartigan (1975, p. 75)), which is renamed as BSAS in Theodoridis and Koutroumbas (2009, p. 633).

All of these authors used k as the integer designating the number of clusters, and most of this early literature referred to *the sequential version* as "k-means," and not "c-means." Duda and Hart's very influential text on pattern classification used k and c more or less interchangeably, and the use of c instead of k probably grew out of that work. The outrage that some show about this issue is pretty amusing – after all, this is just a dummy variable.

There continues to be great confusion about the distinction between the batch and sequential versions of k-means. Here is how Macqueen (1967, p 283) describes the algorithm he analyzed on p. 290 in his 1967 paper:

> *When the sample points are rearranged in a new random order, there is some variation in the grouping which is obtained.*

This makes it clear that the algorithm described and analyzed by MacQueen is NOT the batch version of k-means ascribed to Lloyd (1957), whose basic form is A6.1 (m = 1). You will find this statement about the leader algorithm repeated almost verbatim in Hartigan (1973, p. 106). Johnson and Wichern (2007, p. 696) describe a batch version of k-means but attribute it incorrectly to MacQueen. And *if you queried Google in 2015 with the words "k-means," you would have read this in Wikipedia:*

> *History*
>
> *The term "k-means" was first used by James MacQueen in 1967 [1], though the idea goes back to* Hugo Steinhaus *in 1957 [2].The* standard algorithm *was first proposed by Stuart Lloyd in 1957 as a technique for* pulse-code modulation, *though it wasn't published outside of* Bell Labs *until 1982 [3]. In 1965, E.W. Forgy published essentially the same method, which is why it is sometimes referred to as Lloyd-Forgy [4].*

I believe that these statements in Wiki about k-means are at least misleading and, at worst, incorrect (as they often are). Lloyd developed a batch version, and MacQueen analyzed a sequential version. Many authors in the clustering literature call one or the other of these two algorithms (there is more than one version of both approaches) "k-means." But they are clearly very different algorithms. sHCM is a local, greedy algorithm that looks at just one point at a time, while batch HCM tries to account for the global structure in the data set at each pass through the necessary conditions at Equation (6.25).

Another early relative of HCM worth mentioning is the minimum variance partitioning scheme of Ward (1963), which forms a sort of bridge between the objective function and graph-theoretic approaches to seeking CPOV clusters. This algorithm has the flavor of *sequential, agglomerative hierarchical non-overlapping* (SAHN) algorithms such as the single linkage model and algorithm presented in Chapter 8, but the criterion for choosing the pair of clusters to merge at each step is based on the minimum value of J_1.

Calinski and Harabasz (1974) present an interesting variant of HCM clustering based on cutting a *minimal spanning tree*. They begin by converting $X = \{x_1, \ldots, x_n\} \subset \Re^p$ into D, a (square $n \times n$) *Euclidean distance matrix* (EDM). Then they construct the MST of D. The MST can be cut into (c) clusters by removing (c-1) edges from the MST in $\binom{n-1}{c-1}$ ways. Each of these possible sets of clusters is evaluated using $J_1(U, V; X)$, and the best within groups sum of squared errors partition is chosen using a variance ratio criterion called the VRC index that has become a staple of some cluster validity studies. This is a novel way to cut the finite but very large search space M_{hcn} down to size. Problem 3 below asks you to show that there are approximately $|M_{hcn}| \approx c^n/c!$ crisp c-partitions of n objects. The method of Calinski and Harabasz reduces the size of the search space at each c to $\binom{n-1}{c-1}$, a considerable reduction in the number of HCM partitions that must be (potentially) evaluated.

A different style of k-means algorithms follows a chain of papers typified by Kumar *et al.* (2004), who develop an algorithm that produces an approximation to the global solution of the HCM problem. Their algorithm approximates the centroid of the input data with the centroid of a sample and uses recursion to the base case (c=1) to find the advertised approximation. Other approaches besides the ones discussed in Chapter 4 include simulated annealing (Babu and Murty, 1994), Tabu search (Al-Sultan, 1995), variable neighborhood search (Hansen & Mladenovic, 2001), and constrained optimization (Steinley and Hubert, 2008).

All of these papers attempt to improve on the alternating optimization approach to approximating a solution to HCM problem (6.9). And, yet, as noted by Ostrovsky *et al.* (2012), practitioners continue to use Lloyd's AO (along with EM and hierarchical methods), which they describe as "old standbys with no known performance guarantees."

Another approach to optimizing the HCM model is followed by the devotees of spectral cluster analysis. Partitioning graphs with spectral methods began in 1973 with the work of Donath and Hoffman (1973). The success of this class of methods is due to the very clever method of extracting spectral features from the data *prior to clustering*. Almost all of the current work in this area then turns to HCM to look for the clusters. A more accurate name for this approach would be "spectral feature extraction with c-means clustering," but it is already too late to make such a change. For example, von Luxberg (2007) lists three versions of spectral clustering in his 2007 survey, and all three invoke "k-means" as the last step.

The habit of saying, "we used k-means for clustering," without specifying whether the batch or sequential version is meant, is especially disturbing. von Luxberg mentions several authors who depart from HCM as the preferred clustering algorithm, but k-means (which version?) is the hands-down winner in the spectral clustering community. Dhillon *et al.* (2004) present a very lucid discussion of the relationship between spectral clustering and kernelized c-means.

The bounds on $J_1(U, V; X)$ given in Section 6.8 concern two methods of dimensionality reduction by random projection and low rank approximations. The general idea of finding lower dimensional representations of X is nothing other than feature extraction as discussed in Section 5.5. Finding a function which preserves the cluster structure in X upon passing to Y, $X \subset \Re^p \mapsto Y = \phi[X] \subset \Re^q$ as shown in Figure 5.2, has acquired a sense of urgency in the last decade due to the almost maniacal frenzy caused by (the need to cluster in) Static ***BIG DATA***, which will be discussed in Chapter 11.

As you can see, there are very many papers that have used, and will continue to use, hard c-means (k-means) as their "go-to" clustering method. There are more than a few handfuls of very different algorithms that seek an approximate solution to the HCM optimization problem (6.9). Many authors of papers about HCM focus on getting as close to the (unknown) global minimum of $J_1(U, V; X)$ as possible, but as emphasized and exemplified in Example 4.10, and re-emphasized in this chapter, the global minimum might not be the one you want. Steinley and Hubert (2008) provide an eminently quotable statement about this:

> *To some, it might border on heresy to suggest that being optimal may not be the same as being best, and therefore a sole concern with optimality may be misplaced. But it depends on how "best" is defined, e.g., best recovery of structure or just a minimum value for some loss function.*

Ostrovsky *et al.* (2012) make this provocative statement about HCM algorithm A6.1 (m = 1) as well as the GMD/EM algorithm (A7.1) and the SAHN methods (A8.1) that appear in Chapters 7 and 8. When you read it, please remember what Yogi Berra told us about the difference between theory and practice in the second paragraph of the preface to this volume:

> *There is presently a wide and unsatisfactory gap between the practical and theoretical clustering literatures. For decades, practitioners have been using heuristics of great speed but uncertain merit ... However, in the last few years, algorithms researchers have made considerable innovations, and even obtained polynomial-time approximation schemes (PTAS's) for some of the most popular clustering formulations. Yet these contributions have not had a noticeable impact on practice. Practitioners instead continue to use a variety of heuristics (Lloyd, EM, agglomerative methods, etc.) that have no known performance guarantees. There are two ways to approach this disjuncture. The most obvious is to continue developing new techniques until they are so good—down to the implementations—that they displace entrenched methods. The other is to look toward popular heuristics and ask whether there are reasons that justify their extensive use, but elude the standard theoretical criteria.*

Finally, and sad to say, but you should hear this. There are at least five academic gangs represented in the panoply that is the HCM (aka "k-means") world – the pattern recognition, vector quantization, self-organizing maps, machine learning, and data mining gangs. And, unfortunately, "gangs" is often the right choice for a term to describe them because many members of these gangs don't speak to each other in a scientific sense. And at times, members of these gangs can be very hostile to each other. So, as you read papers and books that explain what HCM (k-means) is and how to use it, please comb through the literature of all the gangs. Each has a somewhat different perspective on HCM, but all will agree that it is the #1 crisp clustering algorithm on planet earth.

6.10 Exercises

1. $\mathbf{D} : \Re^p \mapsto N_{fc}$ is a fuzzy or probabilistic *classifier*. Let $U^{(k)} = \mathbf{D}(x_k)$ be the label vector for the point $x_k \in \Re^p$. Apply $\mathbf{D}$ to each of the vectors in $X = \{x_1, \ldots, x_n\} \subset \Re^p$ and arrange the resultant n label vectors as a $c \times n$ matrix $U = \begin{bmatrix} U^{(1)} & \cdots & U^{(k)} & \cdots & U^{(n)} \end{bmatrix}$. Is $U \in M_{fcn}$? Can an alteration of U guarantee that it will be a non-degenerate partition of the data?

2. Form the Lagrangian of the function $J_m(U, V; X)$, $m > 1$ at Equation (6.10). Argue that you can use the Lagrange multiplier theory of constrained (equality) optimization to derive the FONCs for U^* in Equation (6.19b1). This is the key step in the proof of Theorem 6.2 (cf., Bezdek (1981)).

3. Prove that exact cardinality of M_{hcn} is $|M_{hcn}| = \left(\frac{1}{c!}\right) \sum_{j=1}^{c} \binom{c}{j} (-1)^{c-j} j^n$. For $c << n$, the last term dominates this sum. Show that this yields the approximation $|M_{hcn}| \approx c^n/c!$. What can you say about the exact cardinality $|M_{hcn_0}|$ of the set of degenerate hard partitions? Use your answer to estimate the size of $|M_{hcn_0}|$ when $c << n$.

4. $U \in M_{hcn}$ is an *aligned c-partition* of O when its entries form c contiguous blocks of 1's in U, ordered to begin with the upper left corner, and proceeding down and to the right. The set of all such partitions is

$$M^*_{hcn} = \{U \in M_{hcn} | u_{1k} = 1,\ 1 \le k \le n_1 : u_{ik} = 1, n_{i-1} + 1 \le k \le n_i,\ 2 \le i \le c\}.$$

For example, $\begin{bmatrix} 1 & 1 & 1 & 0 & 0 \\ 0 & 0 & 0 & 1 & 1 \end{bmatrix}$ is aligned, but $\begin{bmatrix} 0 & 0 & 0 & 0 & 1 \\ 1 & 1 & 1 & 0 & 0 \end{bmatrix}$ is not[17]. Prove that the cardinality of M^*_{hcn} is $\left|M^*_{hcn}\right| = \frac{\prod_{j=1}^{c-1}(n-j)}{(c-1)!}$. Show that for $c << n$, $\left|M^*_{hcn}\right| \approx n^{c-1}/(c-1)!$ is a pretty good approximation to the exact value.

5. Using the approximations in P3 and P4, above, prove that when $c << n$, $\frac{\left|M^*_{hcn}\right|}{\left|M_{hcn}\right|} \approx \frac{n^{c-1}}{c^{n-1}}$. Use this result to compare the approximate sizes of the two sets of partitions at $c = 10$, $n = 10,000$. What do you conclude?

6. Compare the cardinality of M_{hcn} in P3 above to $\binom{n-1}{c-1}$, the number of partitions evaluated by the method of Calinski and Harabasz for $c = 10$, $n = 10,000$. Does this make their MST-based approach to minimizing $J_1(U, V; X)$ more attractive than using AO as in A6.1, $m = 1$? Justify your answer.

7. Suppose $U \in \Re^{cn}$ is computed with Equations (6.19b1). Prove that when Equation (6.19b2) has been used, it is still true that $U \in M_{hcn}$. Hint: assume some row of U contains only 0's and then show that this is impossible.

8. Prove the statement made in connection with iteration through conditions (6.25), that J_1 has the descent property for successive iterates t and t + 1, i.e., that $J_1(U_{t+1}, V_{t+1}; X) \leq J_1(U_t, V_t; X)$. Do the same thing for J_m, $m > 1$.

9. Prove the assertion at Equation (6.26c), that as m approaches 1 from above,

$$\lim_{m \to 1} \left\{ \frac{\left(\|x_k - v_i^*\|_A^2\right)^{\frac{1}{m-1}}}{\sum_{j=1}^{c}\left(\left\|x_k - v_j^*\right\|_A^2\right)^{\frac{1}{m-1}}} \right\} = \left\{ \begin{array}{ll} 1; & \|x_k - v_i^*\|_A^2 \leq \left\|x_k - v_j^*\right\|_A^2, j \neq i \\ 0; & \text{otherwise} \end{array} \right\}.$$

10. Prove that $J_2(U^*, V^*; X)$ is the average harmonic mean of the squared errors between the n data vectors in X and the c cluster centers comprising V^*.

Definition 6.14. Inner products, vector norms, and metric distances.

Let $VS(\Re)$ be a real vector space equipped with the operations of vector addition and scalar multiplication. For scalars $\alpha, \beta \in \Re$ and vectors $u, v, w \in VS(\Re)$, the *inner* (aka dot or scalar) *product*, *norm*, and *norm metric* are defined as follows:

$$\textbf{SP1:} \langle u, u \rangle > 0 \text{ if } u \neq 0 \;;\; \langle 0, 0 \rangle = 0 \text{ (positive-definite);} \tag{6.80a}$$

$$\textbf{SP2:} \langle u, v \rangle = \langle v, u \rangle \text{ (symmetric);} \tag{6.80b}$$

$$\textbf{SP3:} \langle \alpha u + \beta v, w \rangle = \alpha \langle u, w \rangle + \beta \langle v, w \rangle \text{ (linear).} \tag{6.80c}$$

[17]Aligned partitions are important to the clusiVAT extension of single linkage clustering discussed in Section 11.7.

$$\mathbf{N1}: \|\mathrm{v}\| > 0 \text{ if } \mathrm{v} \neq \mathbf{0} \; ; \; \|\mathbf{0}\| = 0 \text{ (positive-definite)}; \tag{6.81a}$$

$$\mathbf{N2}: \|\alpha \mathrm{v}\| = |\alpha| \; \|\mathrm{v}\| \text{ (homogeneous)}; \tag{6.81b}$$

$$\mathbf{N3}: \|\mathrm{v} + \mathrm{w}\| \leq \|\mathrm{v}\| + \|\mathrm{w}\| \text{ (triangle inequality)}. \tag{6.81c}$$

$$\mathbf{M1}: \mathrm{d}(\mathrm{u}, \mathrm{v}) > 0 \text{ if } \mathrm{u} \neq \mathrm{v} \; ; \; \mathrm{d}(\mathrm{u}, \mathrm{u}) = 0 \text{ (positive-definite)}; \tag{6.82a}$$

$$\mathbf{M2}: \mathrm{d}(\mathrm{u}, \mathrm{v}) = \mathrm{d}(\mathrm{v}, \mathrm{u}) \text{ (symmetric)}; \tag{6.82b}$$

$$\mathbf{M3}: \mathrm{d}(\mathrm{u}, \mathrm{w}) \leq \mathrm{d}(\mathrm{u}, \mathrm{v}) + \mathrm{d}(\mathrm{v}, \mathrm{w}) \text{ (triangle inequality)}. \tag{6.82c}$$

---

11. Let $A \in PD^{pp}$. Prove that $\forall\ \mathrm{x}, \mathrm{y} \in \Re^p$, $\langle \mathrm{x}, \mathrm{y} \rangle_A = \mathrm{x}^T A \mathrm{y}$ is an inner product on $\Re^p \times \Re^p$ by verifying properties SP1 –SP3. Then prove that $\|\mathrm{x}\|_A = \sqrt{\mathrm{x}^T A \mathrm{x}}$ is a norm on $\Re^p$ by verifying N1–N3. Finally, verify properties M1–M3 for $d_A(\mathrm{x}, \mathrm{y}) = \|\mathrm{x} - \mathrm{y}\|_A$ to conclude that it is a metric on $\Re^p \times \Re^p$.

12. Let A and B denote real $n \times n$ matrices. Show that the function $tr(B)^T = \langle A, B \rangle$ is an inner product on the vector space of real matrices.

13. Prove the assertion in Definition 6. 4 that $A \in PD^{pp} \Leftrightarrow A^{-1} \in PD^{pp}$.

14. All norms are equivalent on finite dimensional normed vector spaces $\boldsymbol{VS}(\Re)$, which means that for any norms $\|*\|$ and $\|*\|_a$, there are constants κ and K so that $\kappa \|\mathrm{x}\| \leq \|\mathrm{x}\|_a \leq K \|\mathrm{x}\|\ \forall\ \mathrm{x} \in \boldsymbol{VS}(\Re)$. This is true for all finite dimensional inner product vector spaces, which are a subset of the finite dimensional normed vector spaces. Does this imply that Theorems 6.5 and 6.6 can be extended to the more general case of any inner product norm?

15. Verify that the reformulation of the HCM objective function given by Zha *et al.* (2001), namely, $J_1(U, \mathbf{V}; X) = tr(X^T X) - tr(P^T X^T X P)$, where P is the orthonormal matrix $P = diag\left(1/\sqrt{n_i}\right)$, $n_i = |X_i|$, is correct.

16. This problem is the basis of MacQueen's 1967 version of sequential hard c-means (or k-means). Suppose the first q points in the data set $X = \{\mathrm{x}_1, \ldots, \mathrm{x}_n\} \subset \Re^p$ belong to c crisp clusters $\{\mathrm{x}_1, \ldots, \mathrm{x}_c\}$ with $n_i = |X_i| : 1 \leq i \leq c$. Let the current c cluster centers (which are the mean vectors of the $\{X_1, \ldots, X_c\}$) be $\{\mathrm{v}_1^{(old)}, \ldots, \mathrm{v}_c^{(old)}\} \subset \Re^p$. Data point x_{q+1} arrives, and the distance from it to the c cluster centers is computed. Let $\mathrm{v}_j^{(old)}$ be the closest mean to this input so that x_{q+1} is to be added to cluster X_j. Prove that the new mean for cluster X_j is given by

$$\mathbf{v}_j^{(new)} = \frac{n_j \mathbf{v}_j^{(old)} + \mathrm{x}_{q+1}}{n_j + 1}. \tag{6.83}$$

17. Write a program to implement A6.1. Test it by verifying Example 6.3 with the Iris data (you can get Iris almost anywhere – just type "Iris Data" into your search browser). You will not get exactly the same results because initializations will vary, but your program should produce very similar results.

18. Probably you will have MATLAB. If you do, submit Iris to the MATLAB implementations of HCM and FCM and compare your outputs from problem 17 to the ones this commercial package produces. They should be very similar. Or, do this problem with a different toolbox, such as pyclus, the one implemented in Python.

19. Does writing your own code for A6.1 provide any advantage over simply running a commercially available version of it? Discuss.

20. There is a generalization of the HCM/FCM models and algorithms called which is defined by adding n positive weights $w = \{w_1, \ldots, w_n\}$ to the objective function at Equation (6.8), obtaining

$$J_{m,w}(U, \mathbf{V}; X, \mathbf{w}) = \sum_{i=1}^{c} \sum_{k=1}^{n} w_k u_{ik}^m \|\mathbf{x}_k - \mathbf{v}_i\|_A^2; \; m \geq 1. \tag{6.84}$$

Note that the weight vector **w** in Equation (6.84) is a user-defined parameter, i.e., this vector is not a part of the optimization variables. Equation (6.84) reduces to Equation (6.8) when all the weights are 1. Prove that the necessary conditions for U* to minimize Equation (6.84) at Equations (6.19b) and (6.25b) are valid for the weighted cases. Then prove that cluster centers for the weighted case must be computed as

$$\mathbf{v}_i^* = \sum_{k=1}^{n} w_k (u_{ik}^*)^m \mathbf{x}_k \Big/ \sum_{j=1}^{n} w_j (u_{ij}^*)^m; \; 1 \leq i \leq c. \tag{6.85}$$

The resultant models and algorithms are called *weighted HCM/FCM* (wHCM/wFCM).

21. Let $U \in M_{fcn}$ be a fuzzy c-partition of $X \subset \Re^p$. Compute a companion centroid $\mathbf{V}$ for this U with Equation (6.19a) at m = 2. Harden U per Equation (3.4), $U \mapsto \mathbf{H}(U)$. Recall that Dunn's index V_{DI} is computed with Equation (6.56). Prove that

$$V_{XB,2}(U, \mathbf{V}; X) \leq \frac{1}{[V_{DI}(\mathbf{H}(U))]^2}. \tag{6.86}$$

From Equation (6.86), we have that $V_{XB,2}(U, \mathbf{V}; X) < 1 \Leftrightarrow V_{DI}(\mathbf{H}(U)) > 1$, that is, if and only if $\mathbf{H}(U)$ is a CS partition of X in the sense of Dunn. This also shows that Dunn's index on $\mathbf{H}(U)$ goes to infinity if and only if the XB index on $U \in M_{fcn}$ goes to zero.

22. The proof of case (ii) in Proposition 6.1 depends on proving that $\left(1 + 2\varepsilon + 2\varepsilon^2 - \frac{2}{c}\right) > \frac{2}{c^2}$ holds for $\varepsilon \in (0, 1)$ and $2 \leq c \leq (n - 1)$. Convince yourself that this is true by comparing graphs of the two functions for a few values of c. What do the graphs tell you for $\varepsilon \in (0, 1)$ (in which case, there is no proof)? Now try to prove this in equality directly.

7

Probabilistic Clustering – GMD/EM

"You haven't told me yet," said Lady Nuttal, "what it is your fiancé does for a living." "He's a statistician," replied Lamia, with an annoying sense of being on the defensive. Lady Nuttal was obviously taken aback. It had not occurred to her that statisticians entered into normal social relationships. The species, she would have surmised, was perpetuated in some collateral manner, like mules. "But Aunt Sara, it's a very interesting profession," said Lamia warmly. "I don't doubt it," said her aunt, who obviously doubted it very much. "To express anything important in mere figures is so plainly impossible that there must be endless scope for well-paid advice on how to do it. But don't you think that life with a statistician would be rather, shall we say, humdrum?" Lamia was silent. She felt reluctant to discuss the surprising depth of emotional possibility which she had discovered below Edward's numerical veneer. "It's not the figures themselves," she said finally, "it's what you do with them that matters."

– The Undoing of Lamia Gurdleneck, K. A. C. Manderville

7.1 Introduction

This chapter is devoted to the most popular and heavily used probabilistic model and algorithm for cluster analysis. The model is called the mixture model, the objective function is a likelihood function to be maximized (hence, *maximum likelihood estimation* (MLE)), and the AO algorithm most often used to approximate solutions to the model is called the *expectation-maximization* (EM) algorithm. Since the EM algorithm *is* AO by definition, we need not specify it as EM/AO. There are other ways, such as the method of moments, to optimize the likelihood function that defines this model, but the EM algorithm is by far the most heavily favored choice.

The EM algorithm is driven by the statistical notion of maximum likelihood. The most important difference between this approach to clustering and the c-means schemes is that mixture analysis makes a number of assumptions about the statistical nature of the process that generates the unlabeled data. The c-means approach, on the other hand, is essentially free of distributional assumptions. In other words, c-means asks very little of the data, whereas clustering with EM asks a lot. Perhaps surprisingly, for many – but not all-data sets, there is little to choose between CPOV clusters found by the three (HCM, FCM, and EM) schemes discussed in Chapters 6 and 7. This is because all three models are at their best when doing the same thing: recognizing and capturing CPOV clusters that are essentially ellipsoidal-shaped "clouds" of points. The explanation for this appears in Section 9.3.1.

7.2 The Mixture Model

There is no lack of excellent texts on this subject, including Everitt and Hand (1981), Titterington *et al.* (1985), and Mclachlan and Peel (2000). The book by McLachlan and Basford (1988) has many applications of mixture models to clustering. This subject leads to some mathematical results that require a pretty deep understanding of the theory of probability and statistics. The treatment presented here remains true to the stated objective of any introductory text – it is basic.

Following the usual tradition in statistics, Greek letters are used for population parameters, and English letters are used for estimates of them. In the context of this chapter, the input data set is regarded as a set of n observations (a sample of size n) which are assumed to be drawn i.i.d. from a mixed population of c vector-valued random variables $\{\boldsymbol{X}_i; 1 \leq i \leq c\}$, that have prior probabilities (sometimes called *mixing proportions*) $\{\pi_i; 1 \leq i \leq c\}$, and class-conditional *probability density functions* (PDF's) $\{g_i; 1 \leq i \leq c\}$. The convex combination of these two sets of quantities is the PDF of the mixed distribution $\boldsymbol{X}$ of the $\{\mathbf{X}_i\}$.

$$f(\mathbf{x}) = \sum_{i=1}^{c} \pi_i g_i(\mathbf{x});\ 0 \leq \pi_i \leq 1; \sum_{i=1}^{c} \pi_i = 1. \tag{7.1}$$

The function at Equation (7.1) is called a *mixture* of the c components $\{\pi_i g_i(\mathbf{x})\}$. The densities $\{g_i(\mathbf{x})\}$ may come from different probability families, but this case is very difficult to handle. In practice, a much more typical situation is that all c densities come from the same family, such as a mixture of binomials or log-normals, etc. And for us, they will come from the simplest and most well-behaved and easy-to-analyze family of all – namely, they will always be Gaussian (aka "normal") distributions. The *posterior probability*, given $\mathbf{x}$, that it came from class i is denoted by $\pi(i|x)$. *Bayes rule* relates the elements of Equation (7.1) to the probabilities $\{\pi(i|\mathbf{x})\}$:

$$\pi(i|\mathbf{x}) = \frac{\pi_i g_i(\mathbf{x})}{f(\mathbf{x})};\ 0 \leq \pi(i|\mathbf{x}) \leq 1\ \forall\ i;\ \sum_{i=1}^{c} \pi(i|\mathbf{x}) = 1. \tag{7.2}$$

Suppose functional forms for the $\{g_i(\mathbf{x})\}$ in the mixture are known (for example, they are all multinomials, they are all Gaussians, some are multinomials, some are log-normals, etc.), but each $g_i(\mathbf{x})$ depends on an unknown vector $\mathbf{q}_i$ of parameters in *parameter space* Ω_i. In this case, the parametric form of Equation (7.1) is

$$f(\mathbf{x}; \mathbf{Q}) = \sum_{i=1}^{c} \pi_i g_i(\mathbf{x}; \mathbf{q}_i);\ \mathbf{Q} = (\mathbf{q}_1, \ldots, \mathbf{q}_c)^T \in \Omega = \Omega_1 \times \cdots \times \Omega_c. \tag{7.3}$$

The prior probabilities $\{\pi_i\}$ are usually also unknown; so, typically, the overall vector of unknown parameters for each component includes an estimate p_i of π_i. The first question that might occur to you is: given an unlabeled data set $X = \{\mathbf{x}_1, \ldots, \mathbf{x}_n\} \subset \Re^p$ drawn i.i.d. from $f(\mathbf{x}; \mathbf{Q})$[18], do these samples contain enough information about the parameters so that $\mathbf{Q}$ can – in principle at least-be recovered from them? This will be the case when the mixture is *identifiable*.

[18]Don't confuse this boldface **Q** used for statistical parameters in Chapter 7 with the unbolded Q which represents a reference comparison matrix in the notation for external cluster validity indices, $V(U|Q)$.

Definition 7.1. Identifiability for a mixture of unknown PDFs.

Assume that the samples in $X = \{\mathbf{x}_1, \ldots \mathbf{x}_n\} \subset \Re^p$ are drawn i.i.d. from a mixture distribution X of c random variables $\{X_i\}$ that have prior *probabilities* $\{\pi_i\}$, class-conditional PDF's $\{g_i(\mathbf{x}; \mathbf{q}_i)\}$, and mixture density $f(\mathbf{x}; \mathbf{Q}) = \sum_{i=1}^{c} \pi_i g_i(\mathbf{x}; \mathbf{q}_i)$; $\mathbf{Q} = (\mathbf{q}_1, \ldots, \mathbf{q}_c)^T \in \Omega = \Omega_1 \times \cdots \times \Omega_c$, the mixture

$$f(\mathbf{x}; \mathbf{Q}) \text{ is identifiable} \Leftrightarrow \forall\, \mathbf{Q}_1, \mathbf{Q}_2 \in \Omega,\ \mathbf{Q}_1 \neq \mathbf{Q}_2,\ \exists \text{ an } \mathbf{x}^* \in \Re^p \ni f(\mathbf{x}^*; \mathbf{Q}_1) \neq f(\mathbf{x}^*; \mathbf{Q}_2). \quad (7.4)$$

In words: $f(\mathbf{x}; \mathbf{Q})$ is identifiable, if and only if, for any two parameter vectors that are different, $\mathbf{Q}_1 \neq \mathbf{Q}_2$, there is a vector $\mathbf{x}^*$ for which the values of the mixture with these two parameters also differ. In other words, distinct values of the parameter vector $\mathbf{Q}$ should determine distinct values of the mixture density.

To verify Equation (7.4), you will need $\mathbf{Q}_1 \neq \mathbf{Q}_2$ and $\mathbf{x}^*$. But you don't know any of these vectors and there is no known general way that can tell you how to find them. Why show Equation (7.4) then? To alert you that the mixture model has some hidden assumptions. This is one of them. In practice, this aspect of the problem is usually ignored since it almost never causes a computational roadblock. Section 3.1 in Titterington *et al.* (1985) contains an excellent discussion and several examples on various aspects of this topic.

There are many principles of statistical inference that can be applied to the problem of parametric estimation-i.e., how to use a data set assumed to be distributed as f for some $\mathbf{Q}$ to find the best-fitting function in a parametric family. One of the most popular ways to estimate $\mathbf{Q}$ using unlabeled data begins with the *maximum likelihood estimation* (MLE) model. Given n samples $X = \{\mathbf{x}_1, \ldots \mathbf{x}_n\} \subset \Re^p$ that are drawn i.i.d. from a mixture $f(\mathbf{x}; \mathbf{Q})$, which is presumed to be identifiable, the likelihood function of $\mathbf{Q}$ is the joint density

$$L(\mathbf{Q}; \mathbf{X}) = \prod_{k=1}^{n} f(\mathbf{x}_k; \mathbf{Q}). \quad (7.5)$$

This equation expresses the likelihood of observing *these* samples X as a function of the parameter vector $\mathbf{Q}$. A maximum likelihood estimate of $\mathbf{Q}$ is a value $\mathbf{Q}^* \in \Omega$ that maximizes $L(\mathbf{Q}; \mathbf{X})$. Intuitively, this method produces a set of parameters that are most likely to have resulted in the observations that were used to secure the estimate. The mixture model that is analogous to the optimization problems for c-means at (Equations (6.9) and (6.10)) is as follows.

Definition 7.2. The Maximum Likelihood (ML) model for finite mixtures.

Let $\mathbf{X} = \{\mathbf{x}_1, \ldots, \mathbf{x}_n\} \subset \Re^p$ be an identifiable sample with respect to the mixture $f(\mathbf{x}; \mathbf{Q}) = \sum_{i=1}^{c} \pi_i g_i(\mathbf{x}; \mathbf{q}_i)$. The *MLE model* for $f(\mathbf{x}; \mathbf{Q})$ is the optimization problem

$$\underbrace{\text{maximize}}_{\mathbf{Q} \in \Omega} \left\{ L(\mathbf{Q}; \mathbf{X}) = \prod_{k=1}^{n} f(\mathbf{x}_k; \mathbf{Q}) \right\}. \quad (7.6)$$

Since logarithms convert products to sums, it is customary to work with the equivalent and much more convenient log-likelihood model

$$\underbrace{\text{maximize}}_{\mathbf{Q} \in \Omega} \left\{ \log L(\mathbf{Q}; \mathbf{X}) = \sum_{k=1}^{n} \log f(\mathbf{x}_k; \mathbf{Q}) \right\}. \quad (7.7)$$

This works because $L(\mathbf{Q}^*;\mathbf{X}) = \max\{L(\mathbf{Q};\mathbf{X})\} \Leftrightarrow L(\mathbf{Q}^*;\mathbf{X}) = \max\{\log L(\mathbf{Q};\mathbf{X})\}$. When the underlying component distributions are all normal (Gaussian) distributions, identification of the parameters in Equation (7.6) or Equation (7.7) is called *Gaussian mixture decomposition* (GMD). $L(\mathbf{Q};\mathbf{X})$ and $\log L(\mathbf{Q};\mathbf{X})$ are the *likelihood* and *log likelihood* functions for X.

$L(\mathbf{Q};\mathbf{X})$ and $\log L(\mathbf{Q};\mathbf{X})$ are analogous to c-means objective function at Equation (6.8). And the MLE model in Definition 7.1 corresponds to the c-means models in Definition 6.5. Solutions for Equation (7.7) can be sought without reference to a specific type of component PDFs, but to get further than very general equations by, for example, zeroing the gradient of $\log L(\mathbf{Q};\mathbf{X})$, specific forms for the $\{g_i(\mathbf{x})\}$ must be chosen. The general case leads to some pretty complicated and deep analysis, which deviates from the premise that this is a basic text. Consequently, the general (i.e., non-Gaussian) case of Equation (7.7) is not pursued further in this book.

The overwhelming majority of papers dealing with probabilistic clustering assume that the underlying component distributions are Gaussian. Finding a $\mathbf{Q}^*$ that solves Equation (7.7) in the GMD case is (in theory) a lot simpler than solving it in the general case. Even in this most favorable case, the maximum likelihood function can be unbounded. Kiefer and Wolfowitz (1956) give a simple example of this type of singularity for a Gaussian mixture of two univariate components with different unknown means and variances. Duda and Hart (1973) mention this, but they also point out that in many instances, finite local maxima of $L(\mathbf{Q};\mathbf{X})$ or $\log L(\mathbf{Q};\mathbf{X})$ can still provide very satisfactory results. The next subsection covers the essential facts about the multivariate normal distribution that we need for probabilistic clustering based on GMD.

7.3 The Multivariate Normal Distribution

Please refer to Chapter 4 of Johnson and Wichern (2007) for a more detailed and excellent treatment of definitions and results concerning the multivariate normal distribution. We are going to sample only those appetizers from the full menu that are needed to understand GMD.

Definition 7.3. The multivariate normal (Gaussian) distribution.

A random variable $\boldsymbol{X} = \{X_1, \ldots, X_p\}$ has a *Gaussian* (or Normal) distribution when its probability density function $\Re^p \rightarrowtail \Re^+$, has the form

$$g(\mathbf{x}) = \frac{e^{-\frac{1}{2}\|\mathbf{x}-\mu\|^2_{\Sigma^{-1}}}}{(2\pi)^{\frac{p}{2}}\sqrt{\det\Sigma}}; \tag{7.8a}$$

$$\mu = (\mu, \ldots, \mu_p)^T \text{ is the population mean vector;} \tag{7.8b}$$

$$\Sigma = [\mathrm{cov}\mathbf{X}] = \begin{bmatrix} \sigma_{11} & \cdots & \sigma_{1p} \\ \vdots & \cdots & \vdots \\ \sigma_{p1} & \cdots & \sigma_{pp} \end{bmatrix} \text{ is the population covariance matrix.} \tag{7.8c}$$

The norm in Equation (7.8a) is the square of the Mahalanobis norm induced by the inverse Σ^{-1} of the population covariance matrix Σ at Equation (7.8c). Recall that the sample-based form of the Mahalanobis norm induced by the sample covariance matrix S^{-1} appeared at Equation (6.50b). Σ (and hence Σ^{-1}) are positive-definite; Σ, $\Sigma^{-1} \in PD^{pp}$. The *i*th diagonal element of Σ is the variance $\sigma_{ii} = \text{var}(X_i)$ of the *i*th component of X, sometimes written as $\sigma_{ii} = \sigma_{ii}^2$. Off the diagonal, $\sigma_{ii} = \text{cov}(X_i, X_j)$ is the covariance between the pair of random variables X_i and X_j. We indicate that X is Gaussian by writing $X \sim N(\mu, \Sigma)$; in words, "X is normally distributed with mean μ and covariance matrix Σ."

This distribution dominates the statistical landscape for several reasons. The most important mathematical reason probably lies with the central limit theorem, which asserts that, for n samples $X = \{\mathbf{x}_1, \ldots, \mathbf{x}_n\} \subset \Re^p$ drawn i.i.d. from $X \sim N(\mu, \Sigma)$, the limiting distribution as n approaches infinity is $\sqrt{n}N(0, \Sigma)$. Many readers take this at face value and assume that almost all data sets are "eventually normally distributed." Of course, $n = 300$ is about the same distance from infinity as $n = 10^{18}$; so this result is psychologically seductive at best. It is a bad reason to assume that data are effectively drawn from the normal distribution.

Much more importantly perhaps, $N(\mu, \Sigma)$ is mathematically tractable. In particular, it is infinitely continuously differentiable, (relatively) easy to integrate, and there are a number of important theorems about it that ease the burden of mathematical manipulations that involve it. Moreover, the geometric properties of this distribution are well-known and are easy to develop. Starting with Equation (7.8a), for some real constants α, β and λ we have

$$g(\mathbf{x}) = \frac{e^{-\frac{1}{2}\|\mathbf{x}-\mu\|^2_{\Sigma^{-1}}}}{(2\pi)^{\frac{p}{2}}\sqrt{\det\Sigma}} = \alpha \Leftrightarrow e^{-\frac{1}{2}\|\mathbf{x}-\mu\|^2_{\Sigma^{-1}}} = \beta \Leftrightarrow \|\mathbf{x} - \mu\|^2_{\Sigma^{-1}} = \lambda^2 \tag{7.9}$$

Thus, level sets (sets of constancy) of the normal density are sets in the domain of g that have the form

$$L_g(\lambda) = \{\mathbf{x} \in \Re^p : \|\mathbf{x} - \mu\|^2_{\Sigma^{-1}} = \lambda^2\} \tag{7.10}$$

Figure 7.1 illustrates how a plot of the PDF $g(\mathbf{x}) = N(\mu, \Sigma)$ *would* look if we could see into p-space (and how it *does* look in the actual 3-space in this diagram). Note that the domain of g (i.e., $\Re^p$) lies in the horizontal plane (2-space) in Figure 7.1.

Level sets of the Mahalanobis norm at Equation (7.10) are an infinite family of hyperellipsoids in $\Re^p$ (ellipses in $\Re^2$) that take their shapes from the eigenvalues and eigenvectors of Σ^{-1}. The eigenvectors of Σ^{-1} determine the directions of the p axes, and the eigenvalues of Σ^{-1} determine the amount of stretching along each axis. Please note that the ellipsis in Figure 6.6 correspond to the topological structure associated with the inner product norm induced by $A = \Sigma^{-1}$. The assumption of normality for the samples in X induces this topology on $\Re^p$.

As λ increases, the concentric hyperellipsoids (ellipsis in Figure 7.1) that are level sets in $\Re^p$, the domain of g, get larger and larger and, thus, capture proportionately more and more of the probability associated with the distribution. Most importantly, if we *sample* a PDF such as the one in Figure 7.1 randomly and without replacement, we expect to gather a cloud of points in the domain space $\Re^p$, centered about the mean vector μ. The spread of the samples is determined by the eigenstructure of Σ^{-1}. Samples of this kind are sometimes called "Gaussian clusters," but a more accurate term would be Gaussian subsets - which might then become CPOV Gaussian clusters after a computational cluster analysis. For small values of p, the cloud of data points will be shaped much like the level sets corresponding to the eigenstructure of Σ^{-1}, but as p increases, the shape of a Gaussian cluster departs from what our intuition about cluster shapes and central tendency suggests. Dasgupta (1999) shows that for a large number of features (p), most of the probability mass of

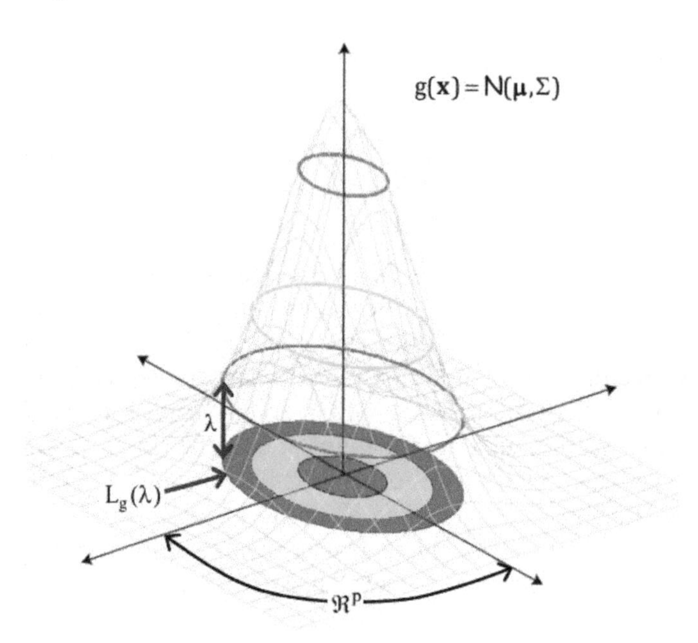

Figure 7.1 The p-variate Gaussian density (shown for p = 2).

$N(0, \sigma^2 I)$ is far away from the mean, concentrated in a thin shell of radius $\approx \sigma\sqrt{p}$. So, the intuitive meaning of the shape of "Gaussian clusters" becomes more and more uncertain as p increases.

When the component densities are all Gaussian, the p-variate mixture at Equation (7.3) becomes

$$f(\mathbf{x}; \mathbf{Q}) = \sum_{i=1}^{c} \pi_i g_i(\mathbf{x}; \mathbf{q}_i), \text{ where, } X_i \sim N(\mu_1, \Sigma_i), \ \ i = 1, \ldots, c, \tag{7.11}$$

where the component density g_i's have form of Equation (7.8a) and $\mathbf{q}_i = (\pi_i, \mu_i, \Sigma_i) \in \Omega_i = (0, 1) \times \Re^p \times PD^{PP}$. The function $f(\mathbf{x}; \mathbf{Q})$ at Equation (7.11) is called the *Gaussian mixture model* (GMM). To get labeled samples $X = \{\mathbf{x}_1, \ldots, \mathbf{x}_n\} \subset \Re^p$ from this mixture when the population parameters are known, we first choose class $i \in \{1, \ldots, c\}$ with probability π_i. Then we sample $g_i(\mathbf{X})$. Each of the component densities is sampled $n_i = \lfloor \pi_i/n \rfloor$ times, and the n_i samples X_i drawn from component i are i.i.d. as $\boldsymbol{X} =\sim N(\mu_i, \Sigma_i)$. The result of this sampling procedure is the n-sample $X = \bigcup_{i=1}^{c} X_i$. Since the samples in each X_i are labeled by class during the draw, if we wanted sample-based estimates of the parameters of each component PDF, the estimation problem for $\mathbf{Q}$ would degenerate into c separate problems of estimating $\mathbf{q}_i$ (not including the priors $\{\pi_i\}$) with the samples in X_i. This sampling procedure provides us with a crisp ground truth partition of X, $U_{GT} = [u_{ik}] \in M_{hcn}$.

Figure 7.2 illustrates samples drawn from a mixture of three 2D normal distributions. You see the samples in 3-space but, again, imagine them in p-space. Look in the horizontal plane (which represents p-space in our imaginations) to see the results of this sampling procedure for the three normal distributions shown there. What you see are three HPOV Gaussian clusters, that is, three clouds of points in the domain of the mixture density. The number of points in each cluster is n_i; the center of each cluster is μ_i. And the shape of each cluster is determined by the eigenstructure of Σ_i^{-1}.

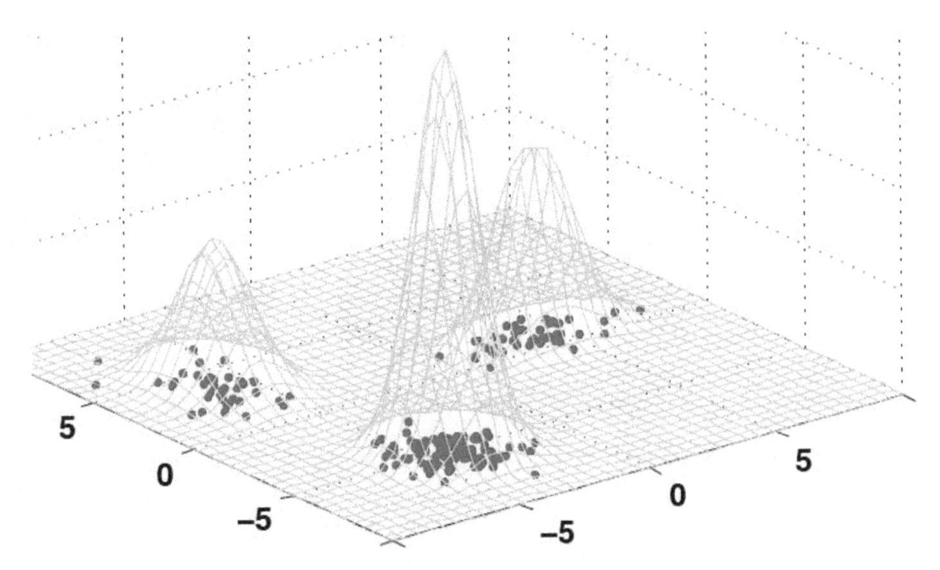

Figure 7.2 Samples drawn from a mixture of three 2D Gaussian distributions.

Figure 7.2 should make the correspondence between "good CPOV clusters" for HCM/FCM and GMD/EM quite transparent. All three models and their AO algorithms are good at finding CPOV clusters of points shaped like these, as long as there is "enough separation" between them. When the means are well-separated, all three algorithms will find roughly the same substructure, but as the means get closer, points from the components will start mixing, which in turn will mix up the algorithms trying to separate them. It is tempting to think that GMD will excel at finding Gaussian subsets, but, surprisingly, for the right norm in the objective function at Equation (6.8), the c-means models often do quite well, or even better, than GMD/EM. And, conversely, all three of these algorithms often *fail* in roughly the same way when the subsets in the data are not cloud-like. An example of this type is presented in Chapter 9.

Example 7.1. Visual display for a Gaussian mixture.

Figure 7.3(a) shows data set $X_{5,2}$, a sample of $n = 5000$ points in $\Re^2$ drawn from a mixture of $c = 5$ two-dimensional Gaussian distributions whose population priors, means, and covariances are shown in Table 7.1. The number of data points $n_i = |X_i|$ actually drawn for each HPOV cluster appears in the rightmost column of Table 7. 1. Note that $n_i/n \approx \pi_i$, but these values are not exactly equal due to the vagaries of sampling.

Samples drawn from each component are shown in different colors. Figures 7.3(b) and (c) are the VAT and iVAT images of the Euclidean distance matrix D_E of $X_{5,2}$ made with Algorithm A4.1. The VAT image $I(D_E^*)$ of the VAT-reordered data (D_E^*) is not very instructive or useful. The iVAT image $I(D_E'^*)$ of the transformed reordered distance data $I(D_E'^*)$ gives a pretty fair idea of the structure (including rough estimates of subset sizes) that seems to underlie the distance data for $X_{5,2}$. Bear in mind that iVAT doesn't *obtain* clusters for us: its job is to suggest how many clusters to seek.

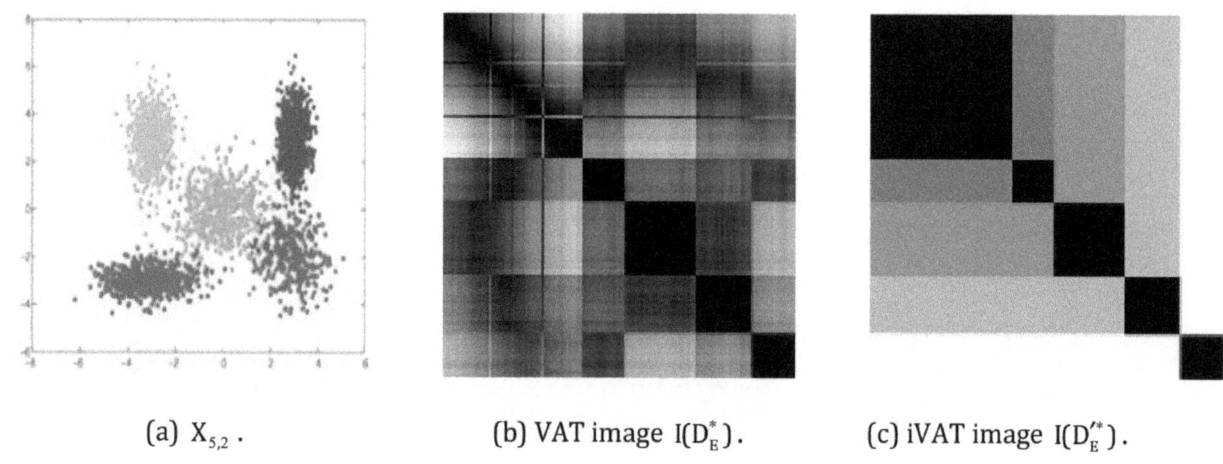

(a) $X_{5,2}$. (b) VAT image $I(D_E^*)$. (c) iVAT image $I(D_E'^*)$.

Figure 7.3 Scatterplot, VAT/iVAT images of c = 5 HPOV Gaussian clusters in $\Re^2$.

Table 7.1 Parameters of the mixture components for data set in Figure 7.3.

i	Prior π_i	Mean μ_i	cov = Σ_i	$n_i = \lvert X_i \rvert$
1	0.2	$(-3, -3)^T$	$\begin{bmatrix} 1 & 0 \\ 1 & 0.2 \end{bmatrix}$	1008
2	0.2	$(0, 0)^T$	$\begin{bmatrix} 1 & 0 \\ 0 & 1 \end{bmatrix}$	959
3	0.3	$(3, 3)^T$	$\begin{bmatrix} 0.1 & 0 \\ 0 & 1 \end{bmatrix}$	1506
4	0.1	$(3, -2)^T$	$\begin{bmatrix} 0.5 & 0 \\ 0 & 1 \end{bmatrix}$	484
5	0.2	$(-3, 3)^T$	$\begin{bmatrix} 0.2 & 0 \\ 0 & 1 \end{bmatrix}$	1043

Please take a moment to correlate the shapes and sizes of these HPOV clusters with the parameters of each component. The various clusters have different (roughly elliptical) shapes determined by the eigenstructure of the inverses of their covariance matrices. For example, the centrally located green cluster corresponds to draws from the essentially circular PDF specified by row 2 in Table 7.1, centered at $(0, 0)^T$ with the identity matrix for its covariance. The red cluster (row 1 in Table 7.1) is stretched in the x-direction because the eigenvalues of the inverse Σ_1^{-1} are in the ratio of 1:5. We will compare clustering outputs obtained with the HCM/FCM/GMD models and algorithms on this data set in Example 7.5. And we will see it's big brother in Figure 11.17 (which is also on the cover of this book).

The iVAT algorithm is one way to convert $X \subset \Re^p$ into a visual display for HPOV analysis (which, of course, is not really needed to see the HPOV clusters in Figure 7.3). Another way to visualize data structure is with scatterplots of low-dimensional projections. Our next example illustrates this approach using linear and non-linear projections of Gaussian subsets which begin in p = 5 dimensions and end up as 2D scatterplots using principal components analysis (PCA) and Sammon's method as given in Chapter 5.

Example 7.2. Feature extraction from a Gaussian mixture.

This example is abstracted from Runkler and Bezdek (2013), who generated c = 5 Gaussian subsets by sampling a mixture of p = 5 dimensional Gaussian distributions. Thirty points were drawn from each of the five Gaussian distributions, comprising a total of n = 150 points for X in the five-dimensional upspace $\Re^5$. All five distributions were circular, with unit variance along each axis. Population means of the five components were: $\mu_1 = (0, 0, 0, 0, 5), \mu_2 = (0, 0, 0, 5, 0), \mu_3 = (0, 0, 5, 0, 0), \mu_4 = (0, 5, 0, 0, 0)$, and $\mu_5 = (5, 0, 0, 0, 0)$.

The generation of this data yields five labeled subsets, which tempts us to believe that there must be five clusters in the upspace that would "look like" the two-dimensional clusters in Figure 7.3(a), but ... we can't see the 5D data; so there is no real HPOV about structure in X. To get a feel for what might be true, X was projected into $Y_{S,2} \subset \Re^2$ by optimizing Sammon's model (5.6) with Newton's method, and $Y_{PC,2} \subset \Re^2$ using the first two principal components of X as in Section 5.5.1. Distance matrices D_X and D_Y were Euclidean distance matrices.

Figure 7.4 is a CPOV scatterplot of the resulting feature vectors in $\Re^2$ (tagged by color). The PCA scatterplot in Figure 7.4(b) is not nearly as attractive as the output of Sammon's model in Figure 7.4(a) because the red, dark blue, light blue, and green points in Figure 7.4(b) are badly intertwined in the PCA scatterplot of $Y_{PC,2}$.

The only points in $Y_{PC,2}$ that are really clearly separated are the cluster of black ones at the bottom of view 7.4(b). The CPOV clusters in the Sammon output $Y_{S,2}$ are *almost* linearly separable. Only the red and green points closest to the origin (0, 0) prevent the five clusters in $Y_{S,2}$ from being linearly separable. But, be careful not to interpret this as "proof" that there are five CPOV clusters in $X \subset \Re^5$ – all we can learn from this display is what might be (or can't be) true, not what *is* certainly true. If we were training a classifier using the extracted 2D features, $Y_{S,2}$ would seemingly provide a design with close to 100% correct resubstitution classification rate. Building a good classifier trained with $Y_{PC,2}$ would be a challenge.

So, this is an example where a non-linear feature extraction model/algorithm seems to yield superior extracted features and a more informative scatterplot than the linear closed-form PCA approach. But this

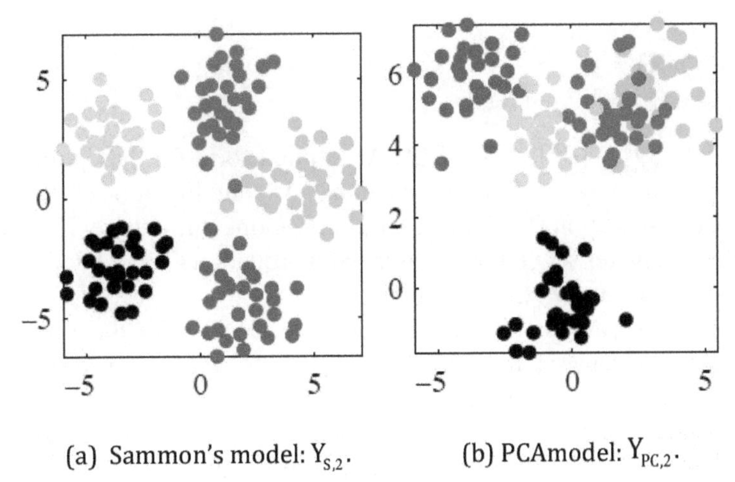

(a) Sammon's model: $Y_{S,2}$. (b) PCAmodel: $Y_{PC,2}$.

Figure 7.4 Sammon and PCA scatterplots of subsets drawn from a 5D Gaussian mixture.

success depends mostly on the data; there will be other data sets that reverse these results – i.e., PCA will render a more satisfactory visual assessment of high-dimensional data than Sammon's model. And for large data sets, Sammon's model becomes computationally unattractive.

7.4 Gaussian Mixture Decomposition

Now we are ready to continue with our development of the EM algorithm for approximating solutions to Equation (7.6) when the component PDF's are (or are assumed to be) Gaussian. If the samples from class i are i.i.d. as $\boldsymbol{X}_i \sim N(\mu, \Sigma_i)$, independence of the draws enables us to write the parameterized class conditional density as the product of the density for each sample:

$$f(X_i; \mathbf{q}_i) = \prod_{\mathbf{x}_k \in X_i} \pi_i g_i(\mathbf{x}_k; \mathbf{q}_i) = \prod_{\mathbf{x}_k \in X_i} \left[\frac{\pi_i e^{-\frac{1}{2}\|\mathbf{x}_k - \mu_i\|^2_{\Sigma^{-1}}}}{(2\pi)^{\frac{p}{2}} \sqrt{\det \Sigma}} \right] \tag{7.12}$$

Denote the natural logarithm to the base e by log. Since logarithms of products are sums of logarithms, we can rewrite Equation (7.12) in the form

$$\log(f(X_i; \mathbf{q}_i)) = \log \left(\prod_{\mathbf{x}_k \in X_i} \left[\frac{\pi_i e^{-\frac{1}{2}\|\mathbf{x}_k - \mu_i\|^2_{\Sigma^{-1}}}}{(2\pi)^{\frac{p}{2}} \sqrt{\det \Sigma_i}} \right] \right) = \sum_{\mathbf{x}_i \in X_i} \log \left[\frac{\pi_i e^{-\frac{1}{2}\|\mathbf{x}_k - \mu_i\|^2_{\Sigma^{-1}}}}{(2\pi)^{\frac{p}{2}} \sqrt{\det \Sigma_i}} \right] \tag{7.13}$$

Recalling that logarithms of quotients are differences of logarithms and that log(exp(z) = z now yields

$$\log(f(X_i; \mathbf{q}_i)) = -\frac{1}{2} \sum_{\mathbf{x}_k \in X_i} \|\mathbf{x}_k - \mu_i\|^2_{\Sigma^{-1}} \log \left(\frac{\pi_i}{\sqrt{\det \Sigma_i}} \right) - \frac{p}{2} \log(2\pi) \tag{7.14}$$

Summing Equation (7.14) over the c components and dropping the constant term leads to the log-likelihood function for the Gaussian mixture,

$$\log L(\mathbf{Q}; \mathbf{X}) = -\frac{1}{2} \sum_{i=1}^{c} \sum_{\mathbf{x}_i \in X_i} \|\mathbf{x}_k - \mu_i\|^2_{\Sigma^{-1}} + \sum_{i=1}^{c} \log \left(\frac{\pi_i}{\sqrt{\det \Sigma_i}} \right) \tag{7.15}$$

where $\mathbf{Q} = (\boldsymbol{\pi}, \mathbf{M}, \{\Sigma_i\}) \in (0,1)^p \times [\Re^p]^c \times [PD^{pp}]^c$. An explicit form for the optimization problem that defines GMD using MLE is obtained by substituting Equation (7.15) into Equation (7.7). The soft clusters we are after -viz., the probabilistic c-partition $P \in M_{fcn}$ of the observations (the data set $X = \{\mathbf{x}, \ldots, x_n\} \subset \Re^p$)-doesn't appear as part of the parameter set $\mathbf{Q}$ that is estimated in Equation (7.15), but the partition P we seek is coupled to $\mathbf{Q}$ through Bayes rule, as we shall see below.

Now we make the crucial assumption that is the basis for GMD with the EM algorithm, namely, that the vectors in the *unlabeled* data set $X = \{\mathbf{x}_i, \ldots, \mathbf{x}_n\} \subset \Re^p$ have been drawn i.i.d. from a normal mixture with all the attendant assumptions this entails. But ... we don't know which sample came from which component of the mixture. In this case, the unknown population parameters for class i are $\mathbf{q}_i = (\pi_i, \mu_i, \Sigma_i) \in \Omega_i = (0,1) \times \Re^p \times PD^{pp}$. Our aim is to use the unlabeled data set X to find *sample-based estimates* $\mathbf{q}_i^* = (p_i^*, \mathbf{m}_i^*, S_i^*)$ for $i = 1, \ldots, c$ of the population parameters for each component.

Wolfe (1970) derived the now very well-known first-order necessary conditions for the ML estimators $\mathbf{q}_i^* = (p_i^*, \mathbf{m}_i^*, S_i^*)$ of $\mathbf{q}_i = (\pi_i, \mu_i, \Sigma_i)$. Let $P^* = [p_{ik}^*] \in M_{fcn}$ denote the ML estimate for the matrix of posteriors and assume that for i = 1, . . . , c, $X_i \sim N(\mu_i, \Sigma_i)$. Wolfe proved the following.

Theorem 7.1. FONCs for GMD (Wolfe, 1970).

Let $X = \{\mathbf{x}_i, \ldots, \mathbf{x}_n\} \subset \Re^p$ be n samples drawn i.i.d. from an identifiable mixture $f(\mathbf{x}; \mathbf{Q}) = \sum_{i=1}^{c} \pi_i g_i(\mathbf{x}; \mathbf{q}_i)$ of c p-variate Gaussian distributions, each $g_i(\mathbf{x}; \mathbf{q})$ distributed as $X_i \sim N(\mu_i, \Sigma_i)$. Then $\mathbf{q}_i^* = (p_i^*, \mathbf{m}_i^*, S_i^*)$ *may be* a local maximum of $\log L(\mathbf{Q}; \mathbf{X}) = -\frac{1}{2}\sum_{i=1}^{c}\sum_{\mathbf{x}_k \in X_i} \|\mathbf{x}_k - \mu_i\|_{\Sigma^{-1}}^2 + \sum_{i=1}^{c} \log\left(\frac{\pi_i}{\sqrt{\det \Sigma_i}}\right)$ only if, for i = 1 to c and k = 1 to n,

$$p_i^* = \sum_{k=1}^{n} p_{ik}^*/n \tag{7.16a}$$

$$\mathbf{m}_i^* = \sum_{k=1}^{n} p_{ik}^* \mathbf{x}_k \Big/ \sum_{k=1}^{n} p_{ik}^* \tag{7.16b}$$

$$S_i^* = \sum_{k=1}^{n} p_{ik}^*(\mathbf{x}_k - \mathbf{m}_i^*)(\mathbf{x}_k - \mathbf{m}_i^*)^T \Big/ \sum_{k=1}^{n} p_{ik}^*(p_i; \mathbf{m}_i; S_i) \tag{7.16c}$$

$$p_{ik}^* = \frac{p_i^* g_i(\mathbf{x}_k; (p_i^*, \mathbf{m}_i^*, S_i^*))}{f(\mathbf{x}_k; (p_i^*, \mathbf{m}_i^*, S_i^*))} \tag{7.16d}$$

What does it mean? Conditions (7.16) are the basis for the AO algorithm called EM that is used to approximately maximize the (log) likelihood function for the optimization problem (7.6) in the special case that the component PDF's are p-variate Gaussian distributions. This theorem is entirely analogous to Theorem 6.2 (HCM/FCM). Similar to the c-means models, these conditions are found by zeroing the gradients of log L (**Q**;X) with respect to each set of unknowns to find first-order necessary conditions at local extrema. Here we will have four sets of uncoupled variables to estimate in the AO loop instead of the two (U and **V**) for the AO of the c-means models

Proof. [Theorem 3, Wolfe, 1970].

Suppose we have *labeled* data $X = \{\mathbf{x}, \ldots, \mathbf{x}_n\} \subset \Re^p$ drawn from the normal mixture at Equation (7.11). The *sample-based* prior probability vector in this case is known and belongs to the set N_{fc} of soft label vectors shown in Equation (3.2b). And for this data set, the matrix P of posterior probabilities, whose *k*th column is $\mathbf{P}^{(k)} = \mathbf{p}(*|k) \in N_{fc}$ for k = 1, . . . , n, is a partition matrix with exactly the constraints that are imposed on fuzzy partitions of X, i.e., $P = [P_{ik}] \in M_{fcn}$. In other words, this posterior probability matrix is a c-partition of X, which is *structurally* identical to fuzzy c-partitions of X (presumably it will be non-degenerate since it is built by sampling each component density at least once). Note especially that this partition P is *not* the crisp c-partition $U_{GT} = [U_{ik}] \in M_{hcn}$ that is connected to X by observation of class origins during the sampling procedure. Bayes rule (7.16d) relates the elements of P to the three sets of parameters $\mathbf{q}_i^* = (p_i^*, \mathbf{m}_i^*, S_i^*)$ for each of the c components of the mixture density.

Probabilistic partitions (P) from EM for GMD and fuzzy partitions (U) from AO for FCM lie in the same set, viz., M_{hcn}. Mathematically then, these two types of partitions are structurally identical. Is there any difference? Sure: the important philosophical difference between the two models is that the values $\{p_{ik} = p_i g(\mathbf{x}_k|i)/f(\mathbf{x}_k)\}$ are interpreted as posterior probabilities instead of fuzzy memberships.

There are also substantial mathematical differences. There is a huge difference in the assumptions laid against solutions of the optimization problems that define these two models. For GMD/EM, we have all of the assumptions for X shown in A7.1; for FCM, there are no assumptions about X other than distinctness of at least c points in the input data. Beyond this, it is fair to ask whether there is any real difference in the numbers p_{ik} and u_{ik}. Yes. This issue was discussed in some detail in Section 2.3. Since this question has been asked and (arguably) answered many times and in many ways since Zadeh (1965) introduced fuzzy sets, a brief extension of Section 2.3 is given here in the context of probabilistic clustering.

The value p_{ik} is the posterior probability that, given $\mathbf{x}_k$, it came from the *i*th mixture component $g_i(\mathbf{x})$. Since it is a probability, upon observation of $\mathbf{x}_k$ (and surreptitiously o_k, the object it represents), this number becomes either 1 (for a yes answer) or 0 (for a no answer). In other words, this number conveys some information about the frequency with which we might expect this event to occur before confirmation, and it always changes to a crisp value after observation (unless it was already 0 or 1).

On the other hand, the value u_{ik} is the membership that $\mathbf{x}_k$ has in cluster i; so it is a *measure of similarity* between $\mathbf{x}_k$ and the properties that define objects belonging to cluster i. Upon observation, not much will happen to u_{ik} if it has been meaningfully assigned. In short: probabilities measure expected frequency of occurrence, whereas memberships measure similarity to fuzzy properties. That's all there is to it!

> ☺ ***Forewarned is forearmed:*** The "soccer ball" structure of cn-partition space depicted in Figure 6.2 for FCM is exactly the same for the posterior probability matrix P at Equation (7.16d). So, don't be too surprised when it turns out that EM's search for good P's sometimes ends up in the same neighborhood as FCM's search for good U's. When P* and U* are close neighbors in M_{fcn}, you might think of them as belonging to different political parties because, sadly, many advocates of these two models are not very neighborly, i.e., not bipartisan at all. ☺

7.5 The Basic EM Algorithm for GMD

Alternating optimization to estimate extrema of the GMD log-likelihood function in the variables (P, **Q**) takes its form from the coupled equations in (7.16). If we know the posterior c-partition P at Equation (7.16d), we can compute $\mathbf{q} = (\pi_i, \mu_i, \Sigma_i) \in \Omega_i = (0,1) \times \Re^p \times PD^{pp}$, the three pieces of **Q** at Equations (7.16a,b,c), denoted as $G_{EM}(P)$; this is the *maximization* or *M step*. Conversely, if we know **Q**, we can calculate P at Equation (7.16d), denoted as $F_{EM}(\mathbf{Q})$: this is the *expectation* or *E-step*. Thus, one complete iteration through Equations (7.16) consists of one E step and then one M step, hence "EM" for this instance of alternating optimization.

Let the *t*th estimate of **Q** be $\mathbf{Q}_t = (\mathbf{q}_{1t}, \ldots, \mathbf{q}_{ct})$. Picard iteration through Equations (7.16) is represented in A7.1 as $\cdots P_t = F_{EM}(\mathbf{Q}_t) \rightarrow \mathbf{Q}_{t+1} = G_{EM}(P_t) \rightarrow \cdots$, where (t) is the iteration counter and the functions F_{EM} (the E step) and G_{EM} (the M step) yield the right-hand sides of Equations (7.16d) and (7.16a, b, c)), respectively. The basic EM algorithm A7.1 for GMD follows almost exactly the same path trodden by c-means algorithm A6.1 in the sense that both schemes are simply Picard iteration through their necessary conditions. Here is a listing for A7.1.

Algorithm A7.1. Basic EM algorithm for GMD.

	Algorithm 7.1 Basic EM for GMD
1	**In** Unlabeled feature vector data set $X = \{\mathbf{x}_1, \ldots, \mathbf{x}_n \subset \Re^p\}$
	Implicit assumptions about X:
	(i) All samples are drawn from $f(\mathbf{X}; \mathbf{Q}) = \Sigma_{i=1}^{c} \pi_i g_i(\mathbf{x}; \mathbf{q}_i)$;
	(ii) Samples drawn i.i.d. from $g_i(\mathbf{x} : \mathbf{q}_i)$ are distributed as $X_i \sim N(\mu_i; \Sigma_i)$;
	(iii) Pick class (i) with probability π_i : sample $g_i(\mathbf{x})$ $n_i = \pi_i/n$ times
	(iv) f is an identifiable mixture
2	***% User choices for model parameters***
3	**Set** Number of clusters: $c \in \{2, \ldots, n-1\}$:
4	***% User choices for implementation parameters***
5	Termination threshold: $\varepsilon \in (0, \infty)$:
6	Termination norm: $\| * \|_{err}$:
7	Iterate limit T: $t \in \{2, \ldots, T\}$
8	Initial guess: choose $\mathbf{Q}_0 = (\mathbf{p}_0, \mathbf{M}_0, \{S_i\}) \in N_{fc} \times [\Re^p]^c \times [PD^{pp}]^c$
9	$P_0 = F_{EM}(\mathbf{Q}_0) : \mathbf{Q}_1 = G_{EM}(P_0) : t = 1$
10	% $\mathbf{M}_t = (\mathbf{m}_{1,t}, \ldots, \mathbf{m}_{c,t})$ is the vector of estimated means at iterate t
11	***While*** $(t \leq T \text{ and } \|\mathbf{M}_t - \mathbf{M}_{t-1}\|_{err} > \varepsilon)$
12	$P_t = F_{EM}(\mathbf{Q}_t)$ % This is the **E** step
13	$\mathbf{Q}_{t+1} = G_{EM}(P_t)$ % This is the **M** step
14	***Endwhile***
15	$P^* = P_t : \mathbf{Q}^* = \mathbf{Q}_t$
16	**Out** (P*, Q*)

Default Choices for A7.1: $\varepsilon = 0.001$; $T = 100$. ***Initialization rule:*** Choose c, then randomly select: (i) a label vector of initial priors $\mathbf{p}_0 \in N_{fc}$: (ii) c *distinct* points in X as $\mathbf{M}_0$; (ii) c distinct matrices $S_i \in PD(\Re^{pp})$. ***Termination rule***: The termination norm is the *squared* Euclidean norm on successive estimates of the mean vectors of the c component densities, $\|\mathbf{M}_{t+1} - \mathbf{M}_t\|_{err} = \|\mathbf{M}_{t+1} - \mathbf{M}_t\|_2^2 \leq \varepsilon$. MATLAB's implementation of this algorithm terminates this way.

When the covariance structure of one or more components of the mixture is arbitrary, L(**Q:** X) doesn't always have a finite maximum. Nonetheless, approximate numerical solutions obtained with A7.1 are known to produce useful estimators in many real problems; so probabilistic clustering (that is, estimates of the matrix P) gotten this way are very popular. While the efficacy of this approach seems to depend on the data satisfying the statistical assumptions given in A7.1, it is virtually impossible to know if a given set of unlabeled feature data actually do satisfy them, even when $p < 3$. Most researchers bent on probabilistic clustering simply run A7.1 anyway. (Indeed, this is probably also true for A6.1, albeit no assumptions other than X having c clusters are made for HCM and FCM.) All three of these algorithms are reliably "well-behaved" in the sense that they almost always terminate quickly and quite independently of the theoretical circumstances governing local and global convergence (cf. Chapter 10). Even if the data are distributed as "Gaussian clouds" (and often they are not), CPOV cluster substructure found by EM for GMD can be very misleading. At the end of the day, exactly like HCM and FCM, we will be forced to rely on *cluster validity indices* (CVIs) to separate the wheat from the chaff in sets of candidate partitions offered by GMD/EM.

What about hardening probabilistic partitions when crisp ones are required? Just after Definition 3.3, we wrote that when P is a probabilistic c-partition, $\mathbf{h}(\mathbf{P}^{(k)})$ at Equation (3.4) is just Bayes rule: that is, decide $o_k \in$ class $i \Leftrightarrow p_{ik} = pr(i|o_k) > p_{jk} = pr(j|o_k) \forall j \neq i$. Ties are resolved arbitrarily. Our computations using P from GMD for partition accuracy relative to U_{GT}, as well as comparisons to crisp versions of FCM partitions, are based on using the function **h** at Equation (3.4) for this purpose.

Now that we have an AO (viz., the EM) algorithm to approximate solutions for GMD, and since GMD and FCM both produce partitions in M_{fcn}, it is natural to wonder if the soft CPOV partitions produced by A7.1 are similar to those produced by FCM, A6.1. Example 7.3 compares the partitions found by A6.1 and A7.1 on the data set X_{30}.

Example 7.3. Comparing soft clusters from FCM and GMD for data set X_{30}.

Recall that data set X_{30} has 3 tiny HPOV clusters of 10 points each (cf., Figure 4.11(a)). These data are not drawn from a mixture of c=3 Gaussians, but the three groups of points are (sort of) cloud-like; so you might expect FCM and GMD to find somewhat similar CPOV clusters (*do you?*). The coordinates of the 30 points are shown in columns 2 and 3 of Table 7.2. Subset means for the three sets of points are shown below the data.

The data was processed with the FCM/AO and GMD/EM algorithms. Parameters for FCM: $c = 3$, $m = 2$, $\mathbf{V}_0 = \{\mathbf{v}_{1,0} = (1.5, 2.5)^T, \mathbf{v}_{2,0} = (1.7, 2.6)^T, \mathbf{v}_{3.0} = (1.2, 2.2)^T\}$, termination occurred when $\|\mathbf{V}_t - \mathbf{V}_{t-1}\|_2 \leq \varepsilon = 0.01$, the model norm was Euclidean. GMD was initialized with c = 3, equal priors $\mathbf{p}_0 = (1/3, 1/3, 1/3)$, $\mathbf{M}_0 = \mathbf{V}_0$, and covariance matrices computed by plugging $(\mathbf{p}_0, \mathbf{M}_0)$ into Equation 7.16(c) to initialize the covariance matrices . Termination of A7.1 occurred when $\|\mathbf{M}_t - \mathbf{M}_{t-1}\|_2 \leq \varepsilon = 0.01$. The results are shown in Table 7.2. To two significant digits, the terminal GMD-AO partition is crisp and equal to the HCM partition U* of X_{30} that is shown in Figure 4.11(b). Does this surprise you? Presumably, GMD finds this data set quite compatible with the mixture assumptions underlying GMD.

The FCM partition in the last two columns of Table 7.2 is properly soft, but only barely so, at $m = 2$. These three HPOV clusters are pretty compact and well-separated; so this result should not come as a great surprise either. Note that hardening the fuzzy partition in Table 7.2 with **H** at Equation (3.4) yields the same results as GMD and HCM. If the fuzzification parameter m is set closer to 1, the few soft values seen for FCM in Table 7.2 become closer to those for GMD and HCM. Of course, at $m = 1$, FCM becomes HCM; so this is consistent with the theory of c-means models.

Table 7.2 Terminal EM and FCM partitions and prototypes for data set X_{30}.

Pt.	Data Pt.		GMD-EM			FCM-AO (m=2)		
	x_1	x_2						
1	1.5	2.5	1.00	0.00	0.00	0.99	0.01	0.00
2	1.7	2.6	1.00	0.00	0.00	0.99	0.01	0.00
3	1.2	2.2	1.00	0.00	0.00	0.99	0.01	0.00
4	2.0	2.0	1.00	0.00	0.00	0.99	0.01	0.00
5	1.7	2.1	1.00	0.00	0.00	1.00	0.00	0.00
6	1.3	2.5	1.00	0.00	0.00	0.99	0.01	0.00
7	2.1	2.0	1.00	0.00	0.00	0.99	0.01	0.00
8	2.3	1.9	1.00	0.00	0.00	0.98	0.02	0.00
9	2.0	2.5	1.00	0.00	0.00	0.99	0.01	0.00
10	1.9	1.9	1.00	0.00	0.00	0.99	0.01	0.00
11	5.0	6.2	0.00	1.00	0.00	0.00	1.00	0.00
12	5.5	6.0	0.00	1.00	0.00	0.00	1.00	0.00
13	4.9	5.9	0.00	1.00	0.00	0.01	0.99	0.00
14	5.3	6.3	0.00	1.00	0.00	0.00	1.00	0.00
15	4.9	6.0	0.00	1.00	0.00	0.01	0.99	0.00
16	5.8	6.0	0.00	1.00	0.00	0.01	0.99	0.00
17	5.5	5.9	0.00	1.00	0.00	0.00	1.00	0.00
18	5.2	6.1	0.00	1.00	0.00	0.00	1.00	0.00
19	6.2	6.2	0.00	1.00	0.00	0.02	0.97	0.01
20	5.6	6.1	0.00	1.00	0.00	0.00	1.00	0.00
21	10.1	12.5	0.00	0.00	1.00	0.01	0.02	0.97
22	11.2	11.5	0.00	0.00	1.00	0.00	0.01	0.99
23	10.5	10.9	0.00	0.00	1.00	0.01	0.04	0.95
24	12.2	12.3	0.00	0.00	1.00	0.00	0.01	0.99
25	10.5	11.5	0.00	0.00	1.00	0.00	0.02	0.98
26	11.0	14.0	0.00	0.00	1.00	0.01	0.02	0.97
27	12.2	12.2	0.00	0.00	1.00	0.00	0.02	0.98
28	10.2	10.9	0.00	0.00	1.00	0.01	0.05	0.94
29	11.9	12.7	0.00	0.00	1.00	0.00	0.01	0.99
30	12.9	12.0	0.00	0.00	1.00	0.01	0.03	0.96
			$\mathbf{m}_1$	$\mathbf{m}_2$	$\mathbf{m}_3$	$\mathbf{v}_1$	$\mathbf{v}_2$	$\mathbf{v}_3$
1.77	5.39	11.3	1.77	5.39	11.27	1.77	5.39	11.28
2.22	6.07	12.0	2.22	6.07	11.99	2.22	6.07	12.00

Example 7.4. Probabilistic clustering of the Iris data.

Refer to Example 5.3 for details of the Iris data set. We again pretend that Iris is unlabeled, and we ignore the assumptions posted in A7.1 about the distributions of the samples. Using the default parameters for A7.1, 100 runs of GMD/EM on the IRIS data using the same initialization scheme as given in Example 6.4 result in the values shown in Table 7.3, where all partitions carry the notation U (including P from GMD/EM). All other protocols were the same as in Example 6.4. Table 7.3 compares these results to HCM/FCM as discussed in Example 6.4.

Recall that the partition accuracy (PA) is max-optimal. It is clear from Table 7.3 that all three algorithms produce crisp partitions of Iris (after hardening in the cases of FCM and GMD) that are in good agreement with U_{GT} and also with each other. However, errors in the overall sample mean estimates produced by EM algorithm A7.1 for Iris yield a prototype match (recall that PM values are min optimal) of 0.59, more than twice the 0.23 prototype error achieved bythe FCM algorithm 6.1 (m=2), and almost twice the 0.35 value from HCM algorithm A6.1 (m = 1). These values are computed from the average estimates of the means shown in Table 7.3.

So, while the crisp 3-partitions have very nearly the same average partition accuracy (Table 7.3), both of the c-means algorithms produce significantly better estimates of the sample means than GMD does (Table 7.4), with the FCM centroids being slightly better than the HCM means. Finally, these solutions were not selected from amongst a group of candidates that include other values of c. That is a job for goodness of fit or cluster validity indices, some of which would have us believe that c = 2 is the best choice for CPOV clusters in Iris. And, since there is no way to find "real" HPOV clusters in Iris, judgment is left to the computational evidence to make a final choice.

Table 7.3 Hundred run averages: (PA) and (PM) on X_{Iris}.

	$PA(U\|U_{GT})(\uparrow)$	$PA(\mathbf{H}\|U_{GT})(\uparrow)$	$PM(\mathbf{V}\|\mathbf{V}_{GT})(\downarrow)$
HCM/AO	0.88	☠	0.35
FCM/AO	☠	0.89	0.23
GMD-EM	☠	0.90	0.59

Table 7.4 Sample means of X_{Iris} and average prototype approximations from HCM/FCM/GMD.

Subspecies	Feature	Sample Means $\mathbf{m}_{GT}$	Average Prototypes $\mathbf{v}_{HCM}$	Average Prototypes $\mathbf{v}_{FCM}$	Average Means $\mathbf{m}_{GMD}$
	SL	5.006	5.025	5.008	5.004
Setosa	SW	3.418	3.429	3.402	3.425
	PL	1.464	1.481	1.495	1.461
	PW	0.244	0.255	0.256	0.244
	SL	5.936	5.985	5.907	6.110
Versicolor	SW	2.770	2.777	2.767	2.817
	PL	4.260	4.518	4.390	4.576
	PW	1.326	1.493	1.411	1.477
	SL	6.588	6.732	6.736	6.371
Virginica	SW	2.974	3.035	3.040	2.934
	PL	5.552	5.562	5.593	5.198
	PW	2.026	1.983	2.028	1.874

7.6 Choosing Model and Execution Parameters for EM

The statistical assumptions in Line 1 of A7.1 add an additional consideration to the GMD clustering model and EM algorithm that is not present with the c-means schemes, which are interpreted entirely in terms of the question: are there good CPOV clusters in the data? This adds a level of complexity to our discussion of how to initialize and terminate A7.1 – are we interested in parametric estimates of **Q**; i.e., estimates of the population parameters $\{\mathbf{q}_i = (\pi_i, \mu_i, \Sigma_i) : 1 \le i \le c\}$ of the underlying component distributions? Or do we just want the CPOV soft clusters that reside in the partition P? Before attempting to cross this slippery slope, let us take a look at an alternative.

7.6.1 Estimating c With iVAT

Algorithms 6.1 and 7.1 share the common problem that the number of clusters (or, equivalently, the number of components of the mixture for GMD) must be specified (line 3 in A6.1 and A7.1) before execution begins. McLachlan and Basford (p. 16, 1988) begin their discussion on starting values for the EM algorithm with the somewhat tongue-in-cheek suggestion that we "initially partition the data into the specified number of groups ... ." This begs the question -how exactly do we know a good value for the number of components in the mixture?

Well, visual methods such as iVAT *do* provide a means for estimating c without recourse to partitioning the data. Figure 7.3(c) is a fine example of how to choose c by visual assessment of an iVAT image which enables you to move on to the next set of choices that are needed in lines 5, 6, 7, and 8 of A7.1. But the data in Figure 7.3(a) are available for HPOV inspection. What if the data reside in a high-dimensional space so that the HPOV is unavailable?

Example 7.5. Estimating the number of Gaussian components with an iVAT image of high-dimensional data.

In this example, we will build a 1000-dimensional data set beginning with the data set $X_{5,2}$ discussed in Example 7.1 whose scatterplot is shown in Figure 7.3(a). The parameters of the five component distributions that contribute the samples for each subset of $X_{5,2}$ are listed in Table 7.1, and repeated in columns 1 and 4 of Table 7.5. Consider a "generalization" of $X_{5,2}$ to a data set just like it (well, sort of just like it), called $X_{5,1000} \subset \Re^{1000}$, which has $n = 5000$ samples drawn from $c = 5$ Gaussian distributions in $p = 1000$ dimensions. There is no logical way to choose parameters for $X_{5,1000}$ which generalize the means and covariance matrices underlying the data set $X_{5,2}$ in any real sense; so we use the parameters shown in Table 7.5 for $X_{5,1000}$.

For example, the 1000×1000 covariance matrix corresponding to the first row in the last column of Table 7.5 is the diagonal matrix that has the two entries 1 and 0.2 repeated along its diagonal 500 times. It is easy to generate $X_{5,1000}$, and you might imagine that, if we could scatterplot this data, it would look something like the 2D Gaussian clouds in Figure 7.3(a). But in all likelihood, your imagination would be very wrong since, as we have noted above, Dasgupta (1999) has shown that the samples in $X_{5,1000}$ can hardly be envisioned as five compact clouds like those in $X_{5,2}$. But suppose you decide to model $X_{5,1000}$ with a Gaussian mixture, not knowing that c = 5. What to do? Well, we can make an iVAT image of the data, and see if it suggests anything useful; perhaps, it will enable us to select a number of components to look for.

Figure 7.5 is the iVAT image of the 5000×5000 Euclidean distance matrix $D_{E,5000}$ of $X_{5,1000}$. It takes some time to compute this matrix, which has 5000(4999)/2 off diagonal distances, each in $\Re^{1000}$-almost

Table 7.5 Parameters for samples in data set $X_{5,1000}$.

p = 2	→	p = 1000	p = 2	→	p = 1000
$(-3,-3)^T$	→	$(-3,-3,\ldots,-3)$	$\begin{bmatrix}1 & 0\\0 & 0.2\end{bmatrix}$	→	diag[1, 0.2,...,1, 0.2]
$(0,0)^T$	→	$(0,0,\ldots,0)$	$\begin{bmatrix}1 & 0\\0 & 1\end{bmatrix}$	→	diag[1]
$(3,3)^T$	→	$(3,3,\ldots,3)$	$\begin{bmatrix}0.1 & 0\\0 & 1\end{bmatrix}$	→	diag[0.1, 1,...,0.1, 1]
$(3,-2)^T$	→	$(3,-2,\ldots,3,-2)$	$\begin{bmatrix}0.5 & 0\\0 & 1\end{bmatrix}$	→	diag[0.5, 1,...,0.5, 1]
$(-3,3)^T$	→	$(-3,3,\ldots,-3,3)$	$\begin{bmatrix}0.2 & 0\\0 & 1\end{bmatrix}$	→	diag[0.2, 1,...,0.2, 1]

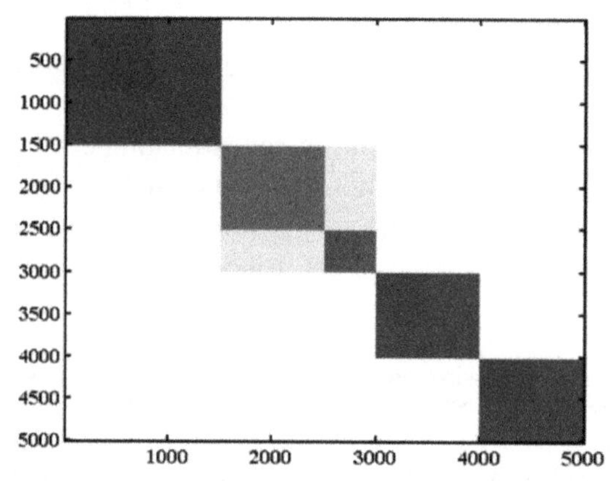

Figure 7.5 iVAT image of $X_{5,1000}$ in upspace $\Re^{1000}$.

more time than it takes to make the 5000×5000 iVAT image from $D_{E,5000}$! This image clearly suggests that $X_{5,1000}$ does indeed seem to have c = 5 clusters; so this affords us a way to begin exploration of the data with GMD algorithm A7.1.

To complete this example, let us have a look at the scatterplots and iVAT images of the best and worst rogue random projections of $X_{5,1000}$ into projected sets, say $Y_{5,2}$, in q = 2 dimensions using the method outlined in Example 5.6 of Section 5.5.2. Here best and worst mean largest and smallest values of the Pearson correlation coefficient at Equation (5.5) in 100 trials. The best and worst sets $Y_{5,2}$ are representations of the upspace data, i.e., $X_{5,1000} \triangleright Y_{5,2}$. At its best, the rogue random projection of $X_{5,1000}$ in the upper left view of Figure 7.6 is not a bad replica of the original data set $X_{5,2}$ in Example 7.1; and this is evident in its iVAT image. At its worst, the projected data look like one big cluster -just ignore the colors to see that the HPOV clusters in Figure 7.6(b) do indeed suggest one big blob. The iVAT image of the worst projection confirms this visual assessment - there is no apparent structure in the iVAT image in view 7.6(b).

This example illustrates two things. First, an iVAT image of an unlabeled data set in any number of dimensions can provide a simple and relatively easy way to estimate the number of clusters that X might contain and, hence, the number c you must provide in line 3 of Algorithms A6.1 and A7.1. And, second, random

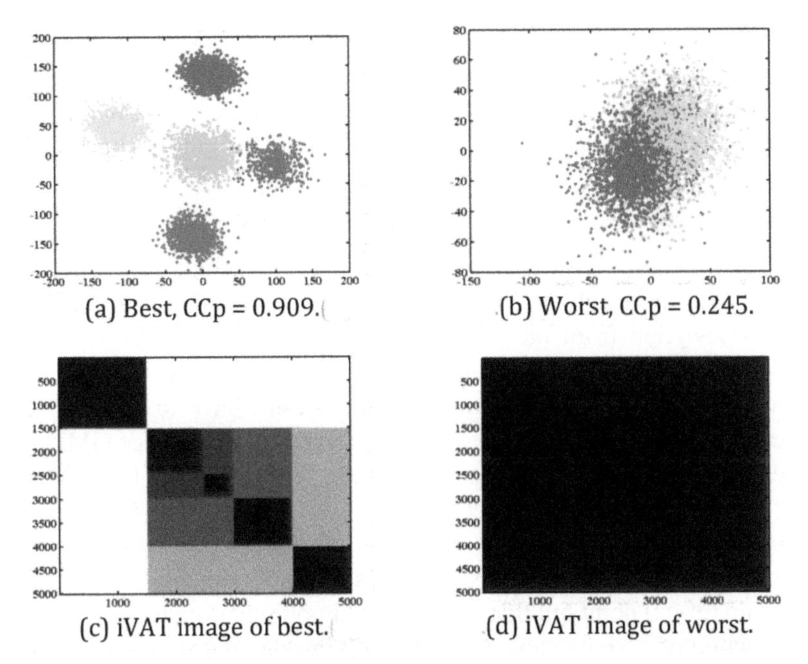

(a) Best, CCp = 0.909. (b) Worst, CCp = 0.245.

(c) iVAT image of best. (d) iVAT image of worst.

Figure 7.6 Scatterplots and iVAT images of RRPs of $X_{5,1000}$ in downspace $\Re^2$.

projection of high-dimensional "Gaussian clusters" can produce good looking Gaussian representations in even two dimensions – ***sometimes***.

7.6.2 Choosing Q_0 or P_0 in GMD

The AO sequence for EM $\cdots \mathbf{Q}_{t-1} \to P_t \to \mathbf{Q}_t \to P_{t+1}$ can be started with either P_0 or $\mathbf{Q}_0$, and as with c-means, the two choices are theoretically equivalent. The economy in time and space realized by starting and stopping with **Q** instead of P is not quite as dramatic as it is for c-means because here **Q** has $c(1 + p + p(p-1)/2)$ parameters to store. But for most data sets, $c(1+p+p(p-1)/2) < n$; so it still seems better to initialize and terminate EM with **Q**. After c is selected by guessing, with iVAT, or by whatever method suits best, you are best served by initializing A7.1 with a set of estimates for the parameters $\mathbf{Q}_0 = (\mathbf{q}_{10}, \ldots, \mathbf{q}_{c0})$ of the Gaussian components. But it is the choice of c that matters most.

In the face of uncertainty about a good value for c, one approach, exactly like the c-means case, is to generate a set of candidate parametric estimates and partitions (cf., Definition 4.4) at various values of c, and then try to choose a best member from amongst the candidates, where "best"is defined as an optimal value of some goodness of fit or cluster validity index, as discussed in the next section. This method acknowledges our ignorance about the true but unknown possible structure in X, including the value of c, the number of components in the mixture. But we still have to start A7.1 somewhere.

There are many studies that propose a scheme for initializing A7.1 that are arguably better than the default given after A7.1, viz., random selection of c vectors in the input data. McLachlan and Basford (p. 16, 1988) suggest finding a crisp partition $U_0 \in M_{hcn}$ of X by some *ad hoc* criterion. Then this partition is used to compute the initial M-step values of the quantities in Equation (7.16a-c), viz., $\mathbf{Q}_0 = G_{EM}(U_0)$.

For example, you could obtain U_0 with HCM algorithm A6.1 or SL algorithm A8.1, but you still need an estimate for c to make this method work. More generally, since $P_t = F_{EM}(\mathbf{Q}_t) \in M_{fcn}$, you could certainly initialize A7.1 with a fuzzy partition $U_0 \in M_{fcn}$ from FCM A6.1 as is done in Example 7.5 below, but again, you are up against the problem that c must be known in order to find U_0 to initialize A7.1 this way.

7.6.3 Implementation Parameters ε, $\|*\|_{err}$,T for GMD With EM

Picking the parameters ε, $\|*\|_{err}$, T, for A7.1 is in very similar to these making these choices for HCM/FCM algorithm A6.1; so begin your thinking about these issues with a look back to Section 6.6.2 for a detailed discussion of these items. Deviation from our default policy is to initialize A7.1 with P instead of $\mathbf{Q}$. This has unfortunate implications for memory requirements, but, sometimes, it is the only way to make a fair comparison with c-means outputs, whose initialization doesn't have to account for the priors and covariances. We will specify the initialization procedure in examples using EM for GMD.

Note that the log likelihood function $L(\mathbf{Q}: X)$ doesn't appear in A7.1, just as $J_m(U, \mathbf{V}; X)$ doesn't appear in A6.1. These objective functions define the models and are used to derive the AO equations, but they are subsequently ignored in the basic implementations given in this volume. Just as for c-means, it is again fair to ask why we don't recommend using successive values of the likelihood function as the termination criterion. Figueiredo and Jain (2002) point out that the log likelihood function, when viewed as a function of (c), is a non-decreasing function of c; so its value is not helpful in selecting the best number of components in a Gaussian mixture for the across-c problem.

Moreover, in the absence of any certain knowledge about what value of $L(\mathbf{Q}: X)$ to shoot for, it is impossible to prescribe a realistic estimate of how much change in this function indicates a good termination point for EM. For the clustering problem, we are after values of the soft partition P of X and possibly of the parameters $\mathbf{Q}$ as well (and not the value of $L(\mathbf{Q}: X)$); so it makes little sense to stop EM iteration unless successive values of one of these sets of unknowns is close. Successive values of $L(\mathbf{Q}: X)$ compress the information about the (cn) values of P or the $c(1+p+p(p-1)/2)$ values of $\mathbf{Q}$ into a single number, much like the average of many real numbers. As previously noted, you can certainly terminate A6.1 or A7.1 on successive values of their objective functions, but, again, it seems clear that too much detail is lost when this is done.

The termination criterion used by A7.1 mimics that of A6.1 by stopping EM when successive sets of estimates of the estimated means (the cluster centers) of the component densities are close. This affords the most direct comparison with the c-means algorithms, which are also terminated this way for the examples in this volume. Convergence theory will be discussed in Chapter 10. Convergence theory for EM is somewhat more complicated than the corresponding theory for FCM because the statistical structure underlying GMD requires consideration of some forms of convergence in probability that don't apply to fuzzy c-means.

The last example in this subsection studies the quality of GMD for extracting partitions from X when the input data matches the assumptions implicit in the GMD model, viz., X is a set of HPOV Gaussian clusters. Specifically, we return to the data shown in Figure 7.3(a).

Example 7.6. Finding probabilistic clusters in $X_{5,2}$.

Recall that the data set used in Example 7.1 and shown in Figure 7.3(a) consists of 5000 points drawn i.i.d. from the mixture of c = 5 Gaussian distributions with parameters as listed in Table 7.1. The draws were made according to the assumptions listed under "inputs" in line 1 of A7.1. This scheme led to the distribution of points with class labels from the draws as shown in Table 7.6. We also relabeled the points by assigning the (possibly new) *Bayes label* k to each sample **x** when the maximum *a posteriori* probability that **x** came from distribution k was maximum as computed with Equation (7.16d). Thus, we have two different "ground truth" crisp partitions of $X_{5,2}$: U_{GT} and U_{GT*} corresponding to the labels in Table 7.6.

Figure 7.7 shows the decision boundaries (quadratic Voronoi diagram, Theodoridis and Koutroumbas, 2009) defined by the Bayes labels. When the components of the mixture are completely known, some writers refer to these decision regions as the "statistically optimal Bayes classifier." Table 7.7 lists the results of processing $X_{5,2}$ averaged over 100 runs of FCM algorithm A6.1 (m = 2) and GMD algorithm A7.1, using the default choices for all parameters for these two algorithms (c = 5 was preset). Table 7.7 also lists the average terminal estimates of the five cluster centers. The notation GMD/HCM in Table 7.7 indicates the average results of 100 runs of GMD when initialized with the cluster centers obtained with hard c-means A6.1.

It is a bit surprising that FCM provides an average relabeling accuracy (PA) that is about 10% better than GMD on $X_{5,2}$. This data set would seem to possess a good match to the assumptions underlying A7.1, but here, at least, GMD doesn't do quite as well as FCM. This happens more often than unwary users might

Table 7.6 Labels for the samples shown in Figure 7.3(a).

Class	Class Labels U_{GT}	Bayes Labels U_{GT*}
1	1008	1018
2	959	946
3	1506	1509
4	484	477
5	1043	1050

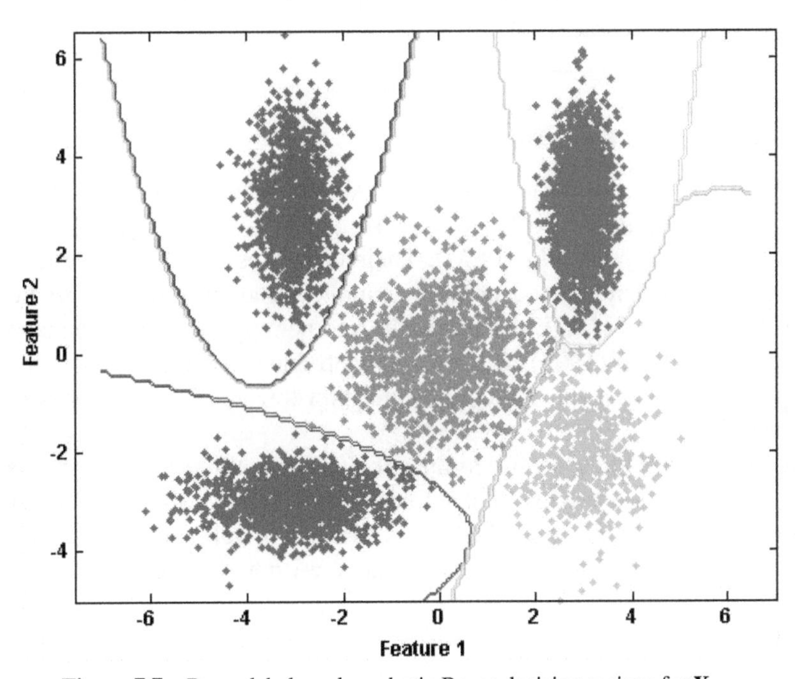

Figure 7.7 Bayes labels and quadratic Bayes decision regions for $\mathbf{X}_{5,2}$.

Table 7.7 Average partition accuracy, prototype matching, and prototype estimates using FCM and GMD on $X_{5,2}$.

	$PA(\mathbf{H}(U)\|U_{GT})(\uparrow)$	$PA(\mathbf{H}(U)\|U_{GT*})(\uparrow)$	$PM(\mathbf{V}\|V_{GT})(\downarrow)$	
FCM	0.976	0.982	0.466	☠
GMD	0.866	0.871	5.461	☠
GMD/HCM	0.918	0.928	2.170	☠
Cluster Center	Population Means μ	Ave. FCM Prototypes $\mathbf{V}_{FCM}$	Ave. GMD Means $\mathbf{M}_{GMD}$	Ave. GMD/HCM Means $\mathbf{M}_{GMD/HCM}$
1	$(-3,-3)^T$	$(-3.05,-2.97)^T$	$(-2.55,-2.69)^T$	$(-3.00,-2.97)^T$
2	$(0,0)^T$	$(-0.05,-0.04)^T$	$(0.96,0.33)^T$	$(0.59,0.55)^T$
3	$(3,3)^T$	$(2.97,3.04)^T$	$(2.88,2.95)^T$	$(2.98,3.12)^T$
4	$(3,-2)^T$	$(2.87,-1.90)^T$	$(2.77,0.27)^T$	$(2.93,-1.06)^T$
5	$(-3,3)^T$	$(-2.97,3.07)^T$	$(-2.30,3.04)^T$	$(-2.96,2.99)^T$

suspect. The GMD results can be substantially improved by initializing A7.1 in a somewhat better way than by simply randomizing the starting point. To see the effects of this strategy, the last column of Table 7.7 reports the results of the algorithm "GMD/HCM," which denotes the use of HCM (A6.1, $m = 1$) to find a good initial guess for starting GMD algorithm A7.1. This strategy is in line with the suggestion of McLachlan and Basford mentioned above.

Specifically, we ran HCM with its default parameters to termination 100 times and then transferred the terminal prototypes $\mathbf{V}^*_{HCM}$ to an initial guess for the means $\mathbf{M}_{GMD,0}$. Assuming initial priors, $p_i = 1/5, i = 1, \ldots, 5$ enables computation of an initial set of covariance matrices for GMD, and AO in A7.1 then proceeds as usual. The final mean values for this hybrid GMD approach listed in Table 7.7 show that it improves the average partition accuracy about 5% and decreases the average prototype error by about 50% -a substantial improvement, but still not quite as accurate as the FCM results.

7.6.4 Acceleration Schemes for GMD With EM

Some initial attempts to accelerate the basic method of A7.1 followed Aitken's (1926) approach. An early departure from Aitken's idea is due to Peters and Walker (1978), who give a direct approach for acceleration of A7.1 called *parametrized EM* (PEM). Their idea is essentially a steepest ascent scheme. Darken and Moody (1992) present an adaptive learning technique based on stochastic gradient ascent that is the basis of one version of the *online EM* (OEM) algorithm. Another approach based on conjugate gradients is offered by Jamshidian and Jennrich (1993). A third approach due to Lange (1995) uses classical quasi-Newton optimization to gradually steer the basic EM algorithm towards the Newton-Raphson algorithm (which converges quadratically).

Ortiz and Kaelbling (1995) provide a useful comparison of these three approaches (seven algorithms based on three methods) and give three numerical examples on small ($n = 2000$) sets of samples from pairs of 2D Gaussian mixtures. The results are mixed: PEM is fastest on well-separated clusters, but speedup based on conjugate gradients is quite good when the clusters are not well-separated. Reported speedups were in the range 1–12:1.

Moore (1998) presents an acceleration method for GMD/EM algorithm A7.1 based on representation of the estimates found at each iteration in a multi-resolution kd-tree he calls *Fast EM*. This method is illustrated using synthetic sets of Gaussian clusters comprising c = 5 to 320 components in p = 2 to six dimensions having n = 12, 500 to 640, 000 samples. Table 1 in Moore (1998) beautifully summarizes the results by comparing Fast EM to basic GMD/EM as each of the parameters n, p, and c varies with the other two fixed in the synthetic data. He observes that as n increases with c and p fixed, the computational advantage of fast EM increases more or less linearly beyond something like n = 50, 000, with speedup of 795:1 at n = 640, 000. As p increases with n and c fixed, the benefits realized by using a kd-tree in fast EM drops from 300:1 at p = 2 to 9:1 at p = 6. Finally, with n and p fixed, fast EM shows a speedup in the range of 100 to 300 to 1 as c runs from 5 to 320.

Thiesson *et al.* (2001) present two schemes for accelerating A7.1 based on partial E-steps (line 12 in A7.1). The first approach uses a version of the incremental EM algorithm, which cycles through data cases in blocks. The number of cases in each block determines the efficiency of the algorithm. This method aims at selecting a near optimal block size. The second approach, called *lazy EM*, is based on the assumption that not all of the data is equally significant for each iteration. This idea is philosophically related to the acceleration methods for HCM discussed in Chapter 6 which use the triangle inequality to avoid some of the cluster center calculations. Thiesson's approach identifies significant cases and then proceeds for several iterations using only the significant cases. Experiments were conducted on two real-world data sets and a number of synthetic data sets which were samples of Gaussian mixtures. Their results for the two real-world data sets (MSNBC and Speech) showed an average speedup in the range 1.7–2.8:1.

Biernacki *et al.* (2003) compare four methods for accelerating GMD/EM A7.1 with better initial guesses: (i) random initialization; (ii) a *classification EM* (CEM) algorithm; (iii) a *stochastic EM* (StEM) algorithm; and (iv) *previous short runs of EM* (SrEM) itself. They conclude that simple random initialization is often outperformed by all three of the experimental methods and that SrEM is probably the best of the three options studied.

Walker and Ni (2011) discuss using the method of Anderson Acceleration (cf., Section 6.6.3) to speed up GMD/EM algorithm A7.1. They report an experiment that used acceleration factor ($\eta_{aa} = 3$), and data sets of n = 100, 000 unlabeled draws from a mixture of three univariate Gaussian distributions with fixed, known prior probabilities (mixing proportions) and variances, specifically, $(p_1, p_2, p_3) = (0.3, 0.3, 0.4)$; $(\sigma_1, \sigma_2, \sigma_3) = (1, 1, 1)$. The experiments examined the rate of convergence for three different sets of means: $\mu_1 = (0, 2, 4)$, $\mu_2 = (0, 1, 2)$, and $\mu_3 = (0, 0.5, 1)$. The first set of means provides pretty well-separated samples, the second set yields intermediate mixing of samples, and the last set results in rather well-mixed samples. Figure 7.8 (Figure 1 in Walker and Ni) shows the results of these experiments for the three cases.

The convergence of the unaccelerated algorithm (the blue dashed lines) is dramatically altered by moving the means closer together. The bottom curve is the well-separated case, and the top curve is the poorly separated case. On the other hand, the Anderson accelerated iterates shown in red converge much more rapidly and are not significantly affected by decreasing the separation of the mixture sub populations.

7.7 Model Selection and Cluster Validity for GMD

The assumption that data are distributed as a normal mixture can be viewed in two ways. First, this may be an appealing framework to model unknown distributional shapes. The GMD/EM algorithm can estimate the parameters $\mathbf{Q}^*$ of the (assumed) Gaussian components as stated in line 1 of A7.1. ***Or*** ... we can interpret the use the mixture model as an expedient to generate probabilistic clusters P* in the data. In the first case,

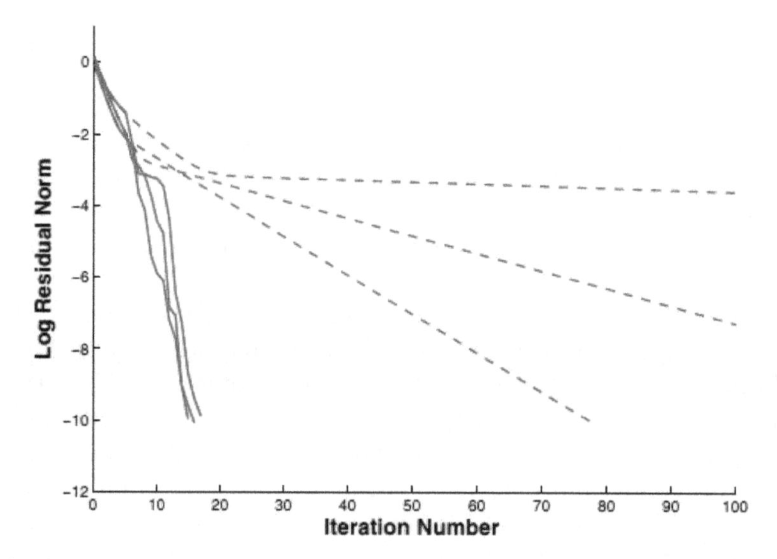

Figure 7.8 Convergence of GMD/EM with (red solid) and without (blue dashed) Anderson Acceleration.

there is the question of how many components to include in the normal mixture model, and, in the second, the question of how many clusters to accept. And, of course, the two cases are inextricably coupled to each other through Bayes rule; so either question automatically begs the other. While many authors simply mix and match these two interpretations, it seems useful to formalize these two points of view about outputs of the GMD/EM model and algorithm.

7.7.1 Two Interpretations of the Objective of GMD

The two interpretations of GMD are given in separate definitions to emphasize their differences.

Definition 7.4. Parametric estimation and Goodness of Fit Indices (GOFIs).

A set of estimates $\mathbf{Q}^* = (\mathbf{p}^*, \mathbf{M}^*, \{S_i^*\})$ corresponding to Equations (7.16a-c) is interpreted as a set of parameters providing the best fit Gaussians to the data in hand. Evaluation of a pool of N candidates $\mathbf{Q}_1^*, \mathbf{Q}_2^*, \ldots, \mathbf{Q}_N^*$ is done with *goodness of fit indices* (GOFIs).

Definition 7.5. Probabilistic clustering and Cluster Validity Indices (CVIs).

The number of components declared for a Gaussian mixture is equal to the number of CPOV clusters that are presumed to be in the data set X, and the posterior probability matrix $P^* \in M_{fcn}$ at Equation (7.16d) produces probabilistic CPOV clusters. Evaluation of a pool of N candidates $\{P_1^*, P_2^*, \ldots, P_N^*\}$ is done with *cluster validity indices* (CVIs).

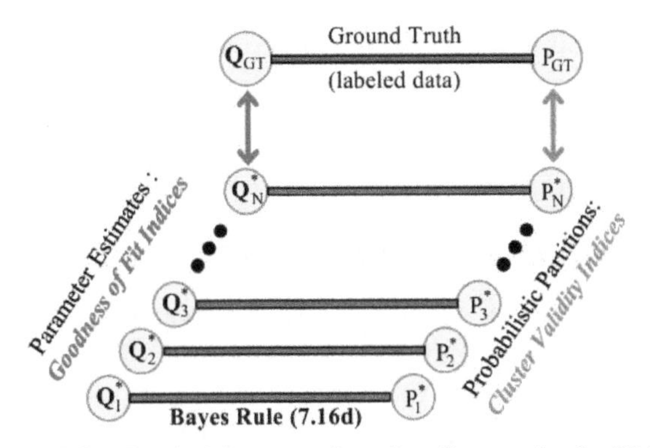

Figure 7.9 The dual interpretation of quality metrics for GMD.

For unlabeled data to be fitted with a mixture of Gaussian components, both interpretations of EM outputs are hopelessly intertwined with the question – how many components comprise the mixture? And, hence, at what value of c shall A7.1 be initialized? It should come as no surprise that, for $p > 3$ at least, since we don't (and never will) know the true but unknown value of c (assuming the data do come from a mixture of Gaussians), we usually resort to post-clustering evaluation of pools of candidates with indices that assess goodness of fit (Definition 7.4) or cluster validity (Definition 7.5). Figure 7.9 illustrates the dual nature of evaluation of the results of applying GMD to unlabeled data.

Interpretation of the implication of Definition 7.4 is shown on the left side of Figure 7.9, where we see GOFIs being used to choose a best set of parametric estimates $\mathbf{Q}^* = (\mathbf{p}^*, \mathbf{M}^*, \{S_i^*\})$ for the Gaussian components of an assumed mixture from N candidates $\mathbf{Q}_1^*, \mathbf{Q}_2^*, \ldots, \mathbf{Q}_N^*$. On the right, per Definition 7.5, CVIs are used to identify a best probabilistic partition P^* of the input data from $\{P_1^*, P_2^*, \ldots, P_N^*\}$. Exercising a choice on either side of this diagram automatically couples it through Bayes rule to the corresponding estimate on the other side of this diagram (think of the coupling like the ends of a rigid barbell as depicted in Figure 7.9). Please keep a clear mind about the distinction between the two types of validation measures.

☺ ***Forewarned is forearmed:*** The literature about what type of post-clustering evaluation is being made on GMD/EM outputs can be confusing since either type automatically makes a choice for the dual structure as well. Hence, you will sometimes see GOFIs such as *Akaike's* (1974) *information criterion* (AIC), which deals explicitly with **Q** and not P, called cluster validity indices. And, conversely, measures such as the soft generalization of *mutual information* (MI; Thomas and Cover, 1991), which deals exclusively with P and not **Q**, are called GOFIs. ☺

Just as with c-means, the question of validation for partitions as in Definition 7.5 comprises two possibilities. First, as we change c, the estimated parameters obviously change, and so does the quality of the proposed solution from either the clustering or parametric estimation viewpoint. This is called *validation across c*. And, second, with c fixed, there may be any number of competitors generated by A7.1 as it ranges across different choices for its control parameters (such as initialization and termination criteria). This is called *validation within c*[19]. These two possibilities were illustrated in Example 3.2. Most users of clustering algorithms concern themselves primarily with the across c problem, for it is, without question, more important to be looking in the right solution space than it is to be looking at competitors in the wrong space.

7.7.2 Choosing the Number of Components Using GMD/EM With GOFIs

This subsection discusses methods aimed at resolving issues connected with the parametric estimation problem in Definition 7.4. There are literally hundreds of papers that discuss the best way to estimate the number of components in a mixture of normals. And the fact that there *are* hundreds tells you that none of them generalize well across a variety of input data sets.

Figueiredo and Jain (2002) subdivide (most) model selection methods into two broad categories: deterministic and stochastic. These authors declare that most stochastic methods are too demanding computationally to be of much practical value. They also assert that virtually all deterministic algorithms for fitting the data with an unknown number of components have their roots in the EM algorithm, which, of course, puts us squarely back into the quagmire that plagues initialization of A7.1 in the first place.

McLachlan and Rathnayake (2014) provide an extensive discussion of the selection of a "best" number of components in an assumed mixture of Gaussians. They define the ***order*** of the mixture as c*, the (true but unknown) smallest value of c for which the Gaussian components have non-zero priors that is compatible with the data. Two basic sets of methods are identified for choosing the order of the mixture: (i) goodness of fit criteria (internal CVIs) that involve some penalized form of the likelihood function that lead to *information criteria* (IC); and (ii) hypothesis testing with a *likelihood ratio test* (LRT). According to McLachlan and Rathnayake, penalized likelihood criteria, like AIC, are less demanding than the LRT, but IC tests produce no scalar that quantifies confidence in the result, such as a *P*-value from an LRT statistic. However, the classical regularity conditions for $-2\log\lambda_{LR} = -2L(\mathbf{Q}) + 2n_{pc}$, where λ_{LR} is the LR test statistic, and n_{pc} is the total number of parameters estimated for a mixture of c components, don't hold; so the ratio fails to have an asymptotic null distribution. We leave further discussion of the LRT method to McLachlan and Rathnayake.

Information-based criteria for validation of mixtures apparently first appeared in Wallace and Boulton (1968). These criteria can be recognized by their first term, which is always $-2L(\mathbf{Q})$. The best-known criterion of this type is Akaike's (1974) information criterion,

$$\mathrm{AIC}(\mathbf{Q}) = -2L(\mathbf{Q}) + 2n_{pc} \tag{7.17}$$

where, again, n_{pc} is the number of parameters estimated for c components. The minimum value of AIC over a set of candidates is taken as the best fit GMD model of the data. According to McLachlan and Rathnayake (2014), the AIC tends to overestimate the number of components in the mixture. Another popular internal CVI for model selection is the *information complexity* (ICOMP) criterion first proposed by Bozdogan (1994).

[19] Statistically oriented users may argue that the within c problem is easily resolved by always taking the largest observed maximum of L(Q). However, it is easy to find examples where this strategy misleads. Indeed, it has been pointed out that a more general curiosity of optimization theory is that the global solution to a well-posed problem may be unsatisfactory on any number of other accounts. This is principally due to the mismatch between the process we humans observe, our models of it, and the data we collect to represent it. See Example 4.7, or p. 220 in Duda and Hart (1973), for examples of this type of failure by the hard c-means objective function.

It is related to the negative of the Hessian of L,which is Fisher's information matrix

$$F(\mathbf{Q}) = -H_L(\mathbf{Q}) = -[\partial L(\mathbf{Q})/\partial(\mathbf{Q})_i\partial(\mathbf{Q})_j] \tag{7.18}$$

The formula for ICOMP is truly intimidating:

$$\begin{aligned} ICOMP(\mathbf{Q}) = {} & -2L(\mathbf{Q}) + cp\left(\frac{p+3}{2}\right)\log\left[\frac{\sum_{i=1}^{c}\left\{\eta + \sum_{j=1}^{p} S_{i,jj}^2\right\}}{cp\left(\frac{p+3}{2}\right)}\right] \\ & - \left\{(p+2)\sum_{i=1}^{c}\log(\det(S_i)) - p\sum_{i=1}^{c}\log(np_i)\right\} - cp\log(2n) \end{aligned} \tag{7.19}$$

where $\eta = \frac{tr(S_i)}{p_1} + \frac{tr(S_i^2)}{2} + \frac{(tr(S_i))^2}{2}$. Cutler and Windham (1994) introduced a modification of ICOMP they called $ICOMP_W$ which is based on their estimate of Fisher's information matrix F at Equation (7.18), viz., $F(\mathbf{Q}) \approx \hat{F}(\mathbf{Q}) = \sum_{k=1}^{n} \nabla L(\mathbf{x}_k; \mathbf{Q})(\nabla L(\mathbf{x}_k; \mathbf{Q}))^T$. Using this estimate, they defined

$$ICOMP_W(\mathbf{Q}) = -2L(\mathbf{Q}) + \left(\frac{n_{pc}}{2}\right)\log(tr(\hat{F}(\mathbf{Q})^{-1}/n_{pc}) - \log(\det(\hat{F}(\mathbf{Q})^{-1}))/2 \tag{7.20}$$

Please note that the three GOFI criteria displayed here all involve estimates of the Gaussian parameters **Q**. The probabilistic partition $P \in M_{fcn}$ coupled to each **Q** through Equation (7.16d) plays no explicit role in these goodness of fit indices. There are many other GOFIs that are used for model selection in Gaussian mixtures. Please see the detailed discussions in Cutler and Windham (1994), Figueiredo and Jain (2002), Chapter 6 of McLachlan and Peel (2000), and especially the excellent recent survey by McLachlan and Rathnayake (2014).

Example 7.7. GMD model selection using internal GOFIs.

The data for this example are drawn from mixtures of p = 2 dimensional Gaussian PDFs having either c = 3 or c = 6 components. Six data sets were generated for each of two cases, resulting in a total of 12 data sets. Table 7.8 summarizes the model parameters used for data generation. The priors are fixed at 1/c, and the means are distributed on rays separated by equal angles, centered equal distances from the origin at radii r = 2 or 3 according to c = 3 or 6. Thus, for example, at c = 3, there are three means on lines separated by 120°, each 2 units from the origin, $\mathbf{0} = (0, 0)^T$. There are three different sets of covariance matrices, shown in Table 7.8 as cases **Sep**, **Med**, and **Over**, which stand for *Separated, Medium, and Overlapping*. The covariance matrices for these cases were scalar multiples of the identity (I) with diagonals (variances) of 0.5, 1, and 2, respectively, and were equal for the components of each mixture.

For each data set in Table 7.8, GMD/EM algorithm A7.1 was run three times, each run being initialized by an output from FCM algorithm A6.1. The range of values for c was $2 \le c \le 6$ for the c = 3 component samples, and $4 \le c \le 8$ for the c = 6 component samples. So, each of the 12 data sets produced a pool of 5(3) = 15 probabilistic candidate partitions. However, the partition matrix P corresponding to Equation (7.16d) is just a by-product of A7.1 in this example – it doesn't play any role in the computation of AIC,

Table 7.8 Parameters for samples in 12 Gaussian data sets.

	c = 3	c = 6
$\{\pi_i\}$	1/3	1/6
$\{\mu_i\}$	r = 2	r = 3
Sep : $\Sigma_i = 0.5 \cdot 1$	SA3: n = 1, 000	SA6: n = 2, 000
	SB3 : n = 10, 000	SB6 : n = 20, 000
Med : $\Sigma_i = 1$	SA3: n = 1, 000	MA6 : n = 2, 000
	SB3 : n = 10, 000	MB6 : n = 20, 000
Over : $\Sigma_i = 2.1$	OA3: n = 1, 000	OA6 : n = 2, 000
	OB3 : n = 10, 000	OB6 : n = 20, 000

Table 7.9 Model selection with three GOFIs: # successes in 12 data sets.

GOFI Index	Best (maxL(**Q**))	Avg. 3 trials
(↓) AIC	9: (3+)	10: (2+)
(↓) $ICOMP_W$	9: (2+, 1−)	9: (2+, 1−)
(↓) ICOMP	2: (10+)	4: (8+)

ICOMP, and $ICOMP_W$. In other words, these three internal indices are being used here in the context of goodness of fit (of parameter estimates) to the "true but unknown" population parameters of the Gaussian components.

Table 7.9 summarizes the number of successes (meaning that the known number of components were chosen from each group of 15 competitors) by each of the three internal GOFIs at recovering a best GMD model from amongst the candidates presented by GMD/EM algorithm A7.1. In this table, and also in Table 7.10, (+ means overestimate of c; -means underestimate of c).

In Table 7.9, the *best run* (= largest value of L(**Q**) in each set of three) appears in the middle column; and the third column contains the result using the *average value* of each index over the outputs of the three runs on each data set. The parenthetical numbers following each entry in the table indicate the number of over (+) and under (−) estimates in the 12 tries. For example, 9: (2+, 1−) for $ICOMP_W$ means 9 successes in 12 tries, with 2 overestimates and 1 underestimate of the correct number of Gaussian components in the mixtures from which the data were drawn. The main point of this example is that different GOFIs will select different models. While ICOMP suffers by comparison here, there will be other data sets that reverse the trends indicated by the values in Table 7.9. Note especially that choosing the mixture components using the maximum attained value of the log-likelihood function leads to less successes than using average values of the objective function for both AIC and ICOMP. This confirms what has been said many times: the most extreme value of an objective function doesn't always point to the "best" result.

7.7.3 Choosing the Number of Clusters Using GMD/EM With CVIs

Just as there are many GOFIs available for model selection, there are many CVIs that can be used to choose a "best" soft partition $P^* \in M_{fcn}$ from a set of probabilistic candidates, $CP = \{P_1, \ldots, P_N\} \subset M_{fcn}$. Since the same structure that underlies fuzzy partitions also carries all the probabilistic ones, there is no reason why any index that will evaluate fuzzy partitions will not also be useful evaluating the quality of probabilistic

Table 7.10 Evaluation with crisp and fuzzy CVIs: # successes in 12 trials.

Eqn.	CVI	Type	Best (max L(**Q**))	Average (3 trials)
(6.51)	(↓) DB_{22}	Crisp	5: (7−)	9: (2+, 1−)
(6.52)	(↑) GDI_{33}	Crisp	10: (2−)	10: (2−)
(6.53)	(↑) PC	Fuzzy	2: (10−)	2: (10−)
(6.57)	(↓) XB_2	Fuzzy	10: (2−)	10: (2−)

ones. Moreover, we can harden P's in CP using the same function **H** for hardening U (Definition 3.3, which in this context amounts to using Bayes rule on each column of P), and then we can apply crisp CVIs to the hardened versions $\{\mathbf{H}(P_1), \ldots, \mathbf{H}(P_N)\} \subset M_{hcn}$ of all candidates in CP = $\{P_1, \ldots, P_N\} \subset M_{fcn}$.

Example 7.8. Selection of a best P* using crisp and fuzzy internal CVIs.

This example is a continuation of Example 7.7 that adds four internal CVIs to the analysis of outputs made by A7.1 as described there for the 12 data sets shown in Table 7.8. The four CVIs used to assess the soft partitions for the 12 data sets were all discussed in Chapter 6. Table 7.10 contains the same type of information as Table 7.9 for *the same GMD/EM partitions* of the data sets described in Example 7.7 whose parameters are listed in Table 7.8. The difference is that, here, we illustrate evaluation of the outputs when they are interpreted as soft clusters of the 12 data sets. The crisp indices DB_{22} and GDI_{33} are applied to hardened candidates, $\{\mathbf{H}(P_1), \ldots, \mathbf{H}(P_N)\} \subset M_{hcn}$. The fuzzy indices are applied to the probabilistic soft partitions in CP = $\{P_1, \ldots, P_N\} \subset M_{fcn}$. Calculation of XB_2 uses the pair (P, **M**) and ignores the priors and covariances of the c estimates.

Please compare the results in Tables 7.9 and 7.10. You will see that the fuzzy Xie-Beni CVI does just as well when applied to the problem on the right side of Figure 7.9 as the goodness of fit index $ICOMP_W$ does in selecting a Gaussian model using the goodness of fit approach depicted on the left side of Figure 7.9. Both enjoy 10 successes in 12 data sets over 3 trial averages. The only real difference is that $ICOMP_W$ overestimates c twice, while XB_2 underestimates c twice. Finally, note that both crisp indices do pretty well on hardened versions of the candidates. Indeed, the generalized Dunn's index GDI_{33} on hardened versions of the probabilistic partitions found by A7.1 on these 12 sets of Gaussian clusters delivers the same number of successes (10 out of 12, on average) as the fuzzy index XB_2 and the probabilistic goodness of fit index $ICOMP_W$. As mentioned in Chapter 6, the partition coefficient doesn't fare so well in this set of experiments.

7.8 Notes and Remarks

Mixture models, MLE, the history of the EM algorithm, and the beginnings of AO are almost hopelessly and certainly non-linearly intertwined. The accuracy and complexity of a history depends on its author, and, ultimately, exact data points in the time line of these topics are fuzzy at best. Interested readers will find no better entrance to this topic than the papers of Butler (1986) and Meng and Van Dyk (1997).

Pearson (1894) used the method of moments to estimate the parameters of a mixture of two univariate normal densities with unequal variances. Several authors subsequently dabbled with the method of moments, but it was computationally intractable prior to the emergence of computers. Tan and Chang (1972) compare moment estimation to MLE and show moments to be inferior to MLE when modern computing power is available. See McLachlan and Basford (1988, p. 4) for a summary of graphical methods that attempted to overcome computational difficulties in the intervening years between the early (moments) and modern (MLE) methods.

Newcomb (1886) is usually credited with the invention of the seeds of the EM/AO algorithm for mixture analysis. Fisher (1925) introduced a method of scoring MLEs that some interpret as EM, while others credit McKendrick (1926) as the starting point of EM for MLE. Important stops along the way for the basic GMD algorithm A7.1 were made by Rao (1948) and Hartley (1958). The early ML methods, however, were so demanding computationally that real interest in MLE by the EM algorithm awaited the power of the modern computer. Hasselblad (1966) showed how to use EM for MLE on a reasonably large (for 1966) mixture of Gaussian distributions.

Arguably, the most important step in the evolution of GMD with EM was Wolfe's (1970) work, which is essentially the basis of the basic GMD algorithm A7.1 which remains so popular today. Meng and Van Dyk (1997) state that no development was as important for EM as the paper by Dempster, Laird, and Rubin (1977). These authors apparently were the first to suggest the name "expectation-maximization"– EM for short – for the method, and they gave a very complete survey of the theory and applications at that time. This paper has about 70,000 citations (and climbing) at Google Scholar! The book by Titterington, Smith, and Makov (1985) tabulates 90 citations for applications of mixture models.

In addition to numerical convergence of AO iteration as presented in Chapter 10, there is, for EM, the additional consideration of convergence in probability to the statistical parameters that underlie the assumed distributions of the components of the mixture. See Wu (1983) and de Leeuw (1994) for discussions on both types of convergence.

As is the case for HCM and FCM, an entire volume could be devoted to the topic of how best to initialize A7.1. According to McLachlan and Basford (1988), one of the most important considerations for good starting values is the vector $\mathbf{p}_0 \in N_{fc}$ of prior probabilities. Recall that our default specification randomly chooses $\mathbf{p}_0 \in N_{fc}$. Bezdek and Dunn (1975) discuss the use of FCM-AO to provide good initializations for these values at significant reductions in computation time for A7.1 by taking $p_{i,0} = \sum_{k=1}^{n} u_{ik,t}$, where U_t is a terminal c-partition of X found by A6.1. See Bezdek *et al.* (1985) for more on this approach. These papers point out the similarities of and differences between estimates produced by FCM/AO and the GMD/EM algorithms. Hathaway and Bezdek (1986) give an example that proves that the point prototypes $\{\mathbf{v}_i\}$ from FCM/AO *can't* be a statistically consistent estimator of the Gaussian mixture means $\{\mathbf{m}_i\}$ from an arbitrary univariate mixture. Another approach to initialization for GMD with EM by clustering with a deterministic competitive learning model is given by McKenzie and Alder (1994).

Figueiredo and Jain (2002) offer a relative of A7.1 that is claimed to be insensitive to bad initializations, and that finds a best fit number of Gaussian components on the fly. A very similar dynamic approach is advocated by Lee *et al.* (2006), who begin with a single component and keep adding components until the mutual relationship between any two mixtures becomes dependent. Biernacki *et al.* (2003) discuss several methods for choosing starting values for the EM algorithm to get maximum likelihood parameter estimates in mixture models. Three methods based on random initialization using a classification EM algorithm, a stochastic EM algorithm, or previous short runs of EM itself are compared. Melnykov and Melnykov (2012) propose initializing A7.1 by choosing points with higher concentrations of neighbors and using a truncated normal distribution for the preliminary estimation of dispersion matrices.

Many fuzzy clustering algorithms have been proposed that are either hybrids of or are related to GMD or that use the maximum likelihood principle in some other way. For example, Gath and Geva (1989) studied the estimation of components of normal mixtures using a fuzzy clustering approach that attempts to marry FCM with GMD. See the text by Kaufman and Rouseeuw (2005) for some other typical algorithms of this type. An extensive numerical comparison of the use of crisp and fuzzy cluster validity indices and probabilistic goodness of fit criteria for choosing the best number of clusters in a Gaussian mixture appears in Bezdek *et al.* (1997). Examples 7.7 and 7.8 are derived from this larger study.

7.9 Exercises

1. Prove Wolfe's Theorem 7.1. Start by forming the Lagrangian of the log likelihood function at Equation (7.15). Use the LaGrange multiplier method to derive the FONCs (7.16). Do this by ignoring the inequality constraints and then show that they are satisfied by the resultant necessary conditions on the posterior probabilities.
2. Let $X = \{x_i, \ldots, x_n\} \subset \Re$ be n observations from a $c = 2$ component univariate Gaussian mixture distributed according to Equation (7.11) as follows:

$$f(\mathbf{x};\mathbf{Q}) = \pi\frac{e^{-\frac{1}{2}\left(\frac{x-\mu_1}{\sigma_1}\right)^2}}{\sqrt{2\pi\sigma_1}} + (1-\pi)\frac{e^{-\frac{1}{2}\left(\frac{x-\mu_2}{\sigma_2}\right)^2}}{\sqrt{2\pi\sigma_2}}$$

   Show that the log likelihood of f is infinite if $\mu_1 = x_1, \sigma_1 = 0$. For p-dimensional data, show that there exists an infinite log likelihood whenever there is a cluster containing less than $(p+1)$ instances (assume arbitrary covariances within clusters).
3. The GMD model in Chapter 7 is based on the assumption that the components of the mixture distribution are Gaussian, i.e., that $f(\mathbf{x};\mathbf{Q}) = \sum_{i=1}^{c} \pi_i g_i(\mathbf{x};\mathbf{q}_i)$; $g_i(\mathbf{x};\mathbf{q}_i) = e^{-\frac{1}{2}\|\mathbf{x}-\mu_i\|^2_{\Sigma_i^{-1}}}/(2\pi)^{\frac{p}{2}}\sqrt{\det\Sigma_i}$; . While this is the most common model, many other mixtures are of interest in pattern recognition. Suppose the component densities are c binomial distributions for inputs $x \in \{0, 1, \ldots, N\}$ so that the mixture density is $f(\mathbf{x};\mathbf{Q}) = \sum_{i=1}^{c} \pi_i \binom{N}{x} (q_i)^N (1-q_i)^{N-x}$. Show that this mixture is not identifiable when the prior probabilities $\{\pi_i\}$ are known and $N < c$.
4. Wolfe's theorem can be proved using the calculus of several variables or, alternatively, with linear algebraic techniques. Rework problem 1 above without calculus.

8

Relational Clustering – The SAHN Models

Mathematical objects are determined by – and understood by – the network of relationships they enjoy with all the other objects of their species.

–Barry Mazur

Two types of numerical data were discussed in Section 3.2, object or feature vector data (X) and relational data (R). Chapters 6 and 7 dealt almost exclusively with feature vector data, the exception being that iVAT images are always based on relational data built from X or directly from R. Many applications depend directly on clustering relational data -e.g., information retrieval, data mining in relational databases, and numerical taxonomy; so methods in this category are important. This chapter is devoted to models and algorithms that produce hierarchies of crisp clusters from *square* relational (usually dissimilarity) data. These algorithms are unquestionably the most heavily used group of relational clustering methods. They can be implemented in either agglomerative (clumping) or divisive (splitting) forms. We study the agglomerative case: the divisive model is in every way equivalent, implemented in "reverse order." The family name of these methods is SAHN, which stands for ***S****equential,* ***A****gglomerative,* ***H****ierarchical, and* ***N****on-overlapping.* The reason for each of these adjectives is discussed shortly.

The basic idea in relational clustering is to group together objects in an object set O that are "closely related pairs" to each other and "not so closely" related to objects in other clusters, as indicated by their relative pairwise relational strengths. The objects are often implicit, but we can still look for groups in O by clustering based on the strength of relationships between pairs of objects. Our discussion begins with a review of crisp relations and similarity measures on the cross product of X (and O) with itself.

8.1 Relations and Similarity Measures

Let the n objects be $O = \{o_1, \ldots, o_n\}$. These objects may be anything – motorcycles, guitars, fish, cigars, beers, patients with heart problems, or whatever. A crisp *binary relation* R in O is a crisp subset $R \subset O \times O$, which has the membership function $\rho = O \times O \rightarrowtail \{0, 1\}$. The n^2 numbers $\rho(o_i, o_j)$ characterize the memberships of pairs of objects (o_i, o_j) in the relation R. We write $\rho(o_i, o_j) = 1 \Leftrightarrow o_i R o_j$, and say o_i is *R-related* to o_j.

For example, if O is a set of fish, ρ (salmon, trout) = 1, ρ (salmon, crab) = 0. Note especially that salmon and trout are not *equal*, but they are *equivalent* under the relation ρ. The relationships in R can be arrayed as the entries of an n × n *relation matrix* $R(\rho; O) = [r_{ij} = \rho(o_i, o_j)]$. We often write $R(\rho : O)$ simply as R and refer to R as "a relation," "a relation matrix," or often call it "the relational data." This abusive terminology is justified by noting that, for crisp relations, the three descriptions of R are equivalent.

When $r_{ij} \notin \{0,1\}$, it is customary to regard r_{ii} as the *strength* of the relationship between o_i and o_j A (square binary) relation R is *reflexive* $(I_n \subset R)$ when every element in O is fully related to itself $r_{ii} = \rho(o_i, o_i) = 1 \,\forall\, o_i \in O$. *Irreflexive* relations satisfy the opposite condition, i.e., $r_{ii} = \rho(o_i, o_i) = 0 \,\forall\, o_i \in O$. Similarity relations are reflexive, and dissimilarity relations are irreflexive. R is *symmetric* $(R = R^T)$ when $r_{jk} = r_{kj} \forall j, k$. This means that whenever o_j is related to o_k at any level, o_k is related to o_j at the same level. The third property that often travels in the company of these two is *transitivity*, but the algorithms we discuss never assume it; so we will not discuss it here.

Many types of proximity relations (similarity or dissimilarity) are found in various applications, but this text deals mainly with *dissimilarity data* D, which exhibit pairwise dissimilarities on n vectors $X = \{\mathbf{x}_1, \ldots, \mathbf{x}_n\} \subset \Re^p$. The elements of D for this case may have four properties:

$$p1 : d_{ij} \geq 0 \,\forall\, i, j \quad \text{(positive);} \tag{8.1a}$$

$$p2 : d_{ij} = 0 \Leftrightarrow \mathbf{x}_i = \mathbf{x}_j \quad \text{(hollow);} \tag{8.1b}$$

$$p3 : d_{ij} = d_{ji} \,\forall\, i \neq j \quad \text{(symmetric);} \tag{8.1c}$$

$$p4 : d_{ij} \leq d_{ik} + d_{kj} \,\forall\, i \neq j \neq k \quad \text{(triangle inequality).} \tag{8.1d}$$

Some authors combine the terms positive and hollow and substitute the term *positive-definite* for these two properties. Please note that property $p2 : d_{ij} = 0 \Leftrightarrow \mathbf{x}_i = \mathbf{x}_j$, automatically guarantees that the diagonal elements of D are zero, $d_{ii} = 0 \,\forall\, i$. Property p1 requires that all of the off-diagonal elements are greater than or equal to zero. If a data set $X = \{\mathbf{x}_1, \ldots, \mathbf{x}_n\} \subset \Re^p$ has n *distinct* points, then p2 asserts that all of the off-diagonal entries of D are positive. But if X has repeated points, there will be zero entries off the diagonal of D. This is the case, for example, with the Iris data: points #102 and #143 have the same coordinates (5.8, 2.7, 5.1, 1.9); so the distance between this pair is zero. This doesn't deter the construction of MSTs in Iris.

Definition 8.1. Metric and Euclidean distance matrices: realizations of X.

(a) $M_n^+ = \{D \in \Re^{nn}; (8.1a, b, c)\}$ is the set of all positive, hollow, symmetric matrices.

(b) $M_n^{met} = \{D = [d_{ij}] \in \Re^{nn}; (8.1a, b, c, d) \text{ hold}\}$ is the set of all *metric* matrices.

(c) $M_n^{euc} = \{D = [d_{ij}] \in \Re^{nn} : d_{ij} = \|\mathbf{x}_i - \mathbf{x}_j\|_2\}$ is the set of *Euclidean distance matrices* (EDMs). D is an EDM if and only if there is a set of vectors $X = \{\mathbf{x}_1, \ldots, \mathbf{x}_n\} \subset \Re^s$ in some Euclidean space $\Re^s$, $s \leq (n-1)$, such that $d_{ij}^2 = (\mathbf{x}_i - \mathbf{x}_j)^T (\mathbf{x}_i - \mathbf{x}_j) = \|\mathbf{x}_i - \mathbf{x}_j\|_2^2$ is *squared* Euclidean distance between pairs of vectors in X. When D is Euclidean, X is called a *realization* of D. Realizations are never unique. The *smallest* s for which there is a Euclidean realization of D is its *embedding dimension.*

(d) Evidently, $M_n^{euc} \subset M_n^{met} \subset M_n^+$. D has metric distances when it is Euclidean but not necessarily conversely.

For ease of reference, the names and definitions of the three most important sets of dissimilarity matrices are:

$$M_n^+ = \{D = [d_{ij}] \in \Re^{nn}; (8, 1a, b, c) \text{ hold}\}; \tag{8.2a}$$

$$M_n^{met} = \{D \in M_n^+; (8.1d) \text{ holds}\} \quad \text{(metric);} \tag{8.2b}$$

$$M_n^{euc} = \{D \in M_n^{met} : [d_{ij}] = \|\mathbf{x}_i - \mathbf{x}_j\|_2\} \quad \text{(Euclidean).} \tag{8.2c}$$

☺ ***Forewarned is forearmed:*** Some authors use *squared* distances for the definition of an EDM (Hathaway *et al.* (1989), Hathaway and Bezdek (1994), Dattorro (2005), Khalilia *et al.* (2014)). But many others use *unsquared distances* (Mardia *et al.* (1979), Gower and LeGendre (1986), Benasseni *et al.* (2007)). It is very important to know which type of D (squared or unsquared) you are dealing with when performing a cluster analysis or multidimensional scaling using D as input[20]. Why? The theory and implementation you use depends on your definition of an EDM. You can find clusters in or derive feature vectors from both unsquared and squared inputs, but they will not necessarily be the same clusters or vectors for the two cases. ☺

The linkage algorithms require input matrices to have the three properties (8.1a, b, c), i.e., $D \in M_n^+$. We will encounter many D's that are neither metric nor Euclidean because they fail to satisfy the triangle inequality (8.1d), but most of these dissimilarity matrices can still be dealt with by the linkage algorithms. And, we will look for clusters in some D's that satisfy only the first two of these properties (zero diagonals, some positive off-diagonal entries but asymmetric).

Figure 8.1 depicts the relationship between the Euclidean and non-Euclidean matrices in Definition 8.1 and sets of feature vectors that may be related to them. The clustering algorithms described in this volume that can be applied to these dissimilarity matrices are shown parenthetically in this figure. The four models and algorithms abbreviated as RHCM/RFCM (*relational HCM/FCM*) and NERHCM/NERFCM (*non-Euclidean RHCM/RFCM*) are not covered in this volume but are shown here for completeness. See Hathaway and Bezdek (1994) or Khalilia *et al.* (2014) for more on these models and algorithms.

Given an object data set $X = \{\mathbf{x}_1, \ldots, \mathbf{x}_n\} \subset \Re^p$, with $\mathbf{x}_i \in \Re^p$ characterizing object o_i, there are many functions (ν, η in Figure 8.1) that can convert X or Y into relational data. When ν is an inner product A-norm on $\Re^p$, $D = [\|\mathbf{x}_i - \mathbf{x}_j\|_A]$ is an EDM, and this is far and away the most typical case. The function ϕ depicts feature extraction as shown in Figure 5.2.

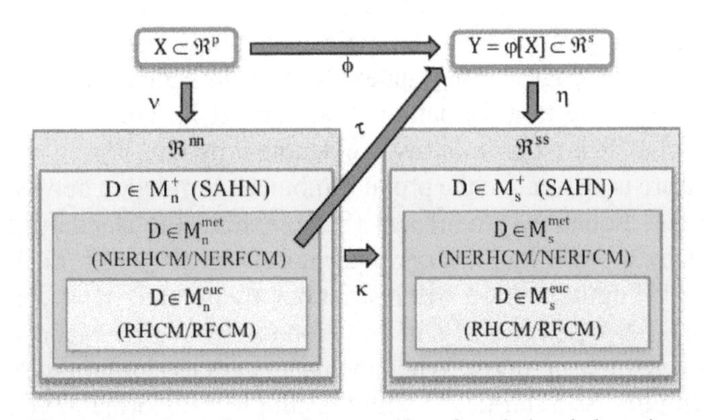

Figure 8.1 Sets of square input matrices for relational clustering.

[20]You will know whether the $[d_{ij}]$ are squared or not if you create D from X with some norm metric. But you may not know which case applies if D is given as input without this specification.

For the first five measures in Table 8.1, the vectors **x** and **y** are in $\{0,1\}^p$: for the last four measures, **x** and **y** are in $\Re^p$. The numbers a, b, c, and d for the first five similarity coefficients in Table 8.1 are computed from binary p-vectors **x**, **y** as follows:

a = # of 1 – 1 matches
b = # of 1 – 0 mismatches
c = # of 0 – 1 mismatches
d = # of 0 – 0 matches

Table 8.1 Some similarity transformations of real and binary data.

Name	Formula	Equation
Simple match	$S_{SM}(\mathbf{x},\mathbf{y}) = (a+d)/p$	(8.3a)
Double match	$S_{DM}(\mathbf{x},\mathbf{y}) = \frac{2(a+d)}{2(a+d)+b+c}$	(8.3b)
Double mismatch	$S_{DM}(\mathbf{x},\mathbf{y}) = \frac{a+d}{a+d+2(b+c)}$	(8.3c)
Ignore 0 – 0 (Jaccard)	$S_J(\mathbf{x},\mathbf{y}) = a/(a+b+c)$	(8.3d)
Yule	$S_Y(\mathbf{x},\mathbf{y}) = (ad-bc)/(ad+bc)$	(8.3e)
cos(**x**, **y**)	$S_{cos}(\mathbf{x},\mathbf{y}) = s_{corr}(\mathbf{x},\mathbf{y}) = \langle\mathbf{x},\mathbf{y}\rangle/\|\mathbf{x}\|\|\mathbf{y}\|$	(8.3f)
Tanimoto	$S_{Tan}(\mathbf{x},\mathbf{y}) = \langle\mathbf{x},\mathbf{y}\rangle/(\langle\mathbf{x},\mathbf{x}\rangle + \langle\mathbf{y},\mathbf{y}\rangle - \langle\mathbf{x},\mathbf{y}\rangle)$	(8.3g)
A-Norm	$d_A(\mathbf{x},\mathbf{y}) = \|\mathbf{x}-\mathbf{y}\|_A = \sqrt{(\mathbf{x}-\mathbf{y})^T A(\mathbf{x}-\mathbf{y})}$	(8.3h)
q-Norm	$d_q(\mathbf{x}-\mathbf{y}) = \|\mathbf{x}-\mathbf{y}\|_q = (\Sigma_{j=1}^{p}\|x_j-y_j\|^q)^{1/q}, \quad q \geq 1$	(8.3i)

Real binary relations in O computed from pairs of vectors in X are variously called coefficients of: similarity, proximity, association, correlation, distance, discrimination, likeness, disagreement, and so on, seemingly forever! Sneath and Sokal (1973, p. 129) stated almost 50 years ago that "There are so many association coefficients in the biological and non-biological literature that any attempt at an exhaustive catalog of them would require many pages." They index 34 measures under the heading "coefficient of." Table 8.1 lists just a few of the many transformations that convert X into R. When the relation R is a similarity, we denote it by S; when R is a dissimilarity, we denote it by D.

The cosine similarity measure is also called the product-moment correlation between the vectors **x** and **y**. The last two rows in Table 8.1, Equations (8.3h) and (8.3i), exhibit the standard dissimilarity relations (distances) induced on $X \times X$ by the two families of vector norms on $\Re^p \times \Re^p$. Each of the inner product and Minkowski norm families is infinite, but we will restrict our attention to the special cases we called the five good norms in Definition 6.10. When $A = \Sigma^{-1}$ in Equation (8.3h), we get Mahalanobis distance. When $q \to \infty$ in Equation (8.3i), we get the sup norm. When the data begin as similarities S with $0 \leq s_{ij} \leq 1$ for all i and j, we can convert them to dissimilarities D with the simple transformation $d_{ij} = 1 - s_{ij}$. When the s_{ij} are not in the unit interval, they are usually normalized with division by their maximum value.

Example 8.1. Similarity and dissimilarity transformations of X_{30}.

We return to the data set X_{30} shown in Figure 4.11(a) that was used in Examples 4.8 and 7.3. X_{30} has c = 3 HPOV clusters of 10 points each. Figure 4.11(b) illustrates that (for most initializations) HCM will

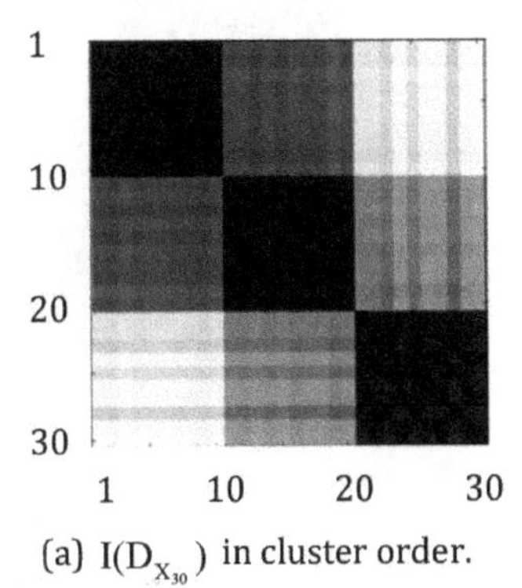

(a) $I(D_{X_{30}})$ in cluster order.

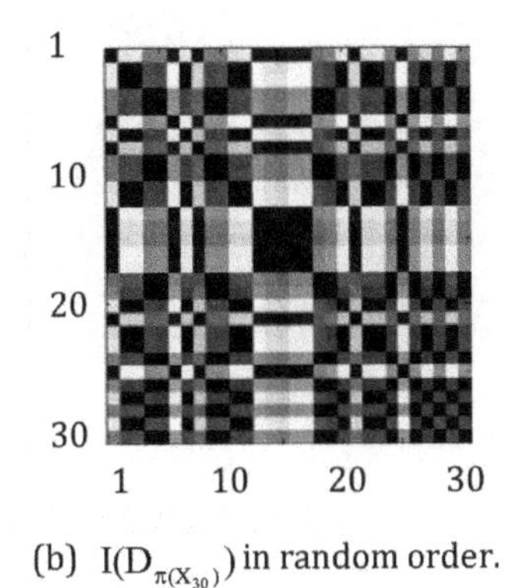

(b) $I(D_{\pi(X_{30})})$ in random order.

Figure 8.2 Images of two Euclidean distance matrices of data set X_{30}.

terminate at these same CPOV clusters when the model norm for J_1 is Euclidean. We can convert this data set to relational dissimilarity data by calculating the 30×30 matrix $D_{X_{30}} = [\|\mathbf{x}_i - \mathbf{x}_j\|_2]$ of (unsquared) Euclidean distances between pairs of points in $D_{X_{30}}$. Figure 8.2(a) is the image $I(D_{X_{30}})$ of X_{30} when it is represented by $D_{X_{30}}$. This matrix and its image correspond to the input order (the cluster order) shown in Figure 4.11(b) that labels the first 10 points 1–10, the middle group of points 11–20, and the cluster on the upper right as 21–30. The three HPOV clusters in Figure 4.11(b) are clearly seen in Figure 8.2(a), appearing as the three dark 10×10 blocks along the diagonal of the image (black is zero, white is max). An iVAT image of $D_{X_{30}}$ would show these three clusters quite clearly.

This is one way we try to "see" geometric properties in relational data. Here, the three HPOV clusters in feature space appear as they do in Figure 8.2(a) in *visual relation space* because the intracluster distances are together in three "*relational clouds*," (the objective of VAT/iVAT reordering).

But .... the image changes if we change the order (the list of labels) of the data. The image in view 8.2(b) is based on the (unsquared) Euclidean distance matrix $D_{\pi(X_{30})} = [\|\mathbf{x}_{(i)} - \mathbf{x}_{(j)}\|_2]$ of a permutation of the labels in X_{30} done with the permutation function π. Reordering the indices in X also reorders the rows and columns of its relational representation. The idea of reordering data images was introduced in Example 4.4.

Let's compare $I(D_{X_{30}})$ to images made by other transformations of X_{30}. Specifically, we calculate the unsquared distance matrices in the same *input order* as Figures 4.11(a) and 8.2(a) corresponding to the other four good norms at (6.50b-e), viz., the Diagonal norm $D_{D^{-1}}$, the Mahalanobis norm, $D_{S^{-1}}$, the 1-norm D_1, and the sup norm D_∞. We also compute the dissimilarity matrices corresponding to the cosine D_{cos}, (8.3f) and Tanimoto D_{tan}, (8.3g) similarity coefficients in Table 8.1. For these last two, we calculate the similarity matrices using the formulas from Table 8.1 and then convert them to dissimilarities with the element-wise transformation $d_{ij} = 1 - (s_{ij}/\max_{k,j}\{s_{kj}\})$. The images of these six D's are shown in Figures 8.3(a)–(g).

The structure of the transformed dissimilarity data (as seen visually) seems very similar to the Euclidean norm in Figure 8.2(a) for the diagonal, city block (1-norm), and sup norms and the Tanimoto dissimilarity. We emphasize that the dark blocks we see in these images are an artifact of building and viewing their distance matrices in the cluster order of the data which is shown in Figure 4.11(b). But the Mahalanobis and correlation transformations seem to obscure the structure of X_{30} quite a bit. You should always bear in mind that the transformation you choose when converting $X \rightarrowtail D(X)$ can distort the original structure of data quite unexpectedly.

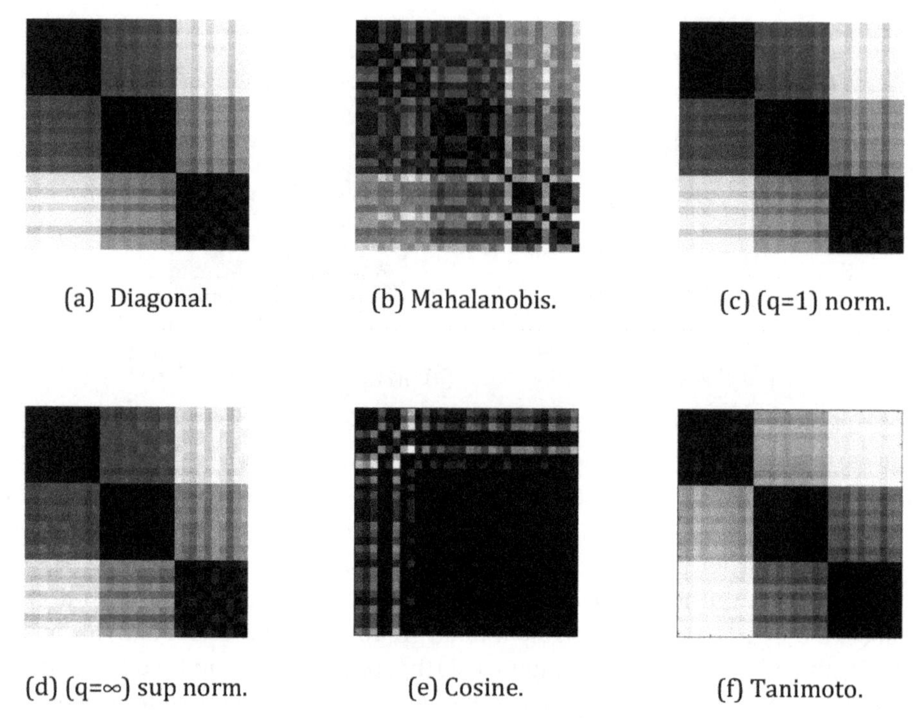

Figure 8.3 Images of some dissimilarity transformations of X_{30}.

Be careful not to read too much into the images in Figures 8.2 and 8.3. All of the feature vector algorithms discussed so far are batch methods – that is, they "see all of the data at once" during each iteration. Consequently, batch methods tend to overlook permutations in the order of presentation and usually find the same clusters when applied to either X_{30} or $\pi(X_{30})$. The correct wording is "usually" instead of "will" since different initializations of the batch algorithms in Chapters 6 and 7 on any data set may send them to different local terminal states, as illustrated in Example 4.8. In contrast, seeking clusters in the six dissimilarity relations extracted from X_{30} (i.e., the six different distance matrices depicted in Figure 8.3), all of which are based on the same cluster order, and all of which represent the same set of input vectors, with relational clustering algorithms such as single linkage that take D as inputs might result in some very different interpretations of X_{30}.

8.2 The SAHN Model and Algorithms

Agglomerative algorithms start with each object in its own singleton cluster (c = n) and, subsequently, merge similar clusters until the procedure terminates at a single cluster (c = 1). In the divisive approach, the processing proceeds the other way, beginning with all the points in a single cluster (c = 1) and then splitting clusters by some rule until every object eventually ends up in its own singleton cluster (c = n). Historically, there have been very few changes to the original models discussed in this section. Most of the recent work in this area concerns itself with more efficient implementations, approximations, and applications to big data sets. Good expositions of many of these methods appear in Sneath and Sokal (1973), Jain and Dubes (1988), Johnson and Wichern (2007), and Theodoridis and Koutrombas (2009).

The heart of the SAHN models is a measure of *set distance* $\delta_{SAHN}(A, B)$ between pairs of finite, crisp subsets A and B in a metric space. A general formulation of SAHN models is provided by Lance and Williams (1967), who call a SAHN procedure *combinatorial* if the (*set*) dissimilarity $\delta_{SAHN}(WV, A)$ between a merged pair WV of clusters W and V, and the remaining clusters $\{A\}$ can be computed from the previous dissimilarities $\delta_{SAHN}(W, V)$, $\delta_{SAHN}(W, A)$, $\delta_{SAHN}(V, A)$ and the cluster sizes $|W|$ and $|V|$[21]. The general formula for combinatorial SAHN distances is

$$\begin{aligned}\delta_{SAHN}(WV, A) = \alpha_w \delta_{SAHN}(W, A) + \alpha_v \delta_{SAHN}(V, A) + \beta \delta_{SAHN}(W, V) \\ + \gamma|\delta_{SAHN}(W, A) - \delta_{SAHN}(V, A)|\end{aligned} \tag{8.4}$$

where $\{\alpha_w, \alpha_v, \beta, \gamma\}$ are coefficients that determine various members of the SAHN family. Table 5.2 in Sneath and Sokal (1973) lists various choices of the four coefficients $\{\alpha_w, \alpha_v, \beta, \gamma\}$ in Equation (8.4). The important point about this is that combinatorial SAHN distances allow the computation of set distances between coarser clusters from those from finer clusters without returning to the input data matrix at each step of the procedure, whereas the original input data must be retained when non-combinatorial SAHN models are employed. There are many possibilities for these choices, but like the five good norms, there are only three SAHN distances that make more than token appearances in the literature – viz., *single linkage* (SL), *complete linkage* (CL), and *average linkage* (AL). In the formulas below, the input dissimilarities are $D = [d_{ij}] \in M_n^+$, and the notation, "$i \in W$" means $o_i \in W \subset O$, and similarly for "$j \in V$":

$$\delta_{SL}(W, V) = \min_{i \in W, j \in V}\{d_{ij}\}; \quad \alpha_W = \alpha_V = \frac{1}{2}; \quad \beta = 0; \quad \gamma = -\frac{1}{2} \tag{8.5a}$$

$$\delta_{CL}(W, V) = \max_{i \in W, j \in V}\{d_{ij}\}; \quad \alpha_W = \alpha_V = \frac{1}{2}; \quad \beta = 0; \quad \gamma = \frac{1}{2} \tag{8.5b}$$

$$\delta_{AL}(W, V) = \sum_{i \in W}\sum_{j \in W} d_{ij}/\{|W| \cdot |V|\}; \quad \alpha_W = \frac{|W|}{|WV|}; \quad \alpha_v = \frac{|V|}{|WV|}; \quad \beta = \gamma = 0 \tag{8.5c}$$

Table 5.2 in Sneath and Sokal (1973) lists four different formulas for the average linkage set distance AL. Formula (8.5c) is the one most often identified as average linkage, and, more to the point, it is the UPGMA or *unweighted paired group mean average* version of AL introduced by Sokal and Michener (1958). Figure 8.4 repeats these formulas and illustrates their geometric meanings for these three cases when D is a Euclidean distance matrix (not all of the distances are illustrated for the AL case). When one of these three set distances is chosen, the corresponding SAHN clustering model and algorithm inherits its name. Without confusion, we speak of SL, CL, and AL clustering, understanding this to mean SAHN clustering using the named set distance. The implications of choosing one of these set distances are discussed after the general idea of the SAHN model and algorithm are given. Table 8.2 gives a simple, informal verbal description of the agglomerative form of SAHN clustering.

This procedure is certainly sequential, and it is **a**gglomerative because it starts at $(c = n)$ and, by merging clusters, ends at $(c = 1)$; the clusters are **h**ierarchical because once objects are joined, they remain joined until termination; and it is **n**on-overlapping because the clusters formed at each step are crisp. The SAHN procedure terminates in at most $(n - 1)$ steps. We say "at most $(n - 1)$" because it is possible that ties will

[21]Another use for the symbol "Vee"! This one is traditionally used in the literature for this topic; so I will leave it here and ask your tolerance for yet another abuse of this non-unique and over-worked notation.

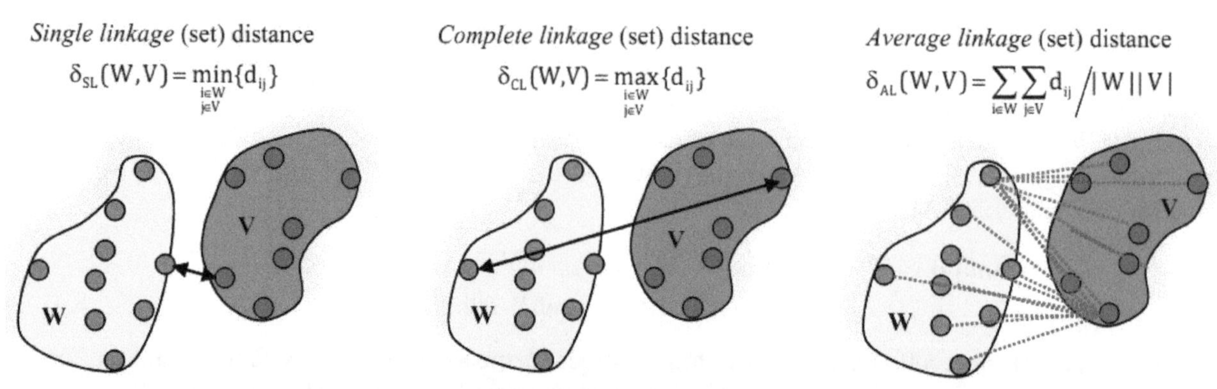

Figure 8.4 Set distances for the SL, CL, and AL SAHN models.

Table 8.2 An informal description of the SAHN clustering algorithms.

Input	$D = [d_{ij}] \in M_n^+$ (a dissimilarity relation on $O \times O$)
SAHN 1	Choose a SAHN distance $\delta_{SAHN}(A, B)$ (Equation (8.4)).
SAHN 2	Start with each of the n objects in O in its own cluster $(c = n)$.
SAHN 3	Find and merge the two closest clusters (W and V) into WV.
SAHN 4	Delete rows and columns in D corresponding to W and V.
SAHN 5	Add a row/column to D with the new distances between WV and the remaining clusters.
SAHN 6	Repeat Steps 3–5 until all n objects are in one cluster $(c = 1)$.

occur in SAHN 3, and care must be taken to specify what to do when this happens. Since the heuristic is to merge closest clusters, if there are several candidates minimally equidistant from the same cluster, the usual strategy is to merge (all of) the tied candidates at the same time. The important result of this is that a SAHN hierarchy may not contain a candidate partition at each value of c from n to 1.

What kind of model is the SAHN approach? It is not wrong to call SAHN clustering a *heuristic model* since its guiding principle and the definition of "good CPOV clusters" are imbedded in the algorithm that finds the clusters. It is hard to separate "the model" from "the algorithm" based on the verbal description in Table 8.2. Section 8.5 provides an alternative description that may be more satisfying to mathematics aficionados – viz., we will see that SAHN clustering is a *graph-theoretic model* and that the most familiar instance of it -single linkage clustering -depends explicitly on the minimal spanning tree. So, the SAHN model is very different from the three clustering models and algorithms presented in Chapters 6 and 7. Here are the major differences:

1. No objective function defines CPOV clusters.
2. No optimization problem leads to the clustering algorithms[22].
3. No initialization, termination, or iteration limit criteria are needed.
4. The only user-defined parameter is the SAHN set distance δ_{SAHN}.
5. The SAHN algorithm is not iterative: it terminates in $\leq (n-1)$ steps.
6. SAHN finds at most one candidate partition at each c, $1 \leq c \leq n$.
7. SAHN clusters are hierarchical and unique (up to ties).

[22]In general this statement is true. However, please see Section 11.7 in Hartigan (1975) for a way to cast the single linkage algorithm as an exact optimization problem.

Algorithm 8.1 contains a pseudocode corresponding to the steps in Table 8.2. The sets of CPOV clusters $\{C_k\}$ and dissimilarity matrices $\{D_k\}$ are indexed for the case where all mergers are unique (note the comment about ties).

Algorithm A8.1. Basic SAHN algorithm.

	Algorithm 8.1 Basic SAHN Algorithm
1	**In** An $n \times n$ dissimilarity (or distance) matrix $D = [d_{ij}] \in M_n^+$
2	***% User choice for model parameters:*** Set distance $\delta_{SAHN}(A, B)$:
3	***% Default choice for initial singleton clusters:*** $C_n = \{\{1\}, \{2\}, \ldots, \{n\}\}$
4	***For c = n-1 to 1, step-1***
5	Find sets to merge: $(W, V) = \underset{A \neq B \subset C_{c+1}}{\arg\min}\{\delta_{SAHN,c}(A, B)\} : G_\delta(c) = \delta_{SAHN,c}(W, V)$
6	% In case of ties, merge all the tied clusters, or any two of them
7	Merge W and V: $WV = W \cup V$
8	Update C: $C_c = C_{c+1} - \{W, V\} \cup WV$
9	Delete rows and columns corresponding to W and V: $D_{c+1} \rightarrowtail D_{c-1}$
10	Insert new row and column distances $\{\delta_{SAHN,c}(WV, S) : WV \neq S \subset C_c\} : D_{c-1} \rightarrowtail D_c$
11	***Next k***
12	**Out:** Crisp data clusters $C_{SAHN} = \{C_n, C_{n-1}, \ldots, C_1\} : C_c \in M_{hcn}$
13	Crisp candidate partitions $CP_{SAHN} = \{U_{SAHN,c} \in M_{hcn} : c = n, n-1, \ldots, 1\}$
14	Merger distances(aka Gap distances) $\{G_\delta(c) = \delta_{SAHN,c}(W, V) : 1 \leq c \leq (n-1)\}$

® ***Forewarned is forearmed:*** Note especially the counter c in line 4 of A8.1 which indexes the steps and number of clusters in hierarchical clustering. The step at which there are c clusters is the *c*-th level in the hierarchy. Thus, level n occurs at c = n; level k at c = n − k; and level 1 at c = 1. There can be confusion about this term. Other authors may use the reverse terminology, calling the level at c = n the first level so that the *n*-th level occurs at c = 1. ®

All of our previous clustering methods can produce sets of candidate partitions by varying the parameters of the model. In contrast, once a SAHN set distance $\delta_{SAHN}(A, B)$ is chosen, the candidates generated by SAHN clustering are a fixed set of (n) hierarchically nested CPOV clusters. The only variable is the number of clusters, c, and it is automatically visited for every c during the execution of A8.1. When the dissimilarity matrix $D = [d_{ij}] \in M_n^+$ is built by computing dissimilarities between pairs of objects (the most common instance being distances between pairs of vectors in $X = \{\mathbf{x}_1, \ldots, \mathbf{x}_n\} \subset \Re^p$), the function chosen to compute the $[d_{ij}]$ is a second user-defined parameter, and, as we have seen, this can alter the outputs of A8.1.

Validation of the candidates produced by a SAHN model corresponds to choosing one of the $(n-2)$ non-trivial partitions for $2 \leq c \leq (n-1)$ as the "best" (most valid) one. Since the candidate partitions $CP_{SAHN} = \{U_{SAHN,c} \in M_{hcn} : c = n, n-1, \ldots, 1\}$ are crisp, they can be evaluated with any crisp cluster validity index. But ... please note that each partition is a different size, so the interpretation of relative CVI values is especially difficult. There is also a heuristic method based on the biggest jump (or gap distance) in the set of merger distances (line 14 in A8.1) that is often advocated as a natural CVI for SAHN candidates. We will have a look at this method in Section 8.7.

8.3 Choosing Model Parameters for SAHN Clustering

One of the most attractive features of the SAHN methods is that, once the input dissimilarity matrix $D = [d_{ij}] \in M_n^+$ is known, there is only one parameter to choose, namely, the set distance $\delta_{SAHN}(A, B)$. Equations (8.5) define the three set distances that are used most often in this approach to clustering. It is easy to describe the geometric circumstances that are favored by each of these choices.

Figure 8.5 compares the behavior of the SAHN clustering model using SL and CL distances when applied to cluster chains and cluster clouds. To put this in perspective, please review the geometry of unit balls for the five good norms in Figure 6.6. The illustrations in Figure 8.5 correspond to $D = [d_{ij}] \in M_n^+$ being a Euclidean distance matrix (EDM) for pairs of vectors in $\Re^2$.

Single Linkage (aka Minimum Distance, Nearest Neighbor). The SL distance (8.5a) favors merging objects that belong to elongated chains or strings such as those depicted in Figure 8.5(a). SL will do quite well at reproducing HPOV clusters of this kind as long as the "gap distance" (intercluster distance G_δ) between the points in separate chains is greater than the intracluster distances $\{d_{ij}\}$ within each chain. Conversely, SL will do poorly when presented with dissimilarities corresponding to objects in cluster clouds, like those illustrated in Figure 8.5(c), because the distance-finding mechanism (which, as we will discover shortly, is nothing more than adding edges to the minimal spanning tree in D) will "jump the intercluster gap" before it completes intracluster connections. Figure 4.6(d) shows how SL is related to the MST of the distance matrix;

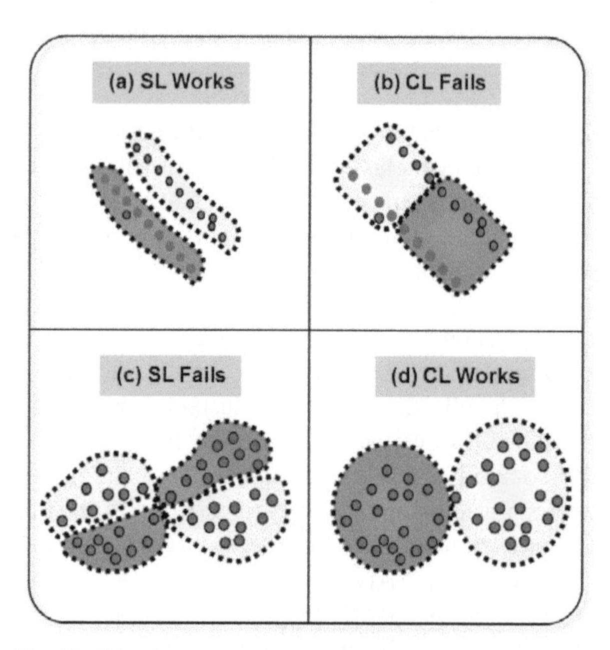

Figure 8.5 The (Euclidean) geometry for success and failure with SL and CL clustering.

Figure 8.22 will illustrate how SL processing jumps the gap when points in-between crisp clusters provide a pathway for the jump.

Complete Linkage (aka Maximum Distance, Furthest Neighbor). SAHN with the CL distance (8.5b) behaves in a manner which is, in some sense, diametrically opposite to the behavior of SL. This distance favors cluster clouds such as that shown in Figure 8.5(d) and fails badly on cluster chains such as those illustrated in Figure 8.5(b). CL is not used nearly as much as SL. Why? Because there are many other clustering methods that fare well in the presence of cluster clouds. Indeed, all three of the methods presented so far (HCM, FCM, and GMD/EM) are good cloud finders but relatively poor chain finders. This is probably the main reason CL doesn't see much action, while SL appears quite often.

Average Linkage (aka Average Distance). We will not use average linkage in the sequel but include it here to complete the presentation of the three main linkage methods. The AL distance at Equation (8.5c) is supposed to mediate between the extremes corresponding to SL chains and CL clouds. Thus, AL is in some ways analogous to GMD/EM, which, in theory anyway, is capable of simultaneously representing clouds for clusters drawn from component densities with almost equal covariance in all p directions, and strings, which correspond to clusters drawn from component densities with (nearly) singular covariance. But AL is not well-regarded by many practitioners because the result of compromising the two extremes is usually to smooth chains and stretch clouds, thereby obscuring clusters of both types that may be present in the data.

Example 8.2. Evolution of SL and CL partitions.

This example illustrates the evolution of SL and CL partitions on the data set X_5 shown in Figure 8.6. Without confusion, we identify X_5 simply as $\{1, 2, 3, 4, 5\}$ with (x, y) coordinates as shown in Figure 8.6. It is arguable whether there *are* HPOV clusters in X_5, but this example exhibits everything you need to know to understand SL and CL; so it has a lot of pedagogical value.

Calculation of the Euclidean distances between the points in X_5. converts the object data to the dissimilarity data

$$D_5 = \begin{bmatrix} 1 & 1.41 & 4.12 & 3.54 & 3.00 \\ & 0 & 3.00 & 2.55 & 2.24 \\ & & 0 & 0.71 & 1.41 \\ & & & 0 & 0.71 \\ & & & & 0 \end{bmatrix} \tag{8.6}$$

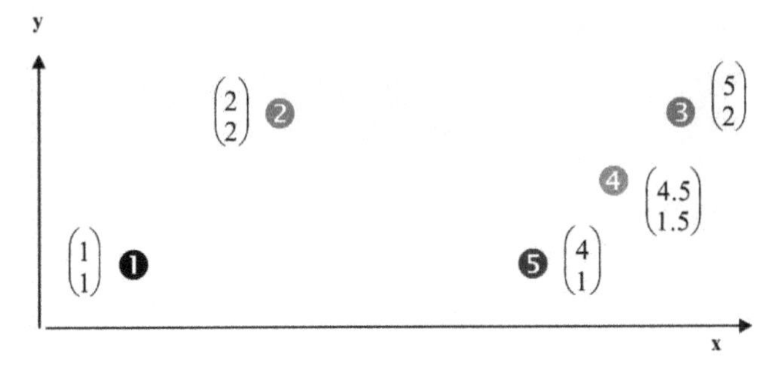

Figure 8.6 Data set X_5.

Single Linkage. Each of the five objects begins in its own cluster; so the initial set of clusters is $C_5 = \{\{1\}, \{2\}, \{3\}, \{4\}, \{5\}\}$. SL finds the *smallest* distance in D_5 and merges the corresponding points. Looking at the data, you can see there is a gap distance tie (value 0.71) between the pairs (3, 4) and (4, 5). What happens? There are two options: (i) you can break the tie arbitrarily and merge either pair, reduce D_5 to D_4, and move on to the next level, where the other point will join this first pair next; or (ii) you can merge all three points now since it will happen at the next step if not done immediately. The choice is an implementation issue with the following consequence: choice (i) yields a partition at $c = 4$, and then the next partition would be produced by choice (ii), which will jump directly from $c = 5$ to $c = 3$. Merging all tied points is always more efficient computationally. Figure 8.7 illustrates the entire processing sequence (the five points in the data set are *not* drawn to scale).

Say you take the second tie-breaking option and merge (3, 4, 5) all at once; so the new clusters are $C_3 = \{\{1\}, \{2\}, \{3, 4, 5\}\}$ with reduced distance matrix D_3. Now find the two clusters in C_3 closest to each other in the sense of SL distance. This search results in the merger of $\{1\}$ and $\{2\}$ since these two singleton sets are closer to each other than either is to $\{3, 4, 5\}$. Hence, $C_2 = \{\{1, 2\}, \{3, 4, 5\}\}$ with distance matrix D_2. Finally, these two clusters will be merged, resulting in the single cluster $C_1 = \{\{1, 2, 3, 4, 5\}\}$.

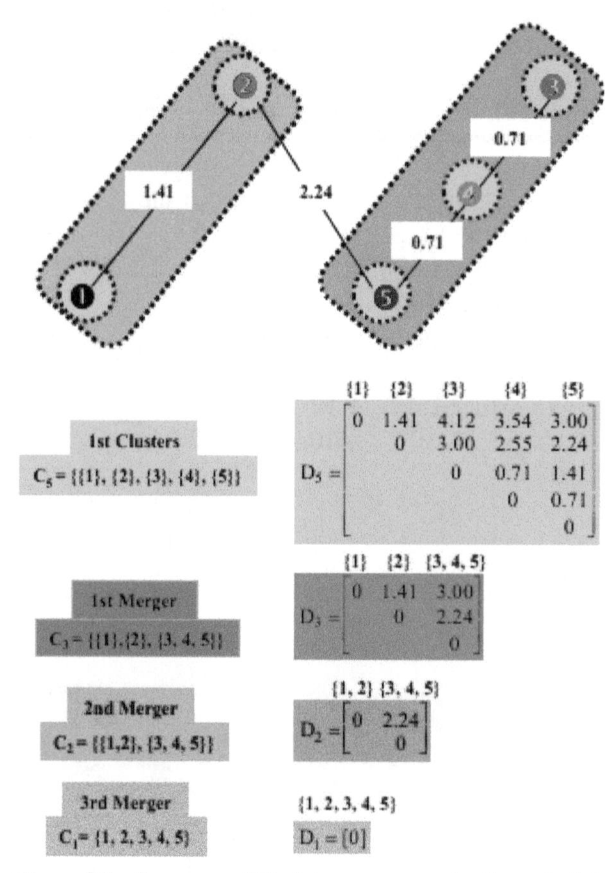

Figure 8.7 Sequence of SL distance mergers for Example 8.2.

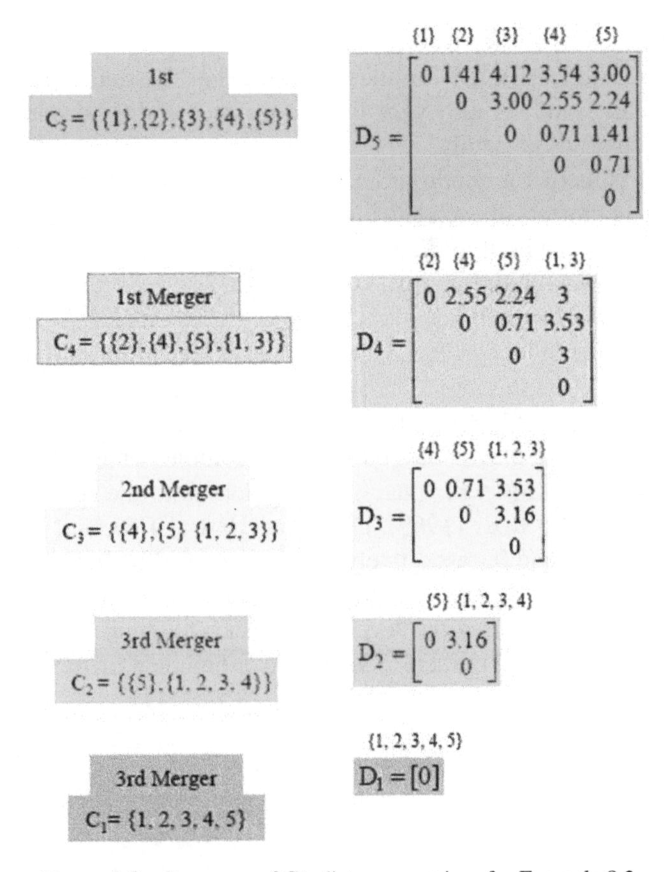

Figure 8.8 Sequence of CL distance matrices for Example 8.2.

Complete Linkage. What will CL do with D_5? Here is the sequence of actions it will take. Beginning again with $C_5 = \{\{1\}, \{2\}, \{3\}, \{4\}, \{5\}\}$, CL finds the *largest* gap distance in D_5 and merges the corresponding points, which, in this case, are the points 1 and 3 with the gap distance value 4.12, resulting in D_4 for the clusters $C_4 = \{\{2\}, \{4\}, \{5\}, \{1,3\}\}$. The next merger adds 2 to (1, 3}, resulting in D_3 for $C_3 = \{\{4\}, \{5\}, \{1,2,3\}\}$. Then CL adds point 4 to the larger cluster, yielding D_2 for $C_2 = \{\{5\}, \{1,2,3,4\}\}$. CL ends with the same terminal cluster as SL, $C_1 = \{1,2,3,4,5\}$ with distance D_1. The sequence of dissimilarity matrices for the CL evolution are shown in Figure 8.8.

Comparing Figures 8.7 and 8.8, the only difference in the two results is that points 1 and 2 are joined to each other by SL before they consider points 3, 4, and 5, whereas 2 joins 3, 4, and 5 before accepting point 1 as a linkage neighbor. There is a convenient way to visually compare the evolution of SL and CL clusters using a dendrogram, the representation we turn to now.

8.4 Dendrogram Representation of SAHN Clusters

A *dendrogram* is a graphical depiction of the results of any hierarchical clustering that portrays the sequential evolution and arrangement of the clusters. This method of visually representing hierarchical clustering is very appealing and is widely seen in earlier studies that used SAHN clustering, especially in numerical taxonomy of biological data. It is quite effective for understanding the "reasoning" underlying the formation of clusters by a SAHN algorithm, but it is also very limited by its inability to portray more than several hundred objects with any degree of visual acuity.

There are various ways to construct a dendrogram, which is a special type of rooted tree. The most common approach is to represent clusters at each level in the tree as nodes or vertices in a tree that is drawn with its leaves at the top and root at the bottom (like a real tree). However, it is not unusual in the literature to see the tree "inverted" – i.e., drawn with the root at the top and leaves at the bottom of the page. Both styles are common, and MATLAB offers a choice of either style.

For a dissimilarity matrix of size n, dendrograms in this text begin at the top of a page, with n leaves at level n corresponding to the $c = n$ singleton clusters $C_n = \{\{1\}, \{2\}, \ldots, \{n\}\}$, and end at the root, where there is just $c = 1$ cluster $C_1 = \{1, 2, \ldots, n\}$ at level 1. The clusters C_k appear at level $c = k$ in the dendrogram. Don't confuse levels in the dendrograms in this volume, which, by our definition, descend from n to 1 as aggregation progresses down the page from the leaves to the root, with other notations in the literature. For example, Gower and Ross (1969), Duda and Hart (1973), and Rosen (2007) all assign different integers, namely, $(n - 1)$, n, and 0, respectively, to the root level in their dendrograms.

The vertical scale of a dendrogram depends on the particular model. For SL as discussed here, the absolute distance down the tree from each previous level corresponds to the distance (the"gap distance" G_δ, computed at line 5 of A8.1) between adjacent merger levels. As a practical matter, the vertical axis for the

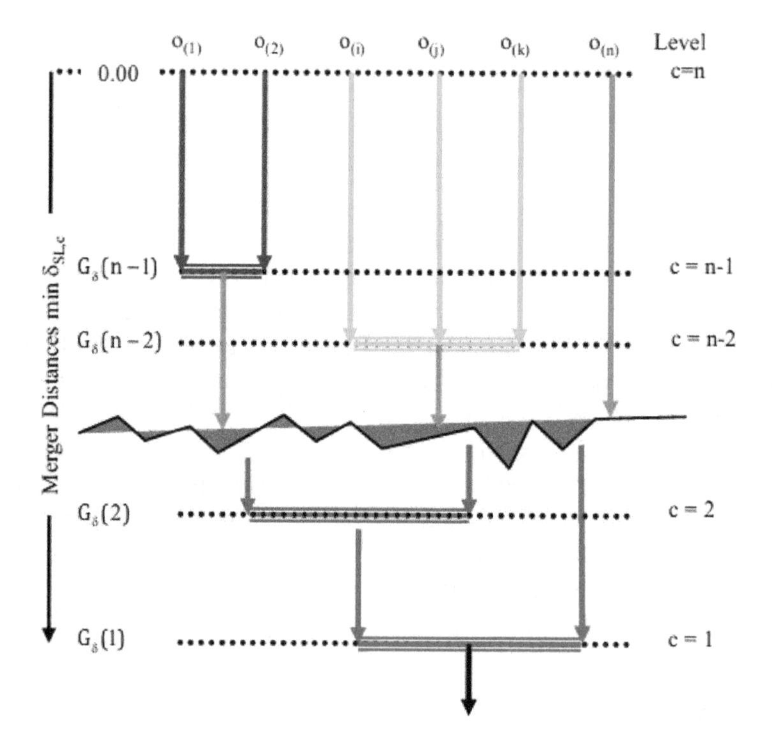

Figure 8.9 Generic dendrogram depicting hierarchical SL clusters.

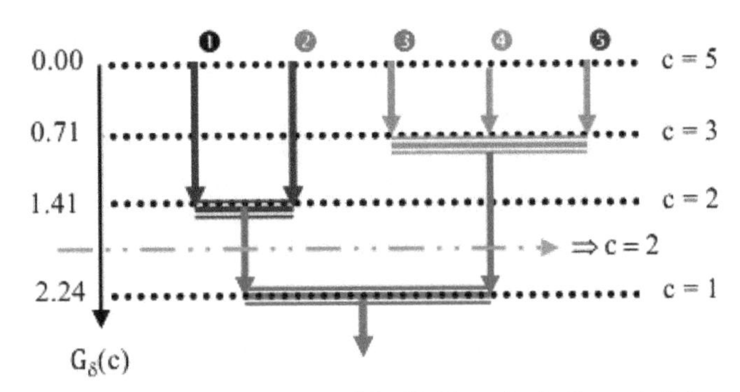

Figure 8.10 Dendrogram of the SL hierarchy in Example 8.2.

dendrogram is often not to scale, but the SL merger distances, which are monotone increasing as c decreases from n to 1, will always be shown. The gap distances can play an important role in determining which set of clusters to validate. Figure 8.9 shows a generic SL dendrogram.

The objects shown at the top (leaves) of the dendrogram are $\Pi(O) = \{o_{(1)}, o_{(2)}, \ldots, o_{(n)}\}$, some permutation of the input object set $O = \{o_1, o_2, \ldots, o_n\}$. It is always possible, after obtaining a hierarchy of clusters by any SAHN method, to rearrange the objects so that the dendrogram appears as in Figure 8.9 – that is, so that there will not be any vertical line crossings. The dendrogram corresponding to the SL hierarchy of X_5 in Example 8.2 is shown in Figure 8.10. A dendrogram for the complete linkage case corresponding to the CL clusters in Example 8.2 is left as an exercise.

Making a horizontal cut through the dendrogram *in-between* merger levels shows the corresponding number of clusters (c is the number of vertical lines cut at the level above the cut). For example, any cut of the tree in Figure 8.10 between the values 1.41 and 2.24 cuts two vertical connectors as shown in the figure; so there are c = 2 clusters at this instant. And so on, until any cut below the merger distance 2.24 results in c = 1 cluster.

You can see in Figure 8.10 that the distance tie between points (3,4) and (4,5) in data set X_5 results in the absence of a set of clusters at c = 4. This is the result of using the tie-breaking strategy: merge all points that yield the same set distance at any level in the hierarchy.

8.5 SL Implemented With Minimal Spanning Trees

8.5.1 The Role of the MST in Single Linkage Clustering

Gower and Ross (1969) pointed out that all the information needed to build an SL hierarchy on a square relation was contained in a *minimal spanning tree* (MST) of the data as described in Definition 4.2 for a weighted graph $G = (V, E, W)$. To see the relationship between the MST and the SL hierarchy, first revisit Figure 4.6(d), which shows how SL connections are related to the MST of the Euclidean distance matrix. To understand this more fully, consider the dendrogram in Figure 8.9. Imagine a "cut" (a horizontal line) through this tree anywhere between the levels c = k and c = (k – 1) in the dendrogram, as shown in Figure 8.11.

All distances "above" the cut (that is, above level k) must be less than or equal to gap distance $G_\delta(k)$, and all the distances below the cut (meaning at level k – 1) must be greater than $G_\delta(k)$. Thus, there must be an edge in the MST of D, depicted in Figure 8.11 as e_{st}, that connects some vertex $\mathbf{v}_s$ in the vertex set

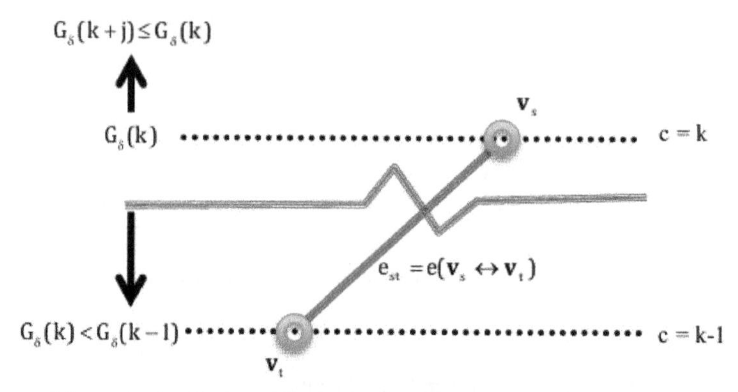

Figure 8.11 Cutting a SL dendrogram between levels k and k − 1.

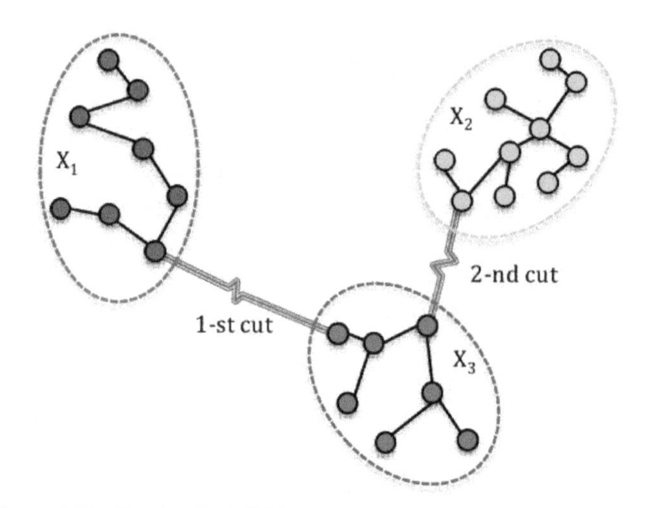

Figure 8.12 Parsing the MST into subtrees which form the SL clusters.

on the nodes "above the cut" to some vertex $\mathbf{v}_t$ "below the cut," and this edge has the minimum weight for all edges that cross the cut. This shows that we can recover the SL clusters produced by applying A8.1 to D by instead constructing the MST of D (i.e., of the graph G = (V, E, W)), followed by making a backpass through the MST to cut edges.

After the MST is built, the backpass ascends up through the tree cutting the sequence of connecting edges in descending order of magnitude. The first cut (is the deepest!) will occur at the longest edge connecting levels c = 1 and c = 2; subsequent cuts are made to the second largest edge, third largest edge, and so on, until the tree is ultimately cut into n leaves. The vertices belonging to the subtrees at any level are the SL clusters at that level. Figure 8.12 illustrates how the SL clusters are formed by this procedure (cf., Figure 4.3(b) to see 11 star clusters formed this way).

At the beginning of the backpass, all of the vertices in Figure 8.12 are connected; so the backpass begins with the SL clusters $C_1 = \{1, 2, \ldots, n\} = X_1 \cup X_2 \cup X_3$. The first cut occurs across the edge with the largest weight (distance) in the tree and isolates two subtrees of vertices; so, after this cut, there are c = 2 SL clusters, $C_2 = \{\{X_1\}, \{X_2 \cup X_3\}\}$. The second cut is made across the next largest edge, resulting in c = 3 SL clusters, $C_3 = \{X_1\}, \{X_2\}, \{X_3\}$. The backpass continues along the sequence of edges, making a (possible) total of n − 1 cuts and, at the end of the pass, arrives at the set of n singletons, $C_n = \{\{1\}, \{2\}, \ldots, n\}$. Here is Algorithm 8.1, recast in the language of the single linkage model:

Algorithm A8.2. MST - single linkage (SL) algorithm.

	Basic (MST-Based) Single Linkage Clustering Algorithm
1	**In:** An $n \times n$ dissimilarity matrix $D = D_n \in M_n^+$
2	**% User choice for model parameter (set distance)**
3	Select set distance $\delta_{SL}(A, B)$ for finite crisp subsets A, B
4	**% Initialization**
5	Initial (singleton) clusters: $I = \emptyset; J = \{1, 2, \ldots, n\}$
6	Pick starting object (vertex) $m \in J : I = \{m\} : J = J - \{m\}$
7	**% Build MST**
8	**For k = 2 to n**
9	% Find a minimum feasible edge e_{ij}: No tie-breaking strategy is employed
10	Select $(i, j) \in \underset{v_p \in I, v_q \in J}{\arg\min}\{d_{pq}\} : I = I \cup \{j\} : J = J - \{j\} : \delta_{SL,k} = d_{ij}$
11	**Next k**
12	**% Extract nested partitions by backpass cuts in MST**
13	**For k = 2 to n**
14	Remove edge link having distance $\delta_{SL,k}$
15	C_K connected subtrees (clusters of vertices) in the MST
16	**Next k**
17	**Out:** Crisp data clusters $C_{SL} = \{C_n, C_{n-1}, \ldots, C_1\} : C_c \in M_{hcn}$
18	Crisp candidate partitions $CP_{SL} = \{U_{SL,c} \in M_{hcn} : c = n, n-1, \ldots, 1\}$
19	Merger distances (aka Gap distances) $\{G_\delta(c) = \delta_{SL,c}(W, N) : 1 \le c \le (n-1)\}$

Default Choices for A8.2: Happy surprise: ***NONE***!

Since the basis of A8.2 is graph theory, the description of the SAHN model for SL can be revised a bit. We stated earlier that this was a heuristic model, but in view of this alternate formulation, it is not wrong to call this instance of SAHN clustering a graph-theoretic model with the formal structure of the MST.

Example 8.3. Extracting SL partitions from the MST of X_5.

This example illustrates the idea of the backpass depicted in Figure 8.12 and implemented in lines 12–16 of A8.2 for the data set used in Example 8.2. Look back at Figures 8.7 and 8.10. Imagine that you have arrived at the bottom of the dendrogram in Figure 8.10 (you are at the bottom of Figure 8.7 too) so that all 5 points in X_5 belong to the cluster C_1. Now start a traversal back up the dendrogram, which corresponds to a backpass through the MST. The first split is made by cutting the edge from point 2 to point 5, and the edge distance is 2.24 as seen in Figure 8.7. This produces c = 2 subtrees in the MST, which are the clusters named C_2 at level 2 in Figure 8.7. Continuing this way results in the following sequence of SL partitions of X_5:

$$\{1,2,3,4,5\} \to \{1,2\} \cup \{3,4,5\} \to \{1\} \cup \{2\} \cup \{3\} \cup \{4\} \cup \{5\}.$$

Thus, the clusters at any level c can be obtained from the MST as the subsets of vertices in the subtrees obtained by deleting all edges of length greater than $\{\delta_{SL}(A,B)\}$.

What happens when the MST is not unique? Example 8.4 illustrates what can happen.

Example 8.4. SL partitions can differ if the MST is not unique.

The top view in Figure 8.13 is Figure 8.13(a), which shows an undirected, weighted, connected graph G_6 which has n = 6 vertices and m = 8 edges with integer weights (recall that a graph is connected if it has no unreachable vertices; all of the nodes of G_6 can be reached by some path). If we imagine the vertices to

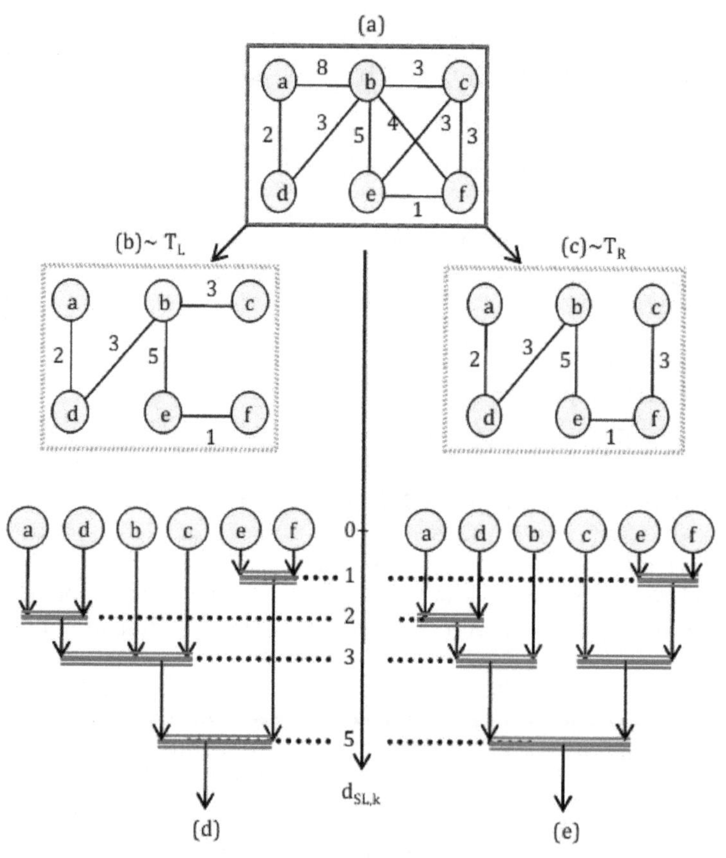

Figure 8.13 SL hierarchies for G_6 from minimal spanning trees T_L and T_R.

Table 8.3 SL hierarchies for MSTs T_L and T_R in Figure 8.13.

	T_L	T_R
C_6	{a}∪{b}∪{c}∪{d}∪{e}∪{f}	{a}∪{b}∪{c}∪{d}∪{e}∪{f}
C_5	{a}∪{b}∪{c}∪{d}∪{e, f}	{a}∪{b}∪{c}∪{d}∪{e, f}
C_4	{a, d}∪{b}∪{c}∪{e, f}}	{a, d}∪{b}∪{c}∪{e, f}}
C_3	(none)	(none)
C_2	**{a, b, c, d}∪{e, f}}**	**{a, d, b}∪{c, e, f}}**
C_1	{a, b, c, d, e, f}	{a, b, c, d, e, f}

represent objects and the weights to represent dissimilarities, we can ask SL to build clusters in this graph. If the weight of each edge in a connected graph is distinct, then the graph contains exactly one (unique) minimum spanning tree. Since 4 of the 8 weights are equal, there may be more than one MST for this graph. Figures 8.13(b) and 8.13(c) show two different MSTs (Left = T_L and Right = T_R) for G_6, both having 14 as their (minimum) sum of weights.

Since the MSTs are different, the SL hierarchies from these two MSTs are also different. Dendrograms from each tree are displayed in Figure 8.13. View 8.13(d) corresponds to tree 8.13(b), and panel 8.13(e) goes with 8.13(c). The SL cluster hierarchies for each tree are shown in Table 8.3.

Because of the distance tie, there are no clusters at level c = 3 in either SL hierarchy. At level c = 2, the two trees yield different sets of clusters, as shown in Table 8.3. Which of these solutions do you expect to see if you submit the dissimilarity matrix D in this example to A8.2?

Remark: The input data to A8.2 is a 6 × 6 matrix D, but this matrix is not in M_n^+ because Equation (8.1a) requires that each off-diagonal distance be greater than 0, but the graphs in Figure 8.13 have six "missing" edges (which are not zero values); so these two graphs are incomplete but connected. Since there are missing edges, you might be tempted to insert 0's into these off-diagonal elements of matrix D for processing with A8.2, but this is unnecessary. This example shows that the MST, iVAT, and SL can function on a slightly more general class of matrices than $D \in M_n^+$. That is, $D \in M_n^+$ is sufficient, but not necessary, for SL clustering.

Next, let's compare the number of label disagreements that SL makes on the Iris data to the solutions offered by our previous clustering methods.

Example 8.5. SL partitions the Iris data.

The SAHN algorithms operate on dissimilarity data; so let's convert X_{Iris} into a distance matrix D of size 150 by computing its *ij*th entry as the Euclidean distance between $\mathbf{x}_i$ and $\mathbf{x}_j$ in Iris. We will refer to this matrix as $D_{E,Iris}$. The $n(n-1)/2$ off-diagonal distances are not all distinct; so we can't conclude that the graph of $D_{E,Iris}$ has a unique MST, but applying algorithm A.8.2 to $D_{E,Iris}$ produces a unique set of crisp candidate partitions of Iris, $CP = \{C_1, C_2, \ldots, C_{150}\}$. Without a change in the minimum edge selection strategy of A8.2, these are the only partitions of Iris that SL has to offer. Which partition should we choose? We will take up the question of a validation strategy that enables us to make an informed choice amongst these many candidates shortly.

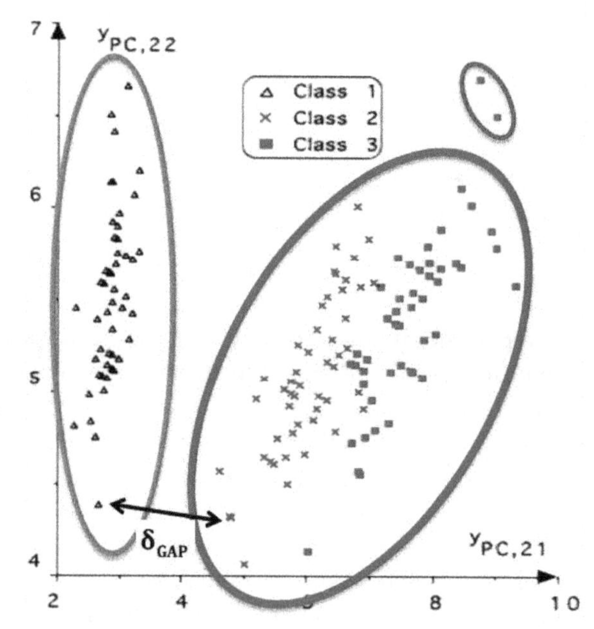

Figure 8.14 SL partition of Iris using X_{Iris} at c = 3 (cf., Figure 5.8).

Here, we let the physical labels for Iris guide us to an examination of the SL partition at c = 3. Visual assessment of this solution is best understood against the background scatterplot of the 2D principal components of Iris that were shown in Figure 5.1 that are repeated here in Figure 8.14. The leftmost 50 plants (the Setosa flowers) are clustered together, but the remaining 100 plants (Versicolor and Virginica) are split into 2 clusters having 98 mixed red and blue points and 2 red points. Thus, SL commits 48 "label mistakes" at c = 3. In terms of the partition accuracy at Equation (6.34), this corresponds to $102/150 = 0.68$, or 68% correct.

To understand how and why this happens, consider the MST of the data. The MST will connect the 50 Setosa plants without "jumping the gap" to the other 100 points since the distance δ_{GAP} shown in Figure 8.14 between Setosa and the remaining 100 plants is greater than any pair of intrapoint distances in the Setosa subset. After A8.2 crosses the gap, the remaining 100 points are almost close enough to each other for a complete connection, but the 2 red Virginica points in the upper right corner of the plot are just far enough away so that they split off to form their own cluster at c = 3.

The partition of Iris found by SL at c = 2 simply joins the 2 point cluster to the 98 point cluster, leaving us with the solution for Iris which seems most satisfactory from the CPOV. Previous examples have argued that Iris probably has c = 2 CPOV clusters when processed with most clustering models, and this result doesn't contradict that assertion. Indeed, this seems like further evidence that Iris doesn't contain three CPOV clusters from almost any reasonable (algorithmic) point of view. We will discuss a way to choose "the best" SL solution shortly.

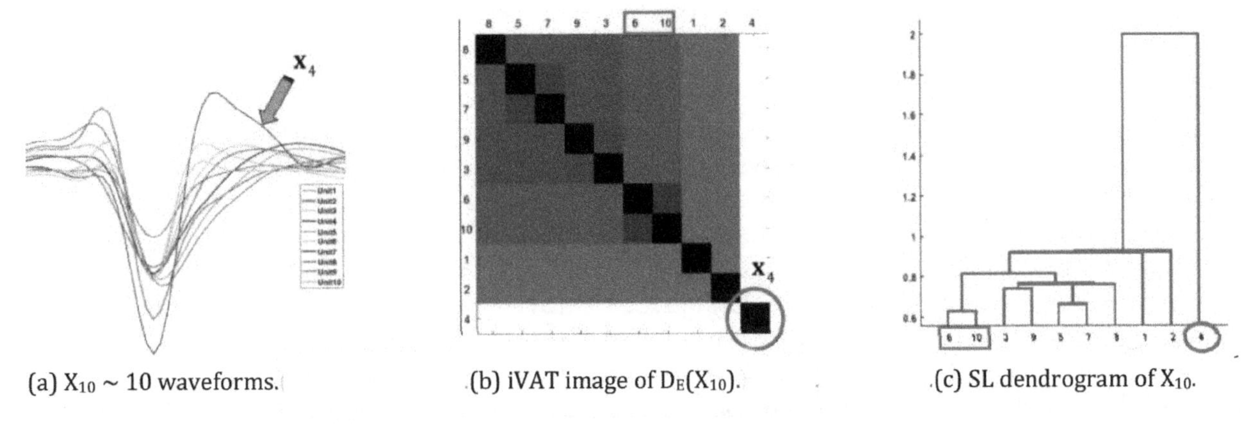

(a) $X_{10} \sim 10$ waveforms. (b) iVAT image of $D_E(X_{10})$. (c) SL dendrogram of X_{10}.

Figure 8.15 iVAT as a visualization of SL clusters and their dendrogram.

Example 8.6. iVAT is a "visual" front end for single linkage.

This example discusses the relationship between SL and the iVAT cluster heat map which is illustrated in Figure 8.15. View 8.15(a) is a set of 10 waveforms, (X_{10}), each represented by a sample vector of p = 72 values, equally distributed along the horizontal axis. The waveform labeled $\mathbf{x}_4$ is (visually) anomalous compared to the general shape trend of the other nine graphs. Figure 8.15(b) is the iVAT image of X_{10} made from the 10×10 matrix D_E of Euclidean distances between pairs of waveform vectors. Integers along the borders of the iVAT image of X_{10} show the identity of each pixel after iVAT reordering. Waveform $\mathbf{x}_4$ is isolated as a 1×1 dark block in the lower right corner of Figure 8.15(b). This illustrates the potential of an iVAT image to suggest anomalies in a set of high-dimensional input data.

The iVAT image also suggests that the other nine waveforms are members of one large cluster. A closer look at the image reveals that this 9×9 diagonal block has some internal hierarchical substructure. For example, intensities in the highlighted (6, 10) pair suggests that these two waveforms are most closely related, followed by the (5,7) pair and then the (3,9) pair. These internal pairings are a bit hard to see in Figure 8.15(b), but if you zoom the view, they are clearly there.

The SL hierarchy of the 10 inputs is easily extracted by applying a backpass that cuts edges in the iVAT MST in reverse order. Figure 8.15(c) is the dendrogram of the clusters produced by extracting the single linkage hierarchy this way. The anomalous waveform $(\mathbf{x}_4)$ is very reluctant to join the SL hierarchy, coming in as the last merger on the dendrogram, more than a full unit above the other nine elements in terms of merger distances along the vertical axis of Figure 8.15(c). The evolution of SL clusters in Figure 8.15(c) is clearly visible in the iVAT image. Figures 8.15(b) and 8.15(c) make the relationship between iVAT and single linkage quite transparent: an iVAT image can be interpreted as a visual front end to SL clustering.

8.5.2 SL Compared to a Fitch-Margoliash Dendrogram

The SAHN models are not the only way to erect cluster trees. Fitch and Margoliash (1967) present a detailed discussion about building a different type of tree for relational data that might reveal possible clusters in the set of 20 organisms listed in Table 8.4. Fitch (1988) recounted the 1967 work in a delightful one page essay that is historically charming "there were 10 published protein sequences." The original work, now more than 50 years old, was prescient about data collection (as gene sequencing) and methods of analyzing it by cluster analysis.

HPOV of the 20 Organisms. If you asked 10 people to list the clusters they think exist in the set of 20 objects in Table 8.4, you would almost certainly get a variety of responses. A first question would be-based on what properties of the objects? Certainly, the responses would depend on this choice. If, for example, you group them according to the number of legs (putting aside some contentiously amusing questions about whether a kangaroo has two legs and two arms or four legs, a penguin has two feet but no legs, and so on), there would be c = 4 clusters: 0 legs: $\{14, 15, 18 - 20\}$; 2 legs: $\{1, 2, 9 - 12\}$; 4 legs: $\{3 - 8, 13\}$; 6 legs: $\{16, 17\}$.

Perhaps, the most natural property for cluster membership would be in terms of their biological classes. There are some oddities in this data set that make the concept of biological clusters in it confusing. For example, while the turtle and rattlesnake are both classified as reptiles, one has four legs, and the other none. And from the HPOV, the tuna can't occupy its own cluster. Most people don't understand the fine distinctions made by systematic zoologists. They are grouped here into what seem (to me!) to be the most natural HPOV clusters at the c = 6 clusters shown in Table 8.5. The point is that the "ground truth" for HPOV clusters in this data depends on (i) the human that constructs it; and (ii) the properties of the objects used in the determination.

CPOV of the 20 Organisms. To ask a computer what clusters exist in these 20 objects, you need a numerical representation of the objects. We could build an object data set to represent these organisms in much the same way as was shown in Example 5.1. Features to nominate might include # of legs, # of arms, # of wings, furred or feathered, and so on. Different choices for the numerical representation would, of course, probably offer different interpretations of CPOV clusters in the object set.

Example 8.7. SL and phylogenetic trees of the mutation relation data.

Fitch and Margoliash (1967) represented the objects in Table 8.4 via the dissimilarity data shown in Table 8.6 (shading in this table is used only to enhance readability). The distances in Table 8.6 are defined by the authors this way: "the *mutation distance* between two cytochromes is defined here as the minimal number of nucleotides that would need to be altered in order for the gene for one cytochrome to code for the other." Since this relational data is distance data, we will call it D_{mut}. There is no set of feature vectors underlying D_{mut} that represents the 20 organisms; so this is "pure" relational data.

Figure 8.16 is a stylized rendition of the phylogenetic tree, which appears in Fitch and Margoliash as their Figure 2. This tree was built by applying their algorithm to the dissimilarity data in Table 8.6, which were further described as average mutation distances between minimum numbers of mutations required to interrelate pairs of cytochromes c. The original figure had distances attached to each edge in this tree.

Table 8.4 Twenty organisms in Fitch and Margoliash (1967).

1	Man	6	Pig	11	Chicken	16	Fly
2	Monkey	7	Rabbit	12	Penguin	17	Moth
3	Dog	8	Kangaroo	13	Turtle	18	Baker's Mould
4	Horse	9	Duck	14	Rattlesnake	19	Bread Yeast
5	Donkey	10	Pigeon	15	Tuna	20	Skin Fungus

Table 8.5 (c = 6) HPOV (zoological similarity) clusters in the data of Table 8.4.

C_1 Mammals		C_2 Birds	C_3 Reptiles	C_4 Fish	C_5 Insects	C_6 Non-sentient
Man	Donkey	Duck	Turtle	Tuna	Fly	Baker's Mould
Monkey	Pig	Pigeon	Rattlesnake		Moth	Bread Yeast
Dog	Rabbit	Chicken				Skin Fungus
Horse	Kangaroo	Penguin				

Table 8.6 D_{mut}: mutation distances between pairs of cytochrome c.

	1	2	3	4	5	6	7	8	9	10	11	12	13	14	15	16	17	18	19	20
1	0																			
2	1	0																		
3	13	12	0																	
4	17	16	10	0																
5	16	15	8	1	0															
6	13	12	4	5	4	0														
7	12	11	6	11	10	6	0													
8	12	13	7	11	12	7	7	0												
9	17	16	12	16	15	13	10	14	0											
10	16	15	12	16	15	13	8	14	3	0										
11	18	17	14	16	15	13	11	15	3	4	0									
12	18	17	14	17	16	14	11	13	3	4	2	0								
13	19	18	13	16	15	13	11	14	7	8	8	8	0							
14	20	21	30	32	31	30	25	30	24	24	28	28	30	0						
15	31	32	29	27	26	25	26	27	27	27	26	27	27	38	0					
16	33	32	24	24	25	26	23	26	26	26	26	28	30	40	34	0				
17	36	35	28	33	32	31	29	31	30	30	31	30	33	41	41	16	0			
18	63	62	64	64	64	64	62	66	59	59	61	62	65	61	72	58	59	0		
19	56	57	61	60	59	59	59	58	62	62	62	61	64	61	66	63	60	57	0	
20	66	65	66	68	67	67	67	68	66	66	66	65	67	69	69	65	61	61	41	0

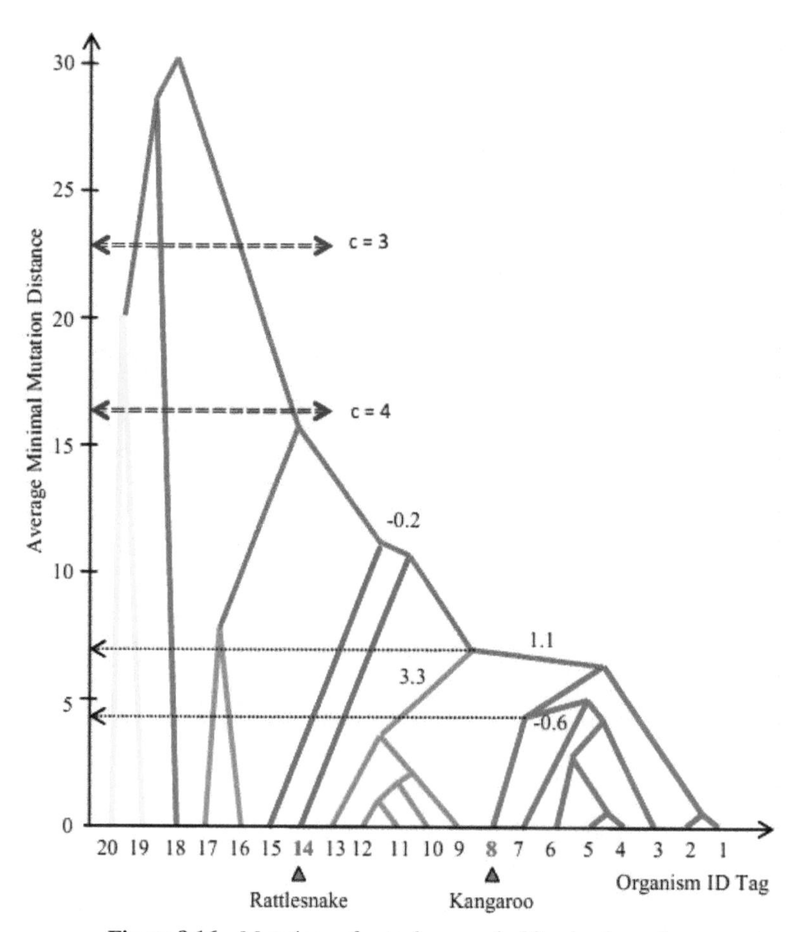

Figure 8.16 Mutations of cytochrome c in 20 animal species.

All but four of these distances are omitted in Figure 8.16. The edge weights 3.3 and 1.1 in Figure 8.16 are an example of an ordinary join. These two edges connect the subgroups (CPOV clusters) $\{1-8\}$ and $\{9-13\}$ at the vertical scale value of about 7 or so, and the values attached to the joining edges denote the "jumps" in the scale from each group below the join required to get to the next merger level. The construction in Figure 8.16 is legitimately regarded as a dendrogram and can be interpreted as showing cluster substructure in the mutation data. However, there are some important differences between the dendrograms we have seen so far and this one. Each join in Figure 8.16 occurs at the intersection of two non-vertical lines. It is a simple exercise to redraw this tree with leaves at the top, root at the bottom, and horizontal lines where clusters merge, in the style of our dendrograms such as in Figure 8.9 – *almost*. But when you try to do this, two problems arise.

First, look at the merger of the subgroup $\{3, 4, 5, 6, 7\}$ to the kangaroo $\{8\}$. The kangaroo makes a *negative jump* of -0.6 to achieve the merger at a level of about 4, which is 0.6 *less than* the merger level of $\{7\}$ to $\{3, 4, 5, 6\}$. This makes it clear that this is not a SL-type clustering. Indeed, no SAHN algorithm could produce this dendrogram since the merger distances are not monotonic. A second instance of reversed merger distance occurs when the tuna $\{15\}$ makes a reverse jump of -0.2 to join with $\{1, \ldots, 14\}$ at about 11 on the vertical scale. What explains these negative jumps?

It is due to the algorithm that builds this tree. In the words of the authors, "each number on the figure is the corrected minimum mutation distance (see their text) *along the line of descent* as determined from the best computer fit so far found. Each apex is placed at an ordinate value representing the average of the sums of all mutations in the lines of descent from that apex." This explains the unusual edge weights and somewhat odd appearance of the tree.

Figure 8.16 is a CPOV of the data. Does it agree with any HPOV? Should we expect the kangaroo $\{8\}$ to occupy the seemingly odd position it takes in the tree? In terms of the HPOV of the clusters, many (zoologists at least) would argue that $\{8\}$ should not join with $\{1-7\}$ at this level. Fitch and Margoliash attribute this, and two other "discrepancies" with respect to conventional HPOV classification of these organisms, to a deficiency of the numerical data, viz., a very small number of mutations. Thus, the rattlesnake $\{14\}$ doesn't join its fellow reptile, the turtle $\{13\}$, until the turtle is already grouped with the birds$\{9-12\}$. See Fitch and Margoliash (1967) for an in-depth discussion of the phylogenetic interpretation offered by this example. While not cast as an explanation or defense of the difference between HPOV and CPOV clusters, this affords you with an excellent chance to think carefully about the point made in Chapters 1 and 3 about the distinction between what humans and computers understand about clusters. It is important. Now let's see what single and complete linkage do with the D_{mut} data.

Submitting D_{mut} to A8.1 using the SL and CL options leads to the dendrograms shown in Figures 8.17 and 8.18. Compare these trees to the tree in Figure 8.16. All three constructions first link $\{1,2\}$ = $\{\text{man, monkey}\}$ and $\{4,5\}$ = $\{\text{horse, donkey}\}$ at the lowest levels of dissimilarity. This seems intuitively plausible from a zoological point of view.

Table 8.7 (F/M stands for Fitch and Margoliash in this table) exhibits the crisp clusters of the 20 organisms produced by cutting the three trees in Figures 8.16-8.18 at values of c from 2 to 6. Note that

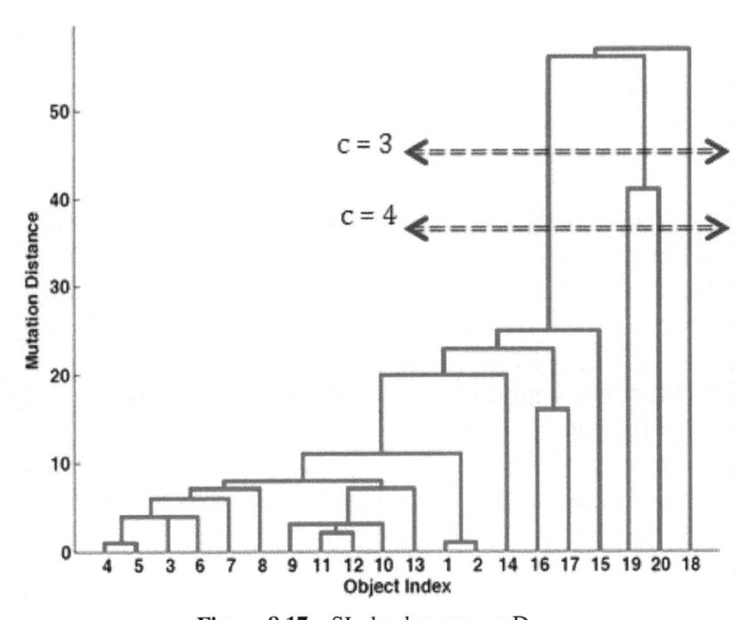

Figure 8.17 SL dendrogram on D_{mut}.

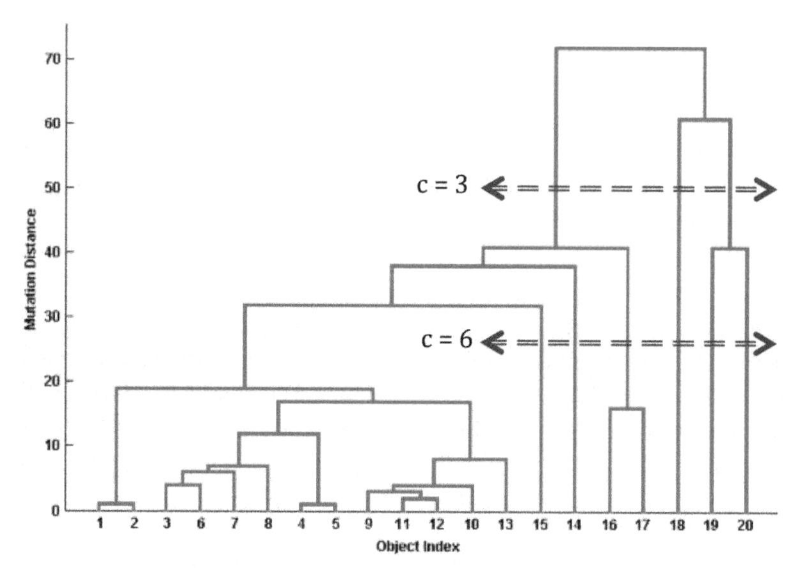

Figure 8.18 CL dendrogram on D_{mut}.

Table 8.7 Clusters in D_{mut} with tree-cutting in Figures 8.16, 8.17, 8.18.

c	F/M	SL	CL
2	{1-17},{18-20}	{1-17, 19-20}, {18}	{1-17},{18-20}
3	{1-17},{18},{19-20}	{1-17},{18},{19-20}	{1-17},{18}, {19-20}
4	{1-17}, {18},{19},{20}	{1- 17}, {18},{19}, {20}	None
5	{1-15}, {16-17}, {18}, {19}, {20}	{1-14, 16-17},{15}, {18}, {19}, {20}	{1-15}, {16-17}, {18}, {19}, {20}
6	{1-14}, {15}, {16-17}, {18},{19},{20}	{1-14}, {15}, {16-17}, {18},{19}, {20}	{1-13, 15}, {14}, {16-17}, {18},{19},{20}

the 20 objects have slightly different locations along the horizontal axis for the three trees. Agreements about clusters are highlighted in different colors. Comparing the partitions in Table 8.7 leads to several observations. SL isolates $\{18\}$ from $\{19, 20\}$ at c = 2, while the other two approaches group 18 with 19 and 20. All three algorithms agree at c = 3. F/M and SL agree at c = 4, whereas CL has a merger distance tie (at value 41) and, hence, jumps directly from c = 3 to c = 5.

The cut lines for F/M and SL at c = 4 group $\{1 - 17\}$ together, and isolate the three singletons $\{18\}$, $\{19\}$, and $\{20\}$. Objects $\{18\}$, $\{19\}$, and $\{20\}$ are the only non-sentient organisms in data, and they are at almost twice the mutation distances that bind the other 17 objects together in the F/M distance data. This explains why these three singletons are isolated first when the trees are cut at c = 4. F/M and CL agree again at c = 5 and differ from SL only in the treatment of object $\{15$ = tuna fish$\}$. At c = 6, F/M agrees with SL, whereas CL isolates $\{14$ = rattlesnake$\}$.

In summary, we have three sets of crisp c-partitions of D_{mut} in Table 8.7 which are, generally speaking, really pretty similar. We will turn the job of selecting a (mathematically) "best" c-partition of these data over to several validation methods in Section 8.7. Whether any one of them is physically "most useful" is a question best argued by zoologists.

The next example examines the behavior of the three SAHN clustering algorithms (SL, CL, and AL) on a data set that is similar to the three rings data in Figure 4.7(a).

Example 8.8. SAHN partitions of the X_{4rings} data.

Figure 8.19(a) is a scatterplot of a two-dimensional data set with n=1000 points called X_{4rings}. The number of points in each ring, beginning with the center ring, is 100, 200, 300, and 400. Each set of points is drawn from circles of different radii and then the coordinates are randomized to produce X_{4rings}. This 2D data is converted to a Euclidean distance matrix D_{4rings} with the 2-norm. Then D_{4rings} is submitted to SAHN algorithm A8.1 for each of the three options for SL, CL, and AL. At n = 1000, visual representation of SAHN partitions as dendrograms becomes intractable. Instead, we resort to a simple visual representation of the clusters at c = 4 in each hierarchical sequence using colors to represent the four different labels.

Refer to Figure 8.5 to refresh your memory about favorable and unfavorable geometries of cluster substructure for each of these clustering algorithms. Looking at the data set in Figure 8.19(a), the HPOV is that X_{4rings} contains four chain clusters in concentric rings, and the intercluster "gap distance" between the rings is much greater than the intracluster connection distance within rings. So, our expectation for CPOV clusters is that SL will do well with this data set (at c = 4), whereas CL will attempt to produce "cluster clouds" in the data. According to the common belief, AL will do something in between these two extremes. The 4-partitions in each of the three sequences of nested crisp clusters are shown in Figures 8.19(b)-8.19(d). The four clusters are color-coded and their interpretation should be quite clear. As you can see, the CPOV 4-partition obtained by SL on X_{4rings} is perfect: each of the four clusters contains only points colored for that cluster. The partition accuracy of SL is 100%.

The complete linkage partition in Figure 8.19(c) divides the data into four sectors that are roughly symmetrical with respect to each other. For example, the red and dark blue clusters seem quite similar.

We can make a rough visual estimate of the accuracy of this partition using the color scheme as a reference. Let the rings be labeled 1–4, inner to outer. The dark green region in view 8.19(c) contains points in all four rings 1–4. The light blue region has points in rings 2, 3, and 4. Correctly labeled points in ring 3 might be red or blue: either choice shows roughly the same numbers of points in rings 3 and 4. So, about $^3/_4$ of the points in rings 2, 3, and 4 are visually incorrect, leading to a partition accuracy of about 25%. Overall, the CL partition performs as expected – the four rings are interpreted by it as four clouds as shown in Figure 8.19(c). In other words, the CL partition behaves as expected and is very bad.

Is the AL solution in view 8.19(d) better than the CL partition? Errors in the outer two rings are about the same. The only real difference is that rings 1 and 2 (and parts of 3 and 4) are grouped together by AL. There is not much difference between these two sets of CPOV clusters: CL and AL both produce pretty bad 4-partitions of this data set. Since each of the SAHN models produces c-partitions from n to 1 (up to ties resolution), we would need some way to decide which partition(s) provide "good" CPOV interpretations of the data.

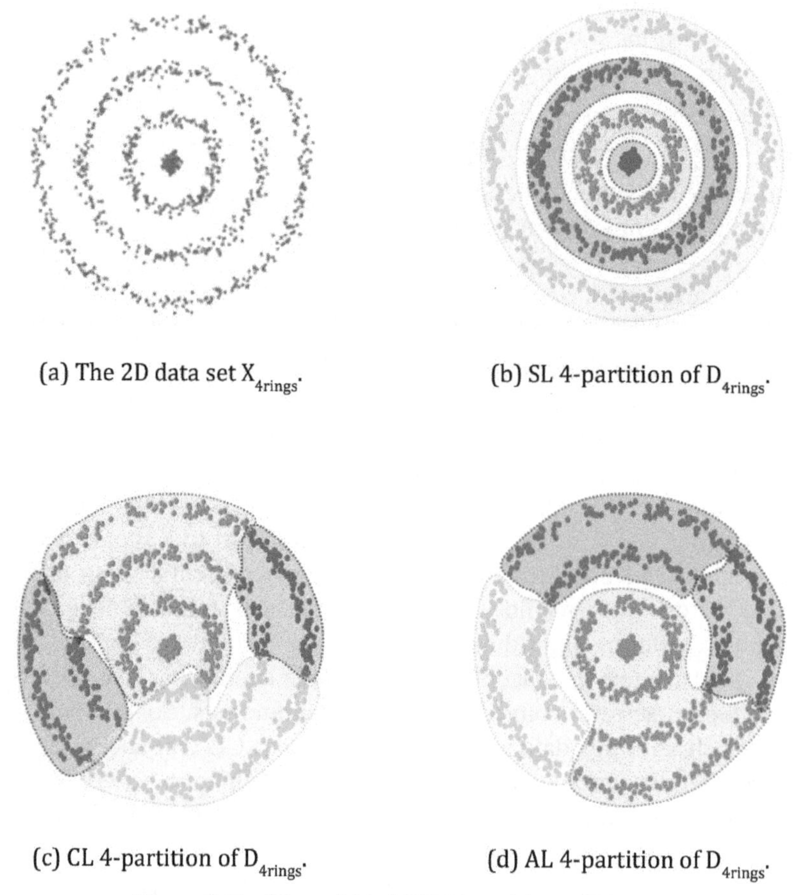

Figure 8.19 Figure 8.19. SAHN 4-partitions of X_{4Rings}.

8.5.3 Repairing SL Sensitivity to Inliers and Bridge Points

The appearance of the MST in A8.2 makes it clear that there is an intimate relationship between the VAT/iVAT images and SL clustering that was visually demonstrated in Figure 8.15. The sequence in which the edges are added to the MST is used to reorder D to D* in A4.1, and then VAT produces the image I(D*) of D. Since the MST is also the basis of single linkage clustering as seen in A8.2, an easy follow-up to the VAT reordering operation is to use the VAT/iVAT MST to obtain SL clusters. Thus, c clusters (c subtrees in the MST) are found by cutting the $c - 1$ largest edges in the MST. The next example, which is abstracted from Ghafoori *et al.* (2018), illustrates the deleterious effect that inliers between otherwise well-separated clusters have on both single linkage clusters and iVAT images of the input data.

Example 8.9. The effect of inliers between well-separated clusters on SL.

Figure 8.20(a) is a scatterplot of a $p = 2$-dimensional data set X_1, comprising $c = 2$ Gaussian clusters which appear to be pretty compact and well-separated from the HPOV. The data has $n = 245$ points; 97 green

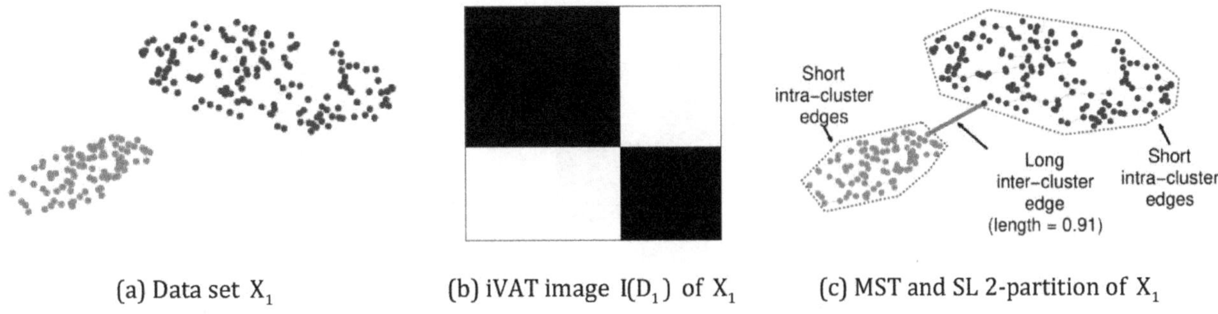

Figure 8.20 Scatterplot, iVAT image, and SL 2-partition of two HPOV Gaussian clusters in X_1.

points and 148 red points. These data are converted to Euclidean distance matrix D_1 (as are each of the four data sets in this example) that is used as input to iVAT algorithm A4.1 and SL algorithm A8.2. The iVAT image $I(D_1)$ is shown in Figure 8.20(b). This image implies that the input data possess two well-separated clusters, in agreement with our visual assessment of the scatterplot.

Figure 8.20(c) is the MST on the vertices (points in X_1) of the data set. The crisp SL 2-partition of X_1 is shown by the dashed magenta cluster boundaries, and this CPOV clustering of X_1 is consistent with our HPOV assessment of cluster structure in X_1. The thick red line in this view is the long intercluster edge, which has an edge length of 0.91. This distance is sufficiently larger than all of the other edge distances in the MST that the intracluster connection sequences are not altered by a "jump across the gap," which, in this case, will occur after all of the green points have been connected.

Figure 8.21(a) shows a new data set $X_2 = X_1 \cup \{\mathbf{y}_1\}$ made by adding the point y_1 that lies along the long red edge in Figure 8.20(c) to X_1. The added point is equidistant between the two vertices it connects. This anomalous point is called a bridge or inlier point since it provides a bridge from the green vertices to the red vertices. The edge length from $\mathbf{y}_1$ to either subset is 0.46, half the length of the original bridge edge in Figure 8.20(c). The new Euclidean distance matrix is denoted as D_2.

The addition of the single inlier $\mathbf{y}_1$ is not enough to change our HPOV opinion about clusters in the data. Nor does it upset the construction the SL 2-partition of the new data set much, which is shown in Figure 8.21(c), where the point $\mathbf{y}_1$ is included (since there is a distance tie), in the green subset. But it is enough to have a somewhat noticeable effect on the iVAT image of D_2, which is shown in Figure 8.21(b).

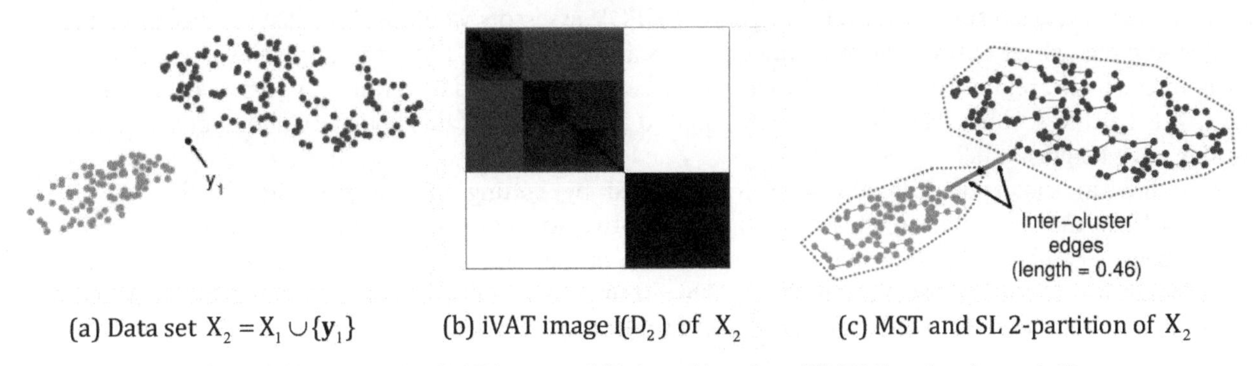

Figure 8.21 Scatterplot, iVAT image, and SL 2-partition of two HPOV Gaussian clusters in X_2.

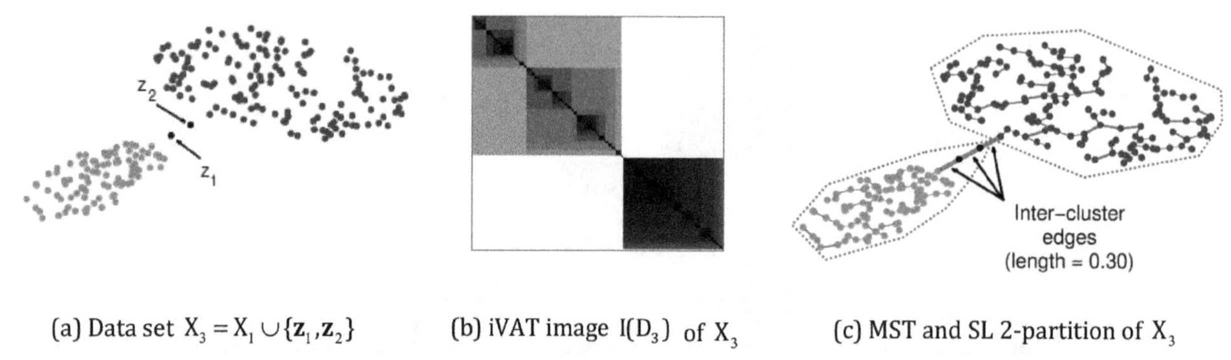

(a) Data set $X_3 = X_1 \cup \{\mathbf{z}_1, \mathbf{z}_2\}$ (b) iVAT image $I(D_3)$ of X_3 (c) MST and SL 2-partition of X_3

Figure 8.22 Scatterplot, iVAT image, and SL 2-partition of two Gaussian clusters X_3.

You may have to zoom on this image to see it clearly, but the upper diagonal block in this image has changed a bit. The uniformly dark upper block in Figure 8.20(b) has been replaced by several smaller substructures within the block in Figure 8.21(b). Just one point! It is like the effect your brother-in-law (who just escaped from prison) would have on the tranquility of your Christmas dinner if he suddenly appeared at the front door while the turkey was being carved.

Continuing in this way, now add two points $\{\mathbf{z}_1, \mathbf{z}_2\}$ to X_1 creating the set $X_3 = X_1 \cup \{\mathbf{z}_1, \mathbf{z}_2\}$. The two new points subdivide the long red edge into thirds of equal edge length 0.30 as shown in Figure 8.22(c). Figure 8.22(c) shows that two inliers are still not enough to imbalance the SL 2-partition of the data set X_2, which still has $c = 2$ apparent HPOV clusters. The two new points belong to the green subset. But what about the iVAT image? Compare the upper diagonal blocks of Figures 8.20(b), 8.21(b), and 8.22(b) to see how the addition of two bridge points affects the original iVAT image of X_1. Not only has the upper block suffered further deterioration but also the integrity of intensities in the lower block is beginning to waver (again, hard to see, but it's there). It is like your brother-in-law is really twins – twice as much chaos for revelers at the Xmas party! Well, things are about to get even worse.

If we add one just one more point to the long edge, it will shorten the intercluster gap distance enough to upset the iVAT image and the corresponding SL partition of the data. Figure 8.23 shows the data set $X_4 = X_1 \cup \{\mathbf{w}_1, \mathbf{w}_2, \mathbf{w}_3\}$. The addition of three equidistant points to the long edge divide the gap distance by 4, and the edge length between the bridge points drops to 0.23.

Now the iVAT image in Figure 8.23(b) gives the misleading impression that the data set X_4 contains three CPOV clusters and a singleton, even though most HPOV assessors would still see just two because the human pattern recognizer will see the three bridge points as insignificant. Figure 8.23(c) shows the MST of X_4. The three shaded regions in view 8.23(c) are the sets of points that iVAT wants us to see as the three clusters. It is hard to see, but there is an edge of length 0.24 that crosses this boundary and is greater than the bridge edge lengths of 0.23.

Figure 8.24(a) shows the 2-partition of X_4 obtained by cutting the longest edge in this MST. Figure 8.24(b) is the 3-partition obtained by cutting the two longest edges in the MST. Both of these partitions bear very little relation to the apparent clusters in the data set. Yikes! It is like your brother-in-law brought his whole cell block to the party! This example should cement your understanding of the relationship between iVAT, SL, MSTs and the effects of inlier noise in data sets.

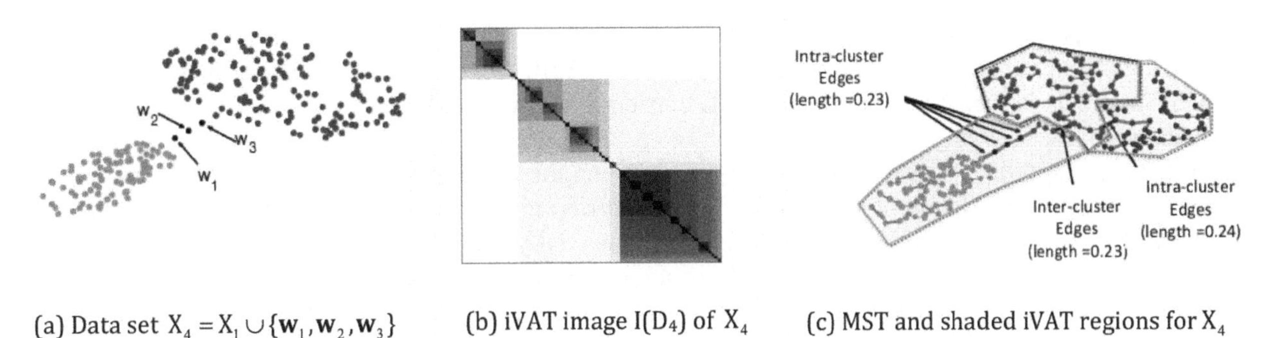

(a) Data set $X_4 = X_1 \cup \{\mathbf{w}_1, \mathbf{w}_2, \mathbf{w}_3\}$ (b) iVAT image $I(D_4)$ of X_4 (c) MST and shaded iVAT regions for X_4

Figure 8.23 Scatterplot, iVAT image, and iVAT Regions for X_4.

(a) c=2. (b) c=3.

Figure 8.24 SL partitions of X_4 at c = 2 and c = 3.

8.5.4 Acceleration of the Single Linkage Algorithm

Since single linkage depends on the minimal spanning tree, any method for accelerating the construction of the MST automatically improves the prospects of using SL for big data. Most studies along these lines confine themselves to MSTs built on Euclidean distance matrices of p-dimensional data (Definition 8.1), which are quite naturally called *Euclidean minimal spanning trees* (EMSTs). Naive construction of a EMST on n vertices from a dissimilarity matrix has $O(n^2)$ time and space complexity. There are a number of algorithms that improve this. Pettie and Ramachandran (2002) give tight upper and lower bounds on the construction. Osipov *et al.* (2009) discuss a method based on filtering Kruskal's MST algorithm.

General algorithms that scale the Euclidean distance matrix are not very useful for scaling up single linkage to big EMSTs. In the Euclidean case, the edge set connects all pairs of points; so linear scaling in the number of vertices (n) corresponds to quadratic scaling in the number of edges $(n(n-1)/2)$. This necessitates consideration of other approaches. For two-dimensional data (p = 2), an EMST can be found in $O(n \ln n)$ time and $O(n)$ space using Delaunay triangulation (Delaunay, 1934). More background on this is available in the excellent classic presentation offered by Preparata and Shamos (1985).

March *et al.* (2010) present an approach for computing the EMST that includes a very interesting discussion about how to compute tighter complexity estimates than the usual approach using properties of the input data as part of the analysis. These authors present a scalable dual-tree Boruvka's algorithm for the EMST problem and give an exact analysis of it for p = 2 dimensions. Four numerical experiments are presented, several of which qualify for big data status.

March *et al.*'s paper is a good one: it deserves a careful reading. For higher dimensions (p = 3), finding an optimal algorithm remains an open problem. Vineet *et al.* (2010) develop a GPU-based implementation of the MST that may be helpful in accelerating SL clustering that is analogous to Anderson *et al.*'s 2008 implementation of FCM.

Müllner (2015) approaches acceleration of SL by changing the way the basic algorithm A8.2 is implemented (cf., Section 11.5). His version of SL is called *fastcluster*, implemented in the Python programming environment. He compares run times of fastcluster to commercially available versions of SL in MATLAB and Mathematica. Two types of runtime comparisons are made: (i) vector data inputs, which must first be converted to a dissimilarity matrix; and (ii) direct inputs of dissimilarity data. Fastcluster compares quite favorably to these commercially available implementations. Mullner reports an average speedup of 10:1.

8.6 Cluster Validity for Single Linkage

The cluster validity problem for single linkage is more straightforward than it is for c-means or GMD because the set of candidate partitions (CP) in the SL case has *at most* one U at each value of c. It is impossible to generate more than one set of SL candidates for a fixed input matrix D; so the validity problem devolves into choosing one of the n candidates in CP. Choosing $c = 1$ or $c = n$ amounts to rejecting the hypothesis that D contains HPOV clusters (it always contains the trivial CPOV clusters at these two values); so examination of the candidates is usually restricted to looking at the partitions corresponding to $c = 2$ to $c = (n - 1)$. However, the partitions U_1 and U_n at the two extremes will be left in because the behavior of various crisp CVIs at the two endpoints is interesting.

Since the U's in CP from SL are crisp partitions, the CVIs used in Example 6.6 for hard c-means candidates are applicable to this problem as well, but interpretation of values is a problem, since each one corresponds to a different value of c. Of course, there are many other internal CVIs that can also be used. Because of the special nature of the SL model, there is an additional and quite popular heuristic CVI that is based on the gap distances at which clusters are merged. The heuristic is based on plotting the sequence of gap distances $\{G(c) : 1 \leq c \leq n\}$ against c. If the user spots a "knee" at c in the corresponding graph of G(c) vs. c, the "correct" value for c is taken to be the *previous value* (that is, the value $c^* = c - 1$) just before the large jump across the knee gap[23]. The rationale for this heuristic is that the clusters found just before jumping the knee gap are more tightly bound than the ones obtained by forcing the merger of the two sets on opposite sides of the knee gap.

Example 8.10. Crisp cluster validity for SL partitions of Iris.

The Iris data was transformed into a Euclidean distance matrix $D_{E,Iris}$ with the Euclidean norm. Then A8.2 was applied to $D_{E,Iris}$, resulting in SL partitions from $c = n$ to $c = 1$. The Iris data doesn't contain 150 distinct vectors, so some values of c were not represented in the SL hierarchy. Then the SL partitions corresponding to $c = 2$ up to $c = 9$ were extracted from the SL hierarchy for this example. Table 8.8 shows values of the three cluster validity indices that were introduced in Section 6.7.1 on these 8 SL partitions of Iris. Also shown in Table 8.8 are the gap distances. Evidently, all three internal CVIs select $c^* = 2$ as the preferred value for the number of CPOV clusters in Iris.

[23]Not to be confused with the knee cap!

Table 8.8 Three crisp CVIs on HCM candidate partitions of Iris.

c	Type	2	3	4	5	6	7	8	9
DB_{22}	(↓)	**0.43**	0.53	0.65	0.65	0.84	0.93	1.56	2.81
GDI_{11}	(↑)	**0.34**	0.17	0.15	0.15	0.15	0.15	0.13	0.13
GDI_{33}	(↑)	**0.43**	0.01	0	0	0	0	0	0
G(c)	"knee"	**1.64**	0.82	0.73	0.65	0.63	0.62	0.56	0.54

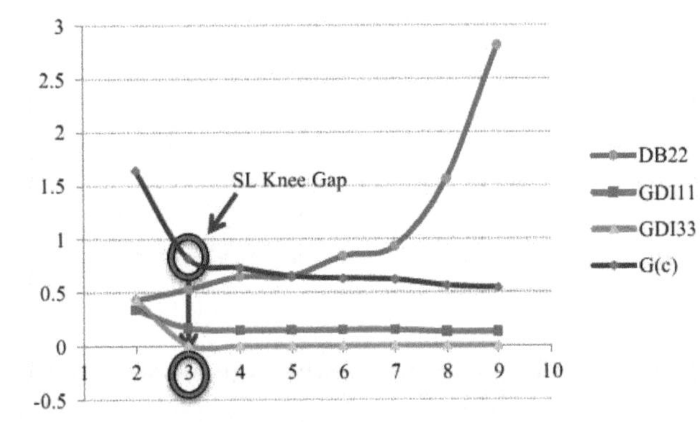

Figure 8.25 Graphs of three crisp CVIs and G(c) on SL partitions of Iris.

Figure 8.25 plots the rows of Table 8.8. The plots of Dunn's index (GDI_{11}) and the generalized Dunn's index GDI_{33} are remarkably similar for these SL partitions. Both maximize at about 0.5 at $c^* = 2$, and then quickly fall way and level off close to 0.0 as c increases. The minimum for DB_{22} at $c = 2$ is not quite as sharp, but clearly indicates for $c^* = 2$. Finally, the plot of G(c) in Figure 8.25 shows that its knee occurs at $c_{knee} = 3$, so the SL validity heuristic also instructs us to choose $c^* = 2$.

You should compare these results to those in Example 6.6, which presents the same type of validation analysis for HCM partitions of the Iris data as shown here for the eight unique SL candidates corresponding to the values $2 \le c \le 9$. Figure 8.14 shows that the 3-partition for Iris built by SL contains 48 label mistakes. This example shows that the knee gap heuristic avoids choosing this bad partition of Iris by pointing to the value of $c^* = 2$, the value prior to the biggest jump. So here, the "best" CPOV SL partition of Iris found by optimizing the knee gap statistic[24] agrees with the "best" CPOV HCM partition of Iris – provided "best" in Example 6.6 means best value of DB_{22} or GDI_{33}. Recall that there, the best GDI_{11} value pointed to $c^* = 3$ for Iris. In practice, you should try to aggregate as many votes for "best partition" as you can with different CVIs. Each additional vote for a particular candidate bolsters your confidence in that candidate – but even when ALL the votes agree, you can never be sure.

---

[24]There is another CVI called the "gap statistic" which is not designed for the SL case. See Tibshirani et al. (2001).

8.7 An Example Using All Four Basic Models

The final example of this chapter is a first look at comparing all four of the basic models in this volume. The data set chosen for this example has some interesting properties that will enable us to understand some differences for the four models and their clustering algorithms.

Example 8.11. The Fantastic Four meet X_{20}.

In this last example of Chapter 8, all four of the clustering models and basic algorithms are applied to the n = 20 point data set X_{20} shown in Figure 8.26(a). Before any results are presented, take a moment to look at these 20 points in the plane. Do you see any HPOV clusters in this little data set? If you do, you might hazard a guess that many observers would see c = 2 clusters, points 1 – 10 in cluster 1, and points 11 – 20 in cluster 2. However, these points are not clouds. Points 1 – 10 seem to be (roughly) points along the boundary of a circular cloud; so they are somewhat like a chain cluster in the four rings data. Points 11 – 20 are a bit harder to describe in simple terms. What, only 20 points in 2D? Well, the idea in this example is for you to look at this tiny data set and imagine what each of the four models will see as CPOV clusters, bearing in mind the properties that you know about for each of them.

The first thing you might think of is to make an iVAT image of X_{20} to see if it suggests any potential cluster structure. That image is shown in Figure 8.26(b), where $D_{E,20}$ represents the Euclidean distance matrix between pairs of points in the data set. This image is pretty suggestive: it shows c = 2 main clusters of 10 points each. The indices of the points after reordering with A4.1 are shown along the side and bottom of the image. Thus, points 1 – 10 occupy the upper left block, while points 11 – 20 are in the 10 × 10 block at the lower right. You can also see the substructure of the data that SL is likely to find. This part of this example is very similar to Example 8.6, which showed how to interpret an iVAT image as a visual front end of SL. To see that here, look at the darkest 2 × 2 sub-blocks of the two main structures in the iVAT image. The 2 × 2 block in the lowest part of the upper 10 × 10 is the pair {3, 9}; the 2 × 2 block in the lowest part of the lower 10 × 10 is the pair {18, 20}.

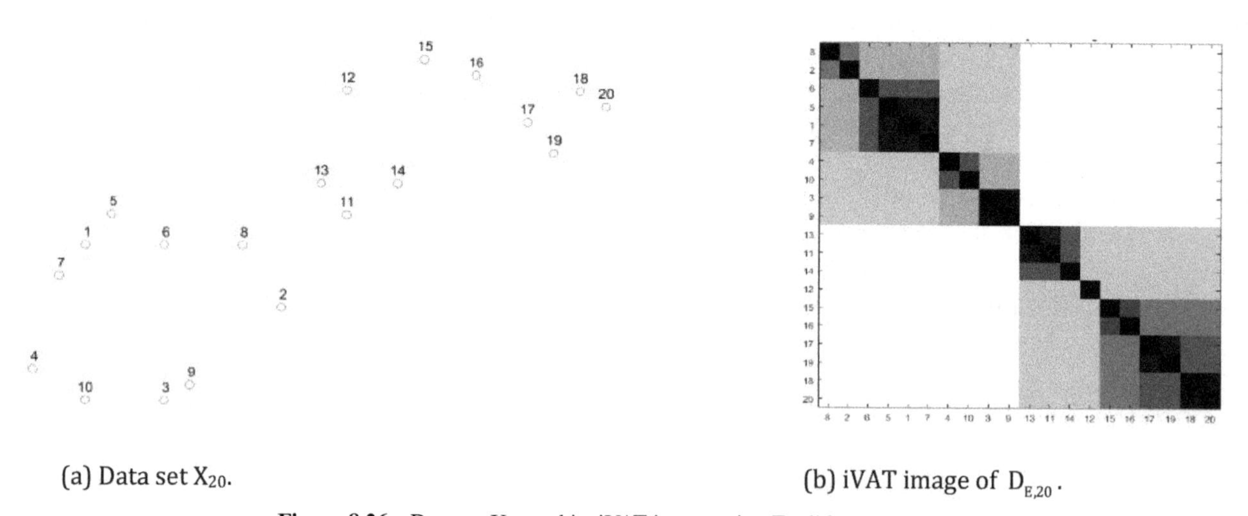

(a) Data set X_{20}. (b) iVAT image of $D_{E,20}$.

Figure 8.26 Data set X_{20} and its iVAT image using Euclidean distances.

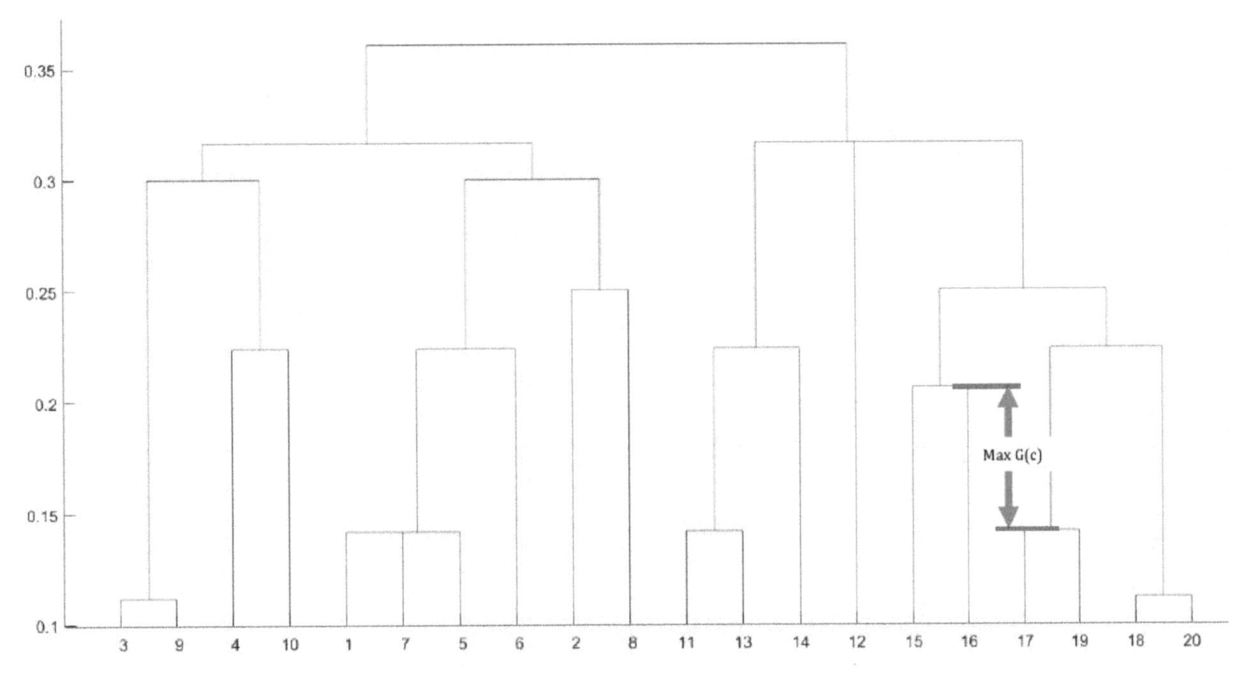

Figure 8.27 Dendrogram of SL hierarchy for data set X_{20}.

Figure 8.27 is the dendrogram of the SL clusters found by MATLAB when applied to this data. First, the pairs {3, 9} and {18, 20} are joined at the level 0.11. At the next level, 0.14, three new clusters are formed simultaneously: {1, 7, 5}, {11,13}, and [17, 19}. You can see the {1, 7, 5} join in the upper 10×10 block, and you can predict from the iVAT image that point {6} will join these three before it is attached to any other set in the hierarchy. But before this happens, {15, 16} are joined at 0.21. Four sets of clusters are formed at 0.225. Two new clusters are formed at 0.25. Then there are two mergers at 0.30. The next-to-last merger occurs at 0.32, and the final merger is made at the gap distance 0.36. Here is the sequence of values of c that are represented by this SL output:

$$c = 20 \rightarrowtail c = 18 \rightarrowtail c = 14 \rightarrowtail c = 13 \rightarrowtail c = 9 \rightarrowtail c = 7 \rightarrowtail c = 5 \rightarrowtail c = 2 \rightarrowtail c = 1 \quad (8.7)$$

There are a lot of distance ties in this little data set because tie resolution is inclusive (all tied points or sets are merged at the same time). The result, seen in Figure 8.27, is that there is no SL partition at the 11 values of c that are missing in Equation (8.7). Looking at the dendrogram shows that at c = 2, the value indicated by the iVAT image of the data, the two SL clusters of X_{20} are the points associated with the two 10×10 blocks in the image.

One final note about the SL result. The heuristic CVI for SL based on the maximum gap distance between levels in the SL tree discussed in the last subsection is shown in Figure 8.27. The biggest jump is 0.07, from 0.14 to 0.21. According to this heuristic then, the "best" SL clusters occur at the level 0.14, viz., c = 18 clusters. This should be warning to you: all CVIs can give very misleading results!

Next, let us apply the three objective function models to X_{20}. The default parameters listed after A6.1 and A7.1 are used, and since the iVAT image in Figure 8.26(b) strongly suggests that there are two major clusters, that parameter is fixed at the value c = 2. Table 8.9 shows the results, with all non-integer values

Table 8.9 SL, HCM, FCM, and GMD initial and terminal CPOV clusters in X_{20} at c = 2.

	x	y	**SL** U^T_{SL}		**Init (I_0)** U^T_{init}		**HCM** U^T_{HCM}		**FCM** U^T_{FCM}		**GMD (I_0)** $P^T_{GMD,1}$		**GMD (I_1)** $P^T_{GMD,2}$	
1	1	0.6	1	0	1	0	1	0	0.97	0.03	1	0	1	0
2	1.75	0.4	1	0	1	0	1	0	0.77	0.23	1	0	1	0
3	1.3	0.1	1	0	1	0	1	0	0.96	0.04	1	0	1	0
4	0.8	0.2	1	0	1	0	1	0	0.94	0.06	1	0	1	0
5	1.1	0.7	1	0	1	0	1	0	0.95	0.05	1	0	1	0
6	1.3	0.6	1	0	1	0	1	0	0.97	0.03	1	0	1	0
7	0.9	0.5	1	0	1	0	1	0	0.96	0.04	1	0	1	0
8	1.6	0.6	1	0	1	0	1	0	0.84	0.16	1	0	0.98	0.02
9	1.4	0.15	1	0	1	0	1	0	0.95	0.05	1	0	1	0
10	1	0.1	1	0	1	0	1	0	0.95	0.05	1	0	1	0
11	2	0.7	0	1	1	0	0	1	0.33	0.67	1	0	0.16	0.84
12	2	1.1	0	1	1	0	0	1	0.19	0.81	1	0	0	1
13	1.9	0.8	0	1	1	0	0	1	0.39	0.61	1	0	0.10	0.90
14	2.2	0.8	0	1	1	0	0	1	0.10	0.90	1	0	0	1
15	2.3	1.2	0	1	1	0	0	1	0.04	0.96	1	0	0	1
16	2.5	1.15	0	1	1	0	0	1	0.01	0.99	1	0	0	1
17	2.7	1	0	1	0	1	0	1	0.01	0.99	0.06	0.94	0	1
18	2.9	1.1	0	1	0	1	0	1	0.05	0.95	0.02	0.98	0	1
19	2.8	0.9	0	1	0	1	0	1	0.03	0.97	0.02	0.98	0	1
20	3	1.05	0	1	0	1	0	1	0.06	0.94	0.01	0.99	0	1

rounded to two significant digits. The second and third columns in Table 8.9 are the x and y (horizontal and vertical) coordinates of the points in X_{20} shown in Figure 8.26(a). All of the remaining sets of column pairs are the transpose of partitions, i.e., the rows of each pair in the table are, in the notation used in this volume, the columns of partitions in M_{fcn}.

The first two columns after the data set show the crisp clusters U_{SL} obtained by SL at c = 2. The fifth and sixth columns, labeled "Init (I_0)" in Table 8.9, are the memberships of the 20 points in the initializing crisp partition U_{Init} for the three objective function models: 16 points in the first subset and 4 points in the second subset. This initial partition is illustrated in Figure 8.28(a). The next two columns in Table 8.9 show the terminal partition U_{HCM} obtained by HCM. Thus, hard c-means agrees with the iVAT portrait of this data set and with SL at c = 2, placing points 1 – 10 in cluster 1 and points 11-20 in cluster 2. This terminal partition is shown in Figure 8.28(b). If you expected SL to behave this way based on your inspection of the geometry of the data, this is your reward!

Columns 10 and 11 of Table 8.9 show the terminal FCM partition U_{FCM} for m = 2 in A6.1. This partition is properly soft, all 20 points having some membership in both clusters. For example, $\mathbf{x}_2$ has the smallest maximum membership (0.77) among the first 10 points. The next smallest membership is 0.84 for $\mathbf{x}_8$. All

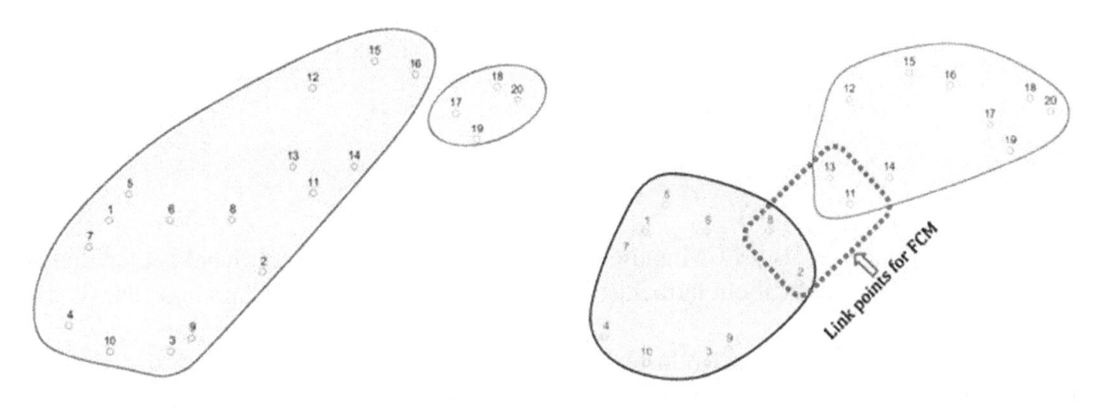

(a) Initialization for HCM, FCM, and GMD. (b) Termination for SL (c = 2), HCM, **H**(FCM), and **H**(GMD).

Figure 8.28 Initial and terminal 2-partitions: HCM, **H**(FCM), and **H**(GMD) for data set X_{20}.

the remaining maximums among the first 10 points are ≥ 0.94. The scatterplot of the data shows that points $\mathbf{x}_2$ and $\mathbf{x}_8$ are closest to the second group of 10. Similarly, points $\mathbf{x}_{11}$ and $\mathbf{x}_{13}$ occupy positions that are more or less equivalent to $\mathbf{x}_2$ and $\mathbf{x}_8$ for the second group of 10, and their memberships are correspondingly lowest among points $11-20$. These four points are highlighted in yellow in Table 8.9 and shown in Figure 8.28(b) as "link points for FCM." This intuitively satisfying result shows how partial memberships in fuzzy partitions can reflect proximity for points in competing clusters. Applying hardening function **H** at Equation (3.4) to this fuzzy partition(m = 2 in A6.1) produces the same terminal partition as HCM (m = 1 in A6.1), and this is also shown in Figure 8.28(b). So, while this data set doesn't have cloud-like HPOV clusters, the CPOV clusters produced by these two models agree with the structure of the data represented by the iVAT image at Figure 8.26(b) and by SL at c = 2.

Next, have a look at the GMD results. There are two sets of results shown for Algorithm A7.1. The next to last pair of columns, labeled GMD(I_0), correspond to the output $P^T_{GMD,1}$ when the initialization is the one specified in Table 8.9. This surprising result shows that GMD settles at a soft partition that is very near to the initializing partition. This terminal partition has 16 crisp points corresponding to the 16 in I_0. Points $17-20$ are slightly soft. Hardening this partition with **H** at Equation (3.4) would produce the initializing partition, which is not very satisfying from a visual HPOV.

The last pair of columns in Table 8.9 are the terminal partition $P^T_{GMD,2}$ from A7.1 when the initialization (I_1) is the default initialization used by the MATLAB implementation. This is an interesting output: 17 of the 20 columns in $P^T_{GMD,2}$ (rows in the table) are listed as crisp label vectors. In the actual outputs, none of these values were integers, and $P^T_{GMD,2}$ was properly soft. For example, point 10 had the values (0.999999999387467, 6.12533040804950e-10). The values in Table 8.9 are formatted to two significant digits; so this value became (1, 0). Three of the four FCM link points, viz., points 8, 11, and 13 highlighted in blue for GMD, are also identified as the most uncertain points in the probabilistic partition $P^T_{GMD,2}$, but the values obtained by GMD are quite a bit stronger (less uncertain) than those produced by FCM. Point 2 is not soft (after formatting) for GMD but is rather soft when seen by FCM. Overall, $P^T_{GMD,2}$ is decidedly less soft than U_{FCM} and is certainly not crisp. As shown in Figure 8.28(b), hardening $P^T_{GMD,2}$ with **H** at Equation (3.4) yields the same crisp 2-partition as the other three algorithms. So, all four basic algorithms produce the same

Table 8.10 Cluster centers for the terminal HCM, FCM, and GMD 2-partitions of X_{20}.

		Initial (I_0) Means	Data Means	HCM Centers	FCM Centers	GMD ($P^T_{GMD,2}$) Means
$\mathbf{x}_1, \ldots, \mathbf{x}_{10}$	V_{1x}	1.57	1.21	1.21	1.21	1.21
	V_{1y}	0.61	0.39	0.39	0.41	0.39
$\mathbf{x}_{11}, \ldots, \mathbf{x}_{20}$	V_{2x}	2.85	2.43	2.43	2.5	2.43
	V_{2y}	1.01	0.98	0.98	1.00	0.98

hard 2-partition on this little data set. But FCM and GMD both supply, via their soft label vectors before being hardened, some additional insight about intracluster CPOV relationships (the link points). This is why soft partitions are useful.

Finally, Table 8.10 compares the means (cluster centers) produced by the three objective function methods for the subsets $\mathbf{x}_1, \ldots, \mathbf{x}_{10}$ and $\mathbf{x}_{11}, \ldots, \mathbf{x}_{20}$. The HCM algorithm estimates of all four values exactly because the terminal HCM 2-partition of X_{20} is $\{\mathbf{x}_1, \ldots, \mathbf{x}_{10}\} \cup \{\mathbf{x}_{11}, \ldots, \mathbf{x}_{20}\}$. The GMD estimates for initialization (I_1) are also exact. FCM doesn't do quite as well: V_{1x} is exact but the other three FCM estimates are all high, by 0.02, 0.07, and 0.02.

Summary

Here is the bottom line for Chapter 8. We don't (and won't ever) know, for the vast majority of object or relational data sets, whether there are cluster clouds or cluster chains (or a mixture of both types with other shape distributions) amongst the objects. The best policy is to always approach clustering in unlabeled data with an open mind. Since we already have a bunch of good cloud finding methods (including complete linkage), the take away method from this chapter is SL, the only method that affords a real alternative to the methods covered in Chapters 6 and 7. However, SL only gives us a way to find *crisp* clusters in relational data. This leaves us a bit shorthanded since there are many validation methods for fuzzy partitions of similarity and dissimilarity data, but we have not developed any way to produce CPs from this type of data. The *relational FCM* (RFCM) and *non-Euclidean RFCM* (NERFCM) clustering algorithms can be used to generate soft partitions from relational data, but these methods are not so basic and are certainly beyond the scope of this volume. Interested readers can refer to Hathaway *et al.* (1989), Hathaway and Bezdek (1994), and Khalilia *et al.* (2014) for more on this subject.

8.8 Notes and Remarks

The notes and remarks section, Section 4.5 of Chapter 4, details the history of the minimal spanning tree and its relationship to SL. Now that the relationship of the MST to the SL model and algorithm have been described, part of that discussion is repeated here. The exact beginning of single linkage *per se* is unknown, but Gower and Ross (1969) and Graham and Hell (1985) both attribute the basic idea underlying this approach to Czekanowski (1909), who used a related procedure as early as 1909 as a basis for visual clustering. As mentioned in Section 4.5 but worth repeating here, Cattell (1944) may have been the first author to use the term *single linkage* when he wrote in 1944:

> *With or without* ***single linkage*** *lists he then manipulates the order of the variables in an attempt to bring linkage correlations alongside the diagonal or as near to it as possible. If the process is successful, the resultant clustering is clearly and strikingly recorded as in Diagram 1.*

The SAHN models, and, in particular, single linkage, became quite popular with the emergence of computer-based numerical taxonomy in the late 1950s. Gower and Ross (1969) credit Sneath (1957) with the introduction of SL to the field of numerical taxonomy in 1957:

> *SINGLE LINKAGE CLUSTER ANALYSIS was put forward by Sneath (1957) as a convenient way of summarizing taxonomic relationships in the form of dendrograms (taxonomic trees).*

But Sneath and Sokal (1973) assert that SL was introduced to numerical taxonomy by Florek *et al.* (1951). This paper, written in French, includes the well-known Polish mathematicians Lukaszewicz and Steinhaus as coauthors, which shows that they were both involved in the development of this method of clustering at a very early stage. We have already mentioned Hugo Steinhaus (1956) in connection with HCM in the notes to Chapter 6. This paper provides compelling evidence that Steinhaus was interested in and made important contributions to several branches of cluster analysis.

It has been pointed out that AL is often regarded as a compromise between the extremes of chaining (SL) and clouding (CL). Lance and Williams (1967) discussed an interesting approach to this problem that is similar in spirit to taking a convex combination of SL and CL. Recall the general formulation of the SAHN model at Equation (8.4), repeated here as

$$\delta_{SAHN}(WV, A) = \alpha_W \delta_{SAHN}(W, A) + \alpha_V \delta_{SAHN}(V, A) + \beta\delta_{SAHN}(W, V) + \gamma|\delta_{SAHN}(W, A) - \delta_{SAHN}(V, A)| \quad (8.8)$$

Lance and Williams present a model they call a flexible SAHN strategy, realized by imposing four constraints on the parameters in Equation (8.8). Altering the value of β effectively distorts the space underlying the data. Specifically, as $\beta \to 1$, the corresponding SAHN model approaches SL. Conversely, as $\beta \to (-1)$, the system resembles CL more and more. Lance and Williams recommend $\beta \to (0.25)$ as a good general compromise between the extremes offered by SL and CL. But to make this interesting idea practical would require extensive experimentation for each new data set; so it has found only limited use amongst SAHN aficionados. The two extremes in this approach can be broadly characterized this way: SL clusters are maximally connected subtrees; CL clusters are maximally complete subtrees.

The dendrogram is a broad term for the diagrammatic representation of a phylogenetic tree such as the one exhibited in Figure 8.16. Representations of branching phylogenetic trees showing relationships among plants and animals appeared as early as 1840 in Hitchcock (1840). According to Mayr (1978), this term was first used in connection with numerical taxonomy in Mayr *et al.* (1953). Constructing dendrograms for data sets that have lots of objects is problematical.

Crowding and occlusion prevent useful visualization of hierarchical cluster structure in the input data even when the number of objects seems quite small. If n is much larger than several hundred, construction of a full exact or literal dendrogram that is visually useful is virtually impossible. Several methods for overcoming this limitation have been proposed. Visualization of a large data dendrogram by a "sketch" or skeleton of the literal graph is in some sense very similar to the skeletal cluster heat map algorithm (A11.3) for static big data sets called siVAT (cf. Chapter 11). The two main approaches for dendrogram skeletons are called (focus + context) and (overview + detail). Figure 8.29 (cf., Figure 1 in Chen *et al.* (2009)) illustrates the idea of the (overview + detail) approach for dendrogram building on larger sets of input objects.

This method apparently extends the utility of dendrogram visualization for data sets with more than a few hundred points by providing something like a summary of the major branches in the cluster tree. Whether this approach is useful for really big static data sets (say, $n \geq 10^6$) remains to be seen.

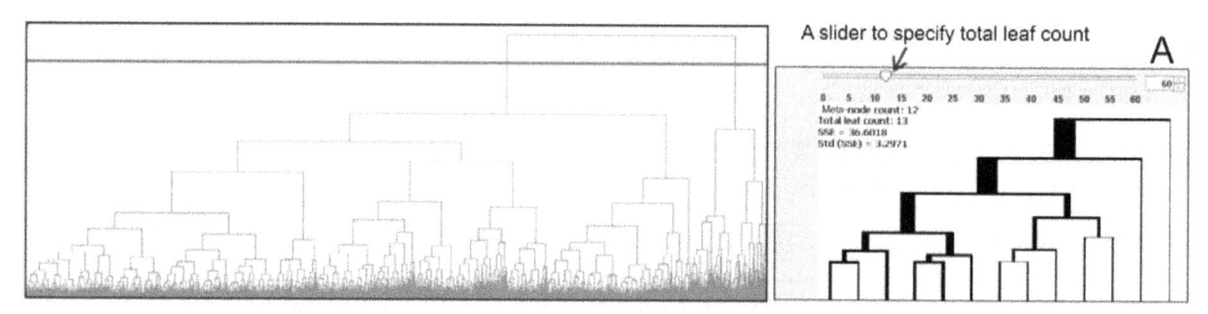

Figure 8.29 Overview (right panel, 12 meta-nodes) and detail (left panel, 3105 nodes) dendrograms.

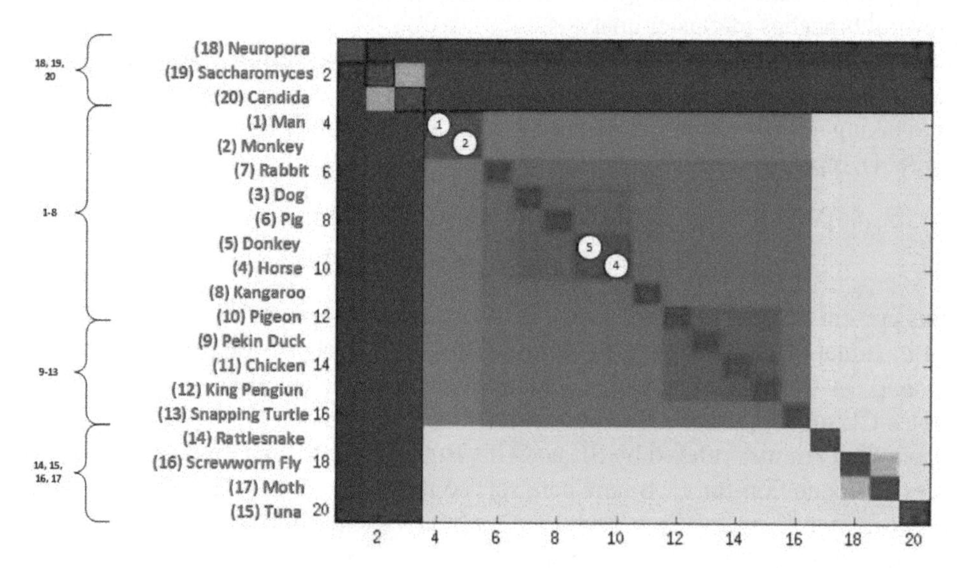

Figure 8.30 iVAT image of the mutation data of Table 8.5.

8.9 Exercises

1. Draw a dendrogram for the CL case in Example 8.2 analogous to the SL case shown in Figure 8.10.
2. Figure 8.30 is the iVAT image made by applying A4.1 to the mutation data of Table 8.6. Compare the visually apparent pre-clustering groups in this image to the clusters reported in Table 8.6 for each of the three clustering algorithms shown there. Which set of clusters best matches the iVAT suggested structure in this unlabeled data set? Discuss.
3. The dark blocks in an iVAT image correspond to groups that cluster together in the SL back pass through the iVAT MST. This could lead to a different method for constructing overview dendrograms, provided you can estimate the jump distances seen in the dendrogram from the iVAT image. Propose a scheme that might do this.
4. Give an example of a square relational data matrix D that might cause iVAT and SL to fail. Can you state any general test for D that would ensure these two models will be usable?

9

Properties of the Fantastic Four: External Cluster Validity

Everything should be made as simple as possible, but not simpler

–Albert Einstein, 1948

This chapter is devoted to comparisons of properties and several examples of the four basic clustering models and algorithms developed in Chapters 6, 7, and 8. We begin with computational complexity. Then we consider the geometric representation of searches for good clusters made by the four algorithms. This is followed by a long and important section on external cluster validity indices. And, finally, we will compare and evaluate the outputs of the four models and algorithms with several numerical examples.

9.1 Introduction

Table 9.1 itemizes 10 properties or descriptors of the Fantastic Four. In Table 9.1, Fcn. = "Function." For clarity, this table has separate columns for hard c-means (HCM, aka "k-means") and fuzzy c-means (FCM). Note especially that, as shown in Table 9.1, HCM algorithm A6.1 ($m = 1$) is *not* regarded

Table 9.1 Some important properties of the four basic methods.

Algorithm	HCM A6.1 ($m = 1$)	FCM A6.1 ($m > 1$)	GMD/EM A7.1	SAHN A8.2
Model type	Objective Fcn.	Objective Fcn.	Objective Fcn.	Graph Theo.
Algorithm type	Iterative: $\approx$ AO	Iterative: AO	Iterative: AO	Non-iterative
Model equation	(6.9)	(6.10)	(7.7) + (7.11)	(8.4)
AO equations	(6.25)	(6.19)	(7.16)	None
Function	$J_1(\mathbf{Q}, X)$	$J_{m>1}(\mathbf{Q}, X)$	$\log L(\mathbf{Q}, X)$	None
Variables	$\mathbf{Q} = (U, \mathbf{V})$	$\mathbf{Q} = (U, \mathbf{V})$	$\mathbf{Q} = (\mathbf{p}, \mathbf{M}, \{S_i\}, P)$	$c \in \{1, \ldots, n\}$
Partition type	Crisp: $U \in M_{hcn}$	Fuzzy: $U \in M_{fcn}$	Prob.: $P \in M_{fcn}$	Crisp: $\{U_c\} \in M_{h(c)n}$
Model choices	$c, m = 1, \|\| * \|\|_A$	$c, m > 1, \|\| * \|\|_A$	$X_i \sim N(\mu_i, \Sigma_i)$	$\delta_{SL}, \delta_{CL}, \delta_{AL}$
Runtime choices (recommended)	$\varepsilon, \|\| * \|\|_{err}, T, \mathbf{V}_0$	$\varepsilon, \|\| * \|\|_{err}, T, \mathbf{V}_0$	$\varepsilon, \|\| * \|\|_{err}, T,$ $Q_0 = (\mathbf{p}_0, \mathbf{M}_0, \{S_{i,0}\})$	None
Input data	$X \subset \Re^p : \|X\| = n$	$X \subset \Re^p : \|X\| = n$	$X \subset \Re^p : \|X\| = n$	Matrix $D \in M_n^+$
# variables *at fixed c*	$c(n + p)$	$c(n + p)$	$c(n + p) + cp(p - 1)/2$	cn

as a true alternating optimization scheme (hence, the designation $\approx$ AO). As was pointed out in Chapter 6, the update equation for the HCM cluster centers is a FONC, but the update equation for HCM memberships is just a NC based on the geometric argument displayed in Figure 6.3. This doesn't alter the fact that Lloyd's algorithm is AO. The technical distinction that underlies this remark will be discussed again in Chapter 10.

One of the most important aspects of the comparison in Table 9.1 is the last row of the table, which shows the *number of unknowns* that each of the four basic algorithms is asked to estimate. The growth in time required to execute the four basic algorithms clearly depends on c, n, and p for the three objective function schemes. The SAHN algorithms don't explicitly recognize the effect of p on their estimate of U, but if D comes from X, p is a hidden variable for this algorithm too. To be fair, there should be two estimates for A8.2, one for each type of input data. For a fixed c and n (discounting the time required to build D from X if this is done), SAHN estimates the fewest parameters of the four, while GMD/EM estimates the most. SAHN models look for $\{U_c\} \in M_{h(c)n}$ at each $c \in \{1, \ldots, n\}$; so the number of variables estimated at any c is (cn).

The dominant parameter for all four algorithms has traditionally been n, the number of objects in the data, because the usual method of accounting for run times uses n for the input size. For input data sets X, the emergence of the "big data" idea has placed a growing emphasis on the importance of p, the number of features possessed by each data vector (more on this in Chapter 11). The number of dimensions (p) and the number of clusters (c) both enter run time calculations based on counting the number of execution steps for n inputs because they determine different numbers of steps per line when run time is estimated; so the traditional approach is followed in the sequel.

9.2 Computational Complexity

The issue of complexity has been treated lightly in previous sections, and in keeping with the idea that this is an elementary text, we will not delve into this topic too deeply. Algorithms A6.1 and A7.1 are given in their most naive forms because the key to understanding how these models operate – their strengths and weaknesses -lies with understanding their basic formulations. For a serious treatment of computational complexity, interested readers are referred to Cormen *et al.* (1996). Rosen (2007) and Stinson (1985) both provide somewhat gentler but also technically correct developments of this important topic. Here, we are content to give a short review of the topics relevant to analysis of the algorithms presented for the Fantastic Four.

9.2.1 Using Big-Oh to Measure the Growth of Functions

Most computer scientists and engineers are comfortable with the notion of big-oh as a measure of "how fast" an algorithm runs, but there will be many readers for which this notion is something of a mystery. The intent of this subsection is to provide a quick overview with enough background information so that Table 9.1 makes sense to you.

With the current focus on extracting information from "big data," there is increasing pressure on the ability of algorithms to *scale up linearly* as the size of problem inputs grows. What does that mean? Linear growth in (x) is represented graphically by the line $y = f(x) = x$. This indicates that y, the output of f (which represents an algorithm here), grows at exactly the same rate as the input x. If $x = 1, y = f(1) = 1$

and if $x = 2, y = f(2) = 2$; so doubling the input also doubles the output of f. What about $y = g(x) = 1000x$? The function g is certainly linear. Does the function g also have linear growth? If $x = 1, g(1) = 1000$; if $x = 2, y = g(2) = 2000$. So, the output for g is still twice as much when its input is doubled, even though the slope of the line $y = g(x)$ is 1000 times steeper than the slope of the line $y = f(x)$. So, the *rate of growth* of f and g is the same, even though the magnitude of the outputs produced by f and g are quite different.

What about $y = h(x) = x + 1000$ – is this also linear growth? Yes. For small values of x, y seems relatively unaffected by changes in x. For example, when $x = 1, h(x) = 1001$, but doubling x doesn't double h(x) since $h(2) = 1002$. But as x grows larger and larger, the effect of the constant term on the output becomes less and less important. For example, $h(100,000) = 101,000$ and $h(200,000) = 201,000$; so doubling this input *nearly* doubles the output. As long as the *rate* of growth behaves in the same way, we are willing to concede that f, g, and h all "grow at the same rate." The notion of "big-oh" growth captures this idea mathematically.

Definition 9.1. Big-oh notation for growth of functions.

Let $f, g : \Re \rightarrowtail \Re$. f(x) is "big-oh of g(x)" if and only if there is a real positive constant $\mathcal{K} > 0$ and integer $n_0 > 0$ so that

$$|f(x)| \leq \mathcal{K}|g(x)| \text{ whenever } x > n_0. \tag{9.1}$$

The constants $\mathcal{K}$ and n_0 are called *witnesses* to the relationship. The technically correct way to indicate this relation is to write $f \in \boldsymbol{O}(g)$, where $\boldsymbol{O}(g) = \{f : \Re \rightarrowtail \Re \mid |f(x)| \leq \mathcal{K}|g(x)| \exists\, \mathcal{K} > 0, x > n_0\}$. However, the popular convention is to write $f = \boldsymbol{O}(g)$ instead. The function g in (9.1) is called a *reference function* for f.

Note that $f \in \boldsymbol{O}(g) \nLeftrightarrow g \in \boldsymbol{O}(f)$. Lecturers in discrete mathematics courses and the analysis of algorithms (I was one of them) delight in having their students agonize over how to find witness pairs for a specified (f, g). It's not so important to be able to do this; some practice in finding witnesses is left to the exercises. The important point about this notion is that it provides a means for comparing the asymptotic growth in time complexity for run times of different algorithms as the problem size grows without limit (i.e., as $n \rightarrow \infty$). The usual hierarchy of reference functions looks like this for real positive constants C and a:

$$C \leq \log x \leq x \leq x \log x \leq x^2 \leq \cdots x^k \leq \cdots a^x \leq \cdots x! \tag{9.2}$$

When you look at Equation (9.2), don't forget that this means *asymptotically.* Most of these reference functions have common names. In order of increasingly faster rates of growth: $\boldsymbol{O}(C)$ is constant; $\boldsymbol{O}(\log x)$ is logarithmic; $\boldsymbol{O}(x^k)$ is polynomial with subcases: $k = 1$ is linear; $k = 2$ is quadratic, $k = 3$ is cubic, etc.; $\boldsymbol{O}(a^x)$ is exponential; and finally, $\boldsymbol{O}(x!)$ is factorial growth. It's a common misconception to look at Equation (9.2) and think: wow, factorial growth is really fast, isn't that what I want? NO! Sure, you want a fast algorithm, but big-oh estimates the *rate* of growth; so what you want your algorithm to have is the opposite of fast growth rate – the *slowest* growth rate it can achieve, because this means its response to larger inputs is less dramatic than if its growth rate were "higher." Thus, constant growth (C) is most desirable, and so on. There are a number of standard theorems about big-oh relationships between functions in the various categories. Here are the most important ones.

Theorem 9.1. Big-oh results for combinations of functions.

Let $f_1, f_2, g_1, g_2 : \Re \rightarrowtail \Re$. Then

$$f(x) = a_n x^n + a_{n-1} x^{n-1} + \cdots a_1 x + a_0 \Rightarrow f \in O(x^n) \tag{9.3a}$$

$$\left\{ \begin{array}{l} f_1 \in O(g_1) \\ f_2 \in O(g_2) \end{array} \right\} \Rightarrow (f_1 + f_2) \in O(|\max\{g_1|, |g_2|\}) \tag{9.3b}$$

$$f_1, f_2 \in O(g) \Rightarrow (f_1 + f_2) \in O(g) \tag{9.3c}$$

$$\left\{ \begin{array}{l} f_1 \in O(g_1) \\ f_2 \in O(g_2) \end{array} \right\} \Rightarrow (f_1 \cdot f_2) \in O(g_1 \cdot g_2) \tag{9.3d}$$

Proof. See Chapter 3 in Rosen (2007).

What does it mean? Result (9.3a) tells us that when $f(x)$ is a polynomial, the growth rate of f is dominated by its highest order term. In particular, we can ignore all the lower order terms. So, for example, $f_1(x) = 2000x^2 + 5000x + 10^7$ and $f_2(x) = x^2$ eventually behave like the same quadratic function of x (i.e., $f_1, f_2 \in O(x^2)$ as x goes to infinity. Equation (9.3c) is a corollary of Equation (9.3b). Results (9.3b) and (9.3d) are very useful when different parts of an algorithmic procedure have different run time functions.

Please remember that the inequalities shown in Equation (9.2) are *asymptotic* inequalities -i.e., this is the *eventual order* for these types of functions for large enough x. For example, $x^2 < x$ for $0 < x < 1$; $x^2 = x$ for $x = 1$; and $x^2 > x$ for $x > 1$. We say that x^2 *overtakes* x as x gets larger, and, for this pair of functions, it doesn't take long. The point at which two functions provide the same (theoretical) growth rate is called the *break even point*: for x and x^2, the break even occurs at $x = 1$.

There are other examples where it takes a good while (i.e., a large x) to restore the asymptotic order in Equation (9.2). This is an important point. Why? It is easy to let asymptotic estimates of growth rates lull you into believing that algorithms with the same big-oh growth rates are really "equivalent" (and they are, ... *in the limit*). But we never reach infinity in real computations; so exact comparisons are very useful, and, often, such calculations separate algorithms into practically useful vs. theoretically equivalent. Example 9.1 will give you an idea of what this statement means.

Example 9.1 Comparison of growth rates for specific functions

In Table 9.2, imagine that you have obtained an expression for T(n), the *running time* associated with a particular algorithm whose *input size* is n. The values in this table make it clear that for the specific instances of run time functions in the table, the "larger" functions in hierarchy (9.2) may have to wait to overtake the "slower" ones. For example, at $n = 10$, all of the functions "above" $T(n) = 2^n/1000$ in the table dominate it, even though, asymptotically, we are sure that it will eventually dominate them. At $n = 100$, this function has blown up and dominates all of the functions above it in the table. But it will never catch $T(n) = n^n/1000$, the function below it in the table, and above it in Equation (9.2).

Table 9.2 Growth rates and run times in ms (cf., Stinson (1985)).

Growth rate (ms)	T(n)	n = 1	n = 10	n = 100
log n	5 log n	0	16.6	33.2
n	2n	2	20	200
n	20, 000n + 1000	30,000	230,000	2,010,000
n log n	$3n \log_2 n$	0	99.6	1993
n^2	$4n^2$	4	400	4000
n^3	$2n^3$	2	2000	$2*10^6$
n^{10}	$n^{10}/1000$	0.001	10^7	10^{17}
2^n	$2^n/1000$	0.002	1.02	10^{27}
n^n	$n^n/1000$	0.001	10^7	10^{200}

The second thing that asymptotic growth rates can't tell you is illustrated by the second and third rows of Table 9.2. Here we see two functions that both have linear growth, $T_1(n) = 2n$ and $T_2(n) = 20,000n+1000$, i.e., $T_1, T_2 \in O(n)$ But the execution time of programs with these as their run time functions is spectacularly different, T_2 taking something like 1000 to 2000 times as much time to run as T_1 for any size input. The only foolproof way to compare algorithmic run times is to implement the methods and compare their actual run times on various inputs.

☺ ***Forewarned is forearmed:*** Example 9.1 is a cautionary tale about big-oh time complexity estimates. It is always nice to say that your algorithm has the same time complexity as a competitor, but this hardly makes them equivalent in real life. At the end of the day, big-oh estimates are like the central limit theorem: they provide some psychological reassurance that things seem to be ok. And, of course, reviewers of your papers will inevitably ask for these estimates, so just make sure you can get them. ☺

How do you get them? There are a number of ways to characterize the ***run time*** or *computational complexity* T(n) of an algorithm. Different operations such as max, min, sum, product, exponentiation, etc., take different amounts of time. But it is not practical to account for these differences even though they do affect the run time because exact accounting makes the analysis quite tedious while not improving its utility much at all. A much better way to proceed is to count the number of times each step in an algorithm is made, make the simplifying assumption that each of these operations takes the same amount of time (unit time), and simply add up the number of steps to arrive at T(n). As long as different algorithms are compared by the same accounting method, we can at least have an idea of their comparative computational complexity, without having to implement them.

The last item needed to complete our discussion of how to estimate the big-oh complexity of an algorithm is the notion of *input size*. For many algorithms, the natural measure of input size is the number of items in the input. This is the traditional way to account for time complexity in clustering algorithms. The number of items in $X = \{\mathbf{x}, \ldots, \mathbf{x}_n\} \subset \Re^p$ is n for A6. 1 and A7.1. And n is also the natural measure of size for inputs $D \in M_n^+$ for A8.1. This is how we approach the estimates in the next section.

9.2.2 Time and Space Complexity for the Fantastic Four

Table 9.3 summarizes time and space complexities for the algorithms given in Chapters 6–8. All operations and storage space are counted as unit costs. We follow convention in using the standard notation "O =" for "big-oh" as equality instead of set containment. The counts in Table 9.3 are "schoolbook" complexity estimates. For example, inverting an $n \times n$ matrix by Gauss-Jordan elimination is accounted for by the big-oh estimate $O(n^3)$. There are somewhat more efficient algorithms for this operation, such as Strassen's algorithm ($O(n^{2.807})$) or William's method ($O(n^{2.373})$) for matrix inversion, but this level of sophistication is not needed here.

Moreover, economies that might be realized by special programming tricks or properties of the equations involved are not taken into consideration. For example, the fact that the covariance matrices for EM are symmetric is not used to reduce various counts from n^2 to $n(n-1)/2$, and we don't assume space economies that might be realized by overwriting of arrays, etc.

Finally, miscellaneous storage of constants, temp files, and the like are not counted. Therefore, these "exact" estimates of time and space complexity are exact only with the assumptions used to make them. Importantly, however, the asymptotic estimates that are shown in Table 9.3 for the growth in time and space are largely unaffected by changes in counting procedures. Some approaches and implementation tricks for improving the runtimes given for these algorithms beyond the acceleration methods already discussed are in the notes and remarks section of Chapters 9 and 11.

In Table 9.3, c = number of clusters, p = number of features (for object vector data X), n = number of objects, and t = *number of iterations to termination*. Complexity of the basic hard c-means (HCM) and fuzzy c-means (FCM) algorithms as implemented in the pseudocode at A6.1 is tc^2pn, but this can be reduced to tcpn using the hidden partition trick that was discussed in Chapter 6. Space consumption shown for the SAHN methods assumes that D is given as input, and all n c-partitions of the $n \times n$ input matrix D are stored inside the program. Much less space would be needed if each partition was removed after its construction. More time is required for SAHN if D is constructed from X with a vector norm on $\Re^p$, but this construction takes $O(n^2)$ time; so the asymptotic time complexity is unaffected. The actual time for this construction can be quite high depending on n, p, and the norm chosen.

The complexity of GMD/EM algorithm A7.1 is relatively high due to the need to compute the inverses and determinants of (c) $p \times p$ matrices in each half-step of the basic algorithm. There are many ways to speed up A7.1, some of which were discussed in Chapter 7. Since computer time keeps decreasing and space limits keep increasing, estimates such as these may seem a bit dated. Perhaps their main utility is that they afford a means for comparing different clustering algorithms on the basis of how much time and space they take (asymptotically) relative to each other.

As Example 9.1 emphasizes, interpreting what complexity estimates really mean for users is fraught with difficulty. Theoretically, two clustering algorithms, say $A_1 = O(n)$ and $A_2 = O(1,000,000n)$ both

Table 9.3 Time and space complexity for the basic (or naïve) Fantastic Four.

	Input	Time	Space
HCM/FCM A6.1	$X \subset \Re^p : \lvert X \rvert = n$	(tc^2pn)	$p(c+n)$
EM/GMD A7.1	$X \subset \Re^p : \lvert X \rvert = n$	(tc^2pn^3)	$2c(p^2+p+1)+cn$
SAHN A8.1	$D \in M_n^+$	SL: (n^2)	$n(3n+5/2)$
		CL: $(n^2 \log n)$	$n(3n+5/2)$
		AL: $(n^2 \log n)$	$n(3n+5/2)$

have linear time complexity; so they will both "reach infinity" in the same amount of time. But, of course, this is not going to happen; so the constants are important – you can be (almost) sure that A_1 will finish clustering a lot faster than A_2 on $n = 100,000$ objects in $p = 1000$ dimensions.

9.3 Customizing the c-Means Models to Account for Cluster Shape

An important limitation of the c-means models at Equations (6.9) and (6.10) is that A6.1 tries to impose an implicitly defined preferred shape on all c clusters which is determined by the model norm chosen. Attempts to match geometric shapes of clusters more exactly than HCM/FCM does is approached either by adjusting the model norm (and, hence, the shapes of the open and closed unit balls in feature space as shown in Figure 6.6) or by changing the fitting prototypes $\mathbf{V}$ from points to some more complex type of geometric objects such as linear varieties or hyperquadrics. These two general approaches both involve a change to the distance term in the c-means models, as shown in Figure 9.1.

9.3.1 Variable Norm Methods

Surprisingly, many users of crisp k-means (HCM) assume that one of its limitations is a preference for circular clusters. But this artifact is due to using the Euclidean norm, which, in many cases, is the default model norm. As shown in Chapter 6, Lloyd's algorithm can be implemented and works fine when any inner product A-norm is used in Equation (6.8). In this section, we consider adjustments to the inner product distance used as the model norm in Equations (6.9) and (6.10).

There have been many studies on the effect of changing A in the distance calculation for $d_{ikA} = \|\mathbf{x}_k - \mathbf{v}_i\|_A^2$ on the assumption that *all* of the clusters in X have roughly the same A-norm geometry and that they are "cloud-like." This problem was first addressed in a very important proceedings paper by Gustafson and Kessel (1979), who introduced the idea of allowing the norm-inducing matrix A to vary from cluster by cluster. Their idea was to let each cluster assume a best-matched shape via individual norms that adapted to the shapes of individual clusters during AO iteration. They put $\mathbf{A} = (A_1, \ldots, A_c)$; $A_1 \in PD^{pp}\ \forall i$, and modified the optimization problem at Equation (6.10) for J_m to include the c matrices $\mathbf{A}$ as part of the unknowns,

$$\underbrace{\min}_{\substack{(U,\mathbf{V},\mathbf{A}) \\ \det(A_i)=\rho_i}} \left\{ J_{m,GK}(U,\mathbf{V},\mathbf{A}) = \sum_{i=1}^{c}\sum_{k=1}^{n} u_{ik}^{m}\|\mathbf{x}_k - \mathbf{v}_i\|_{A_i}^2 \right\} \tag{9.4}$$

The unknowns in the GK model are the triplet $\mathbf{Q} = (U, \mathbf{V}, \mathbf{A})$, where $\mathbf{V}$ is still a vector of point prototypes. Values for these unknowns are estimated using the same AO scheme as in A6.1 but with an added set of FONCs for each of the matrices $\{A_i\}$. Information about the shape of (that is, the points that partially or fully belong to) the *i*th cluster resides in the eigenstructure of A_i. Specifically, the eigenvalues of A_i determine the amount of stretch along each axis of the hyperellipse associated with the A_i-norm. The orientation of the local axes is determined by the eigenvectors of A_i.

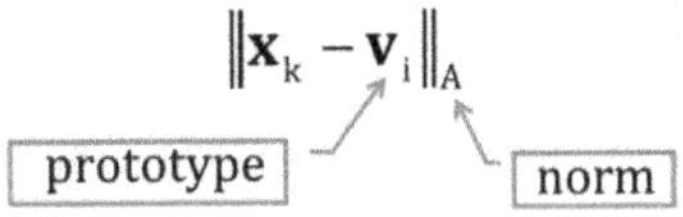

$$\|\mathbf{x}_k - \mathbf{v}_i\|_A$$

Figure 9.1 Potential adjustments to accommodate cluster shapes used in the c-means models.

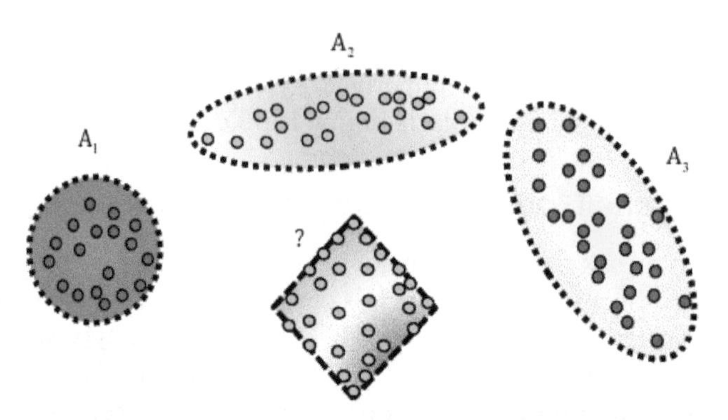

Figure 9.2 Adjustment of inner product norms cluster by cluster.

Figure 9.2 illustrates the geometric idea of the GK model. The three elliptically shaped clusters have *local* inner products induced by the matrices A_1, A_2, and A_3: this is what the GK model is designed to accommodate. The diamond-shaped cluster labeled (?) in Figure 9.2 would be well-matched by the 1-norm (the green diamond in Figure 6.6), but this is not an inner product norm; so it doesn't fit into the GK scheme. This cluster would be approximated by a fourth inner product norm in the GK model if c = 4 clusters were being sought.

Much is known about the GK model. To make the optimization feasible, problem (9.4) adds a set of constraints on the determinants of the norm-inducing weight matrices $\{A_i\}$. Gustafson and Kessel simplified this constraint by requiring the c determinants to be equal, $\rho_i = 1 \forall i$. This constraint is needed to make the differentiation of $J_{m,G,K}(U, \mathbf{V}, \mathbf{A})$ with respect to A_i tractable, and it essentially forces the c clusters to have equal volumes. In terms of the open ball geometry this creates, the GK model loosens the shape requirement that the open balls all have the same A-norm shape but tightens the size requirement in that they all fit into equal volumes. The GK model and its AO algorithm are not exemplified in this volume, but this model is very useful for understanding the overall relationship between inner product norms and how they affect local distances. Moreover, it forms a mathematical bridge of sorts between the c-means models and Gaussian mixture decomposition. See Bezdek *et al.* (1999b) for a fairly complete treatment and an example of the GK model and algorithm.

Please compare the equations at (6.8), (9.4), and the sample-based form of (7.15), which show the FCM, GK, and GMD objective functions, rewritten here in notation to exhibit the similarities and differences between these three functions:

$$J_m(\mathbf{Q}; \mathbf{X}) = \sum_{i=1}^{c}\sum_{k=1}^{n} u_{ik}^m \|\mathbf{x}_k - \mathbf{v}_i\|_A^2; \quad \mathbf{Q} = (U, \mathbf{V}); \quad m \geq 1; \tag{6.8}$$

$$J_{m,GK}(\mathbf{Q}; \mathbf{X}) = \sum_{i=1}^{c}\sum_{k=1}^{n} u_{ik}^m \|\mathbf{x}_k - \mathbf{v}_i\|_{A_i}^2; \quad \mathbf{Q} = (U, \mathbf{V}, \{A_i\}); \quad m \geq 1; \tag{9.4}$$

$$\sum_{i=1}^{c} L(\mathbf{Q}_i; \mathbf{X}) = -\frac{1}{2}\sum_{i=1}^{c}\left(\sum_{k=1}^{n_i} \|\mathbf{x}_k - \mathbf{m}_i\|_{S_i^{-1}}^2\right) + \sum_{i=1}^{c}\left[\log\left(\frac{p_i}{\sqrt{\det S_i}}\right)\right]; \quad \mathbf{Q} = (\mathbf{p}, \mathbf{M}, \{S_i\}, P). \tag{7.15}$$

The number of estimated parameters for the HCM/FCM algorithms is $c(n + p)$. The GK functional estimates $c(n + p) + cp(p - 1)/2$; so GK estimates $cp(p - 1)/2$ more parameters than FCM or HCM. GMD estimates

$c(n + p) + c + cp(p - 1)/2$ parameters, just c more than GK; so these two algorithms are fairly equivalent in terms of the unit work needed to acquire parametric estimates. HCM and FCM bear a slighter computational load than GK or GMD, which increases with c and p but not n.

Next, have a look at the means, norm-inducing matrices, and c-partitions of X. Associate these three sets of variables as follows: $\mathbf{v}_i \leftrightarrow \mathbf{m}_i : \mathbf{A}_i \leftrightarrow S_i^{-1} : U \leftrightarrow P$. The most striking feature which is common to all three objective functions is the double sum of distances between data points and cluster centers. For m = 1, if we choose $\{\mathbf{v}_i = \mathbf{m}_i \forall i\}$, $\mathbf{A} = \mathbf{A}_i \forall i$ in Equation (6.8) and $\{\mathbf{A} = S_i^{-1} \forall i\}$ in Equation (9.4), this part of the three functions is equal up to the constant $-1/2$. What is "left over" in the log-likelihood function at Equation (7.15) is the extra information about the priors and covariances that the Gaussian mixture assumptions entail.

From this, it is easy to see that when the priors are all (roughly) equal to (1/c), and the covariance matrices are all (roughly) equal to the identity matrix, then HCM, GK, and GMD will produce very similar CPOV clusters in X. This is seen in Examples 6.3, 7.3 and 8.11. For m>1 in A6.1, the squared error between data point $\mathbf{x}_k$ and prototype $\mathbf{v}_i$ is weighted by u_{ik}^m; so the sum of squared errors is distributed across the c soft clusters in the data in various ways, depending on the value of m. Thus, when m is close to 1, FCM and GMD will produce similar CPOV clusters. In the Gaussian model, Bayes rule for p_{ik} plays the role taken by the fuzzy weight u_{ik} in FCM and GK. The variable weights due to variation in parameter m in FCM and GK enable flexible soft partitions of the data that are somewhat more rigid due to the fixed nature of the $\{p_{ik}\}$ in GMD.

Another point concerns the geometric distribution of the clusters preferred by each of these three functions. HCM/FCM is at its best when X contains cluster clouds of roughly equal elliptical shapes. GK adds a dimension of flexibility to the shape constraint imposed by FCM (and HCM) by allowing each norm-inducing weight matrix $\{A_i = S_i^{-1} \forall i\}$ to migrate to a best shape for the *i*th cloud. But the geometric price GK pays for this added shape flexibility is that all c clouds must have equal volumes. MLE of CPOV clusters obtained by GMD/EM offers the most shape flexibility since the covariances $\{S_i^{-1} \forall i\}$ can stretch clusters (almost) to infinity, and there is no equi-volume constraint binding the c clusters. In view of these remarks, it is not so surprising that many empirical studies show that the c-means models are quite adept at finding the "correct" clusters in data sets like the Gaussian clusters in Figure 7.3(a). Sometimes, even more adept than GMD/EM. So the supposed advantage of variable norms is often mitigated by the fact that the cluster shapes in samples from Gaussian mixtures are almost always cloud shaped.

When we are dealing with unlabeled data, we are presumably interested in whether or not the objects underlying the data have the property we attribute to the vectors representing the objects. For example, we invent data sets like X_{4rings} in Example 8.7 to illustrate the concept of chain clusters, and there we find that SL is good at recognizing them. Two questions you should ask: (i) what sort of real objects might underlie such data? and (ii) if the data were drawn from a set of concentric hyperspheres in, say, 4-space, how would we know that SL would be a good choice for these 4D chains? Well, we wouldn't, because there would be no HPOV clusters. And, in this case, we are left to infer what we can about (object) geometries from the CPOV clusters we construct using clustering algorithms.

9.3.2 Variable Prototype Methods

The second general method for accommodating clusters that don't have hyperellipsoidal distributions is to alter the fitting prototypes inside the distance measure, as shown in Figure 9.1. So far, the only kind of prototypes that have been used in the three objective function models are point prototypes – i.e., the cluster centers $\mathbf{V} = \{\mathbf{v}_1, \ldots, \mathbf{v}_c\} \subset \Re^p$ for A6.1 or the mean vectors $\mathbf{M} = \{\mathbf{m}_1, \ldots, \mathbf{m}_c\} \subset \Re^p$ for A7.1.

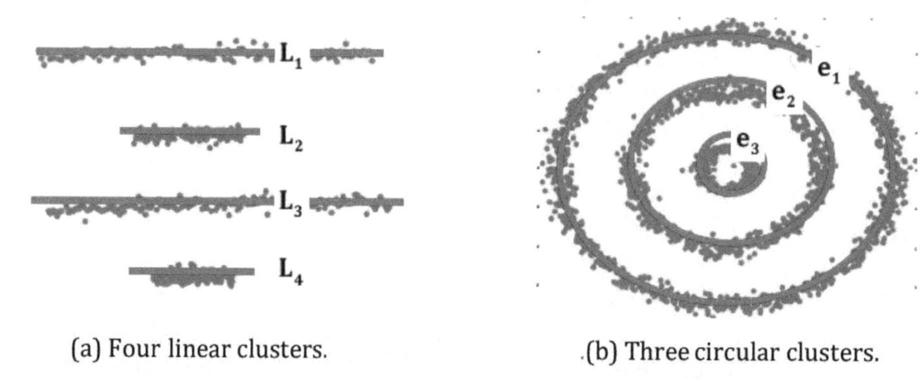

(a) Four linear clusters. (b) Three circular clusters.

Figure 9.3 Some distributions that are best fitted with non-point prototypes.

Some authors interpret the weight matrices $\{A_i \text{ or } S_i^{-1} \forall i\}$ inducing individual norms for each cluster in the GK and GMD algorithms as part of the set of prototypes for the c clusters, but the point of view here is that these matrices are unknowns in the argument of the GK objective function.

There is a large body of literature that is devoted to ways to replace the point prototypes of A6.1 and A7.1 with more general constructs. Some of the structures are new geometric objects, and some are abstract constructs. The simplest type of geometric upgrade is from points to lines. Consider using a linear prototype, rather than a point, to represent line-like clusters. Now let your imagination wander up to the next level of complexity. Suppose the data have circular clusters, e.g., like those in X_{4rings} (cf., Figure 8.19(a)); a circular prototype could be used as a prototype for each cluster in the ring,and so on.

Figure 9.3 illustrates this for several two-dimensional data sets that have HPOV sets of clusters. Figure 9.3(a) shows four linear clusters (in blue) that would be well-matched by the four red line segments $\{\mathbf{L}_1, \mathbf{L}_2, \mathbf{L}_3, \mathbf{L}_4\}$. Figure 9.3(b) contains three elliptical clusters in blue: good prototypes for this data set would be the three red ellipses $\mathbf{e}_1, \mathbf{e}_2, \mathbf{e}_3$. Are there clustering algorithms that can detect structures like this and produce these kinds of prototypes? ***Yes***.

The earliest reference to the explicit use of non-point prototypes in connection with generalizations of FCM was Bezdek *et al.* (1978). These authors discussed a primitive method for fitting fuzzy clusters with lines in two-dimensional data. The *fuzzy c-varieties* (FCV) models (Bezdek *et al.* 1981a,b) grew out of this effort and were the first generalizations of FCM that explicitly used a specific family of non-point prototypes. The prototypes used by FCV are r-dimensional linear varieties, $0 \leq r \leq p-1$. For example, linear varieties of dimension 0 are points, of dimension 1 are lines, of dimension 2 are planes, and so on, up to dimension $(p-1)$, where they are hyperplanes. The mathematics of this type of extension are beyond the scope of this volume, so readers will be spared most of the details of this rather arduous development. But we detour from the main route long enough to explain how it is done.

The FCV adjustment to the objective function at Equation (6.8) begins by replacing the points $\{\mathbf{v}_i\}$ in Equation (6.8) with linear varieties, say $\{\mathbf{B}_i\}$. The simplest case is to use lines $\{\mathbf{L}_i\}$ instead of points. Consider a single point $\mathbf{v}_i$ and a line, say, $\mathbf{L}_i$. Replacing $\mathbf{v}_i$ with $\mathbf{L}_i$ transforms the distance from $\|\mathbf{x}_k - \mathbf{v}_i\|_A^2 \rightarrow \|\mathbf{x}_k - \mathbf{L}_i\|_A^2$. Recall that a line $\mathbf{L}_i$ in any finite dimensional vector space can be specified by finding a point $\mathbf{v}_i$ on the line and direction vector $\mathbf{s}_i$ of the line; so the line can be parametrized as $\mathbf{L}_i = \{\mathbf{v}_i; \mathbf{s}_i\}$.

The next question is: how do we measure the distance from $\mathbf{x}_k$ to $\mathbf{L}_i$? Since the norm being used is an inner product A-norm, we have one of the most powerful tools in mathematical analysis at our disposal – namely, the *orthogonal* (OG) projection theorem. This theorem tells us exactly how to compute the *OG*

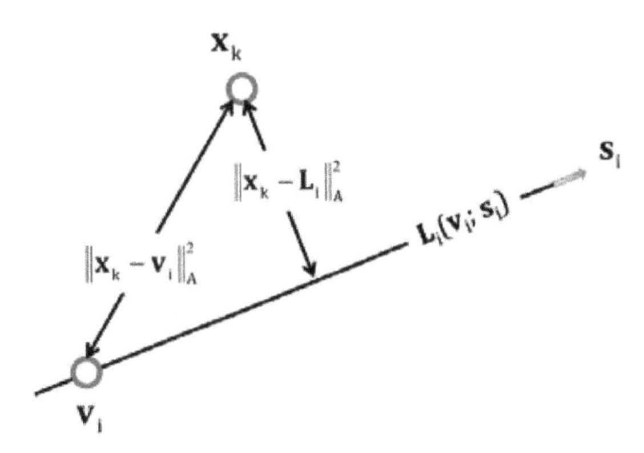

Figure 9.4 Distances from x_k to point v_i and line $L_i(v_i; s_i)$.

distance from $\mathbf{x}_k$ to $\mathbf{L}_i$. Without going into too much detail about how to do this, suffice it to say that this enables us to make the calculation of $\|\mathbf{x}_k - \mathbf{L}_i\|_A^2$. The difference between the distances $\|\mathbf{x}_k - \mathbf{v}_i\|_A^2$ and the OG projection distance $\|\mathbf{x}_k - \mathbf{L}_i\|_A^2$ is illustrated in Figure 9.4 for the case p = 2. Once this principle is grasped, it is a matter of slogging through the calculations to arrive at first order necessary conditions for the unknowns of the fuzzy c-lines objective function, which is just $J_m(\mathbf{U}, \mathbf{L}; \mathbf{X}) = \sum_{k=1}^{n} \sum_{i=1}^{c} (u_{ik})^m \|\mathbf{x}_k - \mathbf{L}_i\|_A^2;\ m \geq 1$, where $\mathbf{L} = (\mathbf{L}_1, \ldots, \mathbf{L}_c)$ is a set of c prototypical lines that best fit the data in the sense of this least squares model. The additional set of unknowns (the direction vectors $\{\mathbf{s}_i\}$ which are part of the fuzzy c-lines model) is a function of the largest eigenvector of the matrix A. The details of these calculations appear in the proof of Theorem 23.1 in Bezdek (1981). Note that the prototypical lines are estimated *simultaneously* with a fuzzy c-partition U of the (potentially) unlabeled data. This is what distinguishes this type of model from something like linear regression, which is done "one line at a time" with labeled data.

The appearance of the Gustafson and Kessel (1978) and FCV papers of Bezdek *et al.* (1981a, b) spawned a wide variety of new clustering models that simultaneously estimated a partition matrix U and sets of non-point prototypes. Delving further into this Pandora's box takes us too far from the objective of the volume; so the rest of this section is presented in the spirit of capturing your interest in exploring the literature in this field.

Perhaps, the biggest drawback of FCV and convex combinations like *fuzzy c-elliptotypes* (FCE) is that these models find c clusters with prototypical "shapes" that are all the same. The fitting prototypes in FCE are mathematical objects that arise from optimizing the FCV model using convex combinations of the distance, viz., $\alpha\|\mathbf{x}_k - \mathbf{v}_i\|_A^2 + (1-\alpha)\|\mathbf{x}_k - \mathbf{L}_i\|_A^2$. These surrogates are not geometric entities; rather, they are mathematical constructs. The reason for the fixed structural elements in FCV is that FCV uses the same real dimension (r) and its convex combinations all use the same "mixture of dimensions" for all c clusters; so cluster substructure having these characteristics is imposed on the data whether they possess it or not. In some sense, this amounts to escaping the structure imposed on the input data by a fixed norm in exchange for imposing the same kinds of (linear variety) prototypes on the input data. This problem resulted in the first locally adaptive fuzzy clustering method (the GK model), and the next generation of locally adaptive clustering methods followed rapidly on the heels of the FCV models.

The earliest scheme for local adaptation in the FCV models was due to Anderson *et al.* (1982). They suggested a scheme called *adaptive fuzzy c-elliptotypes* (AFCE) which proposed that the value of α used in convex combinations of the FCV functionals should be different for each cluster, reflecting a customized distance measure that best represents the shape of each cluster. When convex combinations are used, there

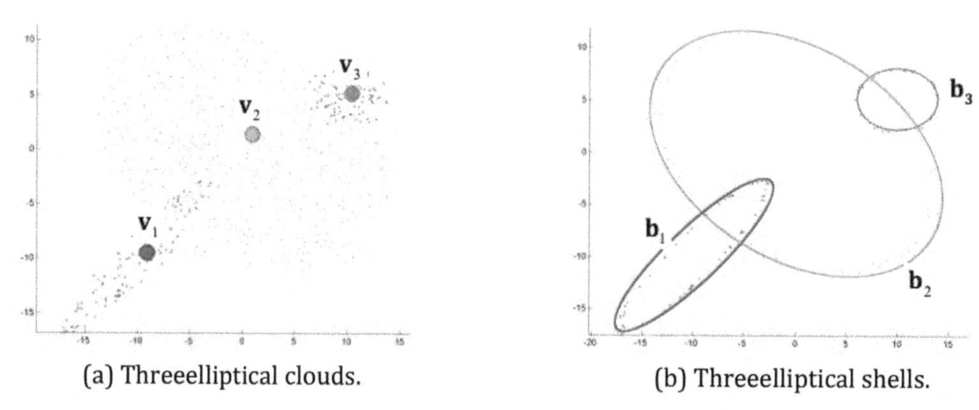

(a) Threeelliptical clouds. (b) Threeelliptical shells.

Figure 9.5 Point prototypes (FCM) and elliptical shells (AFCS).

is no dimensionality of prototypes (it is the *distances* in the FCV objective function that become convex combinations in Bezdek *et al.* (1981a, b), and not the fitting prototypes). The basic idea in FCE is to mediate between the geometric needs for point prototypes (central tendencies) and varietal structure (shape or dispersions). But convex combinations of FCV such as FCE fix the amount by which each factor contributes to the overall representation of all c clusters.

It is arguable whether linear clusters such as those in Figure 9.3(a) are linear clouds or boundary shells. Some writers interpret them as the boundary case between the two types of structures. The use of non-linear shell (curved) prototypes was introduced in Coray (1981), who showed how to optimize a fuzzy c-circles function. This line of research rapidly evolved to the *fuzzy c-shells* (FCS) algorithms (Davé, 1990, 1992) and the *fuzzy c-spherical shells* (FCSS) algorithms of Krishnapuram *et al.* (1992). Davé and Bhaswan (1992) proposed the *adaptive fuzzy c-shells* (AFCS) model for elliptical shells. Figure 9.5 shows the type of outputs that would emanate from using FCM (point prototypes in Figure 9.5(a) and AFCS (elliptical shells) for the data sets in Figure 9.5).

Bezdek *et al.* (1995) introduced a method that generates shell prototypes corresponding to level sets of any norm on $\Re^p$. This model, called the *norm-induced shell prototype* (NISP) model, was optimized using a genetic algorithm instead of alternating optimization. A typical output of the NISP is illustrated in Figure 9.6, where the data points are red circles, and the fitting prototypes are the two yellow diamonds that are obtained after 100 generations of the NISP algorithm using the 1-norm on the input space. This output

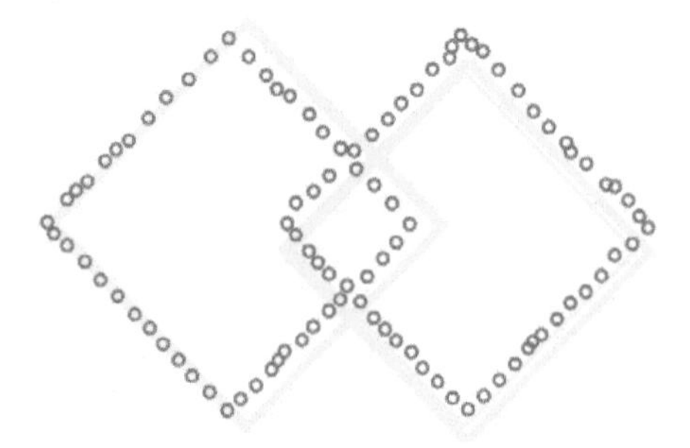

Figure 9.6 Overlapping diamonds fitted to the red data by NISP using the 1-norm.

would be a satisfactory solution for the (?) data shown in Figure 9.2, but using this algorithm on the entire data set in Figure 9.2 would badly misrepresent the other three ellipsoidal clusters seen there.

The last development mentioned here is the family of *fuzzy c-regression models* (FCRM) introduced by Hathaway and Bezdek (1993). The objective functions in this family have regression models (functions, such as linear regression models, quadratic regression models, etc.) as the prototypes. Minimization of a member of this family yields simultaneous estimates for the parameters of c-regression models and a fuzzy c-partitioning of the data.

Well, you can see that these generalizations have a life of their own. All of the algorithms discussed in this subsection might be properly viewed as data fitting models in the sense of regression. Indeed, all clustering models are in fact data fitting models, but this aspect of them usually takes a back seat to the partitions they generate. Not mentioned here, but important, are that many of these models have experienced generalizations to the case where $U \in M_{pcn}$ is a possibilistic c-partition of X (cf., Krishnapuram and Keller, 1993). And at the other extreme, most of these customized prototype models have crisp counterparts that need only slight alterations for the case $m = 1$.

9.4 Traversing the Partition Landscape

This section discusses differences in the way each of the Fantastic Four conducts searches for an optimal partition of the input data. To begin, let us recall how hard bunny and soft bunny deal with the geometric landscape shown in Figure 6.2(c), the soccer ball of soft partitions of n objects.

Recall that in Example 6.2, during the search for a "good" crisp CPOV partition of the input data, HCM bunny can only hop from vertex to vertex during his or her quest, always hoping to land on a c-partition that is good enough to end the trip. Figure 9.7 represents the trip that HCM bunny takes in a slightly different way. The subscript "c" is there to remind you that the values depicted in the diagram all depend on this parameter.

The set $S_{1,c}(X)$ in Figure 9.7, as in Figure 6.11, is the set of values of the objective function $J_{1,c}(U, \mathbf{V}; X)$ associated with optimal pairs for the HCM model, i.e., for each c,

$$S_{1,c}(X) = \{J_{1,c}(U, \mathbf{V}; X) : (\mathbf{V}, X) \text{ satisfy Theorem 6.3 at fixed c}\} \tag{9.5}$$

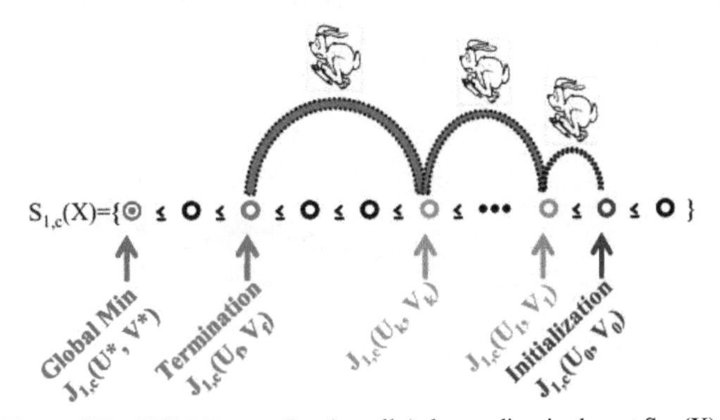

Figure 9.7 HCM Bunny: $J_{1,c}$ (usually) descending in the set $S_{1,c}(X)$.

The values along the horizontal axis in Figure 9.7 are shown in decreasing order from right to left and are in $1-1$ correspondence with the good vertices in Figure 6.2(c), i.e., the non-degenerate crisp c-partitions of X (and also, as we shall see in the next chapter, the red spikes in Figure 10.1(b)). In this figure, hard bunny is progressing nicely toward the global minimum of $J_{1,c}$: each hop improves the estimate (decreases the value of) $J_{1,c}$. But alas, hard bunny either gets trapped at a fixed point, runs out of time, or is simply exhausted (these hops are hard work!) and, so, terminates at some point $(U_f, \mathbf{V}_f)$ before reaching the global minimum. This graphically illustrates the descent property $J_{1,c}(U_{t+1}, \mathbf{V}_{t+1}; X) \leq J_{1,c}(U_t, \mathbf{V}_t; X)$ of HCM iteration. The only exception to this is when HCM gets caught in a limit cycle and stalls (the bunny hops back and forth between two fixed points in $S_{1,c}(X)$) so that no further progress can be made (this happens very rarely but necessitates the word "usually" in the figure caption).

This diagram should make it intuitively clear to you that the concept of a "local minimum" doesn't apply to the crisp HCM (k-means) model. Chapter 10 will explain this from a mathematical point of view. In particular, if this point puzzles you, or intrigues you, page ahead to Figure 10.1(b), which shows the set $S_{1,c}(X)$ in Equation (9.5) from a slightly different perspective.

A diagram like Figure 9.7 for FCM/AO and GMD/EM would exhibit values of the objective function as points on a wavy surface (a nonlinear manifold) in many dimensions. You can imagine this surface looking a little bit like a (smoothed) landscape of the Rocky mountains. An edge view of this surface along a slice of it is depicted Figure 9.8. Of course, the surface varies as a function of c, emphasized again in the subscript for the objective function.

The set $S_{m,c}(X)$ in Figure 9.8 is a subset along this path of the set of values of the objective function $J_{m,c}(U, \mathbf{V}; X)$ associated with optimal pairs for the FCM model, i.e.,

$$S_{m,c}(X) = \{J_{m,c}(U, \mathbf{V}; X) : (U, \mathbf{V}) \text{ satisfy Theorem 6.2 at fixed c}\} \tag{9.6}$$

Recall that soft bunny in Example 6.2 could hop like hard bunny, but soft bunny could also burrow into the interior of the soccer ball in Figure 6.2(c). We see soft bunny in Figure 9.8 beginning at some point in M_{fcn}, either at a vertex or inside the soccer ball, and moving "downhill," in accordance with the descent property of the FCM functional, i.e., $J_{m,c}(U_{t+1}, \mathbf{V}_{t+1}; X) \leq J_{m,c}(U_t, \mathbf{V}_t; X)$. Soft bunny might follow the blue path in Figure 9.8 and stop at or near a local minimum (or, very rarely, a saddle point) of $J_{m,c}(U, \mathbf{V}; X)$. But it is also possible because there is continuous path for this bunny through M_{fcn}, that this rabbit hits a "flat spot" in the iterations which allows it to bridge the gap and take the green bypass path instead, toward a possibly smaller minimum as shown in the diagram.

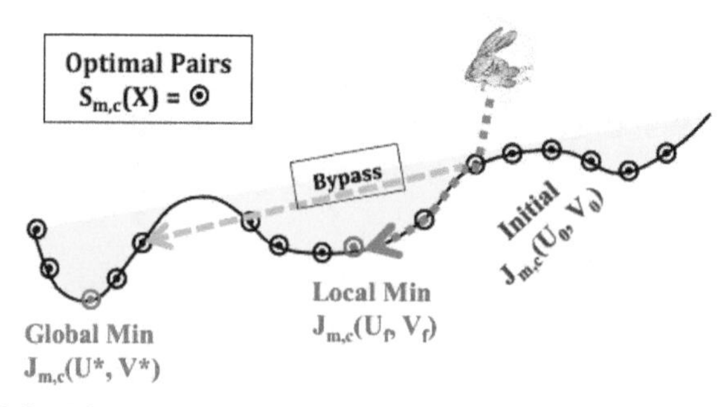

Figure 9.8 FCM Bunny: $J_{m,c}$ (usually) descends in the set $S_{m,c}(X)$ for $m>1$.

☹ ***Forewarned is forearmed:*** Even if hard and soft bunnies get to their global minimums, this doesn't necessarily mean that they have arrived at the very best clustering solution for the problem at hand. That is something you will have to check. Even worse, each time you initialize these little fellows at a different place, they may end up at a new location because they don't have very good memories. And this is just for a single fixed value of c. ☹

What about GMD/EM? Well, just turn Figure 9.8 upside down so that soft bunny is climbing uphill, and you will have a correct mental image of what EM bunny is doing, and how he or she might get caught at a local maximum. The one real difference is that a diagram like Figure 9.8 for GMD/EM may be unbounded above, while values of $J_{m,c}(U, \mathbf{V}; \mathbf{X})$ on the error surface for FCM/AO are always bounded below (by zero).

Next, let us imagine how the SAHN bunny for A8.1, and, in particular, the SL bunny for Algorithm A8.2 moves through the briar patch of vertices in M_{hcn}. This bunny has a very different experience because c decrements from c = n to c = 1. This scheme starts with every point in its own singleton cluster; so SL bunny has no first choice: it begins at the unique vertex M_{hnn}, which is the initial partition I_n = diag[1], the identity matrix for $\Re^n$. What happens next?

The soccer ball that SL bunny sees next is $M_{h,n-1,n}$. Assume, without loss in generality, that the distances in D are all distinct. SL bunny can hop to exactly one of $(n-1)$ vertices amongst the many that exist in $M_{h,n-1,n}$, and then SL algorithm A8.2 merges the two objects, say (i, j), that correspond to the minimum distance in D. The next hop will be to $M_{h,n-2,n}$, where the current $(n-1)$ clusters will become $(n-2)$ clusters after merger. At this stage, SL bunny has a few more choices; there are $(n-2)$ possible joins of some point k to (i, j); or there are $(n-2)(n-3)$ possible joins of a new pair instead of this one, by some other pair, say (s, t). So, the number of potential vertices that SL bunny can visit increases from $(n-1)$ to $(n-2)+(n-2)(n-3)=(n-2)^2$. That is quite a jump if n is large, but, still, it is quite a few less than all of the possible vertices that exist at this stop.

As SL proceeds, the next hop is always to a new soccer ball which has fewer total vertices than the preceding one, but the available number of candidate partitions increases at each step, for a while. The last hop is to $M_{h,1,n}$ at the unique partition $U_{1\times n}$ = [**1**] which assigns all n points to a single cluster. As you can see, it is a lot harder to illustrate the hopping sequence of SL bunny in a simple figure than it is for the three iterative members of the fantastic four, since each SL hop is to a set of partitions in a different space, of which only a (non-constant) few are potential candidates. Figure 9.9 is a rough attempt to capture the way SL bunny hops. The "lattice" shown is a cutout of part of the surface of a pair of successive partition soccer balls about the current centroid of each one.

Moreover, there is not an error surface or objective function that monitors the progress of SL bunny – this bunny has a more or less pre-determined sequence of constrained hops that result in the unique nested sequence of crisp k-partitions $\{U_{SL,k} \in M_{hkn} : n \geq k \geq 1\}$. If SL is implemented as a bottom-up approach, the sequence in Figure 9.9 is executed in reverse (hopping from right to left). Before we turn to some comparative examples, let us continue our exploration with a side trip that will facilitate comparisons of sets of candidate partitions.

9.5 External Cluster Validity With Labeled Data

Table 9.4 summarizes the internal CVIs that have been discussed so far. In this section, some methods of validation based on external cluster validity measures are discussed.

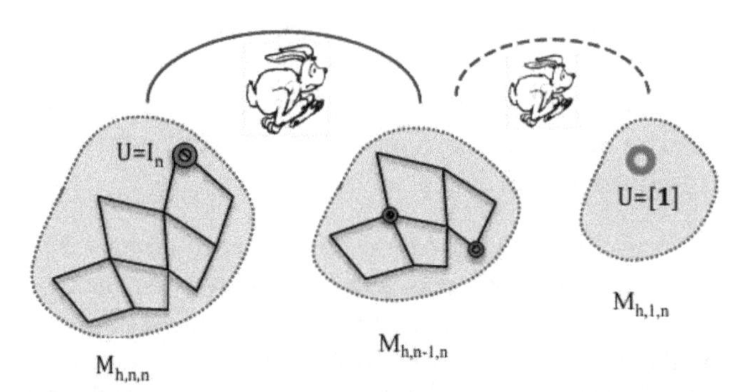

Figure 9.9 Single linkage bunny visits only a few vertices at each stop.

Table 9.4 Internal CVIs/GOFIs that appear in previous examples.

Name	Eqn. #	Notation	Example(s)	Algorithm
Davies-Bouldin	(6.53)	V_{DBqt}	6.6, 7.7, 8.8	HCM
Dunn's index	(6.56)	$V_{DI} = V_{GDI,11}$	6.6, 8.8	HCM
Gen'l. Dunn33	(6.58)	$V_{GDI,33}$	6.6, 7.7, 8.8	HCM
Partition Coeff.	(6.59)	V_{PC}	6.7, 6.8, 7.7	FCM
Partition entropy	(6.60)	V_{PE}	6.8	FCM
Xie-Beni, m=2	(6.64)	$V_{XB,2}$	6.7, 6.8, 7.7	FCM
Xie-Beni, m>1	(6.64)	$V_{XB,m}$	6.7	FCM
Normalized PC	(6.69)	$V_{PC,R}$	6.8	FCM
Normalized PE	(6.70)	$V_{PE,B}$	6.8	FCM
Aikake I. C.	(7.17)	V_{AIC}	7.6	GMD/EM
Inf. comp. (Bozdogan)	(7.19)	V_{IOMP}	7.6	GMD/EM
Inf. comp. (wind./cutler)	(7.20)	V_{IOMPw}	7.6	GMD/EM
Gap distance	None	G(c)	8.10	SL

9.5.1 External Paired-Comparison Cluster Validity Indices

An *external* CVI, say $V_{ext}(U|Q)$, uses the label information available in a reference matrix, $Q = U_t$ or U_{gt} (cf., Section 6.7.2 to refresh your memory on this notation for labeled data; also recall that U_{GT} stands for both kinds of ground truth data) to secure an "independent" opinion of the quality of each candidate. We have (*secretly*) met one external CVI already – viz., the partition accuracy $PA(U|U_{GT})$ defined at Equation (6.34). This measure, used in Examples 6.3, 6.8, and 7.3, is perhaps the canonical external CVI.

The notation $PA(U|U_{GT})$ indicates that U is to be compared to the given structure in the external ground truth partition U_{GT}. In the terminology of cluster validity, $PA(U|U_{GT})$ is a max-optimal (↑) index. In this section, we discuss several other external CVIs that will help with the job of cluster validation *for labeled data*.

If Q is a given *reference matrix*, we write $V_{ext}(U|Q)$. It is almost always the case that $Q = U_t$ or U_{gt}. However, there is a second case, and that is when U and Q are on equal footing. Perhaps, they are two CPOV partitions obtained by the same clustering model and algorithm that differ only by an initialization parameter, e.g., two different initializations of HCM. Or, these two matrices might be partitions obtained by

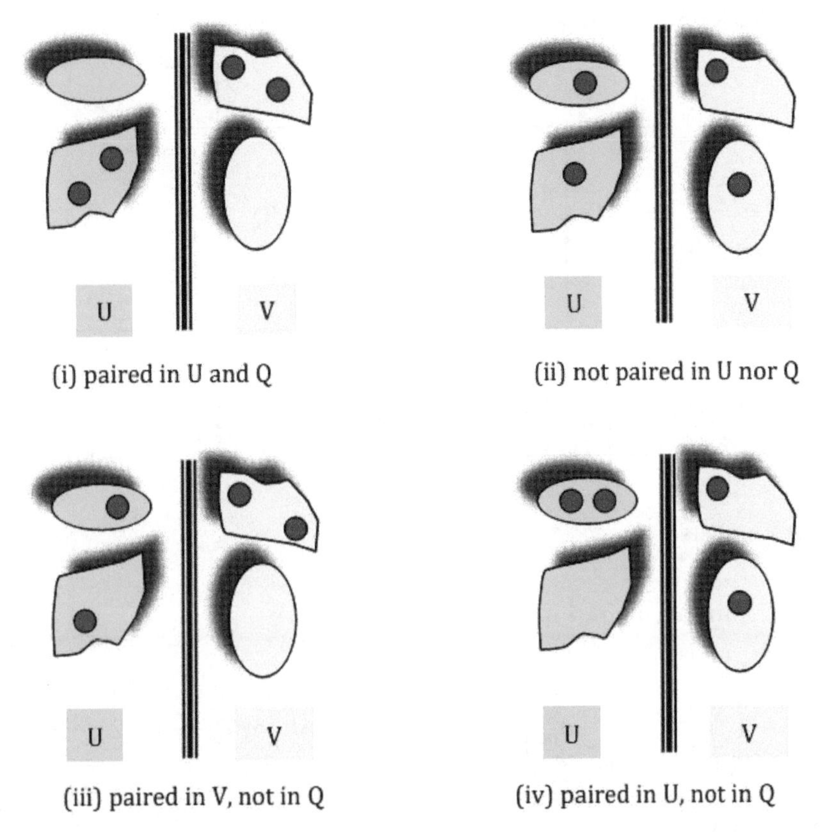

Figure 9.10 Paired comparisons for pairs of points relative to clusters in U and Q.

different clustering models and algorithms applied to the same input data. In this case, external CVIs that compare them are often called *coefficients of partition similarity*. Then the job of the CVI becomes one of assessing the pairwise similarity between U and Q. This change in the use of the index is sometimes signaled by writing $V_{ext}(U, Q)$ instead of $V_{ext}(U|Q)$. This is a pretty subtle distinction, but the idea should be clear. On the one hand, many same-sized U's are compared to a fixed given Q by $V_{ext}(U|Q)$. And, on the other hand, U and Q are simply same-sized partitions compared to each other for similarity by $V_{ext}(U, Q)$. The examples in this chapter all use external indices in the first sense, i.e., $V_{ext}(U|Q)$.

There are lots of ways to use Q in external validation. Here we discuss external CVIs that depend on pair-based comparisons. Let $U \in M_{hrn}$ and $Q \in M_{hcn}$ be two *crisp* partitions of n objects. *Note that U and Q need not possess the same number of clusters (rows),* $r \neq c$. This is one of the great advantages of pair-based comparison indices: U and Q can have different numbers of clusters, so the correspondence problem discussed in Example 6.3 becomes moot. The classical pair-based approach to comparing U and Q begins by considering the four combinations for pairs of objects from the set O in clusters of U and Q as shown in Figure 9.10. The four types are numbered (i)–(iv) following Hubert and Arabie (1985).

The comparison of U to Q begins by representing counts of the four types of pairings in Figure 9.5 in the $r \times c$ contingency Table 9.5 that contains counts of the number of occurrences of each of the four types over the $n(n-1)/2$ distinct, unordered pairs in $O \times O$. The elements $[n_{ij}]$ of the *contingency matrix* $N = UQ^T$

Table 9.5 Contingency table comparing crisp partitions U and Q.

	Partition $Q \in M_{hcn}$ $\mathbf{Q}_{(j)} = j$th row of Q	Row sums →
Partition $U \in M_{hrn}$ $\mathbf{U}_{(i)} = i$th row of U	$N = \begin{bmatrix} n_{11} & n_{12} & \cdots & n_{1c} \\ n_{21} & n_{22} & \cdots & n_{2c} \\ \vdots & \vdots & & \vdots \\ n_{r1} & n_{r2} & \cdots & n_{rc} \end{bmatrix} = UQ^T$	$n_{1\bullet} = \sum_{j=1}^{c} n_{1j}$ $\vdots$ $n_{r\bullet} = \sum_{j=1}^{c} n_{rj}$
Column sums ↓	$n_{\bullet 1} = \sum_{i=1}^{r} n_{i1}$, $n_{\bullet 2} = \sum_{i=1}^{r} n_{i2}$, $\cdots$, $n_{\bullet c} = \sum_{i=1}^{r} n_{ic}$	$n_{\bullet\bullet} = \sum_{j=1}^{c}\sum_{i=1}^{r} n_{ij} = n$

are used to compute four numbers:

$$a = \frac{1}{2}\sum_{i=1}^{r}\sum_{j=1}^{c} n_{ij}(n_{ij} - 1);\ \text{number paired in U and Q; type (1);} \tag{9.7a}$$

$$d = \frac{1}{2}\left(n^2 + \sum_{i=1}^{r}\sum_{j=1}^{c} n_{ij}^2 - \left(\sum_{i=1}^{r} n_{i\bullet}^2 + \sum_{j=1}^{c} n_{j\bullet}^2\right)\right);\ \text{\# not paired U nor Q; type (2);} \tag{9.7d}$$

$$b = \frac{1}{2}\left(\sum_{j=1}^{c} n_{\bullet j}^2 - \sum_{i=1}^{r}\sum_{j=1}^{c} n_{ij}^2\right);\ \text{number paired in Q, not U; type (3);} \tag{9.7b}$$

$$c = \frac{1}{2}\left(\sum_{i=1}^{r} n_{i\bullet}^2 - \sum_{i=1}^{r}\sum_{j=1}^{c} n_{ij}^2\right)\ \text{number paired in U, not Q; type (4).} \tag{9.7c}$$

Equations (9.7a)–(9.7d) and the n_{ij} are the building blocks of many similarity measures that can be used to compare (or match) U and Q. These four equations simply count the number of occurrences amongst the $n(n-1)/2$ pairs of each of the four types of unordered pairs shown in Figure 9.5. Thus, (a) at Equation (9.7a) is the number of pairs which are paired in some cluster of U and some cluster of Q; d is the number of pairs not paired in U or Q, and so on. Equations (9.7) are presented and indexed in the order (a, d), (b, c) to indicate that the sums (a + d) and (b + c) are interpreted, respectively, as (the total number of) *agreements* and *disagreements* between U and Q. For example, Rand's (1971) index, $V_{RI}(U|V) = \frac{(a+d)}{(a+d)+(b+c)}$, is the (number of agreements) divided by the (number of agreements + number of disagreements).

Confusion can arise about the four coefficients in Equations (9.7) because of their similarity to the numbers *a*, *b*, *c*, and *d* which are used to compute the first five similarity coefficients in Table 8.1 from binary p-vectors **x**, **y** in $0, 1^p$. There are many good references on contingency tables. One of the best is Everitt's (1977) book. If you compare Table 1.2 on p. 4 of Everitt's book to Table 9.5 above or to Table 4.5

Table 9.6 Contingency tables for vector variables (X, Y) and crisp partitions (U, Q).

9.6(a): Table 2.3 (Everitt, 1977) = Volume Table 8.1				**9.6(b)**:Table 4.6 (Jain and Dubes, 1988) = Volume Table 9.6			
Comparing (X,Y) for independence or vectors $\mathbf{x},\mathbf{y} \in \{0,1\}^n$ for similarity				Comparing $U \in M_{hrn}$ to $Q \in M_{hcn}$ for similarity			
			Sums				Sums
	a	b	$a+b$		a	b	a+b
	c	d	$c+d$		c	d	c+d
Sums	$a+c$	$b+d$	$a+c+b+d = n$	Sums	a+c	b+d	a+c+b+d = n(n-1)/2
a = # of 1-1 matches $\sim x_i = y_i = 1$				a = # pairs matched in U and Q			
b = # of 1-0 mismatches $\sim x_i = 1,\ y_i = 0$				b = # pairs matched in Q, not U			
c = # of 0-1 mismatches $\sim x_i = 0,\ y_i = 1$				c = # pairs matched in U, not Q			
d = # of 0-0 matches $\sim x_i = y_i = 0$				d = # pairs unmatched in U and Q			

in Jain and Dubes (1988), you will see that all three tables ***appear to be*** identical – but they are not. As above, n_{ij} is the count of the number of items in the *i*th cluster (or category) in the first variable (U or ***X***) that also occur in the *j*th cluster (or category) in the second variable (Q or ***Y***). All three tables can be converted to an alternate 2 × 2 contingency table, but, at this point, great caution needs to be exercised because the entries notated as (a, b, c, d) or (a, b, c, d) in the 2 × 2 tables correspond to different entities in the two cases. The main difference (yellow cells) is the sum of all four terms.

In Everitt's case, the values are used for testing n-samples of variate ***X*** against variate ***Y*** for statistical independence, and our case, they are used for comparing partition U to partition Q for cluster similarity. The constants (a, b, c, d) appear just above Table 8.1 and are shown again on the left side of Table 9.6. Unfortunately, the same variable names (a, b, c, d) are almost always used in both contexts. *In this book the two usages are identified with different fonts as in Table 9.6.*

To add to the confusion, the same functional forms and names are used for various external cluster validity indices. For example, the Jaccard index at equation (8.3d), $s_J(\mathbf{x}, \mathbf{y}) = a/(a + b + c)$, assesses the similarity between binary vectors $\mathbf{x}, \mathbf{y} \in \{0, 1\}^n$ using the coefficients (a, b, c, d) in Table 9.6(a) *over n pairs of values*. This usage of (a, b, c, d) coincides with that made by the values in Table 9.6(a) for contingency analysis of (the independence of) variates ***X*** and ***Y*** for the special case of binary variables with n components. The Jaccard index of *partition similarity* is $s_J(U, Q) = a/(a + b + c)$. This index has the same functional form as the index in Table 8.1, but uses coefficients (a, b, c, d) as in equations (9.7) and Table 9.6(b), which count the four types *over n(n−1)/2 pairs of objects in the clusters in U and Q*. ***Please*** be really careful when you see these four variable names in a book or paper. Find the definitions of them that apply to the case in hand.

The important thing to note about Table 9.5 is that the contingency matrix N can be calculated as $N = UQ^T$ for *any* U and Q that are conformable (cf., Anderson *et al.* 2010). Entry n_{ij} from $N = UQ^T$ is the number of objects common to classes i in U and j in Q. Here is an example that illustrates the counts in N.

Example 9.2. Comparisons in U and Q with N=UQ^T.

Let $U = \begin{bmatrix} 1 & 1 & 1 & 0 & 0 \\ 0 & 0 & 0 & 1 & 1 \end{bmatrix}$ and $Q = \begin{bmatrix} 1 & 1 & 0 & 0 & 0 \\ 0 & 0 & 1 & 0 & 0 \\ 0 & 0 & 0 & 1 & 1 \end{bmatrix}$. Compute the contingency matrix $N = UQ^T = \begin{bmatrix} 2 & 1 & 0 \\ 0 & 0 & 2 \end{bmatrix}$.

Denote the crisp clusters in the rows of U and Q with subscripts; so, for example, $\mathbf{U}_{(1)} = (1, 1, 1, 0, 0)$. Clusters $\mathbf{U}_{(1)}$ and $\mathbf{Q}_{(1)}$ in the first row of U and Q have two objects in common, $\{o_1, o_2\}$, so $n_{11} = 2$; clusters $\mathbf{U}_{(1)}$ and $\mathbf{Q}_{(2)}$ have 1 in common, $\{o_3\}$, so $n_{12} = 1$; and $\mathbf{U}_{(1)}$ and $\mathbf{Q}_{(2)}$ have none in common, so $n_{13} = 0$. And, similarly, for row 2 of N. This is how the entries of N capture each of the four types of pairs illustrated by Figure 9.10.

To compute the four coefficients in Equations (9.7): there are 2 paired in U and Q, i.e., type (i), $\{\{o_1,o_2\},\{o_4,o_5\}\}$, 6 in type (ii), $\{\{o_1,o_4\},\{o_1,o_5\},\{o_2,o_4\},\{o_2,o_5\},\{o_3,o_4\},\{o_3,o_5\}\}$, none in type (iii), and 2 in type 4 (iv), $\{\{o_1,o_3\},\{o_2,o_3\}\}$.

Many well-known crisp comparison indices are functions of the values in Table 9.5. For example, indices of this type include those of Yule (1900), Jaccard (1908), Rand (1971), Fowlkes and Mallow (1983), and Goodman and Kruskal (1954). All of these measures have been used as similarity indices for pairs of partitions; and in the context of cluster validity, they are *pair-based CVIs* (PB-CVIs). In common with all CVIs, these indices have seen success and failure with different algorithms and data sets. One of the most consistent performers and important members in this family is the max-optimal (↑) *adjusted Rand index* (ARI) of Hubert and Arabie (1985).

Definition 9.2. The Adjusted Rand index (ARI) of Hubert and Arabie (1985).

Let $U \in M_{hrn}$ and $Q \in M_{hcn}$ be crisp partitions of n objects, and let $N = UQ^T$. Compute a, b, c, and d with Equations (9.7). The crisp max-optimal (↑) ARI of Hubert and Arabie is

$$V_{ARI}(U|Q) = \frac{2(ad - bc)}{(a + b)(d + b) + (a + c)(d + c)} \tag{9.8}$$

The "adjustment" of the Rand index that makes it the ARI in Equation (9.8) is a correction for chance based on the formula exhibited in Chapter 6 at Equation (6.51), viz., for any CVI V,

$$\widehat{V} = \frac{V - \mu_V}{\max(V) - \mu_V}, \quad \mu_V = E[V] \tag{9.9}$$

The calculation of Equation (9.9) requires a statistical null hypothesis for the computation of the expected value $\mu_V = E[V]$ and applies to the case where V is max-optimal (↑). In this case, V_{ARI} is bounded above by 1, will be 0 when the index is equal to its expected value, and can take negative values in extreme cases. Anderson *et al.* (2010) showed that Equation (9.8) can be used when $U \in M_{pcn}$ is soft by first computing

$N = UQ^T$, followed by the calculations at Equation (9.7) and then Equation (9.8).The resultant external (↑) indices are soft ARIs (fuzzy or possibilistic). The fuzzy case will be called $V_{ARI,s}$ in the sequel. Four other soft ARIs are discussed by Anderson *et al.* (2010). Different extensions of this index are discussed in Hullermeier *et al.* (2011).

Example 9.3. A deficiency of internal best c partition validation.

Here is an example of what is wrong with best c evaluation of candidate partitions using internal CVIs. Suppose we have a synthetic data set with six objects in each of $c_t = 2$ HPOV clusters so that the true ground truth partition of the $n = 12$ objects is

$$U_1 = \begin{bmatrix} 1 & 1 & 1 & 1 & 1 & 1 & 0 & 0 & 0 & 0 & 0 & 0 \\ 0 & 0 & 0 & 0 & 0 & 0 & 1 & 1 & 1 & 1 & 1 & 1 \end{bmatrix} \tag{9.10}$$

Imagine that an internal CVI examines some set of crisp partitions in CP and selects

$$U_2 = \begin{bmatrix} 1 & 0 & 0 & 0 & 0 & 0 & 0 & 0 & 0 & 0 & 0 & 0 \\ 0 & 1 & 1 & 1 & 1 & 1 & 1 & 1 & 1 & 1 & 1 & 1 \end{bmatrix} \tag{9.11}$$

as the best CPOV choice amongst the candidates. Then $c_w = c_t = 2$; so the best c strategy declares this a success for the internal CVI that selected U_2. But look at the cluster structure suggested by U_2: it is very different from the apparent HPOV cluster structure seen in the ground truth partition U_t. What's wrong – how does this happen? The best c strategy concentrates entirely on finding the target number of clusters with no regard to whether the clusters in the partition selected are in any way related to the structure indicated by the ground truth labels. Suppose, for example, the set CP also contained

$$U_3 = \begin{bmatrix} 1 & 1 & 1 & 1 & 1 & 0 & 0 & 0 & 0 & 0 & 0 & 0 \\ 0 & 0 & 0 & 0 & 0 & 1 & 0 & 0 & 0 & 0 & 0 & 0 \\ 0 & 0 & 0 & 0 & 0 & 0 & 1 & 1 & 1 & 1 & 1 & 1 \end{bmatrix} \tag{9.12}$$

U_3 is a much better match to the structure represented by U_t than the selected partition U_2, but it can't be declared a winner for the target value $c_t = 2$; so U_3 could never achieve success under the best c scenario with internal CVIs. Similarly, we might employ FCM or GMD/EM to obtain a set of soft candidate partitions. Suppose, for example, that the CPOV candidates included the soft partitions U_{2f} and U_{3f}:

$$U_3 = \begin{bmatrix} .9 & .1 & .2 & .1 & .2 & .1 & .1 & .2 & .3 & .2 & .2 & .1 \\ .1 & .9 & .8 & .9 & .8 & .9 & .9 & .8 & .7 & .8 & .8 & .9 \end{bmatrix} \tag{9.13}$$

$$U_{3f} = \begin{bmatrix} .9 & .9 & .8 & .9 & .8 & 0 & .05 & 0 & .05 & 0 & .05 & .15 \\ 0 & .05 & .05 & .05 & .1 & .95 & .05 & .1 & .05 & .1 & .25 & .05 \\ .1 & .05 & .15 & .05 & .1 & .05 & .9 & .9 & .9 & .9 & .7 & .8 \end{bmatrix} \tag{9.14}$$

U_{2f} might be the winner of an internal, soft CVI, best c competition for fuzzy or probabilistic candidates, even though U_{3f}, with three clusters, is clearly superior to U_{2f} in terms of matching the HPOV structure implied by U_t. Hardening these two partitions with **H** at Equation (3.4) would produce the partitions at Equations (9.11) and (9.12) so that won't get around the problem.

A great advantage of pair-based comparison indices such as the crisp and soft versions of the adjusted Rand index $V_{ARI}(U|Q)$ is that they can be used to make direct evaluations of the candidates against the reference partition without regard to the numbers of clusters they contain. For these four partitions, the values are

$$\text{Crisp case} \quad V_{ARI}(U_2|U_t) = 0: \qquad V_{ARI}(U_3|U_t) = 0.84$$
$$\text{Soft case} \quad V_{ARI,s}(U_{2f}|U_t) = -0.07: \qquad V_{ARI,s}(U_{3f}|U_t) = 0.46$$

Recalling that the crisp ARI is valued in [0, 1] and is max-optimal, we see that for the crisp case, the ARI regards the 3-partition at Equation (9.12) vastly superior (a jump of 84% in its value) to the 2-partition at Equation (9.11). Similarly, the soft ARI $V_{ARI,s}$ rates the 3-partition U_{3f} at Equation (9.14) as a much better match to U_t at Equation (9.10) than the partition U_{2f} at Equation (9.13), showing a jump of (something like) 50% toward its maximum. This illustrates how *structural matching* using an external CVI can overcome a clear deficiency of the best c strategy that is so often used to rank internal CVIs.

There is another group of paired-comparison external CVIs that are based on information theoretic constructs. For crisp partitions $U \in M_{hrn}$ and $Q \in M_{hcn}$, these indices are defined in terms of the entries of the matrix $N = UQ^T$ as follows:

$$H(U) = -\sum_{i=1}^{r} \frac{n_{i\bullet}}{n} \log \frac{n_{i\bullet}}{n} \quad \text{(probabilistic) entropy}^{25}; \tag{9.15}$$

$$H(U,Q) = -\sum_{i=1}^{r}\sum_{j=1}^{c} \frac{n_{ij}}{n} \log \frac{n_{ij}}{n} \quad \text{(joint) entropy}; \tag{9.16}$$

$$H(U|Q) = \sum_{i=1}^{r}\sum_{j=1}^{c} \frac{n_{ij}}{n} \log \frac{n_{ij}/n}{n_{\bullet j}/n} \quad \text{(conditional entropy)}; \tag{9.17}$$

$$MI(U,Q) = \sum_{i=1}^{r}\sum_{j=1}^{c} \frac{n_{ij}}{n} \log \frac{n_{ij}/n}{n_{i\bullet}\, n_{\bullet j}/n^2} \quad \text{(mutual information)}. \tag{9.18}$$

The intuitive basis for *mutual information* (MI) is that this value tells us how much knowing either of the partitions reduces our uncertainty about the other one. These four equations all depend on the elements of the contingency matrix N. Hence, functions built from these four functions also depend on N. There are many such functions: for example, Nguyen *et al.* (2010) list 22 variants of PB-CVIs built from the functions at Equations (9.15)–(9.18). We will illustrate the use of these three:

$$V_{NMI_{Max}}(U|Q) = \frac{MI(U)}{\max\{H(U), H(Q)\}} \quad \text{(normalized max. MI)}; \tag{9.19}$$

$$V_{NMI_{Min}}(U|Q) = \frac{MI(U)}{\min\{H(U), H(Q)\}} \quad \text{(normalized min. MI)}; \tag{9.20}$$

$$V_{VI}(U|Q) = H(U,Q) - MI(U,Q) \quad \text{(variation of information)}. \tag{9.21}$$

---

[25]Be careful not to confuse H(U), Shannon's entropy of the crisp partition $U \in M_{hcn}$, with $V_{PE}(U)$, the fuzzy partition entropy of $U \in M_{hcn}$ at Equation (6.60), which is constantly 1 on all crisp partitions.

These three external CVIs have the following properties: $V_{NMI_{Max}}$ and $V_{NMI_{Min}}$ are (↑), and both are valued in [0, 1]. The *variation of information* (VI) V_{VI} is (↓) and is valued in [0, log(n)]. The next subsection defines a scheme for using external CVIs in general and in examples to follow. We will use the soft versions of these indices when evaluating soft partitions of data sets that are available through the device of first computing the contingency matrix $N = UQ^T$.

9.5.2 External Best Match (Best U, or Best E) Validation

The type of validation described in this subsection is inspired by the work reported in Arbelaitz *et al.* (2013). These authors argue that when the data are labeled, successfully matching a candidate partition to an external CVI that uses true (U_t) or ground truth labels (U_{gt}) to assess the quality of the candidates is a better way to validate candidates than the best c approach. The general idea is similar to best c evaluation of internal indices in that we again have a race between competing CVIs, but, now, they will be using the structural information possessed by a reference matrix Q as a criterion for successful retrieval of a "best U" from the set CP.

Definition 9.3. Best E (aka best U): $U \in M_{hcn}$ is matched to Q with external, pair-based, CVIs.

Given S *pair-based* external CVIs: $CVI_{ext} = \{V_{ext,j} : 1 \le j \le S\}$. Let the labeled data have ground truth $Q = U_t \in M_{hc_tn}$ (synthetic); or $Q = U_{gt} \in M_{hc_{gt}n}$ (real data).

(i) Generate $CP = \{U \in M_{hcn} : c_m \le c \le c_M\}$ from X or D; (E.1)

(ii) For each j, compute $\{V_{ext,j}(U|Q) : U \in CP\}$; determine a "winner" partition $\widehat{U}_j \in M_{f\hat{c}_jn}$; (E.2)

(iii) If $\widehat{U}_j = Q$ declare this test a "success" for $V_{ext,j}$; (E.3)

(iv) **Optional**: repeat (E.2)–(E.4) K times and aggregate the number of successes for each $V_{ext,j}$. (E.4)

Forewarned is forearmed. Please note carefully that the method outlined in Definition 9.3 assumes that the external CVIs are pair-based indices. If not, you will have to attend to the registration problem (cf., Example 6.3) for each evaluation in Equation (E.3).

If the candidate partitions in CP are fuzzy or probabilistic, there are two choices for generalizing the best E method. An obvious way to use this approach is to harden all of the soft partitions with **H** at Equation (3.4) and then apply the method in Definition 9.3 directly. An arguably better option is to use the soft U and crisp Q to find the entries of the contingency matrix $N = UQ^T$ and then use the soft generalizations of any of the external pair-based CVIs that seem attractive to you.

9.5.3 The Fantastic Four Use Best E Evaluations on Labeled Data

This section contains some examples that compare various aspects of the four basic clustering models and algorithms presented in Chapters 6–8. We begin with the following.

Example 9.4. The Fantastic Four meet the Yang data set.

Problem 4 of Chapter 1 asked you how many HPOV clusters the synthetic "Yang" data set in Figure 1.7 contained. The presumptive answer is $c = 4$. Figure 9.11(a) repeats the data set, and Figure 9.11(b) is the iVAT image of "Yang" made with the Euclidean distance matrix $D_{E,Yang}$ of the 2500 points in the data. The four colors correspond to the labeling of the four blocks in the image. The points in each subset (each letter) are the ground truth target labels of the data, i.e., they comprise the matrix $Q = U_{GT,Yang} \in M_{h,4,2500}$ in E.2 and E.3.

The iVAT image supports the notion that there are indeed four HPOV clusters in "Yang." This is mildly surprising, considering the proximity of the letters "a," "n," and "g" to each other in view 9.11(a), but it is a nice surprise, No? The isolated lowermost dark diagonal block in Figure 9.11(b) corresponds to the letter "Y" in the data, which appears somewhat further away from the other three letters than they are from each other.

Figure 9.12 shows typical results (at $c = 4$) of processing the Yang data set with each of the Fantastic Four using their default parameters. The soft partitions from FCM and GMD/EM are hardened with the function **H** at Equation (3.4) to get the crisp clusters shown by different colors in Figure 9.12. The two

(a) Yang data set. (b) iVAT image of $D_{E,Yang}$.

Figure 9.11 Scatterplot and iVAT image of the data set "Yang."

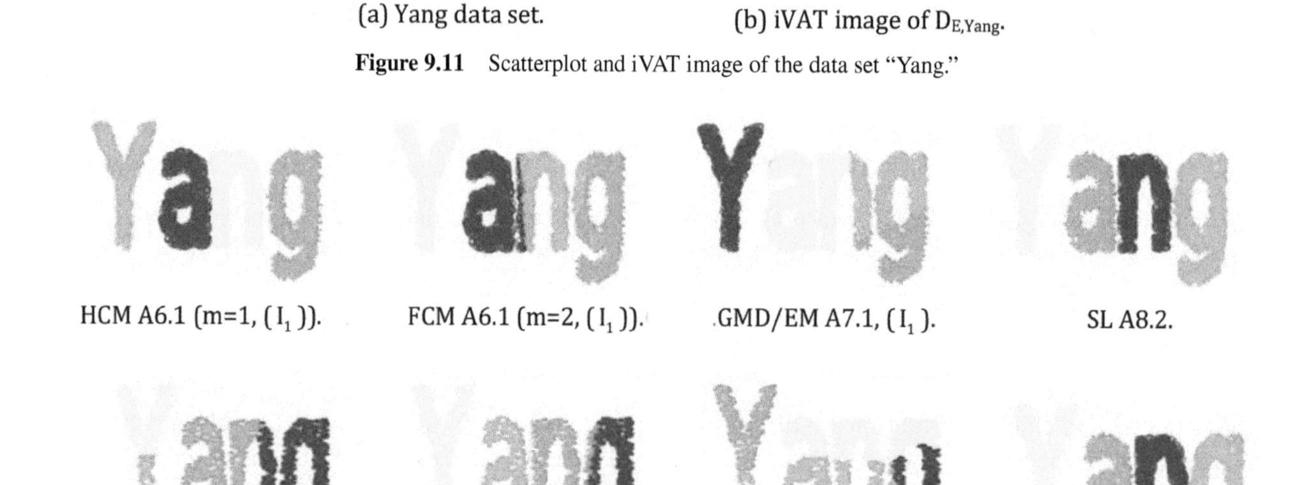

HCM A6.1 (m=1, (I_1)). FCM A6.1 (m=2, (I_1)). GMD/EM A7.1, (I_1). SL A8.2.

HCM A6.1 (m=1, (I_2)). FCM A6.1 (m=2, (I_2)). GMD/EM A7.1, (I_2)). SL A8.2.

Figure 9.12 Typical outputs of the Fantastic Four on the Yang data set.

rows of images correspond to two different initial guesses (I_1 for row 1 and I_2 for row 2) used by all three AO algorithms. SL will produce the same (rightmost) image no matter how many runs are made, and this algorithm does perfectly on the whole data set – a not unexpected result. The first run (upper row) of HCM also produces a perfect result – *very unexpected.* For FCM in the upper row, the left side of the "n" bleeds into the "a," and the left side of the "g" is assigned to the "n." The GMD result in the first row is visually worse than FCM, as you can see by comparing the colors of the clusters for these two outputs.

HCM (c-means) performs much more in line with expectation using the initial guess (I_2) corresponding the terminal partition shown in the second row, by splitting the "Y" and "n." Algorithm 7.1, GMD/EM, does this even more dramatically. The FCM outputs in the upper and lower rows are almost identical. So, in this example, SL behaves as predicted by the geometry shown in Figure 6.7 and provides CPOV clusters that most closely match the presumptive HPOV clusters in the data. The three AO algorithms all split the apparent clusters in different ways in these two views. So, for 100 different initializations, the three objective function methods might provide quite a few different interpretations of CPOV structure in this data set. Let's see if that's true.

What happens if we use various external CVIs per Definition 9.3 and repeat this experiment $K = 100$ times at values of $c = 2, 3, 4, 5$, and 6 on the Yang data (that is, $C_m = 2$, $C_M = 6$ in E.1, $Q = U_{GT,Yang} \in M_{h,4,2500}$ in E.2/3)? Figure 9.13 compares the Fantastic Four this way using four *crisp* external CVIs: V_{ARI} at Equation (9.8); $V_{NMI_{max}}$ at Equation (9.19); $V_{NMI_{min}}$ at Equation (9.20); and V_{VI} at Equation (9.21) with *hardened* partitions for FCM and GMD/EM. The vertical axis in Figures 9.13 and 9.14 is labeled "success rate," whose meaning is defined in E.3. This value is the number of times in $K = 100$ different runs that the CPOV generated clusters $\hat{U}_j \in M_{f\hat{c}_jn}$ chosen by the CVI were the best match to the ground truth partition $U_{GT,Yang}$ of the data. For example, the success rate for c-means matching of $\hat{U}_j \in M_{f\hat{c}_jn}$ to $U_{GT,Yang}$ using $V_{NMI_{max}}$ in Figure 9.13(a) is 0.88. This means that in 88 of the 100 trials with HCM, the best match to the ground truth partition in CP is the hard 4-partition of the input data.

Figure 9.13 invites several observations. First, note that two of the four clustering algorithms (HCM, i.e.,k-means, and SL) are crisp in both sets of experiments; so the success rates for these two algorithms should be identical in views 9.13(a) and 9.13(b), and they are. The four external CVIs rate the candidate partitions quite differently. For example, $V_{NMI_{min}}$ enjoys NO successes for SL or FCM in 100 tries, as can be seen in Figure 9.13(a), where there is no vertical bar for either of these algorithms. The crisp variation of information V_{VI} does match the true partition of Yang in some cases for all four hard versions but has no success (no blue bar) for FCM in the soft case.

Figure 9.13(b) looks at the same experiment using soft versions of these four indices on the soft partitions found by FCM and GMD/EM. This enables us to see how much information -if any -is lost during the hardening procedure for FCM and GMD/EM. The hardened versions of U for the two soft algorithms are made from the soft versions using **H** at Equation (3.4). The crisp and soft versions of V_{ARI} and $V_{NMI_{min}}$ provide the most matches. For example, FCM and SL both enjoy 100 matches to $U_{GT,Yang}$ in 100 trials. HCM (c-means) doesn't lag far behind, at about 88%, and c-means is the only algorithm that gets a good match score with $V_{NMI_{min}}$. Finally, GMD/EM never gets above about 32% for either the hard or soft trials.

What can we conclude from this example? Certainly V_{ARI} and $V_{NMI_{min}}$ rank the candidates against the HPOV ground truth partition of the Yang data more consistently and more accurately than V_{VI} and $V_{NMI_{min}}$. Does this suggest that V_{ARI} and $V_{NMI_{min}}$ are "better" CVIs than V_{VI} and $V_{NMI_{min}}$? Or, does this suggest that the CPOV clusters found by FCM and SL are superior to those found by HCM and GMD/EM?

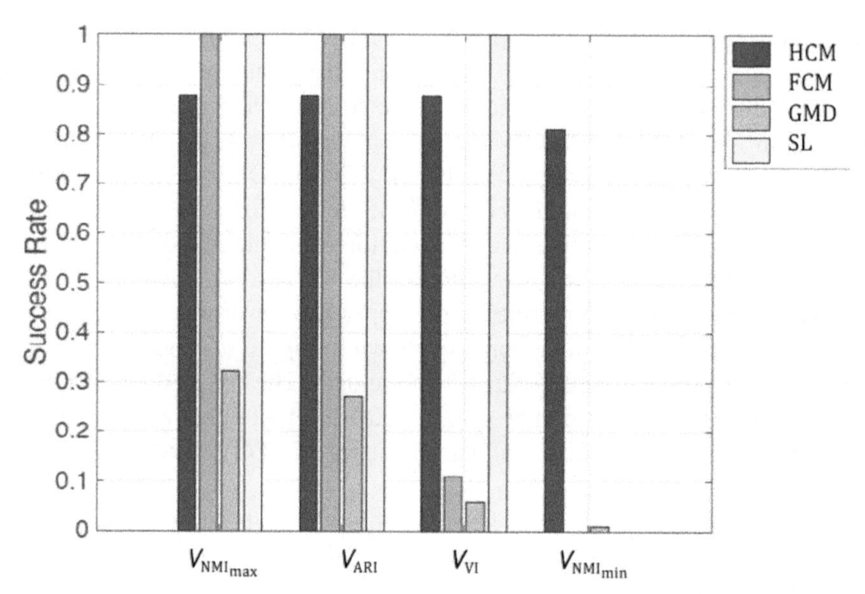

(a) Crisp CVIs on hardened partitions (FCM, GMD/EM) of "Yang."

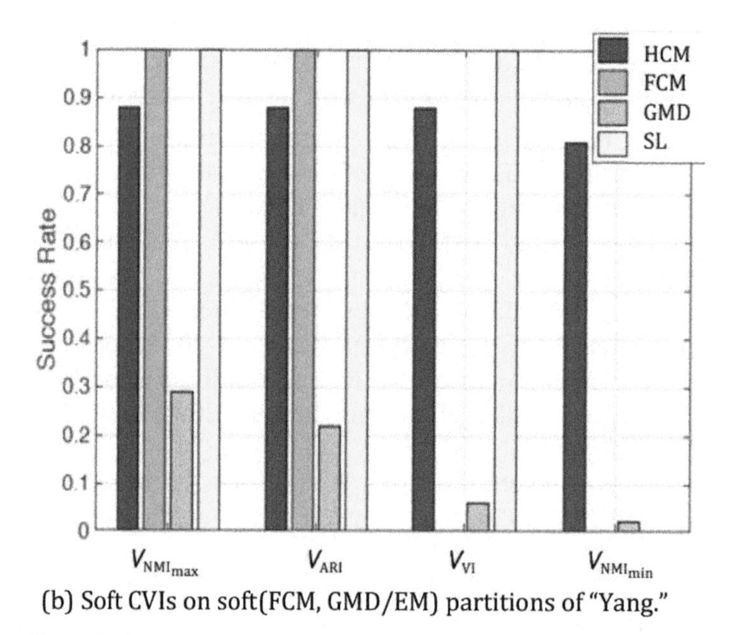

(b) Soft CVIs on soft(FCM, GMD/EM) partitions of "Yang."

Figure 9.13 Success rates for K=100 best-match trials on the Yang data.

How strong are conclusion about these questions? Well, this is one synthetic data set. You would have to process ***hundreds*** of other data sets and obtain roughly similar results before you could be fairly confident in these conclusions.

The next example is identical in spirit and style to Example 9.4. This example involves using best E validation on candidate partitions of the Iris data (Yikes -again?), which should be pretty familiar to you by now.

Example 9.5. The Fantastic Four meet the Iris data.

Figure 9.14 is similar to Figure 9.13 in every way except for the underlying data set, which is Iris for Figure 9.14. Since Iris is a real-world data set, the underlying reference partition for these CVI investigations is $U_{GT,Iris}$, the partition that is displayed at Equation (6.37). For this example, $c_m = 2$, $c_M = 6$ in E.1, $Q = U_{GT,Iris} \in M_{h,3,150}$ in E.2/3, K=100 in E.4.The first thing you may notice is that $V_{NMI_{min}}$ fails to match $U_{GT,Iris}$ to any of the 100 CPOV candidates generated by any of the three objective function clustering algorithms. Now look at the success rate for FCM in Figure 9.14(a), which is the best of the three algorithms (HCM, FCM, and GMD/EM) on hardened partitions of Iris – almost 100% as rated by the three CVIs $V_{NMI_{min}}$, V_{ARI}, and V_{VI}.

Compare the distributions for FCM (blue), EM (green), and k-means (purple) in Figure 9.14(a). They are all identical -so, the CPOV outputs of the three algorithms are quite stable in terms of their CVI matches to these three indices. The GMD/EM method does much better with Iris than it did with "Yang," coming in second at about 83%, followed closely by HCM at about 80%. Believing (actually, *guessing*, since these are 4D data) that the substructure in Iris is roughly cloud shaped confirms what we think is true about all three of these algorithms. Namely, that they do much better on cloud clusters than they do on clusters of more arbitrary shapes such as those possessed by "Yang." You will surely note that the SL algorithm generates NO successes for either case. The SL partition at c = 3 is shown in Figure 8.14, and comparing this partition to the ground truth partition of Iris should tell you that none of these CVIs will find a match for SL at c = 3; so its success rate will be 0 for all of them.

Turning to Figure 9.14(b), the most striking (and pretty surprising) difference between Figures 9.14(a) and 9.14(b) is that, in contradistinction to its behavior in the crisp case (98% success rate), the soft version of the variation of information index $V_{VI,s}$ doesn't match up with $V_{GT,Iris}$ for even one of the 100 candidates generated by FCM (0 successes). But the results for the top two indices, $V_{NMI_{max,s}}$ and $V_{ARI,s}$, are virtually identical to each other in Figures 9.14(a) and 9.14(b). Also note that in both Examples 9.4 and 9.5, there is no significant difference between evaluations made by these two indices before and after hardening of the soft partitions. This speaks well to the stability of these two indices under changes to the set of candidate partitions.

Taken together, Examples 9.4 and 9.5 support our belief that (i) HCM, FCM, and GMD/EM are at their best, and SL is at its worst, with cloud clusters, and (ii) conversely, SL becomes more reliable and the other three clustering algorithms less so as the clusters become less cloud-like in shape. Both the crisp and soft versions of V_{ARI} and $V_{NMI_{max}}$ yield more satisfying results than V_{VI} and $V_{NMI_{min}}$ in these two examples. But we have used just two data sets to compare CPOV partitions generated by the Fantastic Four. We would need *hundreds* of experiments like these with a variety of data sets in order to make statements like this with any real confidence.

The last two examples in this section are partially excerpted and combined from Lie *et al.* (2014, 2017) and don't involve ***100s*** of other data sets, but they do examine 18 sets of labeled object data. These two examples involve only the two soft clustering algorithms (FCM and GMD/EM). The examples are divided

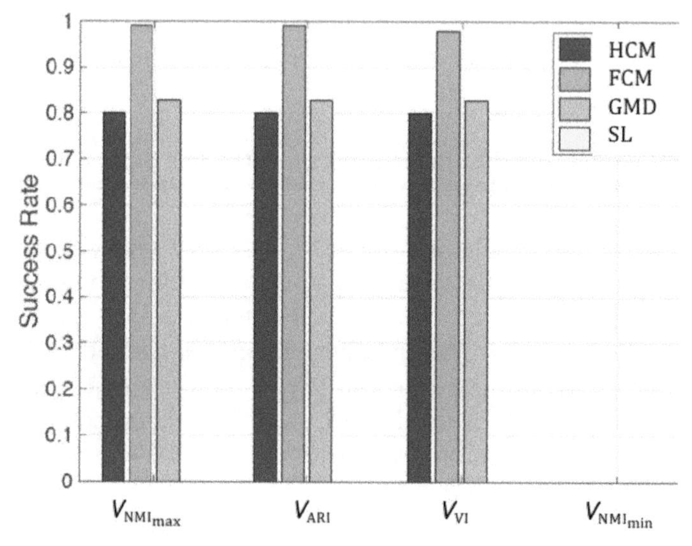

(a) Crisp CVIs on hardened partitions (FCM and GMD/EM) of Iris.

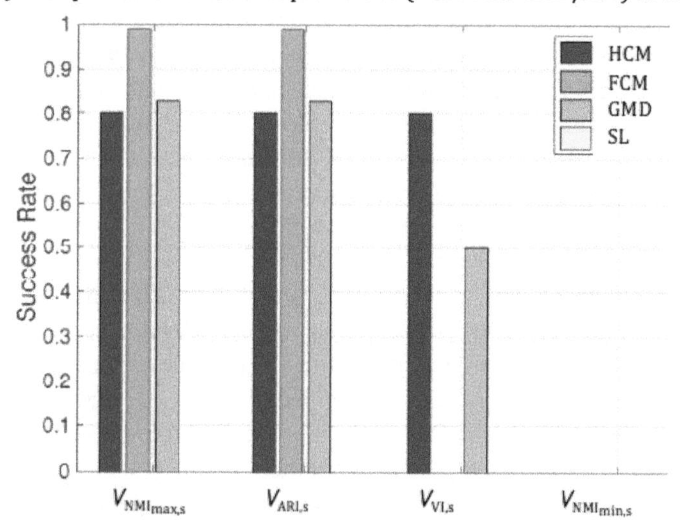

(b) Soft CVIs on soft (FCM and GMD/EM) partitions of Iris.

Figure 9.14 Success rates for K=100 best-match (best E) trials on the Iris data.

this way to emphasize the difference between using synthetic data (labeled by a mathematically constructed partition Q_{U_t}) and real data (labeled by a physical partition $Q = U_{GT}$).

Example 9.6. External CVI validation of FCM and GMD/EM clusters in 18 synthetic labeled data sets.

Figure 9.15 depicts 4 of the 18 synthetic sets of Gaussian clusters studied in Lie *et al.* (2014, 2016). All 18 data sets have a total of n = 1000 points each in two dimensions. Two groups of nine data sets were created from $c = 3$ mixture components, and nine more were created from $c = 5$ mixture components. For each of these two groups, one of two variables was manipulated to control the data. The overall set of 18 sets of HPOV Gaussian clusters is called G_{18}.

The group of nine that have variable separation between clusters but fixed cluster populations are created by distributing the means of the components along rays separated by equal angles according to the different number of clusters, centered equal distances from the origin at various radii r. For example, at $c = 3$, there are three means along lines separated by 120°, each r units from $(0; 0)^T$. Figures 9.15(a) and 9.15(b) show the effect of varying r with $c = 3$ and fixed cluster sizes of $n_i = 333$ points for each of the three clusters. Figures 9.15(c) and 9.15(d) show the data sets corresponding to $c = 5$ and all five clusters have $n_i = 200$ points for $r = 2$ and $r = 6$. The other fixed cluster size data sets are obtained by varying r from 1 to 5 at $c = 3$ and $r = 2$ to 6 at $c = 5$.

Conversely, the other nine synthetic data sets are created by varying the sizes of the clusters with r fixed. Specifically, the number of points in the first cluster, n_1, takes values from 1/6 to 5/6 of $n = 1000$ for $c = 3$ and 1/10 to 5/10 for $c = 5$. The remaining clusters have $n_i = (1 - n_1)/(c - 1); i \neq 1$ in all cases. Some of these data sets have HPOV clusters, e.g., Figures 9.15(b) and 9.15(d). But some of them have very overlapping distributions, and HPOV clusters are hard (if not impossible) to see, e.g., Figures 9.15(a) and 9.15(c). Eliminating the colors, which show the synthetic labels of the draws, makes this very clear: the clusters in panels (b) and (d) are easy to see, but the points in panels (a) and (c) just look like blobs, i.e., $c = 1$.

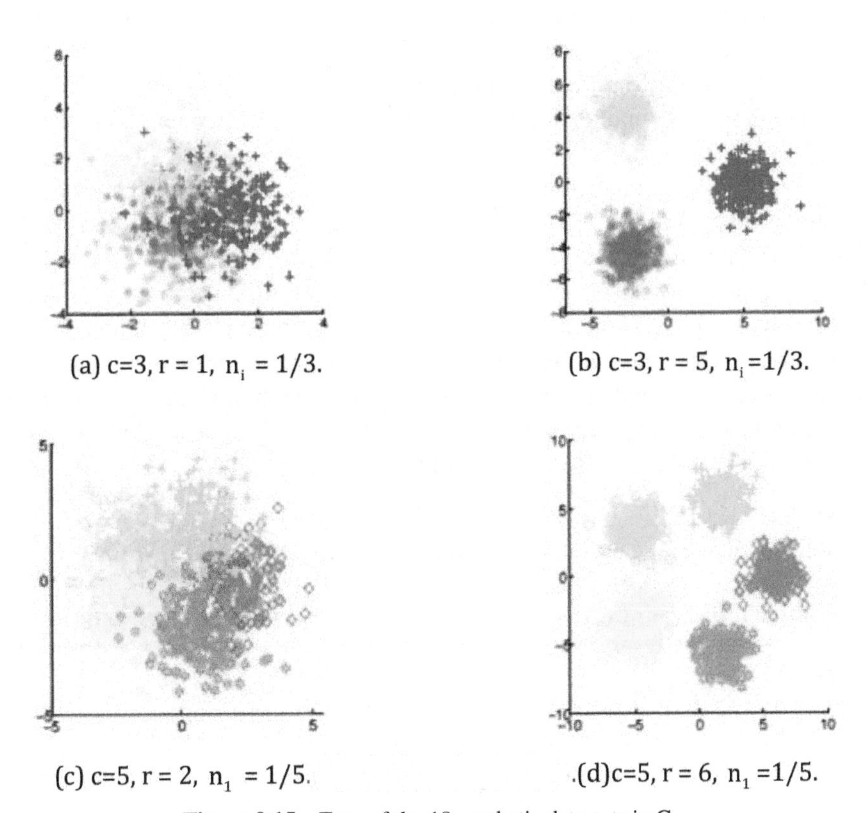

(a) $c=3, r = 1, n_i = 1/3$.

(b) $c=3, r = 5, n_i = 1/3$.

(c) $c=5, r = 2, n_1 = 1/5$.

(d) $c=5, r = 6, n_1 = 1/5$.

Figure 9.15 Four of the 18 synthetic data sets in G_{18}.

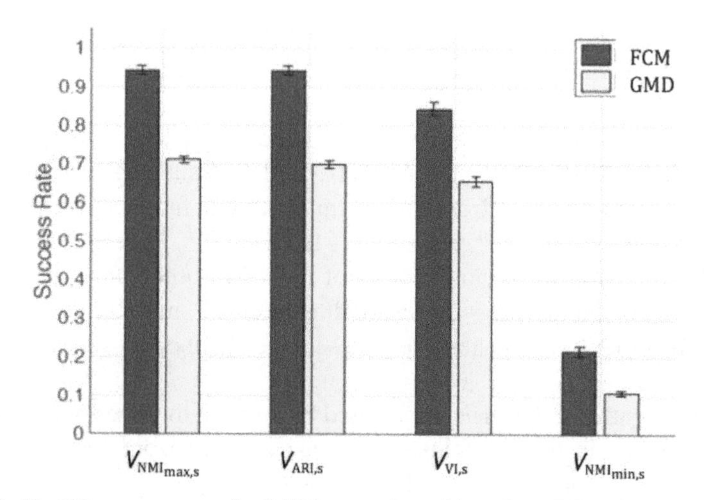

Figure 9.16 Best U success rates of soft CVIs on soft partitions from FCM and GMD/EM on G_{18}.

Figure 9.16 summarizes the evaluation by four soft CVIs of K=100 trials of FCM and GMD/EM on the 18 synthetic data sets in G_{18}. Both algorithms were initialized with the same sets of random vectors for the means in all trials. For GMD/EM, the initial covariance matrices are diagonal matrices whose ith diagonal element is the sample variance of the ith feature vector of each input data set. The initial prior probabilities for GMD/EM are all 1/c. There are 100 trials for each of the 18 sets; so these success rates are for 1800 trials. The error bars indicate the standard deviation across the 1800 trials. So, a value of 0.9 on the vertical axis, for example, means that the algorithm in question recovered $0.9{*}1800 = 1620$ candidates that matched the reference partition.

The indices $V_{NMI_{Max,s}}$ and $V_{ARI,s}$ produce almost identical ratings for the candidates generated by the two clustering algorithms. According to these two CVIs, FCM candidates match U_t in about 94% of the 1800 trials, while GMD/EM candidates match the true partitions in roughly 71% of the trials. The variation of information index $V_{VI,s}$ is somewhat less effective but exhibits roughly the same ratio of success rates for FCM and GMD/EM. Finally, $V_{NMI_{min,s}}$ agrees with U_t on just 21% of the FCM candidates, and 11% of the GMD/EM partitions. So, for the 18 Gaussian data sets, this experiment supports the assertion that $V_{NMI_{Max,s}}$ and $V_{ARI,s}$ are more reliable than $V_{VI,s}$, and, in turn, $V_{VI,s}$ is more reliable than $V_{NMI_{max,s}}$ in matching candidate partitions to the true known labels of these synthetic data sets. Overall, it appears that FCM candidates are somewhat better matches to U_t than GMD/EM partitions for the data sets in G_{18}.

Now let's look at soft validation of soft candidates generated by FCM and GMD/EM for some real data sets.

Example 9.7. External CVIs using nine real-world labeled data sets.

Table 9.7 lists the parameters of nine real data sets which were used by Lie *et al.* (2017). These data sets can be obtained from the UCI machine learning repository (Bache and Lichman, 2013).

Labeled data sets such as these are often used to test clustering algorithms by pretending that the labeled subsets that correspond to the crisp ground truth partition (U_{GT}) of the data are in fact clusters. But to belabor

Table 9.7 Nine real labeled data sets from the UCI ML website.

Name	n	p	C_{GT}
Sonar	208	60	2
Pima Diabetes	768	8	2
Heart-Statlog	270	13	2
Haberman	306	3	2
Wine	178	13	3
Vehicle	846	18	4
Iris	150	4	3
Zoo	101	17	7
Vertebral Column	310	6	3

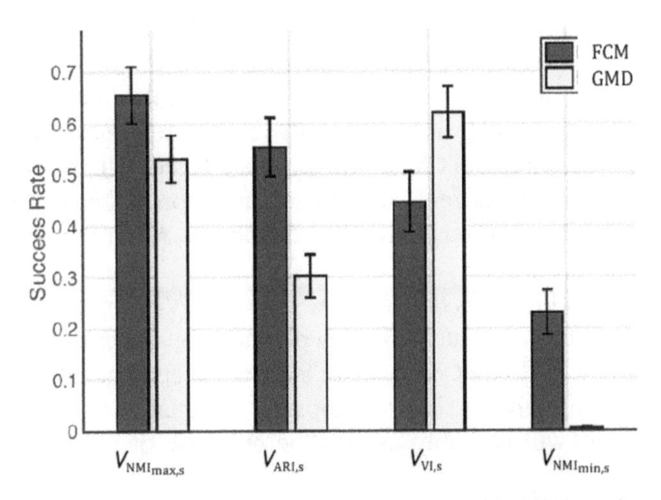

Figure 9.17 Best U success rates of soft CVIs for FCM and GMD/EM on nine real data sets.

the point, these labeled subsets are ***not*** HPOV clusters, and they may or may not correspond to CPOV clusters in the eyes of any clustering model and algorithm. Why use them? Two reasons. First, to compare the efficacy of different clustering algorithms at recovering the labeled subsets on the presumption that equally good clustering algorithms should be able recover similar CPOV clusters. And, more importantly, they can be used in conjunction with an external CVI that uses U_{GT} as its reference matrix to evaluate and rank internal CVIs.

Figure 9.17, rendered in the same way as Figure 9.16, summarizes evaluation of the candidates using the best E method with the four *soft* CVIs $V_{NMI_{Max,s}}$, $V_{ARI,s}$, $V_{VI,s}$, and $V_{NMI_{min,s}}$. The number of trials is again K = 100 trials of FCM and GMD/EM for each of these nine real-world data sets. All other computational protocols mentioned in Example 9.5 hold for these runs too. There are 100 trials for each of the nine sets; so these success rates are for 900 trials. The error bars again indicate the standard deviation across the 900 trials.

Some of the conclusions made in Example 9.6 are valid for this example too, but there are some differences. FCM outputs match the ground truth partitions better than GMD/EM when evaluated with $V_{NMI_{Max,s}}$ and $V_{ARI,s}$, but the reverse is seen for the variation of information index $V_{VI,s}$, which indicates more favorably for GMD/EM than FCM. Finally, $V_{NMI_{min,s}}$ again occupies last place and barely ticks the chart for GMD/EM partitions.

Examples 9.4–9.7 illustrate how you might try to choose an external CVI – without regard to any internal CVI -that provides reliable matches of a known partition for labeled data to CPOV partitions computed by (any) clustering algorithm. The available evidence from the examples so far suggests that the crisp and soft versions of $V_{NMI_{max}}$ and V_{ARI} are high-quality external CVIs. But there are many studies of external CVIs, and as the data and clustering algorithm vary, so do the rankings of the CVIs being tested. It is not too hard to find papers where numerical evidence is exactly the opposite of the conclusions stated here, so always approach what CVIs seem to tell you with a good helping of skepticism. Now we turn to the third type of CVI races, which in some sense is the most important aspect of the topic of cluster validity.

9.6 Choosing an Internal CVI Using Internal/External (Best I/E) Correlation

The best c and best E methods are most realistically regarded as validation of "fake clustering" results because neither of these methods addresses the question of how to evaluate candidate partitions in "real clustering," i.e., when the data are unlabeled. In this case, there is no reference matrix Q; hence, external CVIs are useless – we ***must*** use internal CVIs. Thus, you are faced with the conundrum presented by having to choose an internal CVI from the dozens, and dozens, and more dozens, of internal CVIs on offer in the literature. Each internal CVI has its protagonists and detractors. All of them work some of the time; none of them work all of the time.

For example, look at Table 9.3, which lists the very few internal CVIs discussed in this volume. Which of these would you choose to evaluate candidate partitions produced by a member of the Fantastic Four on unlabeled data? It's not so clear how to do this. Well, there is a way to approach this problem by combining the best c and best E methods using labeled data to choose a "best internal CVI." Once you are happy with your choice, it is the one you will probably use for the real case of unlabeled data that holds your attention.

This type of validation also follows the suggestion of Arbelaitz *et al.* (2013). We pointed out in Section 9.5.2 that these authors argue that matching a candidate partition to an external CVI that uses true (U_t) or ground truth labels (U_{GT}) to assess the quality of the candidates was better than simply optimizing an internal CVI on the candidates. They took this argument one step further and suggested that the relative quality of internal CVIs could be ranked by how often they agreed with the assessment of an external CVI. This is the basis for the following.

Definition 9.4. Best I/E match between internal and external CVIs using labeled (ground truth) data.

Assume S internal indices: $CV_{int} = \{V_{int,j} : 1 \le j \le S\}$: Assume one external index, say $V_{ext}(U|Q)$: finally, let the reference matrix be $Q = U_t \in M_{hc_tn}$ (synthetic data); or $Q = U_{GT} \in M_{hc_{GT}n}$ (real data).

(i) Generate $CP = \{U \in M_{hcn} : c_m \le c \le c_M\}$ from X or D; (I/E.1)

(ii) Compute $\{V_{ext}(U|Q) : U \in CP\}$; determine a "winner" partition $\hat{U}_{\hat{w}} \in M_{fc_{\hat{w}}n}$; (I/E.2)

(iii) Compute for each j: $\{V_{int,j}(U) : U \in CP\}$ determine a "winner" partition $\hat{U}_j^* \in M_{fc_j^*n}$; (I/E.3)

(iv) if $\widehat{U}_j^* = \widehat{U}_{\hat{w}}$ declare this test a "success" for $V_{int,j}$; (I/E.4)

(v) ***Optional*** do (I/E.2)–(I/E.4) K times and aggregate the number of successes for each internal CVI. (I/E.5)

The last example of this section illustrates the use of the I/E approach to select an internal CVI that shows the best match to the soft ARI $V_{ARI,s}$. Let us add two more internal CVIs designed for fuzzy partitions. The first of these is the fuzzy PBM index of Pakhira *et al.* (2004):

$$V_{PBMF,s}(U, \mathbf{V}; X) \left(\frac{\left(\sum_{k=1}^{n} \|\mathbf{x}_k - \bar{\mathbf{m}}\|\right) \left(\underbrace{\max}_{i \neq j}\{\|\mathbf{v}_i - \mathbf{v}_j\|\}\right)}{c \cdot \sum_{i=1}^{c} \left(\sum_{k=1}^{n} (u_{ik})^m \|\mathbf{x}_k - \mathbf{v}_i\|\right)} \right) \tag{9.22}$$

where $U \in M_{fcn}$, $\mathbf{V} = \{\mathbf{v}_1, \ldots, \mathbf{v}_c\} \in \Re^{cp}$, and $\bar{\mathbf{m}} = \sum_{k=1}^{n} \mathbf{x}_k/n$ is the grand mean of the input data set X. The factor $\sum_{k=1}^{n} \|\mathbf{x}_k - \bar{\mathbf{m}}\|$ in the numerator of Equation (9.22) is constant for a given data set, and, according to the authors, its role is to control the range of $V_{PBMF,s}$. For large values of n, this constant can becomes quite large. When the fuzzy clusters in U are compact and well-separated, the numerator should be big, and the denominator should be small; so $V_{PBMF,s}$ is max-optimal (↑), behaving somewhat like the inverse of V_{XB}. Pakhira *et al.* (2004) provide a comprehensive set of numerical examples that compare V_{XB} to $V_{PBMF,s}$

Another internal CVI that is similar to the partition coefficient and partition entropy in that it depends on only $U \in M_{fcn}$ is the Chen-Linkens index (2004). This index has terms that account for both compactness and separation,

$$V_{CL}(U) = \underbrace{\frac{1}{n}\sum_{k=1}^{n} \underbrace{\max_{i}}\{u_{ik}\}}_{\text{compactness}} - \underbrace{\left[\frac{2}{n(n-1)} \sum_{j=i+1}^{c} \sum_{i=1}^{c-1} \left[\frac{1}{n}\sum_{k=1}^{n} \min\{u_{ik}, u_{jk}\}\right]\right]}_{\text{separation}} \tag{9.23}$$

The first term in Equation (9.23) is the average of the maximum memberships in the n columns of U. This term increases as the label vectors in the columns of U become crisper. The presumption is that crisper labels indicate more compactness in the underlying data. The bracketed quantity in the second term is the average membership value in the intersection in the T1-norm of the fuzzy clusters corresponding to the *i*th and *j*th rows of U. This value decreases as the coupling between pairs of rows in U decreases, which, in turn, signifies better separation between the underlying objects. And the entire second term is the average of these $(c)(c-1)/2$ pairwise intersection measures.

The first term in Equation (9.23) is bounded above by 1, the second is bounded below by 0; so when U is crisp, this index is 1. At the other extreme, when every membership is $1/c$, $V_{CL}(\bar{U}) = 0$; so this index has range [0, 1], is independent of c and is max-optimal (↑), very much like the normalized partition coefficient of Roubens at Equation (6.69). Now let us turn to an example condensed from Bezdek *et al.* (2016a) of how to use I/E matching to select an internal CVI.

Example 9.8.I/E matching using the soft ARI as the external CVI.

This example uses the eight data sets shown in Table 9.8. The six real data sets are available from the UCI machine learning website. The synthetic data sets $X_{15,fws}$ and $X_{15,nws}$ are shown in Figure 6.9. Fuzzy c-means CPOV clusters from $c = 2$ to $c = 20$ were generated with MATLAB's implementation of FCM with $m = 2$, $T = 100$, $\varepsilon = 10^{-5}$, and termination was based on the absolute difference between successive values of the FCM objective function because the MATLAB package was used. Each data set was processed ten times. Initialization was done by randomly selecting c distinct vectors in the input data as initial cluster centers for each run.

Table 9.9 contains the results of applying I/E agreement to the eight test data sets in Table 9.8. The values reported in Table 9.9 are the number of times in 10 trials that each internal CVI selected the same hardened partition that the external soft ARI $V_{ARI,s}$ declared a best match to the true (synthetic data) or ground truth (real data) partitions. So, a value of 10 in this table is a perfect score for that index on that data set, while a 0 means that the internal index never recovered the same partition as the one best matched to the reference partition by $V_{ARI,s}$.

Since there are 8 data sets and 10 trials per data set, the total number of trials is 80; so a perfect overall score in this I/E match exercise would be 80. The last row in Table 9.9 shows the column sums (i.e., the overall number of successes in 80 tries) for each of the seven internal CVIs. The best performer is V_{PCR}, the modified partition coefficient of Roubens, which achieves 50 agreements (62.5%) in 80 tries between the internal and external votes. Rouben's CVI is closely followed by $V_{XB,2}$, the Xie-Beni index, which achieves 47 matches in 80 attempts. The worst index in this experiment is the $V_{PBMF,s}$ index, which only recovers the same partition as the soft ARI in 15 of 80 attempts.

By now you should harbor no illusions that the results in this example indicate a "best" CVI in any general sense of the word. Indeed, in the larger example in Bezdek (2016b) using four more synthetic data sets and four more internal CVIs, a fuzzy C index (not discussed in this volume) achieved the top results, 81 successes in 120 tests. The main point of this example is to illustrate how to use the I/E criterion to choose what seems to be a best internal CVI for use when validating CPOV partitions of unlabeled data, by using labeled data that affords ground truth information to an external CVI.

Table 9.8 Eight data sets for Example 9.8.

Synthetic	n	p	c_t
$X_{15,fws}$	5000	2	15
$X_{15,nws}$	5000	2	15
Real	n	p	c_{GT}
Iris	150	4	3
Breast cancer	699	8*	2
Pima diabetes	768	8	2
Vehicle	846	18	4
Wine	178	13	3
Seeds	210	7	3

* The breast cancer data set has nine features per vector, but some of the observations have missing values; so this feature was ignored.

Table 9.9 Number of I/E successes for 10 runs of FCM on 12 data sets.

Synthetic	V_{PE}	V_{PEB}	V_{PC}	V_{PCR}	V_{CL}	$V_{XB,2}$	$V_{PBMF,s}$
$X_{15,fws}$	0	10	9	10	10	8	8
$X_{15,nws}$	0	10	0	10	1	9	1
Real							
Iris	0	0	0	0	1	0	1
Cancer	10	10	10	10	10	10	1
Diabetes	10	0	10	10	10	10	1
Vehicle	0	0	0	0	1	0	1
Wine	10	0	10	10	10	10	1
Seeds	0	10	0	0	1	0	1
Column sum	30	40	39	50	44	47	15

9.7 Notes and Remarks

There are many papers that describe attempts to adjust an objective function so that it accounts for a specific (assumed) distributional property of subsets in feature vector data. There was a great flurry of papers along these lines in the period 1980–2000. The origins of most of these schemes were discussed in Section 9.3. If this topic piques your interest, see Section 2.3 in Bezdek *et al.* (1999), which contains a much deeper treatment of many of the early methods for generalized prototypes. The problem with many of these specialized methods, which are perhaps aptly described as boutique clustering methods, is that they only work well when the data really have subsets distributed in a way that matches the objective function that is designed to recognize them. And for most data sets, there is no way to ever know this, so the application of these schemes in the absence of HPOV intuition is chancy at best.

For example, fuzzy c-lines (FCL) is fine for data sets like the four lines in Figure 9.3(a). Visual HPOV inspection of this scatterplot tells you (i) how many CPOV clusters to seek and (ii) what kind of prototypes are most appropriate. But if you apply this algorithm to the three circular subsets in Figure 9.3(b), FCL will faithfully produce three lines that will have very little appeal as prototypes of the three circular subsets. This problem is greatly exacerbated when the input data are vectors in $\Re^p$ and $p > 3$, for, then, there is no chance of ascertaining, from a HPOV, what kind of prototypes might best match subsets in the data. Indeed, in this case, you will not even know how many to look for. Consequently, most of these papers don't have a very wide range of utility for real problems. The exception of note to this is 2D image data, where the distributions can at least be seen.

The list of external CVIs in Table 9.4 is a short one. Albatineh *et al.* (2006) compile a list of 22 crisp pair-based external comparison indices, which are interpreted as similarity measures on the partition pair (U, Q) in other contexts. Lie *et al.* (2016) extend this to 34 crisp pair-based indices. The method of soft extension using the transformation $N = UQ^T$ discussed by Anderson *et al.* (2010) applies to all of these indices. The Rand index has seen many adjustments, not just the ARI given at Equation (9.8). Hubert and Arabie (1985) have an excellent discussion of this topic. The ARI continues to enjoy success in many fields and has become somewhat of a standard external CVI (Steinley, 2004).

One of the most recent developments in the field of external CVIs is the discovery that some of them can be biased in different ways simply by changing the distribution of the ground truth partition. This surprising result is discussed in Lie *et al.* (2017), who give numerical examples and develop a corresponding theory for ground truth bias in some external pair-based CVIs based on a relationship between Rand's index and the quadratic entropy of the reference partition. This theoretical result carries over to any pair-based external CVI that is a function of the Rand index, including the ARI of Hubert and Arabie, Hubert's index, the quadratic variation of information, and the crisp indices of Rogers and Tanimoto, Mirkin, and Gower and Legendre.

9.8 Problems

1. Without making any computations, state where the location of four-point prototypes will be for the data set X_{4rings} in Figure 8.19(a). Do the same for the three rings data shown in Figure 9.3(b). This should clarify your thinking about the utility of non-point prototypes.

2. Rand's (1971) index is defined as $V_{RI}(U|Q) = \frac{(a+d)}{(a+d)+(b+c)}$ for the coefficients a, b, c, and d at Equation (9.7), crisp $U \in M_{hcn}$ and crisp reference partition $Q \in M_{hrn}$. Compute $V_{RI}(U|Q)$ for $U_1 = \begin{bmatrix} 1 & 1 & 0 & 0 & 0 \\ 0 & 0 & 1 & 1 & 1 \end{bmatrix}$ and $U_2 = \begin{bmatrix} 1 & 1 & 1 & 0 & 0 \\ 0 & 0 & 0 & 1 & 1 \end{bmatrix}$ using the ground truth reference matrix $Q = \begin{bmatrix} 1 & 1 & 0 & 0 & 0 \\ 0 & 0 & 1 & 0 & 0 \\ 0 & 0 & 0 & 1 & 1 \end{bmatrix}$. Which candidate partition matches Q better, U_1 or U_2? Does Rand's index offer any advice?

3. Continue exercise 2 by adding $U_3 = \begin{bmatrix} 1 & 0 & 0 & 0 & 1 \\ 0 & 1 & 1 & 1 & 0 \end{bmatrix}$. Which of the three CPOVs seems best now? Comment on your answer.

4. Recall the ground truth partition for Iris, $U_{GT,Iris}$, displayed at Equation (6.37). The SL 3-partition of Iris is shown in Figure 8.14. Call this partition $Q = U_{SL,Iris}$: it has a first row with 50 adjacent 1's; a second row with 98 adjacent 1's; and the third row contains 2 adjacent 1's for the two point cluster in the upper right corner of Figure 8.14. Compute the match between these two crisp partitions using each of the external CVIs $V_{RI}(U|Q)$, $V_{ARI}(U|Q)$, $V_{NMI_{Max}}(U|Q)$, $V_{NMI_{min}}(U|Q)$, and $V_{VI}(U|Q)$. Which of the five CVIs seems to provide the best information about the match?

5. For the data in Table 9.7, this statement was made: "these labeled subsets are ***not*** HPOV clusters." The basis of this mostly correct statement is these data sets have $p > 3$. But the Haberman data set has $n = 306$ vectors in $p = 3$ dimensions, so you can make a plot of it. Each data point has the label 1 or 2. Retrieve the Haberman data from the UCI data repository and make a plot of it with, and without colors for the labels. What can you say about HPOV clusters in this data set? Do the labeled subsets justify calling them clusters?

10

Alternating Optimization

Premature optimization is the root of all evil

–Donald Knuth (1998)

10.1 Introduction

The theory of alternating optimization presented in this chapter is quite general and pretty technical. Our target is to see how the general theory applies to the basic HCM, FCM, and GMD algorithms that have been discussed in Chapters 6 and 7. This chapter presents two theorems – one for local convergence of AO and one for global convergence of AO – that prescribe the conditions under which we can expect convergence of a very general AO scheme. It is useful and efficient to cover the general theory in some detail for three reasons. The theory in this chapter applies to two of the four clustering algorithms that have been discussed – viz., FCM and GMD/EM. We will see that HCM (k-means) is *not* an AO scheme that is covered exactly by the theory of this chapter, but it is still an AO method that utilizes the general idea of (t, s) splitting. And lastly, the general theory applies to a number of other pattern recognition algorithms that are not discussed in this book. We begin with some notes on the general problem.

10.2 General Considerations on Numerical Optimization

There are dozens and dozens of variations of FCM (e.g., the previously discussed GK, FCV (fuzzy c-varieties), FCQS (fuzzy c-quadric shells), (possibilistic c-means, PCM), (FCRM, fuzzy c-regression models), and many more, including many variations of the basic GMD/EM model and algorithm that use AO algorithms, and the theory in this section, with some slight and usually obvious modifications, covers almost all of them. Moreover, alternating optimization is a widely used method in many other areas of pattern recognition and, more generally, optimization wherever it appears. But the theory of AO is not very accessible in standard texts on optimization theory; hopefully, this chapter will fill in some blanks. Let's begin with some remarks of a general nature about optimization.

10.2.1 Iterative Solution of Optimization Problems

Consider the optimization problem $\min_{\mathbf{X}\in\Psi\subset\Re^s} f(\mathbf{X})$, where $f : \Re^s \rightarrowtail \Re^s$ and Ψ is a constraint set over which the (global) optimization is to be done. This is the *non-linear, constrained optimization* (NCOP) problem for f, which can be recast as a maximization problem by setting $f = (-f)$. The constraint set Ψ specifies side conditions that must be satisfied and serves to delineate a feasible region for solutions of NCOP. There are many types of constraints (linear and non-linear equations and inequalities, ranges of values on the variables, and so on). Please be careful to distinguish between the vector variable $\mathbf{X} \in \Re^s$ in this chapter, and feature vector data sets $X \subset \Re^p$ in previous chapters.

Gill *et al.* (1981) list the following general features of the NCOP optimization problem that bear on the choice of an algorithm for solving $\min_{\mathbf{X} \in \Psi \subset \Re^s} f(\mathbf{X})$:

1. (s), the number of variables or arguments of the function $f : \Re^s \rightarrowtail \Re^s$ and Ψ;
2. whether the variables are confined to specified ranges;
3. whether the problem functions and their derivatives are smooth (continuously differentiable);
4. the highest level of derivatives that can be efficiently coded and evaluated;
5. the proportion of zero elements in the Jacobian and Hessian matrices;
6. the number of general linear constraints compared to (s);
7. the number of constraints that are likely to be active at the solution;
8. whether the problem functions can be evaluated (or are even meaningful) outside of the feasible region

This list was published in 1981 – 40 years ago – so some of these items, e.g., items 4 and 5, will have changed a bit due to improvements in the hardware and software available from implementation, but this list is a valuable place to start when faced with the NCOP problem. Gill *et al.* (1981) is still a highly readable and very useful guide for users faced with solving NCOP and is a great place to start your optimization adventures.

There are many ways to look for a solution, say $\mathbf{X}^*$. Rarely, closed-form optimization via the calculus of scalar fields admits a direct solution (Apostol, 1969). More typically, solutions are approximated with computational methods such as gradient-based descent in all variables (Ortega and Rheinboldt, 1970) and gradient-based alternating descent on subsets of the variables (Luenberger 1969, 1984, Zangwill, 1969). Some interesting evolutionary computation (Fogel, 1995) techniques that don't use optimality conditions from calculus have appeared in recent years. There are many data-driven approaches to solving NCOP such as neural networks (Theodoridos and Koutroumbas, 2009) and rule-based fuzzy systems (Nguyen and Sugeno, 1997) that require input-output pairs of *training data* (as opposed to derivative information) to search for solutions. And last, but certainly not least, there is AO as used by A6.1 and A7.1.

No matter how you decide to approach the NCOP problem, there are some common factors you should recognize. Suppose that iteration begins with an initial guess $\mathbf{X}_0$ and that $\mathbf{X}_k$ is the *k*th approximation to $\mathbf{X}^*$ found by some iterative method. The most general form of iterative procedures to do this is fixed-point iteration in the form $\mathbf{X}_{k+1} = \varphi(\mathbf{X}_k)$, $k = 0, 1, 2, \ldots,$ where $\varphi : \Re^s \rightarrow \Re^s$ is a function dictated by the problem to be solved. For the NCOP problem, the most common scheme is

$$\mathbf{X}_{k+1} = \varphi(\mathbf{X}_k) = \mathbf{X}_k + \alpha_k \mathbf{P}_k;\ k = 0, 1, 2, \ldots, T \tag{10.1}$$

In this equation, the *initial condition* $\mathbf{X}_0$ must be specified. The scalar α_k is the *step length* (of the step from $\mathbf{X}_k \rightarrow \mathbf{X}_{k+1}$); and the vector $\mathbf{P}_k$ is the *direction vector* (the direction in which the step will be taken). Once these three factors are specified, estimates of $\mathbf{X}^*$ can be made with a sequence of approximations $\{\mathbf{X}_k : k = 0, 1, \ldots,\}$ via Equation (10.1). The hope is that this sequence will converge to a solution of NCOP: that is, we want the sequence $\{\mathbf{X}_k \rightarrow \mathbf{X}^*\}$ or, equivalently, that $\|\mathbf{X}_k - \mathbf{X}^*\| \rightarrow 0$ as $k \rightarrow \infty$. The meaning of this type of convergence will be taken up in the next section.

We need some measure of progress that tells us whether the new point is "better" than the previous estimate. If the next value of the objective function satisfies $f(\mathbf{X}_{k+1}) < f(\mathbf{X}_k) \forall k \geq \infty$, the algorithm producing the estimates is said to be a descent method, and we say the algorithm has the *descent property* for f. You will be familiar with several particular instances of this iterative method, such as Newton's method, and the method of steepest descent. Additional material about this is included in the section on Notes and Remarks, and several exercises about this are included at the end of this chapter.

The last thing needed is a termination criterion. The integer T in Equation (10.1) is a fail-safe device which stops the iteration after a specified number of steps at $\mathbf{X}_{T+1}$, independent of its quality. It is much more desirable to stop the iteration before $k = T + 1$, but termination should occur only when sufficient progress toward the solution satisfies some criterion based on successive changes in the estimate. For the iteration in Equation (10.1), there are two obvious choices. We can stop iteration when successive values of the objective function are sufficiently close, i.e., terminate when $|f(\mathbf{X}_{k+1}) - f(\mathbf{X}_k) \leq \varepsilon|$. Or we can stop when successive estimates of $\mathbf{X}^*$ are close; i.e., when $\|\mathbf{X}_{k+1} - \mathbf{X}_k\| \leq \varepsilon$. Which of these two makes more sense to you? Before you answer, look at the problem again:we want an $\mathbf{X}^*$ that solves $\min_{\mathbf{X}\in\Psi\subset\Re^s} f(\mathbf{X})$. We seek the $\mathbf{X}^* \in \Psi$ for which $f(\mathbf{X}^*) \leq f(\mathbf{X}) \forall \mathbf{X} \in \Psi$. The vector $\mathbf{X}^*$ is the objective; so it seems intuitively that $\|\mathbf{X}_{k+1} - \mathbf{X}_k\| \leq \varepsilon$ is a much better measure of progress than $|f(\mathbf{X}_{k+1}) - f(\mathbf{X}_k) \leq \varepsilon|$. This point has been made several times in previous chapters, and it is repeated here one more time: stop when the estimates are close, not when the guide that tracks them is. Now we turn to the theory of the AO scheme.

10.2.2 Iterative Solution of Alternating Optimization with (t, s) Schemes

Sometimes we can't find a solution $\mathbf{X}^*$ of NCOP by direct optimization over the (s) variables $\mathbf{X} \in \Re^s$ (i.e., jointly, in all s variables at once). In this case, an approach that often enables approximation of $\mathbf{X}^*$ is to subdivide the constraint set Ψ into two or more subsets $\{\Psi_i \subset \Re^{p_i}\}$ of constraints and use alternating optimization (AO) iteratively on each subset of (p_i)[26] variables to facilitate an approximation to $\mathbf{X}^*$.

The problems for which AO is a worthy candidate are those for which the simultaneous optimization over all variables is much more difficult than the restricted optimizations over subsets of variables. Five reasons as to why AO *may be* preferable to its competitors when considering a specific instance of NCOP were listed in Bezdek and Hathaway (2002b). Some assertions are based on computational evidence, while others rest solely on intuition about differences between (NCOP) and (AO).

1. When there is a natural division of variables into t subsets for which *explicit* partial minimizer formulae exist. This strong reason is exemplified many times in the pattern recognition literature. Look no further than the three objective function methods (HCM, FCM, and GMD) in this volume, all of which have direct, explicit, and, most importantly, uncoupled components. This case is called *explicit AO*.
2. When there is a division of variables for which explicit partial minimizer formulas exist for *MOST* of the variables. This potentially greatly reduces the number of variables which require application of a numerical solution such as Newton's method. This is the case, for example, for the *fuzzy c-shells* clustering algorithm of Dave (1990). This group of methods is called *implicit AO*.

[26]Don't confuse the notation p_i used here for the dimensions of the t subproblems in a (t, s) scheme with the sample prior probabilities of mixture distributions that appear in Chapter 7.

3. Because AO can be *faster* in some cases. Hu and Hathaway (2002) found that for the FCM functional, AO in some examples is computationally cheaper (in time) than the best standard optimization routines applied to a reformulated version of the FCM objective function.
4. Because of savings in *development time.* When explicit formulas exist for AO, it may be easier to program the AO approach than to try to get a general approach written or adapted to the full NCOP problem. Remember, AO comes with good convergence properties; so no time is wasted trying to pick a good global convergence strategy.
5. Because AO may be more adept at *bypassing local minimizers* than other approaches. Each restricted iteration of AO is typically global and is, therefore, able to "hop" great distances through the reduced variable space in order to find an optimal iterate. On the other hand, Newton's method, and its many descendants, build models based on the current iterate and, in this sense, these schemes are more trapped by local information about the function.

Now suppose $f : \Re^s \rightarrowtail \Re$ is a real-valued scalar field, and let $\mathbf{X} = (x_1, \ldots, x_s)^T \in \Re^s$ be partitioned into t subsets of non-overlapping variables as $\mathbf{X} = (\mathbf{X}_1, \ldots, \mathbf{X}_t)^T$, with $\mathbf{X}_i \in \Re^{p_i}$ for $i = 1, \ldots, t$. It is important to understand the relationship between the dimension (s) of the joint unknowns and the dimensions $\{p_i; 1 \le i \le t\}$ of the partitioned unknowns. Specifically, $\sum_{i=1}^{t} p_i = s$, $1 \le t < s$. And it is equally important to bear in mind that, in this chapter, the vector $\mathbf{X} = (x_1, \ldots, x_s)^T \in \Re^s$ represents the unknowns we seek, while the vector $\mathbf{x} = (x_1, \ldots, x_p)^T \in \Re^p$ of previous chapters is an input feature vector (data vector) for A6.1 and A7.1.

To avoid confusion about the use of the word "partitioning" in this chapter, we refer to the partitioning of the s unknown variables $\mathbf{X}$ into the $\{p_i; 1 \le i \le t\}$ subsets of unknowns $(\mathbf{X}_1, \ldots, \mathbf{X}_t)^T$ as a *(t, s) iteration scheme* or simply a *(t, s) scheme.*

Definition 10.1. Alternating optimization with a (t, s) scheme.

Let $f : \Re^s \rightarrowtail \Re, \mathbf{X} = (x_1, \ldots, x_s)^T \in \Re^s$, be partitioned as a (t, s) scheme having t subsets of non-overlapping variables as $\mathbf{X} = (\mathbf{X}_1, \ldots, \mathbf{X}_t)^T$, with $\mathbf{X}_i \in \Re^{p_i}$ for $i = 1, \ldots, t, \sum_{i=1}^{t} p_i = s,\ 2 \le t < s$. Alternating optimization (AO) is the iterative procedure that begins at an initial point $(\mathbf{X}_1^{(0)}, \mathbf{X}_2^{(0)}, \ldots, \mathbf{X}_t^{(0)})^T$ and generates an iterate sequence $\{(\mathbf{X}_1^{(r)}, \mathbf{X}_2^{(r)}, \ldots, \mathbf{X}_t^{(r)})^T : r = 1, \ldots,\}$ of restricted minimizations of the form

$$\min_{\mathbf{X}_i \in \Psi_i \subset \Re^{p_i}} \{f(\cancel{\mathbf{X}}_1^{(r+1)}, \ldots, \cancel{\mathbf{X}}_{i-1}^{(r+1)},\ \mathbf{X}_i,\ \cancel{\mathbf{X}}_{i+1}^{(r)}, \ldots, \cancel{\mathbf{X}}_t^{(r)})\};\ i = 1, 2, \ldots, t \tag{10.2}$$

The *strikethrough notation* ($\cancel{\mathbf{X}}$) in Equation (10.2) indicates that the $(t - 1)$ sets of struck-through vectors of unknowns are *fixed* with respect to the current *subproblem* at index i.

After a (t, s) scheme is chosen, AO attempts to minimize $f(\mathbf{X}) = f(\mathbf{X}_1, \ldots, \mathbf{X}_t)$ jointly over all (s) variables by instead alternating *sequentially* restricted minimizations over the (t) individual sets of vector variables $\{\mathbf{X}_1 \in \Re^{p_1}, \ldots, \mathbf{X}_t \in \Re^{p_t}\}$. If we put $\Psi = \Psi_1 \times \cdots \times \Psi_t$, iteratively solving the t subproblems in Equation (10.2) is sometimes simpler and is often equivalent to solving NCOP directly. If everything goes just right, a termination point $\mathbf{X}^* = (\mathbf{X}_1^*, \mathbf{X}_2^*, \ldots, \mathbf{X}_t^*) \in \Re^s$ of AO will be an (approximate) solution of the joint NCOP problem. However, we make two sacrifices when we use AO instead of NCOP. First, we may find purely local solutions to NCOP, and, second, we sacrifice joint optimality and a strong convergence theory for a weaker result given by the convergence theory presented in this chapter for FCM and GMD/EM. On the positive side, the two theorems presented in this chapter hold for all AO (t, s) partitions of **X**.

When we base AO algorithms such as A6.1 and A7.1 on zeroing gradients in a (t, s) scheme, we get optimality conditions that identify *all* the stationary points of f, comprising its extrema (local minima and maxima) as well as its saddle points. A tricky point about this is that f can have saddle points that *can't be found by* AO because they don't look like a minimizer when viewed along the coordinate axis restrictions defined by the (t, s) scheme $\mathbf{X} = (\mathbf{X}_1, \ldots, \mathbf{X}_t)^T$. This behavior is illustrated in Example 10.2.

The partitioning of **X** in a (t, s) scheme is probably familiar to many readers for a specific problem such as k-means or GMD/EM, but not in the abstract way it is being presented in this chapter. Here is an example that may make the general idea a bit clearer.

Example 10.1. Solution of a quadratic NCOP problem with a (t, s) scheme.

Consider the quadratic function $f(\mathbf{X}) = f(X_1, X_2, X_3, X_4)$. The NCOP problem for f is

$$\underbrace{\min}_{\mathbf{X}\in\Psi=\Re^4} \{\mathbf{X}^T\mathbf{A}\mathbf{X}\} \text{ with } A = \begin{bmatrix} 100 & 80 & 5 & 1 \\ 80 & 90 & 2 & 1 \\ 5 & 2 & 70 & 40 \\ 1 & 1 & 40 & 80 \end{bmatrix} \quad (10.3)$$

You can imagine what the graph of this function looks like by looking at the graph shown in Figure 10.1(a) below. The quadratic function $f(\mathbf{X}) = \mathbf{X}^T\mathbf{A}\mathbf{X}$ is a hyper-paraboloid in $\Re^4$. The eigenvalues of A are $\{\lambda_1, \lambda_2, \lambda_3, \lambda_4\} = \{14.714, 34.745, 115.009, 175.531\}$ to three significant digits, and A is symmetric. So, A is positive-definite and the exact solution of $\underbrace{\min}_{\mathbf{X}\in\Psi=\Re^4} \{\mathbf{X}^T\mathbf{A}\mathbf{X}\}$ is easily found by setting the gradient of f equal to zero.

$$\mathbf{X}^T\mathbf{A}\mathbf{X} = \|\mathbf{X}\|_A^2 \Rightarrow \nabla_{\mathbf{x}}(\|X\|_A^2) = 2(A + A^T)\mathbf{X} = 2\mathbf{A}\mathbf{X} = \mathbf{0} \Leftrightarrow \mathbf{X}^* = \mathbf{0} \in \Re^4 \quad (10.4)$$

To verify that this a minimum, compute the Hessian matrix of second partial derivatives of f, $\nabla_f^2(\mathbf{X}^T\mathbf{A}\mathbf{X}) = 2A$. This matrix is constant, which verifies minimality. Finally, we know $\mathbf{X}^* = \mathbf{0}$ unique because it is the only solution of $\nabla_{\mathbf{X}}(\|\mathbf{X}\|_A^2) = \mathbf{0}$. Thus, we don't need a numerical method to find the global solution for this simple non-linear unconstrained optimization problem. But to illustrate the AO approach, let us make a (t, s) partitioning of the four unknowns as $(t, s) = (2, 4)$ with the partitioning $\mathbf{X} = (\mathbf{U}, \mathbf{V})$ where $\mathbf{U} = (X_1, X_2)$ and $\mathbf{V} = (X_3, X_4)$. This choice for variable names is quite deliberate because these are the variable names we have used for the hard and fuzzy c-means objective function, and the treatment here will be mirrored more or less exactly in that case.

Next, we need the gradients of f with respect to each subset of partitioned variables. First block the matrix A into four 2 × 2 sub-blocks notated as $A = \begin{bmatrix} B & C \\ D & E \end{bmatrix}$, for example, $B = \begin{bmatrix} 100 & 80 \\ 80 & 90 \end{bmatrix}$ and so on. Next, compute gradients of f restricted to the split variables and set them equal to **0**, resulting in the first-order necessary conditions

$$\nabla_{\mathbf{U}} f(\mathbf{U}, \not{\mathbf{V}}) = 0 \Leftrightarrow \mathbf{B}\mathbf{U} + \mathbf{C}\mathbf{V} = \mathbf{0} \Leftrightarrow \mathbf{U} = -B^{-1}C\mathbf{V}; \quad (10.5a)$$

$$\nabla_{\mathbf{V}} f(\not{\mathbf{U}}, \mathbf{V}) = 0 \Leftrightarrow \mathbf{D}\mathbf{U} + \mathbf{E}\mathbf{V} = \mathbf{0} \Leftrightarrow \mathbf{V} = -D^{-1}E\mathbf{U}. \quad (10.5b)$$

Equations (10.5) express each of the split variables explicitly in terms of the other set of unknowns; so this is an example of explicit AO. Substituting the problem values into (10.5) yields

$$\mathbf{U} = \mathbf{F}(\mathbf{V}) = -B^{-1}C\mathbf{V} = -\begin{bmatrix} 100 & 80 \\ 80 & 90 \end{bmatrix}^{-1}\begin{bmatrix} 5 & 1 \\ 2 & 1 \end{bmatrix}\mathbf{V}; \tag{10.6a}$$

$$\mathbf{V} = \mathbf{G}(\mathbf{U}) = -D^{-1}E\mathbf{U} = -\begin{bmatrix} 70 & 40 \\ 40 & 80 \end{bmatrix}^{-1}\begin{bmatrix} 5 & 2 \\ 1 & 1 \end{bmatrix}\mathbf{U}. \tag{10.6b}$$

If we choose an initial value for either **U** or **V**, we can compute the other value explicitly. Suppose we choose $\mathbf{V}_0 = (1, 1)^T$. Applying Equation (10.6a) to this initial value yields $\mathbf{U}_0 = (-0.1154, 0.0692)^T$. This starts the alternating optimization of f, which takes the iterative form, say, of $\cdots \mathbf{U}_t = \mathbf{F}(\mathbf{V}_{t-1}) \rightarrow \mathbf{V}_t = \mathbf{G}(\mathbf{U}_t) \rightarrow \mathbf{U}_{t+1} = \mathbf{F}(\mathbf{V}_t)\cdots$. For this example, choose the iteration limit $T = 100$ and the termination criterion to be the Euclidean norm of the paired estimates of the split variables, $\|\mathbf{X}_{t+1} - \mathbf{X}_t\|_E \leq \varepsilon = 0.001$. Table 10.1 shows the results of this AO sequence:

Table 10.1 Alternating iterates to minimize $f(\mathbf{X}) = \mathbf{X}^T A\mathbf{X}$ for A at Equation (10.3).

T	$\mathbf{U}_{t+1}$	$\mathbf{V}_{t+1}$	$\|\mathbf{X}_{t+1} - \mathbf{X}_t\|_E$
0	$(-0.1154, 0.0692)^T$	$(0.0083, -0.0036)^T$	2.0251
1	$(-0.0009, 0.0006)^T$	$(0.0000, -0.0000)^T$	0.1337
2	$(-0.0000, 0.0000)^T$	$(0.0000, -0.0000)^T$	0.0011
3	$(-0.0000, 0.0000)^T$	$(0.0000, -0.0000)^T$	0.0000

The AO iterates of f quickly converge (yes, here they do converge!) to the global minimum of f. The speed of convergence depends on the eigenvalues of A and also on the particular (t, s) split chosen. Any other (t, s) split of $\mathbf{X} = (X_1, X_2, X_3, X_4)^T$ will slow down the convergence of AO to the global minimum by about an order of magnitude. Finally, this is not a particularly good example of using AO to solve NCOP since the closed-form solution is easily found directly with the calculus. But it *is* a good example in the sense that all of the elements underlying the derivation of AO for FCM (and also GMD/EM) -except the handling of the constraints for these two NCOP problems -are on display.

Let us apply the concept of the (t, s) scheme to FCM, using the same notational structure as in Example 10.1 and Chapter 6, where the AO iterates appear as Equations (6.25).

Example 10.2.AO iteration of FCM at A6.1, m>1 as a (t, s) scheme.

The iteration in A6.1 is, say, $\cdots \mathbf{U}_t = F_m(\mathbf{V}_{t-1}; \mathbf{X}) \rightarrow \mathbf{V}_t = G_m(\mathbf{U}_t; \mathbf{X}) \rightarrow \mathbf{U}_{t+1} = F_m(\mathbf{V}_t; \mathbf{X})\cdots$. The next $^1/_2$ step for $\mathbf{U}_t$ in terms of the previous pair of $^1/_2$ iterates is $\mathbf{U}_{t-1} = F_m \circ G_m(\mathbf{U}_t)$. In the context of representation as a (t, s) scheme, FCM is a (t = 2, s = cp + cn) scheme. The unknowns are $\mathbf{X} = (\mathbf{U}, \mathbf{V})$, i.e., $(\mathbf{X}_1 = \mathbf{U}, \mathbf{X}_2 = \mathbf{V})$ when the objective function for FCM is written as $f = F_m \circ G_m : \Re^{cp} \times \Re^{cp} \rightarrowtail \Re^+$. The constraints are $\Psi_1 = M_{fcn}; \Psi_2 = \Re^{cp}$, and the subproblems have dimensions $p_1 = cn$ and $p_2 = cp$. Using the other half step representation of FCM instead of this one just swaps the unknowns and dimensions, etc. The (t, s) formulation given here is also correct for the k-means (HCM) objective function at $m = 1$, but the analogy breaks down when we get to the step where gradients for each subproblem are computed as in equations (10.5), for as we know, the operation at (10.5) can't be done for the discrete variables **U**.

Now we are ready to turn to the questions: when can we expect AO to produce a sequence of iterates that approximate a solution to NCOP? And how fast does AO iteration get there?

10.3 Local Convergence Theory for AO

Returning to the NCOP problem $\min_{X \in \Psi \subset \Re^s} \{f(\mathbf{X})\}$, what can we say about convergence of AO to a solution $\mathbf{X}^*$? Convergence theory for iterative algorithms such as FCM (A6.1, m>1) and GMD/EM (A7.1) that approximate solutions to NCOP comprises four main questions:

NCOP (i) *existence*: is there any $\mathbf{X}^*$ that solves NCOP?
NCOP (ii) *uniqueness*: if an $\mathbf{X}^*$ exists, is it the only one?
NCOP (iii) *what is the kind* of convergence to $\mathbf{X}^*$ (i.e., local, global, uniform, pointwise, etc.)?
NCOP (iv) *how fast* does the algorithm approach $\mathbf{X}^*$ (i.e., linearly, superlinearly, quadratic, etc.)?

For this chapter, we will *assume* that the answers to NCOP (i) and NCOP (ii) are yes. We concentrate on how to answer questions NCOP (iii) and NCOP (iv).

Answers to NCOP (iii) are more for psychological reassurance than practical utility. In the first place, the variable under the limit (e.g., n, the number of samples in a feature data set X, or m, the value of the weighing exponent for J_m) can't "get to infinity." We must stop the computer at some point, and what happens up to that point in an infinite sequence may behave quite unpredictably and tell us little about what happens if it could go on forever. Moreover, even with a comprehensive theory in hand, it may be impossible to verify the requisite conditions that support the theory for the problem we are interested in. On the other hand, any definitive answer to question NCOP (iv) – how fast can we get there? -is very good to have. In other words . . .

The *rate* of convergence is much more useful to practitioners than the *type* of convergence because a faster convergence rate usually means fewer iterations are needed to achieve a reasonable approximation to $\mathbf{X}^*$. So, our primary aim in this chapter is to answer question NCOP (iv) for FCM (and as a by-product, for EM as well, and *almost* for HCM). We will discover partial answers to NCOP (iii) – what type of convergence can we expect for these algorithms, at rest stops along the way. We begin with some standard material from numerical analysis. See Ortega (1970) for an in-depth treatment.

Definition 10.1. Orders and rates of convergence.

Let $\mathbf{A} : \Re^s \rightarrowtail \Re^s$ be an iterative algorithm, $\mathbf{Z}_0 \in \Re^s, \mathbf{Z}_{n+1} = \mathbf{A}(\mathbf{Z}_n), n = 0, 1, \ldots,$. When an infinite sequence of distinct iterates of $\mathbf{A}$ converges pointwise to $\mathbf{Z}^*$, i.e., when $\lim_{n\to\infty} \{\mathbf{Z}_0\} = \mathbf{Z}^*(\neq \mathbf{Z}_k \forall k)$, we say $\{\mathbf{Z}_n\} \to \mathbf{Z}^*$ with *order* τ, the exponent in the denominator of the condition exhibited at (10.7),

$$0 \leq \rho \lim_{n\to} \left\{ \frac{\|\mathbf{Z}_{n+1} - \mathbf{Z}^*\|}{\|\mathbf{Z}_n - \mathbf{Z}^*\|^\tau} \right\} < \infty \tag{10.7}$$

The number τ (usually an integer) is the asymptotic convergence rate or *order of convergence*, and ρ is the *asymptotic error constant*. Convergence is *linear* if $\tau = 1$, *quadratic* if $\tau = 2$, *cubic* if $\tau = 3$, and so on.

Rates of convergence given by Equation (10.7) for any τ and ρ are sometimes called "q-convergence" rates, the q standing for the word "quotient." Linear convergence of this kind when $\tau = 1$ is usually called *q-linear convergence*, to distinguish it from another rate of convergence called "root" or r-convergence, which addresses the overall rate of decrease of iteration error. Equation (10.7) shows that q-convergence requires the error to decrease at each iteration. Consequently, q-convergence is stronger than r-convergence (r-convergence implies q-convergence). ***Spoiler alert***: FCM and GMD/EM are locally q-linearly convergent.

One more thing about Definition 10.1 that often causes confusion is the difference between stating that an iterative algorithm has, for example,*runtime complexity* $\boldsymbol{O}(n^2)$ and a *quadratic rate of convergence* $(\tau = 2)$. We want the *lowest* value of $\boldsymbol{O}(n^2)$ we can achieve; runtime complexity $\boldsymbol{O}(n)$ is better (has slower asymptotic growth) than $\boldsymbol{O}(n^2)$. On the other hand, we want the *highest* value of τ we can get; quadratic convergence $(\tau = 2)$ is better (gets there sooner) than linear convergence $(\tau = 1)$.

To make things even more confusing, there are *three types of linear convergence* $(\tau = 1)$ depending on the value of ρ: convergence is said to be *superlinear* if $\rho = 0$, *linear* if $0 < \rho < 1$, and *sublinear* if $\rho = 1$. Any $\tau > 1$ implies superlinear convergence. Think of the limit in Equation (10.1) as describing the behavior of the "tail" of the infinite sequence $\{\mathbf{Z}_n\}$. Larger values of t imply faster convergence in the tail to the limit since the distance from the limit point $\mathbf{Z}^*$ is reduced in the tail by the τth power at each step.

Definition 10.2. Alternate form of Definition 10.1 when $\tau = 1$.

Let $\mathbf{A} : \Re^s \rightarrowtail \Re^s$ be an iterative algorithm, $\mathbf{Z}_0 \in \Re^s, \mathbf{Z}_{n+1} = \mathbf{A}(\mathbf{Z}_n), n = 0, 1, \ldots,$. When an infinite sequence of distinct iterates of $\mathbf{A}$ converges point wise to $\mathbf{Z}^*$ for $\tau = 1$, we have

$$\{\mathbf{Z}_k\} \rightarrow \mathbf{Z}^* \textit{q-linearly} \Leftrightarrow \exists n_0 \in \mathbb{Z}^+ \text{ and } \rho \in [0, 1] \ni \|\mathbf{Z}_{k+1} - \mathbf{Z}^*\| \leq \rho \|\mathbf{Z}_k - \mathbf{Z}^*\| \; \forall k \geq n_0 \qquad (10.7)_{\tau=1}$$

This second form of Equation (10.7) for the special case $\tau = 1$ offers a clearer insight into what q-linear convergence means. When a sequence of iterates has q-linear convergence, the stepwise decrease in $\|\mathbf{Z}_k - \mathbf{Z}^*\|$ depends on the value of ρ. For example, if $\rho = 1/2$, successive approximations to $\mathbf{z}^*$ will be halved at each step (for iterates past that unknown positive integer $n_0 \in \mathbb{Z}^+$!).

Example 10.3. Order of convergence for the sequence $\{\mathbf{z}_n\}$ = *n*th root of $\mathbf{z} > 0$.

Consider the real sequence $\{Z^{2^{-n}}\} = \{Z^{\frac{1}{2^n}}\} = \{Z^{(\frac{1}{2})^n}\} = \{\sqrt[n]{Z}\}$ for any constant $Z > 0$. Each value in this sequence is the square root of the previous iterate. If $Z > 1$, taking square roots will decrease the values of $\{Z^{2-n}\}$ toward 1. If $Z < 1$, square roots will increase the sequence of values of $\{Z^{2-n}\}$ toward 1. So, the powers of Z in this sequence will converge to $Z^* = 1$ as $n \rightarrow \infty$ for any $Z > 0$. Table 10.2 exhibits the first 10 values of this sequence for $Z_0 = 2$ and $Z_0 = 0.5$.

To study the order of convergence, form the quotient $\frac{|Z_{n+1} - Z^*|}{|Z_n - Z^*|^\tau}$, and find τ for which $0 \leq \lim_{n\rightarrow\infty} \left\{ \frac{|Z_{n+1} - Z^*|}{|Z_n - Z^*|^\tau} \right\} = \rho < \infty$. At $\tau = 1$: $\lim_{n\rightarrow\infty} \left\{ \frac{|Z_{n+1} - Z^*|}{|Z_n - Z^*|} \right\} = \lim_{n\rightarrow\infty} \left\{ \frac{Z^{2^{-(n+1)}} - 1}{Z^{2^{-n}} - 1} \right\} = \lim_{n\rightarrow\infty} \left\{ \frac{1}{Z^{2^{-(n+1)}} + 1} \right\} = 1/2$, so this sequence has q-linear convergence. Equation (10.7) looks like this: $|\sqrt[n+1]{Z} - 1| = \frac{1}{2}|\sqrt[z]{Z} = 1|$, which shows that the sequence will rapidly approach the limit point, each iterate reducing the previous estimate by (roughly) $^1/_2$. This can be seen in Table 10.2, where the number of significant digits at each iteration improves by approximately $^1/_2$ of the previous estimate.

Table 10.2 Iterates for the *n*th principal root of z for starting values 2 and 0.5.

n = 0	$Z_0 = 2$	$Z_0 = 0.5$
n = 1	1.414214	0.707107
n = 2	1.189207	0.840896
n = 3	1.090508	0.917004
n = 4	1.044274	0.957603
n = 5	1.021897	0.978572
n = 6	1.010889	0.989228
n = 7	1.00543	0.994599
n = 8	1.002711	0.997296
n = 9	1.001355	0.998647
n = 10	1.00067727	0.999323

Perhaps, the most important mathematical questions associated with (NCOP) and (AO) are: *existence* (does any $\mathbf{X}^*$ exist?) and *uniqueness* (when $\mathbf{X}^*$ exists, is it unique?). It is difficult to resolve these two issues for a general instance of NCOP. Instead, the theory presented here relies on an existence and uniqueness *assumption* for the subproblems comprising the general AO method. Specifically, the notation "=" used in this chapter implies a property of f, $\mathbf{X}_i$ and Ψ_i for $i = 1, \ldots, t$, that we *assume* holds; namely that solutions of the restricted optimizations do *exist* and are *unique*.

Definition 10.3. Existence and uniqueness (EU) assumption.

Let $\Psi = \Psi_1 \times \cdots \times \Psi_t$ with $\Psi_i \in \Re^{p_i}$ for $i = 1, \ldots, t$. Let $\mathbf{X} = (\mathbf{X}_1, \ldots, \mathbf{X}_t)_T \in \Re^s$ be a (t, s) scheme with $\mathbf{X}_i \in \Re^{p_i}$ for $i = 1, \ldots, t$. ***Then*** the NCOP problem $\min\limits_{\mathbf{X} \in \Psi \subset \Re^s} \{f(\mathbf{X})\}$ satisfies the EU assumption (i.e., this NCOP problem has a unique solution) ***if*** for each $\mathbf{X} \in \Psi, i = 1, \ldots, t$,

$$f(\mathbf{X}, \ldots, \mathbf{X}_{i-1}, \mathbf{Y}_i, \mathbf{X}_{i+1}, \ldots, \mathbf{X}_t) \text{ has a unique (global) minimizer over } \mathbf{Y}_i \in \Psi_i. \tag{10.8}$$

We simplify the notation by taking $\Psi_i = \Re^{p_i}$ for $i = 1, \ldots, t$, and, further, we can assume, without loss of generality, that the local minimizer of interest is $\mathbf{X}^* = \mathbf{0} \in \Re^s$. Finally, we assume that the (EU) property (10.8) holds. Define the function $M_i : \Psi_1 \times \cdots \times \Psi_{i-1} \times \Psi_{i+1} \times \Psi_t \rightarrowtail \Psi_i, i = 1, \ldots, t$, as

$$M_i(\hat{\mathbf{X}}_i) = M_i(\mathbf{X}_1, \ldots, \mathbf{X}_{i-1}, \mathbf{X}_{i+1}, \ldots, \mathbf{X}_t) = \underset{\mathbf{X}_i \in \Psi_i \subset \Re^{p_i}}{\arg\min} \left\{ f(\cancel{\mathbf{X}}_1, \ldots, \underbrace{\mathbf{X}_i}_{\text{active}}, \ldots, \cancel{\mathbf{X}}_t) \right\} \tag{10.9}$$

where the vector $\hat{\mathbf{X}}_i = (\mathbf{X}_1, \ldots, \mathbf{X}_{i-1}, \mathbf{X}_{i+1}, \ldots, \mathbf{X}_t)^T$ is missing the active (*i*th) variable. So, the argmin function in Equation (10.9) retrieves the value of the active variable $\mathbf{X}_i$ that minimizes this restriction of f with the other (t − 1) subsets of unknowns fixed. For each M_i in Equation (10.9), define $C_i : \Psi \rightarrowtail \Psi$ as

$$C_i(\mathbf{X}_1, \ldots, \mathbf{X}_t) = (\mathbf{X}_1, \ldots, \mathbf{X}_{i-1}, \mathbf{M}_i(\hat{\mathbf{X}}_i), \mathbf{X}_{i+1}, \ldots, \mathbf{X}_t)^T \tag{10.10}$$

Under fairly mild assumptions, for i = 1, . . . , t the function M_i is continuously differentiable in a neighborhood of $\mathbf{0} = (0, 0, \ldots, 0)^T \in \Re^{p_1} \times \cdots \times \Re^{p_{i-1}} \times \Re^{p_{i+1}} \times \cdots \times \Re^{p_t}$. Under conditions stated below, we have the following.

Lemma 10.1. Differentiability of subproblem functions.

Let $\nabla^2 f(\mathbf{X})$ denote the s × s Hessian matrix of f at **X**. Assume the following four things: ***if***

(i) $f : \Re^s \rightarrowtail \Re$ is twice continuously differentiable (C^2) in some neighborhood of **0**;
(ii) $\nabla^2 f(\mathbf{0})$ is positive-definite at **0**, which is a local minimizer of f;
(iii) there is a neighborhood $\mathbb{N}(\mathbf{0}, \delta)$ of **0** on which f is strictly convex;
(iv) if for i = 1, . . . , t, $\mathbf{Y} = (\mathbf{X}_1, \ldots, \mathbf{X}_{i-1}, \mathbf{Y}_i, \mathbf{X}_{i+1}, \ldots, \mathbf{X}_t)^T \in \mathbb{N}(\mathbf{0}, \delta)$, if $\mathbf{Y}_i^*$ *locally* minimizes $g_i(\mathbf{Y}_i) = f(\mathbf{Y})$, then $\mathbf{Y}_i^*$ is also the unique *global* minimizer of $g_i(\mathbf{Y}_i) = f(\mathbf{Y})$:

Then in some neighborhood of $(\mathbf{0}, \mathbf{0}, \ldots, \mathbf{0})^T \in \Re^{p_2} \times \cdots \Re^{p_t}$ the minimizing function $\mathbf{M}_i$ in Equation (10.9) exists and is continuously differentiable. Similar results hold for all of the $\mathbf{M}_i$'s, i = 2, . . . , t.

What does it mean?

By their construction from the $\{\mathbf{M}_i\}$, Lemma 10.1 guarantees that the $\{C_i\}$ at Equation (10.10) are also continuously differentiable in a neighborhood of **0** and satisfy $C_i(\mathbf{0}) = \mathbf{0}$, for i = 1, . . . , t. Define $\mathbf{S} : (\Re^{p_1} \times \cdots \times \Re^{p_t}) \rightarrowtail (\Re^{p_1} \times \cdots \times \Re^{p_t})$ corresponding to one execution step of all t substeps of AO as the sequential composition of the $\{C_i\}$, that is,

$$\mathbf{S}(\mathbf{X}) = \mathbf{S}(\mathbf{X}_1, \ldots, \mathbf{X}_t) = C_t \circ C_{t-1} \circ \cdots \circ C_1(\mathbf{X}_1, \ldots, \mathbf{X}_t) \tag{10.11}$$

Every component of the composition function $C_t \circ C_{t-1} \circ \cdots \circ C_1$ is continuously differentiable; so Lemma 10.1 implies that S is continuously differentiable in a neighborhood of **0** and that $\mathbf{S}(\mathbf{0}) = \mathbf{0}$. In terms of iteration, note that if we put $\mathbf{X} = \mathbf{X}_k$, then $\mathbf{X}_{k+1} = \mathbf{S}(\mathbf{X})$.

Proof. See Lemma 1 of Bezdek and Hathaway (2003).

Now we take a short detour that helps us understand the general theory of AO a little better. This material is given here in a very brief form: please refer to Noble and Daniel (1987) for a more complete treatment.

Definition 10.4. Spectral norm and spectral radius of a real, square matrix.

Let a real s × s matrix $A \in \Re^{ss}$ have eigenvalues $\{\lambda_1, \ldots, \lambda_s\}$. The *spectral radius* and *spectral norm* of A are

$$\rho(A) = \underbrace{\max}_{1 \le i \le s} \{|\lambda_i|\} \text{ (spectral radius)} \tag{10.12a}$$

$$\sigma(A) = \|A\|_{sp} = \sqrt{(\text{max eigenvalue of } A^T A)} \text{ (spectral norm)} \tag{10.12b}$$

It is usually the case that A is symmetric; so the eigenvalues of A are real. When A is symmetric, $\rho(A) = \sigma(A)$.

Consider solving a linear system $\mathbf{AX} = \mathbf{b}$ by an iterative scheme such as the one at Equation (10.1), $\mathbf{X}_{k+1} = \varphi(\mathbf{X}_k)$; $k = 1, 2, \cdots$. Introduce an auxiliary matrix B, and write the iteration as $\mathrm{B}\mathbf{X}_{k+1} = (\mathrm{B} - \mathrm{A})\mathbf{X}_k + \mathbf{b}$; $k = 1, 2, \cdots$. After k steps, the error vector in the approximation of the solution by the k-th iterate is $\mathbf{E}_k = \mathbf{X}_k - \mathbf{X}$. Substituting this into the iteration yields $\mathrm{B}\mathbf{E}_{k+1} = (\mathrm{B} - \mathrm{A})\mathbf{E}_k + \mathbf{b}$; $k = 1, 2, \ldots,$ or $\mathbf{E}_{k+1} = \mathrm{C}\mathbf{E}_k$, where the matrix $\mathrm{C} = \mathrm{B}^{-1}(\mathrm{B} - \mathrm{A})$. Suppose $\{\mathbf{X}_k\} \to \mathbf{X}^*$, and that $\mathbf{X}^*$ is the solution of the linear system so that $\mathrm{A}\mathbf{X}^* = \mathbf{b}$ and $\mathrm{B}\mathbf{X}^* = (\mathrm{B} - \mathrm{A})\mathbf{X}^* + \mathbf{b}$. Since $\mathbf{E}_k = \mathrm{C}\mathbf{E}_{k-1} = \mathrm{C}(\mathrm{C}\mathbf{E}_{k-2}) \cdots = \mathrm{C}^k\mathbf{E}_0$, the error $\mathbf{E}_k \to \mathbf{0}$ and $\{\mathbf{X}_k\} \to \mathbf{X}^* \Leftrightarrow \mathrm{C}^k \to \mathbf{0}$. Here is a proposition that ties the spectral radius of C to iterative convergence:

Proposition 10.1. Convergence of a general iterative method.

Let $\{\lambda_1, \lambda_2, \ldots, \lambda_s\}$ be the eigenvalues of $\mathrm{C} = \mathrm{B}^{-1}(\mathrm{B} - \mathrm{A})$ for the iteration $\mathbf{X}_{k+1} = \mathrm{B}^{-1}(\mathrm{B} - \mathrm{A})\mathbf{X}_k + \mathbf{b} = \mathrm{C}\mathbf{X}_k + \mathbf{b}$. Then $\mathrm{C}^k \to 0 \Leftrightarrow$ every eigenvalue of C satisfies $|\lambda| < 1$. The rate of convergence is governed by the largest eigenvalue in $\{\lambda_1, \lambda_2, \ldots, \lambda_s\}$, which is just the spectral radius $\rho(\mathrm{C})$ of C.

What it means. The spectral radius of C is a measure of the amount of distortion experienced by stretching along the eigenvector associated with the largest absolute eigenvalue of C. The eigenvalues of C are often related to the convergence of iterative algorithms.

Proof (Strang, 1986). Assume that C has a complete set of eigenvectors, say $\{\mathbf{g}_i : 1 \leq i \leq s\}$. Form the $s \times s$ matrix $\mathrm{G} = [\mathbf{g}_1, \mathbf{g}_2 \cdots \mathbf{g}_s]$ with the eigenvectors as its columns, and let $\mathrm{D} = \mathrm{diag}[\lambda_i]$ be the diagonal matrix of eigenvalues of C. Diagonalize C as $\mathrm{C} = \mathrm{G}^{-1}\mathrm{DG}$.

Then $$\mathrm{C}^k = \underbrace{(\mathrm{G}^{-1}\mathrm{DG})(\mathrm{G}^{-1}\mathrm{DG}) \cdots (\mathrm{G}^{-1}\mathrm{DG})}_{k\text{-times}} = (\mathrm{G}^{-1}\mathrm{D}^k\mathrm{G}) \to 0 \Leftrightarrow |\lambda_i| < 1,\ 1 \leq i \leq s. \quad \blacksquare$$

Our next result utilizes the spectral radius in a similar fashion in the present context. Returning to Equation (10.11), let $\rho = \rho(\mathbf{S}'(\mathbf{0}))$ denote the *spectral radius* of the derivative of $\mathbf{S}$ at 0, $\mathbf{S}'(\mathbf{0})$. This ρ will be the asymptotic error constant for our problem that appears in Definition 10.1. The fundamental property needed for convergence of any AO sequence is contained in Lemma 10.2.

Lemma 10.2. Asymptotic error constant for AO iterates.

Let $f : \Re^s \mapsto \Re$ be differentiable and satisfy the conditions of Lemma 10.1. Let $\mathbf{S}$ be defined by (10.11). Then $\rho = \rho(\mathbf{S}'(\mathbf{0})) < 1$.

What does it mean?

This result tells us that when the spectral radius $\rho = \rho(\mathbf{S}'(\mathbf{0})) < 1$, we can expect good convergence behavior as exemplified by the more general result in Proposition 10.1.

Proof. See Lemma 2 in Bezdek and Hathaway (2003).

Now we are ready to give the main result for local convergence of AO, which is an adaptation of Ostrowski's theorem (Theorem 8.1.7 in Ortega, 1970).

Theorem 10.1. Local convergence of AO.

Let $\mathbf{X}^*$ be a local minimizer of $f : \Re^s \mapsto \Re$ for which the Hessian of f at $\mathbf{X}^*$, $\nabla^2 f(\mathbf{X}^*)$, is positive definite, and let f be twice continuously differentiable in a neighborhood $\mathbb{N}(\mathbf{X}^*, \delta)$ of $\mathbf{X}^*$. Also let assumption (iv) of Lemma 10.1 hold for $\mathbf{X}^*$. Then there is a neighborhood $\mathbb{N}(\mathbf{X}^*, \varepsilon)$ of $\mathbf{X}^*$ such that for any initial guess $\mathbf{X}_0 \in \mathbb{N}(\mathbf{X}^*, \varepsilon)$ the corresponding AO iteration sequence $\{\mathbf{X}_{k+1} = \mathbf{S}(\mathbf{X}_k) : k = 0, 1, \ldots\}$ defined by $\mathbf{S}$ in (10.11) converges q-linearly to $\mathbf{X}^*$.

What does it mean? If the stated assumptions hold, this theorem tells us that AO iteration will move toward a local solution with q-linear speed. This is not very fast, compared to algorithms such as Newton's method that have quadratic convergence. But the actual speed depends on the constants ρ and τ in Definition 10.1 as well as the distance from the initializer to a local minimizer; so for many problems, it's fast enough. The bad news here is that we don't know $\mathbb{N}(\mathbf{X}^*, \varepsilon)$; so we can never be sure that we are actually close enough to any local minimizer of f for this theorem to hold. But there is some good news: this theorem does tell us a lot about what we can expect when we enter AO iteration.

Sketch of Proof [see Theorem 2] in Bezdek and Hathaway (2003) for a full proof]. For notational convenience and without loss of generality, let the local minimizer be $\mathbf{X}^* = \mathbf{0} \in \Re^s$ and $\Psi_i \subset \Re^{p_i}, 1 \leq i \leq t$. We know that $\mathbf{S}$ in (10.11) is continuously differentiable in a neighborhood of $\mathbf{0}$ with $\mathbf{S}(\mathbf{0}) = 0$. Now, with some trouble, the derivative $\mathbf{S}'(\mathbf{0})$ can be calculated as $\mathbf{S}'(\mathbf{0}) = B^{-1}C$, where

$$B = \begin{pmatrix} A_{11} & 0 & \cdots & \cdots & 0 \\ \vdots & \ddots & \ddots & & \vdots \\ \vdots & & \ddots & \ddots & \vdots \\ \vdots & & & \ddots & 0 \\ A_{t1} & \cdots & \cdots & \cdots & A_{tt} \end{pmatrix}; \quad C = \begin{pmatrix} 0 & -A_{12} & \cdots & \cdots & -A_{1t} \\ \vdots & \ddots & \ddots & & \vdots \\ \vdots & & \ddots & \ddots & \vdots \\ \vdots & & & \ddots & -A_{t-1,t} \\ 0 & \cdots & \cdots & \cdots & 0 \end{pmatrix} \tag{10.13}$$

where A is the matrix of second partial derivatives of f at $\mathbf{X}^* = \mathbf{0} \in \Re^s, A_{ij} = f_{X_i X_j}(\mathbf{0}); 1 \leq i, j \leq t$. Theorem 7.1.9 in Ortega (1970) can be applied to show that the spectral radius $\rho(\mathbf{S}'(\mathbf{0})) = \rho(B^{-1}C) = \rho \leq 1$. if $A = B - C$ is a P-regular splitting (B–C is a P-regular splitting if B is nonsingular and B + C is positive definite). It is not too hard to show that B is nonsingular and the symmetric part of B+C is the block diagonal part of A, which is positive definite because A is positive definite. Ostrowski's Theorem is used to finish the proof. Pick $\delta > 0$ such that $(\rho + 2\delta) < 1$. By Theorem 3.8 in Stewart (1973) there is a norm $\|\mathbf{X}_\kappa\|$ on $\Re^s$ and a $\gamma > 0$ such that for all $\mathbf{X}$ in the set $\{\mathbf{X} \in \Re^s : \|\mathbf{X}\|_\kappa \leq \lambda\}$, so

$$\|\mathbf{S}(\mathbf{X})\|_\kappa \leq (\rho + 2\delta)\|\mathbf{X}\|_\kappa \tag{10.14}$$

The result at (10.14) establishes that when initialization of AO is done near enough to $\mathbf{X}^*$, the error is reduced by $(\rho + 2\delta) < 1$ at each iteration, which gives local q-linear convergence of $\{\mathbf{X}_k\}$ to $\mathbf{X}^*$. ■

Theorem 10.1 is not specific to a particular choice of the (t, s) scheme $(\mathbf{X}_1, \mathbf{X}_2, \ldots, \mathbf{X}_t)^T$ for $\mathbf{X}$. A "good" (t, s) scheme is one that gives a substantial error reduction at each iteration. The actual speed of convergence of AO depends on the particular value of ρ in (10.1), and ρ *does* depend on which (t, s) scheme is used. We have already pointed out that FCM is a $(2, c(n+p))$ scheme, $f = J_m, t = 2, \mathbf{X} = (U, \mathbf{V}), s = cn + cp, p_1 = cn, p_2 = cp, \Psi_1 = M_{fcn}, \Psi_2 = \Re^{cp}$. When the assumptions underlying Theorem 10.1 are satisfied, we have

Corollary 10.1. FCM A6.1 (m>1) is locally q-linearly convergent.

Let $X = \{\mathbf{x}_1, \mathbf{x}_2, \ldots, \mathbf{x}_n\} \subset \Re^p$ have $c < n$ distinct points, and let an inner product norm $\|\mathbf{x}_k - \mathbf{v}_i\|_A^2$ be the model norm for FCM. Let $(U^*, \mathbf{V}^*)$ be a local minimizer of $J_m(U, \mathbf{V}; X)$ at which the Hessian $\nabla^2 J_m(U^*, \mathbf{V}^*)$ is positive definite. For fixed $m > 1$, let $\mathbb{N}((U^*, \mathbf{V}^*), \varepsilon)$ be an epsilon neighborhood of this local minimizer. Then if $(U^0, \mathbf{V}^0 = G_m(U^0)) \in \mathbb{N}((U^*, \mathbf{V}^*), \varepsilon)$, the AO iteration sequence $\{(U_{k+1}, \mathbf{V}_{k+1}) = \mathbf{S}((U_k, \mathbf{V}_k)) : k = 0, 1, \ldots\}$ defined by FCM algorithm A6.1 converges q-linearly to $(U^*, \mathbf{V}^*)$.

What does it mean? Just like Newton's method and the method of steepest descent, this tells us that if we start "close enough" to a local minimizer, fixed point iteration with the FCM operator will converge to it q-linearly.

Proof Set $\mathbf{X}^* = (U^*, \mathbf{V}^*)$ in Theorem 10.1. Note that J_m is twice continuously differentiable in (U, **V**). If assumption (iv) of lemma 10.1 holds for J_m, Theorem 10.1 yields the conclusion.

☹ ***Forewarned is forearmed:*** We don't *know* $\mathbf{X}^*$, the limit point which is the presumptive center of the neighborhoods $\mathbb{N}(\mathbf{X}^*, \delta)$ and $\mathbb{N}(\mathbf{X}^*, \varepsilon)$, and even if we did, we can't know the values for δ and ε that will certify the theorem. We only know that if these all things exist, AO will converge q-linearly to this local extrema ☹

At this point, you may be wondering how many local extrema a particular objective function might possess? A function such as J_m, which has $c(p+n)$ unknowns, will usually have more than one – indeed, perhaps many – local minima and (rarely), saddle points. "Strong" local extrema should pull values of J_m toward themselves from a variety of starting positions, but you can never be sure about this subjective assertion. The number of bumps and the size of the local neighborhoods is very much a function of the data set under study. This emphasizes how important it is to use a variety of initializations in algorithms such as A6.1 and A7.1 to avoid stagnation at a "wrong" local solution or saddle point.

Don't confuse the theoretical "speed of convergence" with the "speed of iteration" during computation with an AO algorithm. The rate of convergence depends on the factors identified in Theorem 10.1, whereas the time to complete one iteration clearly depends on p and n, the dimension and number of samples in the input data, the host platform, programming language used, and so on. FCM will have the same rate of convergence whether $(p, n) = (2, 100)$ or $(2{,}000, 1{,}000{,}000)$, but it will obviously take much longer to complete each iteration for the larger input data set. The actual rate of improvement per iteration resides in the value of $\rho = \rho(S'(\mathbf{0}))$, the *spectral radius* of $\mathbf{S}'(\mathbf{0})$; the smaller ρ is, the faster successive iterates will get close to an extrema. In practice, computation of this number, which can be truly intimidating, is not usually worth the trouble, but now you know how to do it if you really want to know.

As a persistent reminder, don't confuse the theoretical notion of convergence with the implementational necessity for termination of AO computations at some point in the iterate sequence. Convergence is a theoretical landmark that is almost always just over the horizon: termination is what you do to reach the nearest point of land. Be aware that many authors use the word convergence (incorrectly) to signify that an algorithm has been terminated by a specified criterion.

The important thing to learn from Theorem 10.1 is that improvement of *any* AO scheme is locally q-linear. The most useful aspect of Theorem 10.1 is that it affords a benchmark of comparison for (possibly) competing algorithms. Thus, FCM/AO and the EM/AO algorithms have the same asymptotic convergence rate as any steepest descent algorithm (even though they are *not* steepest descent algorithms). So, AO is generally slower in the tail than, say, a Newton-Raphson type iteration that typically has quadratic convergence to a local extrema. The actual iteration speed for FCM during computation is much less than it is for EM because of the difference in their computational complexities. Now we turn to the global theory.

10.4 Global Convergence Theory

Global convergence in this context refers to the behavior of iterate sequences started from arbitrary initializations. The global result for AO in Theorem 10.3 establishes that AO sequences converge, at least along a subsequence, to minimizers or (special types of) saddle points of the objective function, starting from an arbitrary point in parameter space. This is *not* the same as saying that an AO algorithm guarantees to find the global extreme point of its objective function. Theorem 10.3 is based on an application of the quite general convergence theory of Zangwill (1969). This result, like the local one, is independent of the (t, s) scheme used to subdivide the input variables. Application of Zangwill's theory to AO for the special case t = s (cyclic coordinate descent) is sketched in both Zangwill (1969) and Luenberger (1969). The first global result for FCM in Bezdek (1980) for $t < s$ used Theorem 10.2.

Theorem 10.2. Zangwill's theorem.

Let $\Psi \subset \Re^s$, and let $P(\Psi)$ be the power set of Ψ. Let the *point-to-set* map $S : \Psi \rightarrowtail P(\Psi)$ represent an algorithm that generates the iterate sequence $\{\mathbf{X}^{(r+1)} \in S(\mathbf{X}^{(r)}) : r = 0, 1, \ldots\}$, beginning at an initial point $\mathbf{X}^{(0)} \in \Psi$. Also, let a *solution set* $\Omega \in \Re^s$ be given. Suppose

(Z1) All points in $\{\mathbf{X}^{(r+1)} \in S(\mathbf{X}^{(r)}) : r = 0, 1, \ldots\}$ are in a compact set $W \in \Re$;

(Z2) There is a continuous function $Z : \Psi \rightarrowtail \Re^s$ such that:

(a) If $\mathbf{X} \notin \Omega$, then $\mathbf{Y} \in S(\mathbf{X}) \Rightarrow Z(\mathbf{Y}) < Z(\mathbf{X})$;

(b) If $\mathbf{X} \in \Omega$, then either the algorithm terminates or $\mathbf{Y} \in S(\mathbf{X}) \Rightarrow Z(\mathbf{Y}) \leq Z(\mathbf{X})$;

(Z3) If $\mathbf{X} \notin \Omega$, then S is *closed* at $\mathbf{X}$.

Then either the algorithm terminates at a solution, or the limit of any convergent subsequence of $\{\mathbf{X}^{(r+1)} \in S(\mathbf{X}^{(r)}) : r = 0, 1, ...\}$ is a solution.

What does it mean? The truth is that in practical terms,it is not so useful to know that there is a subsequence of an iterate sequence which has a limit since you will not usually be able to find the subsequence hidden in the overall set of iterates. And compounding this disclaimer is the fact that the subsequence may itself be infinite; so we will never find such a limit by computational means. But Zangwill's theorem does tell us that AO iteration is well-behaved in a certain sort of weak sense, and it is reassuring to know that algorithms based on this idea are not too chaotic.

Proof. Zangwill (1969, p. 91).

The notion of a *closed* map in (Z3) is an extension for point-to-set functions of the idea of *continuity* for point-to- point functions. Since $\mathbf{S}$ in (10.11) and Theorem 10.3 is a point-to-point function, we will need only to verify continuity of $\mathbf{S}$ at $\mathbf{X} \notin \Omega$ to satisfy (Z3) in Theorem 10.2. With Zangwill's result in hand, we can have a crack at global convergence of AO:

Theorem 10.3. Global convergence of AO.

Assume that the EU property in Definition 10.3 holds for $f : \Re^s \rightarrowtail \Re$. Let $\mathbf{X} = (\mathbf{X}_1, \ldots, \mathbf{X}_t)^T, \Psi = (\Psi_1, \ldots, \Psi_t)$, and assume that $\Psi_i \subset \Re^{p_i}$ is compact $\forall i$. Further assume that *if* $\mathbf{X} \in \Psi$, *then* for $i = 1, \ldots, t$, $f(\mathbf{X}_1, \ldots, \mathbf{X}_{i-1}, \mathbf{Y}_i, \mathbf{X}_{i+1}, \ldots, \mathbf{X}_t)$ has a unique (global) minimizer over $\mathbf{Y}_i \in \Psi_i$. Let $\mathbf{S} : \Psi \rightarrowtail \Re^p$ be defined by (10.11) and define the fixed points of $\mathbf{S}$ as the solution set $\Omega = \{\mathbf{X} \in \Psi : \mathbf{S}(\mathbf{X}) = \mathbf{X}\}$. Finally, represent the iterate sequence as $\{\mathbf{X}^{(r+1)} \in S(\mathbf{X}^{(r)}) : r = 0, 1, \ldots; \mathbf{X}^{(0)} \in \Psi\}$. Then:

$$\mathbf{X}^* = (\mathbf{X}_1^*, \ldots, \mathbf{X}_t^*)^T \in \Omega \Rightarrow \mathbf{X}^* = \underset{\mathbf{X}_i \in \Psi_i}{\text{argmin}}\{f(\mathbf{X}_1^*, \ldots, \mathbf{X}_{i-1}, \mathbf{X}_i, \mathbf{X}_{i+1}, \ldots, \mathbf{X}_t^*)\}; \text{ and} \quad (10.15a)$$

$$\forall r,\ f(\mathbf{X}^{(r+1)}) \leq f(\mathbf{X}^{(r)}) \text{ with equality if and only if } \mathbf{X}^{(r)} \in \Omega; \text{and } \textit{either}: \quad (10.15b)$$

(a) there is an $\mathbf{X}^* \in \Omega$ and number r_0 such that $\mathbf{X}^{(r)} = \mathbf{X}^*$ for all $r \geq r_0$, **or**
(b) the limit of any convergent subsequence $\{\mathbf{X}^{(r_i)}\} \subset \{\mathbf{X}^{(r)}\}$ is in Ω. (10.15c)

What does it mean? Theorem 10.3 tells us when an AO scheme will converge, or have a subsequence that does, to some possibly local solution of NCOP *starting at an arbitrary feasible point* – this is the conventional meaning of the term global convergence in the optimization community. Existence and uniqueness of a global solution are *assumed* to get this result. You would, of course, need to establish that the EU assumption holds before either Theorem 10.1 or 10.3 is certifiable. And as noted, this is not a particularly easy thing to do. If you can verify the mountain of stated assumptions, which is impossible in all but the simplest cases, this theorem guarantees you that AO will find a local extrema of f that is close to the starting guess – from *any starting guess* (this is the "global" part of this result) in the parent space $\Re^s$. Again, please don't confuse the term "global convergence" with the term "convergence to a global solution."

***Sketch of Proof* [see Theorem 3 in Bezdek and Hathaway (2003) for a full proof].** First, we verify (10.15a). Since $(\mathbf{X}_1^*, \ldots, \mathbf{X}_t^*)^T \in \Omega$ is a fixed point of $\mathbf{S}$, it follows that $\mathbf{X}^*$ is also a fixed point of C_i in (10.10), for $i = 1, \ldots, t$, which implies that (10.15a) holds. To show the inequality in (10.15b), we have, using the definition of $S(\mathbf{X})$ at (10.11):

$$\begin{aligned} f(\mathbf{X}^{(r)}) &= f(\mathbf{X}_1^{(r)}, \mathbf{X}_2^{(r)}, \ldots, \mathbf{X}_t^{(r)}) \geq \min_{\mathbf{X}_1 \in \Psi_1} \{f(\mathbf{X}_1, \mathbf{X}_1^{(r)}, \ldots, \mathbf{X}_1^{(r)})\} = f(\mathbf{X}_1^{(r+1)}, \mathbf{X}_2^{(r)}, \ldots, \mathbf{X}_r^{(r)}) \geq \\ &\min_{\mathbf{X}_2 \in \Psi_2} \{f(\mathbf{X}_1^{(r+1)}, \mathbf{X}_2, \ldots, \mathbf{X}_t^{(r)})\} = f(\mathbf{X}_1^{(r+1)}, \mathbf{X}_2^{(r+1)}, \ldots, \mathbf{X}_t^{(r)}) \geq \cdots \geq \\ &f(\mathbf{X}_1^{(r+1)}, \mathbf{X}_2^{(r+1)}, \ldots, \mathbf{X}_t^{(r+1)}) = f(\mathbf{X}^{(r+1)}) \end{aligned} \quad (10.16)$$

Furthermore, if $\mathbf{X}^{(r)} \in \Omega$, then $\mathbf{X}^{(r+1)} = \mathbf{X}^{(r)}$ and equality must hold in the entirety of (10.15). If $\mathbf{X}^{(r)} \notin \Omega$, then for at least one value of i, we have $\mathbf{X}_i^{(r+1)} \notin \mathbf{X}_i^{(r)}$. And in this case, the uniqueness of solutions to $\underset{\mathbf{X}_i \in \Psi_i}{\text{argmin}}\{f(\mathbf{X}_1^{(r+1)}, \ldots, \mathbf{X}_{i-1}^{(r+1)}, \mathbf{X}_i, \mathbf{X}_{i+1}^{(r)}, \ldots, \mathbf{X}_t^{(r)})\}$ implies strictness in the chain of inequalities at (10.16), since $f(\mathbf{X}_1^{(r+l)}, \ldots, \mathbf{X}_{i-1}^{(r+l)}, \mathbf{X}_i^{(r)}, \ldots, \mathbf{X}_t^{(r)}) > f(\mathbf{X}_1^{(r+1)}, \ldots, \mathbf{X}_i^{(r+l)}, \mathbf{X}_{i+1}^{(r)}, \ldots, \mathbf{X}_t^{(r)})$.

Part (10.15c) is exactly the conclusion of Zangwill's convergence Theorem 10.2 applied to alternating optimization. Theorem 10.2 requires verification of the 3 conditions (Z1)–(Z3). Here is a sketch of this verification in the present case. The first condition is that all iterates $\{\mathbf{X}^{(r)}\}$ lie in a compact subset. This is true because of our assumption about each Ψ_i. The second condition follows directly from result (10.15b). The third condition of Theorem 10.2 is that the iteration mapping **S** must be a closed point-to-set mapping. Since in our case **S** is a point to point function, we can establish closedness by demonstrating that **S** is continuous. Continuity of **S** is demonstrated by showing that each of its compositional parts $C_1, \ldots, C_t$ is continuous. This follows by adapting the argument in the proof of Lemma 5.1 in Zangwill (1969, p. 105). ■

In the same way that Theorem 10.1 begets Corollary 10.1 for FCM, we use Theorem 10.3 to secure the global result for FCM.

Corollary 10.2. FCM Algorithm A6.1 (m>1) is globally convergent or has a subsequence that is.

What does it mean? The disclaimer for Theorem 10.3 (in *What does it mean*?) falls squarely onto this corollary. Again, if you can verify the stated assumptions, this corollary guarantees you that FCM will find a local extrema of J_m that is close to the starting guess – from any starting guess in $\Re^s$. Please note again that this corollary doesn't say that the iterate sequence will terminate near the global minimum of J_m.

Proof. Let $X = \{\mathbf{x}_1, \mathbf{x}_2, \ldots, \mathbf{x}_n\} \subset \Re^p$ have $c < n$ distinct points, and let the inner product norm $\|\mathbf{x}_k - \mathbf{v}_i\|_A^2$ be the model norm for FCM. For fixed $m > 1$, let $\Omega = \{(U, \mathbf{V}) \in \Psi = M_{fcn} \times \Re^{cp} : \mathbf{S}((U, \mathbf{V})) = (U, \mathbf{V})\}$, where **S** is defined as in (10.11). Let $\{(U^{(r+1)}, \mathbf{V}^{(r+1)}) = \mathbf{S}(U^{(r)}, \mathbf{V}^{(r)}) | r = 0, 1, \ldots; (U^{(0)}, \mathbf{V}^{(0)}) \in \Psi\}$. If the remaining conditions in Theorem 6.2 apply, we have for FCM algorithm A6.1, that either: (i) there is an $(U^*, \mathbf{V}^*) \in \Omega$ and number r_0 such that $(U^{(r)}, \mathbf{V}^{(r)}) = (U^*, \mathbf{V}^*)$ for all $r \geq r_0$, or (ii) the limit of any convergent subsequence $\{(U^{(r_i)}, \mathbf{V}^{(r_i)})\} \subset \{(U^{(r)}, \mathbf{V}^{(r)})\}$ is in Ω. ■

One more remark. Additional assumptions on the NCOP objective function (f) will yield a much richer theory. For example, if we assume that f is a strictly convex function on $\Re^s$ and that f has a minimizer $\mathbf{X}^*$ at which $\nabla^2 f(X^*)$ is continuous and positive-definite, then AO will converge to the minimizer from any initialization, and the rate of local convergence will be q-linear.

10.5 Impact of the Theory for the c-Means Models

Done with the theorems, let us try to understand what can be learned from them. First, note that HCM doesn't *QUITE* fall under the aegis of Theorems 10.1 and 10.3 because J_1 is not differentiable -indeed, it is not even continuous -in the U variables, and, so, there *can't* be a convergence theory for Lloyd's basic k-means algorithm (m=1 in A6.1) based on the structural requirements for the objective functions covered by the theory in this chapter.

Be careful not to confuse this seemingly contradictory remark with the well-known convergence-in-probability theory for *sequential* k-means (often loosely called k-means or c-means without reference to it being different than the basic AO version given here) presented by MacQueen (1967), which is NOT an AO scheme. Sequential k-means is a competitive learning scheme which is closely related to Kohonen's (1989) self-organizing map algorithm.

The Lloyd iteration (A6.1, m = 1 aka batch "k-means") is AO, but this case is not covered by the theory in this chapter. Indeed, as pointed out in Chapter 6, it is not entirely clear that basic HCM as given in A6.1 *should be* called an AO scheme, at least in the sense of the theory outlined here. Looking back at Definition 10.3, we can ask whether the requirement for restricted minimizations in each set of variables is satisfied for HCM (k-means). Certainly, the next **V** is found from the old U and the FONCs for **V** at Equation (6.25a), and this part of the iteration is covered by the first-order theory. But the next U is found with the last **V** and the 1-np rule (6.25b). Choosing the $\{u_{ik}\}$ with the 1-np rule (cf. Figure 6.3) guarantees that this next step will not increase the value of J_1, but you should tread lightly here. Technically, we might not want to consider the HCM case of A6.1 as AO, but this small difficulty is traditionally overlooked; so most authors continue to regard the implementation of basic HCM in A6.1 with m = 1 as AO, when it should perhaps more properly be referred to as alternating cluster estimation.

Figure 10.1 illustrates the difference between iterates for the two cases. View 10.1(a) shows FCM progressing toward a local minimum along a smooth, continuous path in each set of variables. Figure 10.1(b) is a cartoon depiction of J_1 iterates in (U,**V**) space. Points in this view are (U,**V**) pairs for one complete iteration of HCM. Hence, the appearance of a spiky "telephone" pole function because of the discrete nature of the crisp c-partitions available to HCM.

The colors in the (b) view are keyed to the colors of the soccer ball vertices shown in Figure 6.2 – red for "good U's," and yellow for "bad U's." As HCM proceeds, you can correctly imagine that, for a full cycle in its two subsets of variables, it hops from pole to pole, just like HCM bunny hops from vertex to vertex in Example 6.2. The red spikes in Figure 10.1(b) correspond to values in the set $S_{1,c}(\mathbf{X})$ in Equation (9.5). When HCM gets to a region (in this illustration, if the HCM bunny gets to the origin) where the next

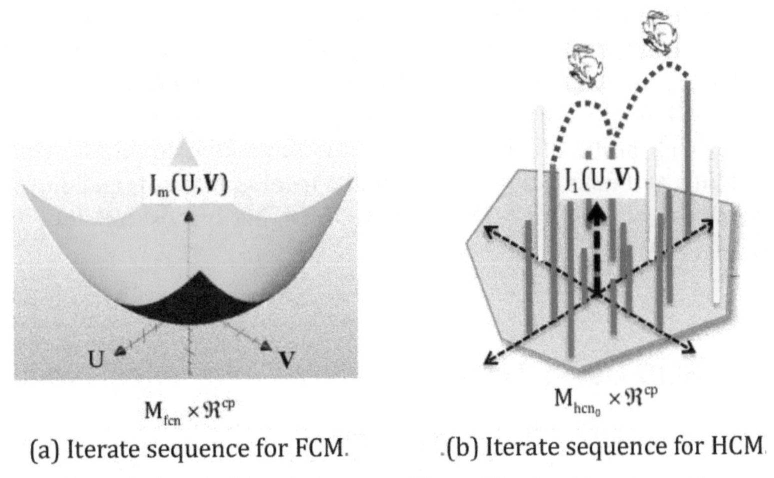

(a) Iterate sequence for FCM. (b) Iterate sequence for HCM.

Figure 10.1 (a): A local minimum of $J_{m>1}$. (b): a local trap state of J_1.

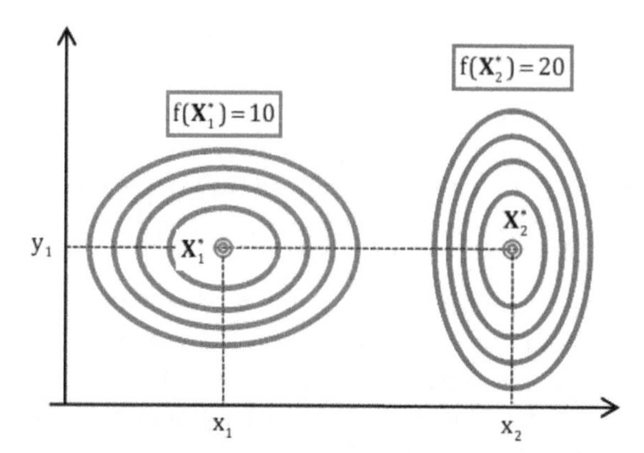

Figure 10.2 AO can't stall at the local minimum $\mathbf{X}_2^* = (x_2, y_1)$.

hop increases J_1, the spike values depicted around the origin in view 10.1(b) all increase; so this is where HCM bunny will stop if the chosen termination criterion is satisfied. Otherwise, A6.1 will continue to hop to another spot, which may or may not be geometrically close (since there is no concept of smooth neighborhood for the domain of J_1), and look some more. Because of this, we should always speak of *local extrema* for FCM but *local trap states* for HCM.

On the other hand, AO is fundamentally different from optimization routines in the joint variables such as those based on Newton's method and gradient descent, which are often trapped by the nearest local minimizer. Example 10.4, which is an expanded version of one in Bezdek and Hathaway (2003), shows this to be the case.

Example 10.4. An extrema that *can't* be *found by* AO.

Figure 10.2 shows some level sets of a continuous function $f(x, y)$ of two variables, partitioned as $\mathbf{X} = (X_1, X_2) = (x, y)$. This function has two local minimizers: $\mathbf{X}_1^* = (x_1, y_1)$ with $f(\mathbf{X}_1^*) = 10$ at the extreme point shown on the left, and $\mathbf{X}_2^* = (x_2, y_1)$ with $f(\mathbf{X}_2^*) = 20$ at the extreme point on the right.

The concentric ellipses about the two local minima are level sets of the function f. If Newton's method, for example, is started close enough to $\mathbf{X}_2^* = (x_2, y_1)$, the gradient will force it to descend quickly to the larger local minimum. Since (x_1, y_1) and (x_2, y_1) lie on a line parallel to the $X_1 = x$ coordinate direction, global minimization at each half-step of AO for X_1 will always converge to $\mathbf{X}_1^* = (x_1, y_1)$. That is, AO can *never* converge to $\mathbf{X}_2^* = (x_2, y_1)$ during this half step, and consequently, not at all. This property of AO sometimes saves it from stalling at uninteresting or undesirable local minimizers and moves it towards more global ones. This good result can happen for FCM and GMD/EM, but is most unlikely for HCM (because HCM is not, strictly speaking, AO)!

10.6 Convergence for GMD Using EM/AO

Convergence theory for GMD algorithm A7.1 is complicated by the fact that there are *two* kinds of convergence questions for this method. First, there are the same questions about numerical convergence for A7.1 that are posed for FCM A6.1 $(m > 1)$: is there any solution? if yes, is the solution unique? Does A7.1 converge locally and/or globally to stationary points of the maximum likelihood function? What is the rate of convergence?

Resolution of this set of questions rests on exactly the same pillar as FCM convergence – viz., Zangwill's theorem. The short answer is that roughly the same theory about *numerical convergence* applies to A7.1 as A6.1 for the special case when the component densities are Gaussian and satisfy the assumptions we have posted for GMD. There is more computation involved, but the AO theory is essentially the same.

However, beyond this (and very much *unlike* FCM!), the EM algorithm A7.1 for GMD estimates the parameters of the component densities of a finite mixture of Gaussian distributions. This adds another set of questions about statistical properties of any limit point of the procedure: are the obtained values good estimates of the unknown parameters of the component distributions? In other words, the probabilistic context of A7.1 adds a second set of questions about (statistical) convergence. Are the estimates $\mathbf{Q}^* = (\mathbf{p}^*, \mathbf{M}^*, \{S_1^*\})$ sufficient? unbiased? efficient? consistent? is convergence "in probability" (pointwise, uniform)? and so on.

This second set of questions don't concern the use of A7.1 to obtain a partition matrix P^* of X – i.e., they don't affect the quality or utility of GMD for probabilistic clustering; so this set of questions lies well outside the scope of this volume and will not be covered in detail here. However, this disclaimer precludes coverage of many interesting results that are connected to relatives of A7.1. Peters and Walker (1978) discuss convergence of Picard iteration through the necessary conditions (7.16a)–(7.16c). Their analysis skips Bayes Rule (7.16d) because this formula is not needed to produce estimates $\mathbf{Q}^* = (\mathbf{p}^*, \mathbf{M}^*, \{S_1^*\})$ of the mixture parameters[27]. For example, they show that with probability 1 as n approaches infinity, iteration through (7.16a)–(7.16c) converges locally to a strong consistent ML estimator of the mixture parameters whenever a certain parameter epsilon lies between 0 and 2. Redner and Walker (1984) update these earlier results and discuss the general case for EM – i.e., without specification of a particular family of PDF's for the components of the mixture. Redner *et al.* (1987) extend the theory just a bit by introducing modal estimators into the model.

The case of interest for us is iteration through all four FONCs (7.16a)–(7.16d) for MLE in the Gaussian mixture model. EM algorithm A7.1 is covered by Definition 10.3, i.e., it is a (t, s) scheme with $t = 4$ sets of explicitly coupled variables, viz., the priors at (7.16a), the means at (7.16b), the covariance matrices at (7.16c), and the posterior probability matrix (the partition matrix) at (7.16d). Consequently, corollaries for A7.1 analogous to 10.1 and 10.2 for A7.1 can be verified in much the same way as was done for FCM. The proofs are left as exercises for the mathematically inclined.

10.7 Notes and Remarks

The field of numerical optimization is much wider than the limited view of it given in this chapter. The field has changed over the years to reflect changes in the theory, and the evolution of the software and hardware available for implementation has greatly altered this branch of mathematics in the last 40 years. But, the underlying basis for much of the modern treatment is quite accessible in many good texts that provide a general introduction to the theory and analysis of iterative algorithms for solving the NCOP problem. In particular, good general references include Gill *et al.* (1981), Dennis and Schnabel (1983), and Luenberger

[27]We need Equation (7.16d) because this is precisely the equation that produces the probabilistic clusters which are the object of interest for clustering in Chapter 7.

(1984). Good advanced treatments that will carry you into the deepest part of this jungle are Collatz (1966) and Luenberger (1969).

The book on non-linear optimization by Zangwill (1969) is the cornerstone for convergence theories for the AO approach given in both (A6.1, m>1) and A7.1 (GMD). The literature on convergence theory for AO has many variations. There are several slightly different versions of the theory presented in this chapter in other papers and books. This type of optimization has other names in the literature (e.g., grouped coordinate descent, block relaxation, etc.). See de Leeuw (1994) for an excellent account of many aspects of this field.

Corollary 2.2 was published in Bezdek (1980) and subsequently improved (actually, corrected just a bit) in Bezdek *et al.* (1987). The general case of AO that pertains to the special cases of algorithms in this volume is covered in some detail in Bezdek and Hathaway (2003). Aficionados of the theory will enjoy reading about conditions that identify the character of a local extrema of J_m , which has been discussed in several papers. For example, Selim and Ismail (1987) proved that the point generated by A6.1 when $m > 1$ is a local minimum provided that the reduced objective function is convex at the point. Kim *et al.* (1988) extended the results of Selim and Ismail to include a fast test of local optimality based on the eigenvalues of a transformation of the Hessian of the reduced function. Their test covers both local minima and saddle points and shows that symmetry of the data set X and $\mathbf{V}_0$ about the origin always results in a saddle point. Wei and Mendel (1994) improved the test for local extrema and showed that if all of the u_{ik} in any row of a terminal partition U* from FCM are less than $(m + 1)/2m$, then U* is a saddle point.

Algorithms A6.1 and A7.1 are examples of *explicit* AO. There is another, more difficult form of AO called *implicit* AO. To see the difference, suppose that we want minimizers $X_i^{(r+1)} = \underset{X_i \in \Psi_i \subset \Re^{p_i}}{\operatorname{argmin}} \{f(\mathbf{X}_1^{(r+1)}, \ldots, \mathbf{X}_{i-1}^{(r+1)}, X_i, \mathbf{X}_{i+1}^{(r)}, \ldots, \mathbf{X}_t^{(r)})\}$ for $i = 1, 2, \ldots, t$. The necessity for the gradients of these t restricted functions to vanish at any candidate solution leads to a set of t coupled equations:

$$\left\{\begin{array}{ll} \nabla_{X_1} f(X_1, \mathbf{X}_2^{(r)}, \ldots, \mathbf{X}_i^{(r)}, \ldots, \mathbf{X}_t^{(r)}) = \mathbf{0} & (\in \Re^{p_1}) \\ \nabla_{X_i} f(\mathbf{X}_1^{(r+1)}, \ldots, \mathbf{X}_{i-1}^{(r+1)}, X_i, \mathbf{X}_{i-1}^{(r)}, \ldots, \mathbf{X}_t^{(r)}) = \mathbf{0} & (\in \Re^{p_i}) \\ \nabla_{X_t} f(\mathbf{X}_1^{(r+1)}, \ldots, \mathbf{X}_2^{(r+1)}, \ldots, \mathbf{X}_{t-1}^{(r+1)}, X_t) = \mathbf{0} & (\in \Re^{p_t}) \end{array}\right\} \quad (10.17)$$

There are two possibilities for the system (10.17). When we can solve each equation explicitly for the active variables as functions of the remaining variables, Equation (10.17) leads to

$$\left\{\begin{array}{l} X_1^{(r+1)} = F_1(X_2^{(r)}, \ldots, X_i^{(r)}, \ldots, X_t^{(r)}) \\ X_i^{(r+1)} = F_i(X_1^{(r+1)}, \ldots, X_{i-1}^{(r+1)}, X_{i+1}^{(r)}, \ldots, X_t^{(r)}) \\ X_t^{(r+1)} = F_t(X_1^{(r+1)}, \ldots, X_i^{(r+1)}, \ldots, X_{t-1}^{(r+1)}) \end{array}\right\} \quad (10.18)$$

System (10.18) is the basis of *Explicit* AO, which allows us to define the iteration function $\mathbf{S} : \Re^s \rightarrowtail \Re^s$ as the vector field $\mathbf{S} = (\mathbf{F}_1, \mathbf{F}_2, \ldots)$, so that $\mathbf{X}^{(r+1)} = \mathbf{S}(\mathbf{X}^{(r)})$, $r = 0, 1, \ldots$. Well known examples of systems of this type include A6.1 (FCM) and A7.1 (GMD/EM), $t = 2$ for FCM, $t = 4$ for GMD/EM. Gray and Neuhoff (1998) are a good source of additional examples of explicit AO algorithms.

Less cooperative problem functions yield a system which *can't* be (completely) solved explicitly as in Equation (10.18). We call this situation *Implicit* AO. In this case, the system can look pretty intimidating since one or more subsets of variables will appear on both sides of the subproblem equations. AO in this harder case is still possible but requires an additional level of effort -namely, *numerical solution* of each implicit necessary condition *at each iteration*. This harder type of alternating optimization is encountered, for example, in the *fuzzy c-shells* clustering algorithm of Dave (1990). This sounds very bad – but, sometimes, it looks worse than it is. Bezdek and Hathaway (1992) proved that each alternation through the implicit FONCs for the fuzzy c-shells model requires but one iteration of Newton's method at each half step for maximum usable accuracy; so things may not be as bad as they seem.

It is also worth noting that the theorems in this chapter are given under a very specific set of assumptions. Theorems 10.1 and 10.3 establish a baseline for AO algorithms that fall with the context described here, but there are AO algorithms for HCM (k-means), for example, that converge faster. There are cases where HCM can converge to a local trap state (most authors incorrectly call it a local minimum) at quadratic speed. For example, Bottou and Bengio (1995) argue that batch k-means (A6.1, $m = 1$) converges quadratically when it is cast as an EM-like algorithm and, further, that it begins close enough to a local extreme point. They assert that, in this case, the quadratic Newton approximation is exact and that batch k-means converges in one step.This is exactly in line with the observation made by Bezdek and Hathaway (1992) for each half-step of the fuzzy c-shells algorithm. All these convergence results about HCM really tell us how quickly we can reach a local trap state when the initial conditions start out very close to an (unknown) terminal state.

Evolution of the theory of convergence for the EM algorithm follows roughly the same time line and path as that for FCM. The papers of Dempster *et al.* (1977), Wu (1983), and Redner and Walker (1984) are the classic references for MLE and the EM algorithm. The importance and contributions of Dempster *et al.* (1977) to the theory were discussed in the notes and remarks for Chapter 7. A glimpse into the history of the emerging theory is found in this quote from p. 4 of their paper:

> *Some of the theory underlying the EM algorithm was presented by Orchard and Woodbury*[28] *(1972), and by Sundberg (1976), and some has remained buried in the literature of special examples, notably in Baum etal. (1970).*

Dempster *et al.* (1977) gave four theorems related to the convergence theory for *generalized EM* (GEM). Subsequently, Wu (1983) showed that their Theorems 2 and 3 were incorrect and provided what is widely regarded as the definitive convergence theory for GEM and EM. Section 2 of Redner and Walker (1984) provides a very thorough summary and historical review of methods for estimating the parameters in a mixture of probability density functions. There have been many papers in the last four decades that update and improve on these basic results, but, for now, these will do.

Finally, don't confuse *alternating optimization* with the iterative optimization algorithm called the *method of gradient descent* (or steepest descent). Property (10.15b) in Theorem 10.3 asserts that all AO algorithms that satisfy the hypotheses have the descent property (that is, successive iterates will continue to decrease the value of f or, at worst, not increase its value). And we know that the FONCs underlying FCM arise by zeroing the gradients of J_m with respect to U and $\mathbf{V}$. From these two facts, it is easy to think that FCM is a gradient descent method, but it is not. Gradient descent methods are similar to AO iteration in the sense that both depend on first derivatives, but the derivatives are used in very different ways.

[28]Max Woodbury was one of the few early specialists in probability theory who also paid serious attention to the evolving area of fuzzy clustering (cf., Woodbury and Clive, 1974).

Specifically, the method of steepest descent assumes that the function f to be optimized has continuous first partial derivatives, and is based on the iterative equation $\mathbf{X}_{k+1} = \mathbf{X}_k - \alpha_k \nabla f(\mathbf{X}_k)$, $k = 0, 1, \ldots,$. The "learning rate" is the positive scalar α_k that minimizes the function $F(\mathbf{X}_k - \alpha_k \nabla f(\mathbf{X}_k))$, which amounts to a line search along the direction of the negative gradient $(-\nabla f(\mathbf{X}_k))$ which begins at the point $\mathbf{X}_k$. The minimizer of F becomes the next value, $\mathbf{X}_{k+1}$. This method has many modern variations and improvements. It makes its way into these notes because it provides a point of reference for the theory of AO given in this chapter. Specifically, steepest descent converges linearly with a convergence ratio $\rho \le \left[\frac{(\lambda_M - \lambda_m)}{(\lambda_M - \lambda_m)}\right]^2$, where λ_M and λ_m are the maximum and minimum eigenvalues of the Hessian matrix $\nabla^2 f(\mathbf{X}^*)$ of f evaluated at the solution. In most cases, there is equality, and many authors that use steepest descent take $\rho = \left[\frac{(\lambda_M - \lambda_m)}{(\lambda_M - \lambda_m)}\right]^2$. When $\lambda_M = \lambda_m$, $\rho = 0$ and gradient descent (in all the variables) is at its best (is fastest): the contours of f are circular and convergence occurs in a single step. Gradient descent becomes slower and slower as the smallest and largest eigenvalues become further apart (the contours become more and more elliptical). This result is the cornerstone of comparative analysis for the convergence theories of many optimization algorithms.

10.8 Exercises

1. Find the limit and order of convergence for each real sequence ($n \in \mathbb{N} = \{1, 2, \ldots, \}$):

1.a $\{z^{2^n} : 0 \le z < 1\}$ ans: $\{z^{2^n}\} \to z^* = 0$; quadratic

1.b $\{z_n = 1 + 0.5^{2^n}\}$ ans: $\{z_n\} \to z^* = 1$; quadratic

1.c $\{z_n = 1 + 0.5^n\}$ ans: $\{z_n\} \to z^* = 1$; q-linear

1.d $\{z_n = n^{-n}\}$ ans: $\{z_n\} \to z^* = 0$; superlinear

1.e $\{z_n = 1 + n^{-n}\}$ ans: $\{z_n\} \to z^* = 1$; superlinear

2. Answer the following questions:

2.a Does it matter that the grouped variables **U** and **V** are switched in the AO formulation in Example 10.1?

2.b Does it matter if you switch the internal variables, for example $(X_1, X_2) \to (X_2, X_1)$, in Example 10.1?

2.c How many (t, s) schemes exist for the problem $\min\limits_{\mathbf{X} \in \Psi \subset \Re^s} \{f(\mathbf{X})\}$?

2.d How many (t, s) schemes exist for problem (6.9), $\min\limits_{\mathbf{M}_{hcn} \times \Re^{cp}} \{J_1(U, \mathbf{V}; X)\}$?

2.e How many (t, s) schemes exist for problem (6.10), $\min\limits_{\mathbf{M}_{fcn} \times \Re^{cp}} \{J_m(U, \mathbf{V}; X)\}$?

2.f How many (t, s) schemes exist for problem (7.7), $\underset{\mathbf{Q} \in \Omega}{\text{maximize}} \{\log L(\mathbf{Q}; X) = \sum_{k=1}^{n} \log f(\mathbf{x}_k; \mathbf{Q})\}$?

3. The general unconstrained quadratic optimization problem is $\min\limits_{\mathbf{X} \in \Re^s} \{f(\mathbf{X}) = \mathbf{X}^T A \mathbf{X} + \langle(\mathbf{B}, \mathbf{X})\rangle + C\}$, where **B** is an s-vector and C is a real constant.

3.a State the conditions for which this problem has a unique solution $\mathbf{X}^*$; then find a general formula for it. State any assumptions you make to secure the answer.

3.b Suppose $A = \begin{bmatrix} 0.78 & -0.02 & -0.12 & -0.14 \\ 0.02 & 0.86 & -0.04 & 0.06 \\ 0.12 & -0.04 & 0.72 & -0.08 \\ 0.14 & 0.06 & -0.08 & 0.74 \end{bmatrix}$ $\mathbf{B} = (0.76, 0.08, 1.12, 0.68)$ and $C = 0$.

Prove that A is positive-definite. Find the exact solution for these values. ans. (to 2 digits) $\mathbf{X}^* = (1.53, 0.12, 1.98, 1.41)^T$

3.c Formulate this problem as a $(t, s) = (2, 2)$ scheme, derive the FONCs for AO to estimate the solution of the problem $\min_{\mathbf{X}\in\Re^s}\{f(\mathbf{X}) = \mathbf{X}^T A\mathbf{X} + \langle(\mathbf{B}, \mathbf{X})\rangle + C\}$ using this split of variables, and then write a program that implements your scheme using the same problem parameters as in Example 10.1 Run your program and compare your AO solution to the one gotten by closed form in 3.b.

4. The EM algorithm A7.1 for Gaussian mixture decomposition is a (t, s) scheme. Write the particulars of this method (i.e., identify t, s, the unknown variables and their dimensions) in the same form as that shown for FCM in Example 10.1.
5. State and prove a corollary for A7.1 that is similar to corollary 10.1 for A6.1.
6. State and prove a corollary for A7.1 that is similar to corollary 10.2 for A6.1.
7. (Refer to the last paragraph of Section 10.6.) Show that the Hessian matrix of $f(\mathbf{X}) = f(X_1, X_2, X_3X_4) = \mathbf{X}^T A\mathbf{X}$ is 2A. Then estimate the convergence rate of gradient descent for the problem $\min_{\mathbf{X}\in\Psi=\Re^4}\{\mathbf{X}^T A\mathbf{X}\}$ with $\begin{bmatrix} 100 & 80 & 5 & 1 \\ 80 & 90 & 2 & 1 \\ 5 & 2 & 70 & 40 \\ 1 & 1 & 40 & 80 \end{bmatrix}$. Compute the first four values of the estimated solution from a starting guess of $\mathbf{X}_0 = (1111)^T$. Compare this to the solution of this problem using AO as shown in Example 10.1.

11

Clustering in Static Big Data

... Curiosity and the urge to solve problems are the emotional hallmarks of our species; and the most characteristically human activities are mathematics

–Carl Sagan

Chapters 1–10 present a more or less comprehensive introduction to the art and science of cluster analysis in batch data by discussing the basics of just four models and algorithms. If you made it this far, you have graduated from the primary school of clustering, and now might be thinking: what is next? Well, almost anything could be in the conversation. Perhaps, the most important topic that was not given the attention it deserves in this volume was *cluster validity*; so this would be an excellent area for you to study in middle school (the second stage of your clustering education).

In this volume, however, I am going to skip a detailed presentation of cluster validity and, instead, briefly discuss what is presumably the most important new topic related to generalizations of the Fantastic Four that you are going to see (wait -that you are *guaranteed* to see!) in the next decade. This chapter takes you on a short safari into the jungle of clustering in static *big data* (aka *massive data, huge data, etc.*). The aim of Chapter 11 is to see how and where the basic algorithms discussed in Chapters 6–8 can (and already have) intersected with the world of clustering in collected batches of big data. No attempt is made to present all the latest wrinkles in this field. Rather, the aim is to get an idea about different ways currently being used to approach the problem of clustering in static big data that can't be processed all at once with the Fantastic Four. Since Chapter 11 is concerned ONLY with *static* data sets, the word "static" is sometimes dropped for the shorter term "big data." Chapter 12 will introduce the field now called "streaming clustering," the latest wrinkle on processing data that stream by in time.

11.1 The Jungle of Big Data

Inarguably, ***BIG DATA*** is the ***BIG BUZZWORD*** of the decade. Unlike real jungles on Planet Earth (many of which are rapidly shrinking or are already gone), the big data jungle is growing like a teenager juiced with steroids. But concerns about how to process big data sets are hardly new. Hill (1990) provided a lucid discussion about scaling up computational methods more than 30 years ago. What *is* new is the current and almost feverish cachet of this term, which has taken a death-like grip on the scientific community in the last 15 years or so.

There are many examples of the phenomenon that a new buzzword creates in science and engineering (in some circles, you would not be wrong to call it a feeding frenzy!). Here are a few familiar examples: neural network, machine learning, evolutionary computation, data mining, fuzzy systems, knowledge discovery, computational intelligence, and so on[29]. For example, machine learning used to be regarded as part of the larger domain called pattern recognition, but the emergence of really useful AI algorithms-principally deep learning in massive neural networks -now gives this term a life of its own. Terms like this come and go for a number of good reasons, but I think this one (big data) is here to stay. Look no further than this statement about unseen efforts by "quants" to secure President Obama's second election taken from Scherer (2012):

> *Data-driven decision-making is another sign that the role of the campaign pros in Washington who make decisions on hunches and experience is rapidly dwindling, being replaced by quants and computer coders who can crack massive data sets for insight. As one official put it, the time of "guys sitting in a back room smoking cigars, saying 'We always buy 60 Minutes"' is over. In politics,* ***the era of big data has arrived****.*

Note 1: [I added this note to the first version of this book on November 20, 2016, about two weeks after Donald Trump won the 2016 election to become the next president of the United States.] From the available evidence, it appears that the Trump campaign used very little of this type of information processing to secure the victory, while Hillary Clinton apparently scaled up the Democratic quant army quite a bit. But to no avail. So? Perhaps, this suggests that the era of big data is here, but it is not so clear that clustering provides a window into "better" results or more useful information.

Note 2: [I added this note on October 16, 2020, three weeks before the 2020 presidential election.] As you will all be aware, it is Trump vs. Biden for the presidency of the USA. In the four years since I wrote Note 1 above, it has become clear that the manipulation of public opinion on social media sites such as Facebook, Twitter, Instagram, Google, and Amazon, using "AI algorithms" to process big data has become a very serious threat to democracy. This prediction was made by O'Neil (2016) in her very accessible book about the good, the bad, and the ugly of big data processing. Please read this little book before you fall in love with big data.

11.1.1 An Overview of Big Data

Well, what *is* big data? That depends on when you ask, who you ask, and what resources they have. Indeed, you can ask a bit more -what *is data*? Here is a definition that appeared not so long ago in Time Magazine (July 6, 2015):

> *Data means any sort of human artifact that can be represented quantitatively and analyzed by a computer.*

While this is a wonderfully general definition, it is a little too broad to be very informative for us. In this book, data is numerical, and, from now on, big data means really big sets of numbers that are already measured and stored (*static data*). The only parameters of big data considered in this chapter are the number of objects (N); and when the data are feature vectors, their dimension (P). So, what is big numerical data? Again, this depends on who, when, and where. This sentence from Cannon *et al.* (1986) is an amusing flashback to the problem of data size when segmenting images with clustering algorithms in 1986:

[29]Two papers by Bezdek (2015, 2016) contain extended discussions about buzzwords - the good, the bad, and the ugly. It is not a trivial topic. Buzzwords are important to science and engineering.

For this study, our images were 256×256 *pixels, thus making* $n = 65,536$. *Larger values of n, say 1 Mbyte, are easily encountered in imaging applications.*

Since the obvious problem with this statement today is that 1 Mb is not so big anymore (indeed, it is relatively quite small), it doesn't seem very useful to even define what big data is. No matter what computing resources you have, there have been and always will be plenty of data sets that will overburden them. It is perfectly clear that technology will advance our ability to collect and store more and more data of higher and higher dimensions until we reach some asymptotic limit of information processing. And there is no doubt that many applications will require exploratory analysis (clustering) in these big data sets.

You won't have any trouble locating information about big data. The IEEE has jumped onto this bandwagon in a big way. There is an *IEEE Transactions on Big Data* sponsored by the IEEE Computer Society. There is the *IEEE Big Data Congress* and an annual *IEEE Big Data Conference*. There are the inevitable (sometimes self-titled) "rock stars of big data." There are dozens of workshops such as the joint IEEE/ACM International Symposium on Big Data Computing. Some very recent textbooks on clustering have begun to give a shout out to big data (cf., Aggarwal and Reddy, 2016). And, of course, you can visit the Internet. I visited http://dataconomy.com/seven-vs-big-data/ on December 17, 2015 and discovered, to my surprise, that the number of ways that data can be "big" has jumped from 3 to 5 to 7, seemingly overnight, also discussed in the paper by Khan *et al.* (2014). I am not sure why the letter "V" has taken the lead in this avalanche of new terms, but, for the record, the list on this day was:

1. **V**olume (big values of c, n, and p in numerical data)
2. **V**elocity (increasing speed of creating and processing data)
3. **V**ariety (data in the generalized sense per the quote above)
4. **V**ariability (data that undergoes changes in meaning)
5. **V**eracity (data that is accurate)
6. **V**isualization (ways to visualize big numerical data)
7. **V**alue (data is worthless; analysis of it adds value)

The most important members of the 7 V's list for clustering in big numerical data are volume and visualization, exacerbated to some extent by velocity. By August 2017, there was (a somewhat whimsical) list of 42 V's! See http://www.elderresearch.com/ company/blog/42-v-of-big-data for the list.

A tacit assumption in cluster analysis is that the data are accurate and that the discovery of CPOV cluster structure has value. Generalizations of the Fantastic Four for big data applications? There are already too many to even catalog in any meaningful way. In one hour of Internet search, you will be able to find more papers about this than you can read in your lifetime. So, only a few references are given here to get you started. Some of these papers have historical importance or introduce a new way to think about clustering in big data that may suggest new models and algorithms to you. Others have already been cited in the text, but it is convenient to relist them here in a collection specifically organized toward clustering in big numerical data. And there are some methods that have already been supplanted by better, faster, or more accurate ways to do the job.

11.1.2 Scalability vs. Acceleration

In this chapter, *N and P are big; n and p are small.* Big static data sets such as $X_N = \{\mathbf{x}_1, \ldots, \mathbf{x}_N\} \subset \Re^P$ are (in principle) collected, stored and processed as sets of vectors. When X_N is unloadable because of the sizes of N and/or P, it is big data, and the question becomes how to look for clusters in it? Static big dissimilarity data ($D_{N\times N} \in M_N^+$, see (8.2a)) impacts the usefulness of the SL algorithm (A8.1/8.2). The parameter that is most often reduced in both cases is N, the number of objects or samples represented in the input data. The reduction of a large number of dimensions (P) to a more manageable number of features (p) by feature extraction or selection is also an important aspect of big data analytics.

Let's introduce the term "*literal clusters*" to mean a partition resulting from successfully running an exact (or literal) implementation of any crisp, fuzzy or probabilistic clustering algorithm (***A***) on any data set. Literal versions of the Fantastic Four are abbreviated in this chapter this way: LHCM is A6.1 (m = 1); LFCM is A6.1 (m > 1); LEM is A7.1; LSL is A8.2. All of the partitions discussed in Chapters 6–9 have been sets of literal clusters. Many approaches to clustering in big data amount to finding a way to approximate literal implementations.

For convenience, let BD_N represent both types (vector or matrix) of big static data, and let SD_n denote small data subsets of BD_N. If we could run the exact, literal version of (***A***) on BD_N the result would be the partition $U_N^{literal} = \boldsymbol{A}[BD_N] \in M_{fcN}$. But we can't. So, we search for ways to scale up clustering algorithms that can produce the exact result on the small data, $U_n^{literal} = \boldsymbol{A}[SD_n] \in M_{fcn}$. Then this result is extended to BD_N in various ways.

Ganti *et al.* (1999a) state that an algorithm is *scalable* if its runtime complexity increases linearly with the number of records (n) in the input data. The hard and fuzzy c-means algorithms (A6.1) are scalable in this sense but are famously slow (q-linear local convergence rates) when processing lots of samples; so scalability in (n) alone is often not enough. And the basic EM algorithm is quite often even slower. What to do?

Scalability seems like an eminently reasonable requirement for clustering algorithms. However, I said in the preface that I don't believe that "if it doesn't work for big data, it's not worth having." Why? Because there will always be applications that can use purpose-built methods that may work quite well for little data (the shallow end of the pool) but don't scale well for big data (the deep end of the pool). Moreover, improving any model and algorithm is best done by starting out small, seeing if it works at all, and then, if it does, pushing it toward a desired performance objective, which might include scaling up to big data. So, it's good to think about how the methods covered in this book -presented only for static (batch) data and illustrated only on small data sets -can be scaled up to bigger and bigger data sets.

Scalability for clustering algorithms applied to big data is often confused with the acceleration of clustering algorithms that are (or were) developed for small data. One approach is to make them run faster with acceleration methods that change their basic rate of convergence. Some methods that try to make the Fantastic Four run faster while not losing too much in terms of approximating the literal result have been discussed in Chapters 6–8. In this chapter, we approach the problem in a different way. Figure 11.1 depicts the relationship between acceleration and scalability in big and little static data sets. We are always on the lookout for ways to make clustering algorithms run faster, but as Figure 11.1 suggests, no amount of acceleration solves the big data problem. There are three data sets in Figure 11.1:

(i) BD_∞ is an infinite population of big data sets.

(ii) BD_N is one member of the population $\sim$ too big to load, so direct calculation of $U_N^{literal} = \boldsymbol{A}[BD_N] \in M_{fcN}$ is not possible. Big data clustering methods enable us to approximate the literal partition, $U_N^{approx} \cong U_N^{literal}$ of the big data. The objective for this type of clustering is *feasibility* (can we find any way to approximately cluster the big data?). Algorithms that can be accelerated for small data may make approximate clustering in big data faster, but acceleration is usually decoupled from scalability issues.

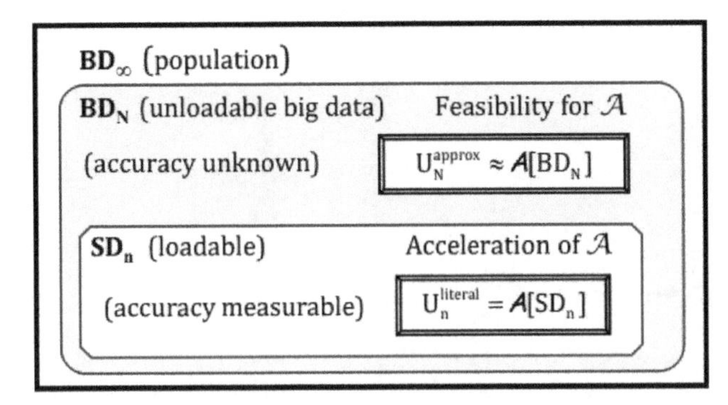

Figure 11.1 Clustering with algorithm (*A*) in big and little static data sets.

(iii) SD_n is a loadable subset of BD_N, which is perhaps constructed by first sampling the big data. Assuming that we can run (*A*) on the sample (well, of course we can; why do this if we can't?), algorithm (*A*) finds literal CPOV clusters $U_n^{literal} = \boldsymbol{A}[SD_n] \in M_{fcn}$ in the small data set. Feasibility for processing SD_n is not an issue ~ we can do it. But acceleration is still on the table ~ can we do it faster? A few of the many schemes for accelerating basic HCM/FCM and GMD/EM were reviewed in Chapters 6 and 7, some are approximations, some speed up the literal algorithm in a clever way. In the end, methods of this kind are inevitably caught in a tradeoff between computation time and accuracy of the approximate partition.

11.2 Methods for Clustering in Big Data

There are lots of ways to classify clustering algorithms that scale up to big static data. For example, we might focus on the style of processing the inputs: streaming, incremental, distributed, and so on. But there is no classification scheme that really captures the variety of methods in the current literature. Figure 11.2 illustrates four common approaches to the problem of clustering in big static data sets. The organizational basis of this illustration is the way the big data is broken into loadable pieces. There are many schemes that don't fit very well into this diagram, but the four styles described here suffice to organize this chapter in a way that introduces a wide variety of approaches. Three of the four methods shown in Figure 11.2 apply to static big data. The fourth method shown in this figure is called "streaming," which is not a method for finding approximate clusters in static big data. This method appears in this diagram because many authors of "streaming clustering" algorithms (cf., Chapter12) describe their approach as an alternative to methods used in static big data clustering, on the presumption that stream processing avoids the collection of batch data altogether. We will see how this unfolds in Chapter 12.

Figure 11.2(a). The left most tier in Figure 11.2 shows one alternative to collection, storage, sampling, and batch processing. In this approach, the points are regarded as *streaming data*; so BD_N is never stored but is replaced by a data stream, say, $\mathbf{x}_1, \ldots, \mathbf{x}_N, \ldots$ or $\{d_{ij,1}, \ldots, d_{ij,N}, \ldots : \forall_{i,j}\}$ that arrives as a time-ordered sequence at times $t_1 \leq \ldots \leq t_N \leq \ldots$. The idea is to build "clusters" (possibly beginning with $\mathbf{x}_1$ as the first cluster center) in the stream, regarding incoming samples as new inputs that are processed incrementally. Processing streaming data thus avoids the big storage and time complexities associated with batch processing of BD_N. However, this is not clustering in any classical sense of the word. So, this method is a hitchhiker for the topic of Chapter 11: i.e., it is not a method for approximate clustering in big static data, so it is crossed out in Figure 11.2.

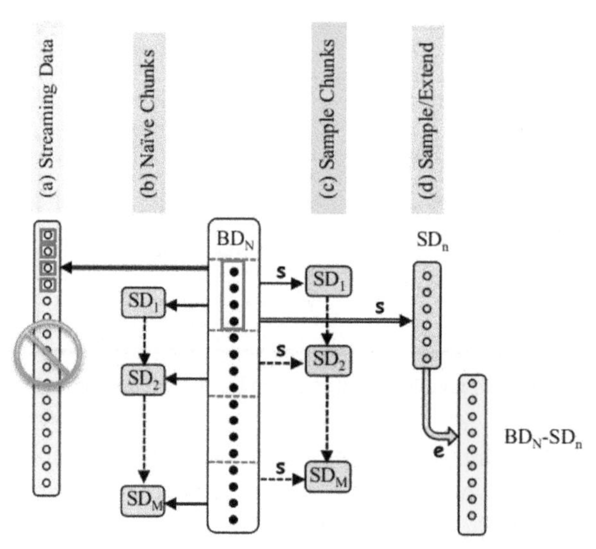

Figure 11.2 Three ways (and a hitchhiker) to make static big data look small.

Figure 11.2(b). The second tier in Figure 11.2(b) depicts clustering in loadable chunks of big static data. Figure 11.3 is a closer look at Figure 11.2(b) which shows three different clustering scenarios that fit under the umbrella of Figure 11.2(b). One style of data handling (upper left in view in Figure 11.3) corresponds to naive chunking of BD_N into M loadable subsets $\{SD_i\}$. The partitioning of BD_N is done by naively cutting successive chunks out of the big data (*naive chunking*)[30]. Once obtained, each chunk is clustered literally by algorithm (***A***), which produces M literal partitions, $\{U_1^{literal}, \ldots, U_M^{literal}\}$. Finally, an aggregation function (***a***) combines the M results, $U_N^{approx} = \boldsymbol{a}\{U_1^{literal}, \ldots, U_M^{literal}\}$. There are two substyles in this approach. Each chunk may be processed in *parallel*, which amounts to complete separation of the chunks. The second way to use naive chunks is to cluster the first chunk with a literal algorithm, summarize the results, and then move along to the next chunk. This method of handling naive chunks is sometimes called *partial data access*, and if only one pass is made through BD_N, the method is called a *single pass* (or single scan) algorithm.

The third situation which is represented in Figure 11.3 is *distributed clustering*. The data underlying distributed (or parallel) clustering is regarded in a somewhat different light than either as chunks of big static data or as streaming data. The Fantastic Four are all centralized algorithms, but all of them have been reconfigured for distributed (or parallel) implementation. Some are direct adaptations of centralized algorithms, which rely on statistics that are easy to compute in a distributed manner. Others generate summaries of local data and transmit them to a central coordinator which then performs the clustering. In this case, the data are imagined as being collected by different sensors at different locations and are not related to BD_N. The objective of clustering depends on the model. Typically, (***A***) is used at each station to find local partitions $\{U_1^{literal}, \ldots, U_M^{literal}\}$. Then these may be passed on to cluster heads (gateway nodes) for aggregation. Often, energy, memory, and/or communication constraints dictate summarizing the literal partitions instead, resulting in M cluster summaries, say, $\{CS_1^{literal}, \ldots, CS_M^{literal}\}$, and then the summaries are passed along and used to build an integrated picture of clusters in the network.

[30]Viewed in this light, the streaming approach in Figure 11.2(a) is the limiting case of naive chunking, with M = N.

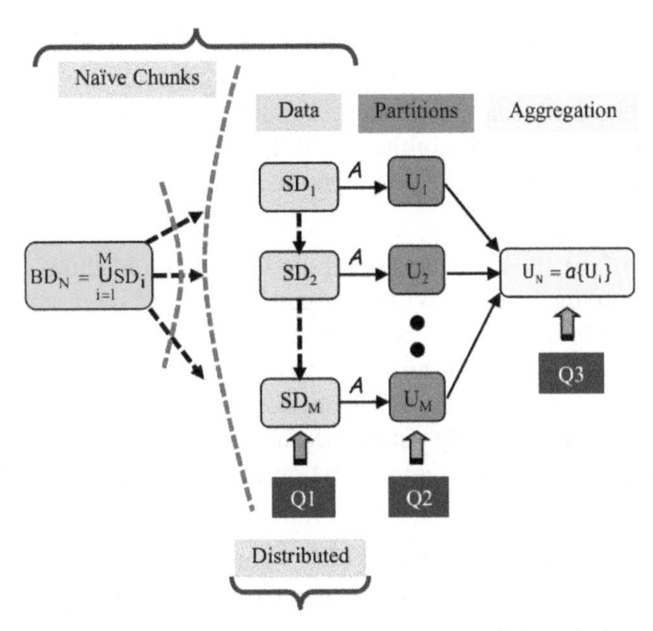

Figure 11.3 Distributed data and naïve chunks of big static data.

In all three cases, each chunk is processed independently, and three questions arise, as shown in Figure 11.3. (Q1): do the M chunks, accessed in parallel or one at a time, all contain representatives of any cluster structure possessed by the data? In the static big data case, this is very important; when collected in the distributed setting, each data (sub)set may legitimately possess quite different cluster structure. (Q2): should the same value of c be used for each literal procedure? And (Q3): how should the M literal partitions be combined into a partition for the big data? Q2 is usually dealt with by using the same value of c for all M literal partitions, which seems very contrary to what might actually occur in the distributed case. The simplest answer for Q3 in the big data case is concatenation to produce $U_N^{approx} \approx U_N^{literal}$. Aggregation in the distributed case varies with the model used.

The most troublesome question about using naive chunks in the static big data case is Q1. Just imagine naively dividing the Iris data into 3 chunks of 50 points, each taken in the standard label order. Then SD_1 contains the 50 Setosa vectors, which are in one physical subset, but unless you knew that, you would probably seek more than 1 cluster in this subset, and you would rapidly go astray. See Hore *et al.* (2009a) for some experiments with Iris along these lines that study this problem. They suggest that the problem of misrepresentation by naive chunking can be partially offset by imposing a non-uniform distribution on the chunks. A popular alternative to independent processing of M chunks in the static big data case is to process the biggest loadable chunk SD_1 with a literal algorithm and then pass a summarization $CS_1^{literal}$ of the results on to the next chunk (the vertical path shown by dashed lines in Figures 11.2(b) and 11.2(c), with or without the data used to build the summarization, which continues until all M chunks are used.

Another field that is seemingly related to the question of how to aggregate $\{U_1^{literal}, \ldots, U_M^{literal}\}$ is often called cluster ensembles or *ensemble clustering*. The general question addressed by researchers in this area is how to integrate M independently obtained literal partitions into one that provides a better idea of cluster structure than any of the individual partitions that contribute to the aggregation. The usual assumption here is that the M partitions are all of the same data set, made by different clustering algorithms, or by independent runs of the same algorithm with different sets of parameters -e.g., different initializations. So, methods of aggregation in this category don't seem applicable to the naive chunking approach for clustering in big static data because the literal partitions don't share a common data set. Ensemble approaches might be useful in the distributed case, depending on the objective of the model. This approach will not be considered here. See Strehl and Ghosh (2002) for an influential early paper on this, and Vega-Pons and Ruiz-Shulcoper (2013) for an update.

Figure 11.2(c). As in Figure 11.2(b), the big data is again subdivided into M chunks, but now each chunk is built "intelligently" with a sampling function (*s*) that hopefully collects representative samples for each chunk from each of the c clusters assumed to be in the big data. Is this possible? Since we probably don't know if the big data even has HPOV clusters, it is not so easy to envision a scheme that guarantees good samples in each and every sample chunk. But there are some limited results in this direction (cf., Theorems 11.1 and 11.2). Once "intelligently" sampled chunks are obtained, both the distributed and summarization models are in play. The only real distinction between Figures 11.2(b) and 11.2(c) is the way the chunks are built.

Figure 11.2(d). The fourth tier in Figure 11.2, Figure 11.2(d), illustrates a method called sampling followed by non-iterative extension. In this approach a loadable sample $SD_n \subset BD_N$ is built by sampling, and a literal partition $U_n^{literal} = \boldsymbol{A}[SD_n] \in M_{fcn}$ is obtained. With the literal result for the small data in hand, $U_n^{literal}$ is extended non-iteratively to the rest of the samples in $BD_n \subset SD_N$, so we end up with U_{N-n}^{approx}, which is usually a partition of the remaining objects in the big data. The word "usually" in the preceding sentence signifies that the row constraint on memberships for the matrix $U_{N-n}^{literal}$ may not be enforced unless precautions are taken to do so. The final step is to combine $U_n^{literal}$ and U_{N-n}^{approx} with an *aggregation function* $\boldsymbol{a}: M_{fcn} \times M_{fc(N-c)} \rightarrowtail M_{fcn}$ that results an approximation to the true but incomputable literal result, $U_N^{literal} \cong U_N^{approx} = \boldsymbol{a}(U_n^{literal}, U_{N-n}^{approx})$. This scheme has been applied to each the Fantastic Four with good results.

Once we have an approximate partition of the big data, how we can evaluate the result? If we could compute $U_N^{literal} = \boldsymbol{A}[BD_N]$, we could compare it to the approximation in any number of ways, such as computing the mean squared error $\|U_N^{literal} - U_N^{approx}\|_{err}^2$ in any convenient matrix norm. But our assumption is that we *can't* compute $U_N^{literal} = \boldsymbol{A}[BD_N]$ on actual big data (hence the "No Error Estimate" in Figure 11.1 for BD_N), so it becomes difficult (well, impossible for unloadable data sets) to verify the quality of the approximation $U_N^{literal} \approx \boldsymbol{a}(U_n^{literal}, U_{N-1}^{approx})$.

One way (indeed, the usual way) to sidestep this difficulty and provide some evidence that the approximate clusters are acceptable is to run algorithm (***A***) on the biggest *labeled* data sets for which $U_n^{literal}$ is computable, and then run the approximation scheme on the same data to get U_n^{approx}. The labels are not used during clustering or the evaluation of the approximation, but now we can compute $\|U_n^{literal} - U_n^{approx}\|_{err}^2$ in any convenient matrix norm. This at least enables comparisons of different approximation schemes against a literal result.

Another path to verification and comparison of different big data clustering schemes is to use the labels supplied with the biggest loadable data sets to compute the partition accuracy at Equation (3.34) of each $U_n^{literal}$ against the ground truth labels. Neither of these methods really verifies how much faith to put in the approximate clusters found in unlabeled big data, but this isn't different than the situation for the CPOV clusters any algorithm (***A***) finds in small (unlabeled) data. The only new wrinkle here is that we have a way to produce the clusters even when we can't find a literal solution directly. Since direct verification of the quality of approximate clustering is not possible, this is the best we can do.

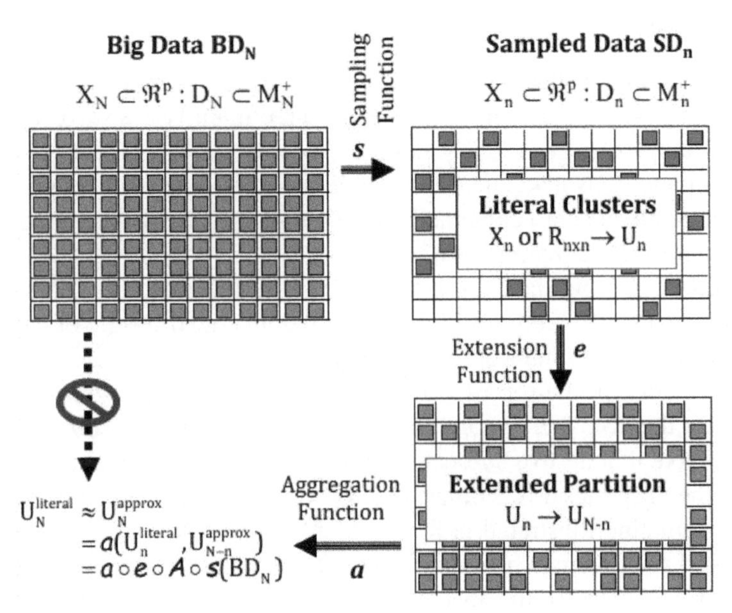

Figure 11.4 Clustering in big static data with *sampling plus extension* (S + E).

Figure 11.4 takes a closer look at the basic architecture for handling big numbers of objects (N) by sampling plus extension. In this diagram the direct vertical path from the big static data to a partition of it constructed with any crisp, fuzzy or probabilistic clustering algorithm (***A***) is inaccessible due to the size of BD_N. Let $\boldsymbol{s}$: $BD_N \rightarrowtail SD_n$ be a *sampling function* as shown in Figures 11.2(d) and 11.4. This function creates a loadable subset of the big data. Since $SD_n \subset BD_N$, $\boldsymbol{A}$: $SD_n \rightarrowtail M_{fcn}$ can be applied to SD_n, which produces the literal result, $U_N^{literal} = \boldsymbol{A}[SD_n] \in M_{fcn}$, for the small data.

Having a literal partition of SD_n in hand, define an *extension function* $\boldsymbol{e}$: $U_n^{literal} \rightarrow U_{N-n}^{approx} \in M_{fc(N-n)}$ as shown in Figure 11.4 that generates labels for all the objects in the set $(BD_N - SD_n)$ using the information contained by the literal solution on the sample. The final step is to combine the two partitions $U_N^{literal}$ and U_{N-n}^{approx} with an *aggregation function* $\boldsymbol{a}$: $M_{fcn} \times M_{fc(N-n)}$ that results in an approximation to the true but incomputable literal result, $U_N^{literal} \approx U_N^{approx} = \boldsymbol{a}(U_n^{literal}, U_{N-n}^{approx})$

Throwing in the clustering algorithm (***A***), the overall result of this procedure can be written as the composition of four functions, $U_N^{approx} = \boldsymbol{a} \circ \boldsymbol{e} \circ \boldsymbol{A} \circ \boldsymbol{s}(BD_N)$. For brevity, the overall construction of U_N^{approx} by the four operations in Figure 11.4 for *sample plus non-iterative extension* will be abbreviated as (S + E). For a fixed (***A***), the sampling function (***s***), the extension function (***e***), and the aggregation function (***a***) all affect the quality of the output. There are lots of choices for the three functions (***a, e, s***).

One popular approach for using samples is based on the *necessary conditions* (NCs) for local extrema of a model's objective function. This approach begins with sampling, followed by literal clustering of the sample, and then non-iteratively extending the partition of the sample to the rest of the data using known necessary conditions for the objective function. This method, illustrated in Figure 11.2(d), can be (and has been) applied to three of our four basic algorithms, viz., HCM/FCM and GMD/EM. It doesn't apply to SAHN algorithms because they don't have necessary conditions, but the overall idea of sampling followed by non-iterative extension can be used in roughly the same way.

11.3 Sampling Functions

The use of samples is a natural way to attack the problem of how to find clusters in big data. There are many ways to get samples and many ways to use them. There are three main issues involved when constructing samples: (i) how to get the samples; (ii) how many samples to collect; and (iii) how to evaluate their quality for a specified purpose. No pretense is made here to provide an exhaustive treatment of sampling theory or of the many ways to use a sample once it is obtained. See Thompson (2012) or the book of chapters edited by Marvasti (2001) for comprehensive treatments of this topic.

In the discussion to follow, the big data set is assumed to be feature vector data, $BD_N = X_N$, so we use these two notations interchangeably. Figure 11.5 depicts the imagined results of applying five different sampling schemes to a set of pretend unlabeled big data ($BD_N = X_N$) having, for the sake of illustration, $N = 28$ points, labeled by colors, in the top row of Figure 11.5. There are 2 green, 11 yellow, 5 blue, 4 orange, and 6 gray points in this X_N. The five colors represent $c = 5$ labeled subsets, i.e., a ground truth partition of the input data. Note that all five subsets have different sizes (cardinalities); so the input data is unbalanced. For example, the gray subset is 3 times, and the yellow subset is 5.5 times, as big as the green one, etc. The five types of sampling illustrated in Figure 11.5 will be called types $(s_1, s_2, s_3, s_4, s_5)$. These five are probably encountered most often in big data clustering, but we will meet several others in the sequel.

There are many other types of sampling that might enter the picture for clustering in big data. For example, a *stratified random sample* is obtained by separating the big data into chunks, $s_1(BD_N) = \{SD_{n_1} \cup SD_{n_2} \cup \ldots \cup SD_{n_r}\}$, and then drawing samples from each chunk with a sampling function of types s_3, s_4 or s_5 prior to clustering in each of the small data sets. Several types of stochastic sampling have been used. Spiral sampling is an exotic sounding possibility. The five types of sampling discussed here represent most of the current work in this area, but certainly other sampling schemes might vastly improve any of these methods. If you are interested in trying out some other methods, start with the excellent text by Thompson (2012).

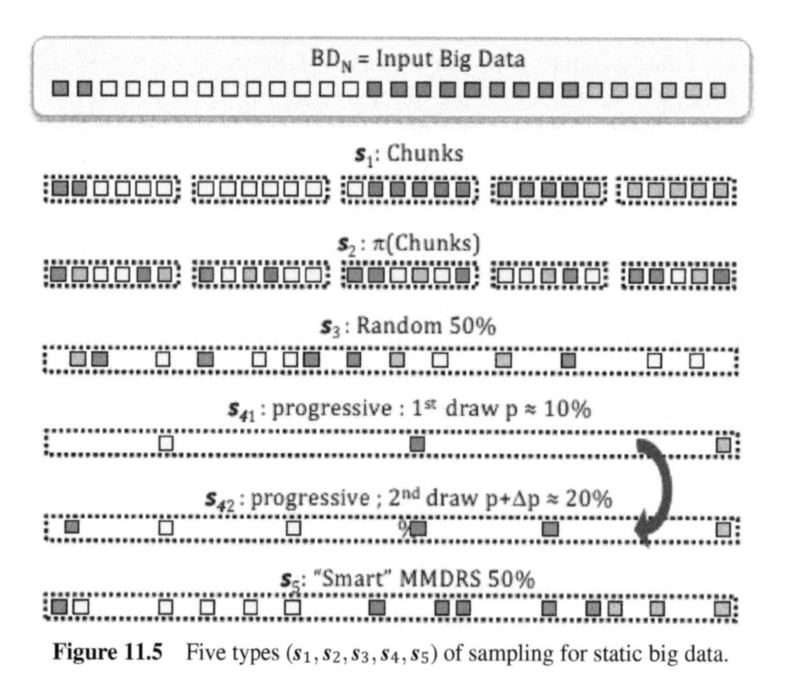

Figure 11.5 Five types $(s_1, s_2, s_3, s_4, s_5)$ of sampling for static big data.

Statisticians sometimes advocate *cluster sampling*, which means get clusters in the data *before* you sample it. This will not be useful for us.

11.3.1 Chunk Sampling

The simplest instance of sampling is type (s_1), which simply divides BD_N into (r) smaller data sets or "chunks," $s_1(BD_N) = \{SD_{n_1} \cup SD_{n_2} \cup \ldots \cup SD_{n_r}\}$ with $N = \sum_{i=1}^{r} n_i$. The chunks may or may not be of equal size, but, presumably, each chunk can be loaded and processed literally. In Figure 11.5, there are three subsets of six points and two subsets of five points. Three of the five subsets contain mixed samples, while two (yellow and grey) are "pure." The divide and conquer approach to clustering in BD_N might send these five smaller data sets off to five processors all implementing algorithm (***A***) (distributed clustering) or it might process them one at a time with (***A***), or (***A***) might use the first one as the initial seed for a compression scheme which then proceeds to the next chunk, and so on. In any case, it is hard to imagine that this type of sampling captures any cluster structure in the input data very accurately. The mixture of c = 5 labels in the source data is hidden by the samples produced by (s_1) in Figure 11.5. Two of the chunks have one uniform cluster, and the other three have two clusters; so the essential structure of BD_N is lost.

The function (s_2) is just a rearranged version of the big static data chunks, $s_2(BD_N) = \pi(s_1(BD_N))$. The five chunks obtained by this function in Figure 11.5 are quite different than those produced by (s_1). This illustrates that any scheme that seeks clusters in big data using the chunking approach may be very susceptible to instability as the input order is changed.

11.3.2 Random Sampling

The third function, (s_3) shown in Figure 11.5, depicts (uniform) random sampling. This is the fastest and most easily understood method of sampling: in a nutshell, all objects are equally likely to be drawn. In Figure 11.5, $SD_n = s_3(BD_N)$ is 50% of BD_N. Note that none of the blue inputs appear in SD_N, illustrating the fact that random sampling may fail to provide a faithful representation of cluster structure in the input data. Several authors have addressed this question theoretically. For example, assuming that each cluster has a certain minimum size, Guha *et al.* (2001) prove the following.

Theorem 11.1. When does random sampling produce points from all c clusters?

Assume that $X_N = BD_N$ has c crisp CPOV clusters, $X_N = \bigcup_{i=1}^{c} X_i; \varnothing = \bigcap_{i \neq j} X_i X_j$, with $|X_i| = N_i$ and $\sum_{i=1}^{c} N_i = N$. Let $SD_n = s_3(BD_N)$ be a random sample of $X_N = BD_N$ of size n. Let $\boldsymbol{X}_j$ denote the random variable for the j-th i.i.d draw, and let $\boldsymbol{X} = \sum_{j=1}^{n} \boldsymbol{X}_j$. Finally, let $0 \leq \lambda \leq 1$ and $0 \leq \delta \leq 1$. Then

$$n \geq \lambda N + \frac{N}{N_i}\left[\log(1/\delta)\sqrt{[\log(1/\delta)]^2 + 2\lambda N_i \log(1+\delta)}\right] \Rightarrow \Pr(\boldsymbol{X} < \lambda N_i) < \delta \text{ for } i = 1, \ldots, c \quad (11.1)$$

What does it mean? This theorem tells us that a random sample of size n, which satisfies the hypotheses of Theorem 11.1 *probably* preserves information about the geometry of the assumed CPOV clusters fairly accurately. The value (n) on the left side of Equation (11.1) is the minimum sample size for which the random sample contains, with the stated probability, at least the fraction λ of the samples from each X_i.

Proof. (Guha *et al.*, 2001).

Given big data set BD_N, can you use this theorem as a rough guide to determine the size of small data set SD_n needed to ensure that it nominally contains some samples from each of the c clusters -the information needed to approximate presumed CPOV clusters in the big static data? The hypotheses in Theorem 11.1 require two user-defined constants, λ and δ. Let us take $\lambda = \delta = 1/2$. Suppose you have ten million points, $N = 10^7$, drawn i.i.d. from a Gaussian mixture of $c = 5$ components with equal priors $p_i = 1/5$, and have an exact draw from each of the five components in the mixture so that $N_i = 2 \times 10^6$ and $N/N_i = 10^7/2 \times 10^6 = 5$. Then

$$n \geq \lambda N + \frac{N}{N_i}\left[\log(1/\delta)\sqrt{[\log(1/\delta)]^2 + 2\lambda N_i \log(1+\delta)}\right]$$
$$= 5 \times 10^6 + 5\left[2 + \sqrt{2^2 + 2 \times 10^6(2)}\right] \approx 5,003,881$$

So for $\lambda = \delta = 1/2$, we need a random sample SD_n of size $n \geq 5,003,881$, more than half of BD_N, to secure a *probability* less than 1/2 that each of the five clusters was sampled 1 million times. This is still a pretty big number of samples. Taking $\lambda = 0.01, \delta = 0.99$ results in $n \geq 100,006 \approx 10^5$ samples. So, if you draw 1% of the data, Theorem 11.1 says that the probability is less than 99% that each of the clusters is sampled fractionally by 1%. How useful is this? Well, since you can adjust the parameters in Theorem 11.1 to produce any value you want by tinkering with λ, and all you get is a probabilistic guarantee, the practical value of Theorem 11.1 is not very high. Compare Theorem 11.1 to the random projection Theorem 5.1. Both of these theorems offer a probabilistic guarantee that the procedure (random projection and random sampling) achieves the desired outcome: certainly, this is psychologically reassuring in both cases. But nothing more. A better way to think about the utility of Theorem 11.1 is in terms of its take-away lesson. If BD_N has cluster structure, and N is really large -say $N \geq 10^9$, it is pretty safe to assume that moderately sized random samples *do* select a representative fraction of each cluster. Studies of various kinds confirm this (cf., Havens *et al.* 2012).

11.3.3 Progressive Sampling

Many researchers assume that a somewhat more goal-directed approach to sampling is needed to "guarantee" that $SD_n = s(BD_N)$ contains "enough" samples from each big data subset to approximate the cluster structure we assume it contains. The nature of the guarantee provided by a particular (s) varies from theoretical results such as Theorem 11.1 that assure proportional representation under certain conditions to intuitive beliefs based on, for example, geometric arguments. The "smart 50%" function (s_5) in Figure 11.5 illustrates the general idea. Thus, $SD_n = s_5(BD_N)$ tries to pick some samples from each of the five subsets, and the target number of representatives of each class is (roughly) the same as their relative proportions in the big data. How do we get a smart sample like this when the data are unlabeled?

One way to get smart (i.e., goal directed) samples is by *progressive sampling*, which is illustrated by the functions s_{41} and s_{42} shown in Figure 11.5. Progressive sampling starts by randomly drawing a small number ($p = 10\%$ of N) of samples: s_{41} produces $n_1 = 3$ samples, $SD_{n_1} = s_{41}(BD_N)$. Then, *if needed*, s_{42} draws and adds three additional samples to the initial draw, $SD_{n_2} = SD_{n_1} \cup s_{42}(BD_N - SD_{n_1}), \triangle X = 10\%)$. The phrase *if needed* is our cue to ask, how is this process terminated?

Some measure of sample quality is used as a *termination function*, say $T_{ps} : P(X_N) \rightarrowtail \Re$ where $P(X_N)$ is the power set (set of all subsets) of X_N. We acquire sample SD_{n_1}, and then use $T_{ps}(SD_{n_1})$ to assess its quality. If the test indicates that this first sample is good enough for the application at hand in the sense of the termination function, the sampling procedure is terminated; otherwise, we draw additional samples, and continue. It is the function T_{ps} that makes the sampling "smart," presumably guiding it towards possessing the information we need to retain to have each cluster properly represented in the sample. The termination function T_{ps} is an important ingredient in the literature of sampling in big static data sets. The upper portion of Figure 11.6 depicts the idea of progressive sampling.

In addition to the question of how to terminate progressive sampling, there is the question of how many samples to draw from the static big data set X_N at each step ~ i.e., what are p and $\triangle X$ in Figure 11.6? A progressive *sampling schedule* specifies how many points in the sample are drawn at each return to the remaining big data and added to the current sample. When p is expressed as a fraction, the initial sample $SD_{n1} = X_{n1} = (pX_N)$ has cardinality $|X_{n1}| = n_1$. If the termination test indicates that more samples are needed, an additional number of samples are drawn from the remaining big data. When the process is terminated after (t) steps, the sample that is accepted is $X_n = \bigcup_{i=1}^{t} X_{n_i}$ with $|X_n| = n = \sum_{i=1}^{t} n_i$.

An *arithmetic schedule* adds a constant number of new samples $\triangle X$ to the current sample at each step: for example, for $\triangle X = 100$ we might have $\{|X_{n_1}| = 100, |X_{n_2}| = 200, \ldots, |X_{n_3}| = 300, \ldots\}$. A geometric schedule increases the number of samples drawn at each step in a geometric progression. For example, if the ratio is 2, the multiplier for the *i*th draw is $2^i \bullet \triangle X$, so we might have $\{|X_{n_1}| = 2^0 \bullet 100 = 100, |X_{n_2}| = 2^1 \bullet 100 = 200, \ldots, |X_{n_3}| = 2^2 \times 100 = 400, \ldots\}$. In general, progressive sampling may be linear, arithmetic, geometric, stochastic, etc.

There is also the question of whether the sampling is with replacement or without replacement. The way $\triangle X$ is used here implies adding more samples to the ones already drawn; so this is sampling without

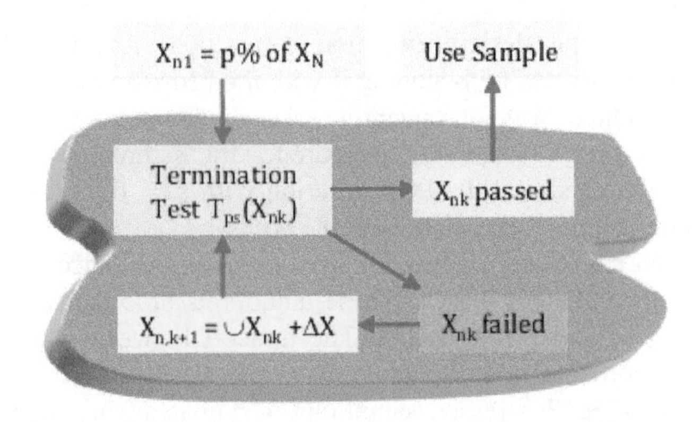

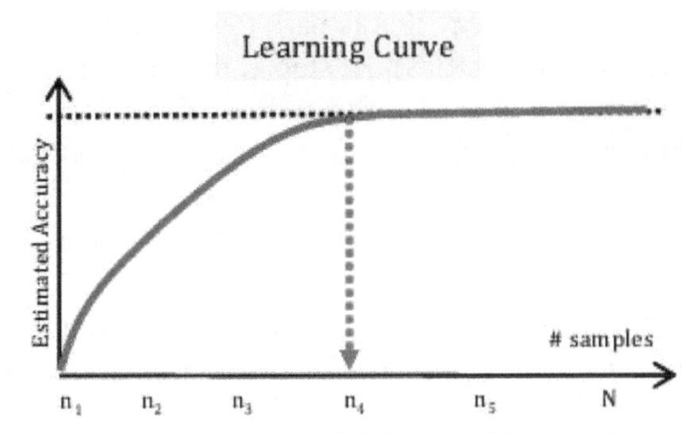

Figure 11.6 Progressive sampling (s_4) for static big data after k steps.

replacement, as shown in Figures 11.5 and 11.6. The statistical theory for the two cases is quite different, but for large numbers of samples, sampling with replacement is not that much different from sampling without replacement. Both types of sampling have been used in various applications: the default assumption here is *without replacement.*

Termination functions for sampling should push the acquired sample toward usefulness in the context of the application envisioned for the sample. In particular, do we want X_n to support classifier design or cluster analysis? These two objectives are different and thus lead to different types of termination functions. One of the earliest measures of sample quality for classifier design was discussed by John and Langley (1996). They suggested the following criterion, which they called the *probably close enough* (PCE) criterion:

$$T_{pce}(X_n, X_N) = Pr(acc(X_N) - acc(X_n) > \varepsilon) \leq \delta. \quad (11.2)$$

where $acc(X_N)$ and $acc(X_n)$ denote the (estimated) classification error that a Naive Bayes classifier could achieve using the sample versus the entire data set. The lower portion of Figure 11.6 shows the basic idea of this kind of termination: accuracy improves as the number samples used for training increases. The aim of T_{pce} is to stop sampling when there is only a small chance that the classifier will do better using the entire data set than when using just the sample. Here is how John and Langley describe the parameters ε and δ in Equation (11.2): ε is a parameter to be specified by a customer describing what "close enough" means, and δ is a parameter describing what "probably" means." (I am pretty sure that neither of these authors is an advocate for fuzzy sets, but you could hardly ask for a better example of linguistic imprecision!)

Sample accuracy in the PCE strategy at (11.2) is estimated using the leave one out cross validation strategy that is very common in classifier design (Evgeniou *et al.*, 2004). Accuracy that might be attained on the entire sample was initially estimated by bounding it with the heuristic that $acc(X_{n+1}) \leq acc(X_n) \Rightarrow acc(X_N) \leq acc(X_n) \therefore stop$. This scheme was subsequently replaced with an estimate found by extrapolation of learning curves. Experiments with 11 data sets compared static sampling to arithmetic progressive sampling; they showed that this method met the PCE criterion with $\varepsilon = 0.02$; and in all cases $acc(X_n)$ was not more than 0.09% higher than the extrapolated sample accuracy.

Provost *et al.* (1999) make an interesting analysis of some general differences between arithmetic and geometric sampling schedules for classifier design. These authors assume that the data are labeled and the pattern recognition objective is a classifier design. The tradeoff between efficiency and accuracy in their paper is measured by the improvement in classifier error rates attained by training a literal induction algorithm at each step during sampling. Progressive sampling is terminated when the computational cost (to train the induction algorithm on SD_n) is equal to the cost of using all N samples in X_N. How is their scheme terminated?

Provost *et al.* advocate a termination strategy that is hard to describe in equation form. Their method is called *linear regression with local sampling* (LRLS). At step k, this scheme samples (j) points in the current sample X_{n_k}. The (j) points are fitted with a linear regression line. If the slope of the regression line is "close enough" to zero, sampling is terminated. This heuristic depends on the fact that the slope of a learning curve decreases as the number of samples increases. So, when the slope of the regression line approaches zero, it is unlikely that further sampling will improve the accuracy of the classifier. The lower part of Figure 11.6 illustrates this idea.

Their analysis concludes that progressive sampling using this termination strategy is more efficient in terms of computational complexity than using all of the samples over a wide range of values for the assumed parameters of the literal algorithm. Provost *et al.* (1999) also show that geometric sampling is more efficient than arithmetic sampling for their application (classifier design) unless the algorithm involved is linear in N (HCM/FCM *are* linear in N) and more than about N/6 −16% of the samples are needed before sampling is terminated. The intended use in their context is to cut off sampling and training when a target error rate is achieved (the unprocessed samples are not subsequently labeled). While the main objective of sampling for us (approximate cluster analysis) has a different target, this paper is well worth a read for its insightful analysis of progressive sampling.

Meek *et al.* (2002) introduce an interesting progressive sampling approach to acquiring enough "good" samples based on a learning curve which monitors increasing computational costs as larger and larger amounts of data are used. This method essentially looks at one chunk of the input data which keeps getting bigger and bigger until processing is terminated without passing through the entire data set. The sampling schedule is arithmetic in p, the percentage of samples drawn at each step, so the sample cardinalities are $\{n_1, 2n_1, 3n_1, \ldots, \}$. The termination function for this scheme is based on the principle of maximum expected utility. Sampling is terminated when future predicted costs outweigh future predicted benefits. The approach is based on evaluating the current GMD clustering model (GMD_i) at step (i) in terms of its benefits and costs, viz., $T_{GMD}(X_n) = U(GMD_i) = B(GMD_i) - C(GMD_i)$, where U, B and C stand for utility, benefit and cost, respectively. As long as this quantity is positive, more samples are drawn.

Sampling and GMD/EM clustering is terminated by Meek *et al.* (2002) when $U(GMD_{i+1}) < 0$. Their approach is illustrated using GMD/EM algorithm A7.1 to find clusters in three real-world data sets (MSNBC, $n = 597,971$; MS.COM, $n = 1,938,877$; and US Census 1990, $n = 2,458,284$). They introduce an interesting modification of the basic algorithm by observing that the learning curves for different levels of convergence (i.e., different values of the termination threshold ε in line 5 of A7.1) are roughly parallel. Based on this, they suggest finding an abbreviated model built with a few iterate steps, a large value of ε, say 0.1, and the initial (smallest) samples, followed by a more extensive procedure with the full model at a smaller value of ε such as 0.00001. Speedups of A7.1 ranged from 5.12:1 for the MSNBC data to 25.05:1 for the US Census 1990 data. This procedure can be viewed as an acceleration method for A7.1, but since sampling and clustering stop at the same time, the full (possibly unloadable) data set may never be accessed. This scheme is a legitimate contender for GMD/EM clustering in big static data sets. This paper is well worth the time spent on a careful reading.

Progressive sampling *for clustering* often follows a slightly different path because the presumption is that we want to produce label vectors for every object in $BD_N = X_N$. In the spirit of Figure 11.4, literal clustering awaits a satisfactory sample of the big static data as determined by some T_{ps}, and then $U_n^{literal} = \boldsymbol{A}[X_n] \in M_{fcn}$ is obtained by applying the literal algorithm to the terminal sample, followed by some type of extension to the rest of the data. Termination functions that are typically used with progressive samples for clustering take the form of statistical hypothesis tests in the general form:

H_0: The sample X_n has been drawn from a population whose distribution is that of X_N; against
H_1: The sample X_n has not been drawn from a population whose distribution is that of X_N.

The exact form of the termination function depends on the nature of the sets X_n and X_N and is usually interpreted as a goodness of fit test between the observations in the two sets, often expressed as relative frequencies in histogram bins. Pal and Bezdek (2002) describe segmentation of images with FCM clustering using arithmetic progressive sampling. They tested two termination functions that compare histograms of the samples to the whole image. The two functions used were the χ^2 test of goodness of fit and the divergence between a pair of probability distributions. The χ^2 test has the following general form,

$$T_{\chi^2}(X_n, X_N) = \sum_{i=1}^{r} \left(\frac{[o_i(f(X_n)) - e_i(f(X_N))]^2}{e_i(f(X_N))} \right) \tag{11.3}$$

where r is the number of non-overlapping categories (bins) that partition X_n and X_N, and $o_i(f(X_n))$, $e_i(f(X_N))$ are the observed and expected frequencies of each of the (i) categories, respectively. The divergence test takes the form

$$T_{div}(X_n, X_N) = n\sum_{i=1}^{r}(p_i(X_N) - q_i(X_n))\log(p_i(X_N)/q_i(X_n)), \tag{11.4}$$

where $p_i(X_N), q_i(X_n)$ are the population and sample probabilities of the i-th bin, and r is the number of non-overlapping categories (bins) that partition X_n and X_N. While the use of these two functions in conjunction with binning of histogram data seemingly limited their application to image processing, Hathaway and Bezdek (2006) used them successfully to illustrate the method of (sampling + extension) for generalized GMD.

11.3.4 Maximin (MM) Sampling

Maximin (MM) sampling (aka the MM rule) has a long and distinguished history. This sampling scheme has some interesting properties and applications, including the clusiVAT algorithm for single linkage clustering of big static data. MM sampling first appeared in Thorndike (1953, p. 270), who described it as a procedure for initializing a set of c prototypes which were then used to build a set of c = 3 clusters in a small dataset using the k-means nearest prototype rule at Equation (6.25b). Here is his description:

> *Our procedure is to assume that the two jobs [objects underlying the data] which are at the greatest distance from one another will axiomatically fall in different families. The third cluster starts with the job which is least near to the other two. Each cluster is built up by adding on that specimen which is nearest to the one which initially defined the cluster.*

Here is an excerpt from Casey and Nagy (1968, p. 495), which summarizes their use of the MM algorithm to construct initial prototypes by the MM rule.

Definition 11.1. Maximin principle (MMP) for sampling.

> *The first sample in the batch to be processed is designated cluster center number one. The distances of the remaining samples from this one are calculated, and the farthest sample is called center number two. The smaller of the two distances from each sample to these two centers is listed, and the sample having the greatest minimum distance is selected. The remaining centers are chosen in turn to have maximum separation from the existing centers. These initial cluster centers are well scattered over the sample space, an intuitively desirable property.*

Note that Casey and Nagy initialize MM sampling with the first sample in the data, while Thorndike began by finding the two samples furthest from each other and designating them as the first two MM samples. Kennard and Stone (1969, p. 140) initialize MM as Thorndike did; and Batchelor and Wilkins (1969, p. 481) initialize in the same way as Casey and Nagy. Gonzalez (1985, pp. 295–296) describes an algorithm on pp. 295–296 and gives pseudocode for it on p. 296, which depends on the MM rule. At first glance, his algorithm looks different than Maximin, but upon closer examination, the two algorithms are identical. Gonzalez's algorithm starts with one cluster that contains all the samples. The center of this cluster is arbitrarily chosen to be one of the samples. Then the sample whose distance to its nearest center is maximal is determined. The second cluster center is chosen to be this sample. All samples in the first cluster that are closer to the second center are moved to the second cluster. Once again, the sample whose distance to its nearest center is maximal is determined. The third cluster is chosen to be this sample. All samples in first two clusters that are closer to the third center are moved to the third cluster. The procedure continues in this manner until c cluster centers are chosen. Tou and Gonzalez (1974, pp. 92–94) use the MM scheme as the basis for a very simplistic clustering algorithm they call the *Maximin distance algorithm*. Their algorithm uses a threshold to decide when to stop generating MM points (as centers) for the clusters being found, but, more typically, the MM algorithm is used to simply generate c′ MM samples (X_{MM}) directly from big feature vector input data set X_N.

The MM sampling procedure is summarized as A11.1 for both input cases (feature vectors or dissimilarity data). Note that c′ is the number of MM samples obtained. When the input data arrives directly in the form of a big N × N dissimilarity matrix $BD_N = D_N \in M_N^+$, the output indices in Line 10 identify c′ *distinguished MM objects* $O_{MM} = (o_{m_1}, \ldots, o_{m_{c'}}) \subset O_N$. In the specification at A11.1, the random initial seed (X_{m_o}) or (m_o) is discarded as soon as the first MM point ((m_1) in line 7) is found. However, this is not always done, and should be considered optional.

Algorithm A11.1. Maximin sampling algorithm.

	Maximin (MM) Sampling (Thorndike, 1953)	
1	**In:** $X_N = \{\mathbf{x}_1, \ldots \mathbf{x}_N\} \subset \Re^p$ and metric d: $\Re^p \times \Re^p \rightarrowtail \Re^+$ *(or)* $D_N \in M_N^+$; c' = desired # of MM samples	MM
2	**Initialize:** $X_{MM} = \varnothing$ *(or)* $O_{MM} = \varnothing$:	
3	If $X_n : \mathbf{x}_{m_0} = rand\{\mathbf{x}_1, \ldots, \mathbf{x}_N\}$ *(or)* If $D_N : m_0 = rand\{1, \ldots, N\}$:	
4	$\mathbf{Z} = (z_1, \ldots, z_N) = (d(\mathbf{x}_{m_o}, 1), \ldots, d(\mathbf{x}_{m_o}, N))$ *(or)* $(d_{m_o1}, \ldots, d_{m_oN})$	
5	**For t ← 1 to c′ do**	
6	$\mathbf{Z} = (\min\{z_1, d(\mathbf{x}_{m_{t-1}}, \mathbf{x}_1)\}, \ldots, \min\{z_N, d(\mathbf{x}_{m_{t-1}}, \mathbf{x}_N)\})$ *(or)* $\mathbf{Z} = (\min\{z_1, d_{m_{t-1},1}\}, \ldots, \min\{z_N, d_{m_{t-1},N}\})$	
7	$m_t = \underbrace{argmax}_{1 \geq j \geq N}\{z_j\}$	
8	$X_{MM} = X_{MM} \cup \{X_{m_t}\}$ *(or)* $O_{MM} = O_{MM} \cup [O_{m_t}]$	
9	**End for**	
10	**Out:** c′ MM indices $M' = \{m_1, \ldots, m_c\}$: c′ MM samples $X_{MM} = \{\mathbf{x}_{m_1}, \ldots, \mathbf{x}_{m_t}\} \subset X_N$ *(or)* c′ MM objects $O_{MM} = \{o_{m_1}, \ldots, o_{m_{c'}}\} \subset O_N$	

Ties in line 6 are broken arbitrarily. Termination of A11.1 is not accomplished by a function such as T_{ps} because MM sampling is guaranteed to stop when c′ MM samples have been obtained. What assurance is there that the MM samples produced by A11.1 represent the distribution of the underlying cluster structure in any meaningful way? For p = 1, 2, or 3 dimensional sets of feature vectors, we can often plot the MM samples against the entire data set, look at them, and decide subjectively whether or not they are truly representative of the set we have sampled. But for $p > 3$ or truly big data sets of any dimension, this method is unavailable. There is one theoretical result that offers a bit more than this about the quality of the distribution of MM samples. This result relates the quality of the objects selected by A11.1 to clusters in X_N by relating the selected objects to compact and separated (CS) clusters as defined by Dunn's (1973) index of cluster validity.

Recall that in Definition 6.12 Dunn asserted that a set X has a crisp CS partition, X is CS $\Leftrightarrow \max_{U \in M_{hcn}} \{V_{DI}(U)\} > 1$. This result, Dunn's CS theorem, plays an important role in the theory of Maximin sampling.

Theorem 11.2. When does an MM sample contain points from all c clusters?

Let $c' \geq c$. Suppose there is crisp CS c-partition of $X_N = \{\mathbf{x}_1, \ldots, \mathbf{x}_N\}$. Then MM Algorithm A11.1 will select at least one object from each of the c clusters.

Proof. Proposition 1, Hathaway *et al.* (2006a).

What does it mean? This theorem tells us that when X_N contains (c) CS clusters, MM algorithm A11.1 guarantees that among the c' DOs, at least one sample is chosen from each of these clusters. Unlike Theorem 11.1, there is no probability associated with the acquisition of a sample that is "good" in this CS sense, but the condition required (that X_N *has* (c) CS clusters) represents a similar degree of uncertainty in the sampling result because the only test for CS clusters in X_N is to compute Dunn's index on all possible crisp c-partitions of X_N, which will be impossible in all but the most trivial instances.

Example 11.1. An MM sample and iVAT image for the Horseshoe data.

The MM algorithm is known to choose samples that faithfully represent the underlying distribution of the input data. Figure 11.7(a) shows the Horseshoe data (X_{hs}). There are 25,000 points in each of the two Horseshoe-shaped HPOV clusters in this data (the green points). The blue points in view (a) are X_{MM}, the $c' = 500$ MM samples extracted from the input data with A11.1. The right view in Figure 11.7(b) is the iVAT image of the Euclidean distance matrix $D_E(X_{MM})$ of the 500 MM points in (a). The iVAT image has an upper block of 375 pixels (samples from the outer Horseshoe) and a lower block of 125 pixels corresponding to the samples in the inner Horseshoe. This image of the MM sample provides a very accurate indication that there are $c = 2$ HPOV clusters with these relative sizes in the sample. It is this good property of MM samples that makes them so useful. Sometimes.

Hathaway *et al.* (2006) appended a second part to MM algorithm A11.1 that is separately displayed here as the DRS algorithm A11.2. First, A11.1 is applied to the big data set to obtain c' MM indices M' and samples X_{MM}. Then c' crisp clusters $\{S_t\}$ in X_N are built using the nearest prototype rule, Equation (6.24), regarding the c' points in X_{MM} as cluster centers. Finally, each of the c' clusters S_t is randomly sampled

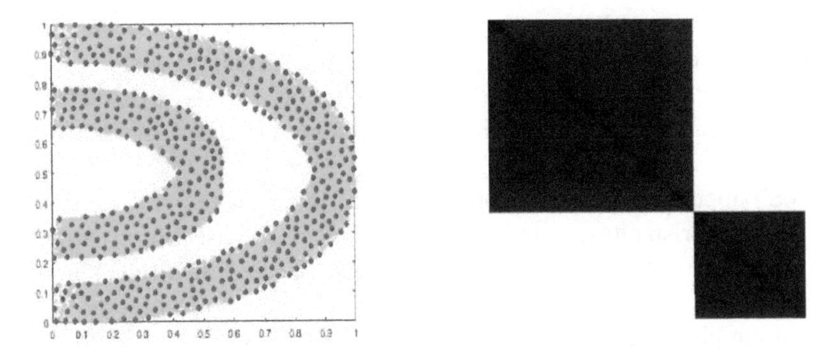

(a) (X_{MM}) = 500 MM samples from A11.1. (b) iVAT image of $D_E(X_{MM})$.

Figure 11.7 MM samples and the iVAT image for data set (X_{hs}).

$n_t = \lceil n(|S_t|/N) \rceil$ times and these samples are aggregated to obtain a data set X_{MMDRS} having a specified number $(|X_{MMDRS}| = n)$ of points.

Algorithm A11.2. (MMDRS) sampling algorithm.

	Maximin Directed Random Sampling Algorithm (Hathaway *et al.*, 2006)	
1	**In:** $X_N = \{\mathbf{x}_1, \ldots, \mathbf{x}_N\} \subset \Re^p$ and metric d: $\Re^p \times \Re^p \mapsto \Re^+$ *(or)* $D_N \in M_N^+$: c′: c′ MM samples M′ : n = desired # of MMDRS samples	DRS
2	**% Initialize:** $S_1 = S_2 = \cdots = S_{c'} = \varnothing$: % $U_{MM} = [0]$	
3	**For t ← 1 to N do**	
4	$q = \underset{1 \le j \le c'}{\operatorname{argmin}}\{d(\mathbf{x}_{m_j}, \mathbf{x}_t)\}$: % $U_{MM}(q, t) = 1$: % For input D_N, $q = \underset{1 \le j \le c'}{\operatorname{argmin}}\{d(m_j, t)\}$	
5	$S_1 = S_q \cup \{t\}$:	
6	**End for**	
7	**For t ← 1 to c′ do**	
8	$n_t = \lceil n(|S_t|/N) \rceil$	
9	Randomly draw n_t unique indices $\{m_t\}$ from S_t	
10	$M = \bigcup_{t=1}^{c'} \{m_t\}$: $n = |M|$	
11	**End for**	
12	**Out:** n MMDRS indices $M = \{m_1, \ldots, m_n\}$: n MMDRS samples $X_n = \{\mathbf{x}_{m_1}, \ldots, \mathbf{x}_{m_n}\} \subset X_N$(or) n MMDRS objects $O_n = \{o_{m_1}, \ldots, o_{m_n}\} \subset O_N$: % MMDRS partition $U_{MM} \in M_{hc'N}$,	

The matrix $U_{MM}(q, t)$ appearing in lines 2, 4, and 12 is commented out since it is not needed to secure the desired MMDRS samples outputted in line 20. It is shown in the specification of A11.2 to help you understand how the partition of X_N into c′ crisp clusters is used to direct the random sampling. Hopefully, this lends some transparency to the DRS scheme. The crisp partition U_{MM} is the "k-means" or *nearest prototype rule* (NPR) partition of X_N built by applying k-means Equation (6.24) to the input data using the c' MM samples as cluster centers. Don't let this notation confuse you: U_{MM} is the MMDRS partition produced by A11.2: $\mathbf{H}(U)$ is the maximum membership matrix produced by the hardening function $\mathbf{H}$ at Equation (3.4).

The word "directed" appears before "random sampling" because the subsets of samples are drawn randomly from each of the c' crisp clusters and not from the entire data set X_N. A careful look at the output of A11.2 shows that the MM input samples (or sample objects) are not themselves in the MMDRS output set. This seemingly counterintuitive result has not shown itself to be very important. It is an easy matter to include these additional samples with the MMDRS output samples if so desired. Here is a comparison of the three sampling algorithms for the horseshoe data set.

Example 11.2. MM, MMDRS, and Random Samples of the horseshoe data.

Figure 11.8 compares samples produced by MM, MMDRS,and random *sampling* (RS) on the horseshoe data X_{hs}. Views (a),(b),and (c) correspond to 50 samples extracted with A11.1 (MM, $c' = 50$), A11.2 (MMDRS, $c' = 50$) and *random sampling* (RS, $n = 50$), respectively. Below, each sample set is the iVAT image of that sample made by applying it to the 50×50 Euclidean distance matrix of the sample. The samples are very dissimilar. The MM sample in view (a) shows that MM samples follow the distribution of the input data pretty well, which corresponds to the same property of MM samples that is illustrated in Figure 11.7.

The MMDRS sample in view (b) looks very different from the MM sample and is not a particularly good replica of the input distribution. However, the ratios of samples drawn which appear above views (a), (b), and (c) tell a different story. Thus, the MM sampling scheme draws twice as many samples (33) from the outer horseshoe as it does from the inner horseshoe (17), even though the two subsets have the same number of points (25,000 each). On the other hand, MMDRS does a much better job of representing the two clusters equally, 23 to 27 being the number of samples drawn from the two subsets. And worst, RS draws 39 points from the inner horseshoe and only 11 from the outer horseshoe, a very misleading imbalance in view of the perfectly balanced input data.

All three of these results will vary with different initializations. But this is a good illustration of differences between the three sampling algorithms. MM is usually better than MMDRS or RS at faithful reproduction of the input distribution, but MMDRS probably produces the most useful samples as a basis for approximate clustering in the input data. But, the computational cost of MMDRS begins to outweigh its advantages as the N increases. For very large values of N, random sampling may well be worth the sacrifice in representational accuracy offered by MMDRS.

The iVAT images in panels (d), (e), and (f) of Figure 11.8 bear this out. The iVAT image of the 50 MM points is essentially devoid of any suggestion of cluster structure (all 50 points are in their own singleton sets along the diagonal). On the other hand, the iVAT image of the MMDRS sample give a pretty good indication that the input data may have two clusters approximately equal in size. The iVAT image of the RS also suggests two clusters, but the sizes of the two main blocks are distorted to agree with the imbalance in the number of samples in the two subsets of X_{RS}. Compare Figures 11.7(b) and 11.8(e): as the number of MM samples increases, so does the quality of the iVAT image of them.

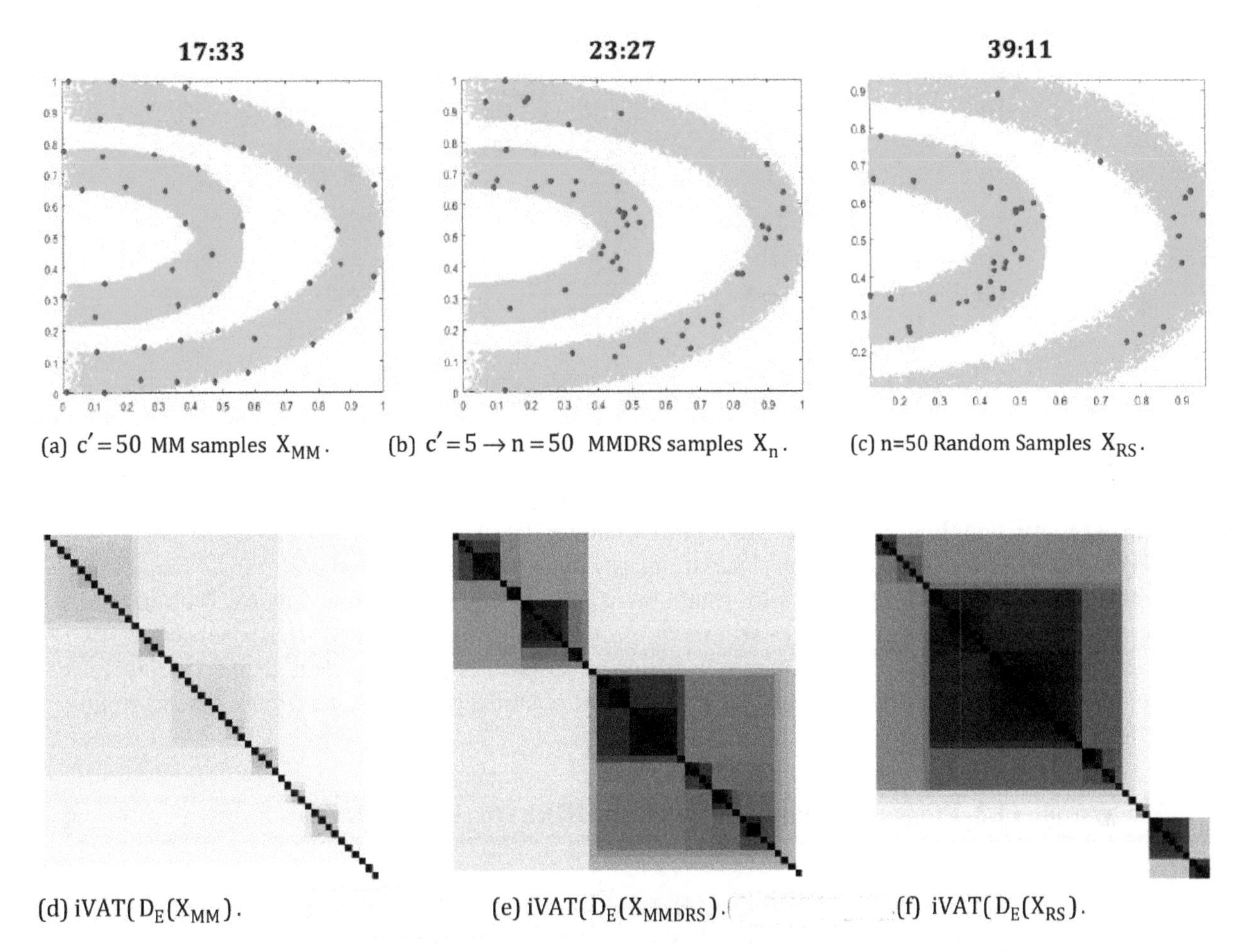

(a) $c' = 50$ MM samples X_{MM}. (b) $c' = 5 \rightarrow n = 50$ MMDRS samples X_n. (c) n=50 Random Samples X_{RS}.

(d) iVAT($D_E(X_{MM})$. (e) iVAT($D_E(X_{MMDRS})$. (f) iVAT($D_E(X_{RS})$.

Figure 11.8 MM, MMDRS, and RS samples and their iVAT images for the Horseshoe data X_{hs}.

Theorem 11.2 shows that Dunn's index at Equation (6.56) plays a central role in the theory of the MM (A11.1) algorithm. This index is also closely related to the MMDRS (A11.2), siVAT (A11.3), and clusiVAT (A11.4) algorithms. Let's recall how this index is computed for a crisp partition $U \in M_{hcn} \leftrightarrow X = \bigcup_{i=1}^{c} X_i; X_i \cap X_j = \varnothing, i \notin j$. The salient feature of Dunn's index for the theory is that its numerator is the (smallest) single linkage distance between any pair of clusters in X, as shown in Figure 11.9.

Please have a closer look at the role played by the individual points in the subsets of X comprising U. The numerator is the distance between the closest pair of points using any metric on the input space between the closest pair of clusters in the partition, as depicted for the Euclidean case in the upper left portion of Figure 11.9. The denominator uses the metric distance between the two points furthest away from each other in the cluster with the largest diameter, as shown in the right upper portion of Figure 11.9.

Now look at the MM samples in Figures 11.7(a) and 11.8(a). These samples hug the boundaries of the two clusters in the horseshoe data because the MM principle drives them toward extreme points in the data. The discussion of progressive sampling in Section 11C.3 never specified how the samples are actually drawn at each step in the sampling schedule. The presumption there is that the draws are random samples. But there

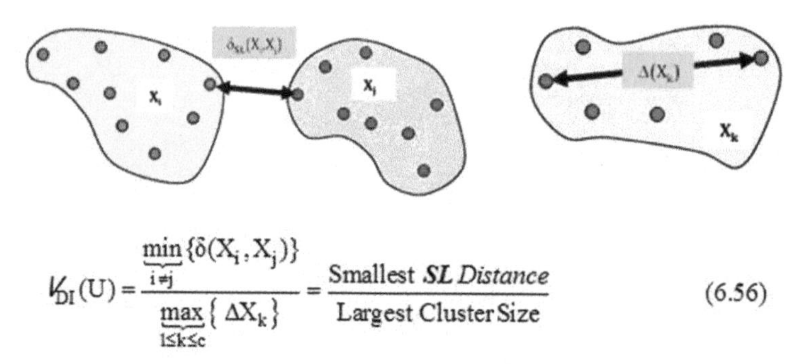

Figure 11.9 The relationship of Dunn's index to single linkage.

is no reason to limit the draws this way. It is perfectly reasonable to use either MM or MMDRS sampling within the progressive sampling cycle in an attempt to combine the good properties of both methods. And we see in Figure 11.9, Dunn's index is driven by extreme points in the clusters in the partition U. Combining these two facts suggests that MMDRS samples might be useful for estimating Dunn's index. Rathore *et al.* (2019a) developed two algorithms that use arithmetic progressive sampling on MMDRS subsets to find samples that provide good estimates of Dunn's index on crisp partitions of big data sets. Example 11.3 abstracts some of the results reported in Rathore *et al.* (2019a) obtained by combining progressive sampling with MMDRS toward this end.

Example 11.3. Progressive Sampling with (MMDRS) to estimate Dunn's index.

The basic idea is to estimate Dunn's index on a very small subset of a big static data set by confining the crisp partition of U on X_N to the sub-partition of U on the MMDRS sample X_n. Rathore *et al.* define four variations of the basic MMDRS algorithm A11.1. Two of the four schemes (iMMDRS and iNMMDRS) use progressive sampling. These two methods are incremental, viz., approximate boundary points are extracted incrementally one at a time (i.e., one per cluster from each labeled cluster) using the MMDRS algorithm. The termination function used by Rathore *et al.* is similar to the one shown in Equation (11.1):

$$\mathbf{T}_{V_{DI}}(X_n, X_N) = V_{DI}(U_{GT}(X_N)) - V_{DI}(U_{GT}(X_n)). \tag{11.5}$$

Note that $V_{DI}(U_{GT}(X_N)) \geq V_{DI}(U_{GT}(X_n))$ for all samples of the big data because the literal value of Dunn's index on the whole data set can't be smaller than an estimate of it made with a subset of the data. The value of the first four columns in Table 11.1 list the parameters of seven labeled data sets used for the experiments reported in Rathore *et al.* (2019a). The iNMMDRS sampling scheme achieved the best results for the methods employed in this paper. The last two columns of Table 11.1 show the exact and estimated values of Dunn's index using the iNMMDRS combination of progressive sampling with A11.2. The estimated values are all exact (to two significant digits). The last row in Table 11.1 shows the average CPU time needed to compute the exact and estimated values of Dunn's index. The estimates based on a combination of MMDRS with progressive sampling yield a speedup of about 1321:1.

Table 11.1 Ten run averages of iNMMDRS estimates of Dunn's index on seven labeled data sets.

Data	N	p	c	Exact Value $V_{DI}(U_{GT}(X_N))$	iNMMDRS approx. $V_{DI}(U_{GT}(X_n))$
XG	55,000	2	3	0.27	0.27
Horseshoe	50,000	2	2	0.07	0.07
ACTR	1,140,000	45	19	0	0
BigX	1,000,000	100	4	1.26	1.26
Forest	581,082	54	7	0.002	0.002
HAR	10,299	561	6	0.09	0.09
MNIST	60,000	784	10	0.15	0.15
Ave CPU time, secs				**185,000**	**140**

There are two other minor results that relate Dunn's notion of CS clusters to MMDRS sampling which provide additional guarantees about MMDRS samples under certain circumstances.

Proposition 11.1. When does MMDRS produce the exact HCM partition of X_N?

Let $X_N = \{\mathbf{x}_1, \ldots, \mathbf{x}_N\} \subset \Re^p$. Let metric $d : \Re^p \times \Re^p \mapsto \Re^+$. If X_N can be partitioned into c compact and separated clusters CS clusters, and $c' = c$, then $U_{MM} \in M_{hcN}$ is the CS partition of X_N.

What does it mean? The meaning of this proposition is pretty straightforward. If the data contain $c' = c$ clusters that are compact and separated in the sense of Dunn, the partition $U_{MM} \in M_{hcN}$ that is commented out that directs the sampling procedure in A11.2 is exactly the partition that c-means would find using the MM samples from A11.1 as cluster centers.

Proof. Theorem 1, Hathaway *et al.* (2006b).

The number of samples drawn from S_t in line 8 of the MMDRS algorithm is $n_t = \lceil n(|S_t|/N) \rceil; 1 \leq t \leq c'$. The fraction $|S_t|/N$ scales the number of desired samples drawn from the *t*th row of U_{MM} by the proportion of samples in that row. Because of the ceiling function, the overall number of samples is approximate. The number of MMDRS samples and the proportions drawn will be *exact* under the extra condition that the sampling proportions are all integers. In this case the ceiling function is not used and $\sum_{t=1}^{c'} n_t = n_s$. This is the content of

Proposition 11.2. When are the MMDRS sample proportions exact?

Let $X_N = \{\mathbf{x}_1, \ldots, \mathbf{x}_N\} \subset \Re^p$. Let metric $\Re^p \times \Re^p \rightarrowtail \Re^+$. Suppose X_N can be partitioned into c CS clusters for $c' \geq c$, and suppose that $|S_t|/N$ is an integer for all t. Then the proportion of objects in the MMDRS sample from subset t equals the proportion of objects in X_N for t = 1 to c'.

What does it mean? Under the hypothesis of this proposition, the proportions of the MMDRS samples are exact.

Proof. Proposition 2, Hathaway *et al.* (2006a).

Do the three results about MM and MMDRS sampling, Theorem 11.2 and Propositions 11.1 and 11.2, have any real practical value? Well, they are pretty weak results. If you could verify the CS hypotheses, you would know from Theorem 11.2 that the MM sampling procedure usually starts off in the right direction, by getting at least one object (i.e., the data representing one object) from each cluster when the data set is compact and separated in the sense of Dunn. And further, when $c' \geq c$, Proposition 11.1 assures you that you would have the exact solution for crisp k-means clustering from MMDRS; so no further analysis would be necessary if you wanted the k-means partition of the input data. And, finally, if the special requirement of Proposition 11.2 is satisfied, you would know that the sample proportions from MMDRS matched the big data proportions exactly.

But remember, even if you have labeled data (in which case you probably will not be clustering it anyway), Dunn's CS criterion is really hard to verify unless the ground truth partition happens to be CS. In consequence, Theorem 11.1 and the three results about MM and MMDRS samples are very similar: none of them guarantees preservation of cluster structure in the samples, but they all carry a certain amount of psychological reassurance about their sampling protocols. If you want to initialize A6.1 with a partition, Proposition 11.2 justifies a modest belief that computing $U_{MM} \in M_{hcn}$ provides a good initialization for HCM or FCM. Or, you can use U_{MM} to compute an initial set of cluster centers with Equation (6.12) or Equation (6.25a). As we shall soon see, a more practical application of MMDRS is its use as a means for scaling up SL to big data.

11.3.5 Aggregation and Non-Iterative Extension of a Literal Partition to the Rest of the Data

With very few exceptions, this idea is limited to vector data inputs (however, see clusiVAT below). After a literal partition of the sample has been obtained, the next step is to extend it to the rest of the input data. Just as there are many types of sampling, there are many types of extension. Distributed approaches often split the static big data into manageable sized subsets (say (r) chunks of type (s_1), $BD_N = \{SD_{n_1} \cup SD_{n_2} \cup \ldots \cup SD_{n_r}\}$), where $N = \sum_{i=1}^{r} n_i$, and then obtain a literal result $\{U_{n_1}^{literal}, U_{n_2}^{literal}, \ldots, U_{n_r}^{literal}\}$ on each chunk. In this scheme the extension to all of the big static data is either already made during the processing, or is found by aggregating the (r) partitions with some type of aggregation function. Some schemes begin by processing (SD_n), the biggest loadable chunk of (BD_N), with the literal algorithm. And then the first data chunk is abandoned, the literal partition of it is represented by some compressed representation, and the algorithm moves on to the next chunk.

Three of our literal algorithms produce $U_n^{literal}$ by processing $(X_n = SD_n)$, along with the associated parameters (HCM/FCM, **V***) for A6.1 or c estimated components (GMD/EM, **p***, **M***, {**S***}) of Gaussian mixtures for A7.1. With the literal parameters obtained by running the algorithm on the sample, a back pass is made through the necessary conditions at (6.19b, FCM) or (6.25b, HCM) or (7.16d, GMD/EM) to generate U_{N-n}^{approx} non-iteratively on the remaining (N-n) objects in the big data $(X_N = BD_N)$. And as we shall see, clusiVAT extends a literal single linkage partition onto the remaining objects in big relational data using a nearest (object) prototype rule similar to Equation (6.25b).This approach leaves you in possession of

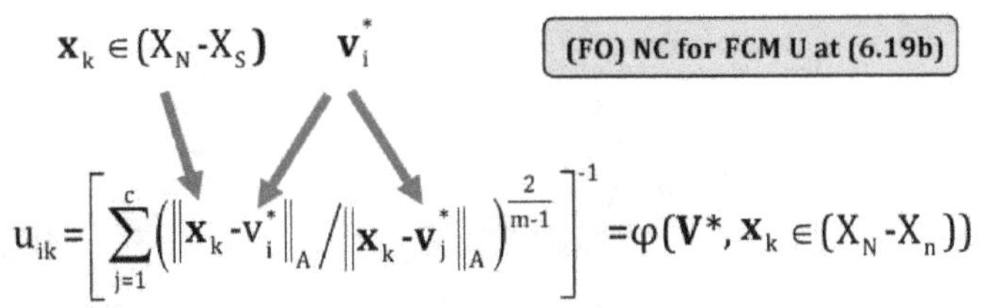

$$u_{ik} = \left[\sum_{j=1}^{c}\left(\|\mathbf{x}_k - \mathbf{v}_i^*\|_A / \|\mathbf{x}_k - \mathbf{v}_j^*\|_A\right)^{\frac{2}{m-1}}\right]^{-1} = \varphi(\mathbf{V}^*, \mathbf{x}_k \in (X_N - X_n))$$

(FO) NC for FCM U at (6.19b); $\mathbf{x}_k \in (X_N - X_S)$; $\mathbf{v}_i^*$

Figure 11.10 Non-iterative extension using literal values for fuzzy c-means.

$(U_n^{literal}, U_{N-n}^{approx})$, which then requires aggregation. Figure 11.10 may help you see exactly how non-iterative extension works for A6.1 in the fuzzy case, m>1. Just plug each unlabeled input point into the FONC for u_{ik} at Equation (6.19b) to get the estimated label for that point. This part of the approximation is quite fast, usually requiring about 1% of the overall CPU time for A6.1 or A7.1.

Once $\{U_{n_1}^{literal}, U_{n_2}^{literal}, \ldots, U_{n_r}^{literal}\}$ or $(U_n^{literal}, U_{N-n}^{approx})$ are in hand, it remains only to combine them. Usually the row size of the partitions to be combined are equal, and in this case they are simply concatenated, In the first case, $U_N^{approx} = \boldsymbol{a}(U_{n_1}^{literal}, U_{n_2}^{literal}, \ldots, U_{n_r}^{literal}) = [U_{n_1}^{literal} \cdots || \cdots U_{n_r}^{literal}]$. In the second case, $U_N^{approx} = \boldsymbol{a}(U_n^{literal}, U_{N-n}^{approx}) = [U_n^{literal}||U_{N-n}^{approx}]$. If you do this, be sure to reorder the columns of the input partitions to have the same order as the input data if comparison to a ground truth partition comes next. The literal partition $U_n^{literal}$ will always satisfy the row sum constraint at (3.3). The extension partition U_{N-n}^{approx} doesn't necessarily enforce the row sum constraint, but there will be at least one value in every row of $U_n^{literal}$, so this constraint will be satisfied after concatenation, even if the approximate clusters in U_{N-n}^{approx} don't.

We did not discuss images in Chapter 1: do you think you can see HPOV clusters in a digital image? Clustering is a very popular method for image segmentation. Clusters in images are regions that display some kind of intra-region similarity such as similar intensities, shapes, textures, colors, etc. Almost all segmentation methods start with feature extraction to convert raw pixel information into some type of feature vectors. And among segmentation methods, k-means (HCM algorithm A6.1, m = 1) is far and away the most popular method, but segmentation with FCM algorithm A6.1, m > 1 and its derivatives has also become very popular (Liew *et al.* (2005)). Our next example illustrates the use of progressive sampling and extension for image segmentation.

Example 11.4. Image segmentation using FCM clustering with progressive sampling.

Pal and Bezdek (2002) describe segmentation of images with FCM clustering using a special type of arithmetic progressive sampling (s_4, with replacement) designed to produce smart samples from large images. FCM algorithm A6.1 (m = 2) is the literal clustering algorithm, and extension function $\boldsymbol{e}$ is through the FONCs at (6.19b), as shown in Figure 11.10. Aggregation $\boldsymbol{a}$ was done by concatenation; $U_N^{approx} = [U_n^{literal}||U_{N-n}^{approx}]$. Two termination functions were used: T_{ps} was either T_{χ^2} at (11.3) or T_{div} at (11.4). These goodness of fit tests were applied to sample histograms of SD_n and the whole image BD_N.

Figure 11.11(a) is a $256 \times 256 = 65,566$ pixels input image (one of four bands from an Indian satellite) with 256 gray-level intensities per band. Two features were extracted from each 3×3 window in the image: the average value of the nine intensities and the "busyness" of the window. Then samples of the 65,536 feature vectors were drawn progressively until termination. The initial percentage was p = 1%, and the arithmetic sampling schedule was also 1%; so the progression of numbers of random samples was: {655,

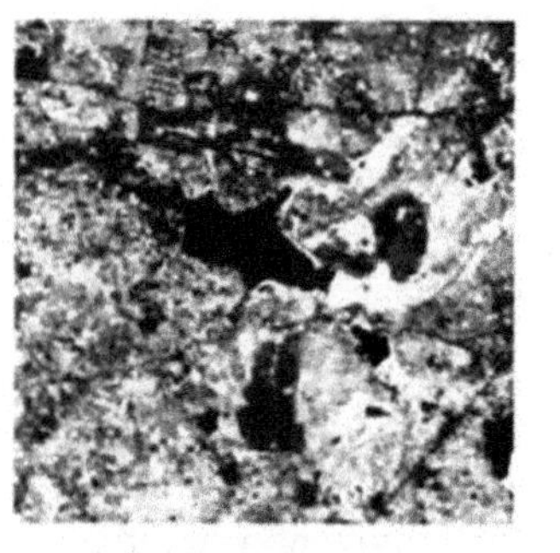

(a) Input image.

(b) LFCM segmentation of X_N.

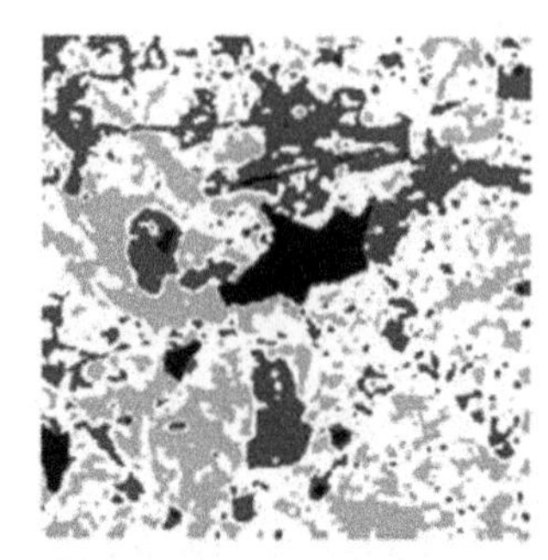

(c) (S+E) Result using X_n = 9% of X_N

Figure 11.11 (a) Indian satellite image. (b) LFCM segmentation. (c) (S + E) FCM segmentation.

1310, 1955, ...}. Termination with the divergence test occurred at 9% of the inputs, viz., 5898 samples. When satisfaction of both functions was required, 29% of the samples were needed.

View 11.11(a) is one band of the input image. Panel 11.11(b) is a segmentation of the input image made by literal FCM applied to all 65,536 pixel vectors. Figure 11.11(c) is the (S + E) approximation to the literal output made by running LFCM on 9% = 5898 samples, and then extending the literal result to the other 91% of the pixels in the image. There is no visible difference between the two segmentations.

The "large images" were not really large since they were all loadable. But the method easily extends to images of arbitrary size since a histogram of any BD_N can be made by paging in and out of a storage device. This algorithm was called eFFCM (*extensible fast fuzzy c-means*).

Hathaway and Bezdek (2006) discuss a similar approach for feature vector data $BD_N = X_N$ using progressive random sampling (s_4, without replacement) on subsets of features chosen by one of four methods. The termination function was either T_{χ^2} at (11.3) or T_{div} at (11.4). These functions assessed the goodness of fit between equi-content histograms for four different ways to build subsets of features in the sample and the big data. This approach extends eFFCM to geFFCM (*generalized fast fuzzy c-means*) and GMD/EM to geFEM (*generalized fast EM*). Numerical experiments are presented as 25 run averages on three loadable data sets having $N = 100,000$ samples from a mixture of $c = 2$ bivariate Gaussian distributions with different covariance structures.

Approximate errors were estimated by computing $E_{approx} = 50 \cdot \|\mathbf{H}(U_N^{approx}) - \mathbf{H}(U_N^{literal})\|/|X_N|$, where $\mathbf{H}$ is the usual hardening function (equation (3.4)) for both FCM and GMD/EM partitions of the data. The acceleration factor due to (S + E) was T_{acc} = (literal time/approx time). Speedups of 2 to 5: 1 (FCM) and 2 to 9: 1 (GMD/EM) were achieved with essentially no loss of accuracy by approximating $U_N^{literal}$ with U_N^{approx} . While none of the data sets in this paper are big (unloadable), the examples show that geFFCM and geFEM achieve both the objectives set for them: (i) they accelerate the literal versions in all cases, and (ii), acceleration is accomplished with very little sacrifice in the quality of extended approximations to literal partitions and prototypes. Finally, since FCM reduces to HCM when $m = 1$ in A6.1, geFFCM reduces to geFHCM, resulting in a way to scale up approximate solutions for k-means clusters using (S + E) in static big data.

Wang *et al.* (2008) point out that using a statistical test for the termination of progressive sampling often results in n samples that are roughly 50% of N. If N is too big, half of N is probably still too big. To avoid over sampling, they introduce a modification of the progressive sampling scheme called selective sampling that combines simple random sampling with MMDRS. This modification uses an intermediate submatrix of D_N which seems to eliminate over sampling; so this is a positive step toward clustering in big data. The acronym that identifies this three-tiered strategy would be something like RS-MM-DRS. Numerical experiments on a (Euclidean) distance matrix of a set of 3,000,000 vectors drawn from a mixture of five bivariate normal distributions in = 2 dimensions demonstrates the feasibility and effectiveness of the modified sampling method. Wang *et al.* (2010) describe a different modification to their 2008 sampling scheme that is discussed in conjunction with the siVAT algorithm A11.3 that utilizes some elements of spectral feature extraction.

This section ends with some observations about how to choose a sampling scheme. The primary requirement for good samples in the context of approximate clustering is that the cluster proportions in the n samples in $(X_n = SD_n)$ be representative of the corresponding proportions for the subsets in $(X_N = BD_N)$. This intuitive objective informs a subjective definition for what constitutes a best set of samples. The main questions are: (i) how to draw the samples; (ii) how many samples to draw; and (iii) how good are they for clustering? If the data are labeled, we can determine how well the samples match the distribution of the labeled subsets in X_N. If the data are unlabeled, as they will be in the case of real clustering, there is no way to ascertain whether this subjective criterion is met. In this case, we can fall back on Theorem 11.1 for RS or Theorem 11.2 and its corollary propositions for MM and MMDRS sampling, but we know that these results are more for psychological reassurance than for constructive information about the sample.

An expectation -fueled by Theorem 11.2 -is that the various DRS methods which begin with MM sampling will produce better distributional representatives of unlabeled data than simple RS in terms of matching proportions of sample and parent. But don't forget: this theory applies only to input data that have CS clusters in the sense of Dunn. How many data sets have CS clusters? Not many. On the other hand, progressive sampling is a style of implementation that controls how many samples to draw with a termination function T_{ps} which presumably relates to capturing faithful samples for the application at hand. There is no stipulation as to how the progressive samples are drawn. Example 11.3 shows that MMDRS and RS can be successfully combined, while Example 11.4 shows that simple random sampling is also effective. So, what to do? There is a lot of numerical evidence to support the following rules of thumb.

1. If the data are small (say, $N < 5000$, $P < 10$), a few MM samples will reproduce the distribution nicely, and a few more MMDRS samples are a good basis for clustering with sampling + extension. But this data is small enough so that any literal algorithm can be used upon it directly. There are some small data applications that can use this good property of MM samples (e.g., boundary estimation), but in the context of sampling big static data sets for approximate clustering, MMDRS will be much more effective than MM sampling. One thing that MM samples do quite well is finding anomalies or noise outliers, because such points will probably be far away from the main structure in the data.
2. If the data are intermediate in size (say, $5000 < N < 100,000, 1 < P < 100$), MMDRS will produce better samples for approximate clustering than random sampling. But ... RS is very fast, while MMDRS is not; so there is a tradeoff in computation time and storage complexity to consider.
3. If the data are big ($N > 100,000, P > 100$), random sampling gets more and more attractive in terms of speed, and as N gets very big ($N > 10,000,000$), RS becomes the only realistic choice. In this case, unless there are hundreds of CPOV clusters, it is likely that all of the CPOV clusters that a clustering algorithm might find will be represented in the random sample. An exception is when there are one or more very small subsets in the input data that are isolated from the main body of points. In this case, RS may miss the small subsets entirely.

11.4 A Sampler of Other Methods: Precursors to Streaming Data Analysis

This section is a collection of early papers about clustering in static big data. Many of these early methods have enjoyed a renaissance of sorts in the world of streaming data analysis. They are included here because most of them were not originally intended for use in "streaming clustering," but many of them have been adapted for the streaming data case as that style of computational analysis has evolved. Most the material in this section could quite logically appear in Chapter 12 instead of here, but I think they belong here because they are a transitional link from clustering in big static data to structural assessment in streaming data.

There are many ways to use chunks or samples if direct substitution into the necessary conditions as described in Section 11.3 is neither feasible nor preferred. Most of the methods in this category depend on the ordering of the input data, whether it be as singletons, naive chunks, or sampled chunks. It is impossible to make a clear distinction between methods based on some type of partial data access or sampling that are categorized by their authors as being incremental (point by point or chunk by chunk), summarization-based, single pass, streaming, and so on. Many of these papers are lumped together in this section without regard to their "type."

There are several approaches of this kind related to c-means that seek either c-medoids or c-generalized prototypes. PAM (*partitioning around medoids*) and CLARA (*clustering large applications*) were introduced by Kaufmann and Rousseeuw (1990) to handle what was considered to be big data in the early 1990s. PAM attempts to find c medoids (points in the data) instead of c means (points built with the data). This algorithm iteratively optimizes the average dissimilarity between crisp clusters. PAM worked well on very small data sets, but its computational complexity is quadratic in N, $\boldsymbol{O}(c(N-c)^2)$. Consequently, PAM's importance was mainly as the basis for upgrading to CLARA.

The original version of CLARA began by randomly drawing $(40 + 2c)$ samples from the big data using uniform random sampling, designated as (s_3) above. PAM then operated on these samples to find the best c-medoids. The remaining objects were crisply labeled with the nearest prototype rule (6.25b) using the medoids as the cluster centers, and the average dissimilarity between crisp clusters was computed. If this improved the objective, these c medoids were retained.

Ng and Han (1994, 2002) generalized CLARA to CLARANS (*clustering large applications based on random search*). Their procedure views the desired medoids as special nodes in a graph representation of the data. Finding the good ones is done by serial randomized search. While CLARA draws a random sample of data points at the beginning of a search, CLARANS draws a random sample of *neighbors* of the data points in each step of its search, which, in some sense makes it a shirt-tail relative of the DRS scheme in A11.2. Wei *et al.* (2000) compare CLARA and CLARANS to several medoid-seeking methods based on genetic search and conclude that CLARANS is the most effective of the four methods they studied. None of these methods extend HCM, but they almost always appear in early references to clustering in big data. See Vijayarani and Nitha (2011) for a study of outlier detection in small data sets using PAM, CLARA, and CLARANS.

Ester *et al.* (1996) developed a very popular clustering method they called DBSCAN (*density based spatial clustering of applications with noise*) and compared runtimes of their approach to CLARANS. They used four types of real-world data sets from the SEQUOIA 2000 database. The only results given in this paper are the relative run times of the two algorithms. The best-case speedup of the ratio of DBSCAN to CLARANS is for N = 10,426, for which the ratio is 1851:1. But these authors don't compare the quality of the clusters in any way; so it is hard to assess the utility of DBSCAN from their numerical experiments. In other words, they show that DBSCAN runs faster -but they don't ask, or answer, the question: does it find better clusters?

A second group of methods find one sample chunk, say SD_1, construct an efficient representation for the literal clusters that are found by processing it (a summarization) and then move on to the next chunk (which may be derived by "intelligent" sampling or naive chunking) with or without retention of the samples. Schemes of this type are represented by the vertical paths of Figures 11.2(b) and 11.2(c). BIRCH (*balanced iterative reducing and clustering using hierarchies*; Zhang *et al.* 1996) was perhaps the first algorithm explicitly designed to handle clustering in big vector data using a compressed representation (summarization) of part of the input data. Interestingly, BIRCH is now also regarded as a "streaming clustering" algorithm: its role in this arena will be discussed in Chapter 12.

In the initial phase of BIRCH as much data as possible is scanned, clustered hierarchically (but not with a SAHN model), and then each cluster is summarized by a *cluster feature* (CF) triple. The *i*th cluster (X_i) is represented by a (2p + 1)-dimensional triple of cluster features (which will be called microclusters in Chapter 12), $\mathbf{CF}_{i,\,\mathrm{Birch}} = (n_i, \mathbf{S}_i, \mathbf{SS}_i)$, where $\mathbf{S}_i = \sum_{k=1}^{n_i} \mathbf{x}_k$ and $\mathbf{SS}_i = (\sum_{i=1}^{n_i} x_{i1}^2, \ldots, \sum_{i=1}^{n_i} x_{in_i}^2)$. Once the CFs are calculated, the data underlying them is deleted, and only the sets $\{\mathbf{CF}_{i,\,\mathrm{Birch}} : 1 \leq i \leq c\}$ which summarize the processing for each chunk of data are retained. The CF's are stored in a height balanced cluster feature tree. As new data arrives, the nodes of the tree are dynamically built and inserted into the tree. Zhang *et al.* state that "BIRCH is shown to perform very well on several large datasets, and is significantly superior to CLARANS (Ng and Han, 1994) in terms of quality, speed and order sensitivity."

Bradley *et al.* (1998) is an influential treatment of the incremental approach for scaling up a variety of algorithms, including sHCM and GMD/EM. These authors posted a list of seven requirements that they think any big data clustering method should satisfy, which they called *data mining desiderata*:

1. Require one scan (or less) of the database if possible: a single data scan is considered costly, early termination if appropriate is highly desirable.
2. Online "anytime" behavior: a "best" answer is always available, with status information on progress, expected remaining time, etc. provided.
3. Suspendable, stoppable, resumable; incremental progress saved to resume a stopped job.
4. Ability to incrementally incorporate additional data with existing models efficiently.
5. Work within confines of a given limited RAM buffer.
6. Utilize variety of possible scan modes: sequential, index, and sampling scans if available.
7. Ability to operate on forward-only cursor over a view of the database.

Bradley *et al.* (1998) suggest using the same compression scheme as BIRCH (i.e., clusters are represented by the same CF triple), but itemize a number of differences between their approach and BIRCH, including: (i) data-compression occurs prior to and independent of clustering; (ii) there is no notion of an allocated memory buffer in BIRCH; (iii) BIRCH requires at least two scans of the data; (iv) simpler statistics are maintained; and (v) three classes of data retention are a generalization of BIRCH's all-discard strategy. Their approach scales up HCM by applying it to primary and secondary compressions of naive chunks. Bradley *et al.* compare their algorithm, scaled HCM (*scaleKM*, aka single pass k-means), *online HCM* (they call this OKM), and random sample HCM (sampKM) using a number of synthetic Gaussian mixtures having $p = 20$, 50, 100; $c = 5$, 50, 100; and two real-world data sets.Experimental outcomes are compared with several measures validity, one of which is based on comparing the ground truth means of the labeled data $\mathbf{M}_{GT}$ with the computed means, $PM(\mathbf{V} \mid \mathbf{M}_{GT})/c = \|\mathbf{V} - \mathbf{M}_{GT}\|_{PM}/c$, a normalized version of Equation (6.41) which they call D_{truth}, the distance between the true and estimated means. Their scaleKM compares favorably to the other algorithms tested, typically achieving estimated means that are much closer to the sample means than the comparison methods. Ganti *et al.* (1999b) generalize the BIRCH framework to one they call BIRCH* by introducing two algorithms called *BUBBLE* and *BUBBLE-FM*. These are single pass algorithms that find "evolving" clusters in big static data in an arbitrary metric space.

Farnstrom *et al.* (2000) assert that scaleKM doesn't scale HCM very well because the summarize/delete operations in the chunk currently residing in the buffer are time-consuming and inefficient. They propose a *simplified single pass k-means*, which, in essence, simply discards all of the input data in each partial data access and moves on to the next chunk. Six variants of scaling k-means, including using HCM on a 1% random sample, are compared on synthetic and real-world data. Naturally, their results support their method, which does look like a substantial improvement on the schemes tested by Bradley *et al.* (1998).

CURE (*clustering using representatives*; Guha *et al.*, 1998, 2001) is another early and important entrant in the race to scale up crisp clustering algorithms that is related to both A6.1 (HCM) and A8.2 (SL). CURE is a type (s_3) sample-based algorithm for large datasets, which performs clustering on the sample, and then extends the results to the entire dataset. CURE seeks a middle ground between a type of hierarchical clustering that is basically a hybrid made by combining a version of SL with some time-saving shortcuts and k-means (HCM). CURE represents clusters in the sample by "well-scattered" points, and, akin to SL, merges nearest clusters until the desired value of c is attained.

CURE can be slow because its time complexity is $O(n^2 \log n)$. Guha *et al.* describe a pre-clustering scheme based on chunking the data with a type (s_1) sampling function that partially overcomes this problem. Once CURE has labeled the sample, the extension function that assigns cluster labels to the remaining data points employs a fraction of randomly selected representative points for each of the final c clusters. Each data point is assigned to the cluster containing the representative point closest to it. But instead of a single centroid, multiple representative points from each cluster are used to label the remainder of the data set; so the extension function is the crisp nearest multiple prototype rule. This is an excellent paper which touches all the bases for this topic. We return to CURE in Section 11.7, where its performance will be compared to big data single linkage (clusiVAT) and several big data versions of HCM (aka k-means).

Single-pass algorithms use all of the data but pass through it just once. There are tons of papers devoted to scaling up GMD/EM A7.1 with methods of this kind. Neal and Hinton (1998) is an oft-cited paper that underlies one online version of EM that is often called incremental EM. Bradley *et al.* (2000) discuss scaling the basic GMD/EM algorithm A7.1 (which they call "classic *Vanilla EM* (VEM)") using a BIRCH-like derivative called *scalable EM* (SEM). Basically, SEM iterates EM algorithm A7.1, applies c-means algorithm A6.1 (m=1) to find small, compact subsets of the data, and finally uses an unspecified member of the SAHN family (cf., Chapter 8) to aggregate summarizations until all of the data have been processed. Here is a thumbnail sketch of SEM: (i) sample $BD_N = X_N \rightarrowtail SD_n = X_n$, where n is the size of the buffer; (ii) run A7.1 on X_n; (iii) identify regions for summarization; (iv) compress these regions with a set of sufficient statistics, i.e., a cluster feature triple, which in this case is $CF_i = (n_i, \sum_{x \in X_i} \mathbf{x}, \sum_{x \in X_i} \mathbf{xx}^T)$; (v) fetch another set of samples and continue until done.

The cluster feature triple of Bradley *et al.* (2000) is easily converted into the representation, $\widehat{CF}_i = (n_i, \mathbf{m}_i, S_i)$, where $(\mathbf{m}_i, S_i)$ are the sample-based estimates of the mean and covariance matrices at equations (7.16b) and (7.16c). These statistics are merged additively, $[(n_i, \mathbf{m}_i, S_i), (n_j, \mathbf{m}_j, S_j)] = (n_i+n_j, \mathbf{m}_i+\mathbf{m}_j, S_i+S_j)$. The SEM approach to scaling is compared to vanilla GMD/EM (VEM algorithm A7.1) and a version of on-line EM (OEM) due to Darken and Moody (1992). Synthetic Gaussian clusters were used for the comparisons, $c = 10, p = 25, N = 10,000$ to 1,000,000. The runtime comparison ratio of SEM:VEM ranges from about 2:1 to 7:1, increasing with the number or records processed, so SEM is faster than A7.1.

Two measures of clustering quality were used by Bradley *et al.* (2000) to compare SEM to VEM and OEM: average values of the estimated log-likelihood and the prototype matching measure shown in Equation (6.41). The Euclidean norm was used for $\|\mathbf{V} - \mathbf{M}_{GT}\|_{PM}$. SEM delivered estimates for the means of the 10 Gaussian distributions that were significantly better than those found by VEM and OEM. Comparisons were also made on three real-world data sets (Reuters Information Retrieval Database, Census Database, REV Digits Recognition Database) that were pretty big. These also showed SEM to be superior to VEM and OEM. Bradley *et al.* (2000) summarize their tests by asserting that SEM outperforms the straightforward alternatives (sample-based VEM and incremental OEM) to scaling existing traditional in-memory implementations to large databases.

Zhang and Forman (2000) present a unified view of sufficient statistics needed for the summarization approach for hard c-means (A6.1, m = 1), hard harmonic c-means, and GMD/EM (A7.1). They assert, per the title of their paper, that distributed data clustering can be efficient and exact. For big static data, they advocate naive chunking and independent (parallel) processing. The analysis is not directed toward clustering accuracy but does discuss one facet of this type of model (relative complexity and efficiency).

Another variant of OEM due to Sato and Ishii (2000) is usually called *stepwise EM* (sEM), not to be confused with the SEM of Bradley *et al.* (2000). Sato and Ishii's sEM can be used for clustering, but their paper is devoted entirely to proving the equivalence of sEM to GMD/EM algorithm 7.1. All of the numerical experiments involve training neural network-based classifiers. They compare sEM to two widely used classifier designs called the Mixture of Experts and the Moody-Darken method. The main difference between the three designs is that they use slightly different log-likelihood functions. Training times are posted for nine classifiers based on these three models. sEM is very competitive. Can sEM be used for big data clustering? Yes. Coming up next.

Liang and Klein (2009) offer a nice treatment that shows how to use incremental OEM (Neal and Hinton, 1998) and sEM (Sato and Ishii, 2000) for clustering. Liang and Klein present two clustering experiments based on speech tagging and document classification. They assert that sEM is 2 – 10 times as fast as GMD/EM algorithm A7.1 and is at least comparable to A7.1 in terms of values of the log-likelihood function as well as accuracy. These authors note that incremental OEM can be realized as a special case of sEM by making the right choices for sEM parameters, (namely, step size reduction power $\alpha = 1$ and mini batch size $\beta = 1$). In view of this, Liang and Klein give a slight nod to sEM but are unequivocal in their assertion that both of these approaches to EM clustering are faster and sometimes more accurate than the basic GMD/EM algorithm A7.1.

Karkkainen and Franti (2007) develop yet another variation of scale up for A7.1 which they call a *gradual model generator* (GMG). The GMG approach is divided into two phases. Phase 1 generates a *Gaussian mixture model* (GMM) from the input data. Stage two employs a hybrid of algorithms A6.1 (c-means, m = 1) and A7.1 (GMD/EM) called *adapted k-means* (AKM) to find clusters in the GMM. This post-processing step doesn't require the original data and is used to reduce the original model to a final form that yields ellipsoidal clusters. The overall procedure is called GMG/AKM.

Example 11.5. Gradual models and adapted c-means to improve A7.1 (GMD/EM).

Figure 11.12 is reproduction of Figure 6 in Karkkainen and Franti (2007) made by GMG/AKM on the data set X_{fws} (cf., Figure 6.9(a) in this volume). The discovered means and cluster ellipsoids shown in Figure 11.12 are pretty convincing: the 15 ellipsoids do seem to capture the central structure of each cluster

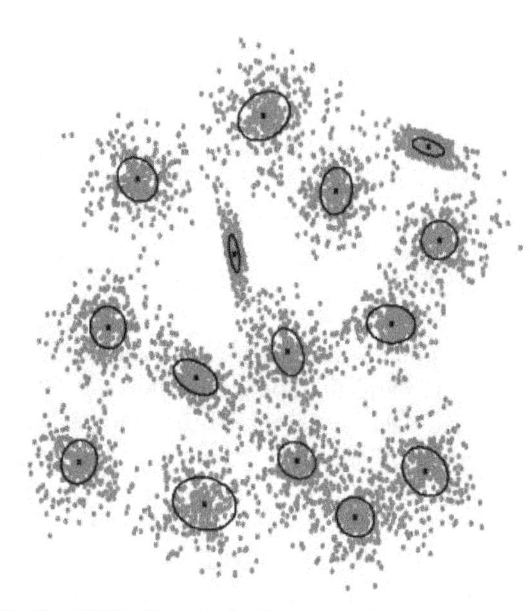

Figure 11.12 GMG ellipsoids for $X_{15,fws}$ (Karkkainen and Franti (2007)).

in this data set. Karkkainen and Franti (2007) compare their GMG/AKM algorithm to two other versions of online EM, viz., SEM (Bradley *et al.*, 2000) and (stepwise) OEM (Sato and Ishii, 2000).

Synthetic data sets used had these ranges: $p = 2$ to 15, $N = 1350$ to 100,000, $c = 9$ to 100. Three real data sets were also tested: $p = 7$ to 16, $N = 68,040$ to 93,945, $c \sim$ unknown. Cluster quality is determined by values of the log-likelihood function $L(\mathbf{Q}; X)$: the larger the value, the better the result. These authors assert that Algorithm A7.1 will produce better results (meaning, higher values of $L(\mathbf{Q}; X)$) than GMG/AKM, SEM, or OEM on any loadable data set. All of the experiments in this paper use loadable data, and the tabulated results bear out their claim. Outputs from GMD/EM algorithm A7.1 are used as the benchmark for comparison. Values of $L(\mathbf{Q}; X)$ from A7.1 were always highest.

Average runtimes are listed for each of the three real data sets for each of four different orderings of the inputs. To get an idea of the results, here are the time ratios of EM algorithm A7.1 to each of the three algorithms tested for the first random submission order on the El Nino data set: GMG/AKM = 18,000:1; SEM = 1.33:1; OEM = 33:1. This approach to online EM is worth a closer look.

A typical approach to scale up HCM/FCM that uses summarization on naive chunks is reported by Hore *et al.* (2009b), who present two single pass methods for scaling up FCM (and by default, HCM as well). *Single pass fuzzy c-means* (spFCM) is discussed in this paper. We will meet the crisp version, *single pass hard c-means* (spHCM) in Section 11.7. Both schemes divide BD_N into $M = \lfloor N/n \rfloor$ naive chunks $\{SD_i\}$ as shown in Figure 11.2(b). The $n = n_1$ points in SD_1 are loaded and processed by wFCM (weighted FCM, cf. Problem 6.21 and Equations (6.84) and (6.85)), with the n initial weights $\mathbf{W}^{(0)} = \mathbf{1}_n = (1, 1, \ldots, 1)^T$. Thus, the initial pass is actually done with LFCM/LHCM (literal fuzzy/hard c-means), resulting in the initial terminal wFCM pair $(U^{(1)}, \mathbf{V}^{(1)})$.

When the first pass is complete, new weights $\mathbf{W}^{(1)} = \left\{\mathbf{w}_i^{(1)} \sum_{k=1}^{n} w_k^{(1)}(u_{ik})^m \mathbf{x}_k / = \sum_{j=1}^{n} w_j^{(1)}(u_{ij})^m;\right.$ $\left. 1 \le i \le c\right\}$ are computed. SD_1 is deleted, $\mathbf{W}^{(1)}$ is merged with SD_2, and wHCM is applied to $\mathbf{W}^{(1)} \cup SD_2$. This procedure continues until all M chunks have been processed. At termination of LHCM on the single pass, $(U^{(M)}, \mathbf{V}^{(M)})$ are available, where $U^{(M)} \in M_{hcn}$ is a crisp c-partition, and $\mathbf{V}^{(M)} \in \Re^{cp}$ is a set of summary cluster centers that presumably encapsulate the clustering history across the M chunks. The final step is to build U_N^{approx} non-iteratively by labeling every $\mathbf{x} \in BD_N$ with $\mathbf{V}^{(M)} \in \Re^{cp}$ and the nearest prototype rule (6.25b) for spFCM, or (6.19b) for spHCM. So, these two single pass methods are $(S + E)$ schemes using type (s_1) sampling.

Single pass FCM/HCM can be used with arbitrarily large input data, but their effectiveness is limited by two facts: (i) each of the M steps suffers from the same limitations and problems as the parent methods (FCM/HCM); and (ii) the output depends on the way BD_N is divided into the M naive chunks. There are two subproblems here: (ii.a) spHCM/spFCM is (chunk by chunk) sequential and, hence, depends on the input ordering of the chunks; and (ii.b) naive chunks may or may not be representative of cluster structure in BD_N, the problem called Q1 in Figure 11.3.

A companion algorithm that appears in Hore *et al.* (2009b) is called *on-line FCM* (oFCM), which by default has a crisp counterpart, *on-line HCM* (oHCM). In this model data might be derived by naively chunking a static big data set BD_N into $M = \lfloor N/n \rfloor$ chunks as is done by spFCM. But the on-line version assumes that each chunk SD_i of size $n = n_i$ (the buffer size) arrives one input at a time and is saved until the buffer is full. Then SD_i is processed by literal FCM. The i-th run of LFCM on SD_i results in the terminal pair $(U^{(i)}, \mathbf{V}^{(i)})$. The row sums of $U^{(i)}$ define a vector of c weights $\mathbf{w}^{(i)} = (w_1^{(1)}, \ldots, w_c^{(i)})^T$, where $w_j^{(i)} = \sum_{k=1}^{n} u_{jk}^{(i)} : 1 \le j \le c$. LFCM clustering in SD_i is summarized by the pair $\{\mathbf{w}^{(i)}, \mathbf{V}^{(i)}\}$, SD_i is deleted from the buffer, and the next chunk is built. After M runs, there are M sets of weights $\{\mathbf{w}^{(i)}\} \subset \Re^p$ and centroids $\{\mathbf{V}^{(i)}\} \subset \Re^p$ which form an ensemble of weighted clustering solutions. Aggregation of the ensemble is done by making one run of wFCM on the union of the centroids, $\mathbf{V}_M = \{\bigcup_{i=1}^{M} \mathbf{V}^{(i)}\}$ using the weights $\mathbf{W} = \{\bigcup_{i=1}^{M} \mathbf{w}^{(i)}\}$. This results in a final set of c cluster centers $\{\mathbf{V}^*\} \subset \Re^p$, which are then used to generate labels for $BD_N = \bigcup_{i=1}^{M} SD_i$ where $N = nM$, with the nearest prototype rule (6.19b) for oFCM, or (6.26b) for oHCM.

The oFCM approach attempts to offset the problem identified by Q1 in Figure 11.2(b) by specifying c to be the maximum expected number of clusters (c_{max}) in the evolving data stream. In the worst case, a given chunk might come from one class only,and, in the best case, a chunk might contain some samples from all *c* classes. See Hore *et al.* (2009b) for arguments about how to interpret this strategy.

Hore *et al.* (2009a) suggest an interesting way to compare approximations to the literal results that any algorithm driven by an objective function can obtain on *loadable* data. This method is applicable to three of our fantastic four viz., HCM, FCM, and GMD/EM. They define the quality of an approximation to the literal result this way:

$$DQ(A_L, A_{approx}) \frac{J_{A_L}(\mathbf{Q}^{literal}) - J_{A_{approx}}(\mathbf{Q}^{literal})}{J_{A_{approx}}(\mathbf{Q}^{literal})} \tag{11.6}$$

This definition is a bit more general than the one used in Hore *et al.* (2009a) because it also encompasses Algorithm A7.1. The notation $\mathbf{Q}$ in Equation (11.6) represents the variables that appear in row 6 of Table 9.1. Specifically, $\mathbf{Q} = (U, \mathbf{V})$ and $J = J_m(U, \mathbf{V} : X_N)$ for HCM $(m = 1)$ and FCM $(m > 1)$; $J = L(\mathbf{Q}; X)$ with $\mathbf{Q} = (\mathbf{p}, \mathbf{M}, \{S_i\}, P)$ for GMD/EM. For hard and fuzzy k-means, Equation (11.6) compares the values of the squared errors committed by the literal and approximate implementations. A negative value indicates that the approximate error is higher than the literal error. Table 2 in Hore *et al.* (2009a) shows values for Equation (11.6) that compare LHCM to spHCM on 10 data sets. The average result is that spHCM is 1.79% better than literal hard k-means in 9 of the 10 trials. The literal algorithm terminated at a smaller error than the approximation just once (on the plankton data). This is a pretty good indication that spHCM is a worthwhile approximation to LHCM, and spHCM can be used on big data that is unloadable for LHCM. Here is an example that combines some of the results in Hore *et al.* (2009a, 2009b) which provides some insight into both of these algorithms.

Example 11.6. Image segmentation with spFCM and oFCM.

In Hore *et al.* (2009b), two sets of unlabeled MR images from the University of Miami and the University of California (San Francisco) were used to test spFCM and oFCM. For example, the Miami images comprised 96 slices, each 512×512 in size. The images all had values for T1-weighted, (weighted) *Proton Density*, and T2-weighted measurements, but only the T1-weighted data were used. Since the images were unlabeled, segmentations made by the FSL routine in the *FMRIB Software Library* (http://fsl.fmrib.ox.ac.uk/fsl/fslwiki/) were chosen as "surrogates for ground truth," i.e., they provided a reference frame that was used a basis for comparison to segmentations made by spFCM and oFCM. This choice was made because FSL was identified in Bouix *et al.* (2007) as the best (probabilistic) segmentation approach in this software library when compared to six other approaches.

Segmentation quality of each of three tissue types (CSF-cerebro spinal fluid, GM = gray matter, and WM = white matter) was evaluated quantitatively on a volume basis. For each voxel, the FSL segmentation was established as the reference label. Then the spFCM and oFCM labels at the same voxel position were compared to the FSL label. When the (FSL, spFCM) or (FSL, oFCM) labels agreed, Hore *et al.* (2009b) counted a match to FSL as if it were a match to all seven segmentation algorithms in the FMRIB library. They report that the average match to the segmentation labels of FSL across all of the images tested as 88.54% for spFCM and 90.33% for oFCM. Since the ground truth (FSL labels) is itself the output of a clustering algorithm, it is incorrect to interpret these percentages as indicants that spFCM and oFCM are not as good as FSL. What can be concluded is that the three segmentation methods are comparable. To complete this example, Figure 11.13 shows the middle row of Figure 1 in Hore *et al.* (2009b), corresponding to the raw T1 data for patient 48. Only three tissue regions are shown: CSF (■), GM (■), and WM (■).

The data for spFCM was randomly arranged (pixel by pixel) and then naively chunked into 10 pieces before processing. The data for oFCM was read from bottom to top in 20 chunks as if it were arriving on

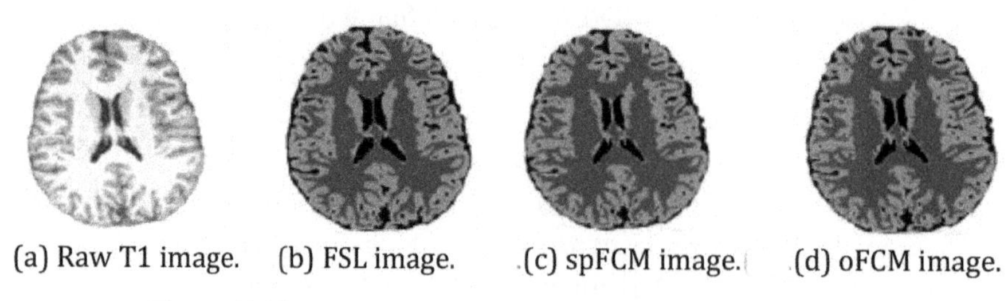

Figure 11.13 Image segmentation using FSL, spFCM and oFCM.

line in a stream. Visual assessment of segmented images that have rich substructure is quite difficult. As you look at Figure 11.13, try to compare each of views 11.13(b,c,d) to 11.13(a), which is the raw data. If you zoom in on this figure, it looks like the spFCM image is slightly better at matching the raw data than the other two. But it is too close to call, and is subjective anyway. The bottom line? spFCM and oFCM are viable options for segmenting images that are too big to load all at once -big data images. And, these two methods are natural candidates for streaming data analysis (cf., Chapter 12).

Havens *et al.* (2012) compared three different techniques that extend FCM algorithm A6.1, $m = 2$, for clustering in big data. The three methods: (i) (S + E) with *random sampling FCM* (rseFCM), which applies FCM to a random sample of the big data, followed by non-iterative extension as illustrated by Figure 11.10; (ii) unweighted and weighted versions of *approximate kernelized FCM* (akFCM), a kernelized version of FCM that begins with random sampling when the kernel matrix K_{NN} is unloadable; and (iii) three incremental techniques that make one sequential pass through subsets of the data, viz., (spFCM), oFCM), and (brFCM). The three sequential techniques more properly belong in Chapter 12 since they are a type of streaming algorithm: they appear here because they were compared to the extension of FCM by (S + E) in Havens *et al.* Time and space complexity analysis is given for 14 different algorithms in Havens *et al.* (2012). Time complexity for the seven vector data forms is linear in N; all but one of the kernelized algorithms are quadratic or cubic in N.

Example 11.7. Clustering in big data with FCM, spFCM, oFCM, brFCM, and akFCM.

Numerical experiments in Havens *et al.* (2012) on loadable and unloadable data sets compare 14 algorithms in terms of speedup and quality of approximations to batch LFCM (for loadable data). The quality of approximate partitions in loadable cases is measured by computing the crisp cluster validity index $V_{ARI}(U|Q)$ at Equation (9.8) on hardened fuzzy c-partitions U made with **H** at Equation (3.4) and ground truth partitions (Q) on three loadable real-world labeled data sets ($X_{15,fws}$ in Figure 6.9(a), the *MNIST* image data, $n = 70,000$ points in $p = 784$ features with $c = 10$ presumed CPOV clusters, and *Forest* ($N = 581,012$, $c = 7$, $p = 54$, Xu and Wunsch, 2009)). Experiments were based on using 1%, 2.5%, 5%, 10%, 25%, 35%, and 50% of the data.

The experiments showed that oFCM was better than rseFCM and spFCM on $X_{15,fws}$; that spFCM performed best on the MNIST data; and that the three algorithms were equally bad (all three averaged about $V_{ARI}(U|Q) \cong 0.03$ in 50 trials for sizes of samples). These experiments don't warrant a strong conclusion other than the usual one: that the results depend quite heavily on the input data, and, of course, that a different CVI might yield very different results. Additional experiments were made by segmenting three large unlabeled MRI images.

The last example in Havens *et al.* is based on an unloadable synthetic data set of $X_N = 5 \times 10^9$ vectors in $\Re^4$ that are randomly drawn with equal probability from three 4D Gaussian distributions, all having the identity as their common covariance matrix, and means at $\mu_1 = (0, 0, 0, 0)^T$, $\mu_2 = (0, 0, 5, 5)^T$, $\mu_3 = (5, 5, 0, 0)^T$ Three algorithms, rseFCM, spFCM and oFCM, operating on samples of 0.001%, 0.01% and 0.1% of the data, are compared to an extrapolated LFCM (based on a loadable sample $|X_n| = 10^6$). Because the data set is so large, the extension step was skipped and the algorithms only returned cluster centers. The quality of the approximations was measured by computing the prototype accuracy $PM(\mathbf{V}|\mathbf{M}_{GT}) = \|\mathbf{V} - \mathbf{M}_{GT}\|_{PM}$. Interestingly, oFCM and spFCM produced the same error (0.28) for every sample size. Random sampling

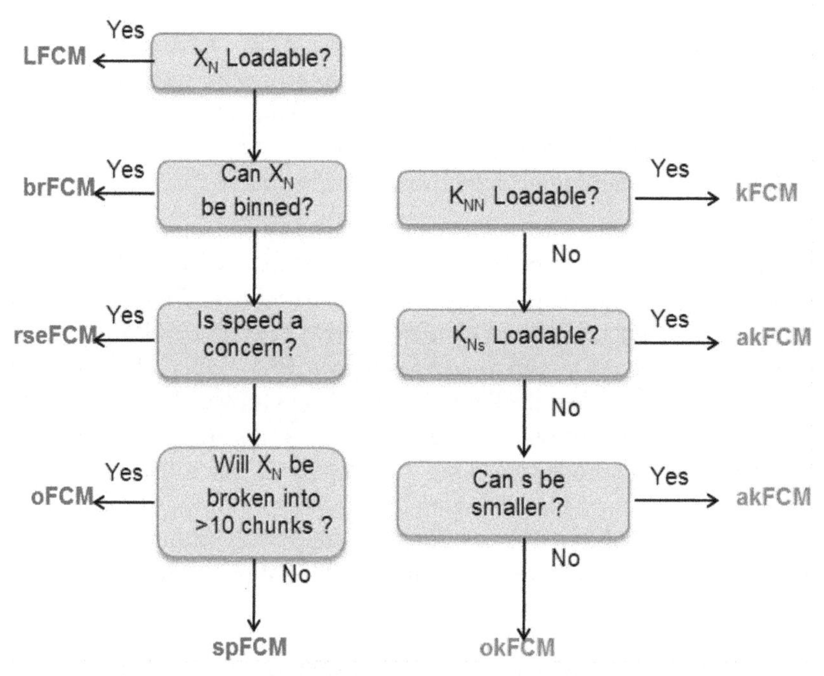

Figure 11.14 Using FCM and kFCM on big data (Havens *et al.* (2012)).

produced PM errors of 0.54, 0.31 and 0.28 for the three samples, corroborating a belief that as the size of the RS increases, so does the approximation accuracy. A conclusion from this example was that streaming FCM schemes such as oFCM and spFCM are competitive with LFCM on big static data sets - this is an important point to remember when we get to Chapter 12.

Figure 11.14 is a reproduction of Figure 7 in Havens *et al.* (2012). Bear in mind, as you look at Figure 11.14, that HCM algorithm A6.1 ($m = 1$) is a special case of FCM; so almost all of the methods shown in this diagram, with small modifications, are applicable to scaling up k-means (HCM) for clustering in big data. The computational evidence is that rseFCM, brFCM, and akFCM are good choices to approximate FCM depending mostly on the type of data being clustered. This paper concludes with the recommendations that are shown in Figure 11.14.

11.5 Visualization of Big Static Data

Often the most effective ways to describe, explore, and summarize a set of numbers - even a very large set - is to look at pictures of those numbers

–E. R. Tufte (2001)

There are a seemingly endless number of approaches that might be useful for reducing big data to some type of visual representation for inspection of clustering tendency assessment (and possibly clustering as well).

Do a Google search for "Textbooks on Visualizing Data" and it will return perhaps a dozen books on this topic that have appeared in the last 10 or 15 years. The classic in this field is Tufte (2001), which provides a great background for modern approaches to data visualization. And, it's a lot of fun to read too.

All of the feature extraction methods discussed in Chapter 5 that produce visual representations of the input data (X or D) have gone under the microscope of scalability -some have failed, but others seem to be quite robust. Since the visual domain is confined to $p \leq 3$ dimensions, most of these methods essentially map upspace data, i.e., batch collections of big data in $\Re^p$, into a visual representation in $\Re^1$, $\Re^2$ or $\Re^3$. The most common target dimension for downspace visualization is $\Re^2$. Some of these methods combine feature extraction with visual representation. The paper of Ferreira de Oliveira and Levkowitz (2003) doesn't explicitly discuss visualization in *big* data, but this is a very worthwhile survey of six approaches to the problem because most, if not all, of these approaches have been or will be scaled up to the big data arena.

Valdes *et al.* (2016) present a very insightful analysis and examples of several visualization techniques, including linear and nonlinear feature extraction, classical projection, and some methods based on various computational intelligence paradigms: principal components, Sammon mapping, Isomap, *locally linear embedding* (LLE), spectral embedding, *t-distributed stochastic neighbor embedding* (t-SNE), generative topographic mapping, neuro-scale, and genetic programming. See Figures 5.17 and 5.18 for projections of the 4D Iris data and 784-dimensional MNIST data with four of these methods. Valdes *et al.* use a virtual reality modeling language to view 3D visualizations that may soon become a standard tool for visualizing big data. This wide-ranging paper deserves a careful reading.

Cluster heat maps (reordered dissimilarity images), which are simple to build and easy to interpret for smallish data sets, are increasingly in demand for visualization of big static data. Hathaway *et al.* (2006) described a version of VAT algorithm A4.1 they called *scalable VAT* (sVAT) which is based on MMDRS. This method generalizes easily to iVAT, thereby becoming siVAT algorithm A11.3. The siVAT algorithm and its relatives have found many applications (Kumar and Bezdek, 2020). In particular, it is the basis of the clusiVAT method discussed in the next section that enables the use of the (S+E) method for single linkage clustering in big static data.

While the term "scalable" was used by Hathaway *et al.* (2006) and has been repeated in many papers and books up to and including Kumar and Bezdek (2020), this adjective is a bit misleading. Recall that an algorithm is *scalable* if its runtime complexity increases linearly with the number of records (n) in the input data. But sVAT/siVAT doesn't produce an $N \times N$ image of the data set: instead, it produces an $n \times n$ image of the dissimilarity matrix of the sample data. So, the sVAT/siVAT image is more properly regarded as a skeleton or core of the incomputable cluster heat map of the big data. A better interpretation of the "s" in the acronym siVAT is to regard it as an abbreviation for the word "sample" instead of scalable. Many papers call sVAT/siVAT cluster heat maps "approximations" of the incomputable big cluster heat maps that are desired. But the word "approximate" suggests that the approximation image is the same size as the literal image, and this is not the case. A11.3 produces a small image that we hope is, in some sense, a good indication of clustering tendency in the big input data that would be made by VAT or iVAT if this were possible.

Hathaway *et al.* (2006) presented MM algorithm A11.1, MMDRS algorithm A11.2, and sVAT/siVAT A11.3 below in a single algorithm they called sVAT which is easily generalized to siVAT. The three steps, MM, MMDRS, and sVAT/siVAT, are presented separately in this chapter because the MM and MMDRS algorithms are valuable in their own right as useful and interesting sampling schemes. The last step of the original method, presented as A11.3 here, simply finds the sample distance (or dissimilarity) matrix $D_n \subset D_N$ corresponding to the MMDRS indices and then constructs the VAT or iVAT image of it. Here is the pseudocode for siVAT (if you want sVAT, call the commented out VAT A4.1, top in line 16).

Algorithm A11.3. sVAT/siVAT visualization algorithm for big data.

	sVAT/siVAT algorithm (Hathaway *et al.*, 2006)
1	**In:** $X_N \subset \Re^p$ and metric $d : \Re^p \times \Re^p \rightarrowtail \Re^+$: *(or)* $D_N \subset M_N^+$: c′:n
2	**Call:** MM (X_N (or) D_N, c′): % return c′ indices $M' = \{m_1, \ldots, m_{c'}\}$
3	**Call:** MMDRS (X_N(or)D_N, c′), % Return: n: $M' = \{m_1, \ldots, m_n\}$: n MMDRS vectors $X_n = \{\mathbf{x}_{m_1}, \ldots, \mathbf{x}_{m_n}\}$ *(or)* n MMDRS objects $O_n = \{o_{m_1}, \ldots, o_{m_n}\} \subset O_N$
4	% Form D_n, the n × n principal submatrix of D_N corresponding to the rows and columns of the indices $\{m_1, \ldots, m_n\}$.
5	$[D_n] = [0]$ For i = 1 to n − 1: For j = i + l to n: $[D_n]_{ij} = [d(\mathbf{x}_{m_i} - \mathbf{x}_{m_j})]$ *(or)* $[d(o_{m_i}, o_{m_j})]$: $[D_n]_{ji} = [D_n]_{ij}$ Next j Next i
6	Apply algorithm [VAT (A4.1, top)] *(or)* iVAT (A4.1, bottom) to D_n
7	**Out**: VAT(D_n) *or* iVAT(D_n)

Figure 11.15 illustrates the relationship between the big data dissimilarity matrix and the output of A11.3. The big dissimilarity matrix D_N comprises the black dots on the left side of the figure. In the vector data case, this matrix isn't needed, but is shown for visualization of the process. If the input data is vectorial, the vectors in X_N represent objects directly. In the object data case, D_N is a known input and the MMDRS indices point to objects underlying the input data. In either case, MMDRS samples $O_n = \{o_1, \ldots, o_n\}$ identify the rows and columns of the submatrix $D_n \subset D_N$ represented by the red dots. Then D_n is passed to A4.1, which produces the small cluster heat map iVAT(D_n) at the right of this figure (this particular image happens to be the 100 × 100 iVAT image that appears in Figure 11.22).

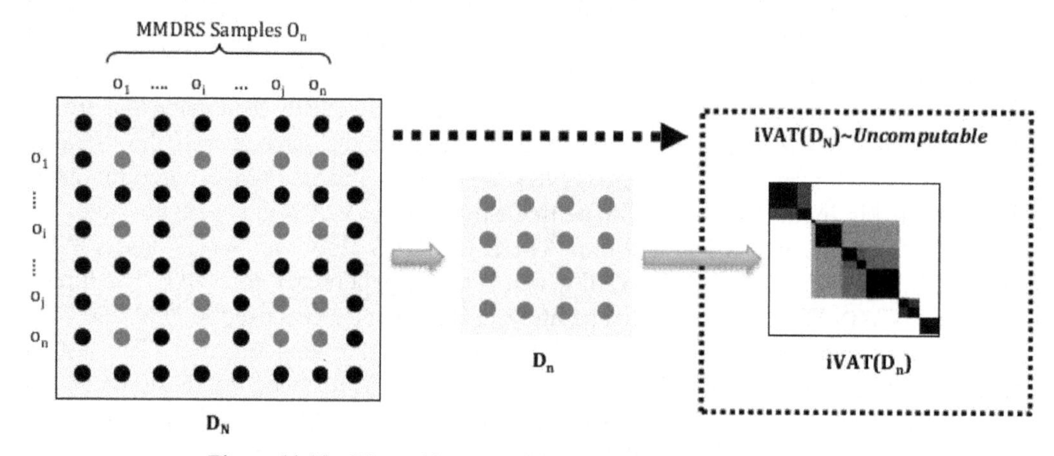

Figure 11.15 The architecture of the sVAT/siVAT sample heat map.

There are many variations of this basic idea which are discussed at length in Kumar and Bezdek (2020). Here is a first example of how the sample cluster heat map mirrors the cluster structure that might be seen if we could build the big data heat map.

Example 11.8. Visual assessment of static big data with siVAT.

This example exhibits sample small data images built by sVAT and siVAT on a moderately big data set (too big for literal VAT or iVAT) using algorithm A11.3. Figure 11.16(a) is a scatterplot of $X_n = 400,000$ points in $\Re^2$ drawn from a mixture of 4 Gaussian distributions which is called $X_{400,000,2}$. There are 100,000 points in each of the c = 4 HPOV Gaussian clusters, and this provides a ground truth partition $U_{GT} \in M_{h4,400,000}$ of the data set. These clusters have various shapes due to different covariance matrix structures, but their means are chosen so that they appear to be compact and well-separated.

However, the maximum diameter in the set of four clusters is 810.48, and the *minimum* distance between a pair of points in the closest (brown and blue) clusters is 174.96. In consequence, Dunn's index $V_{DI}(U_{GT})$ at Equation (6.56) has the value $174.96/810.48 = 0.2159$ on the ground truth partition of the data. This value is surprisingly small but happens because the largest diameter (corresponding to the elongated green cluster) dominates this value of Dunn's index. This doesn't preclude the possibility that there is some as yet undiscovered partition of this data for which Dunn's index is greater than 1, but it is pretty unlikely. There *might be* a crisp partition of $X_{400,000,2}$ in $X_{h4,400,000}$ which is compact and separated in the sense of Dunn as required by Theorem 11.2 to make the sampling guarantee, but the value of Dunn's index for the ground truth partition of this data is less than 1; so the hypothesis in Theorem 11.2 is not confirmed for this partition. Does this mean that MM sampling algorithm A11.1 will fail to find at least one point from each cluster? NO. It just means the guarantee provided by Theorem 11.2 is not in play.

The parameter c' for A11.1 was $c' = 10$; so 10 MM indices will be chosen by A11.1. The target sample size is $n = 300$. Let $X_n = X_{300}$ denote the 300 samples extracted from the data set $X_{400,000,2}$ by Maximin algorithm 11.1. Figure 11.16(b) shows the sVAT image of the Euclidean distance matrix of the pairs in X_{300}. As is often the case, VAT doesn't produce a very useful visual impression of cluster structure in X_{300}. But the siVAT image of X_{300} in view 11.16(c) provides a pretty strong impression that

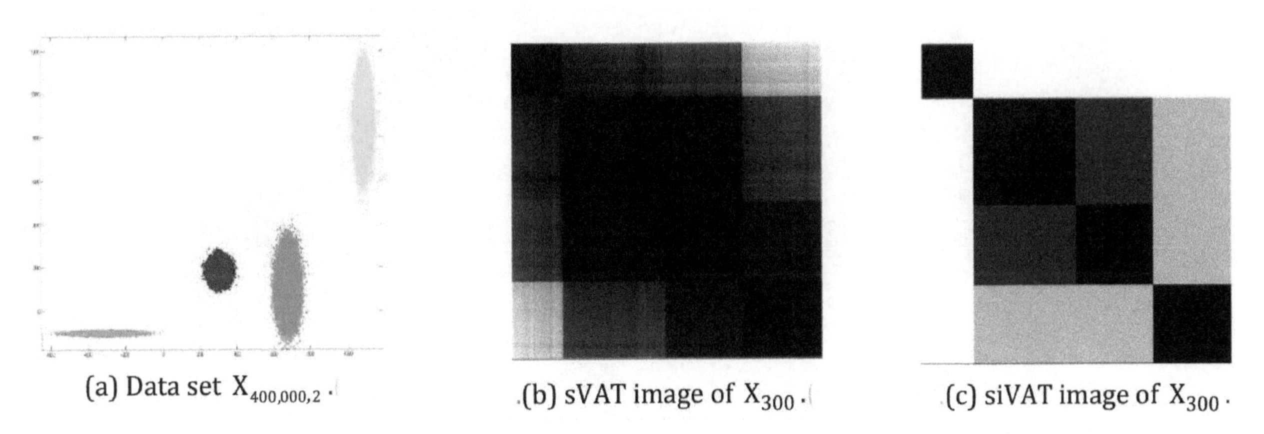

(a) Data set $X_{400,000,2}$. (b) sVAT image of X_{300}. (c) siVAT image of X_{300}.

Figure 11.16 sVAT/siVAT sample heat maps for uncomputable VAT/iVAT images of $X_{400000,2}$.

X_{300} contains c = 4 clusters: there are two (somewhat hard to see) dark blocks within the biggest dark block in this image. Comparing the 300 sample siVAT surrogate of the big data cluster heat map to the HPOV clusters in the data suggests that the iVAT image of all 400,000 points might look very similar. So, the MMDRS sampling scheme has evidently done a good job. The surrogate image in Figure 11.16(c) build by siVAT uses just $300/400,000 = 0.075\%$ of the data, but this is still good enough to get a reasonable idea of clustering tendency in the big input data. This illustrates that siVAT(D_n) can provide a useful estimate of the clustering structure that might be suggested by the incomputable image iVAT(D_N) even when we don't know if Theorem 11.2 is applicable (which we will hardly ever know anyway).

The siVAT algorithm has a sophisticated set of relatives based on spectral feature extraction called specVAT. The approach described by Wang *et al.* (2010) is based on using the spectral decomposition of a normalized Laplacian matrix to reorder D before applying VAT or iVAT to it. Their paper describes three variations of VAT based on this idea: A-specVAT, P-specVAT, and E-specVAT. Here is an example of specVAT, excerpted from Wang *et al.* (2010), that is very similar to the results shown in Example 11.8.

Example 11.9. Visual assessment of static big data with specVAT.

Figure 11.17(a) is a scatterplot of 30,000 samples, $X_{30,000,2}$, from a data set $X_{3,000,000,2}$ that contains $N = 3,000,000$ points drawn from a mixture of c = 5 two-dimensional Gaussians. The data set $X_{3,000,000,2}$ is essentially a big data sibling of the 5,000 points named $X_{5,2}$ that are shown in Figure 7.3(a) – and, the sample $X_{30,000,2}$ of it in view 11.17(a) is the image on the cover of this volume! The sample is shown here only to convey an idea of what the 3,000,000 points would look like in a full scatterplot. The Euclidean distance matrix on the whole data set requires the calculation and storage of $N(N-1)/2 \sim O(10^{12})$ values -way, way, WAY too many! So, making an iVAT image of $X_{3,000,000,2}$ based on the entire dissimilarity matrix is intractable for two reasons: D is too large, and iVAT can't work on a D this big. In fact, it is not possible to

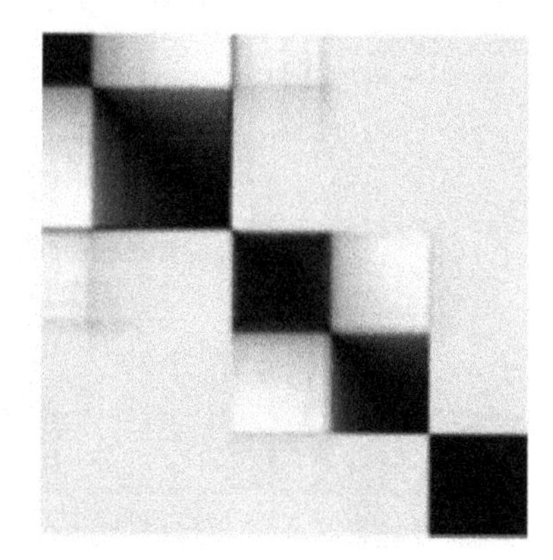

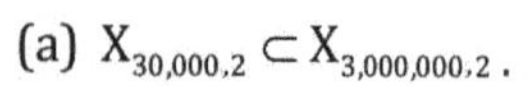

(a) $X_{30,000,2} \subset X_{3,000,000,2}$.

(b) specVAT image of $X_{2,500,2}$.

Figure 11.17 (a) 30,000 points from X_N. (b) specVAT image of $X_n = 2500$ samples of X_N.

make a useful iVAT image of $X_{30,000,2}$ either, even though its Euclidean distance matrix is manageable. This will require some method of downsizing.

The image in view 11.17(b) is made by a version of VAT called *spectral VAT* (specVAT) from $X_n = 2,500$ or 0.083% samples of the original data, $X_n = 2,500 \subset X_{3,000,000,2}$. The visual estimate of c offered by this surrogate image is c = 5, which agrees with the HPOV assessment of the input data that can be made from the sample shown in Figure 11.17(a). To convince yourself that this is useful, imagine an unlabeled data set just like this but in 4-space instead of 2-space. Then no HPOV is available, but images such as this one can still reveal possible cluster structure in the big data.

11.6 Extending Single Linkage for Static Big Data

Algorithm 8.2 is a naive description of basic SL clustering based on Prim's (1957) algorithm for building the MST of D that has its roots in the paper by Gower and Ross (1969). Not much has happened to alter the basic algorithm since then, but Müllner (2013, 2015) presents a very nice update of implementational issues that assumes access to modern software. The 2015 paper discusses theoretical constraints and limitations on implementations of the seven most common SAHN methods: single, complete and average linkage, *weighted pair group mean average* (WPGMA; McQuitty (1967), Ward's method (1963)), unpaired *group mean centroid* (UPGMC), and *weighted paired group median centroid* (WPGMC) linkage (see Sneath and Sokal, 1973 or Everitt *et al.*, 2011). Müllner points out that all of these methods are implemented in standard numerical and statistical software such as R (R Development Core Team, 2011), MATLAB (The MathWorks, Inc., 2011), Mathematica (Wolfram Research, Inc., 2010), and SciPy (Jones *et al.*, 2001). Moreover, Müllner provides comments, explicit instructions, and links to his own implementations in the 2013 paper. Müllner (2015) discusses several alternative implementations of SL, and remarks that

> *For single linkage clustering, [Prim's] MST algorithm is clearly the fastest one. Together with the fact that it has only half the memory requirements of the other algorithms (if the input array is to be preserved), and thus allows the processing of larger data sets, the MST algorithm is clearly the best choice for single linkage clustering.*

According to Müllner (2015), clustering with SAHN algorithms also favors SL when the big data begin as static sets $X = \{\mathbf{x}_1, \ldots, \mathbf{x}_N\} \subset \Re^p$. He asserts that the MST algorithm for single linkage can compute distances on-the-fly. Since every pairwise dissimilarity is read in only once, there is no performance penalty compared to first computing the whole dissimilarity matrix and then applying the MST algorithm. On the contrary, computing pairwise distances in-process can result in faster execution since much less memory must be reserved and accessed. [Prim's] MST algorithm is suitable for any dissimilarity measure which can be computed from vector representations (that is, all scale types are possible, e.g., real valued measurements, binary sequences, and categorical data). Müllner's (2013, 2015) papers offer a very thorough update on single linkage: well worth the effort.

The *sample and extend* (S + E) method discussed for the three objective function algorithms in earlier sections of Chapter 11 also works for single linkage clustering but in a slightly different way. The aggregation step is still $U_N^{approx} = \boldsymbol{a}(U_n^{literal}, U_{N-n}^{approx}) = [U_n^{literal} \| U_{N-n}^{approx}]$, but the construction of U_{N-n}^{approx} from the literal result differs from our previous cases in two ways. First, the extension $U_n^{literal} \rightarrow U_{N-n}^{approx}$ is not carried out by substitution into a FONC for the label using the literal parameters of the small data run, as illustrated in

Figure 11.10 for FCM. And, second, the extension step is not an instantiation of the nearest prototype rule. Instead, the extension uses the *nearest neighbor rule*[31].

The (S + E) strategy for SL is embodied in an algorithm called *clustering with iVAT* (clusiVAT). It is not iVAT that does the clustering -that chore is allocated to single linkage. Implementation of the literal SL algorithm on the small data set SD_n depends on the concept of aligned partitions, which were introduced in Exercises 6.4 and 6.5 but not further discussed until now. Aligned partitions appear in A11.4 and are important to the theory that links SL to clusiVAT, so we take a quick detour to look at this idea.

Definition 11.2. Aligned crisp c-partitions.

$U \in M_{hcn}$ is a crisp *aligned c-partition* of $O = \{o_1, o_2, \ldots o_n\}$ when its entries form c contiguous blocks of 1's in U, ordered to begin with the upper left corner, and proceeding down and to the right. The set of all such partitions is

$$M^*_{hcn} = \{U \in M_{hcn} | u_{ik} = 1, 1 \leq k \leq n_1 : u_{ik} = 1, n_{i-1} + 1 \leq k \leq n_{i-1} + n_i, 2 \leq i \leq c\} \quad (11.7)$$

For example, $U_1 = \begin{bmatrix} 1 & 1 & 0 & 0 & 0 \\ 0 & 0 & 1 & 1 & 1 \end{bmatrix}$ is aligned, but $U_2 = \begin{bmatrix} 0 & 0 & 1 & 1 & 1 \\ 1 & 1 & 0 & 0 & 0 \end{bmatrix}$ is not. There are two useful alternate notations for aligned partitions. First, we can represent an aligned partition with a list of c integers which describes how many 1's occur in each of the blocks in the alignment. For example, the matrix $U_1 = \begin{bmatrix} 1 & 1 & 0 & 0 & 0 \\ 0 & 0 & 1 & 1 & 1 \end{bmatrix}$ is completely specified by the list {2:3}, which denotes that the first two objects are in the first subset, and the second three objects are in the second subset. Second, recall from Chapter 3 that any crisp partition has an alternate representation as a cluster list vector. Using this third notation, we have three representations of the aligned partition: $U_1 = \begin{bmatrix} 1 & 1 & 0 & 0 & 0 \\ 0 & 0 & 1 & 1 & 1 \end{bmatrix} \leftrightarrow [1\ 1\ 2\ 2\ 2] \leftrightarrow \{2 : 3\}$ and the double arrow indicates that they are equivalent, i.e., the forms are reversible or "invertible."

The general notation for $U \in M^*_{hcn}$ is a list with c integers which are the cardinalities of the c crisp clusters in U. Thus, if $X = \bigcup_{i=1}^{c} X_i$ and for all i, $n_i = |X_i|$, we can write $U \in M^*_{hcn} \leftrightarrow \{n_1 : n_2 : \cdots : n_c\}$. Proposition 11. 3 shows why this concept is important for the extension of SL to big static data.

Proposition 11.3. Single linkage partitions extracted from the VAT/iVAT MST are aligned partitions.

Let $O = \{o_1, o_2, \ldots o_n\}$ denote a set of n objects, and let D_n be an $n \times n$ dissimilarity matrix for O. If the dissimilarity values in D_n are all distinct (i.e., the MST is unique), the crisp single linkage clusters in the MST of the VAT reordered dissimilarity matrix D^*_n are aligned c-partitions of O^* for every value of c. An important fact is that the recursive version of iVAT given in the lower block of A4.1 is built from the VAT ordered MST; so iVAT also satisfies this proposition - that is, SL clusters extracted from the iVAT MST are also aligned.

---

[31]Please be careful to note the difference between the nearest prototype (NPR) and nearest neighbor (NNR) rules as used in this primer. The NPR assigns the label of the nearest labeled prototype to each unlabeled object. The NNR assigns the label of the nearest labeled neighbor *in the data* to each unlabeled object. Some authors use the same terminology for both operations.

What does it mean? This proposition gives an explicit link between SL clusters and VAT reordering and is the key to the clusiVAT extension of literal SL in D_n to approximate SL clusters in D_N because it tell us exactly how to find SL clusters in the VAT ordered MST of D_n^*.

Proof. Proposition 1, Havens *et al.* (2009). ■

Recall that the array **d** returned from A4.1 stores the $(n - 1)$ distances between pairs of vertices in the VAT/iVAT ordered MST. These distances are the weights of the edges, stored in the order they are inserted into the MST. To find single-linkage clusters in theVAT/iVAT reordered data, we extract the largest $(c - 1)$ edges from **d**, and then cut them, creating c subtrees. These subtrees are the SL clusters which are represented by the corresponding aligned partition. For example, if we wish to compute the single-linkage 4-partition of D_n, we find the 3 largest values in the array **d**, sort them so that $\{j_1 < j_2 < j_3\}$, store their index values as $\{j_1, j_2, j_3\}$, and form the aligned partition $\{j_1 : j_2 - j_1 : j_3 - j_2, j_4 - j_3\}$, which in turn is re-represented in cluster list vector form. Example 11.10 should make this procedure clear, and it illustrates how the exact SL clusters in D_n are used to label the rest of the data represented by dissimilarities in $(D_N - D_n)$.

Example 11.10. Extracting SL clusters from a VAT reordered MST.

To avoid paging back and forth between Chapter 4 and this example, parts (a) and (d) of Figure 4.6 are repeated here as Figures 11.18(a) and 11.18(b). The $n = 20$ point data set X in view (a) has $c = 4$ HPOV clusters. In Chapter 1, you were alerted to the possibility that SL could regard a single point, such as the orange one at the lower right in view 11.18(a), as a cluster, and that will be the case here. View (b) shows the VAT reordered MST on the data using the initial point indicated by the blue arrow at the lower left, and, in this panel, you can see the three longest edges in the MST. It is these edges that are found and cut to form the SL 4-partition of the input data shown in view 11.18(c). The dashed double-arrowed lines in view (b) indicate the cuts. View (c) shows the SL partition of the 20-point data set unioned with another singleton which was not in the input data, the point $\mathbf{z} \in \Re^2$. This example describes how the point **z** is labeled in the extension step of clusiVAT algorithm A11.4.

To see how the notion of alignment enables the recovery of the SL clusters, Figure 11.19 shows the three alternate ways to represent the SL partition of X extracted from the MST in Figure 11.18(b) by cutting its three longest edges. Figures 11.19(a), 11.19(b), and 11.19(c) are the matrix form, the aligned list form, and

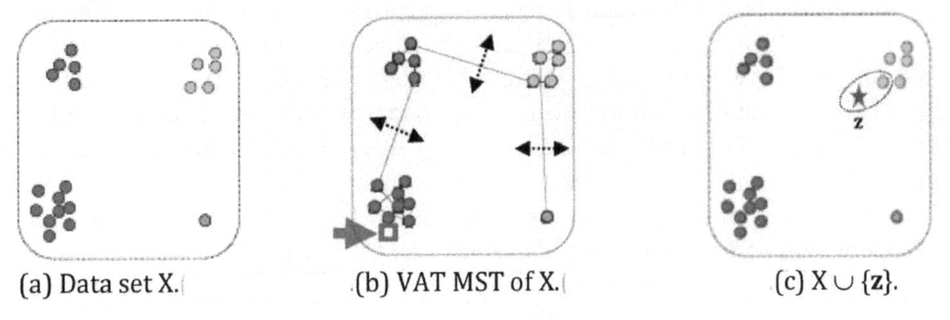

Figure 11.18 Data set X, the iVAT ordered MST of $D_E(X)$, and the augmented data set $X \cup \{z\}$.

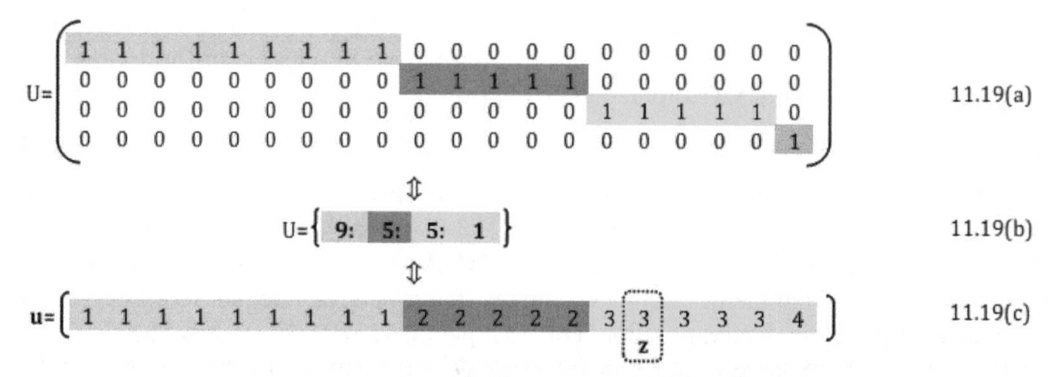

Figure 11.19 Three representations of the SL partition of X.

the cluster list vector form of U, respectively. All three views are color-coded to show the relationship of the memberships in U to the 20 points in X, and all three descriptions are equivalent. So, given any of the three, you can always recover the other two.

The aligned list form of U, viz., {9:5:5:1} and the cluster list vector form (**u**) are both used in clusiVAT A11.4. When you look at this graphic, you probably read it from top to bottom, i.e., given $U \in M_{h,4,20}$, we can write it as {9:5:5:1}. But the double arrow here means "is isomorphic to." For example, we can start at Figure 11.19(b) and proceed to itemize the cn entries in the matrix form of the partition at Figure 11.19(a). Thus, the "9" tells us there are 1's in the first nine columns in the first row of U; the "5" tells us that there are 1's in the next five columns in the second row of U, and so on. And, in turn, the objects corresponding to the columns in U take the indicated cluster labels, as shown in the cluster list vector view at Figure 11.19(c). So, an aligned partition always enables a reconstruction of the crisp c-partition in either matrix or cluster list form.

The partition in cluster list vector form **u** as in Figure 11.19(c), which is built by cutting the $(c-1)$ longest edges in the VAT or iVAT MST of D_n, is an exact SL partition that labels the n objects in the small data set X_n or D_n. So, we can assign cluster labels to all the remaining objects in $(X_N - X_n)$ or $(D_N - D_n)$ using the nearest neighbor rule with these exact (n) SL labels. The point **z** in Figure 11.18(c) is such a point, shown as being nearest to one of the green points in Figure 11.18(c), which bears cluster label 3. The nearest neighbor assignment is seen in Figure 11.19(c), where **z** is assigned to cluster 3 (green).

The proof cited for Proposition 11.3 depends on using Prim's algorithm to build the MST. There is probably a proof if Kruskal's algorithm is used instead, but this is not pursued in the cited paper. Proposition11.3 only applies to relational data that have a unique MST (which, arguably, includes most real-world data sets). A very small perturbation of any data set will transform it into one that does have a unique MST, and such a perturbation will not alter the CPOV cluster structure in the objects underlying the dissimilarity matrix; so this proposition essentially guarantees that SL clusters extracted from a VAT/iVAT re-ordered MST will be aligned. In the alternate notation for aligned partitions, Proposition 11.3 says that the SL partition extracted from the VAT/iVAT reordered MST at each value of c is just $n_1 : n_2 : \cdots n_c$. Now we are ready to combine the elements of MM, MMDRS, siVAT, and SL clustering to make an extension of single linkage to the big static data case. Here is the pseudocode for clusiVAT (the sVAT implementation is easily derived by calling sVAT instead of siVAT).

Algorithm A11.4. Single linkage clustering using (S+E) for static big data X.

	clusiVAT (S + E) **algorithm (Kumar *et al.*, 2016)**
1	**In:** $X_N \subset \Re^p$ *(or)* $D_N \subset M_N^+$; c′: n % (approx. sample size for VAT/iVAT image)
2	**Call:** siVAT(X_N *(or)* D_N, c′, n) **Return:** X_n *(or)* O_n: iVAT(D_n)): index set S: ordering indices {**P**}: edge distances {**d**} % **P** and **d** are returned from A4.1
3	Estimate most likely number of clusters (c) in X_n *(or)* O_n by examination of siVAT(D_n)
4	% Find the indices $\{j_t : 1 \leq i \leq c\}$ of the (c) longest edges in **d**: $\mathbf{d} = (d_1, d_2, \ldots, d_{n-1}) \xrightarrow{\text{sort}\downarrow} \hat{\mathbf{d}} = (d_{\hat{j}_1} > d_{\hat{j}_2} > d_{\hat{j}_c} > \cdots d_{\hat{j}_{n-1}}$ % Sort: $\hat{\mathbf{j}} = \hat{j}_{\hat{j}_1}, \hat{j}_{\hat{j}_2}, \ldots, \hat{j}_{\hat{j}_c} \xrightarrow{\text{sort}\uparrow} \mathbf{j} = (j_1 < j_2 < \cdots < j_c)$
5	% Extract aligned SL clusters from the MST: $U_{SL,n}^{literal} = \mathbf{u}_n = (\underbrace{1, 1, \ldots, 1}_{j_1 \text{ times}}, \underbrace{2, 2, \ldots, 2}_{j_2 - j_1 \text{times}}, \ldots, \underbrace{c, c, \ldots, c}_{j_c - j_{c-1} \text{times}})$:
6	% Extend $U_{SL,n}^{literal} = \mathbf{u}_n$ onto the remaining objects with the nearest-neighbor rule: **For** each $\mathbf{x} \in (X_N - X_n)$ *(or)* each $o \in (O_N - O_n)$ **do**: $j = \underset{\mathbf{x}_i \in X_n}{\text{argmin}}\{d(\mathbf{x}, \mathbf{x}_i)\}$ *(or)* $j = \underset{o_i \in O_n}{\text{argmin}}\{d(o, o_i)\}$ $\mathbf{u}_{N-n}(\mathbf{x}) = \mathbf{u}_n(j)$ *or* $\mathbf{u}_{N-n}(o) = \mathbf{u}_n(j)$
7	**Out:** $U_{SL,N}^{approx} = [U_{SL,n}^{literal} \| U_{SL,N-n}^{approx}] = \mathbf{u}_n \| \mathbf{u}_{N-n}$

Note again that the extension step in clusiVAT uses the nearest neighbor rule, NOT the nearest prototype rule. Do you think that clusiVAT would ever be exact on the big data set BD_N? If you said YES, you would be right! Under certain conditions, clusiVAT does find the literal SL solution on the big data. Perhaps unsurprisingly, this result is again related to Dunn's index. Proposition 11.4 shows that the extension is exact when Dunn's index $V_{DI}(U)$ at Equation (6.96) is greater than 1 for the input data.

Proposition 11.4. When are clusiVAT partitions literal SL partitions?

Let $O = \{o_1, o_2, \ldots o_N\}$ denote a big static data set of N objects, and let D_N be an N × N dissimilarity matrix for O. Let $U_{LSL} \in M_{hcN}$ and $\hat{U}_{SL} \in M_{hcN}$ denote the crisp partitions returned by applying literal SL (LSL) and clusiVAT to D_N. If O can be partitioned into c CS clusters in the sense of Dunn (1973), and $c' \geq c$, then $U_{LSL} = \hat{U}_{SL}$.

What does it mean? When O is a CS data set, clusiVAT will find the literal SL partition of it as long as it runs with $c' \geq c$. This result carries the same weakness as Theorem 11.2 and Propositions 11.1 and 11.2: viz., most data sets don't satisfy the CS requirement, and even when they do, we can hardly ever verify the constraint. For data sets that don't satisfy the Dunn index criterion (i.e., most data sets), clusiVAT still seems to work quite well.

Proof. Proposition 2, Havens *et al.* (2013). ■

The final example in this chapter ties the four algorithms in Chapter 11 together. It is an extension of an example in Kumar *et al.* (2016) that hopefully clarifies all of the steps in the clusiVAT algorithm.

Example 11.11. clusiVAT example using A11.1, A11.2, A11.3, and A11.4.

The eight panels in Figures 11.20–11.22, and 11.23 are abstracted from an example in Kumar *et al.* (2016). Views 11.20(a) and 11.20(b) show a data set X_N comprising $c = 10$ subsets of data drawn from 10 Gaussian distributions. This data set X_N has $N = 1,000,000$ HPOV Gaussian clusters in $p = 2$ dimensions. The shapes and sizes of the 10 subsets in Figure 11.20 are clearly different from each other. Figure 11.20(a) is a scatterplot of the data without colors (labels). How many HPOV clusters do you see in this data? Perhaps this view will tempt you to decide there are $c = 4$ HPOV clusters with a very large bridge between the four subsets and two subsets in the lower part of panel (a)? Or, perhaps, your eye will tell you that the seven subsets in the lower part of the plot are all in one cluster; so $c = 3$. But the shapes of the subsets are quite distinctive, and you can probably see that there are $c = 10$ subsets pretty clearly. The coloring of the labeled samples in Figure 11.20(b) facilitates this notion: it enables you to see 10 HPOV clusters in this data, but this is a trick of the visualization you are being offered -or is it?

Figure 11.21 shows the results of applying Algorithms A11.1 and then A11.2 to this data set. The MM algorithm A11.1 is instructed to find $c' = 20$ MM samples. The 20 samples extracted appear as black dots in Figure 11.21(a). The integers adjacent to the dots indicate the order in which the samples were drawn. Thus, the first sample, $\mathbf{x}_1$, labeled (1) in view (a), is in the brown subset at the bottom left of the figure. If you follow the labels around this view in the order 1, 2, 3,. . . , you can "see" how MM continues choosing the next sample. Thus, sample $\mathbf{x}_2$ at the right of the dark green cluster is the point in the data furthest away from point (1). Sample $\mathbf{x}_3$ at the top of the blue vertical subset on the left is furthest away from $\mathbf{x}_1$ and $\mathbf{x}_2$, and so on. These 20 MM samples are inputted to A11.2 with $n = 100$ as the target number of MMDRS samples. Figure 11.21(b) shows the output. Comparing the (a) and (b) views, you might agree that the MMDRS samples do span all 10 subsets-do you? The 100 points in this view are not colored because, while we do know their labels, this information is not used in the next step.

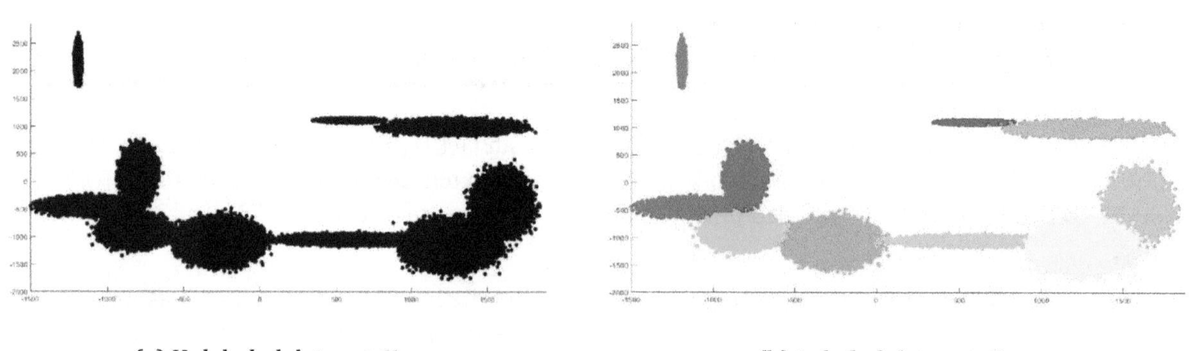

(a) Unlabeled data set X_N. (b) Labeled data set X_N.

Figure 11.20 $X_N = 1,000,000$ points in $p = 2$ dimensions with $c = 10$ labeled subsets.

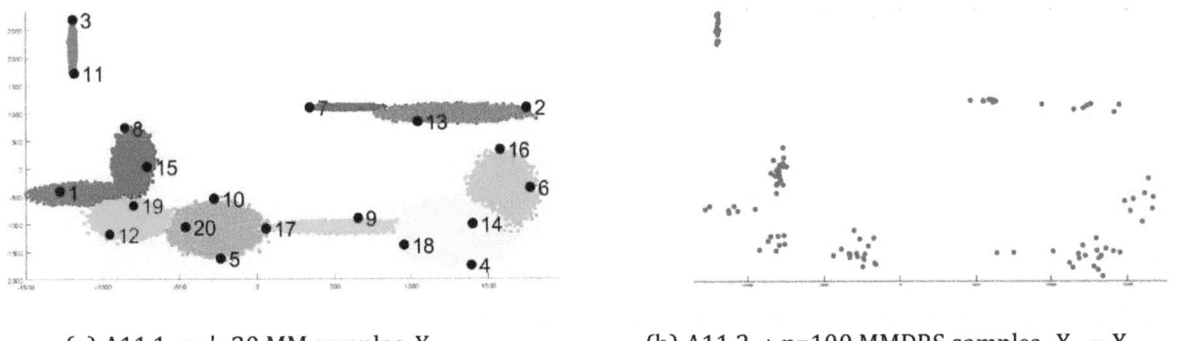

(a) A11.1→ c'=20 MM samples X_{MM}. (b) A11.2→ n=100 MMDRS samples, $X_n = X_{MMDRS}$.

Figure 11.21 MM and MMDRS samples in X_N.

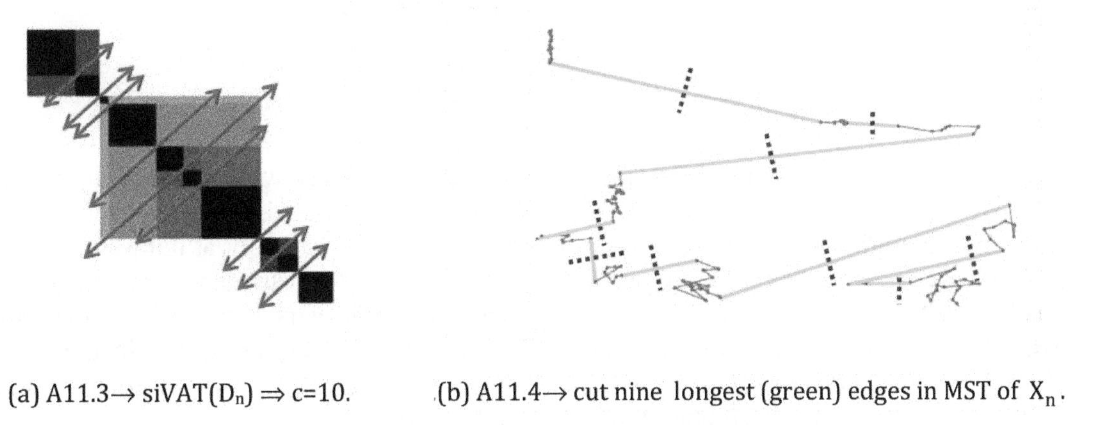

(a) A11.3→ siVAT(D_n) ⇒ c=10. (b) A11.4→ cut nine longest (green) edges in MST of X_n.

Figure 11.22 siVAT image and MST of $D_E(X_n)$.

Figure 11.22(a) shows the results of applying Algorithm A11.3 to the 100 points $X_n = X_{MMDRS}$. Here we see the iVAT image of $D_n = D_E(X_n)$, the 100 × 100 Euclidean distance matrix of X_n. A careful look in a zoomed view of this image shows it to have c = 10 dark blocks along its diagonal; so we conclude that the best SL partition of the 100 MMDRS samples should be a subtree with c = 10 clusters. The nine purple double arrowed lines in Figure 11.22(a) show how this decision was made. The nine green edges in Figure 11.22(b) are the nine longest edges in the MST that clusiVAT algorithm A11.4 finds in the reordered distance array **d** which flows through to A11.4 from the VAT/iVAT algorithm A4.1 that creates the image. The dashed lines in Figure 11.22(b) indicate the cuts.

Figure 11.23(a) shows the partition of X_n built by the siVAT sample. This is the exact SL partition of $X_n = X_{MMDRS}$ that would be built by the literal algorithm for SL that is recorded in A8.2. If we ran SL on the 100 points using A8.2 without the reordering done here, the partition would be the same, but the cluster list vector would not look like the one in Figure 11.19(c). Figure 11.23(b) shows the approximate SL partition $U_{SL,N}^{approx}$ of the big data set X_n built by the "extend" portion of A11.4. If you look compare Figures 11.20(b) and 11.23(b) closely, you will see a few places in the approximate partition, indicated by arrows in panel 11.23(b), where clusiVAT has committed errors. Since this is labeled data, we have a ground truth partition

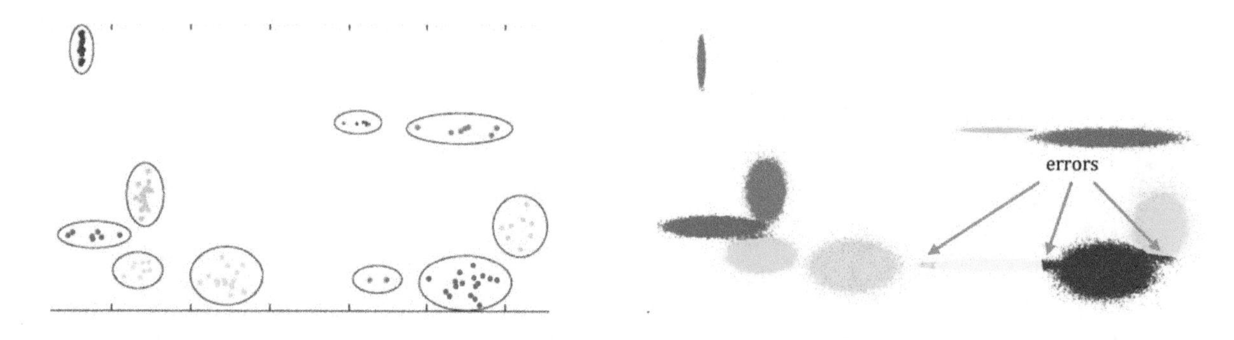

Figure 11.23 clusiVAT partitions of X_n and X_N.

$U_{GT} \in M_{h,10,1000000}$ of the input data; so we can compute the partition accuracy at Equation (6.34) of the clusiVAT partition shown in Figure 11.23(b): $PA(U_{SLN}^{approx}|U_{GT} = 99.92\%)$.

Kumar *et al.* (2016) compared clusiVAT to four other clustering schemes which have been discussed above: HCM algorithm A6.1 (m = 1, aka k-means), spHCM, oHCM, and CURE. Recall that the way spHCM and oHCM are applied to static big data sets is to process the data sequentially, as if it were streaming data. The partition accuracy (Equation (6.34)) and CPU time for these five methods were computed on a variety of small- to medium-sized data sets. For example, in one experiment 12 medium-sized sets of samples with $N = 10^5, 2(10^5), 5(10^5)$, and 10^6 were drawn from fairly well-mixed 2D Gaussian distributions which had $c = 3, 4$, or 5 component distributions whose average size was $N = 450,000$. This experiment yielded the average values shown in Table 11.2 (Table II in Kumar *et al.*, 2016).

The values in Table 11.2 show that clusiVAT had the best overall partition accuracy, was an order of magnitude faster than the three crisp k-means-based approaches, and was about 75 times faster than CURE for these data sets. This seems to corroborate Müllner's statements about the utility of single linkage for (fairly) big static data sets.

Table 11.2 Averages of 25 runs for 12 sets of Gaussian clusters with average size $n = 450,000$, $p = 2$.

	clusiVAT	HCM	spHCM	oHCM	CURE
$\bar{PA}(U\|U_t)$	99.9	75.1	99.2	73.9	95.4
CPU time, s	0.28	4.40	5.16	4.68	20.89

11.7 Notes and Remarks

The topic of acceleration has been of special interest to researchers who use clustering for segmentation of images, dating back to at least Cannon *et al.* (1986) for FCM. A favorite way to modify basic FCM for image segmentation is to attach a spatial constraint in the form of a window function at each pixel in the image (Tolias *et al.* (1998), Liew *et al.* (2000), Ahmed *et al.* (2002), Cai *et al.* (2007), or Szilagi *et al.* (2008)). The use of kernelized versions of HCM and FCM has also made its way into the literature for acceleration when segmenting images by clustering (Liao and Lin, 2007). Many of these techniques are applicable to k-means since it is a special case of FCM. See Phillips (2002) for acceleration methods dedicated to HCM.

Havens *et al.* (2011) accelerate of a version of kernelized FCM (kFCM) by an approximation method which constrains the cluster centers to be linear combinations of q randomly selected members of the n input objects, where $q << n$. Their algorithm requires storage of a $q \times N$ rectangular portion of the full $N \times N$ kernel matrix and the N diagonal values. This economy substantially reduces the computational complexity of literal FCM.

Many studies attempt to accelerate GMD/EM A7.1 by improving its starting point in the search space. One idea is to (quickly) get a good set of clusters centers from HCM and possibly a soft partition from FCM and then use these to initialize A7.1 (Bezdek and Dunn, 1975, Hathaway *et al.*, 1984). Meila and Heckerman (1998) test three ways to improve starting guesses for EM using random initialization, data-dependent MLE from a single distribution, and a SAHN clustering of a random sample of the data. See Ostrovsky *et al.* (2012) for a method of probabilistic seeding that attempts to improve the starting guess for HCM.

The main focus of Chapter 11 is scaling up clustering algorithms to facilitate clustering in big data static data set BD_N by decomposing it in various ways into a set of loadable data sets $\{SD_{i,n}\}$. This chapter has not paid much attention to the other half of the static big data problem, viz., when the number of dimensions (P) is large. A more general approach would begin with BD_{NP} and reduce both N and P into loadable data sets $\{SD_{i,np}\}$. The flip side of the methods discussed and illustrated by Figure 11.2 is to reduce the number of dimensions $(P \rightarrow p)$ while the number of samples (n or N) remains fixed. This style of improvement applies mainly to feature vector data, and includes all of the feature extraction and dimensionality reduction strategies that were discussed in Chapter 5.

There are many other adaptations of the Fantastic Four that have been proposed for clustering in big data. For example, Huang (1998) discusses two algorithms called k-modes and k-prototypes which extend HCM (k-means) to categorical domains and domains with mixed numeric and categorical values without sampling. For categorical data, k-modes replace the means of clusters with modes. The k-prototypes algorithm integrates k-means and k-modes for clustering objects described by mixed numeric and categorical attributes. Experiments on two real-world data sets with half a million objects each are offered to support Huang's claim that these two algorithms are efficient when clustering large data sets having categorical or mixed features.

A simple but sometimes effective alternative to clustering $X = \{\mathbf{x}_1, \ldots, \mathbf{x}_n\} \subset \Re^p$ when P is large is to decompose X into its P one-dimensional feature subsets $\{X_1, \ldots, X_p\} \subset \Re$, where the *i*th subset is composed of the n values of feature i in X. Assuming that each X_i is loadable, literal clustering algorithm $\boldsymbol{A}$ is run on each of the P feature sets, resulting in P literal partitions, $\{U_1^{\text{literal}}, \ldots, U_p^{\text{literal}}\}$. Then an ensemble clustering scheme can be used to aggregate the P results into an approximation of the partition that would ensue by applying the algorithm directly to $X \subset \Re^P$. Cai *et al.* (2013) compare a strategy based on alternating optimization feature by feature, which they call *multi-view hard c-means* (mvHCM), to standard batch c-means (HCM) on six labeled data sets, that have various numbers of features ranging from $P = 2$ to $P = 649$, $k = c$ ranging from $c = 7$ to $c = 397$, and n ranging from $n = 300$ to $n = 30,475$. Some of the data sets are images, and the computational expense of batch k-means can be high. Three measures of agreement (external CVIs) and confusion matrices are used to compare the results. As you might expect, mvHCM is superior to HCM in most their experiments.

The idea of working with samples of big static data is an appealing one, but using a sample always risks the possibility that generalization (usually by non-iterative extension of labels using the nearest prototype rule) will be misleading. Theorem 11.2 is one of the few results that addresses the quality of representation of cluster structure by a sample, but it is a weak guarantee because most unlabeled data sets don't have compact, well-separated (CS) cluster structure in the sense of Dunn's index. Nonetheless, MMDRS sampling (Algorithm 11.2) has been quite effective in many recent studies involving big data. Experiments using it to estimate Dunn's index itself by Rathore *et al.* (2019a) have been discussed. Other variations and applications of MMDRS sampling can be found in Kumar *et al.* (2018) and Rathore *et al.* (2019b).

If random selection of the starting index for MM sampling as shown at line 3 in A11.1 is used, repeated runs of the MM algorithm may lead to different sets of MM samples. The important point is that no matter what initialization is used, the MM objects selected will always satisfy the Maximin rule. Ties can be broken arbitrarily at each step in the procedure if there is not a unique minimizing argument in line 6 at any iterate. The literature contains a number of ways to initialize the MM sampling algorithm in line 3 of A11.1. Different starting points may yield different MM samples, and this will almost certainly be the case, especially if c' is small. Here is a (probably incomplete) list of ways that exact MM sampling has been initialized in the feature vector case:

iMM1.1 $\{\mathbf{x}_{m_0}, \mathbf{x}_{m_1}\}$ = *the two points in X furthest from each other*-Thorndike (1953), Kennard and Stone (1969).

iMM1.2 $\mathbf{x}_{m_0}$ = *the first point in X*-Casey and Nagy (1968), Batchelor and Wilkins (1969), Tou and Gonzalez (1974), Hathaway *et al.* (2006a, b), Wang *et al.* (2008), and many others.

iMM1.3 $\mathbf{x}_{m_0}$ = *a a random point in X*-Steponaviceet *al.* (2016), Rathore *et al.* (2018), and many others.

iMM1.4 $\mathbf{x}_{m_0} = \sum_{j=1}^{N} \mathbf{x}_j / N$ = *the grand mean* of the input data. Gonzalez (1985), Elkan, (2003), Celebi and Kingravi (2015).

iMM1.5 $\mathbf{x}_{m_0} \ni m_0 = \underset{1 \geq j \geq N}{\operatorname{argmax}}\{\|\mathbf{x}_j\|_2^2\}$ = *the vector with maximum squared Euclidean norm* (Katsavounidis *et al.*, 1994).

iMM1.6 MS: $\mathbf{y}_{m_0} \in Y = \phi_i[X] \subset \Re^2$ = *the furthest point from the grand mean of Y*-this complicated method begins by projecting X onto the two coordinate axes $\{i, j\}$ of X for which feature i has the greatest absolute coefficient of variation and feature j has the least Pearson product moment correlation with respect to feature i. Then the grand mean of Y, $\bar{\mathbf{y}} = \sum_{j=1}^{N} \mathbf{y}_i / N$ is computed, and the point $\mathbf{y}_{m_0} \in Y$ that is furthest from $\bar{\mathbf{y}}$ is taken as the initial MS sample in Y. Then a variation of A11.1 which uses the *cumulative distance* instead of the Maximin distance from current MM points to the next MM point is executed on Y -Erisoglu *et al.* (2011).

iMM1.7 MS+ : $\mathbf{x}_{m_0} \in \Re^{ij}$ = *the furthest point from the grand mean of X*- this method computes the samples as in Erisoglu's method using cumulative distance, but no projection is done. The MS samples are built in the input feature space -Celebi and Kingravi (2015).

Methods iMM1.6 and iMM1.7 modify the way distances are used in the MM algorithm and, hence, define a different sampling scheme called the MS (*maximum sum*) algorithm, but both of the MS methods described (in the input and 2D spaces) obtain the first sample as the point *furthest* from the grand mean. In this sense, they are different initialization schemes that could be used at line 3 of A11.1.

Ibrahim *et al.* (2020) compared MM sampling using initializations iMM1.1-iMM1.5 and iMM1.7. They concluded that there is not a significant difference in the MM samples extracted by the six schemes; so the best initialization is, therefore, the cheapest to obtain, viz., initialization iMM1.2 (starting with the first point in the data set as shown in A11.1).

Steponavice *et al.* (2016) proposed an *approximate MM* (AMM) sampling scheme they called "nearly Maximin" sampling[32]. Here is a description of their approach:

Step 1: Randomly choose $\mathbf{x}_1 \in X_N$ as the first element of the MM sample.
Step 2: Randomly draw n_{RS} vectors $Y = \{\mathbf{y}_j\} \subset X_N$ and for all $\mathbf{y}_i \in Y$, calculate the Euclidean distance to the closest element of the existing MM sample.
Step 3: Designate the $\mathbf{y}_j$ having the maximum distance as the next element of the MM sample; repeat steps 2 and 3 until the sample comprises (n) vectors.

The fully implemented AMM scheme constructs each new MM point in $\mathbf{x}_{AMM}$ using a *new* random sample of size n_{RS}. Consider the first two points, $\mathbf{x}_{m_1}$ and $\mathbf{x}_{m_2}$ obtained this way. The distance $d(\mathbf{x}_{m_1}, \mathbf{x}_{m_2})$ is the maximum distance from $\mathbf{x}_{m_1}$ to any one of the n_{RS} random samples chosen on the first pass of the iterative operation. Let $d(\mathbf{x}_{m_1}, \mathbf{x}^*)$ be the maximum distance from $\mathbf{x}_{m_1}$ to $\mathbf{x} \in X_N$. We can't assert that $\mathbf{x}^* = \mathbf{x}_{m_2}$, and the probability that this is the case will decrease with increasing N. Thus, with probability close to 1, the samples extracted by this approximate MM scheme don't satisfy the exact MM principle underlying A11.1.

A simple variation of the AMM scheme is to draw a *single* random sample X_{RS} of size n_{RS} from X_N before entering line 5 in A11.1 and find the set of AMM samples by reusing the same X_{RS} over and over $(c' - 1)$ times. These two variations of AMM have appeared in the literature. If the input to A11.2 is an AMM sample instead of a true MM sample from A11.1, the MMDRS scheme produces different samples. Two variations of this kind have appeared recently. Rathore *et al.* (2018) describe a variation of A11.2 called MMDRS+ which inputs a full implementation of AMM sampling into A11.2. Shao *et al.* (2019) describe another offshoot of MMDRS called MMDRS*, which uses just a single pass of AMM – i.e., one random sample from X_N is used to find all c' AMM samples. The MMDRS+ and MMDRS* algorithms are identical except for this difference in the implementation of Algorithm A11.2. The important difference between the two versions is that Shao *et al.* do random sampling in X_N just once; so the quality of the overall sample obtained by MMDRS* is very dependent on the quality of the single random sample extracted from X_N from which the c' MM samples are built.

Many other techniques for approximate clustering in big data that deserve to be mentioned are not covered in Chapter 11. For example, Feldman *et al.* (2011, 2013) advocate a technique they call *coreset sampling*. Coresetting takes $X_{Np} \subset \Re^p \rightarrowtail X_{np} \subset \Re^p$, so the number of samples decreases from N to n, but the number of dimensions (p) is unchanged. An ε-coreset is a weighted set of points X_{np} whose optimal cost on the coreset (the value of $J_1(U_n^*, \mathbf{V}; X_{np})$ on *any set of centers* $\mathbf{V}$) , is approximately the same as the optimal cost $J_2(U_N^*, \mathbf{V}; X_{Np})$ on the original data X_{Np}, up to a specified accuracy which is some function $\phi(\varepsilon)$ of epsilon.

More specifically, for any X, the HCM objective function is $J_1(U, \mathbf{V}, X) = \sum_{k=1}^{n}\sum_{i=1}^{c} u_{ik}\|\mathbf{x}_k - \mathbf{v}_i\|_A^2$, and the weighted HCM objective function is $J_{1,w}(U, \mathbf{V}, X) = \sum_{k=1}^{n}\sum_{i=1}^{c} w_k u_{ik}\|\mathbf{x}_k - \mathbf{v}_i\|_A^2$. An ε-coreset for $X_{Np} = \{\mathbf{x}_1, \ldots, \mathbf{x}_N\} \subset \Re^p$ with respect to $J_1(U, \mathbf{V}, X_{Np})$ is a set of points $X_{Np} \subset \Re^p$ and a set of weights $\{W(\mathbf{x}) : \mathbf{x} \in X_{np}\}$ such that

$$(1 - \varepsilon)J_1(U_N^*, \mathbf{V}; X_{Np}) \leq J_{1,w}(U_n^*, \mathbf{V}; X_{np}) \leq (1 + \varepsilon)J_1(U_N^*, \mathbf{V}; X_{Np}) \tag{11.8}$$

[32]These authors cite Johnson *et al.* (1990) as a primary reference for MM, but this paper is not about sampling. Rather, it is about MM designs.

Equation (11.8) has the same general appearance as the equation following Theorem 6.5 that relates values of the k-means objective function in the upspace to a randomly projected downspace Y, $X_{Np} \rightarrowtail Y_{Nq}$. In the notation of Chapter 11, $X_{Nq} = Y_{nq}$, Equation (6.71) can be written as

$$J_1(U_P^*, \mathbf{V}_P^*; X_{NP}) \le J_1(U_q^*, \mathbf{V}_q^*; X_{Nq}) \le (2 + \varepsilon) J_1(U_P^*, \mathbf{V}_P^*; X_{NP}) \tag{6.71}$$

Feldman *et al.* (2013) give several $(1 + \varepsilon)$ bounds similar to those shown in Theorem 6.5 when applying HCM to coresets instead of random projections. Both types of results are called "provable guarantees" about approximations to k-means when the data are altered. There are many theorems like Theorem 6.5 for the coreset approach, which aim for provable guarantees like Equation (11.8). Feldman and Langberg (2011) present a very nice summary of many of these results. Their unified approach subsumes many other models, e.g., k-medians, k-lines medians, projective clustering, linear regression, low rank approximation, and subspace approximations. This paper is well worth the time spent on a careful reading. See Rosman *et al.* (2014) for an adaptation of the coreset idea to clustering in streaming data.

Another recent entrant into the use of coreset samples for scalable k-means (HCM) is Bachem *et al.* (2018). These authors also adopt $J_1(U, \mathbf{V}; X)$ as the penultimate measure of cluster quality, and they prove a theorem similar to the result at Equation (11.8) for *lightweight coresets for k-means*. Their main result looks like this:

$$J_1(U^*_{|\hat{C}\hat{S}|}, \mathbf{V}; X_{|\hat{C}\hat{S}|}) \le J_1(U_N^*, \mathbf{V}; X_N) + 4\varepsilon J_1(U_N^*, \bar{\mathbf{V}}; X_N) \tag{11.9}$$

where coreset $\hat{C}\hat{S}$ of X_N has cardinality $|\hat{C}\hat{S}|$, $U^*_{|\hat{C}\hat{S}|}$, is "an optimal solution" of the k-means problem on the lightweight coreset, U_N^* is"an optimal solution" of the k-means problem on the big data set, $\bar{\mathbf{V}}$ is the grand mean of the data, and epsilon is a number greater than 0. Examples given on four data sets compare k-means++ outputs on X_N to those obtained on three types of samples: lightweight coreset $X_{\hat{C}\hat{S}}$, standard coreset X_{CS} and uniform random samples X_{RS}. Average values of $J_1(U, \mathbf{V}; *)$ and CPU times are compared for the four types of input data. Here is their concluding statement about results on the KDD cup data:

> *For KDD with* $k = 100$ *and* $m = 1000$ *[sample size], uniform leads to a speedup of* $2244\times$ *compared to solving on the full data set but also incurs a high relative error of 195.1%. CS reduces the relative error to 16.0% but only obtains a speedup of* $125\times$. *Lightweight coresets capture the best of both worlds — a speedup of* $828\times$ *at a relative error of only 18.5%. In absolute terms, one may compute 100 cluster centers on a 145 751 point data set in 0.42 seconds compared to 345 seconds if one naively uses the full data set.*

This is a pretty good sounding result: the term relative error references the sample values to the value of J_1 on the full data set. However, this method doesn't address the primary question in Chapter 11 of how to extend the sample results to the rest of the data; so this scheme doesn't really scale up k-means, as advertised in the title of the paper. But once the sample results are obtained, it is an easy matter to use them in the same way as illustrated in Figure 11.10; so this might be an excellent improvement to the method for scaling HCM that was illustrated above.

Random projection is, in some sense, the reverse of coreset sampling because it takes $X_{Np} \subset \Re^p \rightarrowtail X_{Nq} \subset \Re^q$; so p, the number of dimensions, is reduced, but the number N of samples is fixed. Cohen *et al.* (2015) provide an excellent summary of results similar to Theorem 6.6 for dimensionality reduction. These authors describe the goal this way: approximate a large data matrix $X_{Np} \subset \Re^p$ with a much smaller data matrix $X_{Nq} \subset \Re^q$ which they call a "sketch" such that solving a given problem on X_{Nq} gives a good approximation to the solution on X_{Np}.

Like the coreset literature, the term "good approximation" in the random projection context means that $J_1(U_q^*, V_q^*; X_{Nq}) = \phi(\varepsilon) \cdot J_1(U_p^*, V_p^*; X_{Np})$ for some function $\phi(\varepsilon)$. The presumption is that when this happens, there has not been an appreciable loss in cluster substructure on passing from X_{Np} to X_{Nq}. Cohen *et al.*'s work includes a $\phi(\varepsilon) = (1 + \varepsilon)$ bound for k-means approximation by random projection that improves the $\phi(\varepsilon)$ bound shown in Theorem 6.5 or the $\phi(\varepsilon) = (2 + \varepsilon)$ bound in Equation (11.8). This paper is an excellent summary of the state of this art. It is a very worthwhile update on many aspects of the evolving theory of low-dimensional approximations and estimates for approximate hard c-means.

> ☺ ***Forewarned is forearmed:*** The point has been made several times in this text that depending of the difference between successive values of any objective function as a sole means for judging cluster quality is dangerous. Close values of such functions on different data sets ($X_{Np} \rightarrowtail X_{np}$ or $X_{NP} \rightarrowtail X_{Nq}$) *might* indicate good structural preservation, but, certainly, a "provable guarantee" in this sense doesn't guarantee that cluster substructure *has been* approximately preserved. Moreover, it is almost impossible to compute the guarantee for actual data because that requires knowledge of the global optima in both spaces, which is hardly (if at all) ever available. All of these results are theoretically appealing (and, for us math freaks, fun to prove too), but none of them have much practical value beyond the psychological reassurance they provide – viz., that clustering in the reduced data sets is well-behaved in the sense of the guarantee. ☺

Distributed clustering is another topic that is inherently related to clustering in big data that was given short shrift in Chapter 11. Younis and Fahmy (2004) is an excellent place to start if distributed clustering sparks your interest. There are tons of papers about converting k-means (HCM algorithm A6.1) and GMD/EM (A7.1) to the distributed case. For example, Balcan *et al.* (2013) is a very nice paper that prescribes a means for using coresets for distributed k-means and k-medians. Each node computes an approximate solution for its local data and then constructs the local portion of a coreset using only its local data and the total cost of each node's approximation. A central coordinator receives contributions from each node, and then clustering is done by the gateway node. Distributed approaches for FCM and SL are also out there, but not so many.

There is an approach to scaling MDS up to big data that has the same distributed flavor as the one Balcan *et al.* (2013) apply to localizing coreset computations. Tzeng *et al.* (2008) discuss an alteration of (metric) *multidimensional scaling* (MDS, time complexity $\boldsymbol{O}(N^3)$) that they call SC-MDS (*split and combine MDS*). These authors show that the time complexity of metric SC-MDS is $\boldsymbol{O}(N)$ when the dimension of the downspace discovered by applying MDS to the Euclidean distance matrix of the input data in the upspace is much less than N. This is a pretty dramatic improvement that makes metric MDS attractive for large data sets. The data is split into overlapping subsets, MDS is applied to each subset, and then the downspace representations are recombined. Their empirical studies using microarray data on the yeast cell cycle showed k-means clusters in the downspace are similar to or slightly better than k-means clusters in the original upspace and are obtained in about one-third of the time.

Finally, since coVAT/coiVAT for rectangular data (Havens and Bezdek, 2012) depend only on being able to scale VAT/iVAT, the rectangular big data case is automatically covered. That is, s-coiVAT produces skeletons of cluster heat maps of big rectangular data such as the microarrays that have become pervasive in biological applications. Make a Google query on "cluster heat maps for big data" to be drawn into this arena, which is very important and useful for bioinformatics and social networks applications. If you do this, you may get a return that "there is no way to make a cluster heat map of big rectangular data." But s-coiVAT can make an approximation on a sample of the big rectangular data that is in every way comparable to the square case depicted in Figure 11.15.

The *non-Euclidean relational hard and fuzzy c-means* (NERHCM/NERFCM) models and algorithms (Hathaway and Bezdek, 1994) are not covered in this volume, but these methods are relational analogs of HCM/FCM for dissimilarity matrix input data. Bezdek *et al.* (2006) discuss a way to scale up these two algorithms for big data using an approach that is quite similar to geFFCM and geFEM. The overall algorithm analogous to geFFCM is called eNERF (*extended NERFCM*). Bezdek *et al.* (2006) apply roughly the same type of progressive sampling that is used by geFFCM to $BD_N = D_N$.

11.8 Exercises

1. Write a program that draws samples from the Iris data. (a) First, draw $n = 3$ random samples from Iris: how many times did you have to do this to get one sample from each of the three labeled subsets? (b) Now draw $c' = 3$ MM samples from Iris using A11.1. Do it until you get one sample from each of the three labeled subsets. Compare the result of (a) and (b). (c) Compute Dunn's index on the ground truth partition of Iris: comments?

2. $U = \begin{bmatrix} 1 & 1 & 0 & 0 & 0 \\ 0 & 0 & 1 & 1 & 1 \end{bmatrix}$ is an aligned 2-partition on $n = 5$ objects. Compute the 5×5 matrix $U^T U$. Relate this to a 5×5 iVAT image for the five objects, where 1 denotes black and 0 denotes white in the image. Now compute the $n \times n$ matrix $U^T U$ for an aligned partition $U \in M^*_{hcn}$ of n objects comprising c clusters (use block matrix multiplication). What can you say about the relationship of this matrix to an iVAT image on the n objects?

12

Structural Assessment in Streaming Data

I would rather have questions that can't be answered than answers that can't be questioned

–Richard Feynman

Feynman's quote here is a very accurate summary of this chapter, for it will end with lots of unanswered questions. The first 11 chapters of this volume are concerned with various topics related to the three basic aspects (assessment, partitioning, and validity) of clustering in fixed (aka static, batch) collections of data. In view of the contents of this last chapter, the first 11 chapters might be called *classical cluster analysis*. Chapter 12 is devoted to a relatively recent emerging topic that *sounds like* it is related to classical cluster analysis because almost all of the algorithms that have appeared in the literature about this are called "stream clustering." The term "sounds like" is chosen to encourage you to think about whether these algorithms should be called clustering algorithms in the same sense as the ones appearing in Chapters 1–11.

But there are many offshoots of the Fantastic Four that have made their way into this domain that owe their existence to these four basic methods; so this is an appropriate finale for our look at the elements of cluster analysis. There will not be many detailed specifics about the algorithms discussed in this final chapter. The objective is to provide a critical look at what this field seems to be about and where it seems to be heading. I will circumvent my academic responsibility to report the details of these algorithms by directing you to the cited papers to read the fine print.

Section 12.1 defines the streaming process, first, in terms of the real world and then in terms of the main computational contribution to this discipline for streaming data-viz., *cluster footprints*. Almost all of the ideas, models and algorithms that we know and love from classical clustering about the three canonical problems defined in Chapter 4 don't apply to this type data. An analogy between the observation of migration patterns of animals and the processing of stream data provides an insightful way to think about this process. Section 12.2 begins with an overview of some notable work in streaming clustering. The terminology that has been established by the research community involved in this activity will be repeated here-but please read it with a critical eye, because much of it is wrong. We need a much more well-defined and sharper semantics for stream processing than the language that appears in Section 12.2. Section 12.3 contains examples of five stream clustering algorithms that should suffice to establish why we need some changes in our thinking about this topic. Section 12.4 discusses some recent research about data visualization and stream monitoring functions in stream processing that forms a sort of bridge between the semantics of classical clustering and the language that needs to make its way into this new field. Section 12.5 summarizes the Chapter 12 position about streaming clustering.

12.1 Streaming Data Analysis

The idea of streaming data is familiar: for example, smart phone transmission, streaming audio and video entertainment, observation of sensor data in real time (weather data, air quality, intrusion detection, global navigation), and so on – the list of examples is virtually endless, no need for citations. Streaming data has not replaced batch (or static) data, but its relative importance is clearly increasing daily-dare I say growing exponentially? The conundrum this creates for the pattern recognition community (machine learning, data mining, classifier design, cluster analysis, etc.) is that most of the techniques that have been developed to date are designed to analyze static data and extract information from it for classification, prediction, and control. These methods are not always easily adapted to the streaming data case. Moreover, streaming data presents its own problems, and some of the methods borrowed from the static case are ill-fitting at best.

Some of you may find the presentation of this material controversial, but this is how science works. Science builds models of the physical world, tests them, rejects those that don't seem to fit, and improves those that are useful and agree with our perception of the real world. Eventually, we arrive at a satisfactory explanation of the process being modeled. So, it is very natural to try to co-opt the techniques and semantics of static data analysis to fit the streaming data case. Natural, but, sometimes, it doesn't work. While it would be foolish to throw away everything connected with static pattern recognition, we need a new way to think about streaming data that will provide a more fertile ground for developing models that extract the information we want from this kind of data.

12.1.1 The Streaming Process

On your dream vacation, a Tanzania photograph safari in Africa, you have seen many animals, both singly and in herds. Just this morning, you saw a large cluster of Ostriches (Figure 12.1(a)). You were startled, however, to see the mixture animals in Figure 12.1(b). Do you see clusters in panel 12.1(b)? Do you see an anomaly? Those three clusters of animals are keenly aware of the anomaly -you wonder, why aren't they running like crazy? Well, maybe the Lion is not hungry, and they know it. Or, maybe Simba is just photoshopped into this picture of giraffes, gnus, and springboks. The clusters in Figure 12.1 are typical of batch data in the real world and in our computational models and algorithms for static cluster analysis: *you see the objects or the data all at once.*

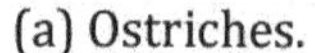

(a) Ostriches.

(b) Giraffes, gnus and springboks with Simba.

Figure 12.1 Clusters of animals: batch data.

Figure 12.2 The Animals all cross the river after dark.

Your objective on the safari is all about pictures, and this particular set of animals is really interesting; so you decide to photograph them. There is a perfect vantage point for your shots across the river. But, by the time you get set up, it is too dark to see them. They all cross the river at night (cf., Figure 12.2).

The next morning, the animals have all passed by. When you go down to the river, all you can find are their footprints in the mud (Figure 12.3.) The word *footprint* (FP) will appear in Chapter 12 many times. In some sense, it is the most important concept in this chapter. Perhaps you recognize the tracks. If so, you can assign a label to each animal that made them, but you certainly can't recover the animals themselves: the animals (the data objects) are gone.

The cluster of ostriches you saw yesterday (Figure 12.1(a)) was really interesting, and you had planned to photograph them too, but they also crossed the river at night. You can imagine them streaming by your observation platform – can you not? Figure 12.4 tries to capture this event. It is impossible to imbed the video (from gifs.com) from which these stills were made into the printed page, but please try to imagine these huge birds moving past your observation platform one at a time -but alas, streaming by after dark. So, the only record of them you have today is their footprints (Figure 12.3). Well, this is streaming data in the real case-all you can ever see are(computational) footprints.

In the pattern recognition sense, a feature vector is associated with object (animal); so the basic data is a possibly infinite sequence of p-dimensional feature vectors, say $X = \{\mathbf{x}_1, \mathbf{x}_2, \ldots, \mathbf{x}_N, \ldots\}, \mathbf{x}_i \in \Re^p \forall i$. Alternatively, this type of data can be thought of a time series (of possibly infinite length). In the computational context, these data pass by an algorithmic station one at a time. The typical stream clustering algorithm maintains a set of cluster footprints. Usually, the algorithm assigns a label to each data point. All four kinds

Figure 12.3 You photograph the footprints: the animals are gone.

Figure 12.4 Ostriches passing by, one at a time: streaming data.

of labels (hard, fuzzy, probabilistic, and possibilistic) are evident in the stream clustering literature. The label and the data point are used to update one (or more) of the cluster footprints, and then, the data (the object) is gone. (If you are one of the animals in Figure 12.1(b), you probably hightailed it to a safer part of the veld.) Most importantly, you hope that your system can spot an anomaly in the tracks that you see (such as), which you may recognize as a danger to the functional capability of your system; an anomaly deserves a special label and footprint of its own.

According to Guha and Mishra (2016), clustering is one of the primary tasks of stream data analysis. In a streaming environment, (so-called) clusters of data are said to emerge, evolve, and mature over time. But in real data streams, the only history of the processing that is usually available to an observer resides in the *cluster footprint*. These – and *not* clusters in a set of data that can be seen (or itemized) -are the entities produced by almost all streaming clustering algorithms.

Let us compare this type of processing to *static cluster analysis* (SCA) as described in Chapters 1–11. First, you don't know how many types of animals you might see (or even if there will be more than one kind); so the need for tendency assessment to estimate a possible number of clusters simply disappears. Second, the animals that pass by your station at night leave a track, and then they are gone. They might collect in clusters on the far side of the river, but you will never see them together; so no partition $U_{cxn} \in M_{pcn}$ is ever built. And if there are no partitions, there is no need to validate them (choose a "best" one); so CVIs are not needed for cluster validity. Figure 12.5 summarizes what was just said: all three canonical problems of SCA that were defined and discussed in Chapter 4 are simply non-existent. So, why call the process of building and maintaining cluster footprints "streaming clustering"? It is a misnomer. There are no clusters. Instead, the cluster footprint serves as the primary (and only) record of processing activity as the data stream past an observer. These algorithms don't form clusters nor do they aspire to form clusters. A much better term for the computational activity that produces cluster footprints is *streaming data analysis* (SDA); so this term will appear often in the sequel.

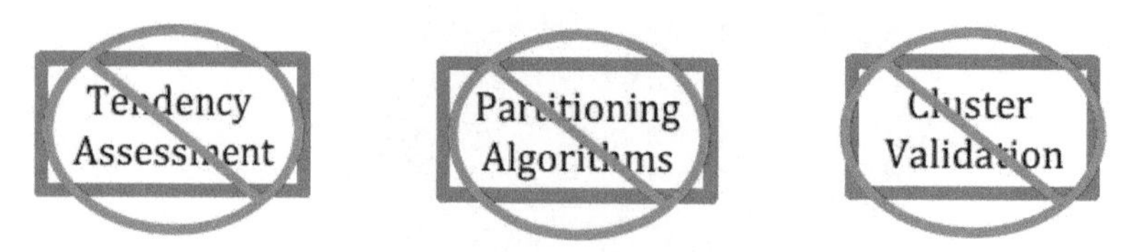

Figure 12.5 The three canonical problems of static cluster analysis: gone.

But, in most of the papers on streaming clustering, the data, which are almost always small and two-dimensional, *ARE* saved and displayed, even though this is impossible in real streaming data analysis. Why save the data? If the data are saved and can be seen visually, then "clusters" can be displayed retroactively, so to speak; and this visual evidence prompts the authors to state that their streaming clustering model "works." This is hindsight evaluation. It is commonplace in both SCA and SDA because, without it, how would we know what these algorithms are doing? Keep this in mind as you continue on your photo safari -in real streaming data, you can't get the data back.Here is an example of what you get from streaming clustering, and how hindsight is used to explain what you see.

Example 12.1. Footprints and hindsight evaluation.

Your graduate student has just run the PrS stream clustering algorithm (cf., Section 12.2.1) on a set of *intrusion detection system* (IDS) streaming data and bursts into your office to show you the results, namely, the set of cluster centers you see in Figure 12.6. The five 2D vectors $\mathbf{V} = \{\mathbf{v}_1, \mathbf{v}_2, \mathbf{v}_3, \mathbf{v}_4, \mathbf{v}_5\} \subset \Re^2$ constitutes the set of footprints left by this stream clustering algorithm. In the real case, this is all you get -the unlabeled data are gone.

What you see in Figure 12.6, a set of computational footprints, is analogous to what the safari passengers see in Figure 12.3 -a set of animal tracks. The safari guide might be able to look at the animal footprints and identify the animals that passed by last night, but you, a passenger on your first trip to Africa, can't. The point is: what would you be able to say about the data stream (the animals) if ALL you had was this set of footprints? Would you say there were five kinds of horned antelopes? Or, perhaps, five herds of giraffes at different spatial locations that were afraid of each other? Or two herds of zebras and three herds of gnus, or what? Well, you could not say any of those things if you could not see the batch data, and without the hindsight provided by seeing the data (a luxury you don't have in the real case), you would really have no idea about the data itself. This is the dilemma we have in real streaming data analysis -what to make of the footprints-and *only* the footprints?

Similarly, confronted only with the computational footprints left by an algorithm that has processed a stream of passing IDS data vectors, you are left with many of the same questions. Are these the footprints of five kinds of attacks (e.g., probe attacks, denial of service (DOS) attacks, users to root (U2R) attacks, remote to local (R2L) attacks, and protocol specific attacks)? Or, is there just one type of attack, with five different spatial locations because of variation in the data collection-e.g., from five sensors in a wireless sensor network? Does each one represent 10 data points or 10,000? And so on. If this algorithm was the subject of a research report, the authors would respond to these questions by showing a picture of the data, as in Figure 12.7.

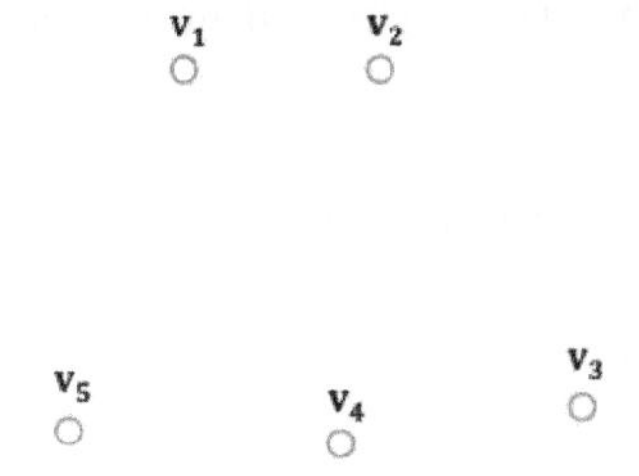

Figure 12.6 Five footprints left by the PrS stream clustering algorithm.

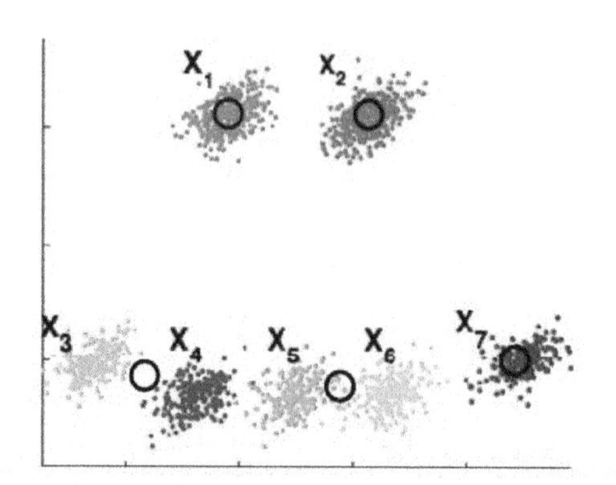

Figure 12.7 The PrS streaming algorithm leaves five footprints for seven HPOV clusters.

Figure 12.7 is a scatterplot of a set X comprising $n = 7000$ labeled data points in $p = 2$ dimensions. The coloring shows the labels of this data set. There are 1000 points in each of the subsets $\{X_j : 1 \leq j \leq 7\}$, which clearly form c = 7 HPOV clusters. The 7000 points were fed into Sebestyen's (1962) PrS algorithm in random order with a size threshold of $T = 15$. When you can see the data, as in Figure 12.7, you can understand a little better what the footprints mean since the five black circles superimposed on it are the locations of the five cluster centers seen in Figure 12.6. Your professor might conclude that you used the wrong parameters for PrS since there should be c = 7 cluster centers, not 5. You tell your professor that this result is easily changed by changing either the threshold T (the cluster size) or the order of sequential feed. And your professor shouts back-what? Well, use your imagination.

Showing the data as in Figure 12.7 is *hindsight evaluation* of the results of the stream clustering algorithm. To do it, you have to save all of the data that passes your station in batch form (or at least a large swath of it, a window of data), and, of course, the data must be viewable (i.e., $p = 1, 2$, or 3 dimensions). If the vectors in the data stream have a dimension greater than 3, you will have to find a means to extract a lower dimensional representation of the batch data and probably of the footprints as well before you can show a figure like Figure 12.7.

> ☺ **Forewarned is forearmed:** You would be hard pressed to find a paper in the streaming cluster literature that doesn't do this. Hindsight evaluation runs counter to the idea of streaming data analysis in real time -so, why do we do it? Well, what are the options? There are none. You want to justify your model and convince others that it does a good job (of what? Certainly not of finding static clusters in the stream), and the only real evidence you can offer is a figure such as Figure 12.7, and, perhaps, values of some hindsight quality metrics such as purity or resubstitution error that use the labels of the data in the stream to benchmark your results. It's not bad to do this[33], but please recognize what hindsight evaluation can tell you about those animals: *not much*-they are gone. ☺

---

[33]Full disclosure: all of the papers I have collaborated on about streaming data analysis use visual hindsight evaluation to show the efficacy of a streaming algorithm. It took me a while to figure out why we were doing this, and to see it's shortcomings. Moreover, we will continue to use this method because - what else is there?

12.1.2 Computational Footprints

Animals leave different kinds of real footprints. Streaming clustering algorithms leave different kinds of computational footprints. Some footprints are pretty simple: cluster centers are almost always part of computational footprints. But are they enough? There are many other mathematical objects in the footprints of various stream clustering algorithms. To describe them, assume that the first N points $\{\mathbf{x}_k : 1 \le k \le N\}$ have been processed and that some points are labeled class (i).The items below are some of the ingredients of the *i*th footprint that are used to represent the points that are active, i.e., are labeled class (i) for the times t in the time period $\{t_{si}, \ldots, t_{fi}\}$. This is a short list -there are other choices. In this list, the terms "in class (i)" and "for class (i)" are shortened to "in (i)" or "for (i)."

$$N = \#\text{ of points processed up to the current time} \quad ; \tag{12.1}$$

$$T_N = (1, 2, \ldots, N) : \text{Times up to the current time} \quad ; \tag{12.2}$$

$$t_{si} \in \mathbb{N} : \text{Time when (i) became active} \quad ; \tag{12.3}$$

$$t_{fi} \in \mathbb{N} : \text{Time when (i) was abandoned (became dormant)} \quad ; \tag{12.4}$$

$$TA_i = \{t_{si}, \ldots, t_{fi}\} : \text{Time set when (i) is active} \quad ; \tag{12.5}$$

$$T_i = \sum_{t \in TA_i} t : \text{Sum of active times for (i)} \quad ; \tag{12.6}$$

$$TT_i = \sum_{t \in TA_i} t^2 : \text{Sum of active times for (i), squared} \quad ; \tag{12.7}$$

$$n_i \in \mathbb{N} : \#\text{ of points in (i)} \quad ; \tag{12.8}$$

$$P_i = n_i/N : \text{Proportion of points labeled (i) in } T_N \quad ; \tag{12.9}$$

$$\mathbf{v}_i = \sum_{\mathbf{x} \in i} \mathbf{x}/n_i : \text{Sample Mean of points in (i)} \quad ; \tag{12.10}$$

$$COV_i = \sum_{\mathbf{x} \in i} \frac{(\mathbf{x} - \mathbf{v}_i)(\mathbf{x} - \mathbf{v}_i)^T}{(n_i - 1)} : \text{Sample Covariance matrix of points in (i)} \quad ; \tag{12.11}$$

$$\mathbf{S}_i = \sum_{k=1}^{n_i} \mathbf{x}_k : \text{Linear sum of p-vectors in (i)} \quad ; \tag{12.12}$$

$$\mathbf{FS}_i = \left(\sum_{k=1}^{n_i} x_{1k}^{(i)}, \ldots, \sum_{k=1}^{n_i} x_{pk}^{(i)} \right) : \text{p-vector of feature sums in (i)} \quad ; \tag{12.13}$$

$$\mathbf{SS}_i = \left(\sum_{k=1}^{n_i} [x_{1k}^{(i)}]^2, \ldots, \sum_{k=1}^{n_i} [x_{pk}^{(i)}]^2 \right) : \text{p-vector of feature sums of squares in (i)} \quad ; \tag{12.14}$$

$$R_i = \underbrace{\max}_{\mathbf{x} \in i} \|\mathbf{x} - \mathbf{v}_i\| : \text{radius of points in (i)} \quad ; \tag{12.15}$$

$$w_i = \left(\sum_{i=1}^{n_i} f(t - T_i)\right) : \text{weight of points in (i)} \quad ; \tag{12.16}$$

$$\mathbf{V}_i = \{\mathbf{V}_{1i}, \mathbf{V}_{2i}, \ldots, \mathbf{V}_{qi}\} : \text{multi-centers for (i)} \quad ; \tag{12.17}$$

$$AN_i : \text{Anomaly list for (i)} \quad . \tag{12.18}$$

These footprint ingredients are very much like a list of ingredients in a cookbook: flour, onions, garlic, eggplant, celery, etc. The ingredients don't represent any particular dish -they are simply available when you decide what to cook. Similarly, the components listed in these equations are combined in many ways in the literature to make footprints with various properties for different stream clustering models and algorithms. Now we turn to some general considerations about streaming clustering algorithms that leave footprints made from various constituents in this list.

12.2 Streaming Clustering Algorithms

Guha *et al.* (2003) provide an extensive review of several styles of clustering in streaming data. According to Guha *et al.*, there are two main approaches to clustering in data streams: online methods and summarization procedures (such as spHCM, oHCM, etc.) applied to windows of the data in the stream. These authors develop a streaming algorithm called LSEARCH based on the k-medians objective function, which chooses cluster centers that minimize the *sum* of distances between the median and the points assigned to each crisp cluster rather than just the largest distance. Figure 6.2 shows how c-means chooses d^*_{ikA}; the c-medians objective is to minimize $\sum_{j=1}^{c} d^*_{jkA}$. Vempala (2007) couples this objective function with low rank approximation. According to Guha *et al.* (2003), k-medoid approaches, including PAM, CLARA, and CLARANS, are not scalable and, thus, are inappropriate for stream analysis.

Online clustering algorithms almost always have two components. The first component is usually a change detection algorithm, which identifies changes in the data stream as potential times for creating a new cluster in the data. There are several recognizable types of changes. For example, drift over time (a battery gradually loses its strength), switching between different phases (temperature, light intensity during daytime, and nighttime), and sudden anomalous events (earthquakes in seismography, heart attacks on ECGs). Different objectives of stream clustering necessitate different types of models to detect and characterize various anticipated types of changes in the data stream. In streaming environments, stream clustering often means: summarize or compress the data stream into a representative model of the data. These algorithms may include an aging component for historical forgetting and/or recovery and return.

The second component in stream clustering is often a technique that builds a model of normal (non-anomalous) clusters that benchmark the expected stable state for the streaming data. This model of normal data then enables anomaly detection, which amounts to departures from the normal background. Many authors have used the BIRCH cluster feature triple, or modified it in some way, to attack anomaly detection in streaming data. Worthwhile papers include the work of Burbeck and Nadjm-Tehrani (2004, 2007) and Assent *et al.* (2012). Zhang (2013) offers a pretty comprehensive survey on outlier detection by various means that includes a section on cluster-based methods. There are a number of very good surveys on this topic, including but not limited to: Silva *et al.* (2013), Nguyen *et al.* (2015), Mousavi *et al.* (2015), Guha and Mishra (2016), Hahsler *et al.* (2017), and, especially, Carnein and Trautmann (2019). The last paper mentioned is a really nice review of 51 streaming clustering algorithms that provides a good overview and comparison of various styles of stream clustering.

The literature of stream clustering is as diverse as any other discipline in pattern recognition, making it virtually impossible to provide an accurate classification of methods into various subgroups. For example, Aggarwal and Reddy (2013) divide data stream clustering algorithms into three broad groups: (i) partitioning methods; (ii) density-based and grid methods; and (iii) model-based methods. Carnein and Trautmann (2019) take a more detailed approach, dividing the 51 algorithms in their survey (which span the years 1987–2018) into four general types: (i) distance-based; (ii) grid-based; (iii) model-based; and (iv) projection methods. They further categorize stream clustering algorithms by the way the various algorithms approach history management of data in the stream. The four approaches are differentiated by the way they deal with windows of data in the stream: specifically, damped, sliding, landmark, and pyramidal time windows. These four styles of stream clustering will be discussed in subsections to follow, but the best way to start this part of our adventure is at the beginning: Sebestyen's algorithm.

12.2.1 Sequential Hard c-Means and Sebestyen's Method

The remarks in Section 6.9 about the origins of the term "k-means" appear there because they deserve a special place in the history of hard c-means (k-means). Those remarks are revisited here because the sequential versions of k-means discussed in Section 6.9 are at the headwaters of the river that is called streaming cluster analysis. Recall MacQueen's (1967) description of his sequential sHCM (or sHKM) procedure:

> "Stated informally, the k-means procedure consists of simply starting with k groups each of which consists of a single random point, and thereafter adding each new point to the group whose mean the new point is nearest. After a point is added to a group, the mean of that group is adjusted in order to take account of the new point. Thus, at each stage the k-means are, in fact, the means of the groups they represent (hence the term k-means). ... When the sample points are rearranged in a new random order, there is some variation in the grouping which is obtained."

MacQueen's sHCM finds clusters in static data. It ends in one pass through a batch data set X with estimates for c crisp clusters and c cluster centers (the c-means), $\mathbf{V} = \{\mathbf{v}_1, \ldots, \mathbf{v}_c\} \subset \Re^{cp}$, which in the streaming context are the cluster footprints. This scheme doesn't adapt well to streaming data since c, the number of (animal) types expected, can't be known in advance but must be specified for implementation of sHCM.

Recall that MacQueen credited Sebestyen (1962) with the invention of sHCM: "Sebestyen (1962) described a procedure called "adaptive sample set construction," which involves the use of what amounts to the sequential c-means process. This is the earliest explicit use of the process with which the author is familiar." Here is a paraphrased version of Sebestyen's algorithm, which will be called PrS:

> [***Procedure S* (PrS)**] "When the first input is introduced, it is assigned to class 1, and this input becomes the mean vector of class 1. The second input arrives. If it is within a threshold distance T of the first mean, it is assigned to class 1, and the mean is updated. Otherwise, it becomes the mean of a new class 2. Continuing, this procedure results in a set of points which represent the locations of clusters of inputs, whose approximate sizes are T."

PrS is easily adapted for use with streaming data by simply continuing it as inputs arrive. Indeed, Sebestyen suggested this -in 1962! -when the idea of streaming data was hardly envisioned. Thus, an argument can be made that PrS is the root node for all stream clustering algorithms. Removal of MacQueen's requirement that c (or k) must be pre-specified is paid for in PrS by having to choose the threshold T instead, which controls the size of the clusters it finds. It was pointed out in Chapter 6 that Hartigan (1975) added an upper limit to the allowable number of clusters found by PrS, and named it the leader algorithm. Theodoridis and Koutroumbas (2009) renamed it BSAS in their wonderful textbook.

The PrS algorithm has enjoyed a resurgence of popularity recently because its inherently sequential nature makes it a natural choice for streaming data. The issue of order dependence when applied to static data is a big problem for its use in batch clustering (cf., Example 12.1), but this is not important for real streaming data since the order in which the data appear can't be altered.

The PrS algorithm leaves only the cluster centers $\mathbf{V} = \{\mathbf{v}_1, \ldots, \mathbf{v}_c\} \subset \Re^{cp}$ as footprints in its wake. The first input is the first mean, $\mathbf{v}_1 = \mathbf{x}_1$, and succeeding points are assigned to class 1 until an incoming point falls outside the Euclidean ball $B_E(\mathbf{v}_1, T)$ centered at $\mathbf{v}_1$ with radius T. This outlier immediately becomes the second cluster center, and so on. More generally, all incoming points that fall within the distance T from cluster center $\mathbf{v}_i$ are labeled class (i), regardless of when they appear in the input sequence. Example 12.1 is based on applying PrS to the 7000 data points shown in Figure 12.7. These two very simple algorithms, PrS and sHCM, are the foundation of the distance-based stream clustering algorithms because the hallmark of almost all of the distance-based SCAs is the use of a distance-based threshold to decide what to do (i.e., how to label) the current point. Sebestyen and MacQueen both used Euclidean distance as the metric of choice, but both of these early schemes will clearly work with any metric on the input space. Now we will embark on a mystery adventure on our safari that, we are told, involves a two-dimensional data set that is known only as $X_?$. Example 12.2 begins with PrS.

Example 12.2. Footprints left by PrS processing of data set $X_?$.

This is the first example in a set of five: Examples 12.2–12.6. Some of the material in these five examples appears in Bezdek and Keller (2021). Figure 12.8 shows two sets of footprints left by processing $X_?$ with Sebestyen's PrS algorithm. The data set has p = 2 dimensions; so these points are vectors in $\{\mathbf{v}_i\} \subset \Re^2$. We know only that these footprints were made by processing $X_?$ in the order that the data appear. The view in Figure 12.8(a) corresponds to the footprints made by PrS when the threshold T in PrS is T = 8.57. This parameter fixes the size of the neighborhoods that capture each of the footprints. Panel 12.8(b) shows the footprints left by PrS when T is cut in half, T = 4.29.

This change in the footprints is exactly what is expected: when the neighborhoods are smaller, we will need more of them to fence in each subset of data that flow past our station. OK. What do you think this set of data looks like? Are there anomalies? Is this one giant herd of zebras spread all over the veld? Or 16 kinds of animals? Or 67 kinds of animals" Or what? This is all you will have in the real streaming case. The identity of data set $X_?$ will be revealed later, when we perform hindsight analysis on Examples 12.2–12.6. Please ask yourself -*what do I want to learn* from processing the stream data?

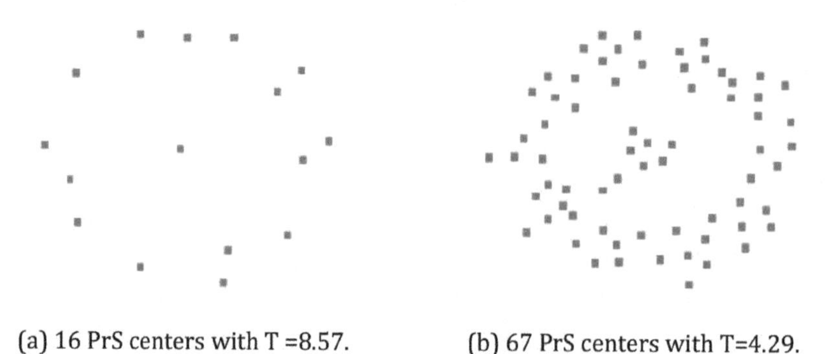

(a) 16 PrS centers with T =8.57. (b) 67 PrS centers with T=4.29.

Figure 12.8 Footprints left by applying PrS to $X_?$ for two choices of threshold T.

12.2.2 Extensions of Sequential Hard c-Means: BIRCH, CluStream, and DenStream

There are many variations of sHCM such as StreamKM++ (Shindler *et al.* (2011); Ackermann *et al.* (2012)). These methods state their objective as construction of a partition of data in the stream into c clusters that minimize an objective function. These methods require c (the number of clusters believed to exist in the stream) *apriori*. It is not clear how the idea of partition-building fits into the idea of SDA as I have described it, but this doesn't mean these methods lack value -only that they are interpreted in a different way.

An interesting approach for scaling up sHCM to static big data is discussed by Domingos and Hulten (2001). Their VFKM (very fast k-means) algorithm employs a progressive sampling scheme whose termination function is an estimated error bound. The approach attempts to assess the adequacy of a given subsample by approximating the difference in the cluster means obtained by sHCM using the current subsample and those that could be obtained with an infinite population. The approximate error bound is used to determine when a sufficiently good sample is identified. They discuss numerical experiments on 20 sets of spherical Gaussian samples, $N = 10,000,000$, $c = 5$, $p = 2, 4, \ldots, 20$. The results are mixed: sometimes, VFKM runs twice as fast as sHCM, but, in other cases, the opposite is true.

A very large family of distance-based methods have evolved from the BIRCH (Zhang *et al.*, 1996) algorithm, shown as the root node for some 26 algorithms in Figure 12.9 (Figure 5 in Carnein and Trautmann (2019)). The most well-known children in this family are *CluStream* (CLU), *DenStream* (DEN), and streamKM++.

Most of the algorithms shown in Figure 12.9 have two phases. In the first (online) phase, they create summaries of data and store it in some type of tree structure. I will call these summaries *minor footprints* (fp). Most of the SCA literature calls them *microclusters*. In the second (offline) phase, crisp clustering of

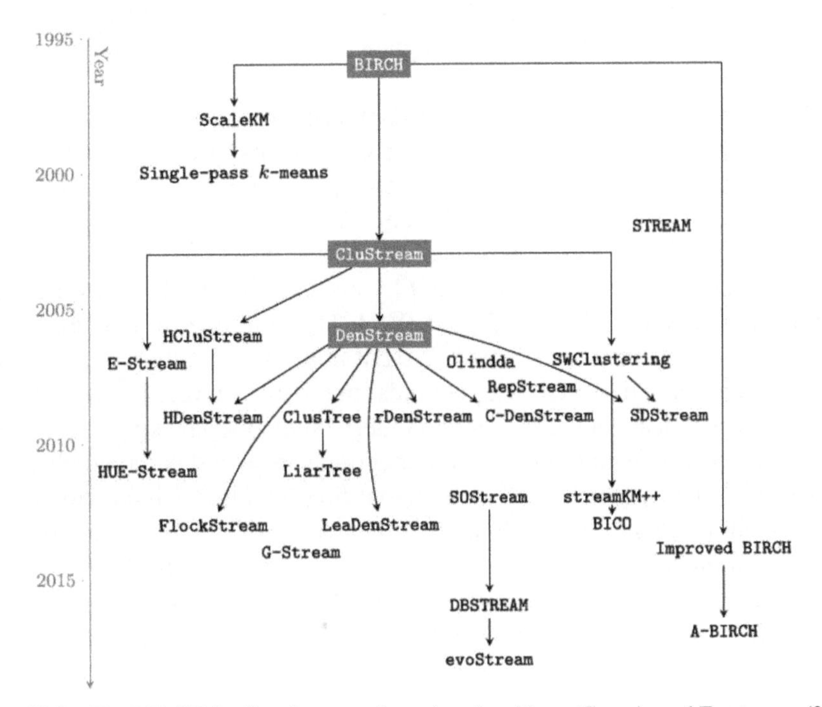

Figure 12.9 The BIRCH family of stream clustering algorithms (Carnein and Trautmann (2019)).

the microclusters is done with algorithm such as single linkage, HCM (k-means), or, perhaps, DBSCAN. On our safari, these resultant higher-level clusters are *major footprints* (FP), usually called *macroclusters*. These algorithms generally require a parameter in the first phase that specifies the radius of clusters, and the number of clusters (c) believed to exist in the data summaries is pre-specified in the second phase. Setting these values is challenging in various application domains, and, as you can imagine, different parameters result in different cluster footprints and, subsequently, different macroclusters. Since there is an offline static clustering phase in what amounts to extracted representations of the streaming data, it is hard to see algorithms of this type as stream data analysis. But this makes it easy to see why they are called stream clustering methods. Let us have a closer look at BIRCH, CLU, and DEN.

Recall from Chapter 11 that BIRCH (*balanced iterative reducing and clustering using hierarchies*, Zhang *et al.* 1996) was initially developed as a means to cluster big static data sets, chunk by chunk. The points identified by BIRCH as belonging to the *i*th structure in the data stream are used to build the parameters of the *i*th microcluster, $\mathbf{fp}_{i,BIRCH} = (n_i, \mathbf{S}_i, \mathbf{SS}_i)$, using the ingredients at Equations (12.8), (12.12), and (12.14). A label value (i) is attached to each footprint and identifies all of the points that have passed through the system that acquire this label. The quantities $\mathbf{S}_i$ and $\mathbf{SS}_i$ are vectors in $\Re^p$; so the *i*th minor footprint has $(2p + 1)$ values[34]. The *i*th and *j*th microclusters merge additively, $\mathbf{fp}_{ij,BIRCH} = (n_i + n_j, \mathbf{S}_i + \mathbf{S}_j, \mathbf{SS}_i + \mathbf{SS}_j)$. This makes footprint mergers easy to manage, if and when required. The microclusters are maintained in a height balanced tree called a CF tree. BIRCH updates the minor footprints after seeing each data point in the stream and, at some point, halts the online phase. Then batch clustering (usually a form of hierarchical clustering) is done on the (parameters of the) set of microclusters, resulting in c macroclusters, $\mathbf{FP}_{BIRCH} = \{\phi(\mathbf{fp}_{i,BIRCH}); 1 \leq i \leq c\}$. Please read the symbol (ϕ) as "some function of," which simply acknowledges the possibility that various crisp clustering algorithms are admissible in the offline phase. BIRCH handles the historical aspect of data flow with a Landmark time window, i.e., a specific time window (the current chunk) is isolated and only the data in this part of the stream are summarized by the minor and major footprints.

Aggrawal *et al.* (2003) developed a summarization approach that is representative of many current streaming models. These authors present an algorithm called *CluStream* (CLU), which consists of an online component that computes statistical information about microclusters and an offline analytical component that performs various tasks related to the evolving clusters. The online algorithm is initialized by HCM clustering of an initializing chunk of data (hence, static clusters in batch data). Each microcluster (a "snapshot" for them, a "footprint" for us) is represented by a set of $(2p + 3)$ values; so these footprints are slightly more complex than the footprints used by BIRCH. The number of microclusters built by CLU is quite a bit higher than the actual (unknown) value of c but much smaller than n, the number of points processed online. The microclusters are stored in a pyramidal tree structure which offers an efficient means for recovering historical information about clusters across time. Computational experiments presented in this paper verify that CLU scales linearly with both c, the number of clusters, and p, the number of dimensions.

CluStream handles history management with a pyramidal time window, which attends most carefully to recent data. The time windows in this approach have variable lengths. You can imagine them as looking similar to equi-content histogram bins. The *i*th minor footprint or microcluster in CLU has the $(2p+3)$ components $\mathbf{fp}_{CLU,i} = (\mathbf{S}_i, \mathbf{SS}_i, TT_i, T_i, N)$. The major footprints of CLU take the form $\mathbf{FP}_{CLU,mc} = \{\phi(\mathbf{fp}_{CLU,i}) : 1 \leq i \leq c\}$. The offline phase of CLU uses (ϕ) = weighted HCM (weighted k-means, Equation (6.85))

[34]There are some versions of BIRCH for which SS_i is a scalar that contains the sum of squares for all data points over all dimensions.

on the minor footprints stored in the leaves of the CF tree. According to these authors, the microclusters represent the current snapshot of clusters, which change over the course of the stream as new points arrive.

Example 12.3. Footprints left by CluStream processing of data set $X_?$.

The two leftmost views in Figure 12.10 show two sets of microclusters left by online CLU processing of our mystery data set $X_? \subset \Re^2$. The processing was done using the packaged version of CluStream software that is available at the MOA website (https://moa.cms.waikato.ac.nz/), described in the book about MOA by Bifet *et al.* (2017) and in even greater detail in the paper by Hahsler *et al.* (2017). Recall that each minor footprint $\mathbf{fp}_{CLU,i} = (\mathbf{S}_i, \mathbf{SS}_i, TT_i, T_i, N)$ has $(2p + 3)$ parameters. Since $X_? \subset \Re^2$, $p = 2$, and each minor footprint is a vector in $\Re^7$. We can't see the 7D microclusters. The red circles shown in Figure 12.10 are 2D visual representations of these microclusters. Thus, the red circles are the footprints available to us for interpretation of the stream clustering. How are the red circles made? According to Hahsler *et al.*, each calculated microcluster for CLU is displayed by calculating an area corresponding to the weight of the microcluster, and each circle encompasses the points in its capture area. This advertises the fact that the captured points are available -that is, stored somewhere in a static data set of crisply labeled clusters. This somewhat mysterious sounding procedure is presented as seven lines of R code in Hahsler *et al.*

The (a) and (b) views show the footprints found by CLU for two user-specified choices: 50 microclusters (panel 12.10(a)) and 100 microclusters (panel 12.10(b)). The plot in Figure 12.10(c) shows the result of offline clustering of the 100 online *microclusters* (mcs) in view 12.10(b) with weighted HCM (weighted k-means) for the specified choice of $c = 11$ macroclusters.

The blue crosses in Figure 12.10(c) show the location of the centers of the 11 macroclusters. Are you wondering why a user of this software would somehow make this particular choice? Does looking at these footprints tell you more about the data that left them behind than the PrS footprints shown in Figure 12.8? Perhaps, you are getting a sense that there is some circular structure in the streaming data. But -50 animals? 100 animals? 11 animals? Are the centrally located microclusters anomalies? Or are these microclusters a terrified group of springboks surrounded by 10 prides of hungry lions? What can you really say now about this streaming data that you were not able to conclude from the PrS footprints? Well, not much. What can

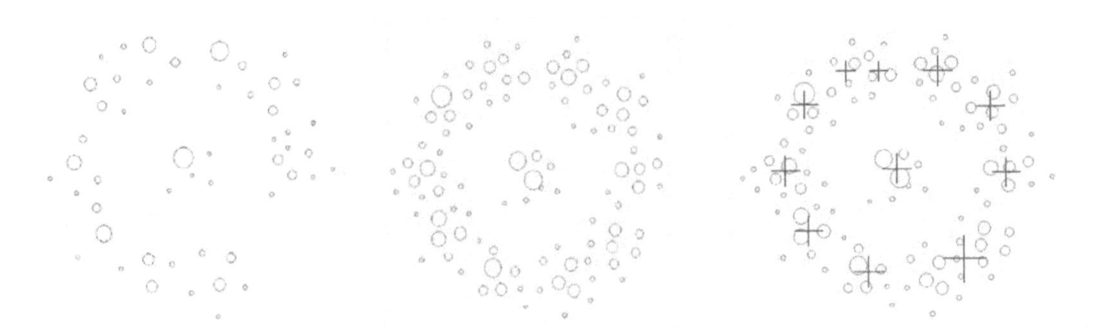

(a) 50 CLU microclusters. (b) 100 CLU microclusters. (c) 100 CLU mcs, 11 MCs for $X_?$.

Figure 12.10 Footprints left by applying CLUto $X_?$.

you say about the stream clustering algorithm that produced these footprints? You can assert, just as in the static case, that they are quite sensitive to their parameters, which are required choices you make to facilitate the processing. You should be asking yourself -how can I make useful choices for the parameters of my stream clustering algorithm? It's a good question.

DEN (Cao *et al.*, 2006) is similar to CLU in some ways, also having both online and offline phases, but the ingredients of its minor footprint set are quite different: $\mathbf{fp}_{DEN,i} = (w_i, \mathbf{v}_i, R_i)$, corresponding to the weight, center, and radius for the points currently identified with the *i*th subset, shown in Equations (12.16), (12.10), and (12.15)[35]. So, the parameter set of each microcluster of DEN contains $(p + 2)$ values. DEN requires a user-specified neighborhood size, and then estimates the density of each microcluster in its neighborhood. DEN organizes microclusters based on their weight as either core and outlier microclusters, in an effort to suppress noise. DEN inserts a newly arrived data point into its nearest microcluster with under certain conditions.

The offline phase of DEN removes minor footprints that have not achieved a sufficient weight, followed by clustering of the remaining footprints, yielding a set of major footprints, $\mathbf{FP}_{DEN,i}\{\phi(\mathbf{fp}_{DEN,i}) : 1 \leq i \leq c\}$, where (ϕ) = DBSCAN (Ester *et al.*, 1996). DEN uses a damped time window for history management. In this approach, each footprint is weighted by the number of points that have contributed to it, and the weight is damped according to a "forgetting factor," typically a function such as λ^j; $0 < \lambda < 1$ where j is the number of steps backwards in time from the current time.

Example 12.4. Footprints left by DenStream processing of data set $X_?$.

Figure 12.11 is entirely analogous to Figure 12.10, with CLU replaced by DEN. Since the data have $p = 2$ dimensions, each microcluster in DEN has four values. View 12.11(a) shows the minor footprints left by processing $X_?$ with DEN as small red circles, produced from its 4D values in much the same way as CLU. This view has 49 microclusters instead of 50 due to parameter choices (it was really hard to pick parameters

(a) 50 DEN microclusters. (b) 100 DEN microclusters. (c) 100 DEN mcs, 11 MCs for $X_?$.

Figure 12.11 Footprints left by applying DEN to $X_?$.

[35] Actually, DEN maintains the sums of squares, $\mathbf{SS}_i$, from which the radius R_i is easily computable.

to guarantee exactly 50). The view in Figure 12.11(b) shows the tracks of the minor footprints left by DEN for 100 microclusters. The rightmost view shows the visual representation of both sets of DEN footprints: the red circle microclusters are the same as the center view. The blue crosses in Figure 12.11(c) are the centers of c = 11 macroclusters found by the offline phase of DEN with DBSCAN. All of the questions posed about CLU footprints in Example 12.3 are in play here as well. The CLU and DEN footprints look similar -both convey a sense that the mystery input data contain c = 11 clusters. The DEN result in Figure 12.11(b) is especially attractive -it certainly looks like the 100 footprints are themselves clusters (of footprints). This might be useful if we knew more about the stream data, but, in the real case, we still wonder about the numbers of different animals, their spatial locations, their sizes, and their anomalies. Please think about what you can learn from the DEN footprints: are they "better" than the PrS or CLU footprints? If so, then how?

The paper by Hahsler *et al.* (2017) is a comprehensive review of the use of MOA in the context of stream clustering. These authors present a very thorough set of examples of not only CLU and DEN but also five additional stream clustering algorithms named: Clustree, sample, window, DStream, and DBStream. Their presentation includes enough snippets of R code to enable you to experiment with these algorithms if so inclined. There are several comparisons of the algorithms when used for static clustering, and then several examples are given for a small set of streaming data. They also discuss how to couple anomaly detection to the streaming algorithms covered and show several examples of doing this. Finally, there are a number of hindsight quality metrics presented that can be used to compare aspects of the various methods. If you are interested in CLU and DEN, this is a very good place to start an exploration on your own. Another good source for this material is Bifet *et al.* (2018). Already mentioned, but worth repeating: implementations of these algorithms are available at the MOA (*massive online algorithms*) website. MOA is an open-source software that allows users to build and run experiments on a great variety of machine learning or data mining algorithms on streaming data. MOA contains several collections of machine learning algorithms besides the ones mentioned here.

The facts that BIRCH, CLU, DEN, and many other "stream clustering" algorithms of similar ilk are (i) also used for static clustering and (ii) utilize some form of static clustering offline to compute the macroclusters from the microcluster parameters makes it very apparent that methods such as these are not wholly committed to the analysis of ***unlabeled*** streaming data that can't be retained in batch form. None of them produce visual interpretations of the data stream that are independent of hindsight verification, and they all require some form of conversion of the computed microclusters and macroclusters to make the footprints visible. Are model-based algorithms different, and possibly better? Well, it is the next destination on our safari, so we will find out.

12.2.3 Model-Based Algorithms

Many (but not all) of the stream clustering algorithms in this category assume that data in the stream are generated by a physical process which can be fitted with a specified model. A typical assumption is that the data emanate from a mixture of unknown statistical distributions. And, naturally, the assumption that the streaming data have a Gaussian mixture model (GMM) as their source occupies the pre-eminent place in this group of algorithms. Consequently, the footprints left by these algorithms are typically populated by prior probabilities, mean vectors, and covariance matrices. Carnein and Trautman (2019) list only six algorithms in this category, beginning with the COBWEB algorithm of Fisher (1987). The literature of model-based stream clustering algorithms almost always uses only p = 2D data for experiments so that visual hindsight verification of model quality is available.

COBWEB operates this way. Each datum in the stream is lodged at a node in a classification tree. Each node in the tree represents a probabilistic concept that summarizes the distribution of attributes attached to data at that node. This model used four operations for footprint management: merger of two nodes, splitting a node, insertion of a node, and passing a datum down the tree. History is managed with a Landmark window, and there is an active quality control metric called category utility that assesses each decision made during processing.

What is the COBWEB footprint? If you examine the classification tree at any instant in time, there will be some current number of nodes that possess various values related to the probabilistic concepts they embody. Each node will have a semantic label. But the ultimate goal of COBWEB is to produce a partition of the data in the classical sense described in Chapters 1–11, and it doesn't leave an easily characterized set of footprints in its wake. In some sense, the COBWEB footprint is the entire tree with values for each concept up to the current time (so, you can somehow see all of the animals that have passed your station simultaneously). While COBWEB can be regarded as a stream clustering model, it was initially developed as a static clustering algorithm which performed hierarchical clustering on batch data by feeding them to COBWEB one at a time -so its application to the streaming data case is exactly like the use of sHCM and PrS for this purpose.

There are quite a lot of model-based stream clustering algorithms in the literature that have recognizable footprints, many which have the general structure of the set of unknowns associated with the GMM which was described in Chapter 7. Some of these algorithms make the assumption that the source of the data is a GMM. Subsequently, different subsets of data in the stream correspond to samples from different components in the mixture. However, there are a number of stream clustering algorithms that leave GMM lookalike footprints that don't make the distributional assumptions of the GMM.

For example, there is a family of streaming models based on elliptical footprints that define anomalies as points exceeding the Mahalanobis distance from the current set of component means. This family began with the *data capture anomaly detection* (DCAD) model for static input data (Rajasegarar *et al.*, 2009). This scheme was generalized for stream clustering by Moshtaghi *et al.* (2012), who developed two incrementally updatable versions of DCAD called IDCAD (*incremental DCAD*) and FFIDCAD (*forgetting factor IDCAD*). Equations for incremental updates to the winning footprint are given for IDCAD and FFIDCAD.

Moshtaghi *et al.* (2012) compare the time and space complexity of these three models and show that IDCAD and FFIDCAD computations for streaming data scale linearly in time and are constant in space. History management for the data stream is accomplished in FFIDCAD using a damped time window model. Three data sets from the domain of wireless sensor networks are used to compare these three models to incremental ARX/RLS learning. The data sets were the Intel Berkeley Research Lab (IBRL) data [IBRL-Web, 2006. http://db.lcs.mit.edu/labdata/labdata. html.]; and the Grand-St-Bernard (GSB) and Le Genepi (LG) data sets available at [SensorScope Web, 2009. http://sensorscope.epfl.ch/index.php]. All three data sets were two-dimensional data, and the method of model verification was hindsight plotting of the data against the discovered footprints. The iDCAD model was extended by Moshtaghi *et al.* (2016) by exponentially weighting its computations, resulting in *exponentially weighted IDCAD* (ewIDCAD). This paper compared the new model to the older ones and also to the well-known multivariate CUSUM algorithm of Crosier (1988) using the same 2D data sets. The ewIDCAD model bases its footprint update on values of the χ^2 probability of misclassification for each point in the stream. The winner of this competition receives the corresponding crisp label.

The parameters of the footprints left by this family of streaming algorithms, in the notation of Equations (7.16), have the same form as those left by a GMM, viz., $\mathbf{fp}_{\text{IDCAD},i} = (n_i, \mathbf{m}_i, S_i^{-1})$, $\mathbf{fp}_{\text{FFIDCAD},i} = (\widehat{n}_i, \widehat{\mathbf{m}}_i, \widehat{S}_i^{-1})$, $\mathbf{fp}_{\text{ewIDCAD},i} = (\hat{n}_i, \hat{\mathbf{m}}_i, \hat{S}_1^{-1})$. Will these footprints be visible? We will be able to see the means and plot elliptical contours of the covariance matrices only for streaming data with two or three dimensions.

The assumption of data emanating from the GMM underlies the Missouri University Stream Clustering (MUSC) algorithms called MUSC1 and MUSC2 (Ibrahim *et al.*, 2018). This family of models assume a GMM as the source of the stream data.The main idea combines possibilistic and fuzzy clustering with GMMs. These algorithms begin by collecting an initial subset of data from the stream which is then processed with possibilistic clustering. This possibilistic partition is used to initialize the GMM, i.e., to compute an initial set of GMM parameters for this subset (i.e., the partition is used to build a single footprint for one animal, viz., $\text{FP}_{\text{MUSC},1} = \{n_1, \mu_1, \Sigma_1, \text{AN}_1 = \varnothing\}$). Note that the anomaly list is initially empty.

In subsequent online processing, the basic idea is very similar to many of the distance-based models that compare distances to a prespecified threshold. An incoming data point, say $\mathbf{x}_{n+1}$, is tested to see if it can be placed with any of the current structures by computing the Mahalanobis distance from $\mathbf{x}_{n+1}$ to each of the current means. If the minimum distance falls within a pre-specified threshold, $\mathbf{x}_{n+1}$ is incorporated into the winning Gaussian cluster, whose mean and covariance matrix are incrementally updated. If $\mathbf{x}_{n+1}$ has low typicality to all of the existing prototypes, it is placed in the anomaly list. The points accumulated in the anomaly list are tested after each update to the footprints to see if they are still unwanted by all of the current structures. These points may or may not eventually be sufficient to warrant the creation of a new cluster of their own. Changes in the anomaly list occur in two different ways. First, if any point in the list can fit in one of the existing clusters by computing the Mahalanobis distance between the outliers and updated cluster centers, the point is moved to that cluster. Second, the outliers themselves are clustered, and this may result in the formation of a new cluster. After updating the Gaussian parameters, $\mathbf{x}_{n+1}$ is discarded.

The *i*th MUSC footprint is $\mathbf{FP}_{\text{MUSC},i} = \{n_i, \mu_i, \Sigma_i, \text{AN}_i : 1 \leq i \leq \#\text{FPs}\}$, composed of the number of points, Gaussian mean, covariance matrix, and outlier anomalies (AN) associated with the *i*th structure. As with all SCAs, the number of footprints that evolve in this scheme depends on the thresholds chosen. Most of the numerical experiments reported using the MUSC approach are for two-dimensional data. If the data is saved, the footprints can be shown with the data: that is, as an aid to hindsight analysis of the MUSC scheme. There is no history management in the MUSC models. The assumption is that data will continue streaming by the observation station, and the update equations for the footprints don't attempt to damp the contributions of past points in time with the current ones. Let's have a look at two footprints left by the MUSC scheme on $X_?$.

Example 12.5. Footprints left by MUSC processing of data set $X_?$.

Figure 12.12(a) shows the footprints (without n_j, the number of points associated with each label) left by processing $X_?$ in input order. That means: the mystery data are ordered by indices 1 to n and are fed into the stream in that order to simulate stream clustering. The algorithm was initialized at the footprint shown by the lightning bolt, and thence proceeded to create new structures as the data evolved. The procedure ended with $c = 11$ footprints. The black dots in both views are the means, and the plotted ellipses are made from the footprint covariance matrices for adaptive values of the Mahalanobis distance threshold. The red squares in both views are data labeled as anomalies that never qualify for inclusion in any of the footprint structures.

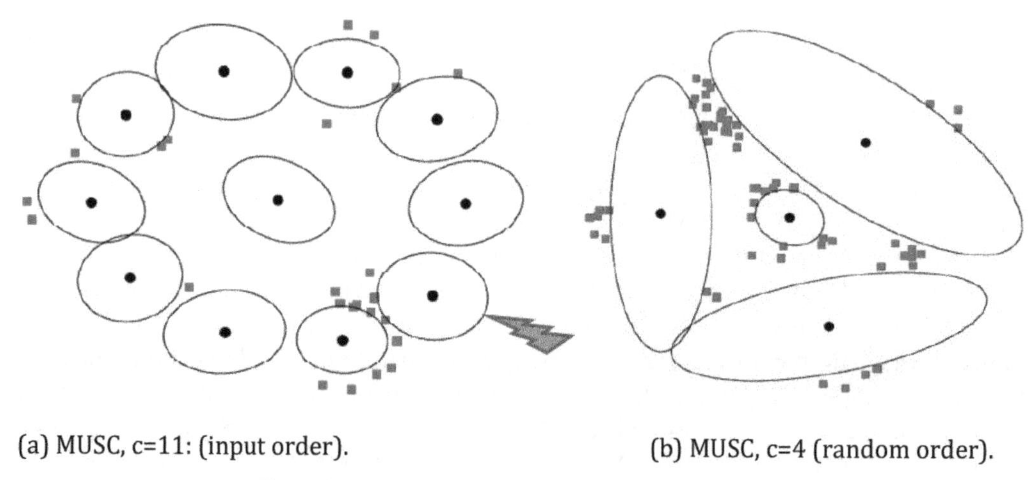

(a) MUSC, c=11: (input order). (b) MUSC, c=4 (random order).

Figure 12.12 Footprints left by applying MUSC to $X_?$.

The view in panel 12.12(a) shows quite clearly which anomalies are just outside the nearest footprints. There are 25 anomalies in this view.

Figure 12.12(b) shows the results of applying the same algorithm to our mystery data, the only change being the order in which the n points appear in the stream. In the (b) view, the points are fed into MUSC in random order. This affects the initial footprint, of course, and all subsequent computations which determine when and where to place additional footprints. In this second run, MUSC finds just c = 4 footprints, and 59 anomalous data that can't qualify for attachment to the subset of any footprint.

This demonstrates several things. First, once again, we see that sequential algorithms are heavily dependent on the order of feed (not a problem in real unlabeled stream data, whose order is not at our convenience). Second, the same stream clustering algorithm can offer radically different impressions of the patterns in the stream by simply altering processing protocols.

The additional information offered by MUSC footprints (some idea of cluster size, the identification of anomalies) is quite welcome of course, but ... are there 11 kinds of animals; or 4; or what? Are there 25 anomalies, or 59? Are the anomalies predators (malware attacks) for the normal data (the peaceful animals inside the elliptical fences)? Do you have a better idea of the mystery data by looking at these views? If you only have 12.12(a) you will have some confidence that there are 11 separate structures in the stream, but the alternate picture of the data painted by 12.12(b) will sow some doubt in your mind as to what really went by last night. Finally, if the data live in a space of dimension $p > 3$, how can you even SEE this type of footprints? Well, this example again leaves us with more questions than answers, but as Richard Feynman pointed out, better to have questions without answers than answers that can't be questioned.

12.2.4 Projection and Grid-Based Methods

Carnein and Trautman (2019) identify just four algorithms in this category: HPStream, Sibling Tree, HDDStream, and PreDeConStream, all damped time window models. This class of stream processing methods has as its raison d'etre the fact that high-dimensional streaming data present a unique set of problems that can't be handled efficiently with methods that are principally designed for low-dimensional data. More specifically, distances in high-dimensional spaces tend to be very similar, clouding an algorithm's judgment about footprint belongingness. Another common problem for the grid-based methods is that their success hinges on choosing the "right" grid, i.e., one whose granularity captures the structure of the data in a meaningful way for clustering. HPStream, the root node for HDDStream and PreDeConStream, is discussed next.

Aggarwal *et al.* (2004) discuss an interesting extension of CluStream called *HPStream* that builds clusters in streaming data in much the same way as CLU but has three very interesting useful additions. First, they add a fading memory structure to HPStream so that the effect of past inputs on evolving clusters is gracefully degraded. Second, they maintain a $c+1st$ cluster when building the microclusters which maintains outliers (an anomaly list, $\cup AN_i$) that are unable to secure membership in any of the current minor footprints. Third, and most interestingly, HPStream has a feature selector attached to each microcluster which identifies the active features (i.e., the most relevant features) for each footprint. All three of these new structures adapt to incoming data and are dynamically maintained to evolve with time using cluster feature triples as in BIRCH that each have $(2p + 1)$ elements.

Aggarwal *et al.* (2004) compared CluStream to HPStream using two real data sets (KDD Cup '99 and the Forest Data, the 10 real variables only) and a number of synthetic labeled data sets generated by predefined probability distributions with n, p, and c as variables. They argue that hindsight evaluation using the sum of squared errors is not a good measure for evaluating projected footprints because it uses all p dimensions, whereas HPStream produces subspace footprints that use fewer dimensions than the base dimension of the streaming data.

Instead, they used the ground truth partitions (are you wondering how streaming data can HAVE a ground truth partition?) to compute cluster purity[36] for the two streaming methods. HPStream produced much higher purity than CluStream on all data sets, often achieving values in the range 90%–100%. Here are the dimensions used by CluStream, compared to the *average* number of dimensions used by HPStream: KDD Cup -34 to 20; Forest -10 to 8; Synthetic -50 to 30. Stream processing rates ranged from 200 to 35,000 points per second. HPStream scales linearly in both c and p. Taken together, the pair of papers about CluStream (Aggarwal *et al.*, 2003) and HPStream (Aggarwal *et al.*, 2004) represent an influential amount of early work about evolving structure in streaming data. They are well worth reading and are much more informative when read together.

There are other partitional methods that are not based on generating labels for the animals (the data objects) based on distances to a prototype animal (mean vector). For example, Table 3 in Carnein and Trautman (2019) lists 13 grid-based algorithms. They identify the fractal clustering algorithm of Barbara and Chen (2000) as the earliest algorithm in this group of methods. These algorithms attempt to capture information about structure in the data stream by subdividing the input space into adjoining cells and recording the density of observations that land in each cell. The set of points in each cell is regarded as a minor footprint. Major footprints (macroclusters) are built by combining adjacent dense cells.

One of the most popular grid-based algorithms is D-Stream (Chen and Tu, 2007), which divides the input dimensions into a number of segments, creating hyper-rectangular grids of fixed and equal size. Each grid cell is used to store synopsis information of data points falling into secondary storage. D-Stream requires an input parameter for identifying dense grids and another input parameter for identifying sparse grids. D-Stream also maintains transitional grids. Finding the right values for these parameters either requires expert knowledge or running the algorithm with various values and choosing the best performer.

[36]The purity of a crisp cluster is the fraction that the largest class assigned to that cluster represents. The overall purity is a max-optimal (↑) external CVI with range in [0, 1]. This measure is exhibited below as Equation (12.19).

History management in D-Stream is accomplished using a damped time window approach, and clustering of the minor footprints is done at specified time intervals, when adjacent dense grids are simply merged.

12.3 Reading the Footprints: Hindsight Evaluation

When a new algorithm (***A***) appears, three questions should always be asked: (i) does (***A***) solve a problem of interest? (ii) If yes to (i), is (***A***) the first algorithm that solves this problem? If yes to (ii), there can't be any comparison; but if the answer to (ii) is no, the third question asked should be (iii) does (***A***) solve the problem at least as well as its competitors? For example, when a new way to invert a matrix comes along, the answer to (i) is yes, to (ii) is no, and for (iii), interested parties will immediately ask -how well does it do on the Hilbert matrix? In essence, inversion of the Hilbert matrix is a litmus test for new matrix inversion routines. If you apply these questions to any stream clustering algorithm, the answer to (i) is maybe, and this informs the answer to (ii). What about (iii)? Well, this is the role of hindsight evaluation -how well did your algorithm do at what you said it would do compared to its rivals? This section discusses how to answer that question.

12.3.1 When You Can See the Data and Footprints

Table 12.1 shows sets of unlabeled footprints from animals (column 1) and from stream clustering algorithms (column 3). The footprints were left by six different animals that crossed the river last night. You can see only one set of tracks made by a single animal - that is just like looking at streaming data -you see one -and only one -piece of data at a time. If you were an expert tracker, you could assign a label (an animal name) to each type of footprint. Column 2 in Table 12.1 contains a silhouette of each animal that made the track in Column 1 for the first five rows. You, as the safari guide, can easily identify them: deer, bear, panther, wolf, and badger. But the sixth track stumps you -what can that be? Is it an animal you have never seen -an anomaly? You try it.

Well, this is exactly the conundrum with streaming data: as each object speeds by, you (that is, algorithm *A*) must decide what it is, how to use it to update one or more computational footprints, and what to do about it once you decide what you have seen. Column 3 in Table 12.1 shows six of the footprints that were built with four different stream clustering algorithms by applying them to the same mystery data set.

What can you say about the data underlying these six computational footprints? Not much. But by now you will have guessed some things about the data. For example, if these algorithms are doing their job (which is what, by the way?), you will be thinking that there are 10 clusters in a ring, surrounding an 11th cluster in the center of the ring. Did you imagine that? Figure 12.13 shows the mystery data set -and -you were right!

The data $X_?$ is a set of $n = 1100$ points in $p = 2$ dimensions which seems to possess $c = 11$ HPOV clusters. The points were created this way. Let $X_? = X_1 \cup X_2 \cup \cdots \cup X_{11}$ with $|X_i| = n_i = 100; 1 \leq i \leq 11$. Each of the 11 subsets are drawn i.i.d. from 11 Gaussian distributions with specified means and covariance structure. The 100 points in X_1 are drawn first. Then 20 points are drawn from the center distribution, X_6. Then 100 points comprising X_2 are drawn, followed by adding another 20 points to X_6 sampled from the center distribution. Continuing this way, the center subset X_6 is completed after the construction of X_5, and the remaining five subsets are then drawn in the order shown.

The footprints shown in Examples 12.2–12.5 were made by pretending this data evolved as a stream around the circle as shown by the dashed circular arc. This is what the term submitted in "input order" meant. The only exception is the footprint shown in Figure 12.12(b), which is captioned as "random order,"

Table 12.1 Footprints of six animals and four stream clustering algorithm.

Unlabeled animal footprints	Animal that made them	Unlabeled SDA footprints	SDA that madethem
			Example 12.2 PrS: 16 centers
			Example 12.2 PrS: 67 centers
			Example 12.3 CLU: 100 mcs + 11 MCs
			Example 12.4 DEN: 100 mcs + 11 MCs
			Example 12.5 MUSC: 11 ellipses + 25 anomalies
	Try labeling this animal from its tracks.		Example 12.5 MUSC: 4 ellipses + 59 anomalies

What can you deduce about the streaming data if all you can see are these footprints?

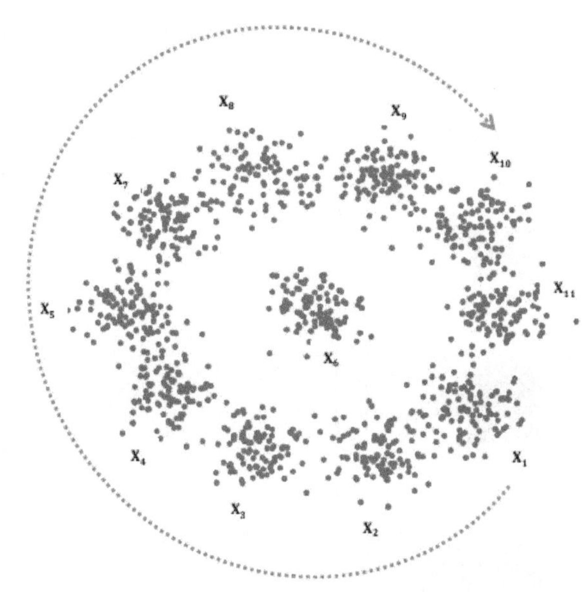

Figure 12.13 Scatterplot of the mystery data set $X_?$.

meaning that the 1100 points were submitted to the MUSC algorithm without regard to the labeled order. And this very misleading footprint should serve as a warning: computational footprints are at the mercy of many implementational choices and user-defined algorithmic parameters. They are much less distinctive than the track of the cat.

Every paper I am aware of (or written!) on this topic finds the footprints, and then, if the data and footprints are in $\Re^2$, there will be figures that show the footprints superposed on the data. Almost all of these papers conclude by saying "you can see how well the algorithm matches the HPOV clusters in the streaming data." The fact that you must HAVE (all of) the streaming data in order to make such a statement is not usually mentioned. This is a type of hindsight evaluation of stream clustering algorithms that essentially asks you to use your knowledge about static cluster analysis to judge the results. And so, you will find as justification a figure like the ones in the second column of Table 12.2, which do exactly this. The data are just blue dots in the two PrS views. They are (hard to see) gray dots for the CLU and DEN views. The data are dots which are colored by the class label assigned to them by MUSC in the two MUSC views.

Ok, now that you can see the data with the footprints, what do you conclude about the stream clustering algorithms? Are you able to say one of them is superior to the others? The first question that should jump into your mind after reading this is: *superior at doing what*? Building static clusters that are well-represented by the footprints? That is not the objective in stream clustering. The lack of a clear set of well-defined objectives is a very real problem with this kind of processing. Various authors mention: anomaly detection, identification of concept drift, building summaries of the data that have appeared in the stream, etc. These good objectives, however, are not so well met by most of the algorithms that appear in the stream clustering literature. A much greater focus of most of these papers is to replicate what is essentially the objective of static cluster analysis.

Looking at the footprints superposed on the mystery data, are you able to say that one of the algorithms has done a better job than the others? If this were static clustering, you would back your answer by noting

Table 12.2 Hindsight evaluation of the footprints shown in Table 12.1.

Unlabeled footprints	Footprints + data set $X_?$	SCA that made the footprint
	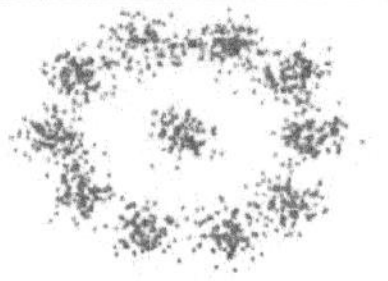	Example 12.2 PrS: 16 centers
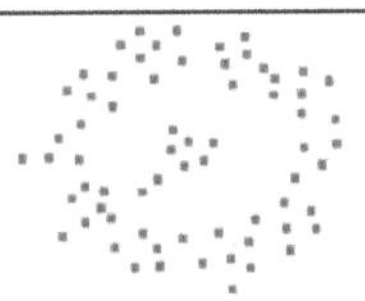	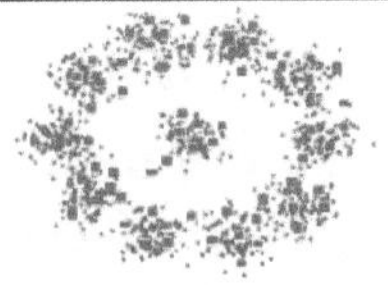	Example 12.2 PrS: 67 centers
		Example 12.3 CLU: 100 mcs + 11 MCs
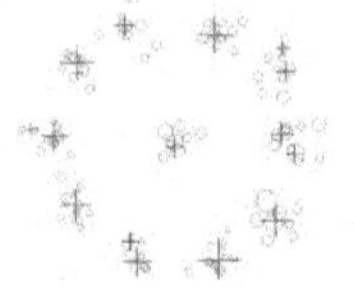	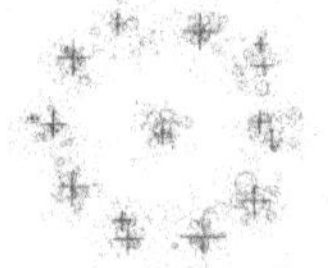	Example 12.4 DEN: 100 mcs + 11 MCs
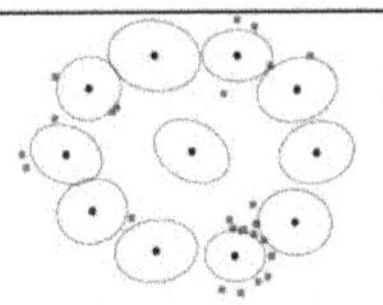	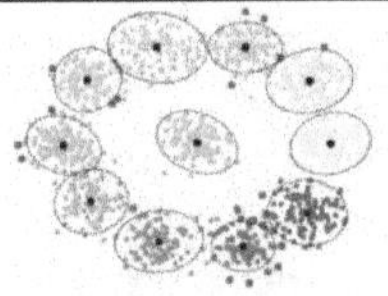	Example 12.5 MUSC: 11 ellipses + 25 anomalies
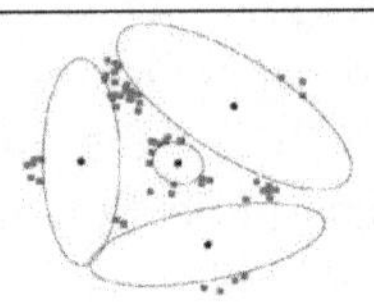	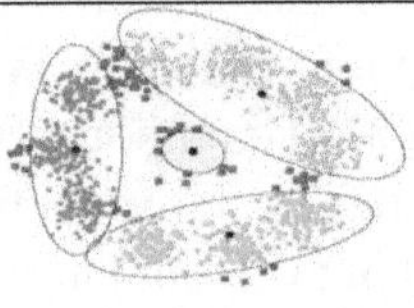	Example 12.5 MUSC: 4 ellipses + 59 anomalies

that you knew something about the data (fake static clustering!). The "best looking" outputs appear to be the DEN and MUSC (11) footprints. The MUSC result is especially appealing because the ellipses capture size and shape (albeit after some judicious parameter tuning) and also display outliers. But to get the 11 cluster footprints shown in rows 3–5 of Table 12.2 required some prior knowledge about the input data.

There is nothing wrong in this approach to the evaluation of SCAs (indeed, what is the alternative?) as long as you remember that what you are seeing is not what an online SCA will be able to present you for very long, if at all. Algorithms which retain a chunk of the data via one of the windowing approaches such as those that are used by BIRCH, CLU, and DEN enable you to see figures like these for the window in use, and this is not without value, but be aware of the limitations of such figures. And, if the data and/or the footprints reside in a space with $p > 3$? Then visualization will be possible, even if you save the data, only after some method of feature extraction is used to project the data into a viewing space. In this more difficult case, researchers in this field co-opt another method from static clustering -viz., the use of cluster validity indices (which, of course, require CPOV clusters you don't have!).

12.3.2 When You Can't See the Data and Footprints

Consider the KDD '99 Cup data (http://kdd.ics.uci.edu/databases/kddcup 99/kddcup99.html, October 2007). This data set was based on the data captured in the DARPA'98 IDS evaluation program. DARPA'98 is about 4 gigabytes of compressed raw (binary) tcpdump data of 7 weeks of network traffic, which can be processed into about 5 million connection records, each with about 100 bytes. The two weeks of test data have around 2 million connection records. The KDD '99 training dataset consists of approximately 4,900,000 single connection vectors each of which contains 41 features (34 continuous, 7 logical) and is labeled as either normal or an attack.

This data has probably been the most widely used data set for the evaluation of anomaly detection methods because it is one of the very few data sets which offers a way to study intrusion detection in streaming data. A number of criticisms have been leveled at KDD '99: there was no validation of false alarm characteristics; some records were not saved, others are duplicated; and the definition of attacks is not precise. There have been a number of papers written about these problems, nicely summarized by Tavallaee *et al.* (2009), who offer a "corrected" KDD '99 data set they call NSL-KDD. You will find many analyses of KDD '99 made by various stream clustering algorithms, and these papers offer various means for evaluating their results. The point here is this: each vector in the stream contains 34 continuous and 7 binary features, i.e., p is either 34 or 41, depending on the features used. Most authors process only the continuous features. Nonetheless, it is clear that there will not be any direct display of the data or computational footprints it leaves that enable us to look at the results of stream clustering algorithms by scatterplots such as those shown in Table 12.2.

One solution to this problem would be to project the high-dimensional data, and footprints made by stream processing, into two or three dimensions. This would provide a visualization of the results, as is done in Table 12.2. But Figure 5.18 should warn you that this approach is fraught with danger. Each CLU microcluster, for example, has $(2p + 3)$ components, which, for the 34 continuous features in KDD '99, would result in a footprint having 71 values. Different projection methods carrying the information from $\Re^{71} \rightarrowtail \Re^2$ will certainly result in different interpretations of what CLU saw in the upspace data stream.

And so, most papers that discuss the use of the KDD '99 cup data to exemplify a stream clustering method and compare it to competitor algorithms use what amounts to a static cluster validity index or clustering model objective function. This is generally true when any other high-dimensional data set such as the MNIST data appears in a stream clustering paper. For example, Aggrawal *et al.* (2003, CLU) and Guha *et al.* (2003, LSEARCH) use a measure they call the *sum of squares quality* (SSQ). The c-means and c-median problems both find c members of the data set, where c, the desired number of clusters, is given as input. For c-means (HCM), the objective is to minimize the overall sum of square errors between the data and the c means, by choosing for each data point the smallest distance, as shown in Figure 6.3. For c-medians, the objective is to minimize the sum of all c distances to each median.

The SSQ measure, as defined by Guha *et al.*, is the sum of all the squared distances to all c medians. For Aggrawal *et al.*, SSQ is the sum of squared distances from the centroids of all the macroclusters in the past time horizon to all the points in the last time horizon. More generally, $SSQ(U,\mathbf{V}) = J_2(U,\mathbf{V})$ is just the c-means objective function. These small differences in the definition of SSQ notwithstanding, one wonders:(i) how are we to specify c, when the number of data types that may appear in the stream is unknown? (ii) How can these computations be made unless all the data are saved and the partition $U \in M_{hcn}$ is built? Well, in the real case, we can't know c, we don't save the data, nor is there a crisp partition of the data. So, the numerical experiments in these papers are conceptualized in terms of entities from static cluster analysis; very unrealistic, but again, maybe it is all we can do?

Aggrawal *et al.* (2004) suggest that SSQ is an inappropriate measure for the quality of their projection method HPStream. Instead, they advocated the use of a measure called purity. There are various definitions of this external CVI in the literature. Assume the availability of the partitions $U, U_{GT} \in M_{hcn}$. The most common equation for purity is calculated by finding the dominant label represented in the *i*th cluster in U, dividing it by n_i, the number of points labeled (i) in the ground truth partition, summing these c values, and then averaging the sum with the multiplier 1/c (cf., Cao *et al.*, 2006). This value is easily calculated from the entries C_{ij} of the confusion matrix $[C(U_{GT}, U)]$ at Equation (6.31):

$$Pur(U|U_{GT}) = \frac{1}{c}\sum_{j=1}^{c}\left(\frac{\underbrace{\max}_{1\leq j\leq c}\{C_{ij}\}}{n_j}\right) : U, U_{GT} \in M_{hcn} \tag{12.19}$$

Some authors replace the multiplier $1/c$ with $1/n$. Purity is 1 when there is perfect agreement between the labels in $U, U_{GT} \in M_{hcn}$, and 0 when the disagreement between them is 100%.The MOA software produces a value for purity when exercising CLU and DEN. For example, the purity of the CLU output in Example 12.3 is 0.889 for 100 microclusters but drops to 0.557 for 50 microclusters of the mystery data set. For DEN, the purity is (curiously) 0.954 for either 100 microclusters (Example 12. 4) or 50 micro-clusters. In any case, the purity is a max-optimal (↑) external CVI that requires two crisp c-partitions $U, U_{GT} \in M_{hcn}$, neither of which is available when creating cluster footprints of streaming data in the real case.

The use of static cluster validity indices to validate the outputs of stream clustering algorithms is hardly limited to SSQ and purity. Many of the internal and external CVIs that have been discussed in Chapters 1–11 appear in Section 6.2 of Hahsler *et al.* (2017), which bears the title "Evaluating clustering of static data streams." The word "static" as used by Hahsler *et al.* is different than the use of the term in the first sentence of this paragraph. Hahsler *et al.* refer to data streams that don't exhibit concept drift as static data streams. This subsection indicates the availability of 13 internal CVIs, including SSQ, silhouette width, max. cluster diameter (denominator of Dunn's index), min. cluster separation (numerator of Dunn's index)), two forms of Dunn's index, Goodman and Kruskal's Gamma index, and the Pearson correlation coefficient. Following this is a list of 16 external crisp CVIS, including precision, recall, F1, purity, Rand's index, the adjusted Rand's index (Definition 9.2), NMI, Fowlkes-Mallow and the variation of information indices. This subsection concludes with one outlier index based on the Jaccard index. The intention of itemizing these is not to cast doubt on the STREAM software; rather, the point is for you to see just how pervasive the use of static clustering concepts is in the context of stream clustering.

Hahsler *et al.* (2017) certainly recognize this fact. Section 6.3, immediately following Section 6.2 on static CVIs, is titled "Evaluation of clustering in dynamic data streams." There is a good discussion here of the great difficulty but clear importance in evaluating stream clustering in the real case. Hahsler *et al.* attribute the concept of horizon windows of saved data that prevails in algorithms such as CLU and DEN to the need for some kind of evaluation when all the data can't be saved. They call this type of evaluation "prequential" and acknowledge that choosing a "bad" horizon can adversely affect the prequential evaluation strategy. Here is an example.

Example 12.6. Hindsight evaluation on data saved in sliding windows.

Hahsler *et al.* (2017) provide a number of excellent examples that utilize various data sets and evaluation strategies to compare four stream clustering algorithms: **biased sample**, **sliding window**, **D-Stream**, and **DBSTREAM**. The examples include this one, which plots the adjusted Rand's index (ARI; Definition 9.2), for the four algorithms in a certain way. Recall from Section 9.4 that $V_{ARI}(U|Q)$ at Equation (9.8) is the (number of agreements) divided by the (number of agreements + number of disagreements) between any pair of partition matrices $U \in M_{hrn}$ and $Q \in M_{hcn}$, corrected for chance by the transformation in Equation (9.9). Hahsler *et al.* call this index cRAND (*corrected Rand index*).

The data for this example are described as a synthetic 2-dimensional data set X having $n = 5000$ points composed of two data streams that cross each other near the middle of the streamas the data evolves. Some of the 5000 points are drawn randomly during evolution of the data, in the same fashion as the center points X_6 in the mystery data at Figure 12.13. Since the number of HPOV clusters in this labeled data is known, $c = 2$ is given to the four algorithms. The four algorithms process the data sequentially, and the labels of points in successive subsets of 250 points are used to build crisp partitions on each chunk. The chunks are saved in a set of edge-to-edge sliding windows, and the saved partitions are used to make the ARI calculations, by comparing each chunk partition to the corresponding subpartition in the ground truth partition. So, there are 20 chunks of 250 points in succession along the time scale at which the ARI evaluation for the four algorithms is made. Let's call the 20 chunks $\{\triangle X_i : 1 \leq i \leq 20\}$.

In this example then, the partitions to be evaluated by the ARI will be 2×250 matrices corresponding to each of the 20 chunks $\triangle X_i$. Let $U(\triangle X_i)$ be the partition built by aggregating the crisp labels produced by one of the four stream clustering algorithms on chunk $\triangle X_i$: and let Q_i be the ground truth sub-partition of the same size, extracted from the overall 2×5000 ground truth partition of the (static) input data. With this setup, we can compute 20 values of the ARI for each algorithm, viz., $\{V_{ARI}(U(\triangle X_i)|Q_i) : 1 \leq i \leq 20\}$. Figure 12.14 is a reproduction of Figure 23 in Hahsler *et al.* (2017) that shows the results. The general idea of this set of graphs is to deduce from a plot of the values $\{V_{ARI}(U(\triangle X_i)|Q_i) : 1 \leq i \leq 20\}$ for each of the four algorithms "how well" that algorithm is doing, where "how well" is defined by good values of $V_{ARI}(U(\triangle X_i)|Q_i)$.

Recall that the best value of the ARI is 1, when there is perfect agreement between the labels of the pair of matrices being compared. Looking at the graphs in Figure 12.14, sub-partitions produced by the DBSTREAM algorithm (dashed blue) return values very close to 1 for (as a guess) the first 10 chunks and the last 5 chunks of data in the stream. For the five chunks from 2500 to 3750 ($\approx$ chunks 11–15), all four algorithms produce their lowest ARI values. Two of the four graphs (the red dashes and solid black line) are

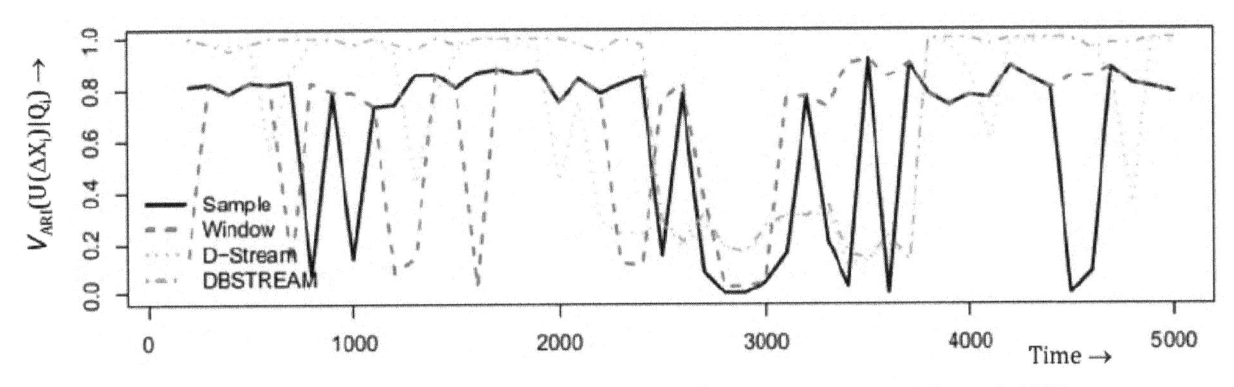

Figure 12.14 Prequential evaluation of four stream clustering algorithms (Hahsler *et al.*, 2017).

very close to zero, indicating that there is not enough separation between points in the stream over each of these five chunks to afford separation of the two subsets in the stream for these five subsets. DBSTREAM and D-Stream do a little better in the crossover area, hovering about values in the range 0.20–0.35 or so. Hahsler *et al.* attribute the higher values of the ARI for D-Stream and DBSTREAM in this area to the fact that the window and sample algorithms don't accommodate noise and must, therefore, place the randomly generated anomaly points into one of the two clusters.

Example 12.6 is a very good step in the right direction -viz., attempting to capture information about a stream processing algorithm by looking at static properties such as the ARI of a partition of a subset of the stream. But this approach stops short of what is really needed -namely, methods for incrementally evaluating what has happened (the historical record), and what is happening at the current input, by looking at the effect of a single data point on the streaming data algorithm. This idea is still in its infancy and is the subject of Section 12.4. Before we move on to this topic, let us have a look at another field which is intimately related to stream clustering -viz., change point detection.

12.3.3 Change Point Detection

When a streaming clustering algorithm $\boldsymbol{A}$ creates a new footprint, it signals that $\boldsymbol{A}$ has detected a change in the essential nature of the data in the stream. In the African animals analogy, a new footprint corresponds to the emergence of a new species in the stream: perhaps, the onset of a herd of giraffes, after a herd of zebras has passed by. This is the underlying idea of change point detection. Most stream clustering papers and texts discuss some aspect of this idea and, in this sense, are related to a very large literature on change detection in time series. See Basseville and Nikiforov (1993) for a nice introduction. This primer can't address the many intricate relationships between the fields of change point detection in time series and streaming clustering, but these two fields are companion technologies; so we take a quick look at a rival camp set up downstream from our own observation platform that might also get a good shot of the animals.

What is a change in a data stream? There are many and often conflicting definitions of this intuitive idea, which are sometimes collected under the general heading of *concept drift*. In the time series literature, a change point is an instance in time at which the statistical properties of points in the stream differ significantly in windows before and after the change point. Here are some questions that Killick (2017) identifies as important for change detection in time series analysis:

> Has a change occurred? If yes, where is the change in time and space? What is the difference between the pre- and post-change data? This may be the type of change and/or the parameter values before and after the change. What is the probability that a change has occurred? How certain are we of the changepoint location? How many changes have occurred (+ all the above for each change)? Why has there been a change?

Bifet *et al.* (2018) classify changes in the streaming clustering case as: sudden, gradual, recurrent, and partial or global. These authors point out that detection of "true changes" that are usually concerned with prediction is aggravated by the existence of transient outliers and persistent noise. These authors don't use the term anomaly, but it is probably the case that a transient outlier corresponds to an anomaly (e.g., do you regard the lion in Figure 12.1(b) as a transient outlier?). Figure 12.15 depicts several of the many types of change points that might occur in a time series.

The data in Figure 12.15(a) is the ts(x) data: it was generated by the R program in Killick (2017). This graph shows a single, sudden jump at the change point (cp), as might happen when the zebras are gone and the giraffes are on the move. The graph in Figure 12.15(b) is the labeled training data set #16 in a set of 85 time series, available at http://timeseriesclassification.com/dataset.php. This illustrates multiple change points and concept drift. The regions in-between the three change points in view 12.15(b) are not characterized by values that experience a more or less fixed variance about a mean value, as shown for the zebra and giraffe regions in view (a). Instead, there is a gradual change or drift from one regime to the next (well, cp2 is certainly pretty abrupt). This might happen, for example, as the voltage in a sensor that generates the time series fluctuates. This type of graph is where the African animals analogy breaks down a bit!

Please see Killick (2017) for an excellent introduction to the difference between online and offline change detection in time series. She points out that in the online version, data arrive one at a time and must be processed on the fly to detect a single change as quickly as possible. This is the type of processing demanded, for example, by intrusion detection. Offline change detection, on the other hand, has the luxury afforded to static cluster analysis: a long time series is collected and can be analyzed for accurate detection of many changes, as might be required in linguistics or genome analysis. And she makes the important point that the changes detected by the two schemes will not often be the same.

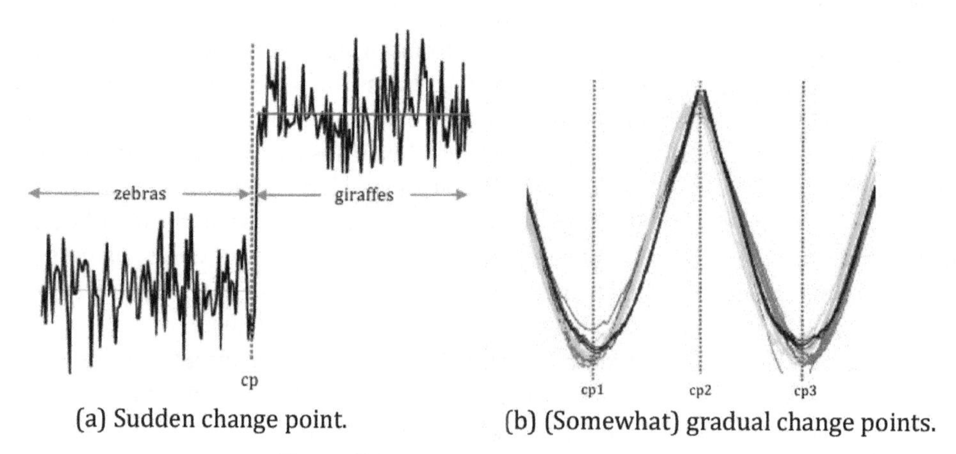

Figure 12.15 Two types of change points.

There is no lack of excellent papers about the nuances of change point detection in time series. Here is a discussion of just one, which is sufficient to illustrate the relationship between change point detection in time series and streaming cluster analysis. Zemeni *et al.* (2020) attack the problem of online change point detection in high-dimensional time series that are presented to an observer in streaming fashion. They give a very good up-to-date review of the state of the art in this area, summarized in their Table 1, which is a useful comparison of 14 change detection methods for the time series problem.

Like Killick (2017), Zemeni *et al.* characterize change point detection methods as ones that subdivide the streaming data into adjoining chunks which are determined on the fly by some statistical criterion of each chunk in the stream. The criterion they introduce is based on the idea that the information gain (drop in entropy) enjoyed by the next evolving chunk in the stream is a good measure upon which to base change point detection. So, their method divides the time series into successive non-overlapping chunks (windows) of variable width that represent different regimes in the data stream. Several theorems are given to support their algorithm, which is subsequently compared numerically to four other recent algorithms on seven (four synthetic and three real) high-dimensional streams of time series. They give the time complexity of their online method as $\boldsymbol{O}(\text{pUB})$, where p is the dimension of the streaming vectors (time series), and UB is an upper bound on memory size.

The Zemeni paper is not about stream clustering but, in fact, accomplishes it *de facto* as illustrated in Figure 12.16 by segmenting the time series data into a number of successive chunks that represent different clusters. The red squares and blue dots in Figure 12.16 represent the crisp partition of data points. In the case of abrupt change points, the labels assigned to the points are analogous to different animals that have streamed by in-between each pair of change points. This illustrates the fact that change point detection algorithms can sometimes be viewed as a special type of streaming clustering. The main difference is that partitioning with change points as shown in Figure 12.16 doesn't leave any footprints. And when concept drift as in Figure 12.15(b) occurs, the distinction between clusters as illustrated in Figure 12.16 becomes quite blurred.

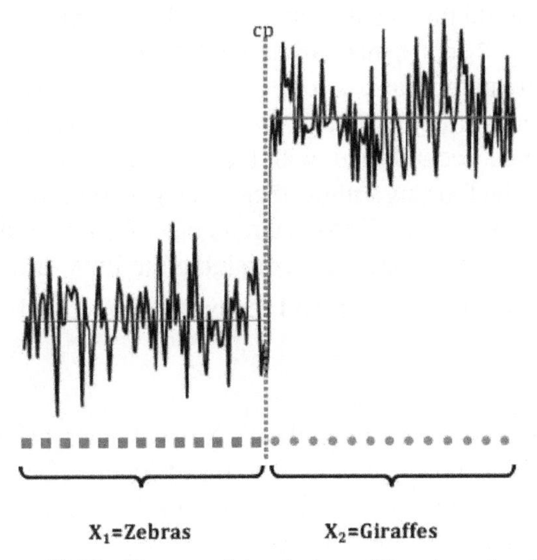

Figure 12.16 Change points crisply partition streaming data.

Zemeni *et al.* (2020) is but one of many papers about change point detection. For now, please note that this paper typifies many methods in the genre: it uses subsets of saved data from the stream to do its job and, hence, again, falls in-between the analysis of static data with classical methods and true streaming data analysis, where the only thing you have to work with is the current animal crossing the river.

12.4 Dynamic Evaluation of Streaming Data Analysis

This section discusses some ways to extend the idea presented in Example 12.6, viz., to create some kind of dynamic online evaluation that assesses the processing done by a stream clustering algorithm as each data point passes by. There are some excellent treatments of the use of sequential evaluation for streaming classifiers (algorithms with the goal of predictive stream learning). For example, Gama *et al.* (2013) give a very accessible treatment of prequential evaluation in this context. They evaluate the ability of a standard change detection algorithm in a set of experiments using three prequential error estimators: the holdout test, prequential error, and prequential error over a sliding window or with forgetting factors. They prove that under suitable conditions, all three of these methods converge to the Bayes error. Their experiments point out that the use of forgetting mechanisms (sliding windows or fading factors) are required for fast and efficient change detection. But this work is in the context of classifier design. Now, let's have a look at what might be done in the context of stream clustering without the benefit of hindsight.

12.4.1 Incremental Stream Monitoring Functions (ISMFs)

We begin this topic with Figure 12.17, which depicts the basic idea. View 12.17(a) corresponds to the situation in Example 12.6 that is illustrated in Figure 12.14, where each chunk $\triangle X_i$ was 250 of the 5000 points, and the chunks were laid out successively. The important difference between Figures 12.14 and 12.17 is the type of measurement that is plotted along the vertical axes. In Figure 12.14, the values plotted correspond to the *external* index $V_{ARI}(U(\triangle X_i)|Q_i)$, whereas the index in Figure 12.17(a) portrays values of a generic *internal* cluster validity index, $V_{int}(U(\triangle X))$. The index used in Example 12.6 was necessarily external because the intention was to do hindsight evaluation with the ground truth partition of the labeled data. The function V_{int} in Figure 12.17 represents an internal CVI because, in the real streaming case, we will not be dealing with labeled streaming data; so our objective is somewhat different. In general, other types of functions might prove equally effective, but let us stick with internal CVIs for now.

Suppose Example 12.6 was done a second time, with the 20 chunks being halved, resulting in twice as many values of the ARI for each of the four algorithms there. The generic form of this operation is depicted in view (b) of Figure 12.17, where $\triangle X \rightarrowtail \triangle X/2$. Do you remember from your first calculus course how the derivative of a real function $y = f(X)$ was defined, by examining the limit of a sequence of finite fractions $\left(\frac{\triangle y}{\triangle x}\right)$ as the denominator got smaller, and smaller, and really smaller? The definition was something like this: $\frac{dy}{dx} = \underbrace{\lim}_{\triangle x \to 0} \left(\frac{\triangle y}{\triangle x}\right)$.

Imagine doing something similar here. But instead of $\triangle X \rightarrowtail 0$, keep halving the chunks until $\triangle X = \{\mathbf{x}_k\}$, a single point in the stream. This is depicted in view (c) of Figure 12.17. Shown there, the first two points in the stream are used to "initialize" the function $\varphi_{ismf}(\mathbf{x})$. Subsequently, this function produces a new value for each point in the stream. The *k*th point $\mathbf{x}_k$ is used for the calculation of $\varphi_{ismf}(\mathbf{x}_k)$ and will probably also be used to update one or more footprints, but after these two chores are finished, $\mathbf{x}_k$ is gone, just like the animals that crossed the river last night. Note that the function in views (a) and (b) is a standard

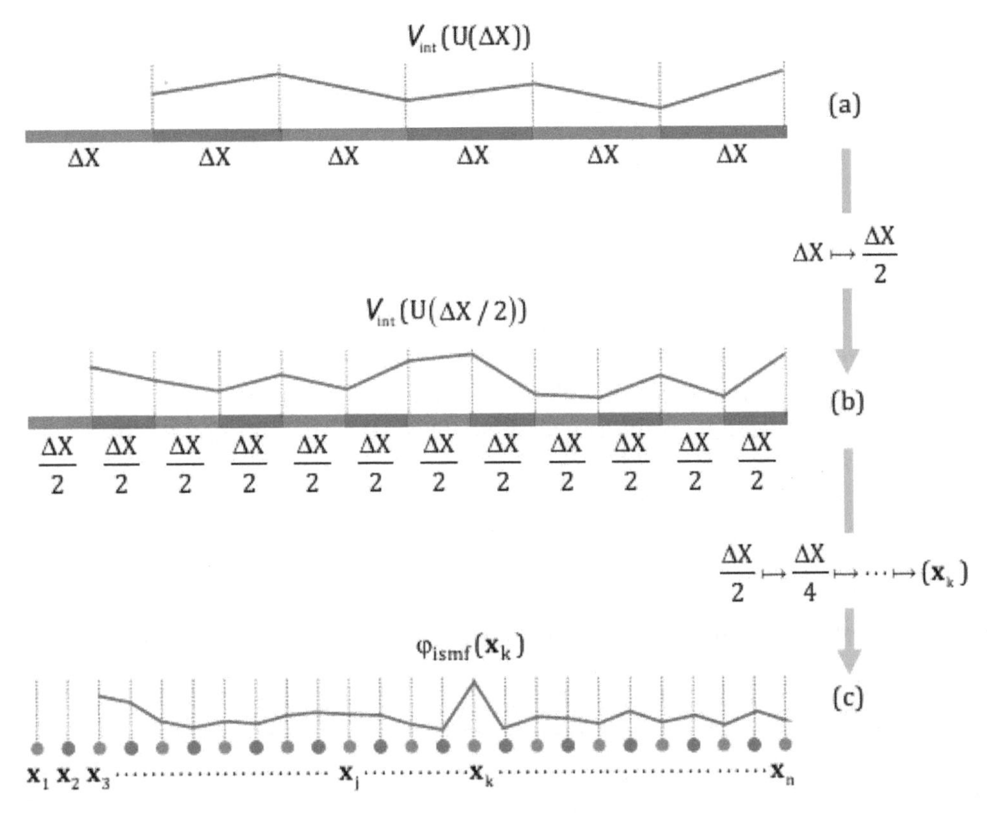

Figure 12.17 The basic idea of incremental stream monitoring functions.

internal CVI, notated V_{int} as in previous chapters, evaluated on a sub-partition of points in the stream. But in the limit, when $\triangle X = \{\mathbf{x}_k\}$, this concept is vacuous.

Since there is no cluster to validate, the name of the function in view (c) is changed from V_{int} to φ_{ismf}. The subscript on φ_{ismf} in view (c) stands for *incremental stream monitoring function* (iSMF). The word incremental is important: it distinguishes the way this function is evaluated -viz., after each input in the data stream is processed by $\boldsymbol{A}$. The objective in this section is to devise functions of this type, which will tell us something about what is being done to the data in the stream *with respect to a specific stream clustering algorithm* $\boldsymbol{A}$.

There are very few schemes of the type illustrated in Figure 12.17(c) that attempt to assess what a stream clustering algorithm $\boldsymbol{A}$ is doing with the data in (near) realtime (i.e., online or instream, as each data point arrives). An early paper of this type appears to be the work of Moshtaghi *et al.* (2018), who derived incremental versions of two static, internal cluster validity indices. One is based on a fuzzy modification of the Davies-Bould in index due to Araki *et al.* (1993); and the other is based on the generalized Xie-Beni index at Equation (6.64):

$$V_{XB_m}(U, \mathbf{V}, X) = \frac{\sum_{i=1}^{c}\sum_{k=1}^{n} u_{ik}^m \|\mathbf{x}_k - \mathbf{v}_i\|_A^2}{n\left(\underbrace{\min}_{i\neq j}\{\|\mathbf{v}_i - \mathbf{v}_j\|_A^2\}\right)}; 1 \leq m < \infty; \tag{6.64}$$

$$V_{DBAr}(U, \mathbf{V}, X) = \left(\frac{1}{c}\right)\sum_{i=1}^{c}\left[\underbrace{\min}_{j\neq i}\left\{(\alpha_1 + \alpha_j)\Big/\left(\|\mathbf{v}_i - \mathbf{v}_j\|_A^2\right)\right\}\right], \text{ where}$$

$$\alpha_i = \sum_{k=1}^{n} u_{ik}^m \left\|\mathbf{x}_j - \mathbf{v}_i\right\|_A^2 \Big/ \sum_{k=1}^{n} u_{ik}^m \tag{12.20}$$

Mostaghi *et al.* (2018) used $m = 2$ and the Euclidean norm $(A = I)$ in both of these formulae, but the more general cases are well-defined. These two indices were originally designed as internal CVIs for cluster validation of partitions generated by FCM in static data; hence, the appearance of the term $u_{ik}^m \left\|\mathbf{x}_j - \mathbf{v}_i\right\|_A^2$, which is part of the FCM objective function, but both of them are well-defined for the crisp case. The Araki modification is not a true generalization of the *Davies-Bouldin index* (DBI) at Equation (6.53) but is very close to it in the crisp case.

Both of these indices are defined in terms of cluster centers and (partition) memberships. Moshtaghi *et al.* derived incremental versions of these two indices that don't require a saved partition of any subset of the data. These indices can be used to monitor the performance of arbitrary stream clustering algorithms (***A***) as long as their footprints contain cluster centers and ***A*** assigns a label vector in N_{fc} to each point in the streaming data. The heart of their approach is the within cluster dispersion in Definition 12.1.

Definition 12.1. Within cluster dispersion (Moshtaghi *et al.*, 2018)[37].

Assume three things:

WCD (i) Streaming algorithm ***A*** processes streaming data, $\mathbf{x}_1, \mathbf{x}_2, \ldots, \mathbf{x}_n \ldots$, where $\mathbf{x}_k \in \Re^p \; \forall k$;
WCD (ii) After processing n points, the footprints of ***A*** include c cluster centers
$\mathbf{V}_n = \{\mathbf{v}_{1,n}, \ldots, \mathbf{v}_{c,n}\} \subset \Re^{cp}$;
WCD (iii) Evaluation of point $\mathbf{x}_k$ by ***A*** generates a set of memberships
$\mathbf{U}^{(k)} = \{u_{ik} : 1 \leq i \leq c\} \in N_{fc}$.

Let $U \in M_{fcn}$ be (conceptually) the current partition that corresponds to the labels assigned to the first n points. The aggregated *within cluster dispersion* (WCD) of the *i*th cluster in $U \in M_{fcn}$ for the first n points is, for any $m \geq 1$,

$$C_{i,n} = \sum_{j=1}^{n} u_{ij}^m \|\mathbf{x}_j - \mathbf{v}_{i,n}\|_A^2;\; 1 \leq i \leq c \tag{12.21}$$

The set of c WCDs in (12.21) is well-defined for crisp, fuzzy, or probabilistic labels. They are a measure of overall compactness of the set of footprints that have been updated one point at a time after n steps.

The quantity in Equation (12.21) appears in many objective functions and CVIs. It is a measure of the compactness of the *i*th cluster about its centroid. The memberships in assumption wcd (iii) are a label vector in N_{fc} for the current input but are shown here in the format of a column of a partition in M_{fcn}

[37]This is the definition from Ibrahim *et al.* (2020). Moshtaghi *et al.*'s original definition was for m = 2.

to make the exposition clear. This set of memberships may be of any type (hard or soft), corresponding to properties of algorithm $\boldsymbol{A}$. For example, if $\boldsymbol{A}$ is sequential hard c-means (sHCM), PrS, or batch c-means (HCM), $\mathbf{U}^{(k)} \in N_{hc}$ will be crisp. If $\boldsymbol{A}$ is FCM or GMD/EM, the memberships will be soft, $\mathbf{U}^{(k)} \in N_{fc}$. The data points are not saved, nor is the partition U actually built or maintained after the incremental evaluation is made, but the concept is needed to understand the derivation of functions that can serve as φ_{ismf}. The objective is to update the WCD by adding an increment based on the current input to the previous value of the WCD:

$$\mathbf{C}_{n+1} = \mathbf{C}_n + \triangle \mathbf{C}_n \tag{12.22}$$

While $\triangle \mathbf{C}_n$ would be easy to compute directly for the next input $(\mathbf{x}_{n+1})$ if we had the previous n points that are shown in the sum at Equation (12.21), we are assuming that we have only the quantities $\{\mathbf{U}^{(n+1)}, \mathbf{V}_{n+1}\}$, just obtained by processing $(\mathbf{x}_{n+1})$ with $\boldsymbol{A}$, and we have the current value of the WCD vector $\mathbf{C}_n = \{\mathbf{C}_{i,n}; i = 1, \ldots, c\}$. Derivation of the next value of the WCD given these inputs is not trivial: see Moshtaghi *et al.* (2018) for complete details. Algorithm A12.1 performs the operations that result in the updated WCD, $\mathbf{C}_{n+1}$.

Algorithm A12.1. Incremental update algorithm for WCD ($\mathbf{C}_n \rightarrow \mathbf{C}_{n+1}$).

	Incremental Update of the WCD (Moshtaghi *et al.*, 2018)	
1	**In:** $\mathbf{U}^{(1)}, \mathbf{V}_1, m \in [1, \infty); c \in \{1, \ldots, n\}$	Initialization
2	**% Initialize**	
3	**For** i =1 to c	
4	$G_{i,1} = (0, 0, \ldots, 0)^T = \mathbf{0} \in \Re^p$	
5	$M_{i,1} = \lvert v_{1,i} \rvert$	
6	$C_{i,1} = 0$	
7	**Next** i	
8	**Data**: m, c, $\mathbf{V}_n, \mathbf{V}_{n+1}, \mathbf{U}^{(n+1)}, \forall i : G_{i,1}, M_{i,1}, C_{i,1}$	Streaming Update
9	**Input**: $\forall i : G_{i,n}, M_{i,n}, C_{i,n}, \mathbf{x}_{n+1}$	
10	**For** i = 1 to c	
11	$Q_{i,n+i} = (\mathbf{v}_{i,n} - \mathbf{v}_{i,n+1})^T G_{i,n}$	
12	$B_{i,n+1} = \lVert \mathbf{v}_{i,n} - \mathbf{v}_{i,n+1} \rVert^2$	
13	$A_{i,n+1} = (u_{i,n+1})^m \lVert \mathbf{x}_{i,n} - \mathbf{v}_{i,n+1} \rVert^2$	
14	$C_{i,n+1} = C_{i,n} + A_{i,n+1} + M_{i,n} B_{i,n+1} + 2Q_{i,n+1}$	
15	$G_{i,n+1} = G_{i,n} + M_{i,n}(\mathbf{v}_{i,n} - \mathbf{v}_{i,n+1}) + (u_{i,n+1})^m (\mathbf{x}_{n+1} - \mathbf{v}_{i,n+1})$	
16	$M_{i,n+1} = M_{i,n} + (u_{i,n+1})^m$	
17	**Next** i	
18	**Output**: $\forall i : G_{i,n+l}, M_{i,n+l}, C_{i,n+l}$	

Stream processing in A12.1 is done one point at a time. The current input enters A12.1 at line 9, and processing continues as long as new points appear in the stream. Algorithm A12.1 requires a first value of c for initialization in line 1, and in the streaming case, this method almost always begins with c = 1. The initial pair $\{\mathbf{U}^{(1)}, \mathbf{V}_1\}$ will depend on streaming algorithm $\boldsymbol{A}$. For example, if $\boldsymbol{A}$ is PrS, $\{\mathbf{U}^{(1)}, \mathbf{V}_1\} = \{[1], \mathbf{x}_1\}$.

The value of c must be incremented in the code every time $\boldsymbol{A}$ creates a new footprint (recognizes a change in the streaming data). The first value of an iSMF that is computable usually requires $c = 2$ because most internal CVIs upon which the iSMF are based are not well-defined for $c = 1$. More complicated footprints, such as those generated by MUSC, may need to be initialized with the first q points in the stream to secure the initial inputs for A12.1. In this case, the first point entering the stream calculations will be $(\mathbf{x}_{q+1})$, but again, most likely the first value of the iSMF will be when $c = 2$.

Algorithm A12.1 produces an incremental update in the value of the WCD for input $\mathbf{x}_{n+1}$. Any internal cluster validity index that has the term (12.21) in its definition can be converted into an incremental stream monitoring function by substitution of the incrementalized value of within cluster dispersion. Equations (12.23) and (12.24) based on the Araki form of the Davies-Bouldin and Xie-Beni internal CVIs were derived in Mostaghi *et al.* (2018):

$$\varphi_{\mathrm{DBAr}_{n+1}}(\boldsymbol{A};\mathbf{x}_{n+1}) = \frac{1}{c}\sum_{i=1}^{c}\underbrace{\max}_{j\neq i}\left\{\frac{\frac{C_{i,n+1}}{M_{i,n+1}}+\frac{C_{j,n+1}}{M_{j,n+1}}}{\|\mathbf{v}_{i,n+1}-\mathbf{v}_{j,n+1}\|^2}\right\} \tag{12.23}$$

$$\varphi_{\mathrm{XB}_{n+1}}(\boldsymbol{A};\mathbf{x}_{n+1}) = \frac{\sum_{i=1}^{c} C_{i,n+1}}{(n+1)\underbrace{\min}_{j\neq i}\left\{\|\mathbf{v}_{i,n+1}-\mathbf{v}_{j,n+1}\|\right\}} \tag{12.24}$$

These two measures were called iCVIs (*incremental cluster validity indices*) in Mostaghi *et al.* (2018) and Ibrahim *et al.* (2020) because they were derived from and reduce to the static cluster analysis forms in the non-streaming case. However, it eventually was recognized that this name is misleading because, in the true streaming case, there are no clusters to validate. Consequently, Bezdek and Keller (2021) suggested changing the name of these incremental measures to iSMFs. The arguments of these functions include the stream clustering algorithm ($\boldsymbol{A}$) to emphasize that the same iSMF will produce different values and, in turn, possibly different interpretations of the action of $\boldsymbol{A}$ on the streaming data for different $\boldsymbol{A}$'s and the same data inputs. Please note that neither of these functions can be computed until $\boldsymbol{A}$ has created $c = 2$ clusters and footprints: both functions are undefined at $c = 1$. A more detailed look at their behavior will explain the rationale for this choice of terminology.

Example 12.7. How iSMF functions are used to monitor a stream clustering algorithm.

This example is based on processing the mystery data set $X_?$ in Figure 12.13 with the MUSC streaming clustering algorithm described in Section 12.2.3. The data set is shown again in the upper left part in Figure 12.18. The 11 footprints (mean vectors and ellipses) created by MUSC are shown here and in Figure 12.12(a), but, here, the final crisp labels of the 1100 points are also plotted by coloring the 11 clusters identified by MUSC. The red squares are unassigned anomalies at the end of MUSC processing. The graph of the incremental DBAr function at Equation (12.23) occupies the right side of Figure 12.18.

To interpret what this graph offers us, please recall how the 1100 data points enter the stream (cf., Figure 12.13). The 100 points in X_1 are drawn first. Then points 101 – 120 are drawn from the center distribution, X_6. Then 100 points comprising X_2 are drawn, followed by adding another 20 points to X_6.

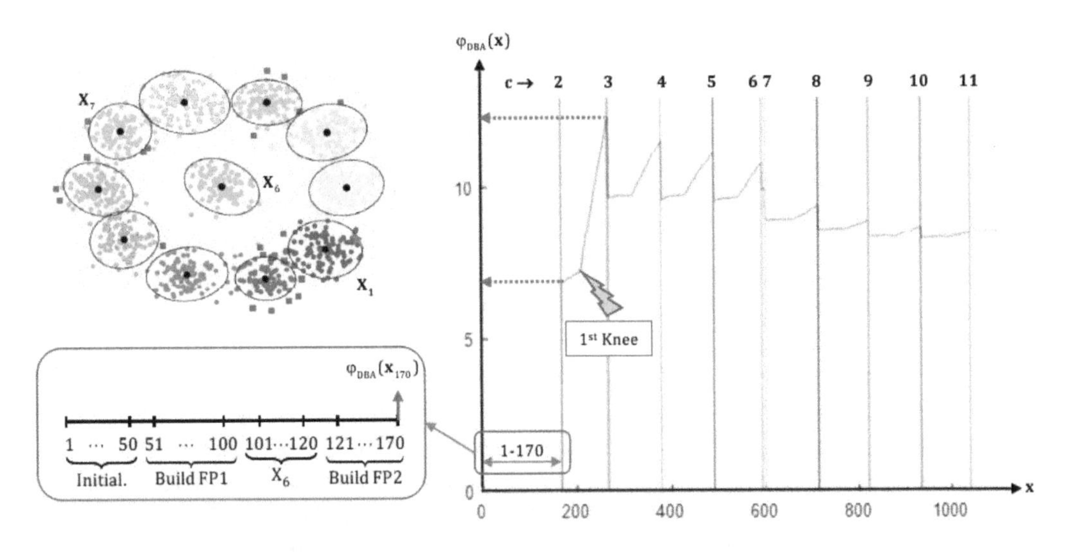

Watch a video of MUSC processing X? with DBAri at: https://www.youtube.com/watch?v=Et1txqXTrd0

Figure 12.18 Graph of φ_{DBAr} watching MUSC process the mystery data.

Continuing this way, the center subset X_6 is completed after the construction of X_5, and the remaining five subsets are then drawn in the order shown. The vertical lines on this graph show the times at which new footprints, corresponding to the number of clusters currently detected, enter the system.

The MUSC algorithm buffers the first 50 points to initialize the GMMs, as shown in the zoom view of the first 170 points in the lower left portion of Figure 12.18. As the data stream in, the remaining 50 points in X_1 (the darkest blue points in the snapshot of the data in the upper left) are used to update the initial footprint (FP1) or are sent to the anomaly tank if they can't qualify for attachment to the first footprint. Now 20 points in the center cluster X_6 arrive and are also sent to the anomaly tank. Now points in subset X_2, drawn from the second distribution, start to appear in the stream. These points can't qualify for use in updating FP1 and, at first, are not numerous enough to justify the formation of a second footprint of their own; so they are also set aside in the anomaly tank. The way MUSC operates, it continues to examine all the points that are temporarily cached in the anomaly tank.

The vertical lines in Figure 12.18 correspond to the jump points (changes detected in the footprint structure determined by algorithm $\boldsymbol{A}$) determined by the DBAr function: be careful to note that these positions are NOT places where the data in the stream experiences a change. The first jump point for $\boldsymbol{A}$ is at position 170, at which time the stream contains 100 points in X_1, 20 points in X_6, and 50 points in X_2; so MUSC has decided that there should be a second footprint ($c = 2$) at this instant in time. The position at which there are two clusters is an artifact of the way that MUSC handles the anomaly list. There are 10 jumps in Figure 12.18, corresponding to creation of clusters at $c = 2, 3, \ldots, 11$. Other stream clustering algorithms may create iSMF graphs with jump points at different locations and, possibly, in different amounts.

® ***Forewarned is forearmed:*** This emphasizes a very important point: the iSMF is not watching the data in the stream -it is watching the algorithm $\boldsymbol{A}$. The jumps in an iSMF graph don't necessarily correspond to changes in the data like change detection methods such as the one developed by Zemeni *et al.* (2020). Instead, an iSMF function detects changes in the footprint (or cluster) structure of the algorithm that supplies the values to it. ⓔ

When the 170th point arrives and is sent to the anomaly tank, the algorithm decides that a second footprint (FP2) is needed; so it collects most of the points currently from X_2 into a second cluster and declares a new structure. Since there are now two structures in the system ($c = 2$), the iSMF fires, and $\varphi_{DBA}(X_{170})$ spikes from 0 to a value of about 7, shown by the lower red dashed line on the graph. Recall that the DB index is min-optimal (lower values point to tighter cohesion and/or greater separation) due to the terms $(\alpha_{i,t} + \alpha_{j,t})/(\|\bar{v}_i - \bar{v}_j\|_q)$ and the same thing holds for the Araki DB index.

What happens after the spike at point 170 in Figure 12.18? Values of $\varphi_{DBA}(X_{170+j})$ at each incoming data point *increase*, indicating a *loss* of cohesion and/or separation in the structure being detected by MUSC. Note the "1st knee" in the graph at about position 210, indicated in Figure 12.18 by the lightning bolt. This is the place where data from the next subset start to arrive, creating a further decrease in the cohesion and separation of the points in the current footprints. The incremental values continue to increase more sharply until MUSC declares a third structure ($c = 3$) at point #270. At this instant, the overall separation increases, the cohesion improves, and there is a sharp vertical drop in the value of the incremental DBA index from about 12 to about 9.5. Then the index begins to climb again as cohesion and separation deteriorate, until MUSC declares that $c = 4$. The jagged graph continues this way out to the end of the data stream, but the peaks and troughs becoming less and less prominent as more and more points arrive. There is a reason for this: without a forgetting factor, an iSMF loses its ability to detect change as more and more points arrive and are used to update its value without regard to their position in the stream.

There is one more thing to see in Figure 12.18 that is important, and this is the almost instantaneous arrival of two spikes at $\mathbf{x}_{600}$. This is due to two things: (i) the way the mystery data evolved; and (ii) the way MUSC buffers and examines the anomalies. When the center subset $\mathbf{x}_6$ is completed and half of the points in subset $\mathbf{x}_7$ have been cached, MUSC declares two new footprints in very close succession, as seen by the very close pair of vertical spikes at $c = 6$ and $c = 7$. A different stream processing algorithm might react quite differently, depending on how it handles anomalies. So, the iSMF in Figure 12.18 provides two important pieces of information about ***A*** that can't be gleaned from the footprint built on the same data set by MUSC that is shown in Figure 12.6: viz., the rapid pair of jumps at $c = 6$ and $c = 7$; and the information provided by the knees in the graph that tell us something about how cumulative cohesion and separation are affecting the decisions made by ***A***. More generally, rapid changes in any iSMF based on the WCD are related to variations in the separation and cohesion of the clusters produced by the stream clustering algorithm ***A***. Therefore, an abrupt change in the value of an iSMF usually signals the appearance of a new structure in the streaming data.

If the iSMF is changed, the information it offers about algorithmic performance may be quite different. Figure 12.19 is a graph of the same experiment underlying Figure 12.18, with the single difference that a different incremental index is being used. The index for Figure 12.19 is the *modified (generalized) Dunn index* (MDI) corresponding to the choices $a = 5$ and $b = 3$ in Equation (6.58). The modification is that distances in the GDI must be squared to make use of the incremental WCD algorithm A12.1. The incremental form of φ_{MDI53} is derived in Ibrahim *et al.* (2020) and is left as an exercise. The MDIs are like the GDIs in the sense that they are max-optimal; so the general behavior of the values in Figure 12.19 is the inverse of their behavior in Figure 12.18. Specifically, after a new footprint is created, the value of φ_{MDI53} decreases, indicating that the new structure has deteriorated the previous separation and/or cohesion in the sense of this modified Dunn's index. The value drops until a "knee" occurs, as shown by the lightning bolt in Figure 12.19.

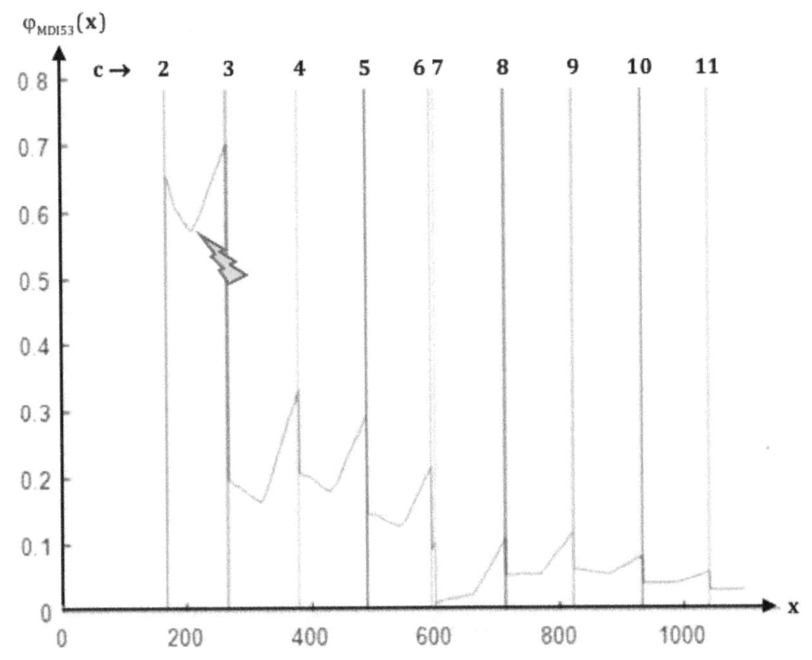

Figure 12.19 Graph of φ_{MDI53} watching MUSC process the mystery data.

Note again that knees lie in-between pairs of spikes in Figure 12.19. At the time of the knee locations, data from the next cluster start to arrive. These data are flagged as anomalies by MUSC until they become dense enough to form a new cluster. While knees do appear in both graphs, φ_{MDI53} shows them much more prominently than φ_{DBAr}, indicating that different iSMFs do provide information about different aspects of the performance of stream algorithm ***A***. Finally, observe that the graph in Figure 12.19 also tends to smaller and smaller values as the data continue to stream: again, this is an artifact due to the lack of a forgetting factor to compensate for the influence of points that passed by long ago.

The functions $\varphi_{\mathrm{DBAr}_{n+1}}$ and $\varphi_{\mathrm{XB}_{n+1}}$ were generalized in Moshtaghi *et al.* (2018) to include a forgetting factor (λ). The general idea is to reduce the effect of past inputs on incremental updates of the iSMFs by degrading their impact as a function of their position in past time. This is important: without a forgetting factor, an iSMF will soon become saturated and will become less and less able to recognize changes in the structure being detected by ***A*** as more and more inputs arrive. The way this is accomplished is to introduce the forgetting factor into the definition of the WCD as follows:

$$C_{\lambda,i,n} = \sum_{j=1}^{n} \lambda^{(n-j)} u_{ij}^{m} \|\mathbf{x}_j - \mathbf{v}_{i,n}\|_A^2;\ 1 \le i \le c;\ 0 < \lambda < 1. \tag{12.25}$$

The formulae for the quantities Q, G, and M in lines 11, 15, and 16 of A12.1 must be changed when the definition of the WCD includes a forgetting factor (cf., Mostaghi *et al.*, 2018). The overall result of incorporating a forgetting factor into these two iSMFs yields the following equations:

$$\varphi_{DBAr\lambda_{n+1}}(\mathbf{A};\mathbf{x}_{n+1}) = \frac{1}{c}\sum_{i=1}^{c}\underbrace{\max}_{j\neq i}\left\{\frac{\dfrac{C_{\lambda,i,n+1}}{\max\{1,M_{\lambda i,n+1}\}}+\dfrac{C_{\lambda,j,n+1}}{\max\{1,M_{\lambda j,n+1}\}}}{\|\mathbf{v}_{i,n+1}-\mathbf{v}_{j,n+1}\|^2}\right\} \tag{12.26}$$

$$\varphi_{XB\lambda_{n+1}}(\mathbf{A};\mathbf{x}_{n+1}) = \frac{(1-\lambda)\sum_{i=1}^{c}C_{\lambda,i,n+1}}{(n+1)\underbrace{\max}_{j\neq i}\{\|\mathbf{v}_{i,n+1}-\mathbf{v}_{j,n+1}\|\}} \tag{12.27}$$

The next example illustrates the effect of adding forgetting factors to $\varphi_{DBAr\lambda_{n+1}}$ and $\varphi_{XB\lambda_{n+1}}$ on a data set that is somewhat similar to our mystery data but has the important difference that there is no "interruption" in the stream to insert a few points into a subset from time to time as was done for subset X_6 in the mystery data. Part of this example is abstracted from Moshtaghi *et al.* (2018), where these two functions had simpler names, viz., $\varphi_{DBAr\lambda_{n+1}} = DB_\lambda(n)$ and $\varphi_{XB\lambda_n} = XB_\lambda(n)$.

Example 12.8. The incremental XB iSMF with and without a forgetting factor.

The data for this example are shown in Figure 12.20(a). This is a set of $n = 2727$ points in $p = 2$ dimensions which are called X_{2727}. There are $c = 11$ labeled subsets named X_1 to X_{11} arrayed in linear fashion as shown in the figure. The first subset, $X_1 = 500$ points, is drawn from a Gaussian distribution with $\mu_1 = (95,75); \Sigma_1 = \begin{bmatrix} 3.84 & -2.64 \\ -2.64 & 4.85 \end{bmatrix}$. Then 9 subsets of 200 points each are drawn from a Gaussian distribution that evolves in equal linear steps to simulate a changing stream of data, terminating in the Gaussian distribution with parameters $\mu_{11} = (5,5); \Sigma_{11} = \begin{bmatrix} 1.52 & -0.54 \\ -0.54 & 1.65 \end{bmatrix}$. from which the samples X_{11} are drawn. After creation of the last distribution, 400 samples are drawn for X_{11}. So, there are $500 + 9(200) + 400 = 2700$ points in the 11 subsets[38]. There are 27 additional blue points in this figure, which are noise generated by adding perturbations of 1% of the samples with draws from the uniform distribution on $[-10,10]$. The streaming aspect of the data set is indicated in Figure 12.20(a), where the time (i.e., position of each input in the stream) starts in the upper right at X_1 and proceeds downward and to the left, culminating in X_{11}. If you were to look at this data set without labels, you would probably say that it was one HPOV cluster. Do the labels (colors) in Figure 12.20(a) encourage you to think that this synthetic data set has 11 HPOV clusters?

The stream clustering algorithm that processed these data was the *online elliptical clustering* (OEC) algorithm of Moshtaghi *et al.* (2016). The footprint ellipses and cluster centers are not easy to see in view (b), and this figure was not included in Moshtaghi *et al.* (2018), but it can be used in the usual way for hindsight evaluation of the OEC algorithm.

[38]This data is described incorrectly in Moshtahgi *et al.* (2018). The description there doesn't show an extra 200 points being drawn for X_{11}, leading to the impression that the data only contains 2525 points. A check of the code verified that the description here is correct.

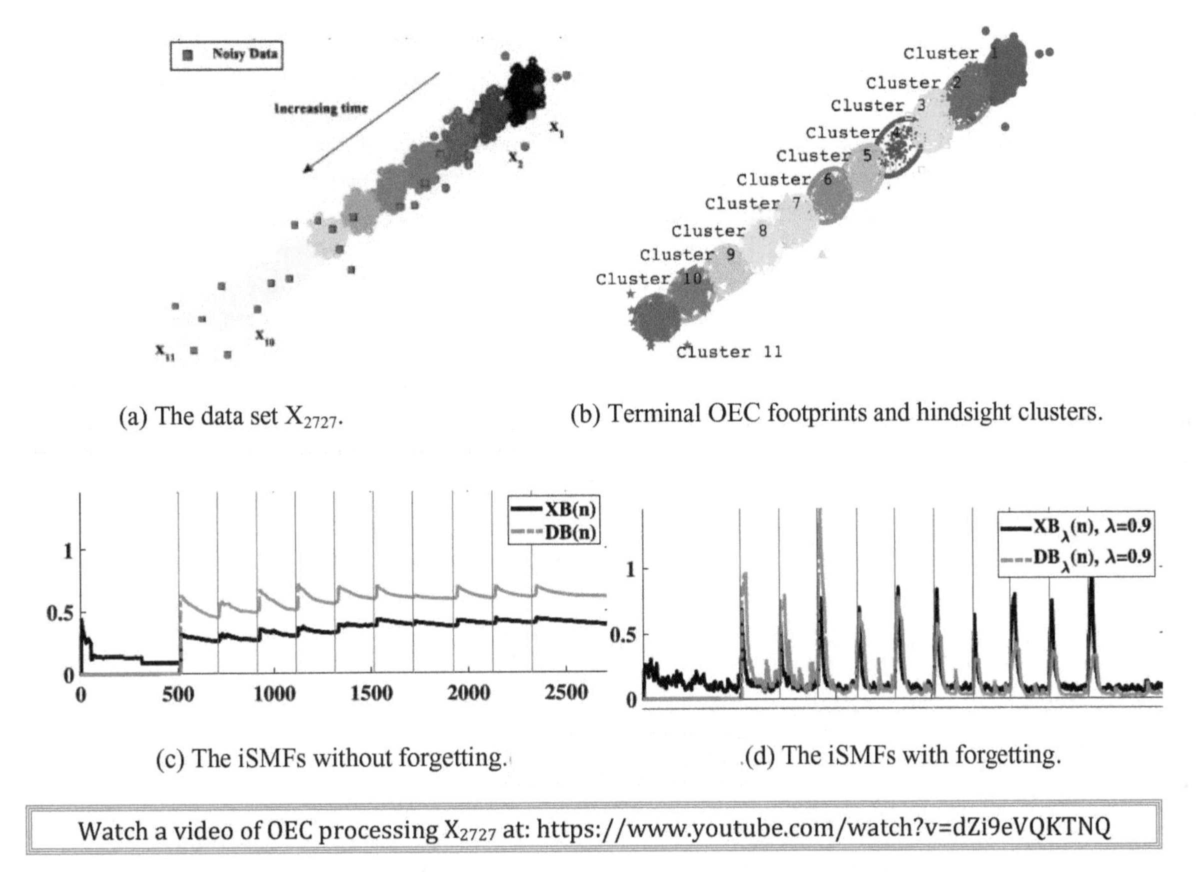

(a) The data set X_{2727}.

(b) Terminal OEC footprints and hindsight clusters.

(c) The iSMFs without forgetting.

(d) The iSMFs with forgetting.

Watch a video of OEC processing X_{2727} at: https://www.youtube.com/watch?v=dZi9eVQKTNQ

Figure 12.20 The incremental XB and DBAr iSMFs with and without forgetting factor (λ).

This algorithm is similar to the MUSC model in that the footprints are cluster centers and ellipsoids, but there is no distributional assumption about GMM, and the algorithm handles anomalies and footprint creation quite differently. The OEC method is initialized with the first (p+1) points in the stream. Since p = 2 for this data, the initialization window uses only the first three points in the stream; so there is not a noticeable lag between startup and declaration of a second cluster by the iSMFs around position 503.The red vertical lines in Figures 12.20(c) and 12.20(d) are the positions at which there is a change in the data stream, and, unlike the long-time delay before the MUSC method, OEC creates a new cluster almost immediately. Since this algorithm only uses the first three points for initialization, the red lines essentially occur when the mean and covariance matrix shift to the next position -in other words, these are the jump points in the data and also are very close to the spikes in the iSMFs because of the way that new clusters emerge almost immediately.

The OEC scheme uses a state-tracker to handle anomalies by monitoring the state of the system. The tracker is a single cluster model with a forgetting factor that accompanies the main processing, but it doesn't contribute to footprint updation. The forgetting factor allows the tracker to respond very quickly to changes in the data stream -if the state tracker is significantly different from the existing clusters, OEC creates a new cluster. The YouTube video of this example shows the tracker following the data, and you can see the footprints being updated as the stream enters each new regime.

Creation of a new cluster by OEC is based on the c-separation between pairs of Gaussians as defined by Dasgupta (1999, 2000) in the context of learning Gaussian mixtures. According to Dasgupta, a 2-separated mixture corresponds to almost completely separated Gaussian distributions. Using this idea, when the tracker becomes 2-separated from the existing clusters, OEC creates a new cluster. Incoming points are candidates for membership in all of the current clusters. The label vectors used by OEC can be soft or hard, but, in this example, hard assignments are made after incrementally updating the iSMFs.

Please compare the graphs in Figures 12.20(c) and 12.20(d). It is clear that without forgetting, both of these iSMFs rapidly lose their ability to provide much information about what ***A*** is doing. The static versions of both of these indices are min-optimal, and both begin at a very low value as points from the first cluster appear in the stream. But as soon as a second cluster appears (OEC recognizes this almost immediately), both incremental indices jump, indicating that the perceived structure has moved to a less cohesive and/or separated state. The XB iSMF is relatively flat thereafter and continues a gradual rise, indicating less and less confidence in the structure being detected. The incremental DB function is a little more sensitive to jumps, but its values are higher than the XB function, indicating that it is not happy with the streaming results either.

In contrast to this, the graphs in Figure 12.20(d) of the functions in Equations (12.26) and (12.27) that have forgetting factors to focus incremental updates on current inputs do quite well at detecting changes in the footprints being built by the OEC algorithm. Very sharp spikes continue to signal the creation of new footprints, and these occur very near to the emergence of points in the next cluster. So, in this case, the OEC algorithm is on a much more equal footing with change detection algorithms such as that discussed by Zemeni *et al.* (2020) that simply identify change points in the data without regard to changes in the footprints created by ***A***.

The two views of the iSMFs in Figures 12.20(c) and 12.20(d) should convince you that the utility of any iSMF is greatly enhanced by adding a forgetting factor that focuses the monitoring on current values in the data stream. As mentioned above, Ibrahim *et al.* (2020) developed and tested two other iSMFs based on modifications of GDIs, but neither of these indices was generalized to the forgetting factor case. This is left to the exercises (or, to a nice MS thesis at your university!).

Please consider the information that can be gleaned from Examples 12.7 and 12.8 about two things: (i) the nature of data flowing past your algorithm in the stream; and (ii) the performance of an ***A*** on the streaming data. Compare what can be learned about these items by the iSMFs in Figures 12.18 and 12.19 to what can be inferred about them from the footprints for the four stream clustering algorithms shown in Table 12.2 and from the four other stream clustering algorithms in Figure 12.14. The footprints alone in Table 12.2 reveal nothing about the performance of the algorithms and very little about the nature of data in the stream. These outputs don't reveal anything about the dynamics of the streaming data, nor the intermediate processing that has been done to arrive at the sets of footprints.

In the case of Figure 12.14, a window of points are saved, a partition is built on them, and a static cluster validity index assesses the performance of four stream clustering algorithms on this offline information. This enables you to ascertain a bit more about how the four algorithms are doing with respect to each other and the data in the stream, but the points of examination are artificial in the sense that they are not data-driven. It is clear that different choices for window position and length will result in different looks at the process. This start and stop procedure is better than hindsight evaluation but falls short of what is needed in real streaming data analysis. In contrast, the iSMFs in Examples 12.7 and 12.8 provide a continuous, point-by-point assessment of what ***A*** seems to be doing.

While the examples given here are tied to the MUSC and OEC stream clustering algorithms, please note that the four iSMFs $\{\varphi_{DBAr}, \varphi_{XB}, \varphi_{MDI43}, \varphi_{MDI51}\}$ can be attached to *any* stream clustering algorithm ***A*** that generates a label vector for the current input and maintains a list of cluster centers. For example, any of the four iSMFs can be used with the PrS and sequential k-means algorithms as well as other methods such as stream k-means++. Finally, the four iSMFs introduced here are by no means an exhaustive accounting of functions which can be incrementalized using the WCD algorithm. There are others that await development. And in all likelihood, there are representations of cohesion and separation other than the WCD in Definition 12.1 that can be used as the basis for other methods of this kind. The use of iSMFs is a new approach, and there is much to learn about it and much room for improvement.

12.4.2 Visualization of Streaming Data

There is another aspect of streaming data analysis that is usually overlooked because there aren't a lot of methods available to accomplish it very well-namely, visualization of the streaming data. Section 11.5 began with a timely quote about the importance of visualizing data by E. R. Tufte. Here is another testimonial:

> *Visualization is critical to data analysis. It provides a front line of attack, revealing intricate structure in data that can't be absorbed in any other way. We discover unimagined effects, and we challenge imagined ones.*
>
> –*W. S. Cleveland, 1993*

Cleveland (1993) is a classic in this field that is devoted to the visual presentation of numerical data. There are three long chapters on univariate, bivariate, and trivariate data, followed by a relatively short chapter about *hypervariate data*. Cleveland says: "For visualization, we need the term "hypervariate" to acknowledge the visual breakpoint between three and four dimensions." This is exactly the case that also distinguishes between the terms introduced in Chapters 1 and 3 for HPOV and CPOV clusters: in a nutshell, we can see them for $p = 1, 2$, or 3; but we can't see them for the hypervariate case, $p > 3$. Many, if not most, real streaming time series are hypervariate data. In this case, we will not be able to employ hindsight analysis to critique and evaluate the performance of stream clustering algorithms. How can we use the advice of Cleveland for this case?

The first thing that probably comes to mind is feature projection. What can we learn about hypervariate streaming data by projection? This idea was explored in detail in Chapter 5, where the static hypervariate ($p = 4$) Iris data was projected orthogonally onto one-and two-dimensional subsets with and without labels. Figures 5.5, 5.6, and 5.7(a, b, and c) should have convinced you that orthogonal projection from a hypervariate upspace to a viewable downspace will reveal very little about the intricate structure that Iris might possess in $\Re^4$. Elmqvuist *et al.* (2008) present an interesting scheme for animating projections in scatterplot matrices for interactive exploration of hypervariate data through various looks at 3D projections of it.

What about feature extraction? Multidimensional scaling and PCA are sometimes advocated for this purpose. Figure 5.17 shows 2D views of Iris made by plotting pairs of extracted features with four popular methods: ISOMAP, Lagrangian Eigenmap, LLE, and t-SNE. This should convince you that plotting extracted features in a viewable downspace is also unreliable. Projection or extraction of a single point in upspace data to a viewable downspace is not useful as a means for understanding the nature of the data in the stream. The value of methods such as these really depends on mapping groups of points that have been collected, so this approach is not viable for point by point interpretation of evolving data.

Finally, Figure 5.21 shows the iVAT image of the 4D Iris data based on its Euclidean distance matrix. The information possessed by this image is different than all of the scatterplots of it discussed in Chapter 5. The iVAT image presents an idea about sets of individuals that form (SL) groups in the data, not locations of individual objects like all of the projection and feature extraction methods. Moreover, the iVAT image is readily available for smallish static hypervariate data of arbitrary dimensionality. The siVAT image affords a way to peek into arbitrarily large sets of static hypervariate data. Can we extend this idea to streaming data? Yes. Example 12.9 previews how this can be done.

Example 12.9. Visualization of the hypervariate Iris data with iVAT.

For a hypervariate time series that terminates after n inputs, with $n \leq 10,000$ or so, the iVAT image can be built point by point as the data arrives. For example, we can do this for the Iris data. The first input will not generate a distance, but we can set D = [1] and display a single black pixel with iVAT. When the second input arrives, an iVAT image of the 2×2 distance matrix can be displayed. When the third input arrives, an iVAT image of the 3×3 distance matrix can be displayed. Continuing this way, the iVAT image of streaming Iris will grow in size as each input arrives, until n = 150, the end of the Iris stream. Figure 12.21 has three snapshots of this procedure at the positions n = 51, n = 102, and n = 150. The iVAT image labeled n = 150 at the right in Figure 12.21 is exactly the image displayed in Figure 5.21 which was constructed all at once by applying iVAT to the 150×150 Euclidean distance matrix of the static Iris data set.

Note the arrow pointing to the single pixel in the lower right corner of the snapshot at n = 51 in Figure 12.21. This is the first input from the second subset of 50 plants (versicolor). At this instant, iVAT has seen the 50 Setosa plants and shows them as the 50×50 square in the n = 51 snapshot. The arrival of point 51 is indicated by the single pixel that iVAT begins as a second structure. At n = 102, there is no breakpoint in the iVAT image, in agreement with all of the evidence we have about Iris that suggests that it has only two CPOV clusters. Continuing out to n = 150, the iVAT sequence makes no distinction for the remaining points in the stream. You can watch this in the YouTube video that is referenced above the caption for Figure 12.21.

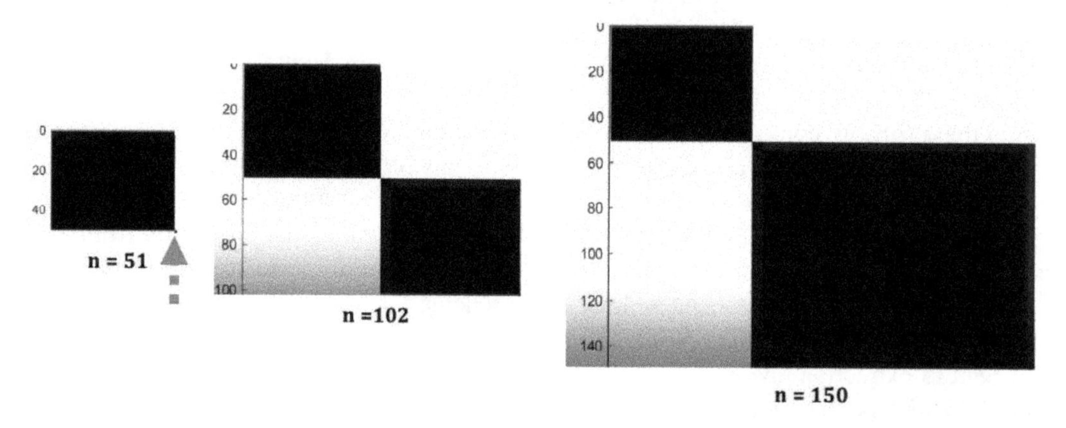

Watch a video of iVAT processing streaming Iris at: https://youtu.be/F4_Nrtf0Ye8

Figure 12.21 Visualization of Iris when streamed to iVAT.

Iris is hypervariate; so it provides a tiny example of how iVAT can be used to watch the evolution of streaming data. Here is another example that uses a slightly larger set of points, but this data set has HPOV cluster structure because the data are three-dimensional.

Example 12.10. Visualization of the Heron Island weather data.

Heron Island is a barrier island in the southern part of the Great Barrier Reef, 80 km north-east of Gladstone, Queensland, Australia. The Heron Island wireless sensor network is part of a large environmental monitoring project associated with the Great Barrier Reef Ocean Observation System. The data for this example are three-dimensional data, comprising (humidity, temperature, and pressure) vectors collected from a single node in a wireless sensor network at the weather station on the island. An observation was recorded every 10 minutes for six hours each day from February 21 to March 22, 2009. So, there are $p = 3$, $n = 1080$ vectors for this little time series. We can see the data and the iVAT image of it evolving in real time; hopefully, this example will provide a bridge in your imagination to leap from the visible to the hypervariate case.

The storm named Hamish formed on March 4 and was declared a category 5 cyclone on March 7. It moved from North to South. There was an unexplained anomaly in the measurements on March 5 from 9.00 AM to 2.10 PM, which significantly altered the values of the measured data from positions 445–476 (this anomaly was probably connected to the initial formation of the cyclone, still far north of Heron Island). Cyclone Hamish crossed the ocean outboard of Heron Island, making its closest approach on March 9, 2009. This event dramatically altered sensor data for that day, most notably, the pressure measurements. The Hamish anomaly occurred between positions 593 and 629. Figure 12.22 (from Kumar *et al.*, 2016) is a 3D scatterplot of the 1080 data vectors as a static batch at the end of the collection period. This figure clearly shows the HPOV clusters in the static data set to be fairly compact and well-separated.

The Heron Island data was collected in a stream in time, and, in this example, it was processed that way by iVAT in exactly the same manner as the Iris data was in Example 12.9. Figure 12.23 shows snapshots of the evolving data (on the left) and the corresponding iVAT image (on the right) at four instants in time taken from stills of the video. The times in the sub-captions were taken from the YouTube video.

The view in panel (a) of Figure 12.23 shows the data and iVAT image just before the March 5 anomaly enters the data stream. You can see that there is a lot of fine substructure suggested in the iVAT image, but the overall impression is that the data to this point forms just one HPOV cluster. Ten seconds later,

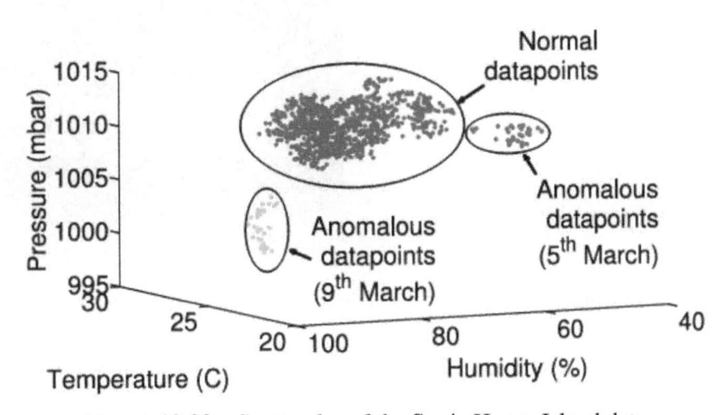

Figure 12.22 Scatterplot of the Static Heron Island data.

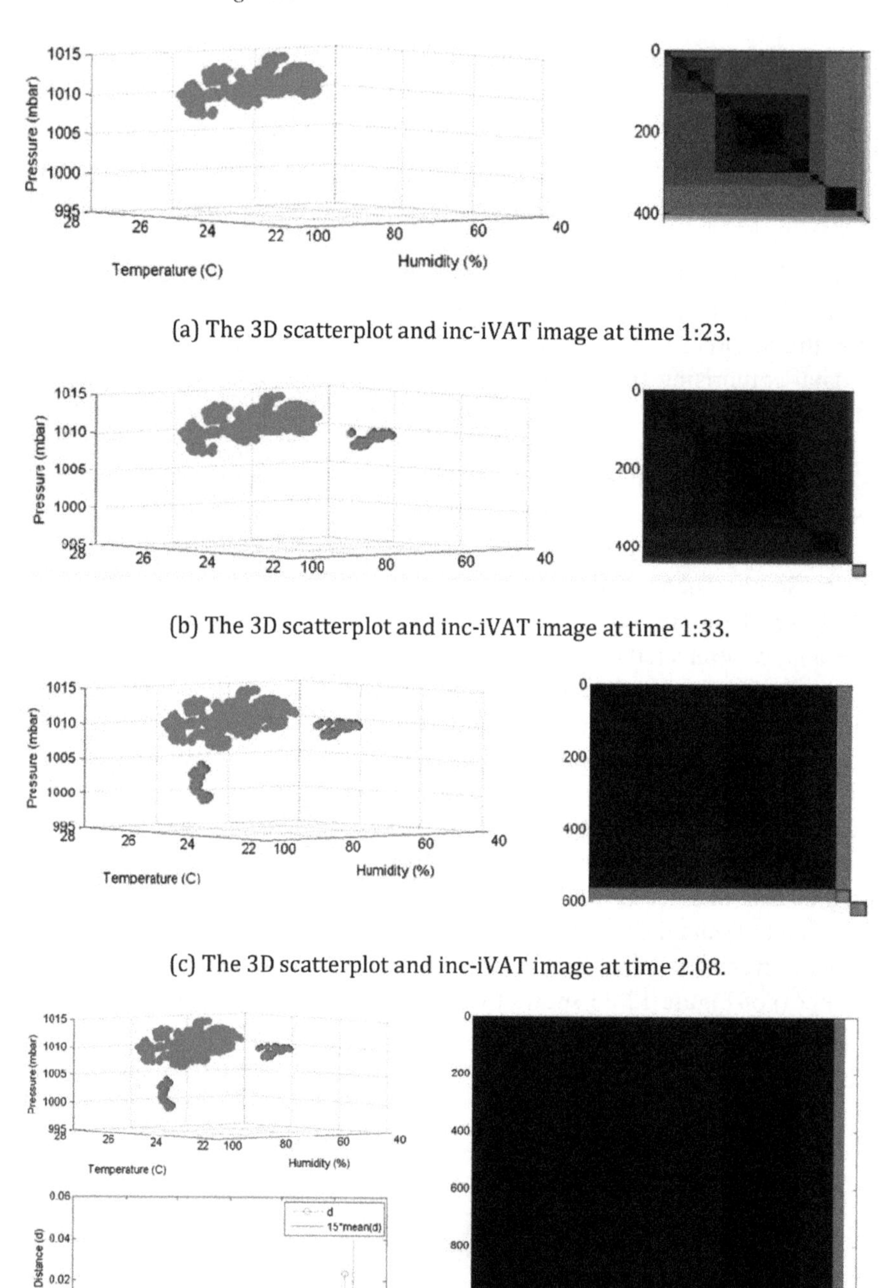

(a) The 3D scatterplot and inc-iVAT image at time 1:23.

(b) The 3D scatterplot and inc-iVAT image at time 1:33.

(c) The 3D scatterplot and inc-iVAT image at time 2.08.

(d) The 3D scatterplot and inc-iVAT image at time 3.42 (termination).

Watch a video of iVAT processing the **Heron Island** data at: https://youtu.be/3c3xK4Wh8r4

Figure 12.23 Visualization of the Heron Island data with iVAT.

some points from the March 5 anomaly have passed by, and panel (b) shows that there are now two HPOV clusters in the scatterplot, and the corresponding iVAT image sees the red points as a second evolving structure. The data set and image grow in size, but the size of the red set of pixels is fixed, corresponding to the number of points in the March 5 anomaly.

Figure 12.23(c) is a snapshot of the scatterplot and iVAT image after the March 9 Hamish anomaly data have entered the stream. Here you can see three HPOV clusters in the current scatterplot, correctly mirrored by the structure in the corresponding iVAT image. After Hamish has passed, the sensor data return to a more normal pattern, which continues out to the end of the sequence. The final scatterplot and iVAT image appear in view 12.23(d). Also seen in the (d) view at the bottom left is a graph of average distances that are used to declare when inputs to iVAT are anomalous. Portions of this example were used by Kumar et al. (2016) to illustrate how clusiVAT can be employed in the service of anomaly detection, but the point here is to see how an iVAT image of a data stream can provide some insight into the nature of the data. Bear in mind that this means with respect to the way iVAT groups data; so it would not be wrong to interpret this as a visualization of what a streaming version of SL clustering might provide. You can see the stream processing for this example at the referenced YouTube site.

Suppose a LOT of animals pass your observation platform -maybe a million zebras, each represented by a hypervariate vector. Can the method of Examples 12.9 and 12.10 be extended to this case? Yes. There are two approaches to iVAT monitoring of streaming data in this more realistic situation: inc-iVAT (*incremental iVAT*, Kumar *et al.*, 2016), and inc-siVAT (*incremental siVAT*, Rathore *et al.*, 2020). Both of these procedures are well-defined for VAT and iVAT, but only the iVAT case will be discussed here.

The memory requirement for the iVAT dissimilarity matrix is $O(n^2)$; so letting the number of points imaged by iVAT grow (as in the Iris and Heron Island Examples 12.9 and 12.10) without bound is impossible. Instead, an inc-iVAT user declares a sliding window of fixed size that is in the range of tractability for iVAT (roughly, $n \leq 10,000$). This concept is very similar to the time horizons used by algorithms like BIRCH, DenStream, and CluStream. Then inc-iVAT is initialized with a static VAT/iVAT image of the data in the sliding window. The inc-iVAT model has a companion algorithm, dec-iVAT (*decremental iVAT*). The two algorithms must be used together but are named separately because the procedures used are very different. This pair of algorithms, introduced in Kumar *et al.* (2016), is hard to summarize in a short form. Both algorithms are described in detail in Kumar and Bezdek (2020): inc-iVAT is algorithm S15; dec-iVAT is algorithm S20. The MATLAB code for both of these algorithms is available at https://github.com/genuine-dheeraj/VAT_family_of_algorithms.

Here is a very brief overview of the general idea. After the initial MST is obtained from the static data in the first window, when a new input arrives from the stream, the current iVAT minimal spanning tree (MST) is updated with an efficient edge insertion scheme. After an edge is added to the MST, dec-iVAT immediately removes a node (and hence edge) from the current MST. As the window moves along, inc-iVAT/dec-iVAT continually operates this edge exchange process, which produces a sequence of fixed-size images that can be used for visual assessment of possible cluster structure in the stream (restricted to the sliding window). This scheme is used in conjunction with clusiVAT for anomaly detection in evolving time series data in Kumar *et al.* (2016).

The inc-iVAT/dec-iVAT algorithms don't lend themselves to easy descriptions in this elementary text. However, it is easy to illustrate the basic idea in a simple example. Figure 12.24 illustrates the basic idea. The nine points are real data in p = 2 dimensions, and the iVAT images below them are also real.

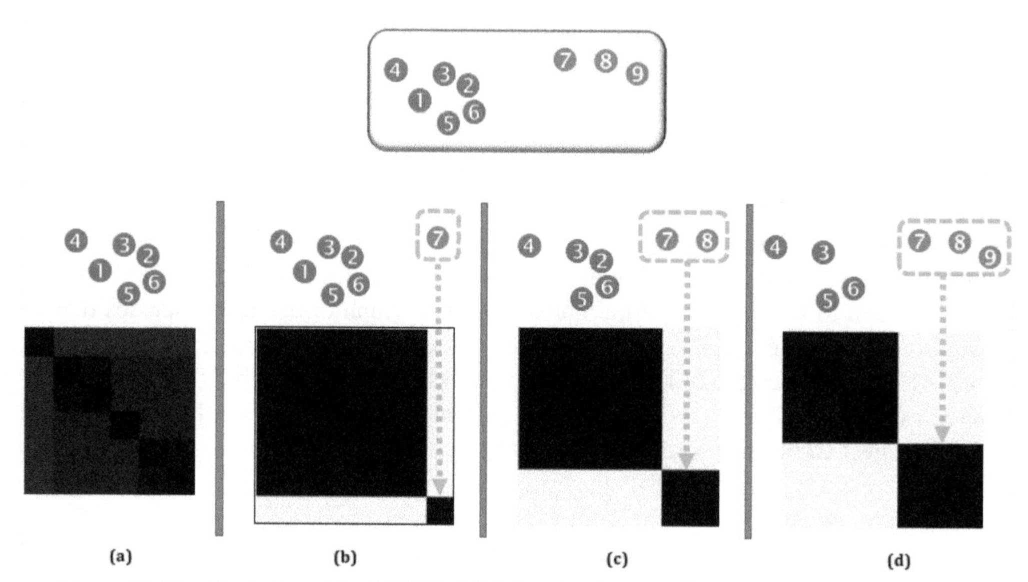

Figure 12.24 Illustration of inc-iVAT/deciVAT for nine inputs and a seven-point sliding window.

The upper panel in Figure 12.24 shows the nine points $\{\mathbf{x}_1, \ldots, \mathbf{x}_9\} \subset \Re^2$ that will arrive at the observation post at times $\{t_1, \ldots, t_9\}$. The points are arranged so that there are two HPOV clusters in the data set: points 1-6 are in the red cluster, and points 7–9 are in the blue cluster.

Choose a sliding window of length 7. As soon as the sliding window is full, the distance matrix will be 7×7; so there will be a continuous row and column exchange in the distance matrix as new points enter via inc-iVAT and old points leave via dec-iVAT. For this little example, the VAT reordering can be recomputed each time there is an exchange. After the first six red points have entered the stream processing, they are grouped together by iVAT in one dark block of size 6×6, as seen in the iVAT image of the Euclidean distance matrix in the (a) view. When the seventh point arrives, since it is in the other HPOV cluster, the (b) view shows that iVAT sees this point as a singleton and creates a 1×1 block in the lower right of the iVAT image. When the eighth point arrives in view (c) since the sliding window can only accommodate seven points, point 1 from the red cluster is removed. Now the iVAT image has two sub-blocks of sizes 5×5 and 2×2 corresponding to the data structure currently seen by iVAT. When the ninth point enters the window in view (d), point 2 is removed, and the iVAT image suggests that there are now two clusters of sizes 4×4 and 3×3 in the stream.

If the sliding window is, say, $n = 5000$, then recalculation of the distance matrix after each exchange, which has $5,000 \times 4,999/2 = 12,497,500$ values, becomes intractable. Suppose D_n is the current distance matrix for a sliding window of length n based on the points $X_n = \{\mathbf{x}_1, \ldots, \mathbf{x}_n\} \subset \Re^p$. Now point $\mathbf{x}_{n+1}$ enters the system, producing the data set $X_n \cup \{\mathbf{x}_{n+1}\}$. To find the new image, we would need to (re)compute the distance matrix D_{n+1} for the new dataset without reusing the results from the previous image. The idea of the inc-VAT algorithm is to circumvent this problem with an incremental version of iVAT that updates the current image by first computing n new distances between the incoming point $\mathbf{x}_{n+1}$ and the previous n points. Following this, one new edge is inserted into the current MST and the new image is displayed. The basis of inc-iVAT/dec-iVAT is to make an exchange (insertion and deletion) of the "right" pair of edges in the current MST so that after the exchange, the new structure is still an MST on the new data $\{\mathbf{x}_2, \ldots, \mathbf{x}_{n+1}\}$ that now reside in the sliding window.

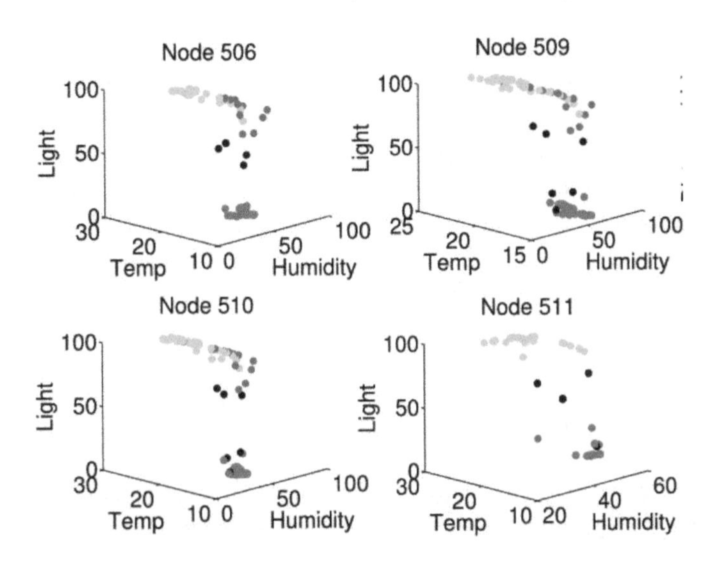

Watch a video of inc-iVAT processing the **Docklands** data at:https://youtu.be/TB8vQOqcPM4

Figure 12.25 Snapshot of Docklands data for two days = 96 measurements.

Figure 12.25 is taken from Kumar *et al.* (2016), where it appears as Figure 22. This shows data collected by a set of four sensors located in the Docklands in Melbourne, Australia. Each sensor collects 3D data (light intensity, temperature, and humidity). The view in Figure 12.25 is a snapshot of the data at each of the four sensors collected over a two-day period, corresponding to 96 samples, and this is the size of the sliding window used by inc-iVAT for the video cited below the Figure. The red, blue, and green dots are normal clusters, and the black dots are anomalies. The overall data comprise $n = 3447$ feature vectors in $p = 12$ dimensions. A video of inc-iVAT processing these 12-dimensional data can be seen at the YouTube link given below Figure 12.25.

The limitation of inc-iVAT/dec-iVAT is that it can only image the stream data in a fairly small sliding window ($n \leq 10,000$); so changes in the streaming data may or may not be seen as the window proceeds across a very long stream. A more comprehensive approach would allow iVAT to continue its inspection of the streaming data as long as there was data moving past the observer without loss of the historical visual record. This necessitates a change from the exchange procedure in the fixed length sliding window of inc-iVAT/deciVAT to a different model.

Rathore *et al.* (2020) recently described a chunk-based model that is somewhat like a wrapper on the inc-iVAT/dec-iVAT model because it uses these algorithms inside each step. This approach is really hard to describe in a page or two. Here is a very brief overview of the algorithms, which are incremental versions of siVAT, designated as inc-siVAT/dec-siVAT. This pair of algorithms deals with the streaming data in chunks. The method begins by buffering the first N_1 points in the stream as the first chunk. Call this chunk X_{N1}. Then, MMDRS algorithm A11.2 is used to obtain a sample S_{n1} of size $n_1 \ll N_1$ from X_{N1}. An iVAT image of S_{n1} is the initial look at the data in the stream, but note carefully that this is the image of a *sample* of points in the stream. After the next chunk X_{N2} passes the observer, the MMDRS samples from the first chunk are incrementally updated using a new sampling scheme called inc-MMDRS (incremental MMDRS), whence $S_{n1} \rightarrowtail S_{n2}$.

The sample update algorithm uses inc-iVAT/dec-iVAT at each stage (after each chunk) to perform data exchanges to obtain the new MMDRS sample S_{n2} without recourse to the data in the preceding chunk. The iVAT image of the updated samples S_{n2} then provides an updated look at the streaming data. This algorithm continues to operate as long as data are available. The image sequence it produces is often more informative about what structure may be in the data than the inc-iVAT/dec-iVAT sequence in the sense that there is less dependence on choosing the "right" window. On the other hand, this upgrade is paid for by looking at iVAT images of samples of each chunk, rather than images updated upon the arrival of each new sample. It is tempting to call the inc-siVAT/dec-siVAT approach a sliding chunk approach, in contrast to inc-iVAT/dec-iVAT, which is a sliding window approach. This knothole view of the inc-siVAT/dec-siVAT approach hardly suffices to answer for its technical details, but the following example will illustrate how both of these approaches provide direct, but admittedly imperfect, peeks into the streaming data in its input space. This example is excerpted from Kumar *et al.* (2020).

Example 12.11. Visualization of the 1-2-3-4-5-4-3 data with inc-iVAT and inc-siVAT.

This example compares inc-iVAT and inc-siVAT on a common data set. The streaming data is a sequence of $N = 100,000$ points in $p = 2$ dimensions. The number of HPOV clusters that are visible in the data varies in time as shown in Figure 12.26. Views (a)-(f) in the figure show six different states. The arrival of points in the stream is as follows: $c = 1$ for the first 12,000 points. Then $c = 2$ as the next 12,000 points arrive. Continuing to add points, there are $c = 4$ clusters at $n = 48,000$ points, each of size 12,000, in view (c). Then 2000 points arrive in a fifth cluster, seen in view (d), where there are $c = 5$ clusters. The next 25,000 points that arrive fill in the gap between the two clusters on the left; so in view (e), there are $c = 4$ clusters. Finally, the last 25,000 points fill in the gap between the two clusters on the right; so when $n = 100,000$, there are $c = 3$ clusters.

For this example, inc-iVAT uses a sliding window of size 5000, and inc-siVAT uses a chunk size of 500. The six snapshots in Figure 12.26 correspond to selected times in the stream where the difference between the two visualization approaches can be easily seen. All six snapshots are across 5000 consecutive points in time. Panel (a) in Figure 12.26 shows the iVAT images made by both algorithms for the 5000 points in the time window $1 - 5000$. At this time, 5000 of the 12,000 points in the lower left (blue) cluster have arrived, there is only one cluster in the data, and both images correctly suggest that this is the case.

The second snapshot is taken across the positions $9500 - 14,500$. At time 14,500, the blue cluster is completed, and 2500 points from the red cluster at upper left have entered the system. From the point of view offered by inc-iVAT, this window contains 2500 points from each of the two clusters; so the inc-iVAT image portrays this by showing us two equal sized blocks of 2500 pixels each. The point is that all inc-iVAT can see is what is in the sliding window. On the other hand, the inc-siVAT image at the bottom of panel (b) reflects the true situation at time 14,500: 12,000 points are imaged in its upper diagonal block, and 2500 points are seen in its lower diagonal block, as indicated by the green lightning bolt pointing to the small block.

The third snapshot is made across the positions from 43,000 to 48,000. At this time, all four subsets of 12,000 each have passed by. The middle view in panel (c) corresponds to the inc-iVAT image of the 5000 points in its current sliding window, ending at time 48,000. Since all 5000 points are in the fourth cluster (the black one at lower right in view (c)), inc-iVAT sees just one cluster; so the image it produces shows just one group of objects. The inc-siVAT image at time 48,000 shows the historical situation correctly: it suggests the stream has had $c = 4$ clusters so far, equal in size. Suppose the first four clusters corresponded to a stream

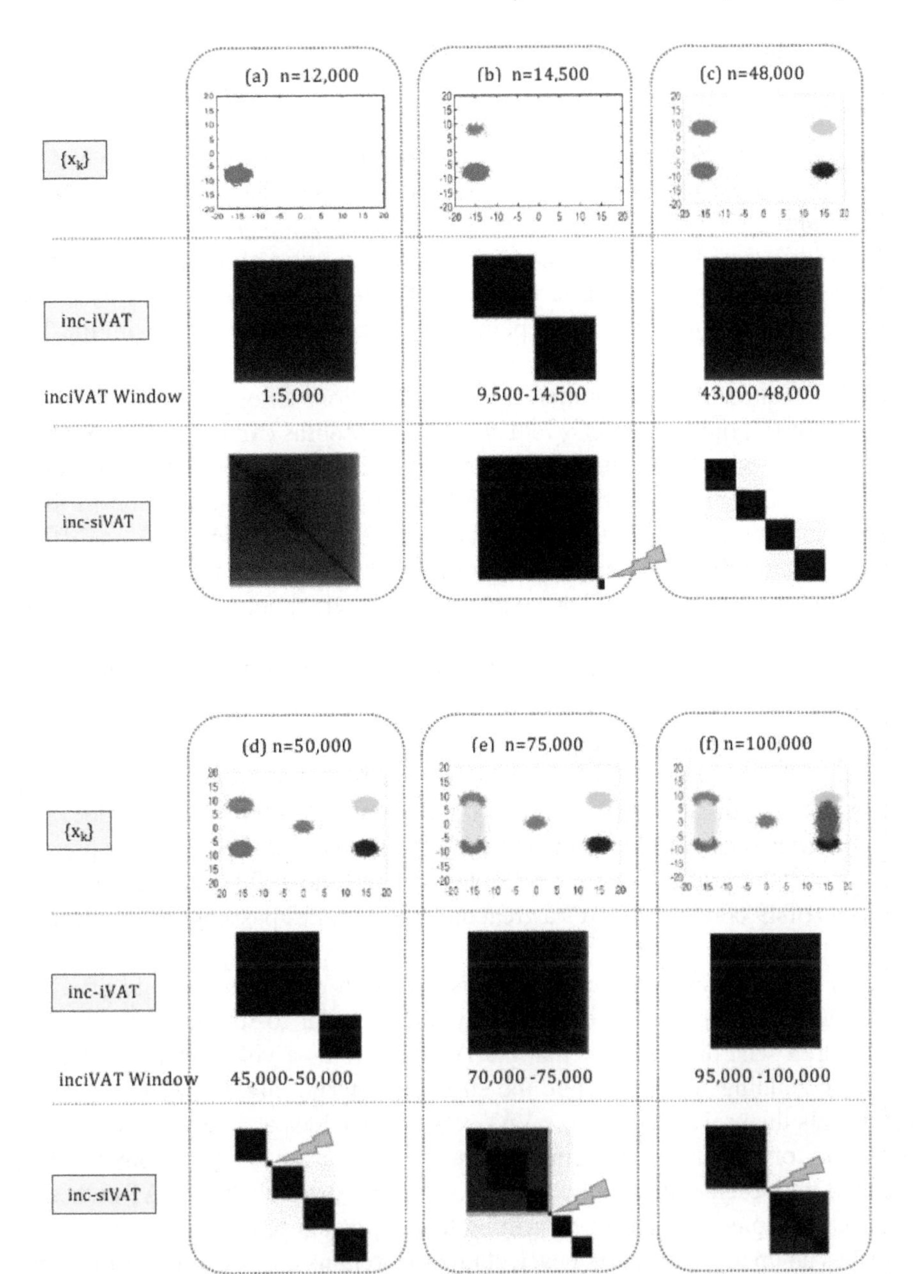

Watch a video of inc-siVAT processing the **1-2-3-4-5-4-3** data at: https://youtu.be/VKbyAA4HO5U
Watch a video of inc-siVAT processing the **5 to 3** data at: https://youtu.be/D3Cq1f4wL10

Watch a video of inc-iVAT and inc-siVAT side by side on this data at: https://youtu.be/UfcY_goNEuk

Figure 12.26 Illustration of inc-iVAT and inc-siVAT for the 1-2-3-4-5-4-3 data set.

of 12,000 zebras, 12,000 giraffes, 12,000 elephants, and 12,000 gnus. The inc-siVAT image in the bottom panel of Figure 12.25(c) accurately records this visually, while inc-iVAT only shows the last 5000 gnus at time 48,000.

The fourth view in Figure 12.26 shows the images for the time spanning 45,000 to 50,000, ending at $n = 50,000$. Recall that the 2000 pink points in the center arrive from 48,001 to 50,000. Now the sliding window of inc-iVAT contains 3000 points from the black cluster and 2000 points from the pink cluster: the inc-iVAT image in the middle panel shows $c = 2$ clusters with size ratio about 3:2. The inc-siVAT image immediately below it, however, depicts the true nature of the data that have passed by, showing $c = 5$ clusters. Note the green lightning bolt in this view, pointing to the pixels associated with the 2000 points in the center. These points are LIONS.

The four herds of 12,000 animals note that there are now 2000 lions amongst them; so they panic. Using the communication skills known only to African veld animals, the Zebras call in 25,000 wild horses (yellow in the top view of panel (e)), who fill in the gap between the zebras and the giraffes on the left side of the data view at $n = 75,000$. This really confuses the lions, who remain in position in the center. The middle view in panel (e) captures the inc-iVAT image made across the time window from 70,000 to 75,000. Since the window contains only the wild horses data from 70,000 to 75,000, inc-iVAT again sees just one cluster, as in panels (a) and (c). But inc-siVAT now suggests that there are $c = 4$ clusters: the biggest block of 49,000 pixels for the (zebras)+(giraffes)+(horses); the two smaller blocks of 12,000 points each for the elephants and gnus on the right side of the data; and the smallest block of 2000 pixels for the lions. When the gnus realizes that the wild horses have intimidated the lions, who will now concentrate on them, they call out 25,000 water buffaloes, the dark green data on the right, and the buffaloes fill in the gap between the elephants and the gnus. The last panel in Figure 12.26 is view (f), taken at $n = 100,000$ points. The inc-iVAT window shown for the times 95,001 to 100,000 now contains 5000 water buffaloes; so this image suggests there is currently one cluster in the data. The inc-siVAT image at the bottom of panel (d) portrays the actual situation: at $n = 100,000$, there are $c = 3$ clusters, and the 2000 lions can be seen trapped in-between the two large sets of 49,000 animals. It's a great day for the safari passengers, seeing all this wildlife on the move.

This example shows the difference in the type of visualization of streaming data offered by the two algorithms. The sliding window approach is conceptually quite similar to streaming clustering algorithms like BIRCH and its children. The difference is that inc-iVAT offers us a visual idea about possible cluster structure of animals in the sliding window, while the stream clustering algorithms produce various kinds of footprints of the animals themselves. The inc-siVAT scheme provides a somewhat more comprehensive picture of the entire history of data flow in the stream because each image is made from the currently saved set of samples associated with each chunk in the stream. So, in this example, inc-iVAT makes a continuous image of the current 5000 samples but forgets everything that happened prior to this window in the stream, whereas inc-siVAT makes an image at the end of each chunk that retains the key information about prior data that have passed by. Thus, inc-iVAT makes 10,000 images in this example (one after each input), whereas inc-siVAT makes 200 (one at the end of each chunk).

Here is a visualization example from the domain of neuroscience that involves hypervariate data in 78 dimensions. The data is extracellular recordings from human cortical tissue *invitro* using a *multielectrode array* (MEA) system (USB-MEA60, Multi-Channel Systems, Germany) collected over a period of about 7 minutes. Each waveform represents a 2.56 ms window of one unit (putative neuron) firing recorded with

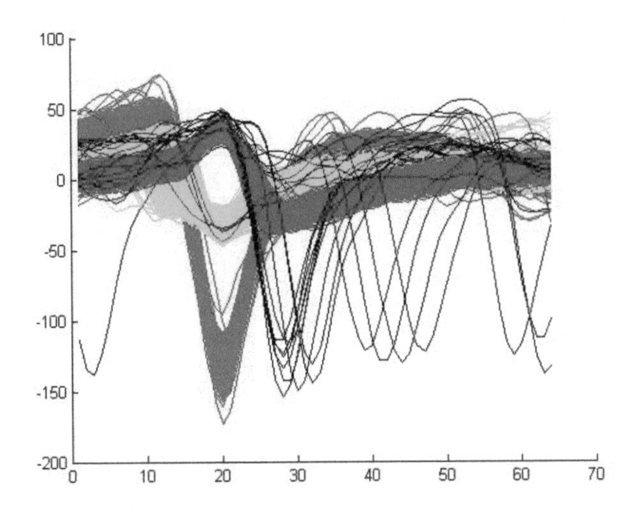

Watch a video of iVAT processing the **Neuron Spikes** data at: https://youtu.be/lbeXQXx8m9E

Figure 12.27 Neuron Spikes data: p = 78 values for each waveform vector.

a sampling frequency of 25 kHz; so each waveform is represented by 78 evenly spaced values which are recorded as one data vector in $\Re^{78}$. The 78D waveform vectors are all registered (aligned) to feature value 22, which corresponds to the peak amplitude for each firing.

The spike waveforms were sorted into c = 3 labeled subsets by a human operator using commercial spike extraction and sorting software (Offline Sorter, Plexon Inc, Dallas, TX, USA) on one channel of the MEA data. The human operator moves a mouse within a black and white image of the 2D PCA projection of the data to capture and label what seem to be the apparent clusters. The result is the c = 3 subsets of labeled waveforms (red, green, and blue) shown in Figure 12.27. There are also a few black waveforms in this graph: these are points in the PCA projection that were outside the capture zones identified during mouse labeling. These waveform vectors were fed into inc-iVAT in a stream. You can watch a video of iVAT processing them at the YouTube link shown below the figure.

The last example mentioned here is inc-siVAT processing of the KDD '99 Cup data. This data set has been discussed at length, and it continues to appear in the literature of stream clustering because although imperfect, it represents an accessible data set that was built for the study of intrusion detection in a data stream. Here is a link to a YouTube video of inc-siVAT processing the 10% data set, N = 493, 500 points in P = 41 dimensions. This is a pretty sizable data set in both N and P: inc-siVAT processes it in 125 seconds. The display above the evolving cluster heat map shows the number of points exchanged (new points added and old points deleted) from the siVAT image for each chunk. The fact that it takes 125 seconds for the processing emphasizes that we are not yet close to a real-time procedure that offers instant visualization of streaming data. But ... it is another step toward that goal.

Watch a video of inc-siVAT processing the **10% KDD** data at:https://youtu.be/M3XEMY27kXg

12.5 What's Next for Streaming Data Analysis?

Other texts present the topic of this chapter in what I would call a more or less standard way. My presentation has a very different intent. The objective of this chapter is to raise some questions about the topic that is currently called "streaming clustering" for you to think about. Here are the important issues that I think need to be discussed. They are all interrelated: a change in one area begets changes to the others.

1. Semantics. The language of static cluster analysis serves no useful purpose but has emerged as a bridge from the past (static clustering) to the future (extraction of useful information from each piece of data in the stream). There is no need to guess whether the stream will contain clusters, and, if so, how many? There are no clusters, no partitions, and no need for choosing a best partition. This was summarized in Figure 12.5: *streaming clustering is not clustering*. Let us abandon the term "streaming clustering" and call this enterprise *streaming data analysis* (SDA). The lexicon we choose for this new field will be driven by answers to the rest of the items in this list.

It is not clear that streaming data really form clusters very often anyway. Suppose 500 zebras stream past your station. The 500 vectors for these objects -seen one at a time -will almost certainly be nearly identical; so the increment to the footprint caused by each animal will probably not change it much. All the zebra data in a static collection will probably form a very tight cluster. If a pride of 11 lions is chasing them and passes by as $\{\mathbf{x}_{501}, \ldots, \mathbf{x}_{511}\}$, the SDA system hopefully sees this abrupt discontinuity (anomaly) in the footprint structure and records these observations as anomalies (or, declares these 11 pieces of data as a new structure). If 300 gazelles are (unwittingly) following the lions, a second footprint should evolve, but it will also be pretty stable (what happens to the zebra footprint if the lions happen to catch some of the zebras?). After the animals cross the river, they will probably pool together in herds, forming clusters (the lions will do this too, after they are full), but the footprints probably will not be very much like clusters in the classical sense of static cluster analysis.

2. Objectives. Channeling Tukey's remark in Section 4.2 (*it is important to understand what you CAN DO before you learn to measure how WELL you seem to have done it*), it is even *more important* to state what you *want to do* before attempting to do it. To answer questions about how well an SDA has done, you need a clearly stated, well-defined objective about what you want the processing to accomplish. The objective can't be "does this algorithm find clusters?" because that question is moot for real stream clustering. So, what do you hope to discover about data in the stream? How many kinds of animals crossed the river? Where were they concentrated? How dense was each herd? Did herds split or coalesce? Did they return? Was there a predator amongst them? Did your algorithm see it?

Surely, the primary goal of SDA is ***anomaly detection***, the problem that plagues voter fraud, financial intrusion, weather data, navigation systems, etc. The SDA must be able to recognize, label, and explain anomalies in the stream. And it needs to be done in real time. That is not a realistically obtainable objective yet, but it should be a target of current research. The most important question that an SDA must be able to answer: *is* this an anomaly? And not, *was that* an anomaly?

The second most important objective is probably ***change detection***. This is a two-part problem. First, there are *changes in the data* in the stream that can be approached with one of the many methods of change detection (Killick, 2017). And, second, there are *changes in the footprints* created by any SDA algorithm ***A*** that is processing the stream data. Different types of change should be well-defined as, e.g., subtle, abrupt, temporary, drift, and so on. Once we have agreed on the characteristics of a type of change, we will be able to design SDA algorithms that are adept at recognizing it. The iSMF functions discussed in Section 12.4.2

are an attempt to "spy on" the SDA algorithm and signal when a change has occurred in the structure being created by $\boldsymbol{A}$. But the current state of iSMF functions is embryonic at best -this is an area that is ripe for development.

3. Footprints. Once the objectives are stated, the next thing to ask is: what type of footprint would help us meet the objectives? How do we choose the ingredients of a footprint to maximize its utility? There are a lot of footprint ingredients on the market today, but there are surely some that have not been defined yet that would be very useful in meeting a stated objective. The unlabeled remnants of the four stream clustering algorithms illustrated in Table 12.2 show that there is a real difference in the information carried by different footprints. At best, the footprints should capture at least: (i) anomalies; (ii) changes in the content of data characteristics such as concept drift; and (iii) data demographics such as size, persistence, times of creation, and deletion. New footprints are emerging all the time.

For example, the StreamsoNG algorithm (Wu *et al.*, 2021) represents each detected structure with a *set* of cluster centers in the input space. Figure 12.28 shows the results of applying this SDA algorithm, streamed in input order, to the mystery data set $X_?$ which is shown in Figure 12.13. This algorithm uses the concept of *neural gas* (NG; Martinetz and Schulten, 1991) to create multi-centers as shown at Equation (12.17). The number of centers (q_i) for each footprint is pre-specified in StreamsoNG, and the algorithm is initialized by pre-clustering a set of buffered data from the initial part of the stream. Then the NG algorithm is applied to the initializing cluster(s) to find multi-centers. In this example, $q_i = 10$ centers (the yellow dots) per cluster were specified; so $\mathbf{V}_i = \{\mathbf{v}_{1,i}, \mathbf{v}_{2,i}, \ldots, \mathbf{v}_{10,i} : 1 \leq i \leq 11\}$ is a set of 110 cluster centers that comprise 11 footprints for this algorithm. The red "x's" in Figure 12.28 are the anomalies declared by StreamsoNG. The initialization accounted for the centers in the first two subsets, and the algorithm finds the remaining structure as the data stream into the algorithm.

Panel (a) of Figure 12.28 shows the footprints and anomalies superposed on the data corresponding to standard hindsight evaluation, as seen above for PrS, CLU, DEN, and MUSC in Table 12.2. The right panel

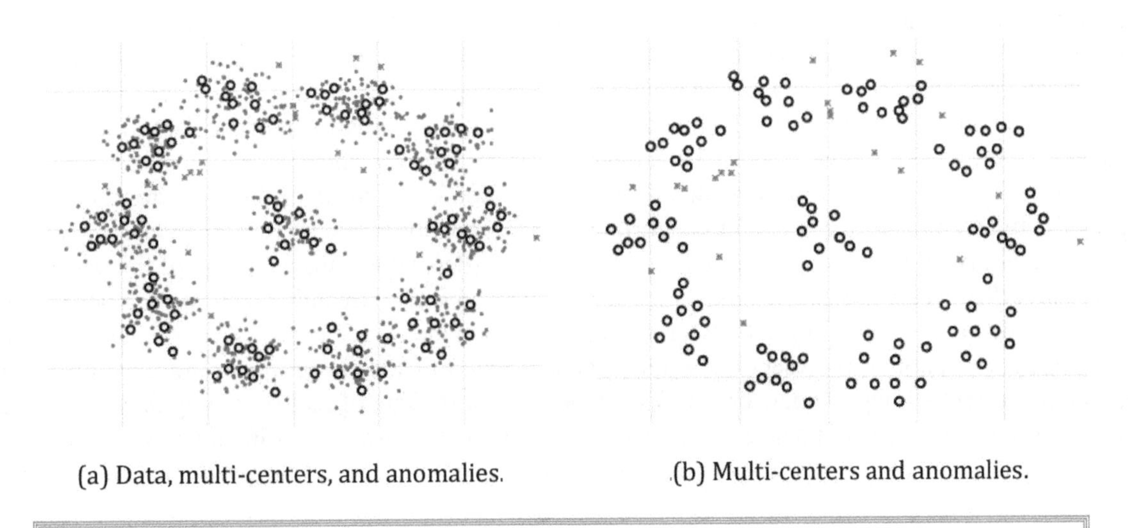

(a) Data, multi-centers, and anomalies. (b) Multi-centers and anomalies.

Watch a video of StreamsoNG processing the $X_?$ data at: https://youtu.be/WX8Zy5rhP1w

Figure 12.28 StreamsoNG meets the mystery data.(a) Hindsight. (b) Footprints only.

(b) of Figure 12.28 shows the footprints and anomalies without the data. Please compare this view with the footprints left by the other four algorithms that appear in column 3 of Table 12.1. Do the multi-centers tell you more about the streaming data than the earlier footprints left by the other four algorithms? This type of representation does convey a sense of compactness (density) of each herd and does show the outliers to each structure; so it has some useful information. But the point here is this: every time you choose an algorithm ***A*** to do SDA, you should ask yourself whether the footprints it leaves really contribute to your stated objective of the processing.

The main idea suggested by this example is that instead of saving all of the animals that reside in a sliding window such as CLU and DEN do, why not save just a few of the most important animals in each herd? Perhaps, SDA algorithms should save a few special animals that provide a "herd summary" for each footprint. For example, we could save a small set of the most representative data points (ones closest to each herd prototype) and update this summary on the fly so that at any instant, a dynamic scatterplot (or heat map) would display the core of the data seen so far. The footprints in Figure 12.28(b) can be used this way. But how will we look at the information maintained by set of multi-center footprints in near real time (especially if the cluster centers are hypervariate)?

4. Hindsight Metrics. Hindsight quality evaluation will remain in place, using labeled data and displaying it (when possible) along with summary statistics, thereby producing visual "clusters" and summary information about the streaming data. Fake streaming data analysis, and its sister enterprise, fake static cluster analysis, have been discussed and are certainly useful. There is not much else we can do to convince ourselves and our readers about the efficacy of an SDA algorithm. We have all been doing this since these two disciplines began. Nothing wrong with either, and we will all continue to do hindsight evaluation for all the reasons stated earlier. But this type of quality assessment hardly satisfies the need for somehow evaluating the utility of the SDA at doing its job in near real time. Prequential "start and stop" evaluation as in Figure 12.14 is a good intermediate step on the way to active evaluation with instream analytics.

5. Instream Analytics. We need methods of instream (online) footprint validation. In the real case, the current footprints, and whatever information they can convey in real time, is all we have. We need instream footprint metrics that can provide a continuous, reliable assessment of whether the SDA is achieving the stated objectives and what the current footprints tell us about the data. Visual assessment of the footprints is not a realistic solution since most of them will not be visible because of data dimensionality.

Spying on the processing algorithm with iSMFs can provide clues to algorithmic performance, but different iSMF functions and different SDA algorithms will probably result in very disparate signals. It is likely that the utility of a ($\boldsymbol{A}$, φ_{iSMF}) pair will be very problem-dependent. Visualization of the streaming data with inc-siVAT provides some insight into structural tendencies in the streaming data and is independent of the dimensionality of the streaming data. Bear in mind that when inc-iVAT or inc-siVAT watches a stream of data, the images essentially represent what a streaming version of SL clustering would have done for the points in the current sliding window or chunk of data. And, at the rate of real streaming data, it will be impossible to watch any point-by-point method of visualization that affords utility in any meaningful way. But don't give up! We should keep improving algorithms such as these to aid in the evaluation of instream processing.

6. History. Some algorithms, e.g., PrS and BSAS, don't provide any facility for historical (data) context. BIRCH and its children use a sliding window (time horizon) to make a snapshot of hindsight quality and summary statistics across the window for "start and stop" hindsight evaluation. And some algorithms, such

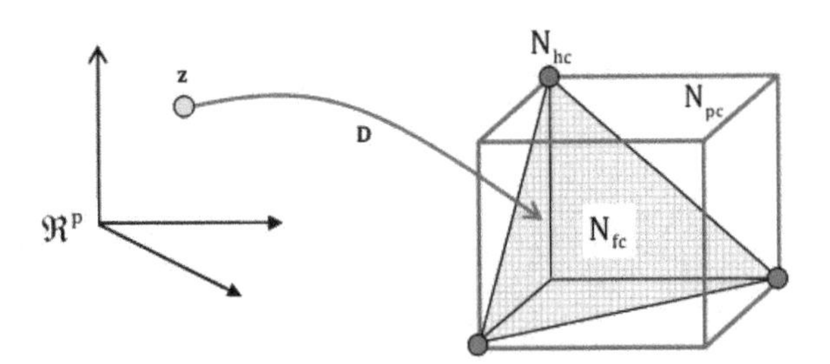

Figure 12.29 Classifier function $\mathbf{D} : \Re^p \rightarrowtail N_{pc}$.

as the XB and DBA iSMFs, utilize a forgetting factor that gradually degrades the importance of data that have already been used for incremental updates. This is an important aspect of SDA for two reasons: (i) incremental computations can be adversely affected by ignoring the effects of past data on the current processing; and (ii) from an informational point of view, knowledge of what has been seen should usefully affect how the current data is being viewed.

7. Clustering or Classification? The view of SDA offered by Bezdek and Keller (2021) concerning the interpretation of what streaming clustering algorithms actually do is that these algorithms are closer to traditional classifiers than to static clustering. Recall the discussion in Section 1.2 about the technical difference between computational clustering and classifier design. Classification concerns itself with find a classifier function, say, $\mathbf{D} : \Re^p \rightarrowtail N_{pc}$ that maps the input space into the set of label vectors (cf., Figure 3.2). Figure 12.29 illustrates this concept.

The best function within a family of **D**-models is constructed or identified using a finite set of "training data." The value **D**(*) is a class label for any object (*) represented by the input data, and once **D** is known, this function provides a way to label *all* the (usually infinitely many) objects in its domain. The type of classifier is defined by the type of labels it produces. So, a crisp classifier is imaged in N_{hc}; a fuzzy or probabilistic classifier is produces label vectors in N_{fc}; and when D yields labels in N_{pc}, it is a possibilistic classifier[39]. Finding the best **D** is called training or learning the function **D**.

There are more ways to design a classifier than you have time to read about in the rest of your life. Traditionally, training data are fed into the system one at a time, and parameters are adjusted until a desired level of correct responses is achieved. Then the system is tested and possibly validated. The classical feed-forward back-propagation neural network is trained this way: the training data are recycled through the system and the parameters of the network usually stabilize at some set of "optimal" values. This type of classifier design is clearly related to streaming clustering in the sense that as each input arrives in SDA, a label vector is generated. The difference is that in classification, the label vector is the end product of a submission, which, in SDA, is then used to update one or more footprints of the stream processing. In operation, both systems discard the label vector.

Stream clustering algorithms do classification in the sense that they assign a label vector to each data (animal) as it streams by. But they do more: they also maintain a set of incrementally updated footprints. Mathematically, a stream clustering algorithm ***A*** might be represented by a composition of two functions: a

[39]This is but one of several ways to define classifier types. Other authors may designate the type of classifier by the type of model used to produce D. For example, when the cluster centers from FCM are used as a basis for a crisp nearest prototype classifier, some authors will call the classifier fuzzy, even if the labels it produces are crisp.

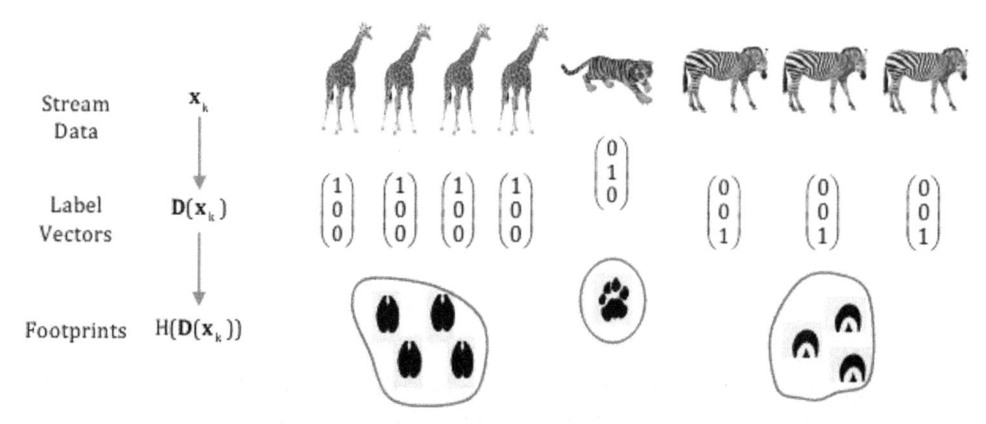

Figure 12.30 An SDA function $\mathrm{H} \circ \mathbf{D} : \Re^p \mapsto \mathrm{FP}$ creating labels and footprints.

classifier function **D** which produces a label vector from the current input, followed by a footprint function H which updates the footprint corresponding to that label, say $\boldsymbol{A}$, where FP is "footprint space." Figure 12.30 illustrates this idea with the animals that have been crossing the river.

There are a lot of missing elements in this characterization of an SDA algorithm: what is the objective of the system? How to train such a system when you don't know which animals may appear? How to test it? How to evaluate it? How to define the function H in a way that renders it agile enough to manipulate (create, delete, and merge) footprints? These questions and others that you will certainly think of-these are the unanswered questions of Feynman. These are the future of this slice of pattern recognition.

12.6 Notes and Remarks

Algorithms like the ones in the BIRCH family (Figure 12.9; Carnein and Trautman, 2019) are available at the MOA website. The STREAM software, presented in Hahsler *et al.*(2019), and methods such as the change detection algorithm of Zemeni *et al.* (2020) all process streaming data, and they all produce clusters in subsets of the data stream. But, none of these methods are truly incremental, and none of them are capable of rendering a (more or less) real-time alert to an anomaly or a change in the stream. Instead, they save part of the stream and process it -usually offline -to get a snapshot of what is happening in the data stream over some subset of time. All of these methods might be called something like *start and stop streaming data analysis.*

Mahmood *et al.* (2007) present an interesting variation on the summarization approaches that characterize footprints in the BIRCH family. These authors present a two-stage sampling scheme for clustering network traffic that uses random sampling in each stage. Their approach is based on the idea that uniform random sampling is heavily biased by cluster density in the streaming data. Streams that have a few large traffic flows (the elephants) dominate uniform samples and tend to overlook smaller patterns (the mice) that may nevertheless be important. They propose a start and stop clustering algorithm called *Echidna* that uses the same cluster feature triple as BIRCH and stores the compressed summarizations in a cluster feature tree. The departure from BIRCH lies in the adaptive two-stage sampling scheme.

In the first stage, input records from a static collection are sampled using uniform random sampling, with a sampling probability p_1. Once a record ($\mathbf{x}$) has been selected, it is tested to see if it belongs to one of the frequently occurring traffic patterns that have been already been identified by matching $\mathbf{x}$ against a buffer B of frequently occurring patterns (like the labeled footprint templates illustrated in Figure 12.30). If $\mathbf{x}$ doesn't match any frequent pattern in B, then it is considered as a new or less frequent pattern and is passed directly to the Echidna clustering algorithm for inclusion into the cluster tree. If $\mathbf{x}$ matches a frequent pattern in B, then $\mathbf{x}$ is considered to be less informative and is passed to a second sampling stage where it would be sampled using uniform random sampling with sampling probability p_2.

If $\mathbf{x}$ passes this second stage, it is sent to Echidna; otherwise, it is discarded. The effect of the second sampling stage is to reduce the rate at which known frequent patterns are clustered so that computational resources can be focused on characterizing the new or less frequent patterns. Thus, random sampling is used; but it is used twice. The way p_1 and p_2 are computed, and how matching inputs are inserted into the CF tree are covered in Mahmood *et al.*

Mahmood *et al.* (2007) tested their approach with the 1998 DARPA Intrusion Detection data set. They compared precision and recall on 25 traces with and without the two-stage scheme and concluded that the accuracy of the underlying clustering algorithm doesn't deteriorate significantly when the sampling rate is reduced using the two-stage scheme. While this paper deals with a specific type of streaming data, the basic principle involved is quite general because many other data streams will have dominant clusters; so this approach warrants a close look.

Ailon *et al.* (2009) present an approach for streaming k-means approximation that modifies the k-means++ algorithm of Arthur and Vassilvitskii (2007) by seeding HCM with a subset of $O(k\log k)$ centers, an algorithm they call k-means#. Their approach uses a divide and conquer strategy that results in a one-pass algorithm with a provable approximation guarantee. Two versions of this scheme are compared to batch HCM (A6.1, m = 1) and online HCM on three data sets with k running from 5–25. In most of their experiments, their method drove the value of $J_1(U, \mathbf{V}; X)$ below the value achieved by A6.1. As with many of these papers, the measure of success is just the value of the objective function. Not much effort is expended verifying the quality of the clusters being found.

The idea of provable guarantees about approximations to k-means solutions on reduced data sets continues to make its way into the literature of clustering in streaming data. Many of these algorithms compute a weighted sketch of the data (a coreset) and then apply k-means to the sample sketch. Schindler *et al.* (2011) describe an algorithm they call StreamKM++ and compare it to k-means#, which itself was an add-on to k-means++. StreamKM++ combines SCHM, HCM, and the so-called "ball k-means" step, which is an additional run of HCM on a subset of the points that are close ("well clustered") to the centers $\mathbf{V}$ found by A6.1 in the HCM step. Another entrant with a similar look that has no experimental results, but a nice theoretical analysis of their single pass version of k-means is the paper by Braverman *et al.* (2011).

You might have the impression from the general tone of Chapter 12 that I regard the current suite of stream clustering algorithms as useless. Nothing could be further from the truth. I view all of the methods described in Chapter 12 as steps along the way toward a more well-defined set of objectives that provide a useful approach to near real-time computational analysis of streaming data. Years ago, I asked Larry Hall (USF, Tampa) what algorithms were favored by the academic community for clustering in big static data sets. He replied: "The big search engines and applications such as Yahoo, Google, Facebook, Twitter, Amazon, etc. all use some kind of clustering, but which methods and in what ways, with what features, etc.

is something we will probably never know. The people who do this for a living at companies like Amazon and Facebook have probably signed non-disclosure agreements in their own blood to never tell - this is how they make money." To give you a better idea of how current methods can be used in the real world, here is an example from one of those secret stockpiles. This description was provided to me by an anonymous colleague who works in algorithm development at one of the big companies that are in the search business.

> "*We subscribe to twitter stream and receive millions of tweets per second. Our goal is to generate summaries of important topics every 15 minutes. We use a streaming clustering algorithm to tag the incoming tweets with a cluster id before persisting the tweet and the id on a data store. A cluster selection algorithm selects a small subset of clusters based on a set of criteria and pushes the tweets in these clusters to a batch summarization model. The summaries are recorded and are presented to interested customers. An incremental stream monitoring algorithm monitors the clustering algorithm to signal whether recent clusters should be used for summarization or not.*
>
> *In this scenario, we need to process tweets one at a time and very fast, maybe a small memory buffer can be used. We don't know the partitioning and the number of partitions but we want to know that the data we put in the datastore is reasonably partitioned without having to load them back in and reprocess so the question is not finding the best partitioning of all the data but knowing the partitions are reasonable.*"

This kind of generic description is all you will probably ever get from people who know how it works, but this example shows that stream clustering, or SDA if you like, is currently important and certainly relevant to many real problems.

Visualization of streaming data is another topic that is really in its infancy at present. There are many excellent papers in various periodicals and books that discuss cluster heat maps and projection methods for static data sets. Perhaps, the most efficient way to peek into what is going on in this area is to go to YouTube and enter "Stream Data Visualization Examples." You will find all of the videos cited in this chapter there, as well as literally hundreds of others, that present many other approaches to this discipline. I have sprinkled quotes about the usefulness of data visualization throughout the text: Ronald Fisher, Brian Everitt, E. R. Tufte, W. S. Cleveland, and John Tukey. I firmly believe that visualization is a key element of mathematics, computer science, and engineering. More generally, visualization is something we depend on to advance our knowledge of the world and its processes. And perhaps most importantly, it's a lot of fun!

12.7 Exercises

1. Recall the generalized Dunn's indices at Equation (6.58): $V_{\mathrm{GDI}_{ab}}(\mathbf{U}) = \frac{\underbrace{\min}_{i \neq j}\{\delta_a(X_i, X_j)\}}{\underbrace{\max}_{1 \le k \le c}\{\triangle_b(X_k)\}}$; $1 \le a \le 6$; $1 \le b \le 3$.

The 43 and 53 cases use the distances $\delta_4(X_i, X_j) = \|\mathbf{v}_i - \mathbf{v}_j\|$; $\delta_5(X_i, X_j) = \frac{\sum_{\mathbf{x} \in X_i} \|\mathbf{x} - \mathbf{v}_i\| + \sum_{\mathbf{y} \in X_j} \|\mathbf{y} - \mathbf{v}_j\|}{n_i + n_j}$; and the diameter $\triangle_3(X_k) = \left(2 \sum_{\mathbf{x} \in X_k} \|\mathbf{x} - \mathbf{v}_k\| / n_k\right)$. The WCD in Definition 12.1 uses the *squared* A-norm.

Modify the 43 and 53 GDIs so that the WCD appears in the formulas. Then derive the following update equations for the iSMFs based on Modified Dunn's indices:

$$(a)\ \triangle(\mathrm{iMDI}_{43,n}) = \frac{\underbrace{\min}_{i \neq j}\{\|\mathbf{v}_{i,n+1} - \mathbf{v}_{j,n+1}\|\}}{\underbrace{\max}_{1 \leq k \leq c}\left\{\frac{2C_{k,n+1}}{n_{k,n+1}}\right\}}; \quad (b)\ \triangle(\mathrm{iMDI}_{43,n}) = \frac{\underbrace{\min}_{i \neq j}\left\{\frac{C_{i,n+1} + C_{j,n+1}}{n_{i,n+1} + n_{j,n+1}}\right\}}{\underbrace{\max}_{1 \leq k \leq c}\left\{\frac{2C_{k,n+1}}{n_{k,n+1}}\right\}}.$$

Well, that's all folks ~ I hope you enjoy clustering as much as I have.

References

Achlioptas, D. (2003). Database-friendly random projections: Johnson Lindenstrauss with binary coins, *J. Comp. and Sys. Sciences*, 66, 2003, 671–687.

Acton, F. S. (1970). *Numerical Methods that Work*, reissued in 1990 by the *Mathematical Association of America*.

Acevedo, M. F. (2017). *Real-Time Environmental Monitoring: Sensors and Systems*. CRC Press.

Aggarwal C. C. (ed., 2007). *Data Streams: Models and Algorithms*, Springer.

Aggarwal, C. C. and Reddy, C. J. (2016). *Data Clustering: Algorithms and Applications*, CRC press, Boca Raton.

Aggarwal, C. C., Han, J., Wang, J. and Yu, P. (2003). A framework for clustering evolving data streams, *Proc. VLDB*, 81–92.

Aggarwal, C. C., Han, J., Wang, J. and Yu, P. (2004). A Framework for Projected Clustering of High Dimensional Data Streams, *Proc. VLDB*, 852–863.

Aggarwal, C. C., Han, J., Wang, J. and Yu, P. S. (2007). On Clustering Massive Data Streams: A Summarization Paradigm, in *Data Streams: Models and Algorithms*, ed. C. C. Aggarwal, 9–38, Springer, NY.

Aharon, M., Elad, M. and A. Bruckstein, A. (2006). k-SVD: An Algorithm for Designing Overcomplete Dictionaries for Sparse Representation," *IEEE Trans. Sig. Proc.*, 54(11), 4311–4322.

Ahmed, M.N., Yamany, S.M., Mohamed, N., Farag, A.A. and T. Moriarty, T. (2002), A modified fuzzy c-means algorithm for bias field estimation and segmentation of MRI data, *IEEE Trans. Med. Imaging* 21,193–199

Ailon, N., Jaiswal, R. and Monteleoni, C. (2009). Streaming k-means approximation. *Proc. NIPS*, 22, 10–18.

Aitken, A. C. (1926). On Bernoulli's solution of algebraic equations, *Proc. Royal Soc. Edinburgh*, 46, 289–305.

Akaike, H. (1974). A new look at the statistical model identification. *IEEE Trans. Automatic Control*, 19(6), 716–723.

Al-Sultan, K. (1995). A tabu search approach to the clustering problem. *Patt. Recog.*, 28, 1443–1451

Albatineh, A. N., Niewiadomska-Bugaj, M. and Mihalko, D. (2006). On similarity indices and correctionfor chance agreement. *J. Classification*, 23(2), 301–313.

Aldenderfer, M. S. and Blashfield, R. K. (1984). *Cluster Analysis*, Sage U. Paper #44, Sage, London.

Altman, D. (1999). Efficient fuzzy clustering of multi-spectral images, *Proc. IGARSS*, 3, 1594–1596.

Anderson, D. G. (1965). Iterative procedures for nonlinear integral equations, JACM, 12, 547–560.

Anderson, E. (1935). The IRISes of the Gaspe peninsula, *Bull. Amer.* ***IRIS*** *Soc.*, 59, 2–5.

Anderson, I. A., Bezdek, J. C. and Davé, R. (1982). Polygonal shape description of plane boundaries, in *Syst. Science and Science*, ed. Len Troncale, SGSR Publ., Louisville, KY, 1, 295–301.

Anderson, D., Luke, R. H. and Keller, J. M. (2008). Speedup of fuzzy clustering through stream processing on graphics processing units, *IEEE TFS.*, 16(4), 1101–1106.

Apostol, T.M. (1969). *Calculus*, Ginn Blaisdell, Waltham, MA.

Araki, S, Nomura, H. and Wakami, N. (1993). Segmentation of thermal images using the fuzzy c-means algorithm, *Proc. FUZZ-IEEE*, 719–724.

Arbelaitz, O., Gurrutxaga, I., Muguerza, J., Perez, J. M. and Perona, I. (2013). An extensive comparative study of cluster validity indices, *Patt. Recog*., 446, 243–256.

Arthur, D. and Vassilvitskii, S. (2007). k-means ++: The advantages of careful seeding, *Proc. SODA*, 1027–1035.

Asrodia P. and Patel, H. (2012), Network traffic analysis using packet sniffer. *Int. Jo. of Engineering Research and Applications*, 2(3), 854–856,

Assent, I., Kranen, P., Baldauf, C. and Seidl, T. (2012). AnyOut: Anytime Outlier Detection on Streaming Data, *Proc. DASFAA*, Part I, LNCS 7238, S.-g. Lee et al. (Eds.), 228–242.

Babu, G. P. and Murty, M. N. (1994). Simulated annealing for selecting optimal initial seeds in the K-means algorithm, *Indian J. of Pure and Applied Math.*, 25, 85–94.

Bache, K. and Lichman, M. (2013). UCI machine learning repository," [Online]. Available @ http://archive.ics.uci.edu/ml

Bachem, O., Lucic, M., Hassani, H., and Krause, A. (2016). Fast and provably good seedings for k-means. *Advances in Neural Information Processing Systems*, 55–63.

Bachem, O., Lucic, M. and Krause, A. (2018). Scalable k-means clustering via lightweight coresets. *Proc. KDD*, 1119 –1127.

Balcan, M., Erlich, S. and Liang, Y. (2013). Distributed k-Means and k-Median Clustering on General Topologies,*Cornell U. Library*, arXiv:1306.0604 [cs.LG].

Ball, G. and Hall, D. A. (1967). A clustering technique for summarizing multivariate data, *Behav. Sci*., 12, 153–155.

Barbara, D. and Chen, P. (2000). Using the fractal dimension to cluster datasets. In: *Proc. SIGKDD*, 260–264.

Basseville, M. and Nikiforov, I. V. (1993). Detection of abrupt changes; Theory and Application, Prentice-Hall.

Batchelor, B. G. and Wilkins, B. R. (1969). Method for location of clusters of patterns to initialize a learning machine, Electronics Letters, 5(20), 481–483.

Baum, L. E., Petrie, T., Soules, G. and Weiss, N. (1970). A maximization technique occurring in the statistical analysis of probabilistic functions of Markov chains. *Ann. Math. Stat.*, 41, 164–171.

Bell, E. T. (1966). *Men of Mathematics*, 5th printing, Simon and Schuster, NY.

Benasseni, J., Dosse, M. B. and Joly, S. (2007). On a General Transformation Making a Dissimilarity Matrix Euclidean, J. Classification, 24, 33–51.

Bengio, Y. (2009). Learning deep architectures for AI, *Foundations and Trends in Machine Learning*, 2(1), 1–127.

Bensaid, A., Hall, L.O., Bezdek, J. C., Clarke, L., Silbiger, M., Arrington, J. and Murtaugh, R. (1996). Validity-guided (re)clustering with applications to image segmentation, *IEEE TFS.*, 4(2), 112–123.

Bezdek, J. C. (1973). *Fuzzy mathematics in pattern classification*, PhD thesis, Cornell U., Ithaca, NY.

Bezdek, J. C. (1974). Cluster validity with fuzzy sets, *J. Cyber*., 3(3), 58–72.

Bezdek, J. C. (1976). A Physical interpretation of fuzzy ISODATA, *IEEE Trans. SMC*, 6(5), 387–389.

Bezdek, J. C. (1980). A Convergence Theorem for the Fuzzy ISODATA Clustering Algorithms , *IEEE Trans. PAMI*, PAMI-2(1), 1–8.

Bezdek, J. C. (1981). *Pattern Recognition with Fuzzy Objective Function Algorithms*, Plenum, NY.

Bezdek, J. C. (1992). On the Relationship between Neural Networks, Pattern Recognition, and Intelligence, *Int. J. Approximate Reasoning*, 6(2), 85–107.

Bezdek, J. C. (1994). The thirsty traveller visits Gamont: A rejoinder to "Comments on 'Editorial: Fuzzy Models - what are they and why?", *IEEE TFS.*, 2(1), 43.

Bezdek, J. C. (2015). The history, philosophy and development of computational intelligence (how a simple tune became a monster hit), Ch. 1 in *Computational Intelligence*, ed. H. Ishibuchi, EOLSS, Oxford, UK, 1–22.

Bezdek, J. C. (2016). [Computational] intelligence: what's in a name? *IEEE SMC Magazine*, 2(2), 4–14.

Bezdek, J. C. and Castelaz, P. (1977). Prototype classification and feature selection with fuzzy sets, *IEEE Trans. SMC*, 7(2), 87–92.

Bezdek, J. C. and Dunn, J. C. (1975). Optimal fuzzy partitions: A heuristic for estimating the parameters in a mixture of normal distributions, *IEEE Trans. Computers*, 24(8), 835–838.

Bezdek, J. C. and Hathaway, R. J. (1992). Numerical convergence and interpretation of the fuzzy c-shells clustering algorithms", *IEEE Trans. Neural Networks*, 3, 787–793.

Bezdek, J. C, and Harris, J. D. (1979). Convex decomposition of fuzzy partitions, *JMAA*, 67, 490–512.

Bezdek, J. C. and Hathaway, R. J. (2002). VAT: A tool for visual assessment of (cluster) tendency, *Proc. IJCNN*, IEEE Press, 2225–2230.

Bezdek, J.C. and Hathaway, R. J. (2002b). Some notes on alternating optimization, *Adv. In Soft Computing* - AFSS 2002, eds. N.R. Pal and M. Sugeno, Springer, New York, 288–300.

Bezdek, J. C. and Hathaway, R. J. (2003). Convergence of alternating optimization, *J. Neural, Parallel and Scientific Computation*, 11(4), 351–368.

Bezdek, J. C. and Keller, J. M. (2021). Streaming data analysis: clustering or classification?, *IEEE Trans. SMC*, 51(1), 91–102.

Bezdek, J. C. and Pal, S. K. (eds, 1992). *Fuzzy Models for Pattern Recognition*, IEEE Press, Piscataway, NJ.

Bezdek, J. C. and Pal, N. R. (1998). Some new indexes of cluster validity, *IEEE Trans. SMC*, 28(3), 301–315.

Bezdek, J. C., Coray, C., Gunderson, R. and Watson, J. (1981a). Detection and characterization of cluster substructure: I. Linear structure: fuzzy c-Lines, *SIAM J. Appl. Math,* 40(2), 339–357.

Bezdek, J. C., Coray, C., Gunderson, R. and Watson, J. (1981b). Detection and characterization of cluster substructure: II. Fuzzy c-varieties and convex combinations thereof, *SIAM J. Appl. Math*, 40(2), 358–372.

Bezdek, J. C., Gunderson, R., Ehrlich, R. and Meloy, T. (1978). On the extension of fuzzy k-means algorithms for the detection of linear clusters, *Proc. IEEE Conf. on Decision and Control*, 1438–1443.

Bezdek, J. C., Hathaway, R. J. and Huggins, V.J. (1985). Parametric estimation for normal mixtures, *Patt. Recog. Lett.*, 3, 79–84.

Bezdek, J. C., Hathaway, R. J. and Pal, N. R. (1995). Norm induced shell prototype (NISP) clustering, *Neural, Parallel and Sci. Comp.*, 3, 431–450.

Bezdek, J. C., Hathaway, R. J., Huband, J. M., Leckie, C. and Ramamohanarao, K. (2006). Approximate clustering in very large relational data, *Int. J. Intell. Sys.*, 21, 817–841.

Bezdek, J. C., Hathaway, R. J., Sabin, M. J. and Tucker, W. T. (1987). Convergence theory for fuzzy c-means: Counterexamples and repairs, *IEEE Trans. SMC*, 17(5), 873–877.

Bezdek, J. C., Keller, J. M., Krishnapuram, R. and Pal, N. R. (1999a). *Fuzzy models and algorithms for Pattern Recognition and Image Processing*, Springer, NY.

Bezdek, J. C., Keller, J. M., Krishnapuram, R., Kuncheva, L. I. and Pal, N. R. (1999b). Will the real Iris data please stand up?, *IEEE TFS.*, **7**(3), 368–9.

Bezdek, J. C., Li, W. Q., Attikiouzel, Y. A. and Windham, M. P. (1997). A geometric approach to cluster validity for normal mixtures, *Soft Computing*, 1, 166–179.

Bezdek, J. C., Moshtaghi, M., Runkler, T. and Leckie, C. (2016a). The Generalized C Index for Internal Fuzzy Cluster Validity, *IEEE Trans. Fuzzy Systems*, 24(6), 1500–1512.

Bezdek, J. C., Ye, X., Keller, J. M., Popescu, M. and Zare, A. (2016b). Random projection below the JL Limit, *Proc. IJCNN*, 2414–2423.

Biernacki, C., Celeux, G. and Covaert, G. (2003). Choosing starting values for the EM algorithm for getting the highest likelihood in multivariate Gaussian mixture models, *Comp. Stat. and Data Analysis*, 41, 561–575

Bifet, A., Gavaldà, R., Holmes, G. and Pfahringer, B. (2018). Machine Learning for Data Streams with Practical Examples in MOA, MIT Press.

Black, M. (1937). Vagueness: An exercise in logical analysis, Philosophy of Science 4: 427–455.

Bobrowski, L. and Bezdek, J. C. (1991). c-Means Clustering with the Norms, *IEEE Trans. SMC.*, 21(3), 545–554.

Borgelt, C. and Kruse, R. (2003). Speeding up fuzzy clustering with neural network techniques, *Proc. FUZZ-IEEE*, 2, 852–856.

Boruvka, O. (1926). 0 jistem problemu minimalnim. *Prace Mor. Prirodoved. Spol. v Brne, Acta Societ. Scient. Natur. Moravicae*, 3, 37–58.

Bottou, L. and Bengio, Y. (1995). Convergence properties of the k-means algorithms, *Proc. NIPS*, 585–592.

Bouix, S., Martin-Fernandez, M., Ungar, L., Nakamura, M., Koo, M.-S., McCarley, R. W., et al. (2007). On evaluating brain tissue classifiers without a ground truth, *NeuroImage*, 36, 1207–1224.

Bourlard, H. and Kamp, Y. (1988). Auto-association by multilayer perceptrons and singular value decomposition, *Biological Cybernetics*, 59 (4–5): 291–294.

Boutsidis, C., Zouzias, A. and Drineas, P. (2010). Random projections for k-means clustering, *Proc. NIPS*, 1–9.

Bozdogan, H. (1994). Mixture-Model Cluster Analysis Using Model Selection Criteria and A New Informational Measure of Complexity, *Proc. First US/Japan Conference on the Frontiers of Statistical Modeling: An Information Approach*, Bozdogan, H. (ed.), 69–113.

Bradley, P. S. and Fayyad, U. M. (1998). Refining initial points for k-means clustering. Proc. ICML, 98, 91–99.

Bradley, P. S., Fayyad, U. and Reina, C. (1998). Scaling clustering algorithms to large databases, *Proc. KDD*, 9–15.

Bradley, P. S., Fayyad, U. and Reina, C. (2000). Scaling EM (expectation-maximization) clustering to very large databases, *Proc. 15th ICPR* (2), 76–80.

Braverman, V., Meyerson, A., Ostrosky, R., Roytman, A., Schindler, M. and Takigu, B. (2011). Streaming k-means on Well-Clusterable Data, *Proc. SODA*, 26–40.

Brennan, R. L. and Light, R. J. (1974). Measuring agreement when two observers classify people into categories not defined in advance," *Brit. J. Math. and Stat. Pschology*, 27, 154–163.

Burbeck, K. and Nadjm-Tehrani, S. (2004). ADWICE - Anomaly Detection with Real-time Incremental Clustering, *Proc Int. Conf. on Information Security and Cryptology*, 407–424

Burbeck, K. and Nadjm-Tehrani, S. (2007). Adaptive real-time anomaly detection with incremental clustering, *Information Security Technical Report*, 12, 56–67.

Butler, R. (1986). Predictive likelihood inference with applications, *J. Royal Stat. Soc.* series B, 48(1), 1–38.

Cai, D., He, X. and Han, J. (2005). Document Clustering Using Locality Preserving Indexing, *IEEE Trans. Knowledge and Data Eng.*, 17(2), 1624–1637.

Cai, W., Chen, S. and Zhang, D. (2007). Fast and robust fuzzy c-means clustering algorithms incorporating local information for image segmentation, *Patt. Recog.*, 40, 825–838.

Cai, X., Nie, F. and Huang, H. (2013). Multi-view k-Means clustering on big data, *Proc. IJCAI*, 2598–26–4.

Calinski, T. and Harabasz, J. (1974). A dendrite method for cluster analysis, *Comm. in Stat.*, 3, 1–27.

Can, F. (1993). Incremental clustering for dynamic information processing, *ACM Trans. Inf. Syst.*, 11(2)2, 143–164.

Can, F., Fox, E., Snavely, C. and France, R. (1995). Incremental clustering for very large document databases: Initial MARIAN experience, *Inf. Sci.*, 84 (1–2), 101–114.

Cannon, R., Davé, J. and Bezdek, J. C. (1986). Efficient implementation of the fuzzy c-means clustering algorithms, *IEEE Trans. PAMI*, 8(2), 248–255.

Cao. F., Ester, M., Qian, W. and Zhou, A. (2006). Density-based clustering over an evolving data stream with noise. *Proc. SIAM*, 328–339.

Capra, F. (1975). *The Tao of Physics*, Shambhala Publications.

Carnein, M. and Trautmann, H. (2019). Optimizing Data Stream Representation: An Extensive Survey on Stream Clustering Algorithms, *Bus. Inf. Syst. Eng.*, 61(3), 277–297.

Casey, R. G. and Nagy, G. (1968). An autonomous reading machine, *IEEE Trans. Computers*, C-17(5), 492–503.

Cattell, R. (1944). A note on correlation clusters and cluster search methods, *Psychometrika*, 9, 69–184.

Celebi, M. E. (ed) (2015). *Partitional clustering algorithms*, Springer.

Celebi, M. E. and Kingravi, H. A. (2015). Linear, Deterministic, and Order-Invariant Initialization Methods for the K-Means Clustering Algorithm, Ch. 3 in Celebi (2015), (ed) (2015). *Partitional clustering algorithms*, Springer.

Chakraborty, D. and Pal, N. R. (2001). Integrated feature analysisand fuzzy rule-based system identification in a neuro-fuzzy paradigm, *IEEE Trans. SMC*, B31(3), 391–400.

Chakraborty, D. and Pal, N. R. (2008). Selecting useful groups of features (Sensors) in a connectionist framework, *IEEE Trans. Neural Networks*, 19(3), 381–396.

Cheeseman, P. (1986). Probabilistic versus Fuzzy Reasoning, in *Uncertainty in Artificial Intelligence,* ed. L. Kanal and J. Lemmer, Elsevier, NY, 85–102.

Cheeseman, P. (1988). An Inquiry into Computer Understanding, *Comp. Intell.*, 4, 57–142 (with 22 commentaries/replies).

Chen, M. Y. and Linkens, D. A. (2004). Rule-base generation and simplification for data-driven fuzzy models, *Fuzzy Sets and Sys.*, 142, 243–265.

Chen, J., MacEachren, A. M. and Peuquet, D. J. (2009). Constructing Overview + Detail Dendrogram – Matrix Views, *IEEE Trans. Visualization and Computer Graphics*, 15(6), 889–896.

Cheng, T. W., Goldgof, D. B. and Hall, L. O. (1998). Fast fuzzy clustering. *Fuzzy Sets and Sys.*, 93, 49–56.

Chernoff, H. (1973). The use of faces to represent points in k-dimensional space, *J. Amer. Stat. Assoc.*, 68, 361–368.

Chinrungrueng, C. and Sequin, C. H. (1995). Optimal adaptive k-means algorithm with dynamic adjustment of learning rate, *IEEE Trans. Neural Networks,* 6(1), 157–169.

Cleveland, W. S. (1993). *Visualizing Data*, Hobart Press, Summit, NJ.

Cohen, M. B., Elder, S., Musco, Ca., Musco, Ch. and Persu, M. (2015). Dimensionality reduction for k-means clustering and low rank approximation, *Proc. STOC 2015*, 163–172.

Collatz, L. (1966). *Functional Analysis and Numerical Mathematics*, Academic Press.

Coray, C. (1981). Clustering algorithms with prototype selection, *Proc. of Hawaii Int. Conf. on Syst. Sci.,* II, Western Periodicals Co.,945–955.

Cormen, T. H., Leiserson, C. E. and Rivest, R. L. (1996). *Introduction to Algorithms*, MIT Press, Cambridge.

Cottrell, G.W., Munroe, P. and Zipser, D. (1989). Image compression by back-propagation; An example of extensional programming, in *Models of Cognition: A Review of Cognitive Science*, ed. N. Sharkey, Norwood, NJ, 208–240.

Cover, T. M. and Thomas, J. A. (1991). *Elements of Information Theory*, Wiley.

Cox, D. R. (1957). Note on grouping, *JASA*, 52(280), 543–547.

Cox, T. F. and Cox, M. A. A. (2000). Multidimensional Scaling, Chapman and Hall.

Coxon, A. P. M. (1982). The user's guide to multidimensional scaling, Heineman, Exeter, NH.

Crosier, R. B. (1988). *Technometrics*, 30, 291–303.

Cutler, A. and Windham, M.P. (1994). Information-based validity functionals for mixture analysis, in Proc. 1st US/Japan Conf. on the Frontiers of Statistical Modeling, ed. H. Bozdogan, Kluwer, 149–170.

Czekanowski, J. (1909). "Zur differentialdiagnose der neandertalgruppe," Korrespondenzblatt der Deutschen Gesellschaft fr Anthropologie, Ethnologie und Urgeschichte, 40, 44–47.

Darken, C. and Moody, J. (1990).Fast adaptive k-means clustering: some empirical results, *Proc. 1JCNN.*

Darken, C. and Moody, J. (1992). Towards Faster Stochastic Gradient Search. *Proc. NIPS*, 1–7.

Dasgupta S. and Gupta,A. (2002). An elementary proof of the Johnson Lindenstrauss lemma, *Proc. ACM SIGMOD-SIGACT-SIGART*, 274–281.

Dasgupta, S. (1999). Learning mixtures of Gaussians, Tech. Report UCB/CSD-99-1047, UC Berkeley, 1–22.

Dasgupta, S. (2000). Experiments with random projections, *Proc. Uncertainty in AI*, 2000, 143–151.

Datta, S., Bezdek, J. C. and Palaniswami, M. (2019). Experiments with Dissimilarity Measures for Clustering Waveform Data from Wearable Sensors, *Proc. IEEE SSCI.*

Dattorro, J. (2005). *Convex optimization and Euclidean distance geometry*, Menbo Press.

Davé, R. N. (1990). Fuzzy shell-clustering and applications to circle detection in digital images, *Int. Journal of General Sys.*, 16, 343–355.

Davé, R. N. (1992). Generalized fuzzy c-shells clustering anddetection of circular and elliptical boundaries, *Patt. Recog.,* 25(7),713–721.

Davé, R. N. (1996).Validating fuzzy partitions obtained through c-shells clustering, Patt. Recog. Lett., 17, 613–623.

Davé, R. N. and Bhaswan, K. (1992). Adaptive fuzzy c-shells clustering and detection of ellipses, *IEEE Trans. Neural Networks*, 3(5), 643–662.

Davenport, J. W., Bezdek, J. C. and Hathaway, R. (1988). Parameter estimation for finite mixture distributions, *Int. J. Comp. and Math. with Applications*, 15(10), 819–828.

Davies, D. L and Bouldin, D.W. (1979). A cluster separation measure, *IEEE Trans. Patt. Anal. and Machine Intell.*, 1(4), 224–227.

Davison, M. L. (1983). *Multidimensional Scaling*, J. Wiley and Sons, NY.

de Leeuw, J. (1994). Block relaxation algorithms in statistics, in *Information Sys. and Data Analysis*, eds. H. H. Bock, W. Lenski and M. M. Richter, Springer, 308–324.

Delaunay, B. (1934). Sur la sphère vide. A la mémoire de Georges Voronoï. Bulletin de l'Académie des Sciences de l'URSS, *Classe des sciences mathématiques et naturelles*, 6 (1934), 793–800.

Dembele, D. and Castner, P. (2003). "Fuzzy c-means method for clustering microarray data," *Bioinformatics*, 19, 973–980.

Dempster, A. P., Laird, N. M. and Rubin, D. B. (1977). Maximum likelihood from incomplete data via the EM algorithm, *J. Royal Statist. Soc. B*, 39, 1–38.

Deng, Y., Kayarat, D., Elasri, M. O. and Brown, S. J. (2005). Microarray Data Clustering Using ParticleSwarm Optimization K-means Algorithm, *Proc. JCIS*, 1730–1734.

Dennis, J. E. and Schnabel, R. B. (1983). *Numerical Methods for Unconstrained Optimization and Nonlinear Equations*, Prentice-Hall.

Devijver, P. and Kittler, J. (1982). *Pattern Recognition: A Statistical Approach*, Prentice-Hall, Englewood Cliffs, NJ.

Dhillon, I. S., Guan, Y. and Kulis, B. (2004). Kernel k-means, spectral clustering and normalized cuts, *Proc. ACM SIGKDD*, 551–556.

Dimitriadou, E., Dolnicar, S. and Weingessel, A. (2002). An examination of indexes for determining the number of clusters in binary data sets, *Psychometrika*, 67(3), 2002, 137–160.

Domingos, P. and Hulten, G. (2001). A general method for scaling up machine learning algorithms and its application to clustering, *Proc. Int. Conf. Machine Learning*, 106–113.

Donath, W. E. and Hoffman, A. J. (1973). Lower bounds for the partitioning of graphs. *J. Res. Develop.*, 17, 420–425.

Drake, J. and Hamerly, G. (2012). Accelerated k-means with adaptive distance bounds. *Proc. 5th NIPS workshop on optimization for machine learning*, paper #13.

Drienas, P., Frieze, A., Kannan, R., Vempala, S. and Vinay, V. (2004). Clustering large graphs via the singular value decomposition, *Machine Learning*, 56, 9–33.

Dubes, R. and Jain, A. (1980). Validity studies in clustering methodology, *Patt. Recog.*, 11, 235–254.

Duda, R. and Hart, P. (1973). *Pattern Classification and Scene Analysis*, Wiley Interscience, NY.

Dunn, J. C. (1973). A fuzzy relative of the ISODATA process and its use in detecting compact well-separated clusters, *J. Cyberns.*, 3(3), 32–57.

Elkan, C. (1993). The paradoxical success of fuzzy logic, *Proc. AAAI*, MIT Press, 698–703.

Elkan, C. (2003). Using the triangle inequality to accelerate k-means, *Proc. ICML*, 147–153.

Elmqvist, N., Dragicevic, P. and Fekete, J. D. (2008). Rolling the dice: Multidimensional visual exploration using scatterplot matrix navigation. *IEEE Trans. Vis. Comput. Graphics,* 14(6), 1539–1548.

Erisoglu, M., Calis, N. and Sakallioglu, S. (2011). A new algorithm for initial cluster centers in k-means algorithm, *Pattern Recognition Letters*, 32(14), 1701–1705.

Eschrich, S., Ke, J., Hall, L. O. and Goldgof, D.B (2003). Fast accurate fuzzy clustering through data reduction. *IEEE TFS.*, 11(2), 262–270.

Ester, M., Kriegel, H. P., Sander, J. and Xu, X. (1996). A density-based algorithm for discovering clusters in large spatial databases with noise, *Proc. KDD*, 226–231.

Everitt, B. S. (1977). *The Analysis of Contingency Tables*, Chapman and Hall.

Everitt, B. S. (1978). *Graphical Techniques for Multivariate Data*, North Holland, NY.

Everitt, B. S. and Hand, D. J. (1981). *Finite Mixture Distributions*, Chapman and Hall, NY.

Everitt, B. S., Landau, S., Leese, M. and Stahl, D. (2011). *Cluster Analysis*, John Wiley & Sons, 5th edition.

Evgeniou T., Pontil M., Elisseeff A. (2004).Leave one out error, stability, and generalization of voting combinations of classifiers, *Machine Learning*, 55, 71–97.

Farnstrom, F., Lewis, J. and Elkan, C. (2000). Scalability for clustering algorithms revisited, *Proc.SIGKDD*, 2(1), 51–57.

Feng, Q., Chen, L., Chen, C. L. P. and Guo, L. (2019). Deep Fuzzy Clustering—A Representation Learning Approach, *IEEE Trans. Fuzzy Systems*, 28(7), 1420–1433.deep learning

Feldman, D. and Langberg, M. (2011). A unified framework for approximating and clustering data, *Proc. STOC*, 569–578.

Feldman, D., Schmidt, M. and Sohler, C. (2013). Turning Big data into tiny data: Constant-size coresets for k-means, PCA and projective clustering, *Proc. ACM-SIAM Symposium on Discrete Algorithms* (SODA '13), 1434–1453.

Feller, W. (1959). *An Introduction to Probability Theory and its Applications*, 2nd Ed. Wiley, NY, 1, 58.

Fern, X. Z. and Brodley, C. (2003). Random projection for high dimensional data clustering: a cluster ensemble approach, *Proc. ICML*, 2003, 186–193.

Ferreira de Oliveira, M. C. and Levkowitz, H. (2003). From visual data exploration to visual data mining: A survey, *IEEE Trans. Visualization and Computer Graphics*, 9(3), 378–394.

Feynman, R. P. (1989). On the Reliability of the [Challenger] Shuttle, in *What Do You Care What Other People Think?* Bantam, NY, 224–225.

Figueiredo, M. and Jain, A. K. (2002). Unsupervised learning of finite mixture models, *IEEE Trans. PAMI*, 24(3), 1–16.

Fisher, D. H. (1987). Knowledge acquisition via incremental conceptual clustering. *Machine Learning*, 2(2):139–172.

Fisher, R. A. (1925). Theory of statistical estimation, *Proc. Camb. Phil. Soc.*, 22, 700–725.

Fisher, R. A. (1936). The use of multiple measurements in taxonomic problems, *Ann. Eugenics*, 7(2), 179–188.

Fitch W. M. and Margoliash, E. (1988). This week's citation classic, *Current Contents*, (27), ISI.

Fitch, W. M. and Margoliash, E. (1967). Construction of phylogenetic trees, *Science*, 155 (3760), 279–284.

Floodgate, G. D. and Hayes, P. R. (1963). The Adansonian taxonomy of some yellow pigmented marine bacteria, *J. Gen. Microbiol.*, 30, 237–244.

Florek, K., Lukaszewicz, J. Perkal, J., Steinhaus, H. and Zubrzycki, S. (1951). Sur la liason et la division des points d'ensemble fini, *Coll. Math.*, 2, 282–285.

Fogle, D. B. (1995). *Evolutionary Computation*, IEEE Press, Piscataway, NJ.

Fowlkes, E.B. and Mallows, C. L. (1983). A method for comparing two hierarchical clusterings, *JASA*, 78, 553–569.

Fränti, P. and Virmajoki, O. (2006). Iterative shrinking method for clustering problems, *Patt. Recog.*, 39 (5), 761–765.

Friedman, A., Keselman, M. D., Gibb, L. G. and Graybiel, A, M. (2015). A multistage mathematical approach to automated clustering of high-dimensional noisy data, *Proc. Nat. Acad. Science*, DOI: 10.1073/pnas.1503940112.

Friendly, M. and Denis, D. (2005). The early origins and development of the scatterplot, *J. Hist. Behav. Sci.*, 41(2), 103–130.

Frigui, H. and Gader, G. (2009). Detection and Discrimination of Land Mines in Ground-Penetrating Radar Based on Edge Histogram Descriptors and a Possibilistic K-Nearest Neighbor Classifier, *IEEE Transactions on Fuzzy Systems*, 17(1), 185–199.N

Futschik, M. E. and Carlisle, B. (2005). "Noise-robust soft clustering of gene expression time-course data, *J. Bioinform. Comp. Bio.*, 3, 965–988.

Gaines, B. R. (1978). Fuzzy and Probability Uncertainty Logics, *Information and Control*, 38, 154–169.

Gama, J., Sebastiao, R. and Rodrigues, P. P. (2013). On evaluating stream learning algorithms, *Machine Learning*, 90, 317–346.

Ganti V., Ramakrishnan, R., Gehrke, J., Powell, A. L. and French, J. C. (1999b). Clustering large data sets in arbitrary metric spaces, *Proc. ICDE*, 502–511.

Ganti, V., Gehrke, J. and R. Ramakrishnan (1999a). Mining very large databases, *Computer*, 32(8), 38–45.

Gath, I. and Geva, A. B. (1989). Unsupervised optimal fuzzy clustering, *IEEE Trans.PAMI,* 11, 773–781.

George A. and Liu J. (1981). *Computer solution of large sparse positive definite Sys.*, Prentice-Hall. Englewood Cliffs, NJ.

Ghafoori, Z., Erfani, S., Bezdek, J. C., Karunasekera, S. and Leckie, C. (2020), LN-SNE: Log-Normal Distributed Stochastic Neighbor Embedding for Anomaly Detection, *IEEE TKDE*, 32(4), 815–820.

Gill, P. E., Murray, W. and Wright, M. H. (1981). *Practical Optimization*, Academic Press.

Goldman, A. Hain, R. and Liben, S. (2012). Oxford textbook of palliative care for children, 2nd Edition, Oxford U. Press.

Gonzalez, T. F (1985). Clustering to Minimize the Maximum Intercluster Distance. Theoretical Computer Science, 38(2–3):293–306

Goodman, L. A. and Kruskal, W. H. (1954). Measures of association for cross-classifications, *JASA*, 49, 732–764.

Gower, J. C. and Legendre, P. Metric and Euclidean Properties of Dissimilarity Coefficients, *J. Classification*, 3, 5–48, 1986.

Gower, J. C. and Ross, G. (1969). Minimum spanning trees and single linkage cluster analysis. *Appl. Stat.* 18, 54–64.

Graham, R. L. and Hell, P. (1985). On the history of the minimum spanning tree problem, *Annals of the History of Computing*, 7(1), 43–57.

Gray, R. M. and Neuhoff, D. L. (1998). Quantization, *IEEE Trans. IT*, 44(6), 2325–2383.

Guha, S. and Mishra, N. (2016). Clustering data streams. In: *Data Stream Management*, Springer, 169–187.

Guha, S., Meyerson, A., Mishra, N., Motwani, R. and O'Callaghan,L. (2003). Clustering data streams: theory and practice, *IEEE TKDE.*, 15(3), 515–528.

Guha, S., Rastogi, R. and Shim, K. (1998). CURE: An efficient clustering algorithm for large databases, *Proc. ACM SIGMOD*, 73–84.

Guha, S., Rastogi, R. and Shim,K. (2001). CURE: An efficient clustering algorithm for large databases, *Inf. Syst.*, 26(1), 35–58.

Guo, X., Xu, E., Liu, X. and Yin, J. (2017). Deep Clustering with Convolutional Autoencoders, Proc. ICONIP, 373–382.

Gurrutxaga, I., Muguerza, J., Arbelaitz, O., Perez, J. M. and Martin, J. I. (2011). Towards a standard methodology to evaluate internal cluster validity indices, *Patt. Recog. Letters*, 32, 505–515.

Gustafson, E. E. and Kessel, W. (1979). Fuzzy clustering with a fuzzy covariance matrix, *Proc. IEEE Conf. on Decision and Control,* IEEE Press, Piscataway, NJ, 761–766.

Guyon, I., Gunn, S., Nikravesh, M. and Zadeh, L. A. (eds., 2006). *Feature Extraction: Foundations and Applications*, Springer.

Hahsler, M., Bolaños, M., and Forrest, J. (2017). Introduction to stream: An extensible framework for data stream clustering research with R., *J. of Statistical Software,* 76(14), 1–52.

Hall, L. O., Ozyurt, I. B. and Bezdek, J. C. (1999). Clustering with a genetically optimized approach, *IEEE Trans. Evolutionary Computation*, 3(2), 103–112.

Hamerly, G. (2010). Making k-means even faster, *Proc SIAM ICDM*, 130–140.

Hamerly, G. and Drake, J. (2015). Accelerating Lloyd's Algorithm for k-Means Clustering, Ch. 2 in Celebri, M. E. (ed) (2015),*Partitional clustering algorithms*, Springer.

Hamerly, G. and C. Elkan (2003). Learning the k in k-means, *Proc. NIPS*, 1–8.

Hansen, P., & Mladenovic, N. (2001). J -Means: A new local search heuristic for minimum sum of squares clustering, *Patt. Recog.*, 34, 405–413.

Hardoon D. R., Szedmak, S. and Shawe-Taylor J. (2004). Canonical correlation analysis: an overview with application to learning methods, *Neural Comput.*, 16(12), 2639–2664.

Hartigan, J. (1975). *Clustering Algorithms*, Wiley, NY.

Hartley H. O. (1958). Maximum likelihood estimation from incomplete-data, *Biometrics* 14, 174–194.

Hasselblad, V. (1966). Estimation of parameters for a mixture of normal distributions. *Technometrics*, 8, 431–444.

Hathaway, R. J. and Bezdek, J. C. (1986). On the asymptotic properties of fuzzy c-means cluster prototypes as estimators of mixture subpopulations, *Comm. Stat.*, 5(2), 505–513.

Hathaway, R. J. and Bezdek, J. C. (1993). Switching regression models and fuzzy clustering, *IEEE Trans. Fuzzy Syst.,* 1(3), 195–204.

Hathaway, R. J. and Bezdek, J. C. (1994). NERF c-Means : Non-Euclidean relational fuzzy clustering, *Patt. Recogition*, 27(3), 429–437

Hathaway, R. J. and Bezdek, J. C. (2006). Extending fuzzy and probabilistic clustering to very large data sets, *Comp. Stat. And Analysis*, 51, 215–234.

Hathaway, R. J., Bezdek, J. C. and Huband, J. M. (2006a). Scalable visual assessment of cluster tendency for large data sets, *Patt. Recog.*, 39, 1315–1324.

Hathaway, R. J., Bezdek, J. C. and Huband, J. M. (2006b). Maximin Initialization for Cluster Analysis, in *Progress in Pattern Recognition, Image Analysis and Applications*, LNCS, Springer, 14–26.

Hathaway, R. J., Davenport, J. W. and Bezdek, J. C. (1989). Relational Duals of the c-Means Clustering Algorithms, *Patt. Recog.*, 22(2), 205–212.

Hathaway, R. J., Huggins, V. and Bezdek, J. C. (1984). A Comparison of Methods for Computing Parameter Estimates for a Mixture of Normal Distributions, eds. W. G. Vogt and M. H. Mickle, *Proc. Modeling and Simulation*, 15, 1853–1860.

Havens, T. C. and Bezdek, J. C. (2011). An Efficient Formulation of the Improved Visual Assessment of Cluster Tendency (iVAT) Algorithm, *IEEE TKDE*, 23(2), 568–584.

Havens, T. C. and Bezdek, J. C. (2012). A new formulation of the coVAT algorithm for visual assessment of clustering tendency in rectangular data. *Int. J.Intell. Sys.* 27(6), 590–612.

Havens, T. C., Bezdek, J. C. and Palaniswami, M. (2013). Scalable single linkage clustering for big data, *Proc. IEEE ISSNIP*, 396–401.

Havens, T. C., Bezdek, J. C., Keller, J. M. and M. Popescu (2008). Clustering in ordered dissimilarity data, *Int. J. Intell. Sys.*, 24(5), 504–528.

Havens, T. C., Bezdek, J. C., Keller, J. M., Popescu, M. and Huband, J. (2009). Is VAT really single linkage in disguise?, *Ann. Math. Artificial Intelligence*, 55(3–4), 237–251.

Havens, T. C., Bezdek, J. C., Leckie, C., Hall, L. O. and Palaniswami, M. (2012). Fuzzy *c*-Means Algorithms for Very Large Data, *IEEE TFS,* 20(6), 1130–1146.

Havens, T. C., Chitta, R., Jain, A. K. and Jin R. (2011), Speedup of fuzzy and possibilistic kernel c-means for large-scale clustering, *Proc FUZZ-IEEE*, 463–470.

Highleyman, W. (1962). Linear decision functions, with applications to pattern recognition, *Proc. IRE*, 50, 1501–1514.

Hill, M. D. (1990). What is scalability? *ACM SIGARCH Computer Architecture Newsletter*, 18(4), 18–21.

Hitchcock, E. (1840). *Elementary Geology*, Dayton and Saxton, New York.

Hodgson, M. E. (1988). Reducing computational requirements of the minimum-distance classifier, *Remote Sensing of Environments*, 25, 117–128.

Hoppner, F., Klawonn, F, Kruse, R. and Runkler, T. (1999). *Fuzzy Cluster Analysis*, Wiley and Sons, Chichester, UK.

Hore, P., Hall, L., Goldgof, D., Gu, Y., Maudsley, A. and Darkazanli, A. (2009b). A scalable framework for segmenting magnetic resonance images, *J. Sig. Proc. Syst*, 54(1–3), 183–203.

Hore, P., Hall, L.O. and Goldgof, D.B. (2009a). A Scalable framework For cluster ensembles, *Patt. Recog.*, 42, 676–688.

Hotelling, H. (1936). Relations between two sets of variates. *Biometrika*, 28, 312–377.

Hu, Y., and R.J. Hathaway (2002). On efficiency of optimization in fuzzy c-means, *Neural, Parallel and Scientific Comp.*, 10(2).

Hu, Y., and R. J. Hathaway (2008). An algorithm for clustering tendency assessment, *WSEAS Trans. Math.*, 7(7), 441–450.

Huang, Z. (1998). Extensions to the k-Means Algorithm for Clustering Large Data Sets with Categorical Values, *Data Mining and Knowledge Discovery*, 2, 283–304.

Hubert, L. (1974). Some applications of graph theory to clustering, *Psychometrica*, 39, 283–309.

Hubert, L. J. and Arabie, P. (1985). Comparing partitions, *J. Classification*, 2, 193–218.

Hüllermeier, E., Rifqi, M., Henzgen, S. and Senge, R. (2011). Comparing Fuzzy Partitions: A Generalization of the Rand index and related measures, *IEEE TFS*, 20(3), 546–556.

Hung, M. C. and Yang, D. L. (2001). An efficient fuzzy c-means clustering algorithm, *Data Mining, 2001, Proc. IEEE ICDM 2001*, 225–232.

Hurvich, C. M. and Tsai, C. L. (1989). Regression and time series model selection in small samples, *Biometrika*, 76(2), 297–307.

Ibrahim, O., Du, Y. and Keller, J. M. (2018). Robust On-Line Streaming Clustering," Proc. Int. Conf. on Inf. Processing and Management of Uncertainty in Knowledge-Based Systems, 467–478, 2018.

Ibrahim, O., Keller, J. M. and Bezdek, J. C. (2020). Evaluating Evolving Structure in Streaming Data with ModifiedDunn's Indices, *IEEE Trans. Emerging Topics in Comp. Intelligence,* 5(2), 262–273.

Indyk, P. and Motwani, R. (1998). Approximate nearest neighbors: Towards removing the curse of dimensionality. *Proc. ACM Symposium on the Theory of Computing*, 604–613.

Indyk, P., Amir, A. Efrat, A. and Samet, H. (1999). Efficient regular data structures and algorithms for location and proximity problems, Proc.40th Symposium on Foundations of Computer Science.

Jaccard, P. (1908). Novelles recherches sur la distribution florale, Bull. Soc. Vaud. Sci. Nat., 44, 223–270.

JafariAsbagh M., Ferrara, E., Varol, O., Menczer. F. and Flammini, A. (2014). Clustering memes in social media streams. *Social Network Analysis and Mining*, 4, 237–262.

Jain, P. and Buksh, B. (2016). Accelerated k-means clustering algorithm, *Int. J. IT and CS*, 10, 39–46.

Jain, A. K. and Dubes, R. C. (1988). Algorithms for Clustering Data, Prentice Hall, Englewood Cliffs, NJ.

Jamshidian, M. and Jennrich, R. I. (1993). Conjugate gradient acceleration of the EM algorithm, *JASA*, 88, 221–228.

Jarník. V. (1930). O jistém problému minimálním. *Práca Moravské Pćírodovćedecké Spolećcnosti*, 6:57–63. In Czech.

John, G. H. and Langley, P. (1996). Static vs Dynamic Sampling for Data Mining, *Proc. KDD*, 367–370).

Johnson W. and Lindenstrauss, J. (1984). Extensions of Lipschitz mappings into a Hilbert space, *Contemporary Mathematics*, 26, 189–206.

Johnson, R. A. and Wichern, D. W. (2007). *Applied Multivariate Statistical Analysis*, 6th ed., Prentice Hall, Englewood Cliffs, NJ.

Johnson, M. E., Moore L. M. and Ylvisaker, D. (1990).Minimax and maximin distance designs, *J. Stat. Planning and Inference*, 26(2), 131–148.

Kamel, M. S and Selim, S.Z. (1994). A new algorithm for solving the fuzzy clustering problem, *Patt. Recog.*, 27(3), 421–428

Karkkainen, I. and Franti, P. (2007). Gradual model generator for single-pass clustering, *Patt. Recog.*, 40, 784–795.

Kärkkäinen, I. and Fränti, P. (2007). Gradual model generator for single-pass clustering, *Patt. Recog.*, 40 (3), 784–795.

Karypis, G. (2003). CLUTO: A clustering toolkit, Dept. Comput. Sci., U. Minnesota, Tech. Rep. 02–017.

Kanungo, T., Mount, D. M., Netanyahu, N. S., Piatko, C. D., Silverman, R. and Wu, A. Y. (2002) An efficient k-means clustering algorithm: analysis and implementation. *IEEE Trans PAMI*, 24, 881–892

Katsavounidis I, Kuo C. C. J. and Zhang Z (1994). A new initialization technique for generalized Lloyd iteration. *IEEE Signal Processing Letters*, 1(10):144–146.

Kaufman, L. and Rousseeuw, P. (1990). *Finding Groups in Data: An Introduction to Cluster Analysis*, Wiley-Blackwell, New York.

Keller, J. and Sledge, I. (2007). A cluster by any other name, *Proc. NAFIPS,* 427–432.

Kennard, R. W. and Stone, L. A. (1969). Computer aided design of experiments, *Technometrics*, 11(1), 137–148.

Khalilia, M., Bezdek, J. C., Popescu, M. and Keller, J. M. (2014). Improvements to the relational fuzzy c-means clustering algorithm, *Patt. Recog.*, 47(12), 3920–3930.

Khan, M. A., Uddin, M. F. and Gupta, N. (2014). Seven V's of big data: Understanding big data to extract value, *Proc. Zone 1 ASEE*, 1–5.

Kiefer, J. and Wolfowitz, J. (1956). Consistency of the maximum likelihood estimates in the presence of infinitely many incidental parameters, *Ann. Math. Stat.*, 27, 887–906.

Killick, R. (2017). Introduction to optimal changepoint detection algorithms, http://members.cbio.mines-paristech.fr/~thocking/change-tutorial/RK-CptWorkshop.html.

Kim, J. and Park, H. (2008). Sparse non-negative matrix factorization for clustering, CSE Technical Reports ; GT-CSE-08–01, SMARTech, Ga. Tech, Library, 1–15.

Kim, T., Bezdek, J. C. and Hathaway, R. (1988). Optimality tests for fixed points of the fuzzy c-means algorithm, *Patt. Recog.*, 21(6), 651–663.

King, I. P. (1970). An automatic reordering scheme for simultaneous equations derived from network analysis. *Int. J. Numerical Methods in Eng.*, 2, 523–533.

Knuth, D. E, (1998). *The Art of Computer Programming, The: Volume 1: Fundamental Algorithms*, 3rd Edition. Addison-Wesley.

Kohonen, T. (1989). Self-Organization and Associative Memory, 3rd Edition, Springer-Verlag, Berlin.

Kolen, J. and Hutcheson, T. (2002). Reducing the time complexity of the fuzzy c-means algorithm, *IEEE TFS*, 10(2), 263–267, 1076.

Kosko, B. (1990). Fuzziness versus Probability, *Int. J. General Sys.*, 17(2–3), 211–240.

Kraskov, A., Stogbauer, H., Andrzejak, R. G., and Grassberger, P. (2005). Hierarchical clustering using mutual information," *Europhysics Letters*, 70(2), 278–284.

Krishnapuram, R. and Keller, J. M. (1993). A possibilistic approach to clustering, *IEEE TFS*, 1(2), 98–110.

Krishnapuram, R., Nasraoui, O. and Frigui, H. (1992). The fuzzy c-spherical shells algorithms: A new approach, *IEEE Trans. NeuralNetworks,* 3, 663–671.

Kruskal, J. (1956). On the shortest spanning subtree of a graph and the traveling salesman problem, *Proc. Am. Math. Soc.*, 7, 48–50.

Kumar, A., Sabharwal, Y. and Sen, S. (2004). A simple linear time (1+e) approximation algorithm for k-means clustering in any dimensions, *Proc. IEEE FOCS*, 454–462.

Kumar, D. and Bezdek, J. C. (2020). Visual approaches for exploratory data analysis: A survey of the VAT family of algorithms, *IEEE SMC Magazine*, 6(2), 10–48.

Kumar, D., Ghafoori, Z, Bezdek, J. C., Leckie, C. and Kotagiri, R. (2018). Dealing with Inliers in Feature Vector Data, *Int. Jo. of Uncertainty, Fuzziness and Knowledge-Based Systems*, 26(2), 25–45.

Kumar, D., Bezdek, J. C., Palaniswami, M., Rajasegarar, S., Leckie, C. and Havens, T. C. (2016). A hybrid approach to clustering in big data, *IEEE Trans. Cybernetics*, 46(10), 2372–2385.

Kumar, D., Bezdek, J. C., Palaniswami, M., Rajasegarar, S.,., Palaniswami, M., Leckie, C., Chan, J. and Gubbi, J. (2016b). Adaptive Cluster Tendency Visualisation and Anomaly Detection for Streaming Data, *ACM Trans. KDD*, 11(2), article #24.

Lance, G. N. and Williams, W. T. (1967). A general theory of classificatory sorting strategies, *Computer Journal*, 9(4), 373–380.

Lange, K. (1995). A quasi-Newton acceleration of the EM algorithm, *Stat. Sinica*, 5, 1995, 1–18.

Laviolette M. and Seaman, J. W. (1994). The efficacy of fuzzy representations of uncertainty, *IEEE TFS,* 2(1), 4–15.

Lee, Y., Lee, K. Y. and Lee, J. (2006). Estimating the optimal number of Gaussian mixtures based on incremental k-means for speaker identification, *Int. J. Inf. Tech.*, 12(7), 13–21.

Legendre, A. M. (1805). Uvelles methods pour la determination des orbites des cometes. Courcier, Paris.

Liang, P. and Klein, D. (2009). Online EM for Unsupervised Models, *Proc. NAACL*, 611–619.

Liao, L. and Lin, T. (2007)., A fast constrained fuzzy kernel clustering algorithm for MRI brain image segmentation, *Proc. ICWAPR*, 82–87.

Lie, Y., Bezdek, J. C., Chan, J., Nguyen, N. X., Romano, S. and Bailey, J. (2014), Generalized information theoretic cluster validity indices for soft clusterings, *Proc. IEEE SSCI*, 24–31.

Lie, Y., Bezdek, J. C., Chan, J., Nguyen, X. V, Romano, S. and Bailey, J. (2016), Ground truth bias in external cluster validity indices, *Pattern Recognition*, in review: arXIV 1606.05596v1.

Lie, Y., Bezdek, J. C., Chan, J., Nguyen, X. V, Romano, S. and Bailey, J. (2017), Extending Information-Theoretic Validity Indices for Fuzzy Clustering, IEEE TFS, 25(4), 1013–1018.

Liew, A.W.C., Leung, S.H. and Lau,W.H. (2000). Fuzzy image clustering incorporating spatial continuity, *IEEE Vis. Image Signal Process.*, 147, 185–192.

Liew, A.W.C., Yan, H. and Law, N. F. (2005). Image Segmentation Based on Adaptive Cluster Prototype Estimation, *IEEE Trans. Fuzzy Systems*, 13(4), 444–453.

Liiv, I. (2010). Seriation and Matrix Reordering Methods: An Historical Overview. *Stat. Anal. and Data Mining,*3(2), 70–91.

Linde, Y., Buzo, A. and Gray, R. M. (1980). An Algorithm for Vector Quantizer Design, *IEEE Trans. Comm,* 28, 84–94.

Lindley, B. V. (1982). Scoring Rules and the Inevitability of Probability, *Int. Stat. Review*, 50, 1–26 (with 7 commentaries/replies).

Ling, R. F. (1973). A computer generated aid for cluster analysis, *CACM*, 16, 355–361.

Liu, H. (ed., 1998). *Feature Extraction, Construction and Selection: A Data Mining Perspective*, Springer.

Liu, H. and Motoda, H. (eds., 2007), *Computational Methods of Feature Selection*, Chapman-Hall/CRC Data Mining and Knowledge Discovery Series.

Lloyd, S.P. (1957). Least squares quantization of PCM, 129–137. (originally an unpublished Bell Labs technical note, reprinted in *IEEE Trans. IT*, 28, March, 1982).

Loua, T. (1873), Atlas statistique de la population de Paris, Paris: J. Dejey.

Lovasz, L. and M. Plummer, M. (1986). *Matching Theory*, Elsevier Science, publ. B.V. and Akadet'miai Kiadot', Budapest.

Luenberger, D. G. (1969). *Optimization by Vector Space Methods*, Wiley, NY.

Luenberger, D. G. (1984). *Linear and non-linear programming*, Addison-Wesley.

Lukaszewicz, J. and Steinhaus, H. (1955). On measuring by comparison, *Zastos. Mat.*, 2, 225–231, 1955; *Math. Reviews*,17, 757, 1956.

Maaten, L. V. D. and Hinton, G. (2008). Visualizing data with t-SNE, *J. Machine Learning Research*, 1, 1–48.

MacQueen, J. (1967). Some methods for classification and analysis of multivariate observations, *Proc. Berkeley Symp. Math. Stat. and Prob.*, 1, eds. L. M. LeCam and J. Neyman, U. California Press, Berkeley, 281–297.

Mahallati, S., Bezdek, J. C., Popovic, M. R. and Valiante, T.A. (2019). Cluster tendency in Neuronal Spike Data, PLOS ONE, https://doi.org/10.1371/journal.pone.0224547.

Mahmood, A., Leckie, C. and Udaya, P. (2007). A scalable sampling scheme for clustering in network traffic analysis, *Proc. InfoScale*, Article 38.

March, W. B., Ram, P. and Gray, A. G. (2010). Fast Euclidean Minimum Spanning Tree: Algorithm, Analysis and Applications, *Proc. ACM SIGKDD*, 603–612.

Mardia, K. V., Kent, J. T. and J. M. Bibby, J.T. (1979). *Multivariate Analysis*. Academic Press, New York.

Marriott, F. H. C. (1974). *The Interpretation of Multivariate Observations*, Academic Press, London, 89.

Martinetz. T. and Schulten, K. (1991). A "neural-gas" network learns topologies, *Artificial Neural Networks*, 397–402.

Marvasti, F. (ed., 2001). *Nonuniform sampling: Theory and Practice*, Springer, London.

Mayr, E. (1978). Origin and History of Some Terms in Systematic and Evolutionary Biology, *Systematic Zoology*, 27(1), 83–88.

Mayr, E., Linsley, E. G. and Usinger, R. L, (1953). *Methods and Principles of Systematic Zoology*, McGraw-Hill, New York.

McBratney, A.B. and Moore, A.W. (1985). Application of fuzzy sets to climatic classification, *Ag. and Forest Meteor*, 35, 165–185.

McKendrick A. G. (1926). Applications of mathematics to medical problems, *Proc. Edinburgh Math.* Soc., 44: 98–130.

McKenzie, P. and Alder, M. (1994). Initializing the EM algorithm for use in Gaussian mixture modeling, in *Pattern Recognition in Practice IV ; Multiple Paradigms, Comparative Studies and Hybrid Syst.*, eds. E.S. Gelsema and L. N. Kanal, Elsevier, NY, 91–105.

McLachlan, G. J. and Basford, K. E. (1988). *Mixture Models : Inference and Applications to Clustering*, Marcel Dekker, NY.

McLachlan, G. J. and Peel, D. (2000). *Finite Mixture Models*, John Wiley & Sons, New York.

McLachlan, G. M. and and Rathnayake, S. (2014). On the number of components in a Gaussian mixture model, *WIREs Data Mining Knowledge Discovery*. doi: 10.1002/widm.1135.

McQuitty, L. L. (1967). A mutual development of some typological theories and pattern analytical methods, *Educ. and Psychology Meas.*, 27, 21–48.

Meek, C, Thiesson, B, and Heckerman, D. (2002). The learning curve sampling method applied to model based clustering, *J. Machine Learning Research*, 2, 397–418.

Meila, M. and Heckerman, D. (1998). An experimental comparison of several clustering and initialization methods, *Proc. UAI*, 386–395.

Melnykov, V. and Melnykov, I. (2012). Initializing the EM algorithm in Gaussian mixture models with an unknown number of components, *Comp. Stat. and Data Anal.*, 56, 1381–1395.

Meng, X.-L., Van Dyk, J. R. (1997). The EM Algorithm - An Old Folk song Sung to a Fast New Tune, *J. Royal Statist. Soc. B*, 59(3), 511–567.

Milligan, G. W. and Cooper, M. C. (1985). An examination of procedures for determining the number of clusters in a data set, *Psychometrika*, 50(2), 159–179.

Moore, A. W. (1998). Very fast EM-based mixture model clustering using multi-resolution kd-trees, *Proc. NIPS*, 543–549.

Moore A. W. (2000). The anchors hierarchy: using the triangle inequality to survive high dimensional data, *Proc. 12th Uncertainty in Artificial Intelligence Conf.*, 397–405.

Moshtaghi, M., Bezdek, J. C., Erfani, S. M., Leckie, C. and Bailey, J. (2018). Online cluster validity indices for performance monitoring of streaming data clustering. *Int. Jo. Intelligent Systems*, 34(4), 541–563.

Moshtaghi, M., Bezdek, J. C., Havens, T. C., Leckie, C., Karunasekera, S., Rajasegarar, S. and Palaniswami, M. (2012), Streaming Analysis in Wireless Sensor Networks, *Wireless Communications and Mobile Computing*, published on line: DOI: 10.1002/wcm.2248.

Moshtahgi, M., Erfani, S., Leckie, C. and Bezdek, J. C. (2016). Exponentially Weighted Ellipsoidal Model for Anomaly Detection, *Int. Jo. Intell. Systems*, 32, 881–899.

Moshtaghi, M., Havens, T. C., Bezdek, J. C., Park, L., Leckie, C., Rajasegarar, S., Keller, J. M. and Palaniswami, M. (2011). Clustering ellipses for anomaly detection, *Patt. Recog*. 44, 55–69.

Mostaghi, M., Leckie, C. and Bezdek, J. C. (2016). On line clustering of multivariate time series, *Proc. SIAM International conference on Data Mining*, 360–368.

Moulton, F. R. (1939). The Velocity of Light. *Scientific Monthly*, 48, 481–484.

Mousavi, M., Bakar A. A. and Vakilian, M. (2015). Data stream clustering algorithms: a review, Int. J. Adv. Soft Comp. Appl., 7, 1–15, 2015.

Müllner, D. (2011). Modern hierarchical, agglomerative clusteringalgorithms, ArXiv:1109.2378, URL http://arxiv.org/abs/1109.2378.

Müllner, D. (2013). fastcluster: Fast Hierarchical, Agglomerative Clustering Routines for R and Python, *J. Stat. Software*, 53(9), 1–18.

Neal, R. M. and Hinton, G. E. (1998). A view of the EM algorithm that justifies incremental, sparse, and other variants, in *Learning in Graphical Models*, M. I. Jordan (ed.), 355–368.

Nesetril, J. (2001). Otakar Boruvka on minimum spanning tree problem: Translation of both the 1926 papers, comments, history, *Discrete Math.*, 233:3–36, 2001.

Newcomb, S. (1886). A generalized theory of the combination of observations so as to obtain the best result. Amer. J. Math., 8, 343–366.

Newling, J. and Fleuret, F. (2017). K-medoids for k-means seeding. In *Advances in Neural Information Processing Systems*, 5201–5209.

Newsarama (2008). http://www.newsarama.com/15339-all-time-top-ten-members-of-the-fantastic-four.html.

Ng R. and Han, J. (1994). Efficient and effective clustering methods for spatial data mining, *Proc. Int. Conf. on Very Large Data Bases*, 144–155.

Ng, R. and Han, J. (2002). CLARANS: A method for clustering objects for spatial data mining, *IEEE TKDE*, 14(5), 1003–1016.

Nguyen, V. X., Epps, J. and Bailey, J. (2010). Information theoretic measures for clusterings comparisons: variants, properties, normalization and correction for chance, *J. Mach. Learning Research*, 11, 2837–2854.

Nguyen, H. L. Woon, y-K. and W-K. Ng, W-K.(2015). A survey on data stream clustering and classification, *Knowl. and Inf. Syst.*, 45(3), 535–569.

Nguyen, H.T. and Sugeno, M. (eds.) (1997). *Fuzzy Systems: Modeling and Control*, Kluwer, Norwell, MA.

Noble B. and Daniel, J. W. (1987). *Applied Linear Algebra*, 3rd edition, Pearson.

O'Neil, C. (2016). *Weapons of Math Destruction: How big data increases inequality and threatens democracy*, Crown Publ.

Orchard, T. and Woodbury, M . A. (1972). A missing information principle: theory and applications. *Proc. 6th Berkeley Symposium on Math. Statist. and Prob.* 1, 697–715.

Orlandia, R. Lai, Y. and Lee, W. (2005). Clustering high-dimensional data usingan efficient and effective data space reduction, *Proc. ACM Conf. Inf. Knowl. Manag.*, 201–208.

Ortega, J. M. (1970). *Numerical Analysis: A Second Course*. New York: Academic Press.

Ortega, J. M. and W.C. Rheinboldt (1970). *Iterative Solution of Nonlinear Equations in Several Variables*. New York: Academic Press.

Ortiz, L. E. and Kaelbling, L. P. (1995). Accelerating EM: An empirical study, *Proc. UAI*, 512–521.

Osipov, P. Sanders and J. Singler (2009). The Filter-Kruskal Minimum Spanning Tree Algorithm, *Proc. ALENEX*, 52–61.

Ostrovsky, R., Rabani, Y., Schulman, L. J. and Swamy, C. (2012). The effectiveness of Lloyd-type methods for the k-means problem, *JACM*, 59(6), Article 28.

Paatero, P. and Tapper, U. (1994). Positive matrix factorization: A non-negative factor model with optimal utilization of error estimates of data values. *Environmetrics* 5(1):111–126, 1994.

Pacheco, J., & Valencia, O. (2003). Design of hybrids for the minimum sum-of-squares clustering problem. *Comp. Stat. and Data Analysis*, 43, 235–248.

Pakhira, M. K., Bandyopadhyay, S. and Maulik, U. (2004). Validity index for crisp and fuzzy clusters, *Patt. Recog*. 37, 487–501.

Pakhira, M. K. and Dutta, A. (2011). Determination of number of clusters using VAT images and genetic algorithms, *Proc. Int. Conf. Emerging Applications of Information Technology*, 357–360

Pal, N. R. and Bezdek, J. C. (1995). On cluster validity for the fuzzy c-means model, *IEEE TFS*, 3(3), 370–379.

Pal, N. R. and Bezdek, J. C. (2002). Complexity reduction for large image processing, *IEEE Trans. SMC*, B-32(5), 598–611.

Pal, N. R. and Chintalapudi, K. K. (1997), A connectionist system for feature selection, *Neural, Parallel and Scientific Computation*, 5, 359–382.

Pearson, K. (1894). Contributions to the mathematical theory of evolution, *Phil. Trans. Royal Soc*. London, 185, 71–110.

Pearson, K. (1901). On lines and planes of closest fit to systems of points in space, *Philosophical Magazine*, 2(6), 559–572.

Pelleg, D. and Moore, A. (1999). Accelerating exact k-means algorithms with geometric reasoning, *Proc. KDD*, 277–281.

Peters, B. C. and Walker, H. F. (1978). An iterative procedure for obtaining maximum-likelihood estimates of the parameters for a mixture of normal distributions, *SIAM J. Appl. Math*., 35(2), 362–378.

Petrie, W. M. F. (1899). Sequences in prehistoric remains, *J. Anthropological Inst. of Great Britain and Ireland*, 29 (3/4), 295–301.

Pettie, S. and Ramachandran, V. (2002). An optimal minimum spanning tree algorithm. *JACM*, 49(1), 16–34.

Phillips, S. J. (2002). Acceleration of k-means and related clustering problems, *Proc. ALENEX*, 166–177.

Poteras, C. M., Mihaescu, M. C. and Mocanu, M. (2014). An optimized version of the k-means clustering algorithm, *Proc. FCCSIS*, 685–699.

Plackett, R. L. (1972). The discovery of the method of least squares, *Biometrika*, 59(2), 239–251.

Preparata, F. P and Shamos, M. I. (1985). *Computational Geometry: An Introduction*. Springer-Verlag New York, Inc., New York, NY.

Prim, R. (1957). Shortest connection networks and some generalisations. *Bell Syst. Tech. J*. 36, 1389–1401.

Provost, F, Jensen, D. and Oates, T. (1999). Efficient progressive sampling, *Proc. 5th KDDM*, 23–32.

Rajasegarar, S., Bezdek, J. C., Leckie, C.A. and Palaniswami, M. (2009). Elliptical Anomalies in Wireless Sensor Networks, *ACM TOSN*, 6(1), 1550–1579.

Rajasegarar, S., Shilton, A., Leckie, C., Kotagiri, R. and Palaniswami, M. (2010). Distributed training of multiclass conic-segmentation support vector machines on communication constrained networks, *Proc. ISSNIP*, 211–216.

Ramasubramanian, V., & Paliwal, K. K. (1990). A generalized optimization of the tree for fast nearest-neighbour search, *Proc. TENCON*, 565–568.

Rand, W. M. (1971). Objective criteria for the evaluation of clustering methods, *JASA*, 66(336),846–850.

Rathore, P., Bezdek, J. C., Kumar, D., Rajasegarar, S, and Palaniswami, M. (2018). Approximate Cluster Heat Maps of Large High-Dimensional Data, *Proc. Int. Conf. on Pattern Recog*. (ICPR), 195–200.

Rathore, P., Ghafoori, Z., Bezdek, J. C., Palaniswami, M., and C. Leckie. (2019a). Approximating Dunn's Cluster Validity Indices for Partitions of Big Data, *IEEE Trans. Cybernetics*, 49(5), 1629–1641.

Rathore, P., Bezdek, J. C. and Palani, M. (2019b), Fast Cluster Tendency Assessment for Big, High-dimensional Data, in *A fuzzy dictionary of fuzzy modeling. Common concepts and perspectives*, Springer.

Rathore, P., Kumar, D., Bezdek, J. C., Rajasegarar, S. and Palani, M. (2019c). A Rapid Hybrid Clustering Algorithm for Large Volumes of High Dimensional Data, *IEEE TKDE*, 31(4), 641–654.

Rathore, P., Kumar, D., Bezdek, J. C., Rajasegarar, S, and Palaniswami, M. (2020). Inc-siVAT: Visual Structural Assessment and Anomaly Detection for High-Velocity Streaming Data, *IEEE Trans. Cybernetics*, 1–14.

Rao, C. R. (1948). The utilization of multiple measurements in problems of biological classification, *J. Royal Statist. Soc.* B, 10, 159–203.

Redner R. A., Hathaway, R. J. and Bezdek, J. C. (1987). Estimating the Parameters of Mixture Models with Modal Estimators, *Comm. in Stat. (A)*, 16(9), 2639–2660.

Redner, R. A. and Walker, H. F. (1984). Mixture densities, maximum likelihood, and the EM algorithm, *SIAM review*, 26(2), 195–239.

Rivera-Galicia, L. F. (2013). Graphical Determination of Groups and Outliers in Distance-Based Cluster Analysis, https://www.researchgate.net/figure/Scatterplot-matrix-for-Iris-Data_fig1_236656495.

Romano, S., Bailey, J., Vinh, N. X. and K. Verspoor K. (2014). Standardized Mutual Information for Clustering Comparisons: One Step Further in Adjustment for Chance," *Proc. 31st ICML*, 1143–1151.

Rosen, K. H. (2007). *Discrete Mathematics and its Applications*, 6th Ed., McGraw-Hill, New York.

Rosman, G., Volkov, M., Feldman, D. Fisher, J. W. and Rus, D. (2014). Coresets for k-Segmentation of Streaming Data, *Proc. NIPS*, 559–567.

Roubens, M. (1978). Pattern classification problems with fuzzy sets. *Fuzzy Sets and Sys.*, I, 1978, 239–253.

Roweis, S. T. and Saul, L. K. (2000). Nonlinear Dimensionality Reduction by Locally Linear Embedding, *Science*, 290 (issue 5500), 2323–2326.

Rumelhart D. E, McClelland J. L. and the PDP Research Group (1986). Parallel distributed processing. *Exploration in the microstructure of cognition*, 1/2, MIT Press, Cambridge

Runkler, T. A. and Bezdek, J. C. (2013). Topology preserving feature extraction with multiswarm optimization, *Proc. IEEE SMC*, 2997–3002.

Ruspini, E. H. (1969). A new approach to clustering, *Inf. and Control*, 15, 22–32.

Ruspini, E. H., Bezdek, J. C. and Keller, J. M. (2019). Fuzzy Clustering: A historical perspective, *Computational Intelligence Magazine*, 45–55.

Sammon, J. W. (1969). A nonlinear mapping for data structure analysis, *IEEE Trans. Computers,* 18, 401–409.

Sato. M. and Ishii, S. (2000). On-line EM algorithm for the normalized Gaussian network, *Neural Computation*,12, 407–432.

Scherer, M. (2012). Inside the Secret World of Quants and Data Crunchers who helped Obama Win, *Time Magazine*, Nov. 19, 2012, 56–60

Scherer, M. (2012). Inside the Secret World of Quants and Data Crunchers who helped Obama Win, *Time Magazine*, November 19, 2012, 56–60.

Schoenberg, I. J. (1935). Remarks to Maurice Frechet's article "Sur la definition axiomatic d'une classe d'espace distancies vector-iellement applicable sur l'espace de Hilbert," *Ann. Math*, 36(3), 724–732.

Schwammle V. and Jensen, O. N. (2010). A simple and fast method to determine the parameters for fuzzy c-means cluster analysis," *Bioinformatics*, 26(22), 2841–2848.

Sebestyen, G. S. (1962). *Decision-Making Processes in Pattern Recognition*, Macmillan, NY.
Seising, R. (2015). On the history of fuzzy clustering, *IEEE SMC Magazine,* 1(1), 20–27.
Seising, R. Trillas, E. Termini, S. and Moraga, C. (eds., 2013). *On Fuzziness: A Homage to Lotfi A. Zadeh.* Springer-Verlag Studies in Fuzziness and Soft Computing, vol. 298, 2 volumes.
Selim, S. Z. and Ismail, M. A. (1986). On the local optimality of the fuzzy ISODATA clustering algorithm, *IEEE Trans. PAMI*, 8(2), 284–288.
Shafer, G. (1992). Rejoinder to comments on "Perspectives on the theory and practice of belief functions," *Int. J. Approximate Reasoning*, 6, 445–480.
Shao, H., Zhang, P., Chen, X., Li, F. and Du, G. (2019). A Hybrid and Parameter-Free Clustering Algorithm for Large Data Sets, *IEEE Access*.
Shannon, Claude E. (1948). A mathematical theory of communication, *Bell System Technical Journal* 27 (3): 379–423.
Shindler, M., Wong, A. and Meyerson, A. (2011). Fast and accurate k-means for large datasets. *Proc. NIPS*, 2375–2383
Silva, J. A., Faria, E. R., Barros, R. C. Hrushka, E. R. de Carvalho and J. Gama, J. (2013). Data Stream Clustering: a survey, *ACM Comp. Surveys*, 46(1), 1–37, 2013.
Simmons, G. F. (1963). Introduction to Topology and Modern Analysis, New York, McGraw-Hill.
Simmons, G. F. (1981). Precalculus Mathematics in a Nutshell: Geometry, Algebra and Trigonometry, Los Altos, CA., Kaufmann Inc.
Siirtola, H. and Makinen, E. (2005). Constructing and reconstructing the reorderable matrix, *Information Visualization*, 4. 32–48.
Sloan, S. W. (1986). An algorithm for profile and wavefront reduction of sparse matrices. *Int. J. Numerical Methods in Eng.*, 23, 239–251.
Sneath, P. H. A. and Sokal, R. (1973). *Numerical Taxonomy*, Freeman, San Francisco, CA.
Sneath, P.H.A. (1957). A computer approach to numerical taxonomy, *J. Gen. Microbiol.* 17, 201–226.
Sokal, R. R. and Michener, C. D. (1958). A statistical method for evaluating systematic relationships, *U. Kansas Sci. Bull*, 38, 1409–1438.
Soukup, T and Davidson, I. (2002). *Visual Data Mining: Techniques and Tools for Data Visualization and Mining*, Wiley, New York, NY, 2002.
Stallings, W. (1977). Fuzzy Set Theory versus Bayesian Statistics, *IEEE Trans. SMC*, 7(3), 216–219.
Steinhaus, H. (1956). "Sur la division des corps matériels en parties" (in French). *Bull. Acad. Polon. Sci.***4** (12), 801–804.
Steinley, D. (2004). Properties of the Hubert-Arabie adjusted Rand index. *Psychol Methods*, 9(3),386–396.
Steponavice, I., Shirazi-Manesh, M., Hyndman, R. J., Smith-Miles, K. and Villanova, L. (2016). On Sampling Methods for Costly Multi-objective Black-box Optimization, *Adv. Stochastic and Deterministic Global Optimization*, Springer, 273–296.
Stewart, G. W. (1973). *Introduction to matrix computations*, Academic Press, NY.
Stigler, S. M. (1981). Gauss and the invention of least squares, *Annals of Statistics*, 9(3), 465–474.
Stinson, D. R. (1985). *An introduction to the design and analysis of algorithms*, Charles Babbage Research Centre, Canada.
Strang, G. (1986). *Introduction to Applied Mathematics*. Wellesley-Cambridge Press,. Wellesley, MA, 1986
Strehl, A. and Ghosh, J. (2002). Cluster ensembles - a knowledge reuse framework for combining multiple partitions, *J. Machine Learning Research*, 3(3), 583–617.

Sundberg, R., (1974). Maximum likelihood theory for incomplete data from an exponential family, *Scand. J. Stat.*, 1, 49–58.

Szilágyi, L, Benyó, Z., Szilágyii, S. M. and H.S. Adam, H. S. (2003). MR brain image segmentation using an enhanced fuzzy c-means algorithm, in: *Proc. IEEE EMBS*, 17–21.

Szilágyi L, Szilágyi S. M. and Benyó Z: (2008). Fast and Robust Fuzzy c-Means Algorithms for Automated Brain MR Image Segmentation, *Encycl. Health care Information Sys.*,eds. Wickramasinghe N and Geisler E, New York, 578–586.

Tan, W. Y. and Chang, W. C. (1972). Some comparisons of the method of moments and the method of maximum likelihoodin estimating parameters of a mixture of two normal densities, *JASA*, 67, 702–708.

Tavallaee, M., Bagheri, E., Lu, W. and Ghorbani, A. A. (2009). A detailed analysis of the KDD 99 data set, *Proc. CISDA*, 53–58.

Theodoridis, S. and Koutroumbas, K. (2009). *Pattern Recognition*, 4th ed., Academic Press, NY.

Thiesson, B. Meek, C. and Heckerman, D. (2001). Acclerating EM for large databases, *Machine Learning*, 45, 279–299.

Thompson, S. K. (2012). *Sampling*, Wiley and Sons, Hoboken, NJ.

Thorndike, R. L. (1953). Who belongs in the family?, *Psychometrika*, 18(4), 267–278.

Tibshirani, R., Walther, G. and Hastie, T. (2001). Estimating the number of clusters in a data set via the gap statistic, *J. Royal Stat. Soc.* (B), 63(2), 411–423.

Titterington, D., Smith, A. and Makov, U. (1985). *Statistical Analysis of Finite Mixture Distributions*, Wiley, NY.

Tolias, Y.A. and Panas, S.M (1998). Image segmentation by a fuzzy clustering algorithm using adaptive spatially constrained membership functions, *IEEE Trans. SMC* A, 28, 359–369.

Torgerson, W. F. (1952). Multi-dimensional Scaling: I Theory and Method, *Psychometrika*, 17(4), 401–419, 1952.

Tou, J. T. and Gonzalez, R. (1974). *Pattern Recognition Principles,* Addison-Wesley, Reading, MA.

Tran-Luu T.D. (1996). Mathematical concepts and novel heuristic methods for data clustering and visualization. *Ph.D. Dissertation*, U. Maryland.

Tribus, M. (1979). Comments on “Fuzzy Sets, Fuzzy Algebra, and Fuzzy Statistics,” *Proc. IEEE*, 67(8), p. 1168.

Tryon, R. C. (1939). *Cluster Analysis*, Edwards Bros., Ann Arbor, MI. (still available at Amazon.com.).

Tufte E. R. (2001), *The Visual Display of Quantitative Information*, 2nd ed, Graphics Press, Cheshire, CT, 2001.

Tukey, J. (1977). *Exploratory Data Analysis,* Addison-Wesley, Reading, MA.

Tzeng, J., Lu, H. H-S., and Li, W-H. (2008). Multidimensional scaling for large genomic data sets, *BMC Bioinformatics*, 9, 179–196.

van der Matten, L. and Hinton, G. (2008). Visualizing data using t-SNE, *J. Machine Learning Res.*, 9, 2579–2605.

Valdes, J. J., Alsulaiman, F. A. and El Saddik, A. (2016). Visualization of handwritten signatures based on haptic information, in *Recent Advances in Computational Intelligence in Defence and Security*, Springer, 2016.

Valente de Oliveira, J. and Pedrycz, W., eds. (2007). *Advances in Fuzzy Clustering and its Applications*, Wiley, West Sussex, England.

Van Rijsbergen, C. J. (1970). A clustering algorithm, *Comp. J.*, 13, 113–115.

Vega-Pons, S. and Ruiz-Shulcoper (2013). A survey of clustering ensemble algorithms, *Int. J. Patt. Recog. and Art. Intell.*, 25(3), 337–372.

Vempala, S. S. (2004). *The Random Projection Method*, v65 in DIMACS, the AMS series in Discrete Mathematics and Computer Science.

Ventkatasubramanian, S. and Wang, Q. (2011). The Johnson-Lindenstrauss transform: An empirical study, *Proc. 2011 ALENEX*, SIAM, 164–173.

Vijayarani, S. and Nitha, S. (2011). An efficient clustering algorithm for outlier detection, *Int. J. Comp. App.*, 32(7), 22–28.

Vineet, V., Harish, P. Patidar, S. and Narayanan, P. J. (2010). Fast Minimum Spanning Tree for Large Graphs on the GPU, *Proc. HPD '09*, 167–171.

von Luxberg, U. (2007). A tutorial on spectral clustering, *Stat. Comp.*, 17, 395–416.

Walker, H. and Ni, P. (2011). Anderson acceleration for fixed point iterations, *SIAM J. Numerical Analysis*, 49(4), 1715–1735.

Wallace, C. S. and Boulton, D. M. (1968). An information measure for classification, *Computer Journal*, 11(2), 185–194.

Wang, L., Bezdek, J. C., Leckie, C. and Ramamohanarao, K. (2008). Selective sampling for approximate clustering of very large data sets, *Int. J. Intell. Sys.*, 23(3), 313–331.

Wang, L., Geng, X., Bezdek, J. C., Leckie, C. and Ramamohanarao, K. (2010). Enhanced visual analysis for cluster tendency assessment and data partitioning, *IEEE TKDE*, 22(10), 1401–1414.

Wang, W. and Zhang, Y. (2007). On fuzzy cluster validity indices, *Fuzzy Sets and Sys.*, 158, 2095–2117.

Ward, J. H., Jr. (1963). Hierarchical Grouping to Optimize an Objective Function, *JASA*, 48, 236–244.

Wei, C.-P., Lee, Y.-H. and Hsu, C.-M. (2000). Empirical comparison of fast clustering algorithms for large data sets, *Proc. Int. Conf. Sys. Sci.*, 1–10.

Wei, W. and Mendel, J. M. (1994). Optimality tests for the fuzzy c-means algorithm, *Patt. Recog.*, 27(11), 1567–1573.

West D. B. (2001). *Introduction to graph theory*, Prentice-Hall, Englewood Cliffs, NJ.

Wilkinson L. and Friendly M. (2009). The history of the cluster heat map, *American Statistician*, 63, 179–184.

Wilks, S. S. (1962). *Mathematical Statistics*, John Wiley, NY.

Williams, K. 2019). *Oddball Gallery*, https://www.youtube.com/watch?v=8UeH1aTBy7k.

Witten, I. H. and Frank, E. (2005). *Data Mining: Practical Machine Learning Tools and Techniques*, 2nd ed, Morgan Kaufmann, San Francisco, CA, 2005.

Wolfe, J. H. (1970). Pattern clustering by multivariate mixture analysis, *Multivariate Behavioral Research*, 5, 329–350.

Wong-Baker FACES Foundation (2015). Wong-Baker FACES§Pain Rating Scale. Retrieved Oct. 9, 2015 with permission from http://www.WongBakerFACES.org.

Woodall W. H and Davis, R. E. (1994). Comments on Editorial: Fuzzy Models - what are they and why?, *IEEE TFS.*, **2**(1), p. 43.

Woodbury, M. A. and Clive, J. A. (1974). Clinical pure types as a fuzzy partition, *J. Cybern.*, 4(3), 111–121.

Wu, C. F. J. (1983). On the convergence properties of the EM algorithm, *Annals of Stat.*, 11(1), 95–103.

Wu, X. and Kumar, V. (2009, eds.). *Top 10 algorithms in data mining*, Chapman and Hall/CRC.

Wu, W. Keller, J. M., Dale, J. and Bezdek, J. C. (2021) StreamSoNG: A Soft Streaming Classification Approach, https://arxiv.org/abs/2010.00635.

Xie, J, Girschick, R. and Farhadi, A. (2016). Unsupervised deep embedding for clustering analysis, *Proc ICML*, 48.

Xie, X.L. and Beni, G.A. (1991). Validity Measure for Fuzzy Clustering, *IEEE Trans. PAMI*, 3(8), 841–846.

Xiong, X., Chan, K. L. and Tan, K. L. (2004). Similarity driven cluster merging method for unsupervised fuzzy clustering, *Proc. UAI*, AUAI Press, Arlington, 1–8.

Xu, R. and Wunsch, D. C. (2009). *Clustering*, IEEE Press, Piscataway, NJ.

Yang, B., Fu, X., Sidiropoulos, N. D. and Hong, M. (2017). Towards K-means-friendly Spaces: Simultaneous Deep Learning and Clustering, arXiv:1610.04794.

Young, G. and Householder, A. S. (1938). Discussion of a set of points in terms of their mutual distances, *Psychometrika*, 3:19–22.

Younis, O. and Fahmy, S. (2004). HEED: a hybrid, energy efficient, distributed clustering approac for ad hoc sensor networks, *IEEE Trans. Mobile Computing*, 3(4), 366–379.

Yu, J., Cheng, Q. and Huang, H. (2004). Analysis of the Weighting Exponent in the FCM, *IEEE Trans. SMC-B*, 34(1), 634–639.

Yule, G. U. (1900). *On the association of attributes in statistics, Phil.Trans. A*, 194, 257–319.

Zadeh, L. A. (1965). Fuzzy Sets, *Inf. and Control,* 8, 448–454.

Zangwill, W. (1969). *Nonlinear Programming*: A Unified Approach. Englewood Cliffs, NJ: Prentice-Hall.

Zha, H., He, X., Ding, C., Simon, H. and Gu, M. (2001). Spectral relaxation for k-means clustering, *Proc. NIPS*, 1–4.

Zhang, B. and Forman, G. (2000). *ACM SIGKDD Newsletter on Scalable data mining algorithms*, 2(2), 34–38

Zhang, D., Ramamohanorao, Versteg, K. S. and Zhang, R. (2009). RoleVAT: Visual Assessment of Practical Need for Role Based Access Control, *Proc. IEEE Conf. on Security Apps*, 13–22, 2009.

Zhang, J. (2013). Advancements of Outlier Detection: A Survey, *ICST Trans. Scalable Inf. Sys.*, 13 (1–3), article e2.

Zhang, T., Ramakrishnan, R. and Livny, M. (1996). BIRCH: An efficient data clustering method for very large databases, *Proc. SIGMOD*, 103–114.

Zhang, J., Yao, Y., Peng, Y. and Yu, H. (2018). Fast K-means clustering with Anderson Acceleration, *arXiv.org>cs>arXiv:1805.10638.*

Zemini, M., Sadri, A., Ghafoori, Z., Moshtaghi, M., Salim, F. D., Leckie, C. and Ramamohanaroa, K. (2020). Unsupervised onlie change point detection in high-dimensional time series, *Know. and Inf. Syst.*, 62, 719–750.

Index

E

F

G

About the Author

3 Redfish, Milton, FL 2014 (on my dock)

Jim received the Ph.D.degree in applied mathematics from Cornell University in 1973. He is the past president of NAFIPS (North American Fuzzy Information Processing Society), IFSA (International Fuzzy Systems Association), and the IEEE CIS (Computational Intelligence Society); founding editor of the *International Journal Approximate Reasoning* and *IEEE Transactions on Fuzzy Systems*; Life fellow of the IEEE and IFSA; and a recipient of the IEEE 3rd Millennium, CIS Fuzzy Systems Pioneer, and Technical Field Award Rosenblatt medals, and the IPMU Kempe de Feret Award. Jim retired in 2007 and will be coming to a university near you soon (especially if there is fishing nearby).

Jim's interests: woodworking, optimization, motorcycles, pattern recognition, cigars, clustering in very large data, fishing, streaming data analysis, blues music, clustering in static big data, gardening, poker, and visual clustering.

Jim's advice: Never wear a hat that has more attitude than you do.

CPSIA information can be obtained
at www.ICGtesting.com
Printed in the USA
LVHW060820110522
718477LV00007B/516